:T DIRECTOR

ı Garfunkel,
ım for Mathematics and its Applications

IBUTING AUTHORS

Management Science

ʌalkevitch, York College, CUNY

I Statistics: The Science of Data

e M. Lesser, The University of Texas at El Paso
. Moore, Purdue University

II Voting and Social Choice

Taylor, Union College
Conrad, Temple University
. Brams, New York University

V Fairness and Game Theory

Taylor, Union College
Conrad, Temple University
. Brams, New York University

/ The Digital Revolution

\. Gallian, University of Minnesota Duluth

/I On Size and Growth

Campbell, Beloit College

/II Your Money and Resources

Campbell, Beloit College

For All
Practical Purp

PROJ

Solom

Conso

CONT

PART

Joseph

PART

Lawre

David

PART

Alan I

Bruce

Steven

PART

Alan I

Bruce

Steven

PART

Joseph

PART

Paul J.

PART

Paul J.

For All Practical Purposes

Mathematical Literacy In Today's World

EIGHTH EDITION

W. H. FREEMAN AND COMPANY
New York

Senior Publisher: Craig Bleyer
Publisher: Ruth Baruth
Senior Acquisitions Editor: Terri Ward
Executive Marketing Manager: Jennifer Somerville
Freelance Development Editor: Lisa Collette
Associate Editor: Laura Capuano
Senior Media Editor: Roland Cheyney
Assistant Editor: Brian Tedesco
Editorial Assistant: Katrina Wilhelm
Photo Editor and Researcher: Christine Buese, Rae Grant
Cover Designer: Blake Logan
Project Editor: Vivien Weiss
Director of Production: Ellen Cash
Printing and Binding: Quebecor

COMAP Production Department:
Production Manager: George Ward
Copy Editor: Joyce Barnes
Text Design and Composition: Daiva Chauhan

Library of Congress Control Number: 2008936372

ISBN-13: 978-1-4292-0900-7
ISBN-10: 1-4292-0900-3

Chapter Openers 1: Angelo Cavalli/age footstock; 2: Reuters/Corbis; 3: Jerzy Dabrowski/dpa/Corbis; 4: Mark E. Gibson/Corbis; 5: Studio DL/Corbis; 6: Robert Michael/Corbis; 7: Charles Gupton/Corbis; 8: D. Hurst/Alamy; 9: William Manning/www.williammanning.com/Corbis; 10: Getty Images; 11: Getty Images; 12: Paul J. Richards/AFP/Getty Images; 13: Corbis Super RF/Alamy; 14: Achim Scheidemann/Corbis; 15: Jerry Cooke/Corbis; 16: Steve Krongard/STONE/Getty Images; 17: NASA/Corby Waste; 18: Juniors Bildarchiv/Alamy; 19: Erik Von Weber/The Image Bank/Getty Images; 20: Michos Tzovaras/Art Resource, NY; 21: Premium Stock/Corbis; 22: Yellow Dog Productions/The Image Bank/Getty Images; 23: James L. Stanfield/Getty Images

Printed in the United States of America
First printing

W. H. Freeman and Company
41 Madison Avenue
New York, NY 10010
Houndmills, Basingstoke RG21 6XS, England

www.whfreeman.com

Photo by Rich Pilling/ MLB Photos via Getty Images

Mark Hirsch/Getty Images

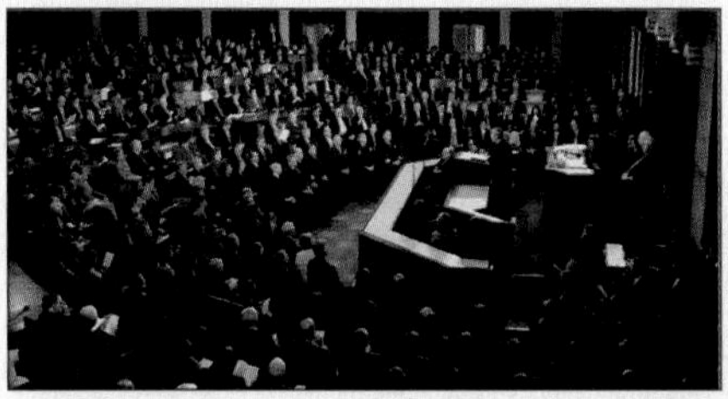

Brooks Kraft/Corbis

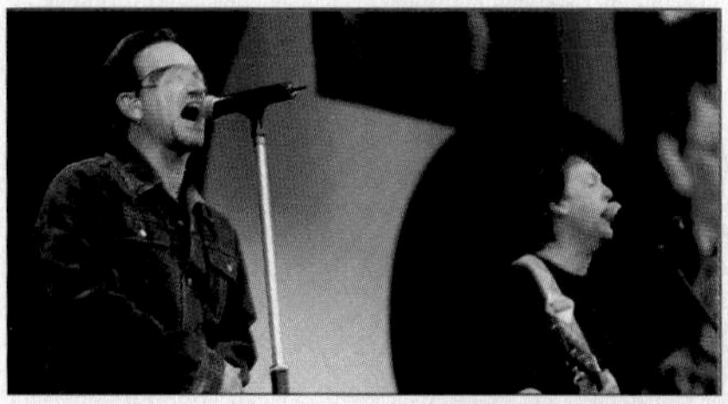

Getty Images

Craig Tuttle/Corbis

Dan Lamont/Corbis

PREFACE

To the Student

For All Practical Purposes, Eighth Edition, continues our effort to bring the excitement of contemporary mathematical thinking to the nonspecialist. In science and industry, mathematical models are the main tools for analyzing and solving problems that arise. In this book, our goal is to convey the power of mathematics by showing you the great variety of problems that can be modeled and solved by quantitative means. An extensive supplements package designed to make study time supremely effective complements the eighth edition. Highlights of the supplements package include the *Student Study Guide* and *Student Solutions Manual.* Between the text and the available resources, *For All Practical Purposes* offers you the tools to succeed in the course and apply your new knowledge to daily life experiences.

There are many ways to talk about why mathematics and its applications matter. You will hear expressions such as "mathematical literacy" or "quantitative literacy." They mean, essentially, that math is important. It is important because knowing it can make your life easier. In other words, it can help to explain how your world works. We created this course and this book because we know that not everyone looks at mathematics in this way.

In school, you spent a great deal of time learning the tools of mathematics–how to manipulate symbols, how to solve equations. In this course, you will spend time learning the uses of mathematics and the power of mathematics to help us to understand so many different parts of our everyday lives and the world itself. We hope that this exploration will give you a broader sense of what our subject is about and why we wanted you to take a math course every year you were in school. It's "for all practical purposes," because, in a sense, you've learned to hammer nails and saw wood. Now we're going to build houses.

Enjoy!

To the Instructor

Because *For All Practical Purposes* stresses the connections between contemporary mathematics and modern society, our text must be flexible enough to accommodate new ideas in mathematics and their new applications to our daily lives. We maintain this flexibility in the eighth edition.

Our primary goal for this edition was to further improve the ease of use for instructors and students alike. An extensive supplements package is available, including the new MathPortal, available packaged with the text or sold separately. This innovative online resource brings together the complete text and its media in one easy-to-use learning space. From the eBook and Assignment Center to the full array of Resources, including practice quizzes, exercise solutions, interactive applets, flashcards, video clips, and much more, it's new, it's innovative, it's a must have!

NEW TO THE EIGHTH EDITION

Improved Pedagogical Structure

The enhanced pedagogy makes it easier to navigate the text. More examples are called out for students, more key terms are in definition boxes, and more key theorems, procedures, and rules have been identified.

New Examples

Each chapter offers two to three new examples, all based on real-world scenarios of particular interest to students, such as:

- Does Running Lead to Winning in Football? (Chapter 6, Example 9).
- Simple Interest on a Student Loan (Chapter 21, Example 1).
- Chaos in Manhattan (Chapter 23, Example 8).

Examples provide new topics for class discussion and new ways of relating to essential concepts.

New Exercises

- **All exercise sets** have been updated and refreshed.
- **End-of-chapter exercises** now include the correlating section number, making it easier to assign homework and create tests.
- The **skills check exercises** have been updated and now include multiple choice and fill-in-the-blank questions, providing a greater variation in assessment formats.

New MathPortal

Available packaged with *For All Practical Purposes*, Eighth Edition, or for purchase online. **MathPortal** brings together *For All Practical Purposes*, Eighth Edition, and its media in one affordable, easy-to-use learning space that offers a range of assessment and course management features. It is organized into three main areas:

- **The FAPP interactive eBook**
- **Resources**
- **Assignment Center**

Content Update Highlights

PART I Management Science

- **Enhanced** presentation of the material related to the Four Color Theorem in Chapter 3.

PART II Statistics

- **New** Technology Spotlights were added to aid in calculating standard deviation, the 5-number summary, correlation, line of best fit, combinatorics, and factorials (covering graphing, scientific, and nonscientific calculators).
- **Revised:** Key formulas are now provided in both computational and conceptual versions (Standard Deviation in Chapter 5, and Correlation in Chapter 6).

- **Expanded coverage** of sample space (e.g., tree diagrams), combinatorics (e.g., counting when order does not matter), probability rules (e.g., multiplication rule for independent events) and descriptive statistics (e.g., mode change).
- Notation used is **more explicitly** explained and is aligned with already-familiar notation, e.g., line of fit uses notation from algebra class: $y = mx + b$.

PART III Voting and Social Choice

- Arrow's Impossibility Theorem to Organ Transplant Policy example has been **expanded.**
- **New** discussion of the National Popular Vote law.
- **Expansion of coverage** on the discussion of positioning in presidential primaries.

PART IV Fairness and Game Theory

- **New section** on Fair Division and Organ Transplant Policies.
- **New** discussion of the work of the winners of the 2005 and 2007 Nobel Prize in Economics.

PART V The Digital Revolution

- **New** discussion of 13-digit ISBN number and new postal Bar Code.
- **New** spotlights on Bar Coding DNA and on Morse Code.
- **New** sub-section on Message Routing.
- **New** error detection method.
- **Updated** material on Web Searches.

PART VI On Size and Growth

- **New** Spotlight: Fitness Test.
- **New** Discussions of rotation symmetry and of vertex type.
- **New** Example: A Group of Non-Numbers.

PART VII Your Money and Resources

- **Updated** treatment of Simple Interest on a Student Loan.
- **Updated** formulae for compound interest, saving and payment.
- **New** sub-section on Real Growth Under Inflation.
- **Updated** real world data.
- **New** Spotlight on The Mortgage Crisis.
- **Expanded** treatment of Chaos.

Focus on Accuracy

For this edition we once again implemented a detailed accuracy checking plan to sustain the quality of the exercises and improve the solutions. To this end, we are very grateful to **John Samons of Florida Community College at Jacksonville.** He tirelessly worked with the authors to ensure accuracy in this edition of the text. John once again collaborated with the supplements author, **Heidi Howard of Florida Community College at Jacksonville**, to ensure both accuracy and consistency between the text and supplements package.

We are also grateful to **Scott Inch of Bloomsburg University, Rosalie Abraham of Florida Community College at Jacksonville**, and **Paul Lorczak** for their participation in a detailed line edit review of the eighth edition.

NEW CUSTOM OPTIONS

In addition to the extensive topics covered in the text, more traditional chapters (including **Problem Solving, Sets, Logic, Geometry, Counting and Probability, Numeration Systems and Personal Finance**) are available with *FAPP* through custom publishing. For more information, please contact your W. H. Freeman representative or go to www.whfreeman.com/fapp8e. Restrictions apply.

MEDIA AND SUPPLEMENTS

The media and supplements package for the eighth edition has been updated to reflect changes in the book. Both instructors and students will benefit from the innovative materials available to them.

Student Resources

NEW! **MathPortal**

http://courses.bfwpub.com/fapp8e (Access code required. Available packaged with *For All Practical Purposes,* Eighth Edition, or for purchase online.)
MathPortal brings together the text *For All Practical Purposes,* Eighth Edition, and its media in one affordable, easy-to-use learning space that offers a range of assessment and course management features. It is organized into three main areas:

- **The FAPP interactive eBook:** integrates a complete and customizable online version of the text with all of its media resources. Students can quickly search the text, and can personalize the eBook with highlighting, bookmarking, and note-taking features. Instructors can add, hide, and reorder content, integrate their own material, and highlight key text.
- **Resources:** organizes all student and instructor resources in one easily searchable location. Resources include self quizzes, interactive applets, flashcards, and games as well as video clips, news feeds, projects, and PowerPoint® sets.
- **Assignment Center:** organizes assignments and guides instructors through an easy-to-create assignment process. Exercises come from the Test Bank, Web Quizzes, and the text, and include many algorithmic problems.

NEW! ***For All Practical Purposes*, Eighth Edition, eBook**

The complete eBook is also available stand-alone, outside of MathPortal, at approximately one-half the cost of the printed textbook.

NEW! **Online Study Center**

www.whfreeman.com/osc/fapp8e (Access code or online purchase required.)
This premium Web-based study alternative helps make study time supremely efficient. Students take a pre-chapter Self-Test that generates a Personalized Study

Plan linking to the online resources relevant to the questions they missed. Instructors have access to an easy-to-manage gradebook and all media resources to help them track student progress and prepare lectures or course Web pages.

Student Study Guide, ISBN: 1-4292-2650-1
Heidi Howard, Florida Community College at Jacksonville
Offers study tips and tools to help students gain a better understanding of course material.

Student Solutions Manual, ISBN: 1-4292-2646-3
Heidi Howard, Florida Community College at Jacksonville
Contains full, worked solutions to the odd-numbered problems in the text.

Book Companion Site: www.whfreeman.com/fapp8e
The complimentary site provides students with access to study tools and instructors a range of assessment, presentation, and course management resources.

Instructor Resources

Instructor's Manual with Full Solutions, ISBN: 1-4292-2649-8
Heidi Howard, Florida Community College at Jacksonville
Includes teaching support for each chapter *and* full solutions for all problems in the text.

Teaching Guide for First-Time Instructors, ISBN: 1-4292-2645-5
Heidi Howard, Florida Community College at Jacksonville
This guide for new instructors, adjuncts, and teaching assistants will help make planning your course and teaching with *FAPP* easier and more effective. Ideas set forth in this guide also offer fresh perspective and ideas to experienced instructors.

Enhanced Instructor's Resource CD-ROM (IRCD), ISBN: 1-4292-2654-4
Created to help instructors develop lecture presentations, course Web sites, and other resources, this CD-ROM allows instructors to **search** and **export** all the resources contained below by key term or chapter:

- All text images.
- Applets, movies, flashcards, spreadsheet projects, self-quizzes available on the Web site.
- Instructor's Manual with Full Solutions.

Assessment

Test Bank

CD-ROM (Windows and Macintosh): 1-4292-2653-6
Printed: 1-4292-2651-X
John Emert, Ball State University
The *Test Bank* offers 75 multiple-choice and fill-in-the-blank questions and 35 short-answer questions per chapter. The easy-to-use CD includes Windows and Macintosh versions on a single disc, in a format that lets you add, edit, and re-sequence questions to suit your needs.

Course Management

WebCT and Blackboard
All the book's Web and testing materials are compatible with WebCT and Blackboard. We offer the electronic content as a service to adopters; please contact your local sales representative.

ACKNOWLEDGEMENTS

For All Practical Purposes continues to evolve in great part due to our many friends and colleagues who have offered suggestions, comments, and corrections. We are grateful to them all.

Rosalie Abraham, *Florida Community College at Jacksonville*
Alison Ahlgren, *University of Illinois at Urbana-Champaign*
Scott Balcomb, *St. Joseph's University*
Nancy Balle, *Ball State University*
Richard Bedient, *Hamilton College*
Rebecca Bergs, *Ball State University*
Terence R. Blows, *Northern Arizona University*
Raouf N. Boules, *Towson University*
Kristina K. Bowers, *Florida State University*
Terry Boyd, *University of Indianapolis*
Linda Braddy, *East Central University*
Barry Brunson, *Western Kentucky University*
Paul Buckelew, *Oklahoma City Community College*
Annette M. Burden, *Youngstown State University*
Shana Calaway, *Shoreline Community College*
Tim Carroll, *Eastern Michigan University*
G. Andy Chang, *Youngstown State University*
Yi Cheng, *Indiana University South Bend*
Leo Chouinard, *University of Nebraska–Lincoln*
Karen Clark, *Tacoma Community College*
Valerie Morgan-Crick, *Tacoma Community College*
Greg Crow, *Point Loma Nazarene University*
Sloan Despeaux, *Western Carolina University*
Rob Donnelly, *Murray State University*
Daniel Dreibelbis, *University of North Florida*
Gina Poore Dunn, *Lander University*
Nancy Eaton, *University of Rhode Island*
Kristy J. Eisenhart, *Western Michigan University*
John W. Emert, *Ball State University*
Sandra Fillebrown, *Saint Joseph's University*
Joseph Fox, *Salem State College*
W. Bart Frye, *Ball State University*
Martha Gady, *Whitworth College*
Monica Pierri-Galvao, *Gannon University*
Steve Gendler, *Clarion University*
Marty Getz, *University of Alaska, Fairbanks*
Carol E. Gibbons, *Salve Regina University*
T. R. Hamlett, *East Central University*
Geoffrey Hagopian, *College of the Desert*
Mohammad Halim, *Ball State University*
Frederick Hoffman, *Florida Atlantic University*
Michael Hull, *Northern Arizona University*
Scott Inch, *Bloomsburg University of Pennsylvania*
Peter Johnson, *Auburn University*
W. T. Kiley, *George Mason University*
Julie Killingbeck, *Ball State University*
Nancy Kitt, *Ball State University*
Samuel Kohn, *Thomas Edison State College*
Kathy Lewis, *State University of New York, Oswego*
Monica Liddle, *State University of New York, Delhi*
Jay Malmstrom, *Oklahoma City Community College*
Barbara Margoulius, *Cleveland State University*
Vania Mascioni, *Ball State University*
Mary T. McMahon, *North Central College*
Christopher McCord, *University of Cincinnati*
Ricardo Moena, *University of Cincinnati*
Steve Morics, *University of Redlands*
Dean Morrow, *Washington and Jefferson College*
Anne Marie Mosher, *St. Louis Community College*
Ellen Mulqueeny, *Cleveland State University*
Mika Munakata, *Montclair State University*
Chris Oehrlein, *Oklahoma City Community College*
Steven Ohs, *Western Michigan University*
Patricia Parkison, *Ball State University*
Deb Pearson, *Ball State University*
Andrew B. Perry, *Springfield College*
Marilyn Reba, *Clemson University*
Leo Robinson, *Ball State University*
Chris Rodger, *Auburn University*
Jennifer Marie Rodin, *University of South Carolina Aiken*
Robin Ruffato, *Ball State University*
Daniel Russow, *Arizona Western University*
Steven Schecter, *North Carolina State*
Brian Siebenaler, *Ball State University*
Debora J. Simonson, *University of North Florida*
Samuel Bruce Smith, *St. Joseph's University*
Patricia Stanley, *Ball State University*
James D. Stoops, *Ball State University*
William R. Stout, *Salve Regina University*
Tamas Szabo, *Weber State University*
Robert Terrell, *Cornell University*
Helen Thorwarth, *Northern Kentucky University*
Aaron K. Trautwein, *Carthage College*
David Urion, *Winona State University*
Bonnie Wachhaus, *Messiah College*
W. D. Wallis, *Southern Illinois University*
Kim Ward, *Eastern Connecticut State University*
John Weglarz, *Kirkwood Community College*
Gideon Weinstein, *Montclair State University*
Cheryl Whitelaw, *Southern Utah University*
Liz Whittern, *Ball State University*
Scott Wilde, *Baylor University*
Meredith Wort, *East Central University*
Mingqing Xiao, *Southern Illinois University*
Christian Yankov, *Eastern Connecticut State University*
Janet Yi, *Ball State University*
Laurie Margaret Zack, *High Point University*
John Zerger, *Catawba College*
Cathleen M. Zucco-Teveloff, *Rowan University*

We owe our appreciation to the people at W. H. Freeman and Company who participated in the development and production of this edition. We wish especially to thank the editorial staff for their tireless efforts and support. Among them are Craig Bleyer, Senior Publisher; Ruth Baruth, Publisher; Terri Ward, Senior Acquisitions Editor; Laura Capuano, Associate Editor; Vivien Weiss, Project Editor; Blake Logan, Cover Designer; Christine Buese, Photo Editor; Roland Cheney, Senior Media Editor; and Ellen Cash, Director of Production. We would also like to extend our appreciation to outside Development Editor, Lisa Collette.

The efforts of the COMAP staff must also be recognized. We thank our production staff, George Ward, Joyce Barnes, and Daiva Chauhan.

To everyone who helped make our purposes practical, we offer our appreciation for an exciting and exhilarating time.

Solomon Garfunkel, COMAP

Management Science PART I

On November 7, 2007 the space shuttle Discovery returned home from a highly successful, if not exactly trouble free 15-day mission. The National Aeronautics and Space Administration (NASA) had reason to be proud of the crew's accomplishments. This shuttle flight was supporting one of mankind's most ambitious ventures on the last frontier of human exploration–the construction of a space station in Earth orbit. This space station is not only a laboratory for new technologies that are being pioneered for ventures into space: It also tests the ability of human beings to live and work in space for extended periods of time. The mission made the history books when a team of women astronauts served as commanders of Discovery and the Space Station for the first time. The cool-headed leaders of this mission, supported by a team of experts back on Earth, scheduled a successful (but not initially planned) space walk to deal with a tear in one of the two "solar wings" designed for the space station.

This flexibility to handle the unexpected illustrated a hidden mathematical story behind this mission. Missions of this complexity require planning, scheduling, resource allocation, and cost minimization on a vast scale. To get a feel for the size of this project, some estimate its cost at about $100 billion. The branch of mathematics concerned with helping governments, businesses, and individuals operate as efficiently as possible is known as **management science** or **operations research (OR)**. The tools of this subject (graph theory, linear algebra, probability theory, and so forth) have evolved over more than a 100 years and often build on simple but clever mathematical ideas. However, it was only about 50 years ago that OR was identified as a distinct branch of knowledge and the systematic study of the subject began.

The chapters that follow chart a wide variety of ways that mathematical ideas make our lives more enjoyable and satisfying. At the same time these ideas challenge us with "simple puzzles" that are just fun to think about. ■

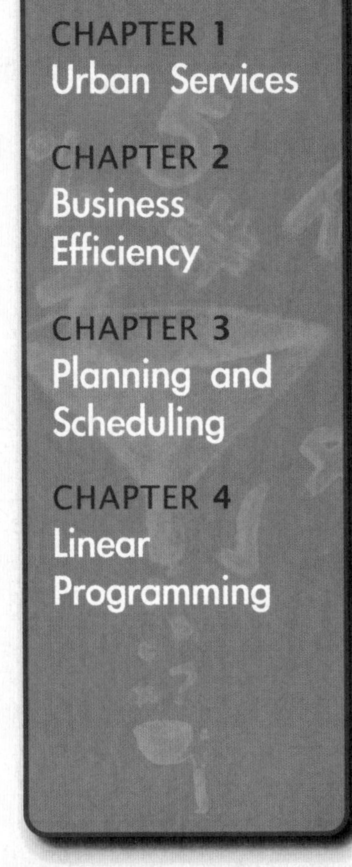

Urban Services

The underlying theme of management science, also called **operations research**, is finding the best method for solving some problem–what mathematicians call the **optimal solution**. In some cases, the goal may be to finish a job as quickly as possible. In other situations, the objective might be to maximize profit or minimize cost. In this chapter, our goal is to save time in traversing a street network while checking parking meters, delivering mail, removing snow, collecting bottles for recycling, or inspecting for potholes.

Let's begin by assisting the parking department of a city government. Most cities and many small towns have parking meters that must be regularly checked for parking violations or emptied of coins. We will use an imaginary town to show how management science techniques can help to make parking control more efficient.

1.1 Euler Circuits

The street map in Figure 1.1 is typical of many towns across the United States, with streets, residential blocks, and a town park. Our job, or that of the commissioner of parking, is to find the most efficient route for the parking-control officer, who travels on foot, to check the meters in an area. Efficient routes save money. Our map shows only a small area, allowing us to start with an easy problem. But the problem occurs on a larger scale in all cities and towns and for larger areas. The bigger the region involved, the greater the potential for cost savings.

The commissioner has two goals in mind: (1) The parking-control officer must cover all the sidewalks that have parking meters without retracing any more steps than are necessary; and (2) the route should end at the same point at which it began, perhaps where the officer's patrol car is parked. To be specific, suppose there are only two blocks that have parking meters, the two lightly shaded blocks that are side by side toward the top of Figure 1.1. Suppose further that the parking-control officer must start and end at the upper left corner of the left-hand

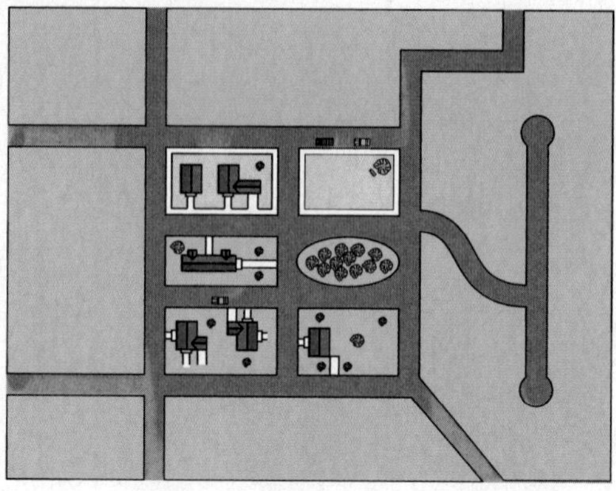

FIGURE 1.1 A street map for part of a town.

block. You might enjoy working out some routes by trial and error and evaluating their good and bad features. We are going to leave this problem for the moment and establish some concepts that will give us a better method than trial and error to deal with this problem.

Describing a Graph DEFINITION

A **graph** is a finite set of dots and connecting links. The dots are called **vertices** (a single dot is called a **vertex**), and the links are called **edges**. Each edge must connect two different vertices. A **path** is a connected sequence of edges showing a route on the graph that starts at a vertex and ends at a vertex; a path is usually described by naming in turn the vertices visited in traversing it. A path that starts and ends at the same vertex is called a **circuit**. A graph can represent our city map, a communications network, a system of air routes, or electrical power lines.

EXAMPLE 1 ▪ Parts of a Graph

We can use the graph in Figure 1.2 to help explain these technical terms. The graph shown has 5 vertices and 8 edges. The vertices represent cities, and the edges represent nonstop airline routes between them. We see that there is a nonstop flight between Berlin and Rome, but no such flight between New York and Berlin. There are several paths that describe how a person might travel with this airline from New York to Berlin. The path that seems most direct is New York, London, Berlin, but New York, Miami, Rome, Berlin is also a path. We can describe these two paths as *NLB* and *NMRB*. Another path would be New York, Miami, Rome, London, Berlin, which can be written as *NMRLB*. An example of a circuit is Miami, Rome, London, Miami. It is a circuit because the path starts and ends at the same vertex. This circuit can best be described in symbols by *MRLM*. Another example of a circuit in this graph would be *LRBL*, which is the circuit involving the cities London, Rome, Berlin, and back to London. In this chapter, we are especially interested in circuits, just as we are in real life. Most of us end our day in the same location where we start it–at home!

Notice that the edges *MB* (which could also be denoted *BM*) and *RL* shown in Figure 1.2 meet at a point that has no label. Furthermore, this point does not have a dark dot. This is because this point does not represent a vertex of our graph; it does not represent a city. It arises as an "accidental" consequence of the way this diagram has been drawn. We could join *M* and *B* with a curved line segment so that

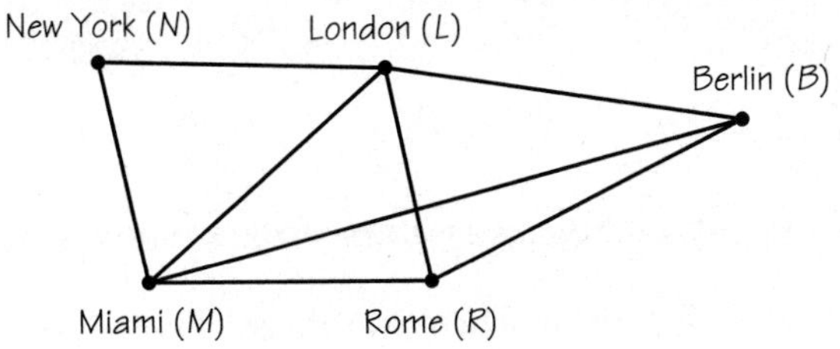

FIGURE 1.2 The edges of the graph show nonstop routes that an airline might offer.

the edges *LR* and *MB* do not cross, or redraw the diagram so as to avoid a crossing in this case (but not in all graphs we might wish to draw). We will be working often in situations where graphs can be drawn without accidental crossings and we will try to avoid such crossings when it is convenient to do so.

Returning to the case of parking control in Figure 1.1, we can use a graph to represent the whole territory to be patrolled: Think of each street intersection as a vertex and each sidewalk that contains meters as an edge, as in Figure 1.3. Notice in Figure 1.3b that the width of the street separating the blocks is not explicitly represented; it has been shrunk to nothing. In effect, we are simplifying our problem by ignoring any distance traveled in crossing streets.

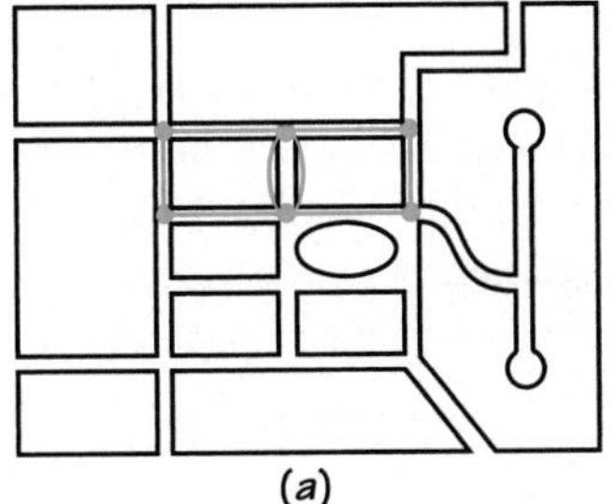

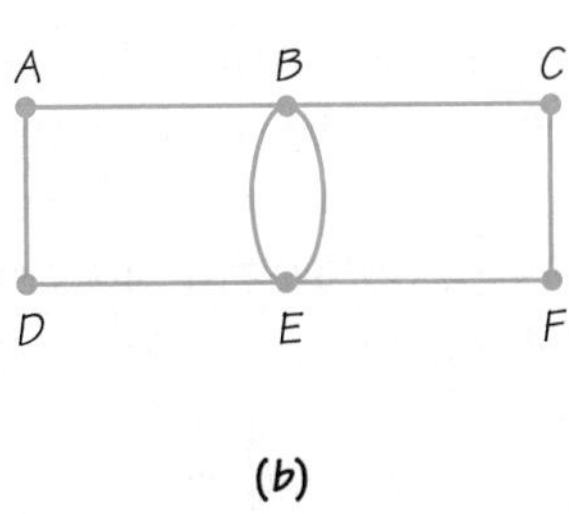

FIGURE 1.3 (a) A graph superimposed upon a street map. The edges show which sidewalks have parking meters. (b) The same graph enlarged.

The numbered sequence of edges in Figure 1.4a shows one circuit that covers all the meters (note that it is a circuit because its path returns to its starting point). However, one edge is traversed three times. Figure 1.4b shows another solution that is better because its circuit covers every edge (sidewalk) exactly once. In Figure 1.4b, no edge is covered more than once, or *deadheaded* (a term borrowed from shipping, which means making a return trip without a load).

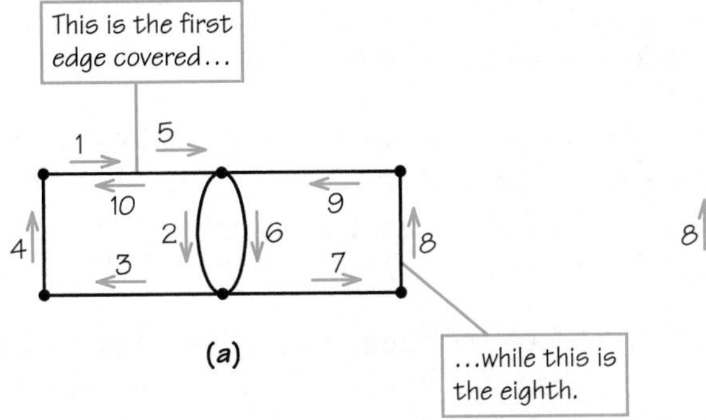

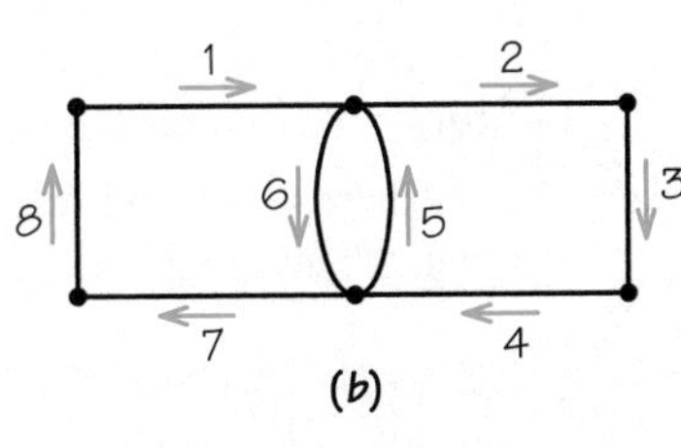

FIGURE 1.4 (a) A circuit and (b) an Euler circuit.

Euler Circuit DEFINITION

A circuit that covers each edge of a graph once but not more than once is called an **Euler circuit**.

Figure 1.4b shows an Euler circuit. These circuits get their name from the great eighteenth-century mathematician Leonhard Euler (pronounced *oy' lur*), who first studied them (see Spotlight 1.1). Euler was the founder of the theory of graphs, or graph theory. One of his first discoveries was that some graphs have no Euler circuits at all. For example, in the graph in Figure 1.5b, it would be impossible to start at one point, return to that starting vertex and cover all the edges without retracing some steps: If we try to start a circuit at the leftmost vertex, we discover that once we have left the vertex, we have "used up" the only edge meeting it. We have no way to return to our starting point except to reuse that edge. But this is not allowed in an Euler circuit. If we try to start a circuit at one of the other two vertices, we likewise can't complete it to form an Euler circuit.

As mentioned in Spotlight 1.2, realistic problems of this type will involve larger neighborhoods that might require the use of a computer. In addition, there may be other complications that might take us beyond the simple mathematics we want to stick to.

Because we are interested in finding circuits, and Euler circuits are the most efficient ones, we will want to know how to find them. If a graph has no Euler circuit, we will want to develop the next best circuits, those having minimum deadheading. These topics make up the rest of this chapter.

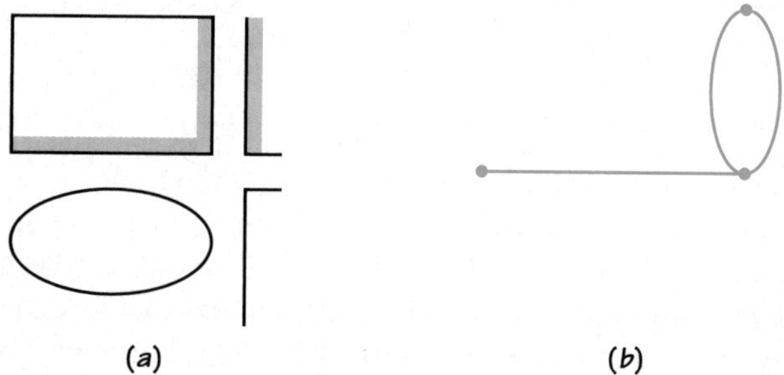

FIGURE 1.5 (a) The three shaded sidewalks cannot be covered by an Euler circuit. (b) The graph of the shaded sidewalks in part (a).

1.2 Finding Euler Circuits

Now that we know what an Euler circuit is, we are faced with two obvious questions:

1. Is there a way to tell by calculation or logical reasoning, not by trial and error, if a graph has an Euler circuit?
2. Is there a method, other than trial and error, for finding an Euler circuit when one exists and finding it quickly?

Loosely speaking, the first question lies within the concerns of mathematicians because it asks whether or not a certain problem admits a solution. Typically, the second question lies in the domain of computer science because it concerns finding the actual answer to a complex version of a problem in a short enough time to be useful.

SPOTLIGHT 1.1 Leonhard Euler

Leonhard Euler (1707–1783) was one of those rare individuals who was remarkable in many ways. He was extremely prolific, publishing over 500 works in his lifetime. But he wasn't devoted just to mathematics; he was a people person, too. He was extremely fond of children and had thirteen of his own, of whom only five survived childhood. It is said that he often wrote difficult mathematical works with a child or two in his lap.

Human interest stories about Euler have been handed down through two centuries. He was a prodigy at doing complex mathematical calculations under less than ideal conditions, and he continued to do them even after he became totally blind later in life. His blindness diminished neither the quantity nor the quality of his output. Throughout his life, he was able to mentally calculate in a short time what would have taken ordinary mathematicians hours of pencil-and-paper work. A contemporary claimed that Euler could calculate effortlessly, "just as men breathe, as eagles sustain themselves in the air." His collected works are not yet fully published.

Leonhard Euler

(Portrait by Emanuel Handmann, Bildnis des Mathematikers, 1753, Oeffentliche Kunstsammlung Basel, Kunstmuseum.)

Euler invented the idea of a graph in 1736 when he solved a problem in "recreational mathematics." He showed that it was impossible to stroll a route visiting the seven bridges of the German town of Königsberg exactly once. Ironically, in 1752 he discovered that three-dimensional polyhedra obey the remarkable formula $V - E + F = 2$ (that is, number of vertices − number of edges + number of faces = 2) but failed to give a proof because he did not analyze the situation using graph theory methods.

Euler investigated these questions in 1735 by using the concepts of **valence** and **connectedness**.

Valence — DEFINITION

The **valence** of a vertex in a graph is the number of edges meeting at the vertex.

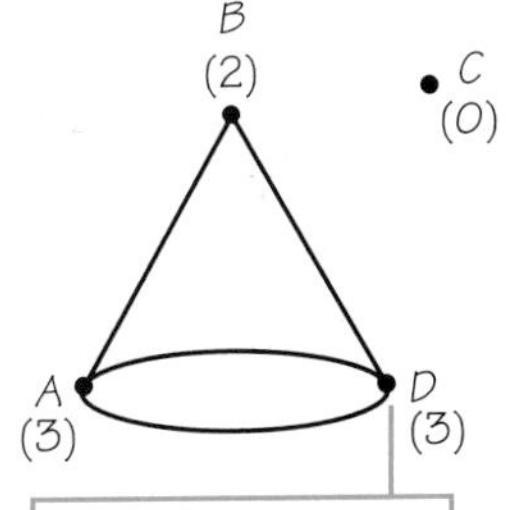
B
(2)
C
(0)
A
(3)
D
(3)
The valence is 3 because three edges meet at D.

FIGURE 1.6
Valences of vertices.

Figure 1.6 illustrates the concept of valence, with vertices A and D having valence 3, vertex B having valence 2, and vertex C having valence 0. Isolated vertices such as vertex C are an annoyance in Euler circuit theory. Because they don't occur in typical applications, we henceforth assume that our graphs have no vertices of valence 0.

Figure 1.3b has four vertices of valence 2, namely, A, C, F, and D. This graph also has two vertices, B and E, of valence 4. Notice that each vertex has a valence that is an even number. We'll soon see that this is very significant.

Connected Graph — DEFINITION

A graph is said to be **connected** if for every pair of its vertices there is at least one path connecting the two vertices.

SPOTLIGHT 1.2 The Human Aspect of Problem Solving

Thomas Magnanti, professor of operations research and management, heads the Department of Management Science at MIT's Sloan School of Management. Here are some of his observations:

Thomas Magnanti
(Courtesy of Thomas Magnanti.)

Typically, a management science approach has several different ingredients. One is just structuring the problem—understanding that the problem is an Euler circuit problem or a related management science problem. After that, one has to develop the solution methods.

But one should also recognize that you don't just push a button and get the answer. In using these underlying mathematical tools, we never want to lose sight of our common sense, of understanding, intuition, and judgment. The computer provides certain kinds of insights. It deals with some of the combinatorial complexities of these problems very nicely. But a model such as an Euler circuit can never capture the full essence of a decision-making problem.

Typically, when we solve the mathematical problem, we see that it doesn't quite correspond to the real problem we want to solve. So we make modifications in the underlying model. It is an interactive approach, using the best of what computers and mathematics have to offer and the best of what we, as human beings, with our own decision-making capabilities, have to offer.

Given a graph, if we can find even one pair of vertices not connected by a path, then we say that the graph is not connected. For example, the graph in Figure 1.7 is not connected because we are unable to join *A* to *D* with a path of edges. However, the graph does consist of two "pieces" or connected components, one containing the vertices *A, B, F,* and *G,* the other containing *C, D,* and *E.* A connected graph will contain a single connected component. Notice that the parking-control graph of Figure 1.3b is connected.

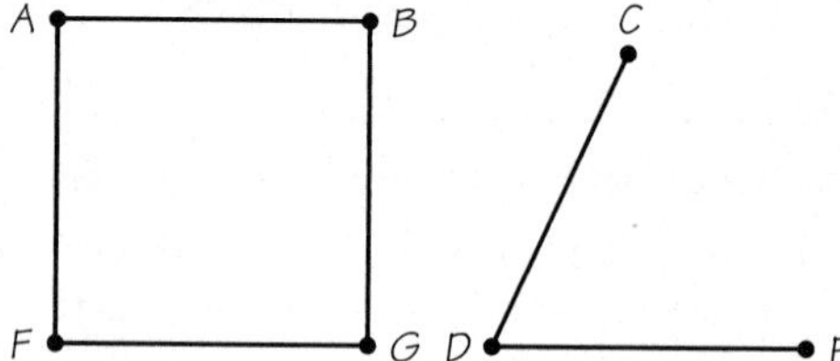

FIGURE 1.7 A nonconnected graph.

We can now state Euler's theorem, his simple answer to the problem of detecting when a graph *G* has an Euler circuit.

Euler Circuit Theorem — THEOREM

1. If *G* is connected and has all valences even, then *G* has an Euler circuit.
2. Conversely, if *G* has an Euler circuit, then *G* must be connected and all its valences must be even numbers.

Because the parking-control graph of Figure 1.3b conforms to the connectedness and even-valence conditions, Euler's theorem tells us that it has an Euler circuit. We already have found an Euler circuit for the graph shown in Figure 1.4b by trial and error. For a very large graph, however, trial and error may take a long time. It is usually quicker to check whether the graph is connected and even-valent than to find out if it has an Euler circuit.

Once we know there is an Euler circuit in a certain graph, how do we find it? Many people find that, after a little practice, they can find Euler circuits by trial and error, and they don't need detailed instructions on how to proceed. At this point you should see if you can develop this skill by trying to find Euler circuits in Figure 1.8a, Figure 1.9a, and Figure 1.10. In doing your experiments, draw your graph in ink and the circuit in pencil so you can erase if necessary.

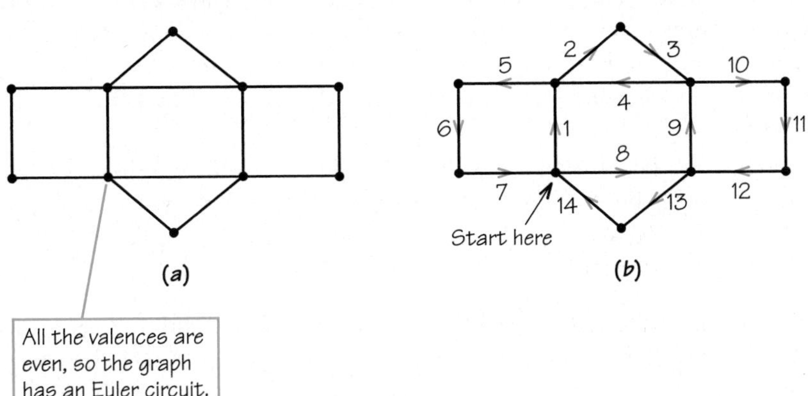

FIGURE 1.8 (a) A graph having (b) an Euler circuit.

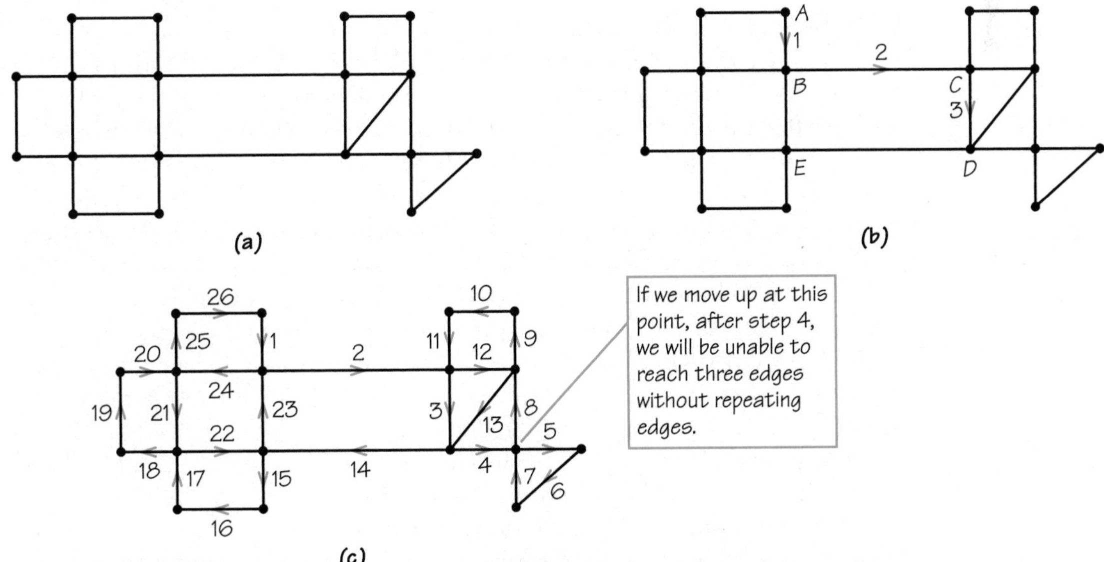

FIGURE 1.9 (a) A graph that has an Euler circuit. (b) A critical junction in finding an Euler circuit in this graph, starting from vertex *A*. (c) A description of a full Euler circuit for this graph.

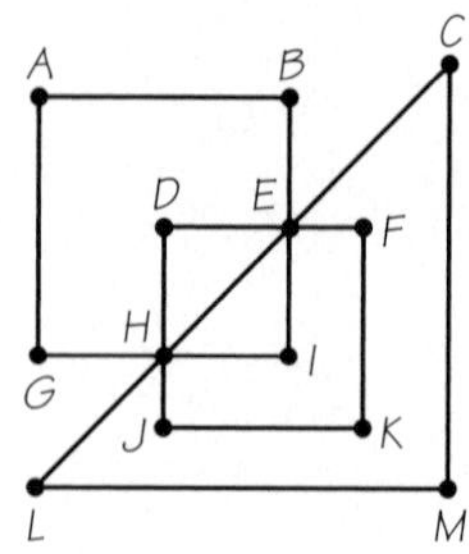

FIGURE 1.10 A graph with an Euler circuit.

If you would like more guidance on how to find an Euler circuit without trial and error, here is a method that works: Never use an edge that is the only link between two parts of the graph that still need to be covered. Figure 1.9b illustrates this. Here we have started the circuit at *A* and gotten to *D* via *B* and *C*, and we want to know what to do next. Going to *E* would be a bad idea because the uncovered part of the graph would then be disconnected into left and right portions. You will never be able to get from the left part back to the right part because you have just used the last remaining link between these parts. Therefore, you should stay on the right side and finish that before using the edge from *D* to *E*. This kind of thinking needs to be applied every time you need to choose a new edge.

Let's see how this works, starting at the beginning at *A*. From vertex *A* there are two possible edges, and neither of them disconnects the unused portion of the graph. Thus, we could have gone either to the left or down. Having gone down to *B*, we now have three choices, none of which disconnects the unused part of the graph. After choosing to go from *B* to *C*, we find that any of the three choices at *C* is acceptable. Can you complete the Euler circuit? Figure 1.9c shows one of many ways to do this.

The method just described leaves many edge choices up to you. When there are many acceptable edges for your next step, you can pick one at random.

EXAMPLE 2 ■ Finding an Euler Circuit

Check the valences of the vertices and the connectivity of the graph in Figure 1.8a to verify that the graph does have an Euler circuit. Now try to find an Euler circuit for that graph. You can start at any vertex. When you are done, compare your solution with the Euler circuit given in Figure 1.8b. If your path covers each edge exactly once and returns to its original vertex (is a circuit), then it is an Euler circuit, even if it is not the same as the one we give. ■

Proving Euler's Theorem

We'll start by proving that if a graph has an Euler circuit *R*, then it must have only even-valent vertices and it must be connected. Let *X* be any vertex of the graph. We will show that the edges at *X* can be paired up, and this will prove that the valence is even. Every edge at *X* is used by *R* as an outgoing edge (leaving from *X*) or an incoming edge (arriving at *X*). If the Euler circuit starts at *X*, then pair up the first edge used by *R* with the last one (when the circuit returns to *X* for the last time). In addition, each other edge at *X* that is used by the circuit as an incoming edge will be paired with the outgoing edge that is used next. Because all edges at *X* are used by the Euler circuit, none more than once, this pairs up the edges.

But what if *X* is not the start of the Euler circuit? Then do the pairing like this: The first incoming edge at *X* is paired with the outgoing one used next, the second incoming edge at *X* is paired with the outgoing one used next, and so on. For example, in Figure 1.11 at vertex *B*, we would pair up edges 2 and 3 and edges 9 and 10. At vertex *C*, we would pair up edges 4 and 5 and edges 8 and 9. Can you see how the pairings would work at *D*? How about vertex *A*?

In studying this particular example, you might think it would be simpler to count the edges at a vertex to see that the valence is even. True, but our pairing method works for a graph about which we know nothing except that it has an Euler circuit.

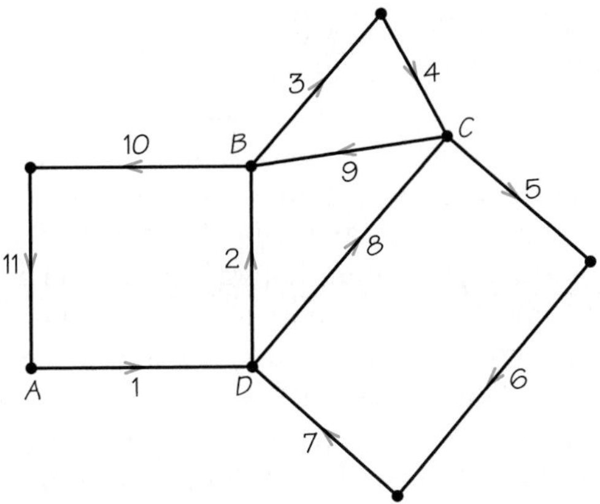

FIGURE 1.11 An Euler circuit starting and ending at A.

To see that a graph with an Euler circuit is connected, note that by following the Euler circuit around we can get from any edge to any other edge (it covers them all) using a portion of the Euler circuit. Because every vertex is on an edge (there are no vertices of valence 0), we can get from any vertex to any other using a portion of the Euler circuit.

So far, this is not a complete proof of Euler's theorem. To complete the proof we would need to prove that if a graph has all vertices even-valent and is connected, then an Euler circuit can be found for it.

1.3 Beyond Euler Circuits

Now let's see what Euler's theorem tells us about the three-block neighborhood with parking meters, represented by dots in Figure 1.12a. Figure 1.12b shows the corresponding graph. (Because we use edges to represent only sidewalks along which the officer must walk, the sidewalk with no meters is not represented by any edge in the graph.) This graph has vertices with odd valences (at vertices C and G), so Euler's theorem tells us that there is no Euler circuit for this graph.

Because we must reuse some edges in this graph to cover all edges in a circuit, for efficiency we need to keep the total length of reused edges to a minimum. This type of problem, in which we want to minimize the length of a circuit by carefully choosing which edges to retrace, is often called the **Chinese postman problem** (like parking-control routes, mail routes need to be efficient). The problem was first studied by the Chinese mathematician Meigu Guan in 1962–hence the name. The remainder of this chapter is dedicated to solving the Chinese postman problem and discussing applications beyond parking control.

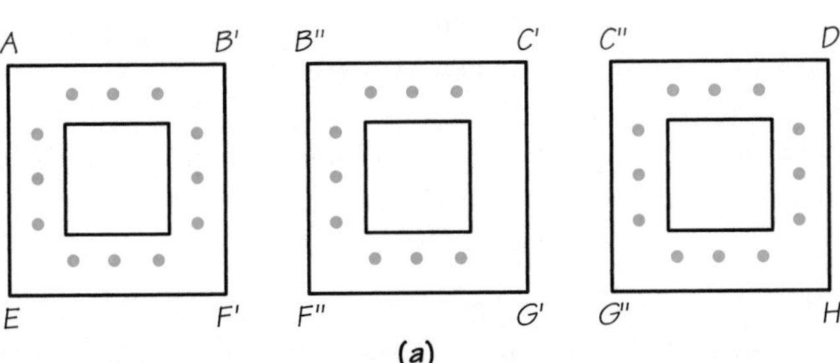

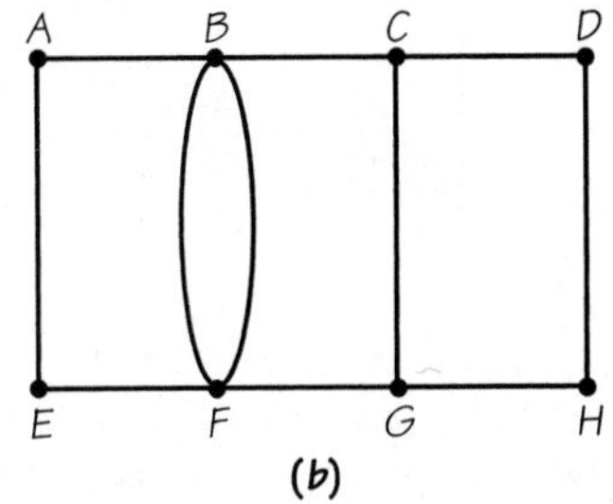

FIGURE 1.12 (a) A street network and (b) its graphic representation. Locations such as B' and B'', C' and C'', F' and F'', and G' and G'' are merged to form the vertices B, C, F, and G. The dots shown represent parking meters.

Solving the Chinese Postman Problem

In a realistic Chinese postman problem, we need to consider the lengths of the sidewalks, streets, or whatever the edges represent, because we want to minimize the total length of the reused edges. However, to simplify things at the start, we can suppose that all edges represent the same length. (This is often called the *simplified* Chinese postman problem.) In this case, we need only count reused edges and need not add up their lengths. To solve the problem, we want to find a circuit that covers each edge and that has the minimal number of reuses of edges already covered.

To follow the procedure we are going to develop, look at the graph in Figure 1.13a, which is the same graph as in Figure 1.12b. This graph has no Euler circuit, but there is a circuit that has only one reuse of an edge (*CG*), namely, *ABCD-HGCGFBFEA*. Let's draw this circuit so that when edge *CG* is about to be reused, we install a new, extra, blue edge in the graph for the circuit to use. By duplicating edge *CG,* we can avoid reusing the edge. To duplicate an edge, we must add an edge that joins the two vertices that are already joined by the edge we want to duplicate. (It makes no sense to join vertices that are not already connected by an edge, because such edges would not represent sidewalk sections with meters; see Figure 1.15.)

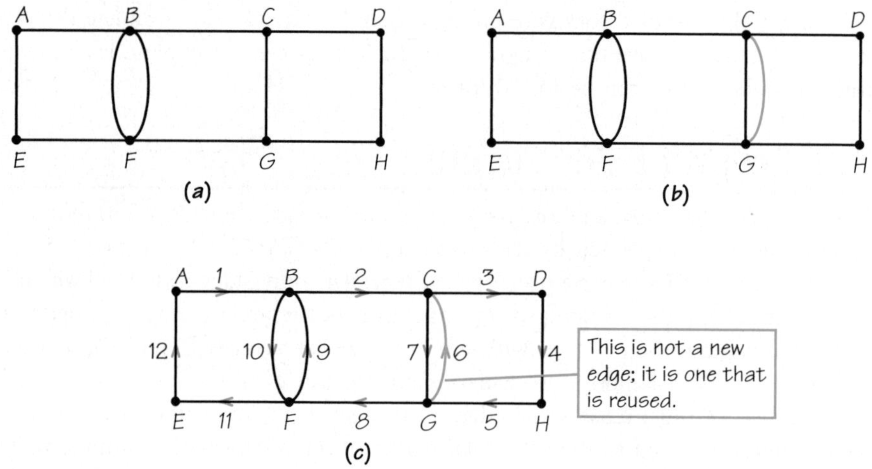

FIGURE 1.13 Making a circuit by reusing an edge.

We have now created the graph of Figure 1.13b. In the graphs we draw, the edges that are added will be shown in color to distinguish them from the original edges, which are shown in black. (You may want to use a similar scheme to help you remember which edges are the originals and which are duplicated in the graphs you draw.) In the graph of Figure 1.13b, the original circuit can be traced as an Euler circuit, using the new edge when needed. The circuit is shown in Figure 1.13c. Our theory will be based on using this idea in reverse, as follows:

1. Take the given graph and add edges by duplicating existing edges, until you arrive at a graph that is connected and even-valent. Note that after a graph is *eulerized,* the new graph produced will have an Euler circuit.
2. Find an Euler circuit on the eulerized graph.
3. "Squeeze" this Euler circuit from the eulerized graph onto the original graph by reusing an edge of the original graph each time the circuit on the eulerized graph uses an added edge.

Eulerizing a Graph DEFINITION

Adding edges that duplicate existing edges to a connected graph to make all valences even is called **eulerizing** the graph.

EXAMPLE 3 ■ Eulerizing a Graph

Suppose we want to eulerize the graph of Figure 1.14a. When we eulerize a graph, we first locate the vertices with odd valence. The graph in Figure 1.14a has two, *B* and *C*. Next, we add one end of an edge at each such vertex, matching up the new edge with an existing edge in the original graph. Figure 1.14b shows one way to eulerize the graph. Note that *B* and *C* have even valence in the second graph. After eulerization, each vertex has even valence. To see an Euler circuit on the eulerized graph in Figure 1.14c, simply follow the edges in numerical order and in the direction of the arrows, beginning and ending at vertex *A*. The final step, shown in Figure 1.14d, is to "squeeze" our Euler circuit onto the original graph. There are two reuses of previously covered edges. Notice that each reuse of an edge corresponds to an added edge.

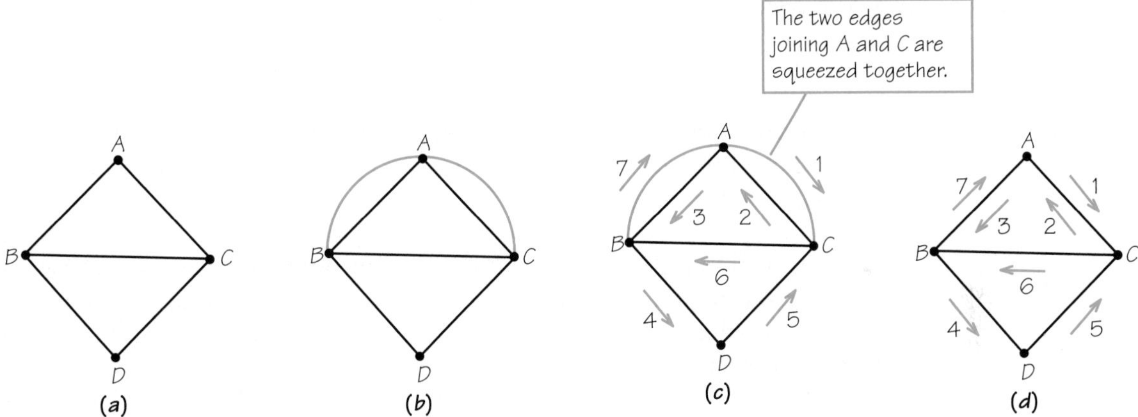

FIGURE 1.14 Eulerizing a graph.

In the previous example, we noticed that we could count how many reuses we needed by counting added edges. This is generally true in this type of problem: *If you add the new edges correctly, the number of reuses of edges equals the number of edges added during eulerization.*

Adding new edges correctly means adding only edges that are duplicates of existing edges. Doing this makes the rule, just stated in italics, always true, and so it is easy to count the needed reuses.

To see why we add only duplicate edges, examine Figure 1.15a. We need to alter the valences of vertices *X* and *Y* by adding edges so that they become even-valent. Adding one long edge from *X* to *Y* (Figure 1.15b) might seem like an attractive idea, but adding this edge is equivalent to asking a snowplow, say, to get from *X* to *Y* without moving along existing streets. At times it is necessary to traverse sections of the graph that have been previously traversed. This is the significance of the duplicated edges. Here the structure of the graph forces us to repeat some edges.

We cannot get away with fewer than three repeats–the three edges *XU, UV,* and *VY* (Figure 1.15c). The duplicated edges are shown in color.

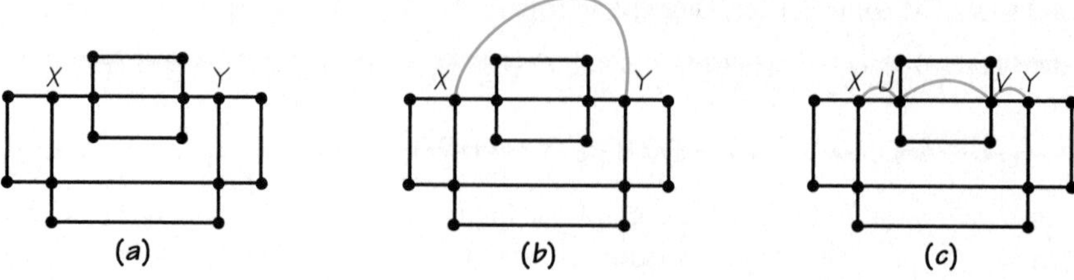

FIGURE 1.15 Eulerizing when the vertices are more than one edge apart.

Now that we have learned to eulerize, the next step is to try to get a best eulerization we can–one with the fewest added edges. It turns out that there are many ways to eulerize a graph. It is even possible that the smallest number of added edges can be achieved with two different eulerizations. This is the reason we use the phrase "a best eulerization" rather than "the best eulerization." Remember, we want a best eulerization because this enables us to find the circuit for the original graph that has the minimum number of reuses of edges.

EXAMPLE 4 ▪ A Better Eulerization

In Figure 1.16a, we begin with the same graph as in Figure 1.14, but we eulerize it in a different way–by adding only one edge (see Figure 1.16b). Figure 1.16c shows an Euler circuit on the eulerized graph, and in Figure 1.14d we see how it is squeezed onto the original graph. There is only one reuse of an edge, because we added one edge during eulerization.

The solution in Figure 1.16 is better than the solution in Figure 1.14 because one reuse is better than two. These examples suggest the following addition to our solution procedure: Try to find an eulerization with the smallest number of added edges. This extra requirement makes the problem both more interesting and more difficult. For large graphs, a best eulerization may not be obvious. We can try out a few and pick the best among the ones we find, but there may be an even better one that our haphazard search does not turn up.

A systematic procedure for finding a best eulerization does exist, but the process is complicated. There is an especially easy technique for eulerizing the following special category of networks often found in our neighborhoods.

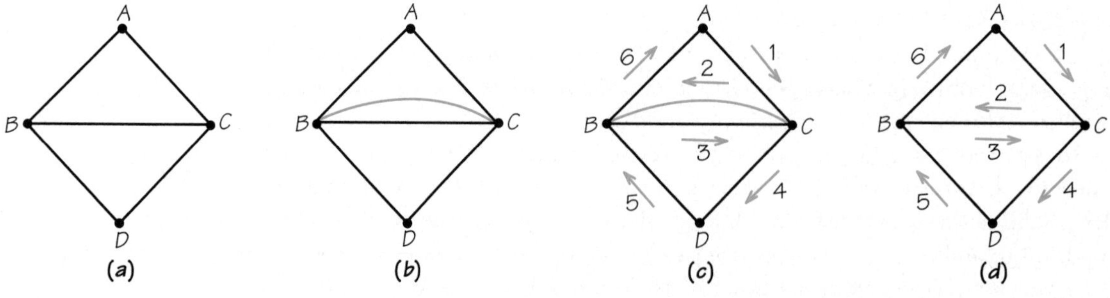

FIGURE 1.16 A better eulerization of Figure 1.14.

Rectangular Network

DEFINITION

If a street network is composed of a series of rectangular blocks that form a large rectangle a certain number of blocks high and a certain number of blocks wide, the network is called **rectangular**.

Examples of rectangular street networks (a 3-by-3, a 3-by-4, and a 4-by-4) are shown in Figure 1.17. The graph on the right in each pair shows a best eulerization for the rectangular street network on the left. There appear to be three different eulerization patterns, depending upon whether the rectangle height and width in the original graph are odd or even numbers. In Figure 1.17a, both lengths are 3, both odd; in Figure 1.17b, one length is odd (3) and one is even (4); in Figure 1.17c, both lengths are 4, an even number.

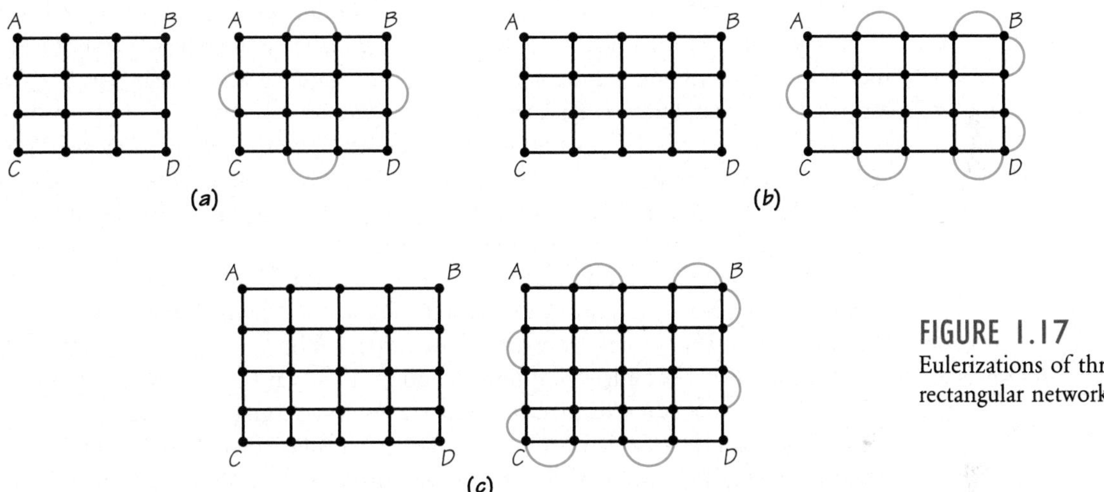

FIGURE 1.17 Eulerizations of three rectangular networks.

Although the patterns appear different, one technique can be used to create all of them. This technique can be thought of as involving an "edge walker" who walks around the outer boundary of the large rectangle in some direction, say, clockwise. He starts at any corner, say, the upper left corner. As he goes around, he adds edges by the following rules. When he comes to an odd-valent vertex, he links it to the next vertex with an added edge. This next vertex now becomes either even or odd. If it became even, he skips it and continues around, looking for an odd vertex. If it became odd (this could happen only at a corner of the big rectangle), the edge walker links it to the next vertex and then checks this vertex to see whether it is even or odd. Each of the three parts of Figure 1.17 has been eulerized by this method.

In a street network that is not rectangular, the eulerization process is started by locating all the vertices with odd valence and then pairing these vertices with each other and finding the length of the shortest path between each pair. We look for the shortest paths because each edge on the connecting paths will be duplicated. The idea is to choose the pairings cleverly so that the sum of the lengths of those paths is the smallest it can be. With a little practice, most people can find a best or nearly best eulerization using only this idea together with trial and error and some ingenuity.

Finding Good Eulerizations

Suppose we want a perfect procedure for eulerizing a graph. What theoretical ideas and methods could we use to build such a tool?

One building block we could use is a method for finding the shortest path between two given vertices of a graph. For example, let us focus on vertices *X* and *Y* in Figure 1.18a. Both have odd valence. We can connect them with a pattern of duplicate edges, as in Figure 1.18b. The cost of this is the length of the path we duplicated from *X* to *Y*. A shorter path from *X* to *Y*, such as the one shown in Figure 1.18c, would be better. Fortunately, the *shortest-path problem* has been well studied, and we have many good procedures for solving it exactly, even in large, complex graphs.

But there is more to eulerizing the graph in Figure 1.18a than dealing with *X* and *Y*. Notice that we have odd valences at *Z* and *W*. Should we connect *X* and *Y* with a path, and then connect *Z* and *W*, as in Figure 1.18d? Or should we connect *X* to *Z* and *Y* to *W*, as in Figure 1.18e? Another alternative is to use connections *X* to *W* and *Y* to *Z*, as in Figure 1.18f. It turns out that the alternatives in both Figures 1.18e and 1.18f are preferable to the one in Figure 1.18d, because they involve seven added edges, whereas Figure 1.18d uses nine.

We know there is a simple way to test whether a connected graph has an Euler circuit: Check to see if the graph is even-valent. Is there a very easy way to compute the number of edges in a best eulerization of a graph? Unfortunately, the answer is "no." However, there is a simple observation that often saves a lot of work. Suppose we count the number of odd-valent vertices in a graph. This number must always be an even number. When we duplicate an existing edge we can never change more than two odd-valent vertices to even-valent vertices. Thus, in a best eulerization of a graph, the number of edges that must be duplicated is at least the number of odd-valent vertices divided by two. If, for example, we have a graph with ten odd-valent vertices and we find an eulerization with five added edges, there may be other eulerizations which also have five duplicated edges, but there can be no eulerization with fewer than five duplicated edges.

Remember that when an unweighted graph is eulerized in an optimal way, then the total cost of traversing each edge at least once can be found by adding the total number of edges in the graph to the number of edges that are reused (duplicated). Small problems involving eulerization can be carried out by trial-and-error methods. Unfortunately, although there is an algorithm that can be applied to find the best eulerization for large problems, the details of this algorithm are quite complex. However, the procedure works quickly not only for graphs without weights but also for graphs with weights on the edges.

1.4 Urban Graph Traversal Problems

Euler circuits and eulerizing have many more practical applications than just checking parking meters. Almost anytime services must be delivered along streets or roads, our theory can make the job more efficient. Examples include collecting garbage, salting icy roads, plowing snow, inspecting railroad tracks for flaws, and reading electric meters (see Spotlight 1.3).

Each of these problems has its own special requirements that may call for modifications in the theory. For example, in the case of garbage collection, the edges of our graph will represent streets, not sidewalks. If some of the streets are one-way, we need to put arrows on the corresponding edges, resulting in a directed graph, or

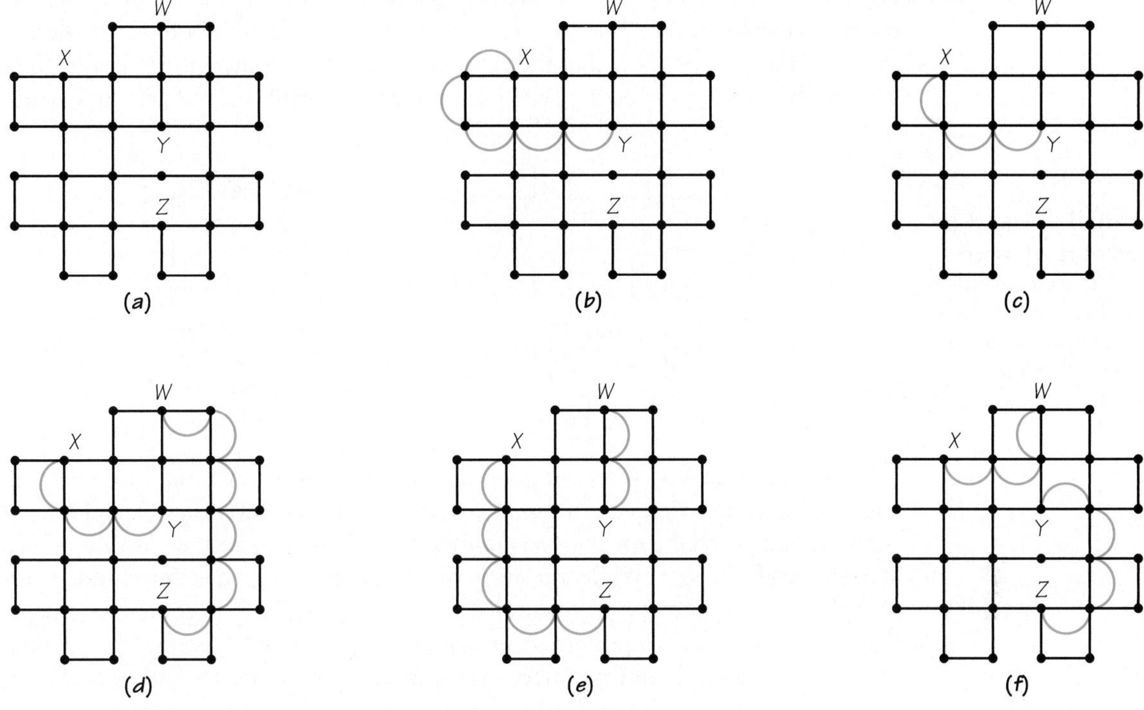

FIGURE 1.18 Choosing among eulerizations.

SPOTLIGHT 1.3 Israel Electric Company Reduces Meter-Reading Task

The Beersheba branch of Israel's major electric company wanted to make the job of meter reading more efficient. When the branch managers decided to minimize the number of people required to read the electric meters in the houses of one particular neighborhood, they set a precedent by applying management science. Formerly, each person's route had been worked out by trial and error and intuition, with no help from mathematics. The whole job required 24 people, each doing a part of the neighborhood in a five-hour shift.

At first, it looks as though one would find a more efficient way of doing the work the same way as in the Chinese postman problem, but there are two important differences. First, the neighborhood was big enough to negate any possibility of having only one route assigned to one person. Instead, it was necessary to find a number of routes that, taken together, covered all the edges (sidewalks). Second, a meter reader who was done with a route was allowed to return home directly. Thus, there was no reason for the individual routes to return to their starting points; therefore, routes could be paths instead of circuits.

The Beersheba researchers found solutions to these problems by modifying the basic ideas we have described in this chapter. They managed to cover the neighborhood with 15 five-hour routes, a 40% reduction of the original 24 five-hour routes. Altogether, these routes involve a total of 4338 minutes of walking time, of which 41 minutes (less than 1%) is deadheading.

digraph. The circuits we seek will have to obey these arrows. In the case of salt spreaders and snowplows, each lane of a street needs to be modeled as a directed edge, as shown in Figure 1.19. Note that the arrows on the map and digraph are not in color because these arrows denote restrictions in traversal possibilities, not parts of circuits.

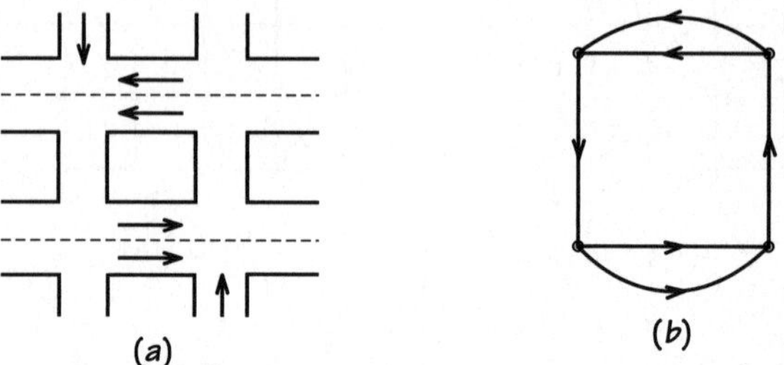

FIGURE 1.19 (a) Salt-spreading route, where each west–west street has two traffic lanes in the same direction, and (b) an appropriate digraph model.

Like salt spreaders, street-sweeping trucks can travel in only one lane at a time and must obey the direction of traffic. Street sweepers, however, have an additional complication: parked cars. It is very difficult to clean the street if cars are parked along the curb. Yet for overall efficiency, those who are responsible for routing street sweepers want to interfere with parking as little as possible. The common solution is to post signs specifying times when parking is prohibited. Because the parking-time factor is a constraint on street sweeping, it is important to find not only an Euler circuit, or a circuit with very few duplications, but also a circuit that visits streets when they are free of cars. Once again, mathematicians have developed techniques to handle this constraint.

Finally, because towns and cities of any size will have more than one street sweeper, parking officer, or garbage truck, a single best route may not suffice. Instead, it becomes necessary to divide the territory into multiple routes. The general goal is to find optimal solutions while taking into account traffic direction, number of lanes, parking-time restrictions, and divided routes (see Figure 1.20).

Management science makes all this possible. For example, a pilot study done in the 1970s in New York City showed that applying these techniques to street sweepers in just one district could save about $30,000 per year. With 57 sanitation districts in New York, this would amount to a savings of more than $1.5 million in a single year. This translates to about $5 million in 2008 dollars. In addition, the same principles could be extended to garbage collection, parking control, and other services carried out on street networks.

This plan was not adopted when first proposed. Because city services take place in a political context, several other factors come into play. For example, union leaders try to protect the jobs of city workers, bureaucrats might try to keep their departmental budgets high, and elected politicians rarely want to be accused of cutting the jobs of their constituents. Thus, political obstacles can overrule management science. As mentioned in Spotlight 1.2, such human factors often arise when applying management science. Perhaps a more acceptable street-sweeping plan would have been devised for New York City if more attention had been paid to the human factors earlier.

Despite the complications of real-world problems, management science principles provide ways to understand these problems by using graphs as models. We can reason about the graphs and then return to the real-world problem with a workable solution. The results we get can have a lasting effect on the efficiency and economic well-being of any organization or community.

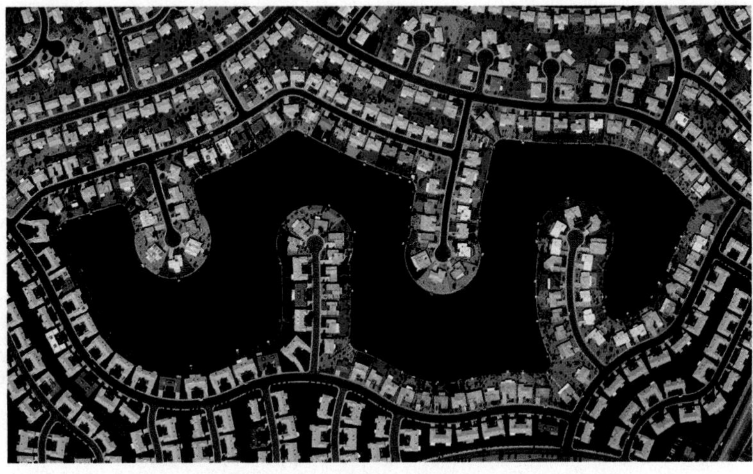

(a)

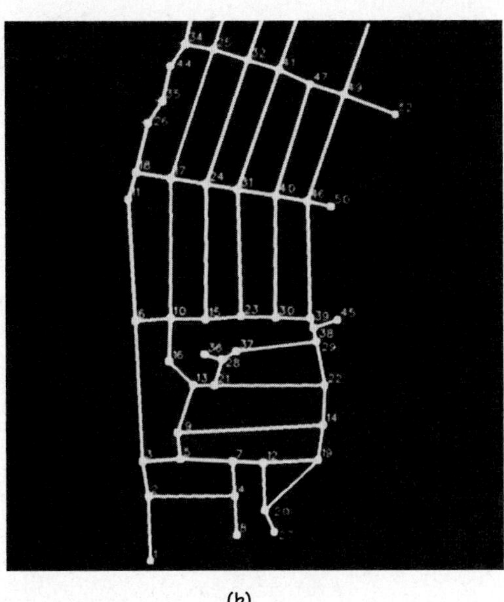

(b)

FIGURE 1.20 (a) Residential neighborhoods, whether they be in cities or the suburbs, require many services such as mail delivery, garbage collection, street sweeping, meter reading, or sewage systems. The mathematical techniques of operations research make it possible to provide these services as cheaply as possible. When optimal solutions to providing such services can be found, everyone is a winner. *(Brand X Pictures/ Picturequest.)* (b) Computers can be used to extract the essential information needed to solve routing problems from photographs.

REVIEW VOCABULARY

Chinese postman problem The problem of finding a circuit on a graph that covers every edge of the graph at least once and that has the shortest possible length. (p. 11)

Circuit A path that starts and ends at the same vertex. (p. 4)

Connected graph A graph is connected if it is possible to reach any vertex from any specified starting vertex by traversing edges. (p. 7)

Digraph A graph in which each edge has an arrow indicating the direction of the edge. Such directed edges are appropriate when the relationship is "one-sided" rather than symmetric (for instance, one-way streets as opposed to regular streets). (p. 18)

Edge A link joining two vertices in a graph. (p. 4)

Euler circuit A circuit that traverses each edge of a graph exactly once. (p. 6)

Eulerizing Adding new edges, which duplicate existing edges, to a connected graph so as to make a graph that possesses an Euler circuit. (p. 13)

Graph A mathematical structure in which points (called vertices) are used to represent things of interest and in

which links (called edges) are used to connect vertices, denoting that the connected vertices have a certain relationship. (p. 4)
Management science A discipline in which mathematical methods are applied to management problems in pursuit of optimal solutions that cannot readily be obtained by common sense. (p. 1)
Operations research (OR) Another name for management science. (p. 1)
Optimal solution When a problem has various solutions that can be ranked in preference order (perhaps according to some numerical measure of "goodness"), the optimal solution is the best-ranking solution. (p. 3)
Path A connected sequence of edges in a graph. (p. 4)
Valence (of a vertex) The number of edges touching that vertex. (p. 7)
Vertex A point in a graph where one or more edges end. (p. 4)

✓ SKILLS CHECK

1. What is the valence of vertex A in the graph below?

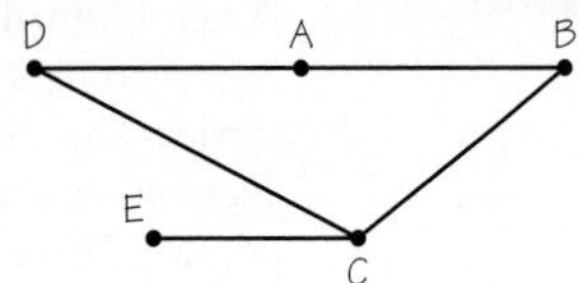

(a) 2
(b) 1
(c) 3

2. The number of vertices in the graph below is _____ , while the number of edges in this graph is _________ .

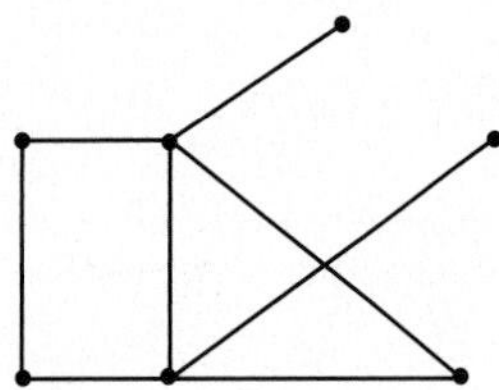

3. The valences of the vertices in the accompanying graph listed in non-increasing order are

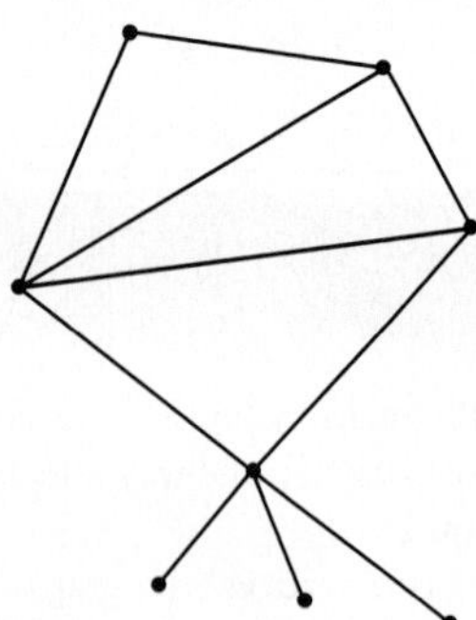

(a) 5, 4, 3, 3, 2, 1, 1, 1.
(b) 1, 3, 4, 4, 5, 5.
(c) 5, 5, 4, 3, 3, 1

4. The graph shown below is not connected because it consists of ________ parts.

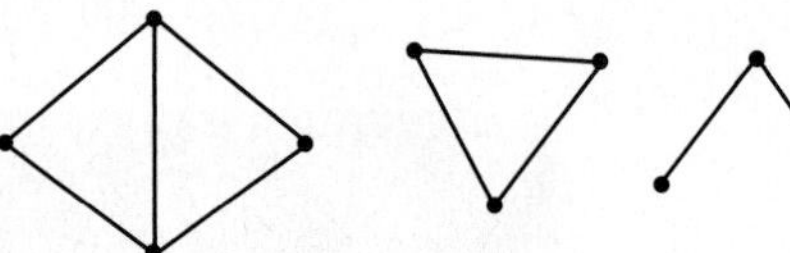

5. The graph below has

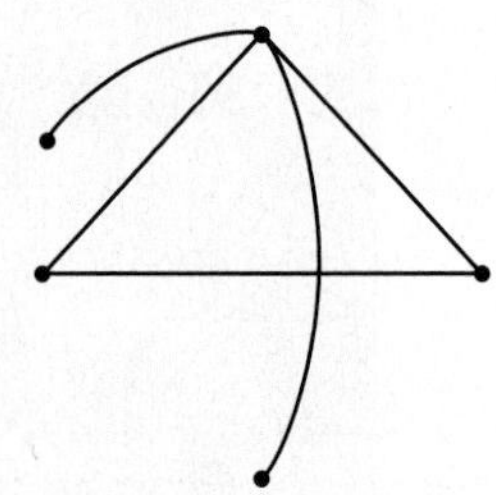

(a) four vertices and six edges.
(b) four vertices and four edges.
(c) five vertices and five edges.

6. The alphabetically ordered list of even-valent vertices of the graph below is ______ , ______ .

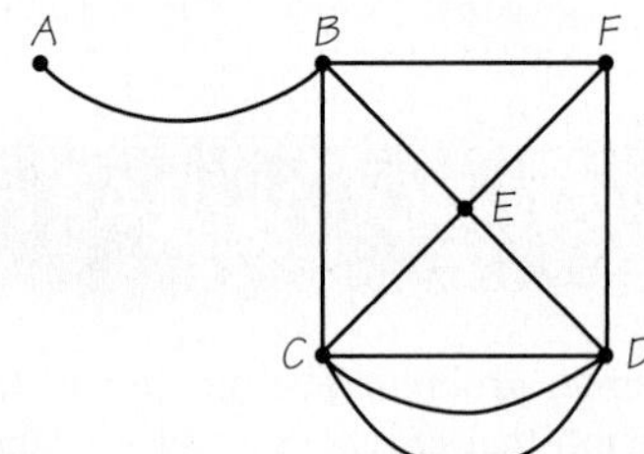

7. Which of the following statements is true about a *path*?

(a) A path always forms a circuit.
(b) A path is always connected.
(c) A path can visit any vertex only once.

8. If a graph consists of four vertices and every pair of vertices is connected by a single edge, the number of edges in the graph is exactly ________ .

9. It is not possible for a graph to have five vertices of valence 3 and six vertices of valence 4 because

(a) there are no graphs with exactly 11 vertices.
(b) a graph cannot have an even number of 4-valent vertices.
(c) a graph cannot have an odd number of odd-valent vertices.

10. If a graph is connected and has seven vertices, the graph must have at least __________ edges.

11. For which of the situations below is it most desirable to find an Euler circuit or an efficient eulerization of the graph?

(a) Sweeping the sidewalks of a small town
(b) Planning a new highway
(c) Planning a parade route in Muncie, Indiana

12. The minimum number of edges which must be duplicated to create a best possible eulerization of the following graph is _______ .

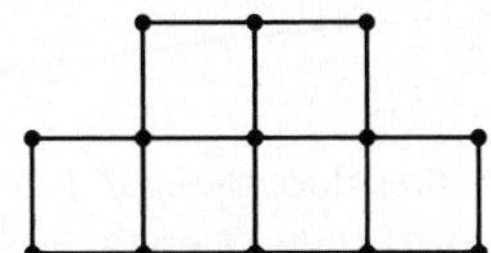

13. Consider the path represented by the sequence of numbered edges on the graph below. Which statement is correct?

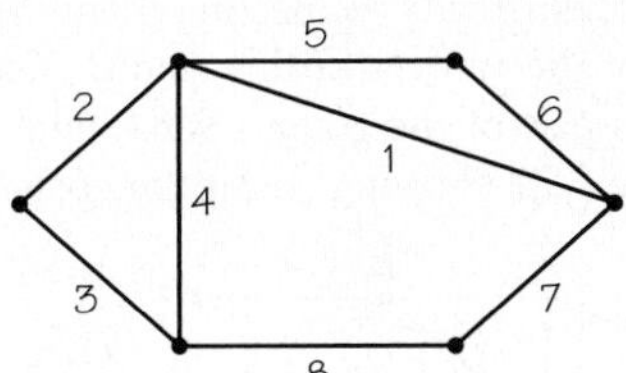

(a) The sequence of numbered edges forms an Euler circuit.
(b) The sequence of numbered edges traverses each edge exactly once but is not an Euler circuit.
(c) The sequence of numbered edges forms a circuit but not an Euler circuit.

14. For the graph below, the minimum *total* number of edges which constitutes a tour of the graph, starting and ending at the same vertex, and which visits each edge at least once, is _______ .

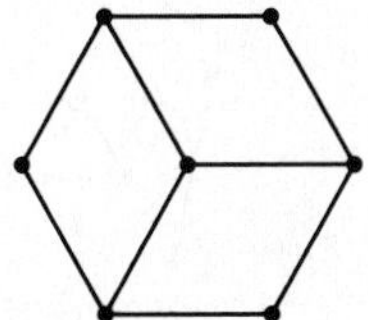

15. Suppose each vertex of a graph represents a baseball team and each edge represents a game played by two baseball teams. If the resulting graph is not connected, which of the following statements must be true?

(a) At least one pair of teams never played a game.
(b) At least one team played every other team.
(c) The teams play in distinct leagues.

16. If a graph has six vertices of odd valence, the absolute minimum number of edges that must be added (duplicated) to eulerize the graph is _______ .

17. Suppose the edges of a graph represent streets that must be plowed after a snowstorm. To eulerize the graph, four edges must be added. The real-world interpretation of this is that

(a) four streets will not be plowed.
(b) four streets will be traversed twice.
(c) four new streets would be built.

18. For each of the following situations, decide whether a graph or a digraph seems a more reasonable model:

(a) A system of hiking trails: __________________ .
(b) An electrical wiring plan for a home: __________ .
(c) A bus route map: ___________________________ .

19. Suppose a civic club offers several craft courses, and each club member can choose to participate in up to two different courses. Let each vertex of a graph represent one of these courses and each edge represent a club member who wants to take the two courses represented by the vertices at its endpoints. What can be said about the vertices in the resulting graph whose valence is zero?

(a) There are no vertices whose valence is zero.
(b) These vertices represent courses that can occur at the same time without displeasing any club member.
(c) These vertices represent the least popular courses.

20. If the valences of the vertices of a graph G are: 5, 4, 4, 4, 3, 2, 2, and 2, the number of vertices of G is _______ and the number of edges of G is _______ .

CHAPTER 1 EXERCISES

■ Challenge ◆ Discussion

1.1 Euler Circuits

1.2 Finding Euler Circuits

1. In the graph below, the vertices represent houses and two vertices are joined by an edge if it is possible to drive between the two houses in under 10 minutes.

(a) How many vertices does the graph have?

(b) How many edges does the graph have?

(c) What are the valences of the vertices in this graph?

(d) Based on the information given by the graph below, for which houses, if any, is it possible to drive to all the other houses in less than 20 minutes?

(e) Based on the graph below, from house *B* which houses require a trip of longer than 20 minutes?

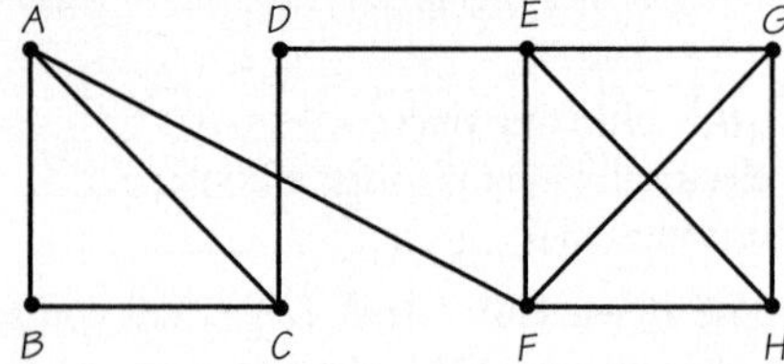

2. (a) Redraw the graph in Figure 1.2 to obtain a graph which has the same information where the edges only meet other edges at vertices.

(b) List all the routes that start on the U.S. side of the Atlantic Ocean and cross the ocean once and immediately.

3. (a) Is the figure below a graph? Explain your answer.

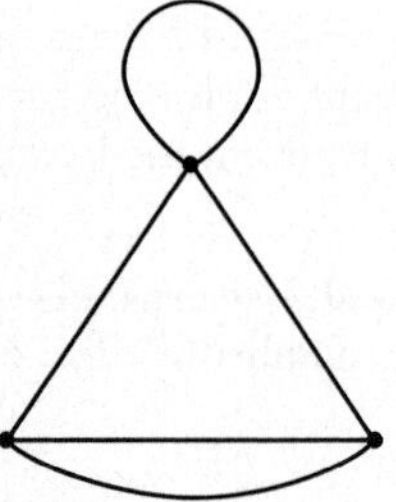

(b) The graph below has edges that "cross" at points that are not vertices of the graph. Which edges are these?

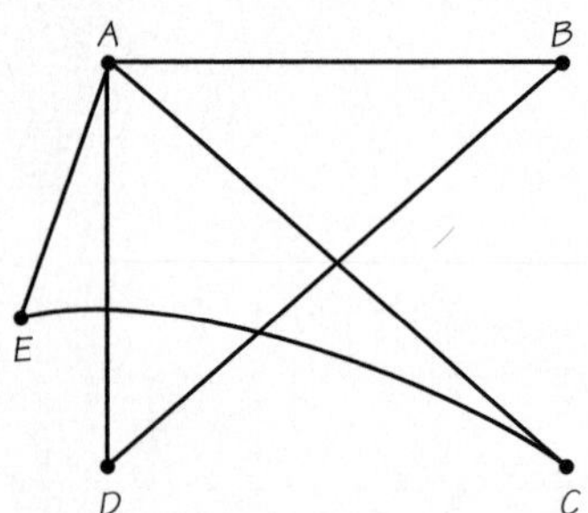

(c) How many vertices and edges are there in the prededing graph?

4. The graph below shows the stores and roads connecting them in a small shopping mall.

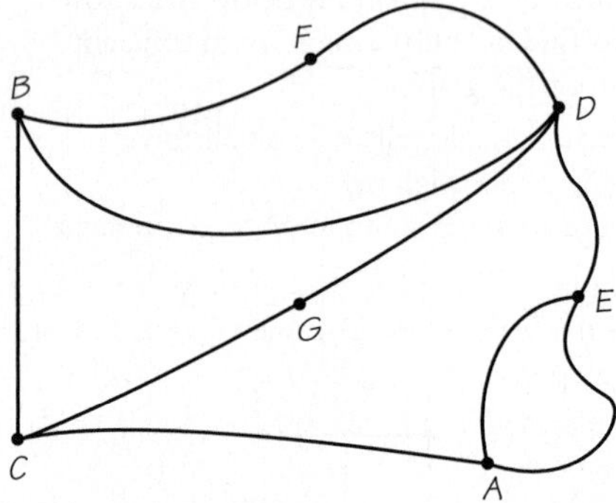

(a) How many stores does the mall have?

(b) How many roads connect up the stores in the mall?

(c) Write down a path from *C* to *F*.

(d) Write down a path from *E* to *B*.

5. In the graph below, the vertices represent cities and the edges represent roads connecting them. What are the valences of the vertices in this graph? (Keep in mind that *E* is part of the graph.) What might the valence of city *E* be showing about the geography?

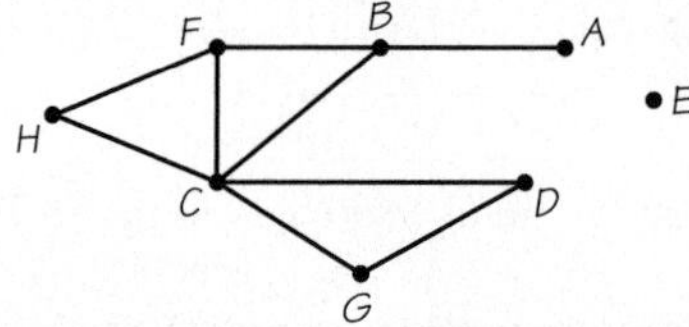

6. In the two graphs below, the vertices represent cities and the edges represent roads connecting them. In which graphs could a person located in city *A* choose any other city and then find a sequence of roads to get from *A* to that other city?

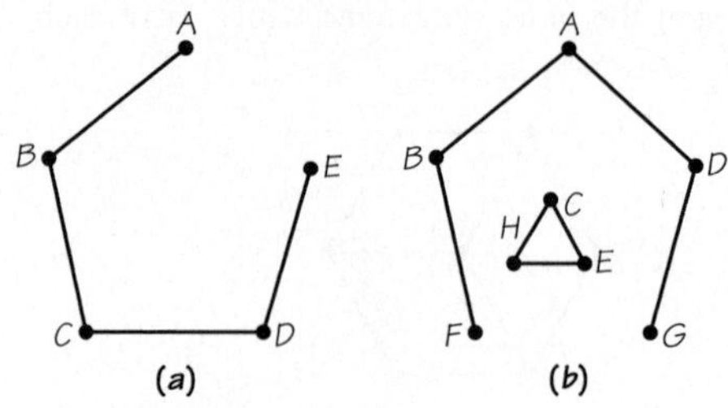

7. Refer to the figure in Exercise 4.

(a) Write down a circuit that includes the vertices *C* and *D* but does not start or end at either of these vertices.
(b) If two paths are considered different if they use different edges, write down:

(i) two different paths from *B* to *D*.
(ii) three different paths from *C* to *F*.
(iii) a circuit that has four edges.

8. Jack and Jill are located in Miami and want to fly to Berlin (see Figure 1.2).

(a) Find three paths for them to carry out this trip.
(b) What is the largest number of paths that can be used to carry out this trip that do not repeat a vertex (city)?
(c) Explain why it is reasonable not to want to repeat a vertex in this situation.

9. (a) How many vertices and edges does the graph in Figure 1.6 have?

(b) How many vertices and edges does the graph in Figure 1.7 have?
(c) How many vertices and edges does the graph in Figure 1.8a have?

10. (a) Add up the numbers you get for the valences of the vertices in Figure 1.6.

(b) Add up the numbers you get for the valences of the vertices in Figure 1.7.
(c) Add up the numbers you get for the valences of the vertices in Figure 1.8a.
(d) Describe the pattern you see in the answers you got for parts (a) through (c).
(e) Show that the pattern describes a fact that is true for any graph. (*Hint:* How many endpoints does an edge have?)

11. In the graph in Figure 1.8a, find the smallest possible number of edges you could remove that would disconnect the graph.

12. In the graphs in Figure 1.17, find the smallest possible number of edges you could remove that would disconnect the graph.

13. Draw a graph with eight vertices that is connected where

(a) each vertex has valence 3.
(b) each vertex has valence 4.
(c) Do all graphs with eight vertices having valence 2 have the same number of edges?

14. Is it possible that a street network gives rise to a disconnected graph? If so, draw such a network of blocks and streets and parking meters (in the style of Figure 1.12a). Then draw the disconnected graph it gives rise to.

15. (a) Draw a connected graph with six vertices, all of whose vertices have valence 2.

(b) Draw a disconnected graph with 6 vertices, all of whose vertices have valence 2.

16. (a) Draw a graph where every vertex has valence of at least 3 but where removing a single edge disconnects the graph.

(b) In what urban settings might a road network be represented by a graph that has an edge whose removal would disconnect the graph?

17. (a) Find a graph where the valences of the seven vertices of the graph are 1, 2, 2, 3, 3, 3, 4.

(b) Find another graph with the same valences as above that is "different" from the one you found for part (a).

18. For some services provided along streets, it may matter whether the roads are one-way or two-way. Give some examples where the street directions do and do not matter for our graph model analysis.

◆ **19.** A postal worker is supposed to deliver mail on all streets represented by edges in the graph below by traversing each edge exactly once. The first day the worker traverses the numbered edges in the order shown in (a), but the supervisor is not satisfied–why? The second day the worker follows the path indicated in (b), and the worker is unhappy–why? Is the original job description realistic? Why?

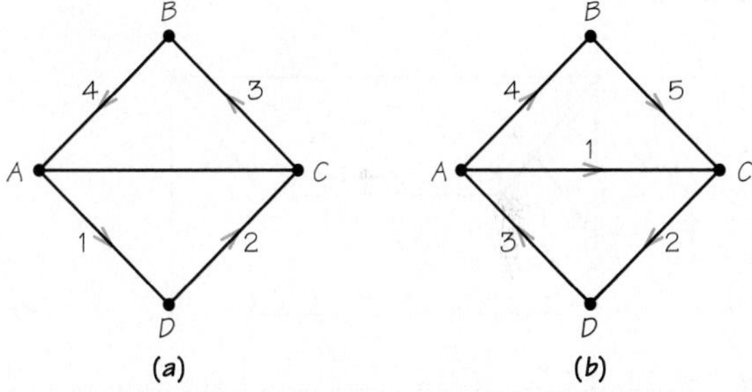

20. For the street network in Exercise 19, draw the graph that would be useful for routing a snowplow. Assume that all streets are two-way, one lane in each direction, and that you need to pass down each lane separately.

21. Find an efficient route for the snowplow to follow in the graph you drew in Exercise 20.

22. (a) Give examples of services that could be performed by a vehicle that moved in the direction of traffic down either lane of a two-way street.

(b) Give examples of services that would probably require a vehicle to travel down each of the lanes of a two-way street (in the direction of traffic for that lane) to perform the service.

23. For the street network shown below, draw the graph that would be useful for finding an efficient route for checking parking meters. (*Hint:* Notice that not every sidewalk has a meter; see Figure 1.12.)

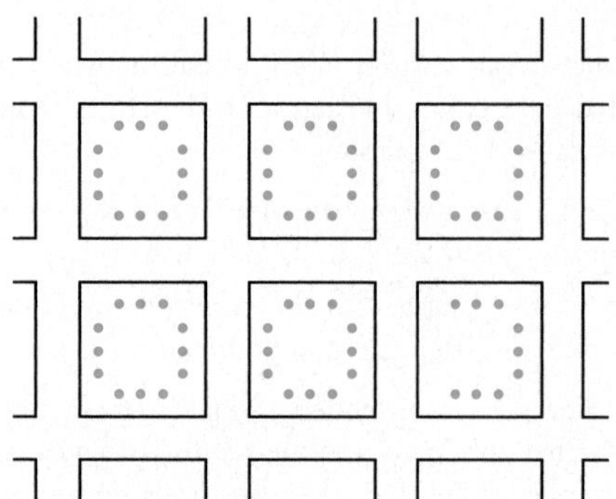

24. (a) For the street network in Exercise 23, draw a graph that would be useful for routing a garbage truck. Assume that all streets are two-way and that passing once down a street suffices to collect from both sides.

(b) Do the same problem on the assumption that one pass down the street suffices to collect from only one side.

25. (a) In the graph below, find the largest number of paths from A to F that do not have any edges in common.

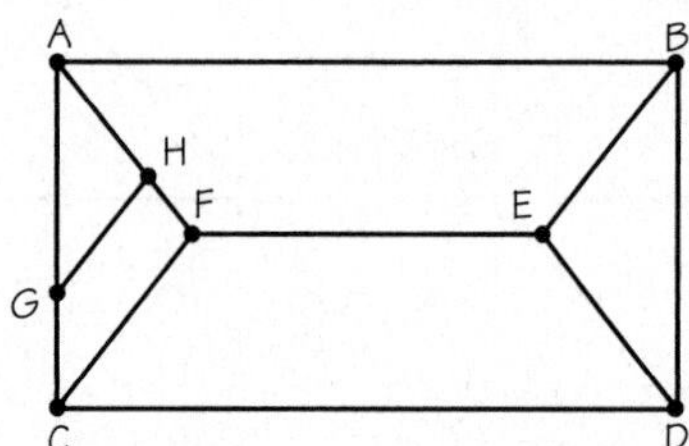

(b) Verify that the largest number of paths with no edges in common between any pair of vertices in this graph is the same.

(c) Why might one want to be able to design graphs such that one can move between two vertices of the graph using paths that have no edges in common?

26. Examine the paths represented by the numbered sequences of edges in both parts of the figure below. Determine whether each path is a circuit. If it is a circuit, determine if it is an Euler circuit.

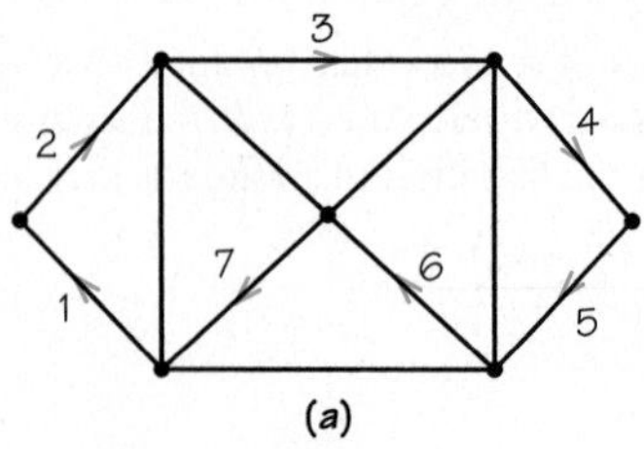

(*a*)

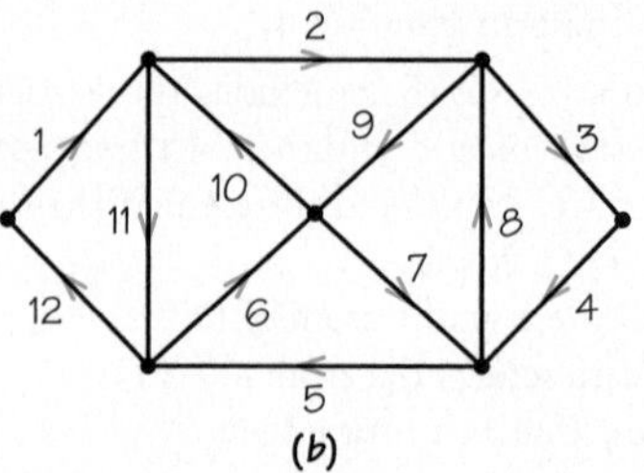

(*b*)

27. In Figure 1.13c, suppose we started an Euler circuit using this sequence of edges: 6, 7, 8, 9 (ignore existing arrows on the edges). What does our guideline for finding Euler circuits tell you *not* to do next?

28. In Figure 1.8b, suppose we started an Euler circuit using this sequence of edges: 14, 13, 8, 1, 4 (ignore existing arrows on the edges). What does our guideline for finding Euler circuits tell you *not* to do next?

29. Find an Euler circuit on the graph of Figure 1.15c (including the blue edges).

30. Find Euler circuits in the right-hand graphs in Figures 1.17a and 1.17b.

31. In the following graph, we see a territory for a parking-control officer that has no Euler circuit. How many sidewalks (edges) need to be omitted in order to enable us to find an Euler circuit? What effect would this have in the associated real-world situation?

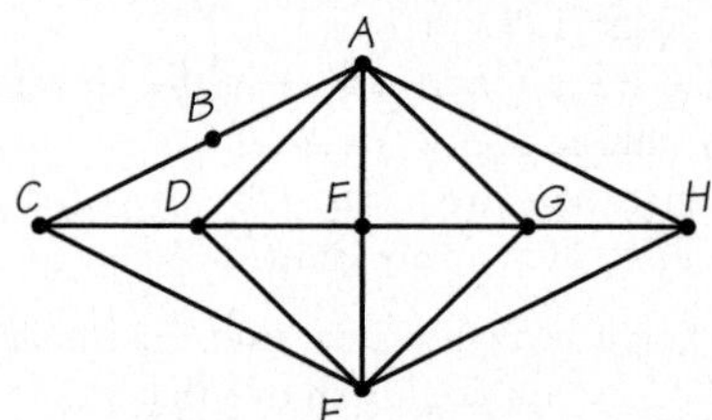

32. An Euler circuit visits a four-valent vertex X, such as the one in the accompanying graph, by using the edges AX and XB consecutively, and then using CX and XD consecutively. When this happens, we say that the Euler circuit cuts through at X.

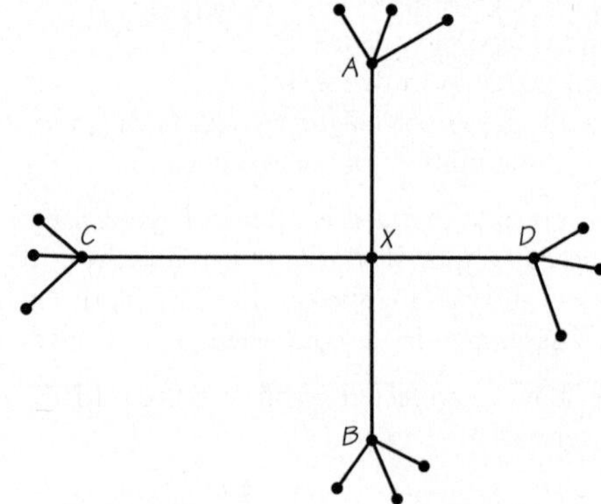

Suppose G is a four-valent graph such as that in the diagram below. Is it possible to find an Euler circuit of this graph that never cuts through any vertex? Explain why it might be desirable to find an Euler circuit of this special kind in an applied situation.

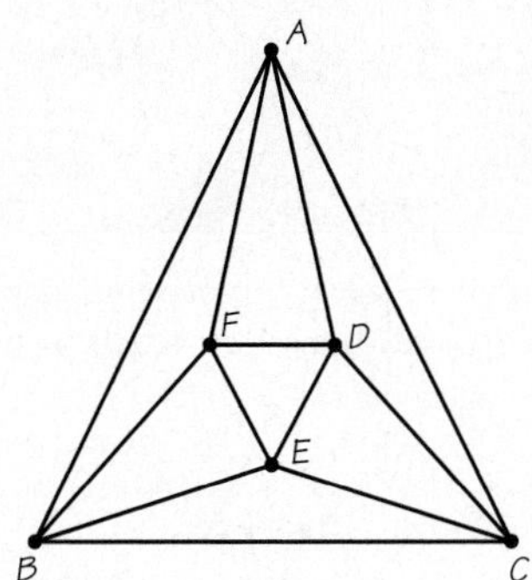

1.3 Beyond Euler Circuits

1.4 Urban Graph Traversal Problems

33. Find an Euler circuit on the eulerized graph (b) of the following figure. Use it to find a circuit on the original graph (a) that covers all edges and reuses edges only five times. Can fewer than five reused edges be achieved?

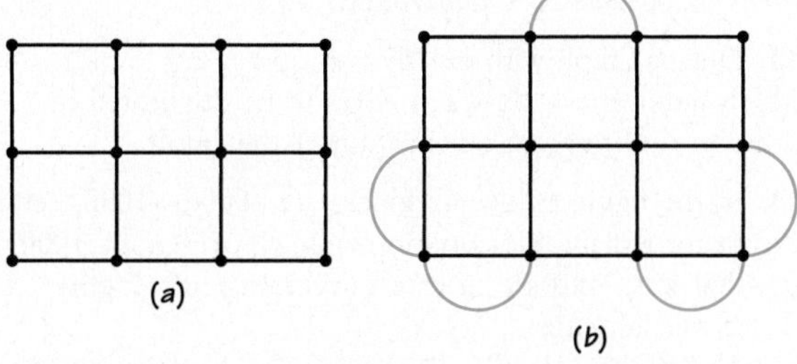

34. Squeeze the circuit shown in graph (a) below onto graph (b). Show your answers by writing numbered arrows on the edges and by listing a sequence of vertices (for example, $ABEB \dots A$).

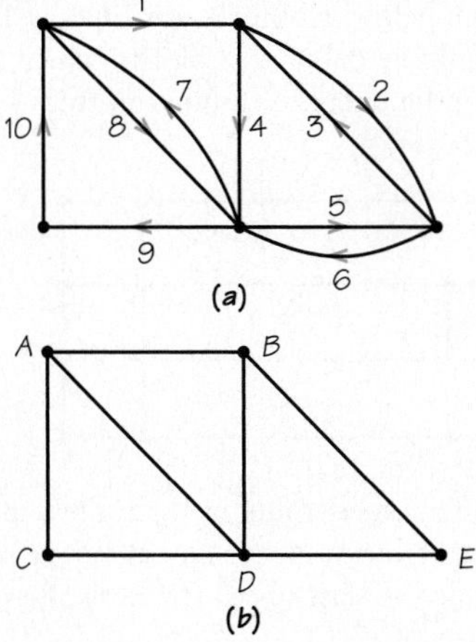

Then squeeze the circuit shown in graph (c) onto graph (d). Show your answers by writing numbered arrows on the edges and by listing a sequence of vertices.

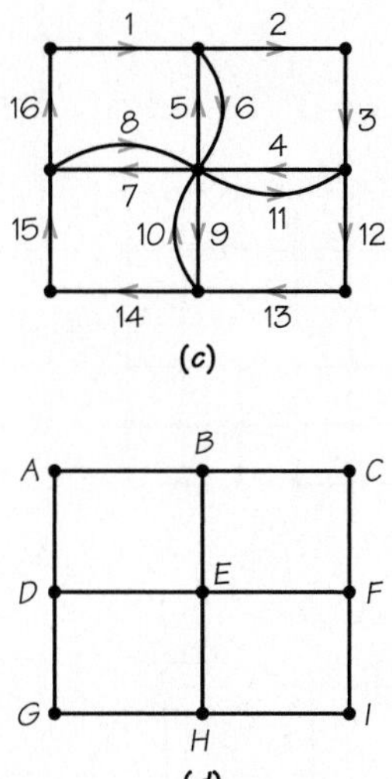

35. A college campus has a central square with sides arranged as shown by the edges in the graph below. Show how all these sidewalks can be traversed at least once in a tour that starts and ends at the same vertex.

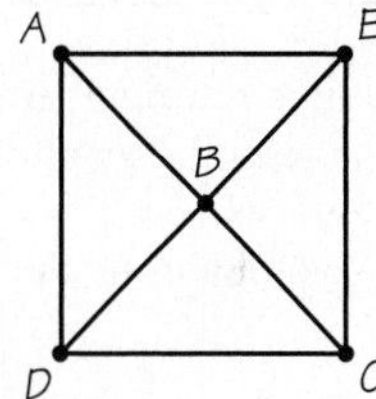

36. In the graph below, add one or more edges to produce a graph that has an Euler circuit.

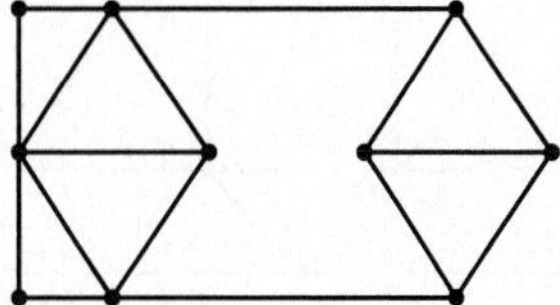

37. Eulerize these rectangular street networks using the same patterns that would be used by the edge walker described in the text.

(a) A 5×5 rectangle

(b) A 4×5 rectangle

(c) A 6×6 rectangle

(d) Can you find an eulerization with nine added edges for a 2-by-7-block rectangular street network? Can you do better than nine added edges?

38. Find good eulerizations for the following graphs, using as few duplicated edges as you can. See "Finding Good Eulerizations" for hints.

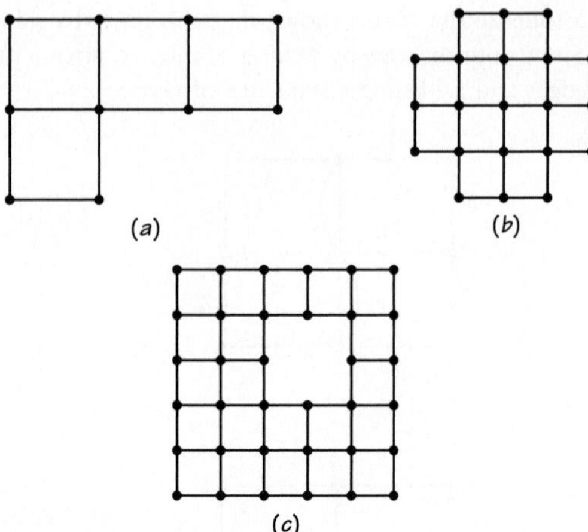

39. For the following graph:

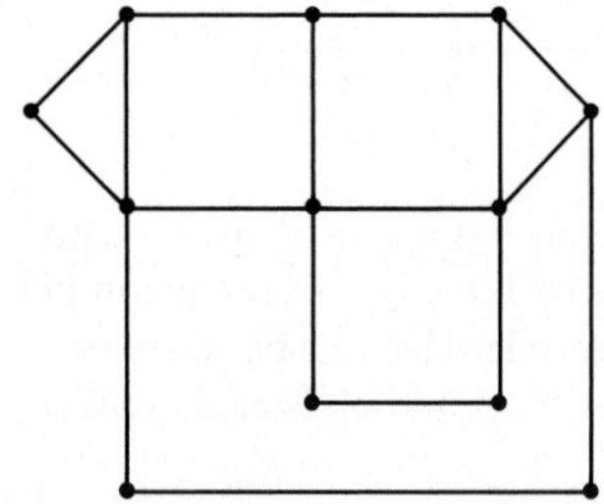

(a) Determine the minimum number of edges that have to be removed for the resulting graph to have all even-valent vertices.
(b) Does the graph you obtain in part (a) have an Euler circuit?

For the graph below:

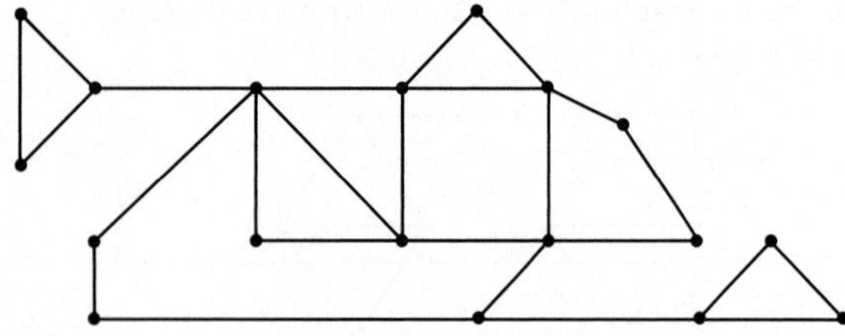

(c) Determine the minimum number of edges that have to be removed for the resulting graph to have all even-valent vertices.
(d) Does the graph you obtain in part (c) have an Euler circuit?

40. The following figure shows a river, some islands, and bridges connecting the islands and riverbanks. A charity is sponsoring a race in which entrants have to start at *A*, go over each bridge at least once, and end at *A*. Draw a graph that would be useful for finding a route that requires the least recrossing of bridges. Show what that route would be. (*Historical note:* This situation resembles the one that inspired Leonhard Euler's 1736 "recreational mathematics" problem that resulted in the first work in graph theory.)

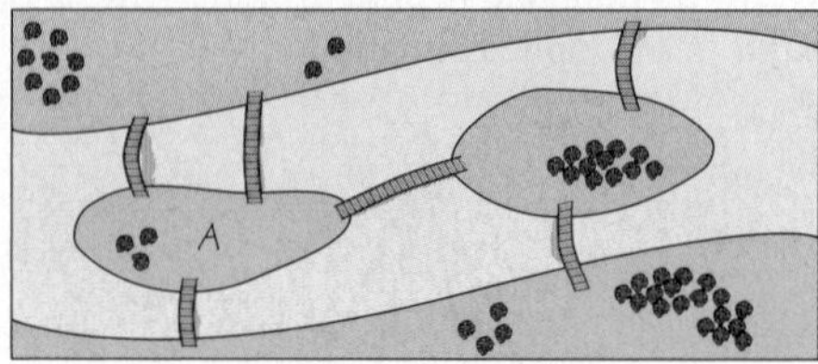

41. Find a circuit in the accompanying graph that covers every edge and has as few reuses as possible.

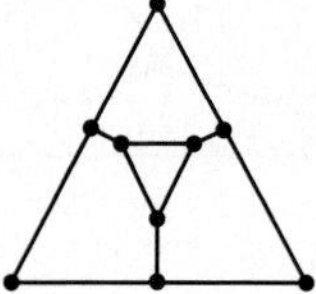

42. **(a)** Discuss the difference between the problem of:
(i) Adding the minimum number of edges to a graph to make all its vertices even-valent, and
(ii) Finding the best eulerization of a connected graph.
(b) In (i) must the graph that results from adding a minimum number of edges to make all the vertices even-valent have an Euler circuit?

43. Draw a graph with exactly two odd-valent vertices which requires exactly seven edges to be duplicated in order to find the best eulerization of the graph.

44. In the figure below, all blocks are 1000-by-1000 feet, except for the middle column of blocks, which are 1000-by-4000 feet. Find a circuit of minimum total length that covers all edges.

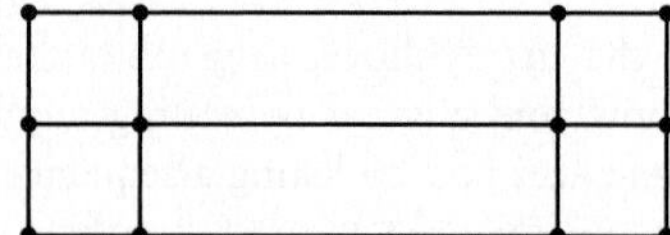

45. In the figure below, all blocks are 1000-by-1000 feet, except for the middle column of blocks, which are 1000-by-4000 feet. Find a circuit of minimum total length that covers all edges.

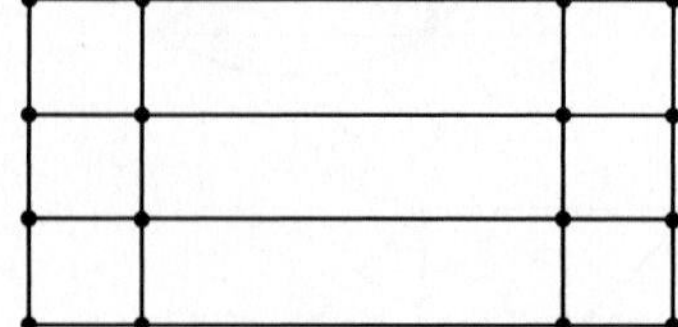

46. **(a)** Find the cheapest route in the following graph, where one starts at vertex *A*, finishes at vertex *A*, and traverses each edge at least once. The cost of a route is computed by summing the numbers along the edges that one uses.

(b) How many edges are repeated in the minimal-cost route?

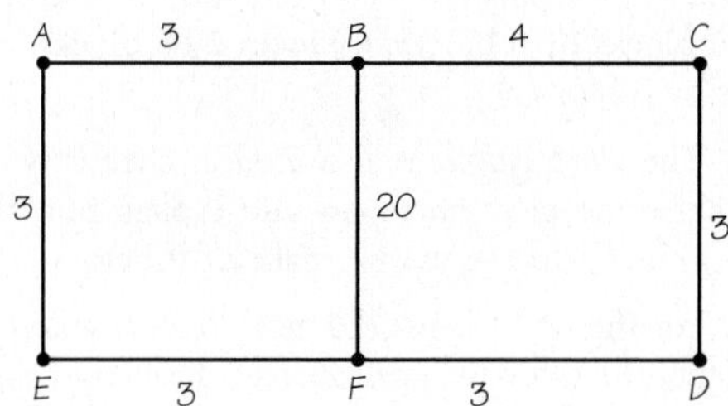

(c) Discuss the implications of this example for the relation between finding good eulerizations of graphs and the problem of finding cheap routes that start and end at the same vertex and traverse each edge at least once.

(d) The physical edge with cost 20 in the diagram is not physically longer than other edges with lower costs attached to them. Explain why in an urban setting it might make sense to assign two stretches of street of similar length very different "costs" for traversing them.

(e) What are some different meanings that "weights" (for example, traffic volume) potentially assigned to edges in a graph might have in an urban setting?

47. Which graphs (see figures below) have Euler circuits? In the ones that do, find the Euler circuits by numbering the edges in the order the Euler circuit uses them. For the ones that don't, explain why no Euler circuit is possible.

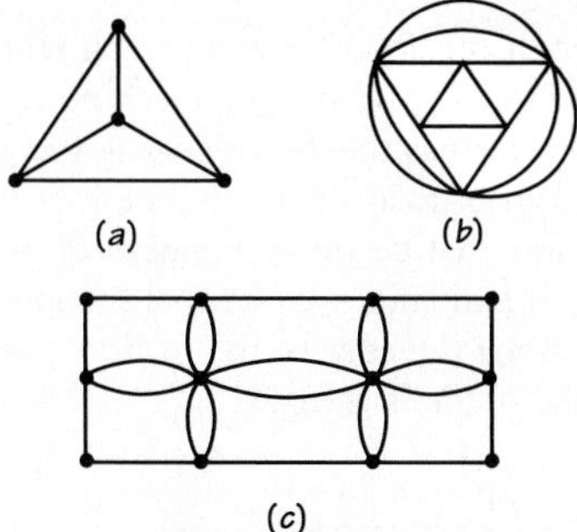

48. Eulerize the graph below by using four new edges. Find an Euler circuit in the eulerized graph and use that circuit to find a circuit of the original graph that covers all edges but reuses edges only four times. How many different ways can the four edges be chosen?

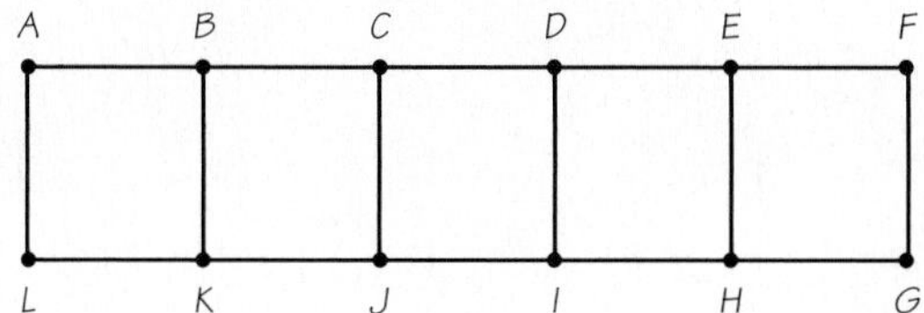

■ **49.** A graph G represents a street network to be traveled by a postal worker who must traverse every street twice, once for each side of the street. In graph G, the edges represent sidewalks. Does such a graph always have an Euler circuit? Explain your answer.

50. In the graph below, find a circuit that covers every edge and has as few reuses as possible.

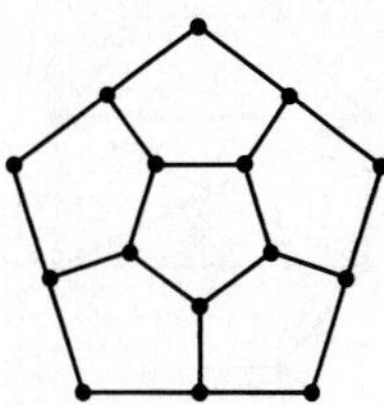

51. (a) Find the best eulerizations you can for the two graphs below.

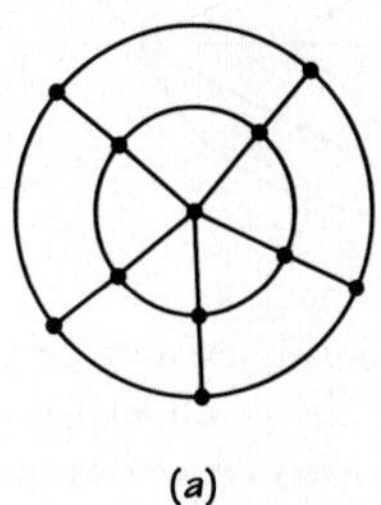
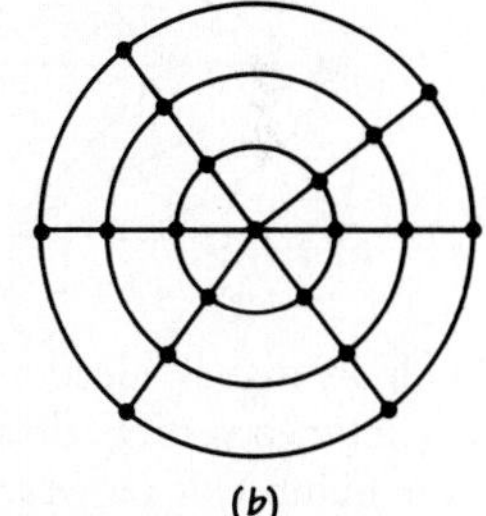

(a) (b)

(b) Graph (a) can be thought of as having five rays and two circles, and graph (b) as having six rays and three circles. Draw a graph with four rays and four circles and find the best eulerization you can for this graph.

(c) Find a "formula" involving r and s for the smallest number of edges needed to eulerize a graph of this type having r rays and s circles.

■ **52.** Suppose that for a certain connected graph it is possible to disconnect it by removing one edge. Explain why such a graph (before the edge is removed) must have at least one vertex of odd valence. (*Hint:* Show that it cannot have an Euler circuit.)

53. Can you draw a graph with six vertices where the valence of each vertex is 5?

54. Each of the following graphs represent the sidewalks to be cleaned in a fancy garden (one pass over a sidewalk will clean it). Can the cleaning be done using an Euler circuit? If so, show the circuit in each graph by numbering the edges in the order the Euler circuit uses them. If not, explain why no Euler circuit is possible.

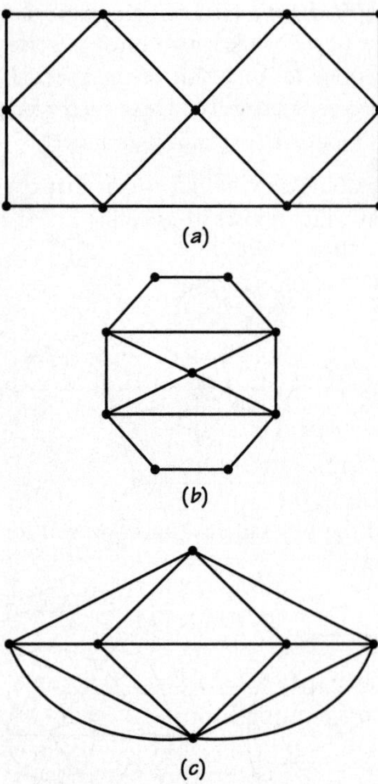

■ **55.** If an edge is added to an already existing graph, connecting two vertices already in the graph, explain why the number of vertices with odd valence has the same parity before and after. (This means if it was even before, it is even after, while if it was odd before, it remains odd.)

■ **56.** Any graph can be built in the following fashion: Put down dots for the vertices, then add edges connecting the dots as needed. When you have put down the dots, and before any edges have been added, is the number of vertices with odd valence an even number or an odd number? What is the number of vertices with odd valence when all the edges have been added (see Exercise 55)?

57. Draw the graph for the parking-control territory shown in the figure below. Label each vertex with its valence and determine if the graph is connected.

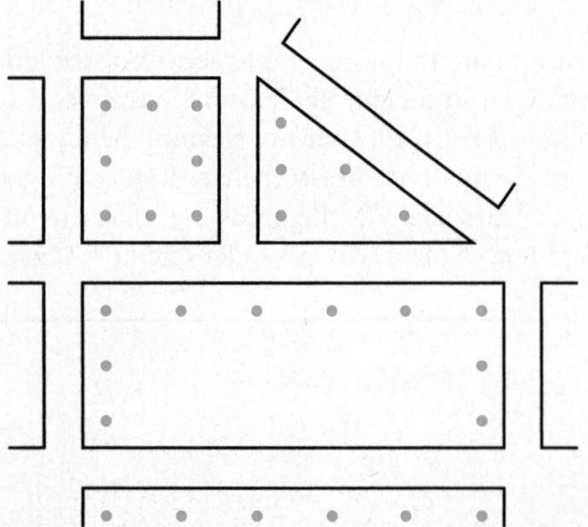

58. If a rectangular street network is r blocks by s blocks, find a formula for the minimum number of edges that must be added to eulerize a graph representing the network in terms of r and s. (*Hint:* Treat the case $r = 1$ separately. Test your formula with the cases 6 blocks by 5 blocks, 6 blocks by 6 blocks, and 5 blocks by 3 blocks.)

◆ **59.** The word *valence* is also used in chemistry. Find out what it means in chemistry and explain how this usage is similar to the use we make of it here.

■ **60.** For the street network below, draw a graph that represents the sidewalks with meters. Then find the minimum-length circuit that covers all sidewalks with meters. If you drew the graph as we recommended, you would find that the shortest circuit has length 18 (it reuses every edge).

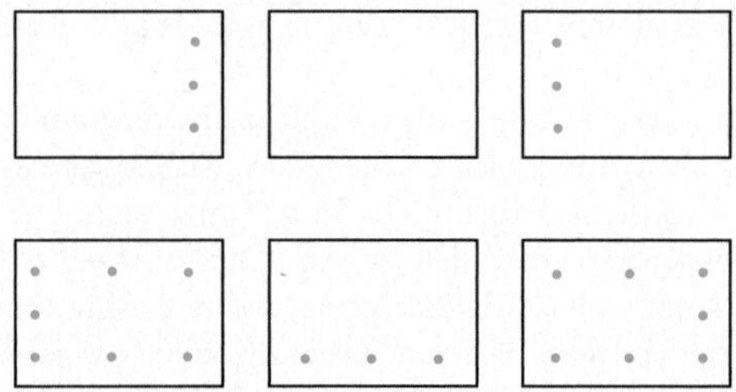

But the meter checker comes to you and says: "I don't know anything about your theories, but I have found a way to cover the sidewalks with meters using a circuit of length 10. My trick is that I don't rule out walking on sidewalks with no meters." Explain what he means and discuss whether his strategy can be used in other problems.

61. Each edge of the accompanying graph represents a two-lane highway. A grass-mowing machine is located at A, and its operator has the job of cutting the grass along each of the edges of road shown. Find a tour for the mowing machine that begins and ends at A. Find such a tour that begins and ends at A and, as the mowing is done, moves along the edge of the road in the same direction as the traffic is going.

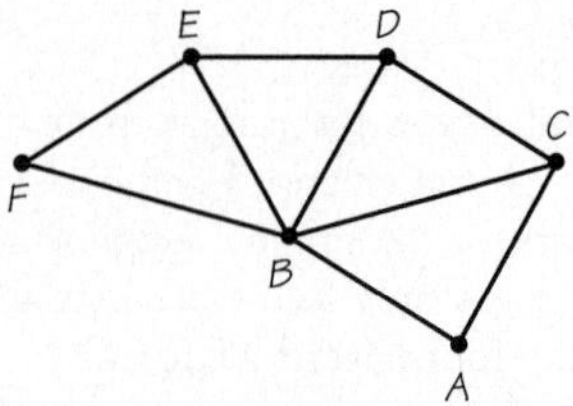

APPLET EXERCISES

To do these exercises, go to www.whfreeman.com/fapp8e.

Eulerizing a Graph

We learned that if a graph has exactly two vertices with odd valences, then an Euler circuit does not exist–but an Euler path does. It is also possible to produce an Euler circuit through the process of eulerization, by duplicating certain edges of the graph. But how many duplications are necessary to obtain an Euler circuit? Investigate this problem and more general related topics using the *Eulerizing a Graph* applet.

Euler Circuits

We know that if all the vertices have even valence, then an Euler circuit exists. Try your hand at finding such circuits in the *Euler Circuit* applet.

WRITING PROJECTS

1. Write a memo to your local department of parking control (or police department) in which you suggest that management science techniques like the ones in this chapter be used to plan routes. Assume that the person to whom you are writing is not extensively trained in mathematics but is willing to read through some technical material, provided you make it seem worth the trouble.

2. Do the same as in Writing Project 1, but to the department in charge of spreading salt on roads after snowstorms.

3. If you were making a recommendation to the mayor of New York City concerning proposed new street-sweeping routes, designed using the theory of this chapter, would you recommend that the changes be adopted or not? Write a memo that outlines the pros and cons as fairly as you can, and then conclude with your recommendation.

SUGGESTED READINGS

BELTRAMI, EDWARD J., *Models for Public Systems Analysis,* Academic Press, New York, 1977. This book gives a good overview of the way that operations research has provided and continues to provide new tools for solving societal problems. Among the ideas discussed are police patrol tactics, organization of emergency services, and scheduling. Some of the mathematics used is advanced.

MALKEVITCH, JOSEPH, and WALTER MEYER, *Graphs, Models and Finite Mathematics,* Prentice-Hall, Englewood Cliffs, NJ, 1974. This introductory book includes much of the same material as presented here but provides more details of the proofs and uses somewhat different algorithms for solving the problems involved.

The following books treat many of the topics discussed here as well as shortest-path problems and matching problems, and they formulate some problems in more realistic terms:

ROBERTS, FRED S., AND BARRY TESMAN, *Applied Combinatorics,* Second Edition, Pearson Prentice Hall, Upper Saddle River, NJ, 2004.

TUCKER, ALAN. *Applied Combinatorics,* Third Edition, Wiley, New York, 1995.

SUGGESTED WEB SITES

www.hsor.org/what_is_or.cfm This site discusses the history of operations research (OR) and some of the areas where OR is being applied. Be sure to follow the "Networks Routing" link to see applications of the Chinese postman problem.

www.geom.uiuc.edu/~doty/applications This Web page provides some examples of how to apply Euler circuits.

www-gap.dcs.st-and.ac.uk/~history/Mathematicians/Euler.html This essay discusses the numerous contributions that Euler made to mathematics, and provides biographical information about him.

www.ams.org/featurecolumn/archive/urban-geom.html This Web page includes an introduction to how graph theory has provided tools for urban operations research.

CHAPTER 2

Business Efficiency

In the previous chapter, we saw that there was an easy way of telling whether a connected graph has a circuit that traverses each of the edges of a graph exactly once–for example, a route for a snowplow that covers the streets of a section of a town. However, the situation changes radically if we make a seemingly small change in the problem: When is it possible to find a route along distinct edges of a graph that visits each *vertex* once and only once in a simple circuit? Perhaps there has been a hurricane and it is important to check whether or not the storm sewers at every corner in town are clogged.

This problem is called the *Hamiltonian circuit problem,* and, like the Euler circuit problem, it is a graph theory problem. The Hamiltonian circuit problem has many applications. Suppose inspections or deliveries need to be made at each vertex (rather than along each edge) of a graph. An "efficient" tour of the graph would be a route that started and ended at the same vertex and passed through all the vertices without reuse, or repetition; that is, the route would be a **Hamiltonian circuit**. Such routes would be useful for inspecting traffic signals or for delivering mail to drop-off boxes, which hold heavy loads of mail so that urban postal carriers do not have to carry them long distances, or delivering Meals on Wheels to the elderly.

Hamiltonian Circuit DEFINITION

A tour, like the ones marked by wiggly edges in Figure 2.1, that starts at a vertex of a graph and visits each vertex once and only once, returning to where it started, is called a **Hamiltonian circuit**.

For example, the wiggly line in Figure 2.1a shows a circuit we can take to tour that graph, visiting each vertex once and only once. This tour can be written *ABDGIHFECA.* Note that another way of writing the same circuit would be *EFHIGDBACE.* A different circuit visiting each vertex once and only once would be *CDBIGFEHAC* (Figure 2.1b). Do not be confused because *C* is written twice when we write down this list of vertices. We can think of the circuit as starting at any of its vertices, but we do start and end at the same vertex.

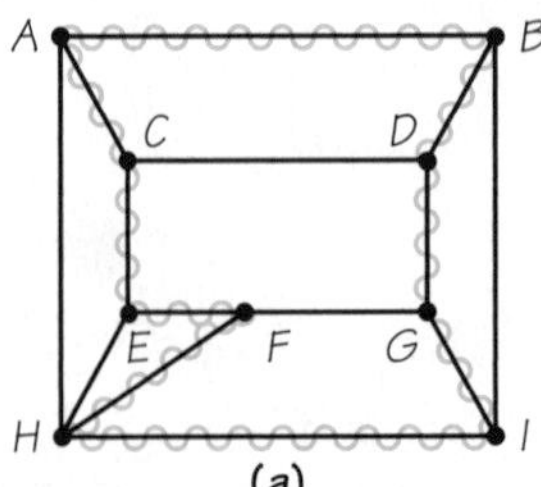

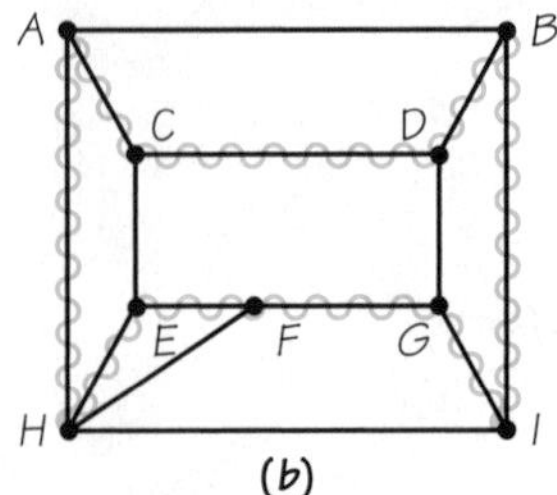

FIGURE 2.1 Wiggly edges illustrate Hamiltonian circuits.

2.1 Hamiltonian Circuits

The concept is named for the Irish mathematician William Rowan Hamilton (1805–1865), who was one of the first to study it. We now know that the concept was discovered somewhat earlier by Thomas Kirkman (1806–1895), a British minister with a penchant for mathematics.

The concepts of Euler and Hamiltonian circuits are similar in that both forbid reuse. An Euler circuit forbids the reuse of edges, while a Hamiltonian circuit forbids the reuse of vertices. However, it is far more difficult to determine which connected graphs possess a Hamiltonian circuit than to determine which connected graphs have Euler circuits. As we saw in Chapter 1, looking at the valences of vertices tells us whether a connected graph has an Euler circuit, but we have no such simple method for telling whether or not a graph has a Hamiltonian circuit.

It may seem that delivering mail in a cheap and timely manner should not be that hard. However, finding the optimal way to deliver mail over a variety of environments, rural, suburban, and urban, is very complex. How should a large geographic area be divided into smaller sections? Should each mail carrier use a truck as a "depot" to resupply small amounts of mail for delivery or should there be deposit boxes on street corners? Mathematics can be used to find answers to such questions. (*Lawrence Migdale/Photo Researchers, Inc.*)

Some special classes of graphs are known to have Hamiltonian circuits, and some special classes of graphs are known to lack them. For example, here is a method for constructing an infinite family of graphs where each graph in the family cannot have a Hamiltonian circuit. Construct a vertical column of m vertices and a parallel column of n vertices, where m is bigger than n, as shown in Figure 2.2a. The figure illustrates a typical case where $m = 4$ and $n = 2$. Now join each vertex on the left in the figure to every vertex on the right. As m and n vary, we get a family of different graphs.

No graph obtained in this manner can have a Hamiltonian circuit. If a Hamiltonian circuit existed, it would have to alternately include vertices on the left and right of the figure. This is not possible because the number of vertices on the left and right are not the same. It is unlikely that a method will ever be found to easily determine whether or not a graph has a Hamiltonian circuit. If Hamiltonian circuits were easy to find in any graph at all, many applied problems could be solved in a less costly way.

In many urban operations research situations "grid graphs" such as the one in Figure 2.2b are of interest. If we wanted an efficient route (circuit) to inspect traffic surveillance cameras located at urban street intersections, we would need to find a Hamiltonian circuit for the graph in Figure 2.2b. Note that in going from one vertex to another we move from a vertex of one color to a vertex of the other color. Since colors would alternate in a Hamiltonian circuit, it follows that the number of vertices of each color would have to be the same if there is a Hamiltonian circuit in this graph. Since the number of vertices of the two colors is *not* the same, there is no Hamiltonian circuit and, hence, no fully efficient route for inspecting the traffic control cameras.

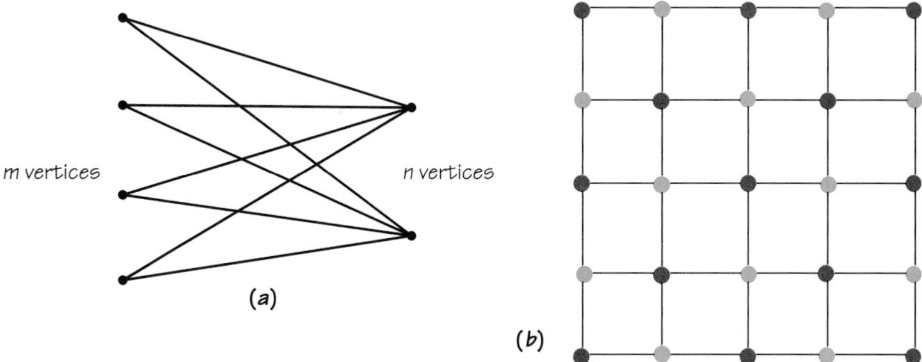

FIGURE 2.2 (a) An example of one graph from a family of graphs that has no Hamiltonian circuit. The number of vertices m on the left is chosen to be greater than the number of vertices n on the right. The case $m = 4$ and $n = 2$ is shown. (b) A graph used to model a portion of a city. Since the graph reflects the block structure of the city, it is known as a "grid graph."

The Hamiltonian circuit problem itself has many applications. This is not unusual in mathematics. Often mathematics used to solve a particular real-world problem leads to new mathematics that suggests applications to other real-world situations. One class of problems to which we can apply Hamiltonian circuits is vacation planning.

EXAMPLE 1 ▪ Vacation Planning

Let's imagine that you are a college student studying in Chicago. During spring break you and a group of friends have decided to take a car trip to visit other friends in Minneapolis, Cleveland, and St. Louis. There are many choices as to the order of visiting the cities and returning to Chicago, but you want to design a route that minimizes the distance you have to travel. Presumably, you also want a route that cuts costs, and you know that minimizing distance will minimize the cost of gasoline for the trip. Similar problems with different complications would arise for bus, railroad, or airplane trips.

Express mail and parcel post delivery companies need to make complicated patterns of deliveries and pick-ups. To do this they need to know driving distances between the various geographical locations involved. Using this information, together with driving times, they can use mathematics to cut costs and to make the pick-ups and deliveries on time. (© *Rhoda Sidney/PhotoEdit.*)

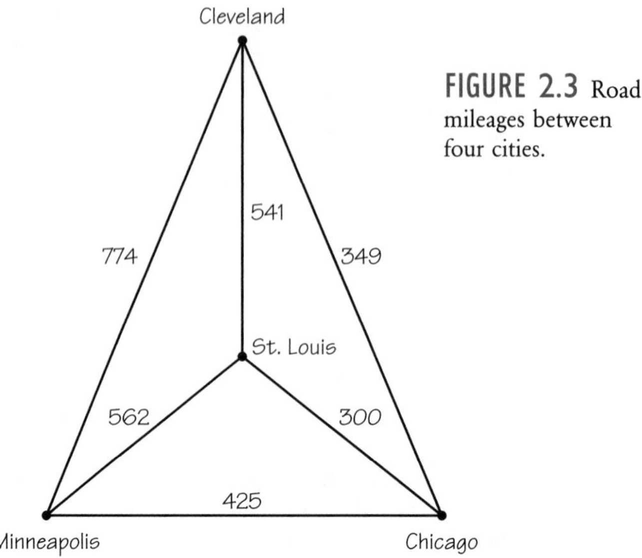

FIGURE 2.3 Road mileages between four cities.

Imagine now that the local automobile club has provided you with the intercity driving distances between Chicago, Minneapolis, Cleveland, and St. Louis. We can construct a graph model with this information, representing each city by a vertex and the legs of the journey between the cities by edges joining the vertices. To complete the model, we add a number called a **weight** to each graph edge, as in Figure 2.3. In this example, the weights represent the distances between the cities, each of which corresponds to one of the endpoints of the edges in the graph. (In other examples the weight might represent a cost, time, satisfaction rating, or profit.) We want to find a minimum-cost tour that starts and ends in Chicago and visits each other city only once. Using our earlier terminology, what we wish to find is a **minimum-cost Hamiltonian circuit**–a Hamiltonian circuit with the lowest possible sum of the weights of its edges.

Finding a Minimum-Cost Hamiltonian Circuit — PROCEDURE

How can we determine which Hamiltonian circuit has minimum cost? There is a conceptually easy **algorithm**, or mechanical step-by-step process, for solving this problem:

1. Generate all possible Hamiltonian tours (starting from Chicago).
2. Add up the distances on the edges of each tour.
3. Choose a tour with total distance being a minimum, that is, as small as possible.

Steps 2 and 3 of the algorithm are straightforward. Thus, we need worry only about Step 1, generating all the possible Hamiltonian circuits in a systematic way. To find the Hamiltonian tours, we will use the **method of trees**, as follows. Starting from Chicago, we can choose any of the three cities to visit after leaving Chicago. The first stage of the enumeration tree is shown in Figure 2.4. If Minneapolis is chosen as the first city to visit, then there are two possible cities to visit next, Cleveland and St. Louis. The possible branchings of the **tree** at this stage are shown in Figure 2.5. In this second stage, however, for each choice of first city to visit, there are two choices from this city to the second city to visit. This would lead to the diagram in Figure 2.6.

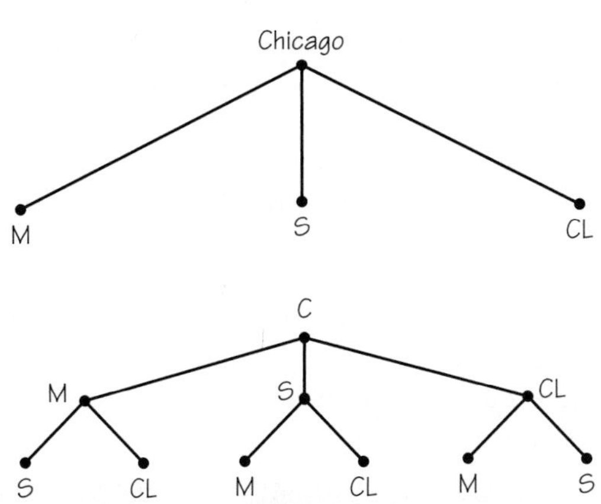

FIGURE 2.4 First stage in finding vacation-planning routes.

FIGURE 2.6 Complete second stage in finding vacation-planning routes.

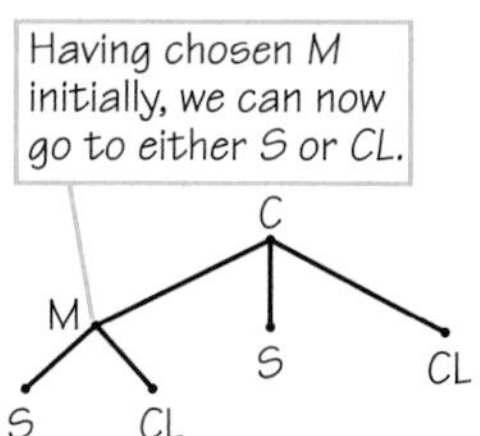

FIGURE 2.5 Part of the second stage in finding vacation-planning routes.

Having chosen the order of the first two cities to visit, and knowing that no revisits (reuses) can occur in a Hamiltonian circuit, there is only one choice left for the next city. From this city we return to Chicago. The complete tree diagram showing the third and fourth stages for these routes is given in Figure 2.7. Notice, however, that because we can traverse a circular tour in either of two directions, the paths shown in the tree diagram of Figure 2.7 do not correspond to different Hamiltonian circuits. For example, the leftmost path (C–M–S–CL–C) and the rightmost path (C–CL–S–M–C) represent the same Hamiltonian circuit. Thus, among what appear to be six different paths in the tree diagram, only three in fact correspond to different Hamiltonian circuits. These three distinct Hamiltonian circuits are shown in Figure 2.8.

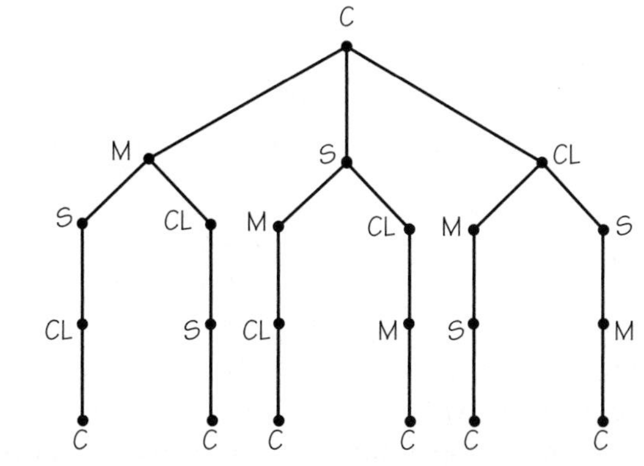

FIGURE 2.7 Completed enumeration of routes using the method of trees for the vacation-planning problem.

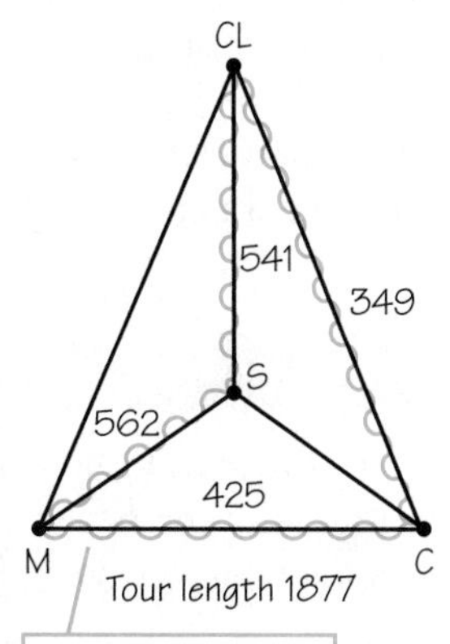

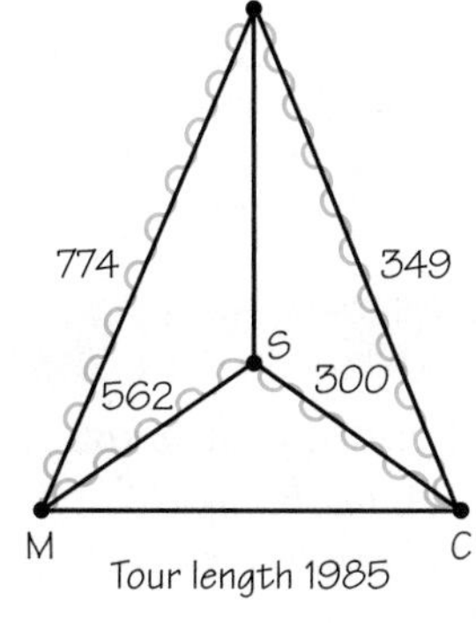

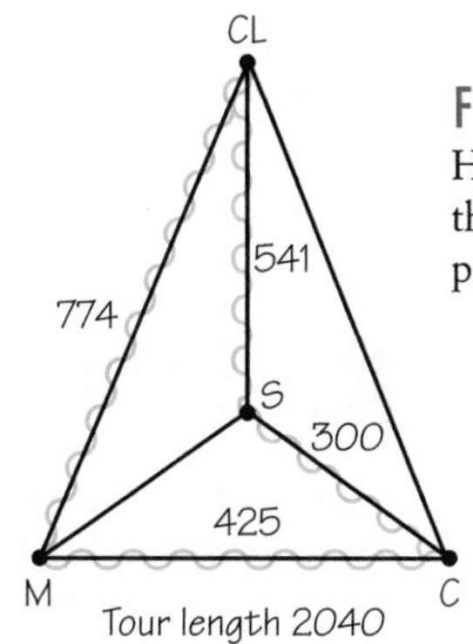

FIGURE 2.8 The three Hamiltonian circuits for the vacation-planning problem of Figure 2.3.

Note that in generating the Hamiltonian circuits we disregard the distances involved. We are concerned only with the different patterns of carrying out the visits. To find the optimal route, however, we must add up the distances on the edges to get each tour's length. Figure 2.8 shows that the optimal tour is Chicago, Minneapolis, St. Louis, Cleveland, Chicago. The length of this tour is 1877 miles, which saves 163 miles over the longest choice of tour.

The method of trees is not always as easy to use as our example suggests. Instead of doing our analysis for four cities, consider the general case of n cities. The graph model similar to that in Figure 2.3 would consist of a weighted graph with n vertices, with every pair of vertices joined by an edge.

Complete Graph DEFINITION

A graph is called **complete** if there is exactly one edge between each pair of vertices in the graph.

A complete graph with five vertices is illustrated in Figure 2.9. The graph in Figure 2.3 is a weighted complete graph with four vertices.

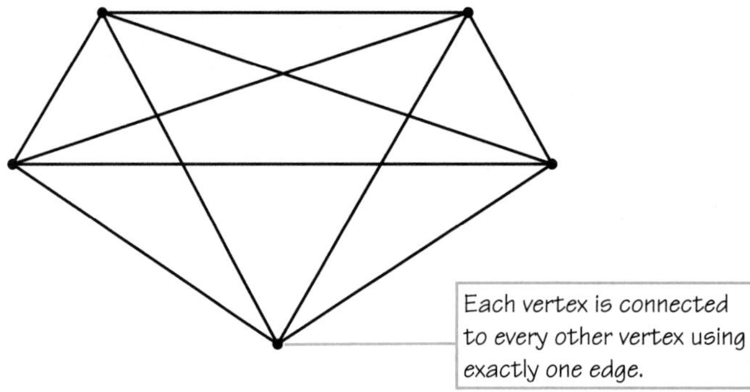

Each vertex is connected to every other vertex using exactly one edge.

FIGURE 2.9 A complete graph with five vertices. Every pair of vertices is joined by an edge.

Fundamental Principle of Counting

How many Hamiltonian circuits are in a complete graph of n vertices? We can solve this problem by using the same type of analysis that we used in the method of trees. The method of trees is a visual application of the **fundamental principle of counting**, a procedure for counting outcomes in multistage processes. Using this procedure, we can count how many patterns occur in a situation by looking at the number of ways in which the component parts can occur. For example, if Jack has 10 shirts and 4 pairs of trousers, he can wear $10 \times 4 = 40$ shirt–pants outfits. Each shirt can be worn with any of the pants. (This can be verified by drawing a tree diagram, but such a diagram is cumbersome for big numbers.)

The Fundamental Principle of Counting DEFINITION

In general, the **fundamental principle of counting** can be stated this way. If there are a ways of choosing one thing, b ways of choosing a second after the first is chosen, . . . , and z ways of choosing the last item after the earlier choices, then the total number of choice patterns is $a \times b \times c \times \cdots \times z$.

EXAMPLE 2 ■ Counting

Here are some examples of how to use the fundamental principle of counting:

1. In a restaurant there are 4 kinds of soup, 12 entrees, 6 desserts, and 3 drinks. How many different four-course meals can a patron choose? The four choices can be made in 4, 12, 6, and 3 ways, respectively. Hence, applying the fundamental principle of counting, there are $4 \times 12 \times 6 \times 3 = 864$ possible meals.
2. In a state lottery a contestant gets to pick a four-digit number that does not contain a zero followed by an uppercase or lowercase letter. How many such sequences of digits and a letter are there? Each of the four digits can be chosen in 9 ways (that is, 1, 2, . . . , 9), and the letter can be chosen in 52 ways (that is, A, B, . . . , Z plus a, b, . . . , z). Hence, there are $9 \times 9 \times 9 \times 9 \times 52 =$ 341,172 possible patterns.
3. A corporation is planning a musical logo consisting of four different ordered notes from the scale C, D, E, F, G, A, and B. How many logos are there to choose from? The first note can be chosen in 7 ways, but because reuse is not allowed, the next note can be chosen in only 6 ways. The remaining two notes can be chosen in 5 and 4 ways, respectively. Using the fundamental principle of counting, $7 \times 6 \times 5 \times 4 = 840$ musical logos are possible. If reuse of notes is allowed, $7 \times 7 \times 7 \times 7 = 2401$ logos are possible.

■

Let's now return to the problem of enumerating Hamiltonian circuits for the complete graph with n vertices. The city visited first after the home city can be chosen in $n - 1$ ways, the next city in $n - 2$ ways, and so on, until only one choice remains. Using the fundamental principle of counting, there are $(n - 1)! = (n - 1)(n - 2) \times \cdots \times 3 \times 2 \times 1$ routes. The exclamation mark in $(n - 1)!$ is read "factorial" and is shorthand notation for the product $(n - 1)(n - 2) \times \cdots \times 3 \times 2 \times 1$. For example, $5! = 5 \times 4 \times 3 \times 2 \times 1 = 120$.

As we saw in Figure 2.7, pairs of routes correspond to the same Hamiltonian circuit because one route can be obtained from the other by traversing the cities in reverse order. Thus, although there are $(n - 1)!$ possible routes, there are only half as many, or $(n - 1)!/2$, different Hamiltonian circuits. Now, if we have only a few cities to visit, $(n - 1)!/2$ Hamiltonian circuits can be listed and examined in a reasonable amount of time. Analysis of a six-city problem would require generation of $(6 - 1)!/2 = 5!/2 = 120/2 = 60$ tours. But for, say, 25 cities, $24!/2$ is approximately 3×10^{23}. Even if these tours could be generated at the rate of 1 million a second, it would take 10 billion years to generate them all. Because it would take so long to solve large vacation-planning problems using this method, it is sometimes referred to as a **brute force method** (that is, trying all the possibilities). Computer scientists and engineers have made it possible to market faster and faster computers. However, governments and businesses need to solve larger scale problems; say, for example, finding a Hamiltonian circuit in a graph with 10,000 vertices. If the methods one knows for solving such problems are not much better than brute force, then it's unlikely that even these faster computers can solve large versions of such problems. Mathematicians and computer scientists are actively seeking procedures that will significantly improve our ability to solve large versions of important problems.

2.2 Traveling Salesman Problem

If the only benefit were saving money and time in vacation planning, the difficulty of finding a minimum-cost Hamiltonian circuit in a complete graph with *n* vertices for large values of *n* would not be of great concern. However, the problem we are discussing is one of the most common in *operations research,* the branch of mathematics concerned with getting governments and businesses to operate more efficiently. This problem is usually called the **traveling salesman problem (TSP)** because of its early formulation: Determine the trip of minimum cost that a salesperson can make to visit the cities in a sales territory, starting and ending the trip in the same city.

Many situations require solving a TSP:

1. A lobster fisherman has set out traps at various locations and wishes to pick up his catch.
2. The telephone company wishes to pick up the coins from its pay telephone booths. (To avoid the high cost of picking up these coins, phone companies in many countries have adopted a system that uses prepurchased phone cards to operate phones. This means that there are no coins to collect.) Due to the increased use of cell phones, fewer pay phones are available.
3. The electric (or gas) company needs to design a route for its meter readers.
4. A minibus must pick up six day campers, deliver them to camp, and return them home later in the day.
5. In drilling holes in a series of plates, the drill press operator (perhaps a robot) must drill the holes in a predetermined order.
6. Physical records generated at automated teller machine (ATM) locations–as backup in case of failure of the electronic systems–must be picked up periodically.
7. A limousine service with a van located at an airport must pick up five customers and deliver them to the airport in time to catch their flights.

Perhaps surprisingly, TSP problems are also solved regularly in the design of computer chips. The components must be located so that the machines involved in the assembly can insert them on the chips as efficiently as possible. Because many chips are manufactured, even a small improvement in the time needed to make a chip can save a lot of money.

The meaning of *cost* can vary from one formulation of a TSP to another. We can measure cost as distance, airplane ticket prices, time, or any other factor that is to be optimized. In many situations, the TSP arises as a subproblem of a more complicated problem. For example, a supermarket chain may have a very large number of stores to be served from a single large warehouse. If there are fewer trucks than stores, the stores must be grouped into clusters so that one truck can serve each cluster. If we then solve the TSP for every truck, we can minimize total costs for the supermarket chain. Similar vehicle-routing problems–for dial-a-ride services that transport senior citizens to activity centers, for example, or that deliver children to their schools or camps–often involve solving the TSP as a subproblem.

2.3 Helping Traveling Salesmen

Because the traveling salesman problem arises so often in situations where the associated complete graphs would be very large, we must find a faster method than the brute force method we have described. We need to look at our original problem in Figure 2.3 and try to find an alternative algorithm for solving it. Recall that our goal is to find the minimum-cost Hamiltonian circuit.

Nearest-Neighbor Algorithm — PROCEDURE

Starting from the home city, first visit the nearest city, then visit the nearest city that has not already been visited. We return to the start city when no other choice is available. This approach is called the **nearest-neighbor algorithm**.

EXAMPLE 3 ■ Applying the Nearest-Neighbor Algorithm

Applying this algorithm to the TSP in Figure 2.3 quickly leads to the tour of Chicago, St. Louis, Cleveland, Minneapolis, and Chicago, with a length of 2040 miles. Here is how this tour is determined. Because we are starting in Chicago, there is a choice of going to a city that is 425, 300, or 349 miles away. Because the smallest of these numbers is 300, we next visit St. Louis, which is the nearest neighbor of Chicago not already visited. At St. Louis, we have a choice of visiting next cities that are 541 or 562 miles away. Hence, Cleveland, which is nearer (541 miles), is visited. To complete the tour, we visit Minneapolis and return to Chicago, thereby adding 774 and 425 miles to the length of the tour. ■

The nearest-neighbor algorithm is an example of a **greedy algorithm**, because at each stage a best (greedy) choice, based on an appropriate criterion, is made. Unfortunately, this is not the optimal tour, which we saw was C–M–S–CL–C, for a total length of 1877 miles. Making the best choice at each stage may not yield the best "global" solution. However, even for a large TSP, one can always find a nearest-neighbor route quickly.

EXAMPLE 4 ■ Applying the Nearest-Neighbor Algorithm Revisited

Figure 2.10 again illustrates the ease of applying the nearest-neighbor algorithm, this time to a weighted complete graph with five vertices. Starting at vertex *A*, we get the tour *ADECBA* (cost 2800) (Figure 2.10a). Note that the nearest-neighbor algorithm starting at vertex *B* yields the tour *BCADEB* (cost 3050) (Figure 2.10b). ■

This example illustrates that a nearest-neighbor tour can be computed for each vertex of the complete graph being considered and that different nearest-neighbor tours can be obtained starting at different vertices. Thus, even though we may seek a tour starting at a particular vertex–say, *A* in Figure 2.10–because a Hamiltonian circuit can be thought of as starting at any of its vertices, we can just as easily apply the nearest-neighbor procedure starting at vertex *B* (rather than at *A*). The Hamiltonian circuit we get can still be thought of as beginning at vertex *A* rather than *B*. Even for complete graphs with a large number of vertices, it would still be faster to

Recently, mathematical researchers have adopted a somewhat different strategy for dealing with TSP problems. If finding a fast algorithm to generate optimal solutions for large problems is unlikely, perhaps we can show that the quick-and-dirty methods, usually called **heuristic algorithms**, come close enough to giving optimal solutions to be important for practical use. For example, suppose we could prove that the nearest-neighbor heuristic was never off by more than 25% in the worst case or by more than 15% in the average case. For a medium-sized TSP, we would then have to choose whether to spend a lot of time (or money) to find an optimal solution or instead to use a heuristic algorithm to obtain a fairly good solution. Investigators at AT&T Research have developed many remarkably good heuristic algorithms. The best-known guarantee for a heuristic algorithm for a TSP is that it yields a cost no worse than one and a half times the optimal cost. Interestingly, this heuristic algorithm involves solving a Chinese postman problem (see Chapter 1), for which a "fast" algorithm is known to exist.

Throughout our discussion of the TSP, we have concentrated on the goal of minimizing the cost (or time) of a tour that visited each of a variety of sites once and only once. However, the subtle issues that arise in specific real-world situations (or that provide a contrast between seemingly similar situations) are the things that make mathematical modeling exciting. For example, suppose the TSP situation is to pick up day campers and take them to and from the camp. The camp wants to minimize the total length of time that the bus needs to pick up the campers. The parents of the campers, however, may want to minimize the time their children spend on the bus. For some problems, the tour that minimizes the mean (average) time that a child spends on the bus may not be the same tour that minimizes the total time of the tour. (Specifically, if the bus first picks up the child who lives the farthest from the camp, and then picks up the other children, this may yield a relatively short time on the bus for the kids but a relatively long time for the tour itself.) Mathematicians return to examine these subtleties between problems at a later time, after the basic structure of the main problem itself is well understood. It is in this way that mathematics continues to grow, explore new ideas, and find new applications.

2.4 Minimum-Cost Spanning Trees

The traveling salesman problem is but one of many graph theory optimization problems that have grown out of real-world problems in both government and industry. Here is another.

Some videophone (pictaphone) and video conferencing requires the creation of specialized networks for which the techniques of finding a minimum-cost spanning tree are required. (*Digital Vision/Punchstock.*)

EXAMPLE 7 ▪ Pictaphone Service

Imagine that Pictaphone service (telephone service that provides a video image of the callers) will be set up on an experimental basis among five cities. The graph in Figure 2.12 shows the possible links that might be included in the Pictaphone network, with each edge showing the cost in millions of dollars to create that particular link. To send a Pictaphone message between two cities, a direct communication link is not necessary because it is possible to send a message indirectly via another city. Thus, in Figure 2.12, sending a message from A to C could be achieved by sending the message from A to B, from B to E, and from E to C, provided the links AB, BE, and EC are part of the network. We assume that the cost of relaying a message, compared with the cost of the direct communication link, is so small that we can neglect this amount. The problem that concerns us, therefore, is to provide service between any pair of cities in a way that minimizes the total cost of the links.

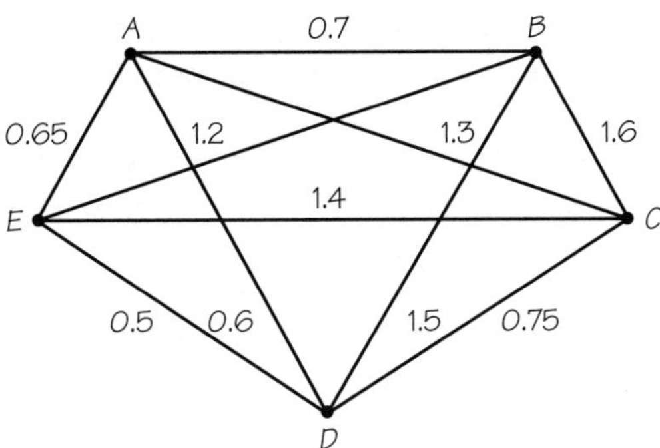

FIGURE 2.12 Costs (in millions of dollars) of installing Pictaphone service among five cities.

Our first guess at a solution is to put in the cheapest possible links between cities first, until all cities could send messages to any other city. Such an approach is analogous to the sorted-edges method that was used to study the traveling salesman problem. In our example, if the cheapest links are added until all cities are joined, we obtain the connections shown in Figure 2.13a.

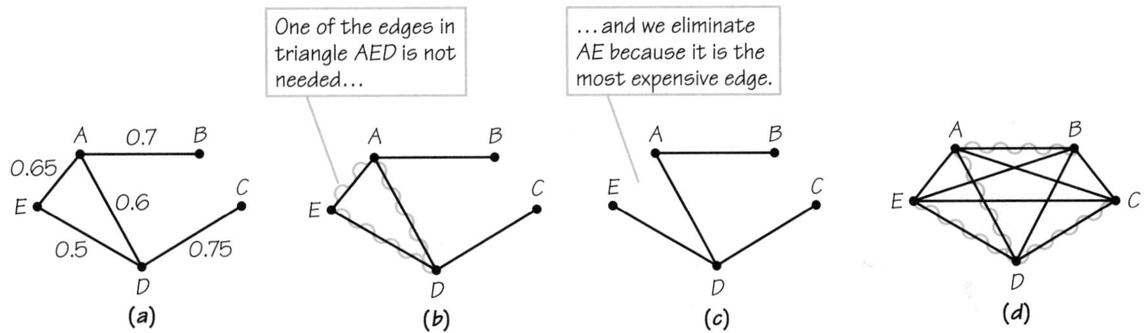

FIGURE 2.13 (a) Cities are linked in order of increasing cost until all cities are connected. (b) Circuit in part (a) highlighted. (c) Most expensive link in circuit in part (a) deleted. (d) Highlighted edges show, as a subgraph of the original graph, those links connecting the cities with minimum cost, obtained using Kruskal's algorithm.

The links were added in the order *ED, AD, AE, AB, DC*. However, because this graph contains the circuit *ADEA* (wiggly edges in Figure 2.13b), it has redundant edges: We can still send messages between any pair of cities using relays after omitting the most expensive edge in the circuit–*AE*. After deleting an edge of a circuit, a message can still be relayed among the cities of the circuit by sending signals the long way around. After *AE* is deleted, messages from *A* to *E* can be sent via *D* (Figure 2.13c). These ideas constitute a procedure developed by Joseph Kruskal (AT&T Research) in 1956.

Kruskal's Algorithm

PROCEDURE

Kruskal's algorithm: Add links in order of cheapest cost so that no circuits form and so that every vertex belongs to some link added (Figure 2.13d).

In Kruskal's procedure, as in the sorted-edges method for the TSP, the edges that are added need not be connected to each other until the end. A subgraph formed in this way will be a **tree;** that is, it will consist of one piece and contain no circuits. It will also include all the vertices of the original graph. A subgraph that is a tree and that contains all the vertices of the original graph is called a **spanning tree** of the original graph.

To understand these concepts better, consider the graph *G* in Figure 2.14a. The wiggly edges in Figure 2.14b would constitute a subgraph of *G* that is a tree (because it is connected and has no circuit), but this tree would not be a spanning tree of *G* because the vertices *D* and *E* would not be included. On the other hand, the wiggly edges in Figure 2.14c and 2.14d show subgraphs of *G* that include all the vertices of *G* but are not trees because the first is not connected and the second contains a circuit. Figure 2.14e shows a spanning tree of *G;* the wiggly edges are connected and contain no circuit, and every vertex of the original graph is an endpoint of some wiggly edge.

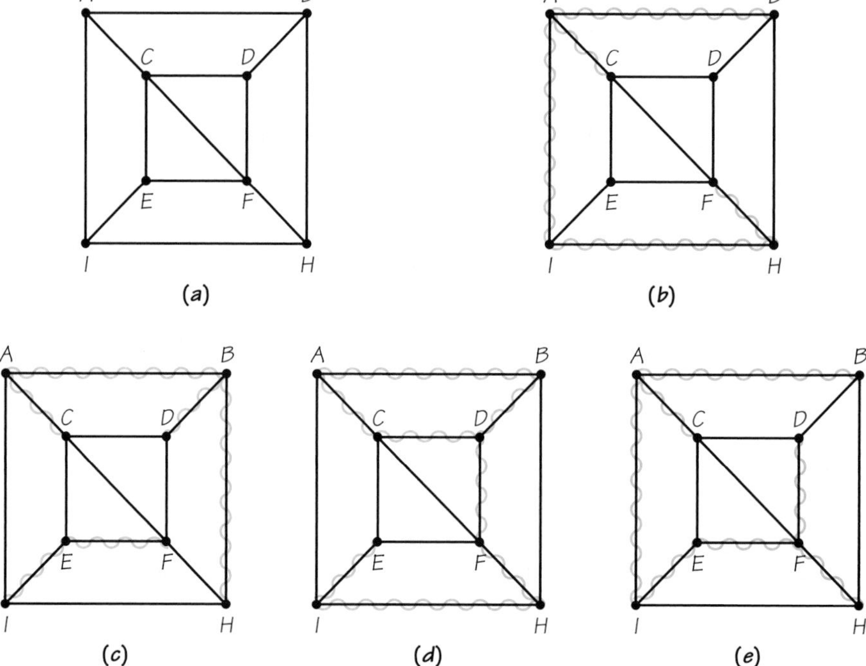

FIGURE 2.14 (a) A graph to help illustrate the concept of a spanning tree. (b) The wiggly edges are a tree, but not a spanning tree, because vertices *D* and *E* are not part of the tree. (c) The wiggly edges are not a tree, because they are not connected. All of the vertices of the graph are, however, end points of wiggly edges. (d) The wiggly edges are not a tree, because they contain the edges of the circuit *BDCAB.* All the vertices of the graph are, however, endpoints of wiggly edges. (e) The wiggly edges form a tree and include all of the vertices of the graph as endpoints of wiggly edges. Thus, the wiggly edges are a spanning tree.

Finding a **minimum-cost spanning tree**–that is, a spanning tree whose edge weights sum to a minimum value–solves the Pictaphone problem. Note that having a different goal in the Pictaphone problem led to a different mathematical question from that of finding a Chinese postman tour or TSP tour. This application required that we find a minimum-cost spanning tree. In Figure 2.15a we have a graph model showing the costs of putting in roads to connect new houses in a suburban land-development project. Applying Kruskal's algorithm–adding the edges in the order of increasing cost, but avoiding the creation of a circuit–yields a minimum-cost spanning tree, indicated by the wiggly edges in Figure 2.15b. This tree is the cheapest one that makes it possible to drive between any pair of homes, though the driving distance between some of the homes will be relatively large, because only roads corresponding to wiggly edges will be built.

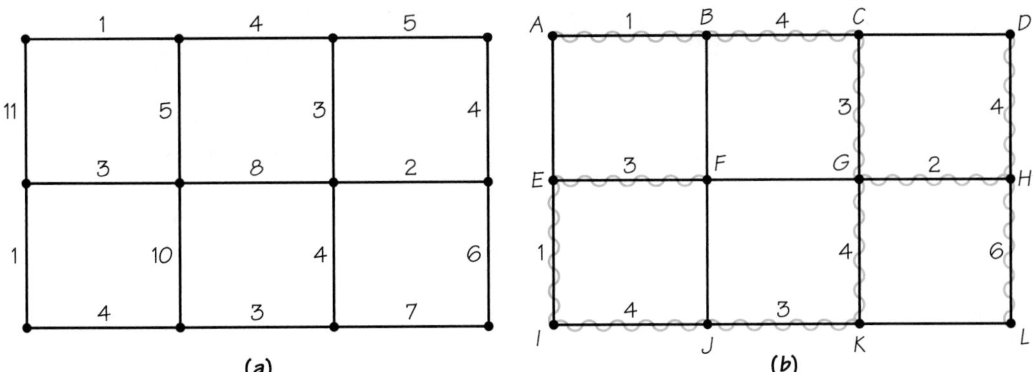

FIGURE 2.15 (a) A graph showing costs for construction of roads between houses. (b) Wiggly edges show a minimum-cost spanning tree for the graph in part (a).

Remember that the weights on the edges of the graph in Figure 2.15a represent the costs of building roads, not the driving distances between the houses. Note that Figure 2.15a is not a complete graph, one in which all possible edges are included. Edges that correspond to roads that would be economically prohibitive to build have not been shown in the graph model. Also, in Figure 2.15b, the two edges of weight 5 (shown in Figure 2.15a) do not become part of the minimum-cost spanning tree, because they would create circuits with edges already chosen.

Although Kruskal's algorithm worked in our example, how do we know that the spanning tree found by this algorithm will always achieve the minimum possible cost? While this sounds very plausible, our experience with the TSP should suggest caution. Remember that for the TSP, the sorted-edges algorithm did not necessarily give an optimal solution even though it is a greedy algorithm like Kruskal's. On what basis should we have more faith in Kruskal's algorithm?

Kruskal proposed his algorithm as a way to solve a pure mathematics problem put forward by Czechoslovakian mathematician Otakar Borůvka. In mathematics it is surprising but not uncommon to find that ideas used to solve problems with no apparent application often turn out to have many real-world uses. Kruskal's solution to the problem of finding a minimum-cost spanning tree in a graph with weights is a good example of this phenomenon.

Kruskal showed that the greedy algorithm described does yield the minimum answer, and his work led to applications of these and related ideas in designing minimum-cost computer networks, phone connections, sewer systems, and road and railway systems. For additional discussion of operations research in the communications industry, see Spotlight 2.2. To explore how one can reconstruct full information from partial information using the tree concept, see Spotlight 2.3.

In our discussion of routing problems in graphs, we have not touched on one of the most obvious ones: finding the path between two specified, distinct vertices while keeping the sum of the weights of the edges in the path as small as possible. (Here there is no need to cover all vertices or to cover all edges.) We have seen that the weights on the edges have many possible interpretations, including time, distance, and cost. The following are some of the many possible applications:

1. Design routes to be used by an ambulance, police car, or fire engine to get to an emergency as quickly as possible.
2. Design delivery routes that minimize gasoline use.
3. Design routes to bring soldiers to the front as quickly as possible.
4. Design a route for a truck carrying nuclear waste.

SPOTLIGHT 2.2 AT&T Manager Explains How Long-Distance Calls Run Smoothly

Although long-distance calls are now routine, it takes great expertise and careful planning for a company like AT&T to handle its vast amounts of telephone traffic. Rich Wetmore was district manager of AT&T's Communications Network Operations Center in Bedminster, New Jersey. Here are his responses to questions about how AT&T handles its huge volumes of long-distance traffic and how it tracks its operations to keep things running smoothly.

How do you make sure that a customer doesn't run into a delayed signal when attempting a long-distance call?

We monitor the performance of our AT&T network by displaying data collected from all over the country on a special wallboard. The wallboard is configured to tell us if a customer's call is not going through because the network doesn't have enough capacity to handle it.

That's when we step in and take control to correct the problem. The typical control we use is to reroute the call. Instead of sending the customer's call directly to its destination, we'll route it via a third city—to someplace else in the country that has the capacity to complete the call.

It would seem that routing via another city would take longer. Is the customer aware of this process?

Routing a call via a third city is entirely transparent [imperceptible] to the customer. I'm an expert about the network, and even when I make a phone call, I have no idea how that individual call was routed. It's transparent both in terms of how far away the other person sounds and in how quickly the telephone call gets set up. With the signaling network we use, it takes milliseconds for switching systems to "talk" to each other to set up a call. So the fact that you are involved in a third switch in some distant city is something you would never know.

You want to be sure to keep costs down while supplying enough service to customers. So how do you balance company benefits with customer benefit?

In terms of making the network efficient, we want to do two things. First, we want our customers to be happy with our service and for all their calls to go through, which means we must build enough capacity in the network to allow that to happen. Second, we want to be efficient for stockholders and not spend more money than we need to for the network to be at the optimum size.

There are basically two costs in terms of building the network. There is the cost of switching systems and the cost of the circuits that connect the switching systems. Basically, you can use operations-research techniques and mathematics to determine cost trade-offs. It may make sense to build direct routing between two switching systems and use a lot of circuits, or maybe to involve three switching systems, with fewer circuits between the main two, and so on.

Many people find it increasingly convenient to use the Web or software installed in their cars to get driving directions and driving time estimates to a place they wish to visit. The software that provides this information relies on algorithms that compute the shortest-path route in an appropriate weighted graph, which involves distances or times.

The need to find shortest paths seems natural. Next we investigate a situation in which finding a *longest* path is the right tool.

2.5 Critical-Path Analysis

Mathematics can confirm the obvious in certain situations while showing that our intuition is wrong in other circumstances. Our next group of applications will illustrate this point.

SPOTLIGHT 2.3 Common Ancestors?

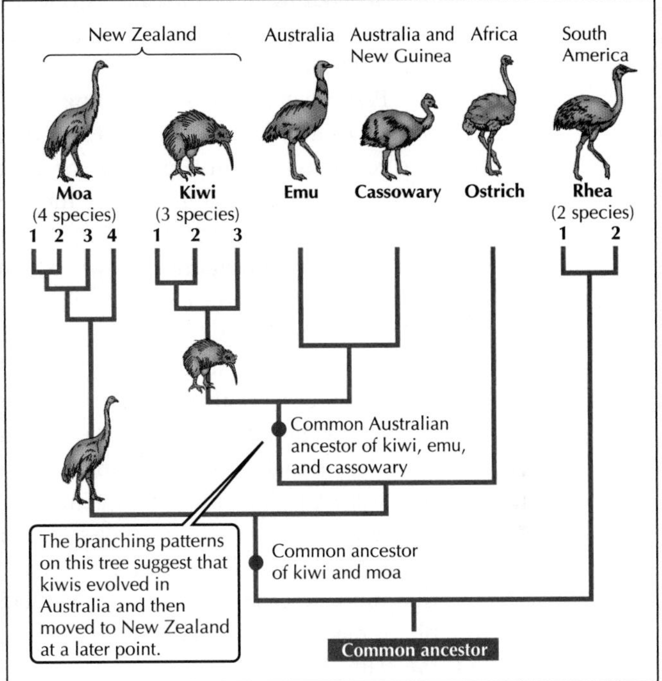

In the study of ancient manuscripts, different manuscripts of the same book are available, even though the original manuscript upon which they are based has been lost. Examples of this include Euclid's *Elements* and Chaucer's *Canterbury Tales*. What interests scholars is reconstructing the relationships between the manuscripts and the common ancestors of the manuscripts, even when some of the ancestors are now missing.

Similarly, perceptual psychologists may be interested in which colors people perceive as being closely related and comparing these perceptions with those of people who are color-blind. Linguists are interested in the connections between languages that seem very different today, but have some words that are similar. Finally, in studying different species, biologists are interested in determining which species are more closely related to each other, including species known only in fossil form, and constructing a "tree" of life that shows which species were ancestors of others.

Reconstructions of this kind are made possible by using graph theory, specifically using the graph theory concept of a tree. The value of the graph theory in these and many other situations lies in using the distance between pairs of vertices in the tree as a way of reflecting the closeness of the relationships that pairs of manuscripts, pairs of colors, or pairs of species have. The distance between two vertices in a tree is the sum of the weights along the one path that joins the two vertices. If there are no weights on the edges, the distance is the number of edges in the path. In some reconstruction problems, a special vertex of the tree called the *root* is singled out. This root plays the role of the original common ancestors, and distances to the root are of critical interest.

In the case of species, trees of family relatedness were traditionally constructed based on similarities of bones and physical appearance. With the discovery of molecular biology, many new avenues have been opened up. We can now draw trees of relatedness based on an organism's genetic material, DNA, or the proteins that the DNA codes for. The traditional trees based on physical traits often show different species as being more closely related than trees based on newer molecular biological approaches. These differences focus scholars on how to resolve the discrepancies and thereby reach a deeper understanding of the unity of life.

A characteristic of American life is its fast pace. People are interested in getting things done quickly and efficiently. This means that when you take your car in to be repaired before going to work, you want to know for sure that the repairs will be done when you pick the car up. You want the trains and the bus that take you to your doctor's appointment to run on time. When you arrive at the doctor's office, you want a technician to be free to take a blood sample and a throat culture. You want your outpatient appointment for an X-ray at the local hospital to occur on schedule. You want the X-ray to be interpreted quickly and the results reported back to your internist.

Scheduling machines and people is a big part of modern life. It is involved in running a school, a hospital, an airline, or in landing a person on Mars, and modern mathematics plays a big part in solving scheduling problems.

Part of what makes scheduling complicated is that the tasks that make up a job usually cannot be done in a random order. For example, to make Thanksgiving dinner you must buy and prepare the turkey before putting it in the oven, and you must set the table before serving the food.

If the tasks cannot be performed in a random order, we can specify the order in an **order-requirement digraph**. The term *digraph* is short for "directed graph." A digraph is a geometrical tool similar to a graph except that each edge has an arrow on it to indicate a direction for that edge. Digraphs can be used to illustrate that traffic on a street must go in one direction or that certain tasks in a job must be completed before other tasks. A typical example of an order-requirement digraph is shown in Figure 2.16. There is a vertex in this digraph for each task. If one task must be done immediately before another, we draw a directed edge, or arrow, from the prerequisite task to the subsequent task. The numbers within the circles representing vertices are the times it takes to complete the tasks. In Figure 2.16 there is no arrow from T_1 to T_5 because task T_2 intervenes. Also, T_1, T_7, and T_8 have no tasks that must precede them. Hence, if there are at least three processors (such as people or machines) available, tasks T_1, T_7, and T_8 can be worked on simultaneously at the start of the job.

Let's investigate a typical scheduling problem faced by a business.

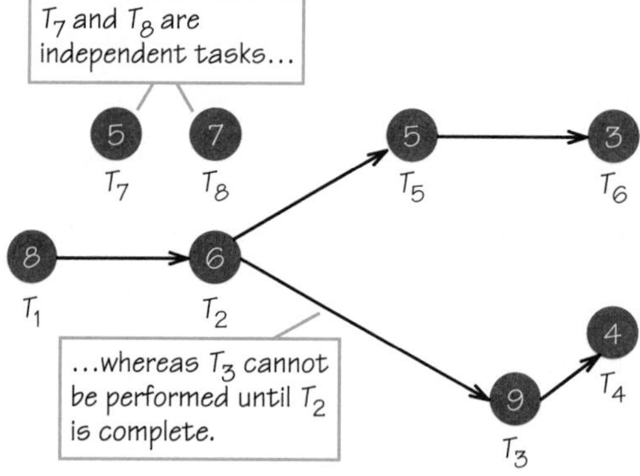

FIGURE 2.16 A typical order-requirement digraph.

EXAMPLE 8 ■ Turning a Plane Around

Consider an airplane that carries both freight and passengers. The plane must have its passengers and freight unloaded and new passengers and cargo loaded before it can take off again. Also, the cabin must be cleaned before departure can occur. Thus, the job of "turning the plane around" requires completion of five tasks:

TASK *A*	Unload passengers	13 minutes
TASK *B*	Unload cargo	25 minutes
TASK *C*	Clean cabin	15 minutes
TASK *D*	Load new cargo	22 minutes
TASK *E*	Load new passengers	27 minutes

Turning a plane around, which involves such tasks as refueling, unloading, and then reloading cargo and passengers, entails very careful scheduling to avoid time slippage. (*David Butow/Corbis Saba.*)

The order-requirement digraph for the problem of turning an airplane around is shown in Figure 2.17. The presence or absence of an edge in the order-requirement digraph depends on the analysis made as part of the modeling process for the problem. It seems natural that we need an arrow between task A and task C, because the passengers have to be unloaded before the cabin is cleaned. Other arrows may not seem natural–say, perhaps the arrow from task B (unload the cargo) to task E (load new passengers). This arrow may be due to government rules or union requirements.

What matters is that the mathematics of solving the problem does not depend on the reason that the order-requirement digraph looks the way that it does. The person solving the problem constructs the order-requirement digraph and then the mathematical techniques we will develop can be applied, regardless of whether or not another business faced with a similar problem might model the problem in a different way. Because we want to find the earliest completion time, it might seem that finding the shortest path in the digraph (path BD with time length $25 + 22 = 47$) would solve the problem. But this approach shows the danger of ignoring the relationship between the mathematical model (the digraph) and the original problem.

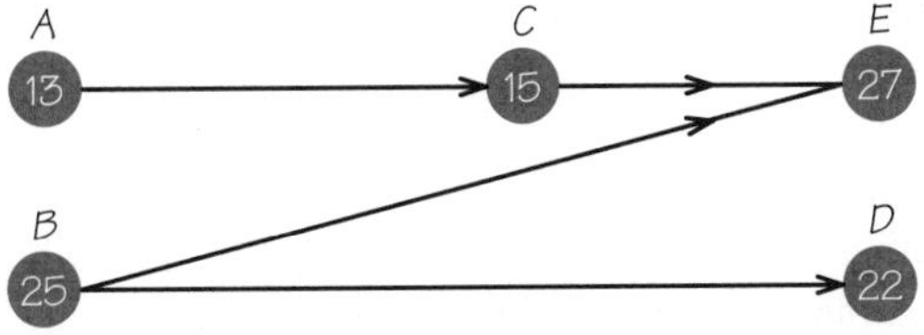

FIGURE 2.17 An order-requirement digraph for turning an airplane around after landing.

The time required to complete all the tasks, A through E, must be at least as long as the time necessary to do the tasks on any particular path. Consider the path BD, which has length $25 + 22 = 47$. Recall that here *length* of a path refers to the sum of the times of the tasks that lie along the path. Because task B must be done before task D can begin, the two tasks B and D cannot be completed before time 47. Hence, even if work on other tasks (such as A, C, and E) proceeds during this period, all the tasks cannot be finished before the tasks on path BD are finished. The same statement is true for every other path in the order-requirement digraph. Thus, the earliest completion time actually corresponds to the length of the longest path. In the airplane example, this earliest completion time is 55 ($= 13 + 15 + 27$) minutes, corresponding to the path ACE. We call ACE the **critical path** because the times of the tasks on this path determine the earliest completion time.

Critical Path

DEFINITION

A **critical path** in an order-requirement digraph is a longest path. The length is measured in terms of summing the task times of the tasks making up the path.

Note that if none of these tasks could go on simultaneously, the time to complete all the tasks would be $13 + 25 + 15 + 22 + 27 = 102$ minutes. However, even though some tasks may be performed simultaneously, the fact that the length of the critical path is 55 means that completion of the tasks in less than 55 minutes is not possible. Only by speeding up the times to complete the critical-path tasks themselves can a completion time less than 55 minutes be achieved.

Suppose it were desirable to speed the turnaround of the plane to less than 55 minutes. One way to do this might be to build a second jetway to help unload passengers more quickly. For example, we could unload passengers (task A) in 7 minutes instead of 13. However, reducing task A to 7 minutes does not reduce the completion time by 6 minutes, because in the new digraph (Figure 2.18) ACE is no longer the critical (longest) path. The longest path is now BE, which has a length of 52 minutes. Thus, shortening task A by 6 minutes results in only a 3-minute saving in completion time. This may mean that building a new jetway is uneconomical.

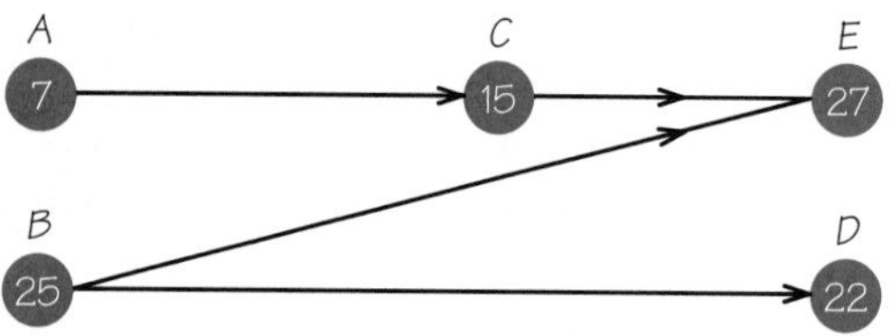

FIGURE 2.18 An order-requirement digraph for turning an airplane around in reduced time due to construction of a new jetway.

Note also that shortening the time to complete tasks that are not on the original critical path ACE will not shorten the completion time at all. Speeding tasks on the critical path will shorten completion time of the job only up to the point where a new critical path is created. Also note that a digraph may have more than one longest path.

Not all order-requirement digraphs are as simple as the one shown in Figure 2.17. The order-requirement digraph in Figure 2.19 has 12 paths, which can be found by exhaustive search. Examples of such paths are $T_1T_2T_3$, $T_1T_5T_9$, $T_4T_5T_9$, and $T_7T_5T_3$. (Although we have not discussed them here, fast algorithms for finding longest and shortest paths in graphs are known.) The critical path is $T_7T_8T_6$ (length 21), and the earliest completion time for all nine tasks is time 21, though the actual completion time may be later than time 21 depending on the resources available to carry out the tasks. Completing the tasks by time 21 depends on having sufficient resources available so that some of the tasks can be worked on simultaneously.

These examples are typical of many scheduling problems that occur in practice (see Spotlight 2.4). Perhaps the most dramatic use of critical-path analysis is in the construction trades. No major new building project is now carried out without a critical-path analysis first being performed to ensure that the proper personnel and materials are available at the right times in order to have the project finished as quickly as possible. Many such problems are too large and complicated to be solved without the aid of computers.

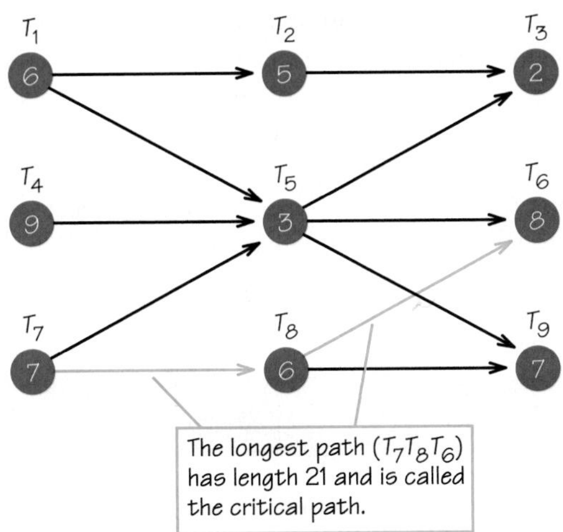

FIGURE 2.19 An order-requirement digraph with 12 paths, to examine how to find the length of the longest path.

SPOTLIGHT 2.4 Every Moment Counts in Rigorous Airline Scheduling

When people think of airline scheduling, the first thing that comes to mind is how quickly a particular plane can safely reach its destination. But using ground time efficiently is just as important to an airline's timetable as the time spent in flight. Bill Rodenhizer was the manager of control operations for an airline that provided shuttle service between Boston and New York. He is considered to be an expert on airplane turnaround time, the process by which an airplane is prepared for almost immediate takeoff once it has landed. He tells us how this well-orchestrated effort works:

> Scheduling, to the airline, is just about the whole ball game. Everything is scheduled right to the minute. The whole fleet operates on a strict schedule. Each of the departments responsible for turning around an aircraft has an allotted period of time in which to perform its function. Manpower is geared to the amount of ground time scheduled for that aircraft. This would be adjusted during off-weather or bad-weather days or during heavy air-traffic delays.
>
> Most of our aircraft in Boston are scheduled for a 42- to 65-minute ground time. Boston is the end of the line, so it is a "terminating and originating station." In plain talk, that means almost every aircraft that comes in must be fully unloaded, refueled, serviced, and dispatched within roughly an hour's time.
>
> This is how the process works: In the larger aircraft, it takes passengers roughly 20 minutes to load and 20 minutes to unload. During this period, we will have completely cleaned the aircraft and unloaded the cargo, and the caterers will have taken care of the food. The ramp service may take 20 to 30 minutes to unload the baggage, mail, and cargo from underneath the plane, and it will take the same amount of time to load it up again. We double-crew those aircraft with heavier weights so that the workload will fit the time it takes passengers to load and unload upstairs.
>
> While this has been going on, the fueler has fueled the aircraft. As to repairs, most major maintenance is done during the midnight shift, when [most of our] several hundred aircraft are inactive. We all work under a very strict time frame.

New security requirements in the wake of the World Trade Center attack (9/11/2001) have increased the difficulty of adhering to timetables in operating shuttle services between East Coast cities such as New York and Boston. This makes it even more important to use analytical tools in keeping operations on schedule.

The critical-path method was popularized and came into wider use as a consequence of the *Apollo* project. This project, which aimed at landing a man on the moon within 10 years of 1960, was one of the most sophisticated projects in planning and scheduling ever attempted. The dramatic success of the project can be attributed partly to the use of critical-path ideas and the related program evaluation and review technique (PERT), which helped keep the project on schedule.

In Chapter 3, we will see how mathematical ideas drawn from outside of graph theory can be used to gain insight into scheduling problems.

REVIEW VOCABULARY

Algorithm A step-by-step description of how to solve a problem. (p. 34)

Brute force method The method that solves the traveling salesman problem (TSP) by enumerating all the Hamiltonian circuits and then selecting the one with minimum cost. (p. 37)

Complete graph A graph in which every pair of vertices is joined by an edge. (p. 36)

Critical path The longest path in an order-requirement digraph. The length of this path gives the earliest completion time for all the tasks making up the job consisting of the tasks in the digraph. (p. 49)

Fundamental principle of counting A method for counting outcomes of multistage processes. (p. 36)

Greedy algorithm An approach for solving an optimization problem, where at each stage of the algorithm the best (or cheapest) action is taken. Unfortunately, greedy algorithms do not always lead to optimal solutions. (p. 39)

Hamiltonian circuit A circuit using distinct edges of a graph that starts and ends at a particular vertex of the graph and visits each vertex once and only once. A Hamiltonian circuit can start at any one of its vertices. (p. 31)

Heuristic algorithm A method of solving an optimization problem that is "fast" but does not guarantee an optimal answer to the problem. (p. 42)

Kruskal's algorithm An algorithm developed by Joseph Kruskal (AT&T Research) that solves the minimum-cost spanning-tree problem by selecting edges in order of increasing cost, but in such a way that no edge forms a circuit with edges chosen earlier. It can be proved that this algorithm always produces an optimal solution. (p. 43)

Method of trees A visual method of carrying out the fundamental principle of counting. (p. 34)

Minimum-cost Hamiltonian circuit A Hamiltonian circuit in a graph with weights on the edges, for which the sum of the weights of the edges of the Hamiltonian circuit is as small as possible. (p. 34)

Minimum-cost spanning tree A spanning tree of a weighted connected graph having minimum cost. The cost of a tree is the sum of the weights on the edges of the tree. (p. 44)

Nearest-neighbor algorithm An algorithm for attempting to solve the TSP that begins at a "home" vertex and visits next that vertex not already visited that can be reached most cheaply. When all other vertices have been visited, the tour returns to home. This method may not give an optimal answer. (p. 39)

NP-complete problems A collection of problems, which includes the TSP, that appear to be very hard to solve quickly for an optimal solution. (p. 41)

Order-requirement digraph A directed graph that shows which tasks precede other tasks among the collection of tasks making up a job. (p. 48)

Sorted-edges algorithm An algorithm for attempting to solve the TSP where the edges added to the circuit being built up are selected in order of increasing cost, but no edge is chosen that would prevent a Hamiltonian circuit from forming. These edges must all be connected at the end, but not necessarily at earlier stages. The tour obtained may not have the lowest possible cost. (p. 40)

Spanning tree A subgraph of a connected graph that is a tree and includes all the vertices of the original graph. (p. 44)

Traveling salesman problem (TSP) The problem of finding a minimum-cost Hamiltonian circuit in a complete graph where each edge has been assigned a cost (or weight). (p. 38)

Tree A connected graph with no circuits. (p. 34)

Weight A number assigned to an edge of a graph that can be thought of as a cost, distance, or time associated with that edge. (p. 34)

✓ SKILLS CHECK

1. Which of the following describes a Hamiltonian circuit for the graph below?

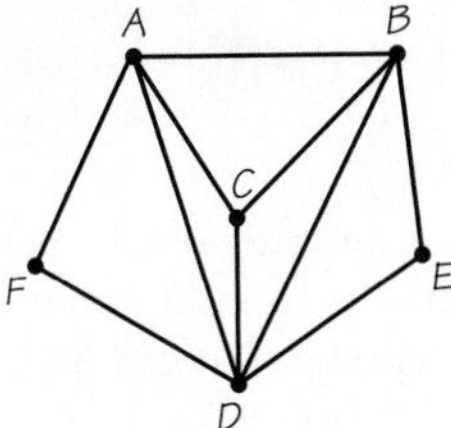

(a) *ABCDFA*
(b) *AFDCBE*
(c) *ACBEDFA*
(d) *ACEBDFA*

2. The cost of the nearest-neighbor tour (Hamiltonian circuit) that starts at vertex *A* for the graph below is ______________.

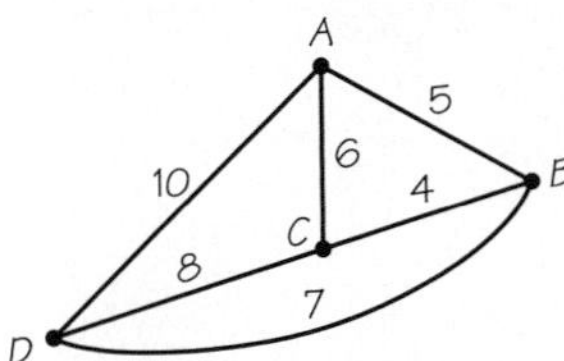

3. Suppose that after a hurricane, a van is dispatched to pick up five nurses at their homes and bring them to work at the local hospital. Which of these techniques is most likely to be useful in solving this problem?

(a) Finding an Euler circuit in a graph
(b) Solving a TSP (traveling salesman problem)
(c) Finding a minimum-cost spanning tree in a graph

4. The cost of the sorted-edges tour (Hamiltonian circuit) for the graph below is ______________.

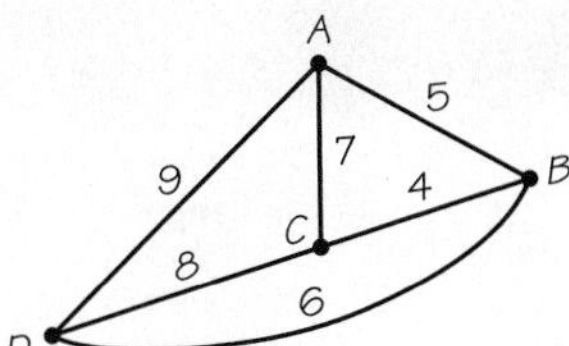

5. The graph shown below has

(a) no Hamiltonian circuit and no Euler circuit.
(b) an Euler circuit and a Hamiltonian circuit.
(c) no Hamiltonian circuit, but it has an Euler circuit.

6. The cost of the nearest-neighbor traveling salesman tour that starts at *B* for the following graph is ________.

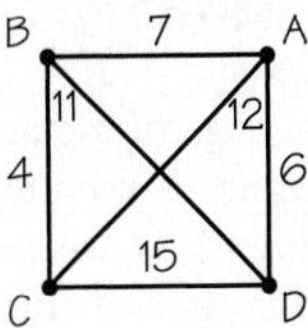

7. When the sorted-edges method and nearest-neighbor method are applied to a complete graph on seven vertices with nonnegative weights,

(a) both methods always give the same optimal answer.
(b) both methods always give the same answer but that answer may not be optimal.
(c) neither method may give an optimal answer.

8. If a graph has *E* edges and *V* vertices as well as a Hamiltonian circuit, then the number of edges in the Hamiltonian circuit is ______________.

9. Paul has packed four ties, three shirts, and two pairs of pants for a trip. How many different outfits can he create if he never wears a tie?

(a) Fewer than 10
(b) Between 10 and 25
(c) More than 25

10. The number of different lunches that Jules can design by selecting one of three meats, one of three salads, and one of six vegetables is exactly ______________.

11. An ice-cream shop offers 3 types of cones, 20 flavors, and 4 different toppings (crushed peanuts, crushed almonds, chocolate bits, or corn flakes). If a customer is allergic to nuts, how many different choices can she choose from?

(a) 240
(b) 120
(c) 25

12. If a three-character password system must begin with a lowercase letter of the English alphabet followed by two decimal digits that may be repeated, the number of different possible passwords is ______________.

13. Assuming a graph with *E* edges and *V* vertices has a minimum-cost spanning tree *T*, which of the following statements *must* be true?

(a) The tree *T* has exactly *V* edges.
(b) The tree *T* includes every minimum-cost edge.
(c) The graph is connected.

14. When arranged in increasing order, the weights of the edges in the following graph that are not part of the minimum-cost spanning tree selected when Kruskal's algorithm is applied are ________, ________, ________.

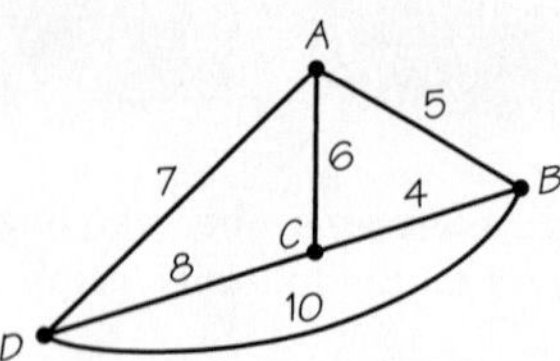

15. Assume that every edge of a graph G has a different cost. If Kruskal's algorithm is used to find the minimum-cost spanning tree T for graph G, which of the following statements *must* be true?

(a) Any other spanning tree for graph G will have more edges than T.
(b) Any other spanning tree for graph G will have a greater cost than T.
(c) The edge of graph G having greatest weight is included in T.

16. The smallest positive integer valued weight that x can have in the graph below so that it could not be selected by Kruskal's algorithm as an edge of a minimum-cost spanning tree is ______________________ .

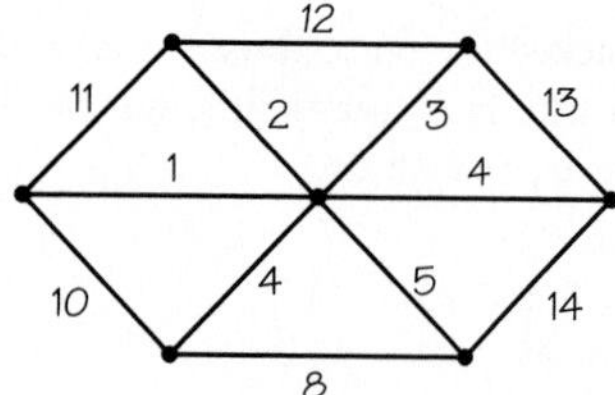

17. If a graph contains a circuit, which of the following statements is true?

(a) The graph cannot be a tree.
(b) The graph must have the same number of vertices as edges.
(c) The graph is not connected.

18. The earliest completion time (in minutes) for a job with the following order-requirement digraph is ______ .

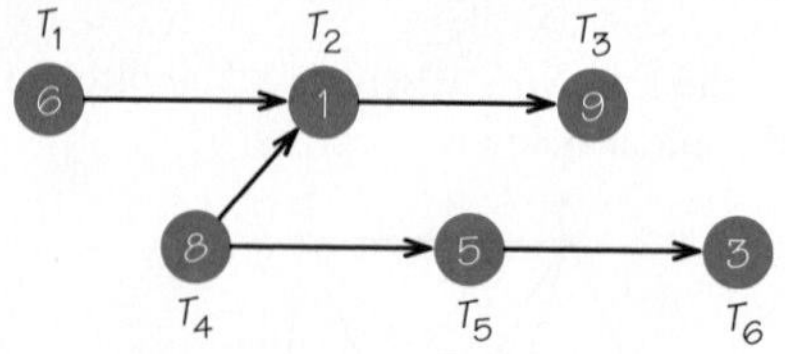

19. Assume a job has an order-requirement digraph with five tasks whose critical path is 25 minutes in length. Based on this information, what can be said about the tasks?

(a) Each task takes exactly 5 minutes.
(b) Some task takes 25 minutes.
(c) The five tasks in total take at least 25 minutes.

20. The length of the critical path in the order-requirement digraph below is ______________________ minutes.

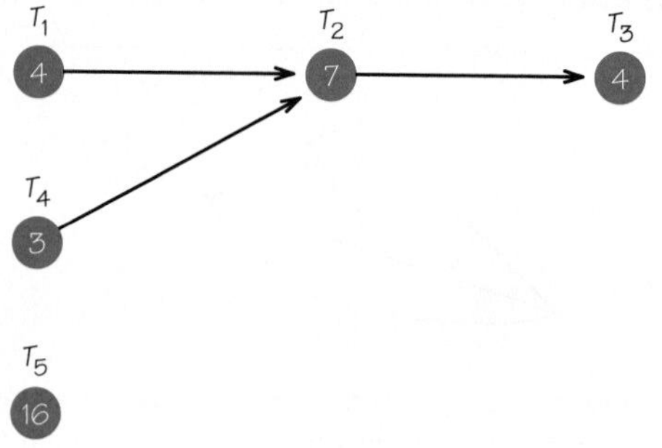

CHAPTER 2 EXERCISES

■ Challenge ◆ Discussion

2.1 Hamiltonian Circuits

1. For the accompanying graphs (a) through (c), write a Hamiltonian circuit starting at X_5.

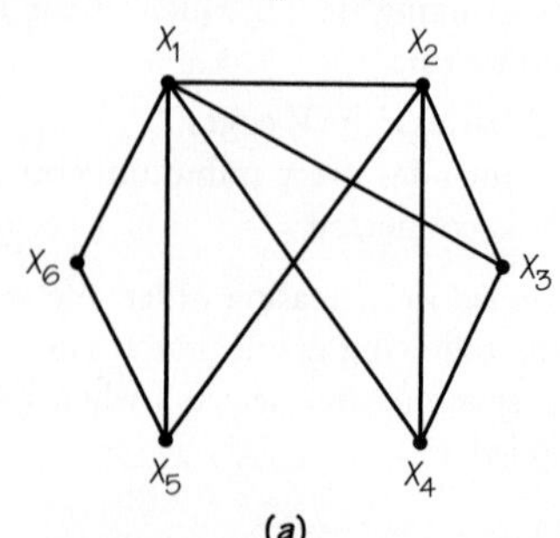

(a)

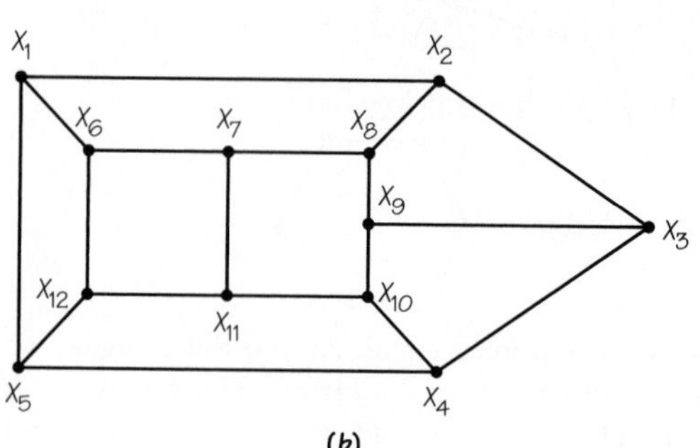

(b)

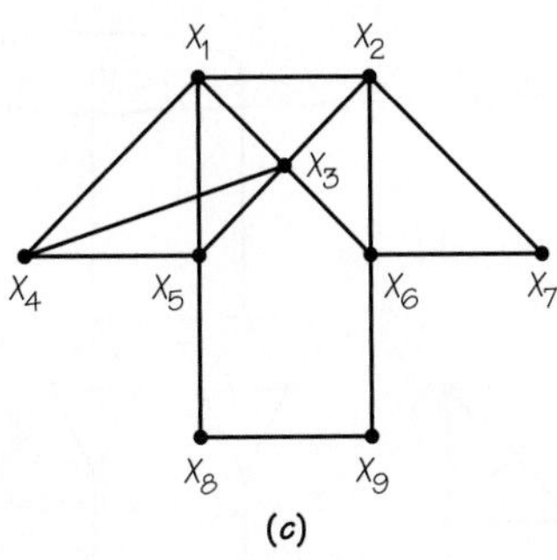

(c)

2. For the accompanying graphs (a) through (c) write down a Hamiltonian circuit starting at X_5.

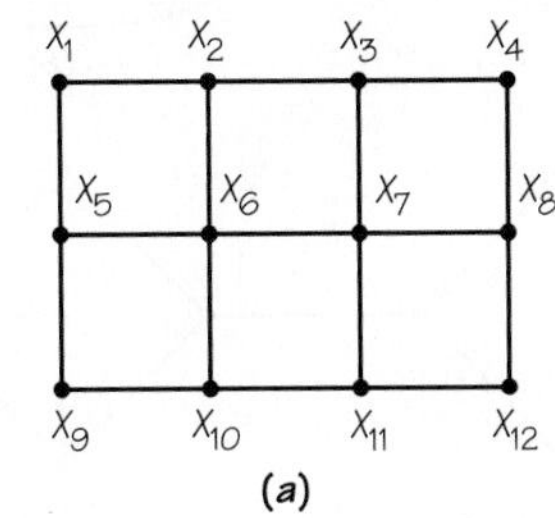

(a)

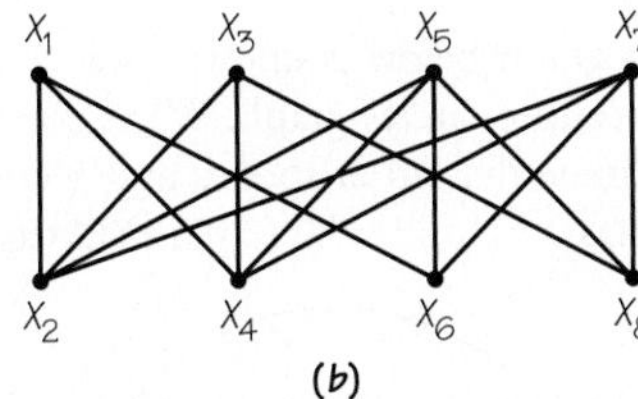

(b)

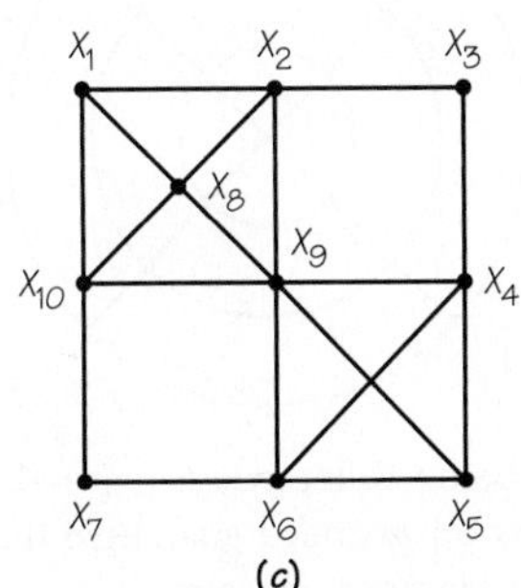

(c)

3. If the edge X_2X_3 is erased from each of the graphs in Exercise 1, does the resulting graph still have a Hamiltonian circuit?

4. **(a)** If the vertex X_6 and the edges attached to X_6 are removed from the graphs in Exercise 1, do the new graphs that result still have Hamiltonian circuits?
(b) If you think of the graphs in Exercise 1 as communications networks, what interpretation might be given to the "removal" of a vertex and the edges attached as described in part (a)?

5. **(a)** If the edge X_6X_7 is removed (erased) from each of the graphs in Exercise 2, do the new graphs that result still have Hamiltonian circuits?
(b) If you think of the graphs in Exercise 2 as communications networks, what interpretation might be given to the "removal" of an edge as described in part (a)?

6. **(a)** Give examples of real-world situations that can be modeled using a graph and for which finding a Hamiltonian circuit in the graph would be of interest.
(b) For each of the examples you mention in part (a), can you adapt the question about the real-world situation involved so that finding an Eulerian circuit in the same graph would be of interest?

7. Suppose two Hamiltonian circuits are considered different if the collections of edges that they use are different. How many other Hamiltonian circuits can you find in the graph in Figure 2.1 that are different from the two discussed?

8. For each of the following graphs, add wiggly edges to indicate a Hamiltonian circuit.

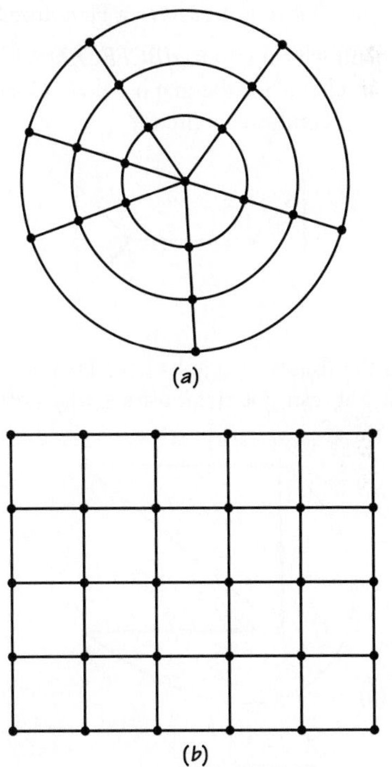
(a)

(b)

9. **(a)** Neither of the following graphs has a Hamiltonian circuit. Is it possible to add a single new edge to these graphs to obtain a new graph that has a Hamiltonian circuit?

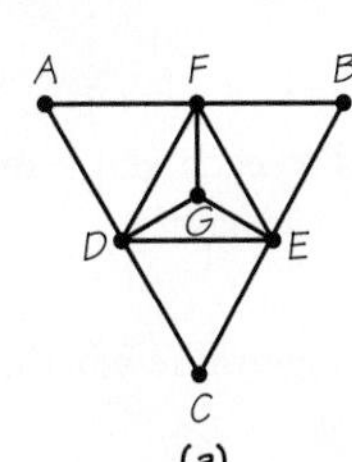

(a)

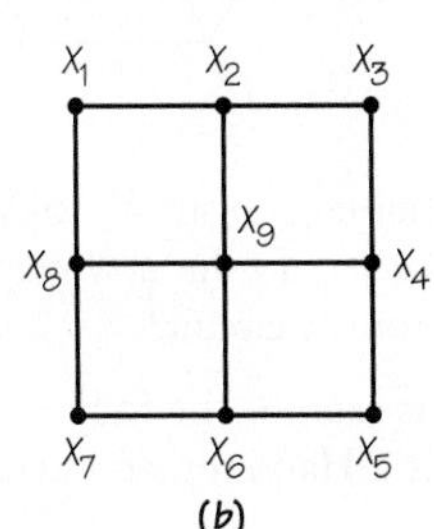

(b)

(b) Find an example of a graph that has no Hamiltonian circuit and will still have no Hamiltonian circuit no matter what single edge is added to it.
(c) Show that it is possible to add 4 additional edges to the graph diagram in part (b) above so that the resulting new graph will still have no Hamiltonian circuit.

10. Explain why the graph below has no Hamiltonian circuit.

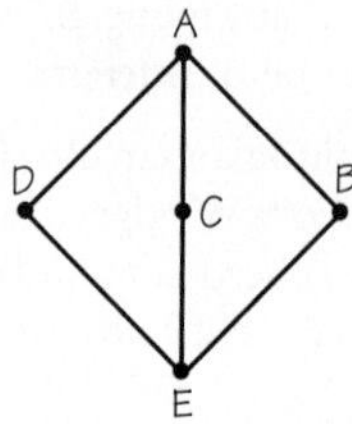

11. Use the graph shown in Exercise 10 to help you construct a connected graph for which every vertex has valence 3 and that does not have a Hamiltonian circuit.

■ 12. Explain why the tour *ABCFECBDA* is not a Hamiltonian circuit for the graph below. Does this graph have a Hamiltonian circuit?

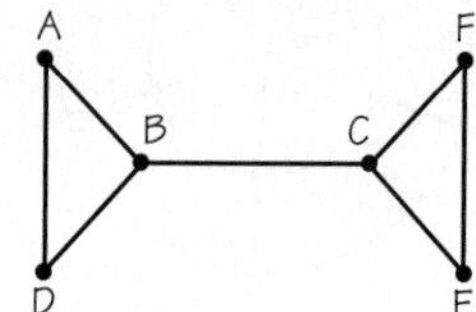

■ 13. Do the following graphs have Hamiltonian circuits? If not, can you demonstrate why not?

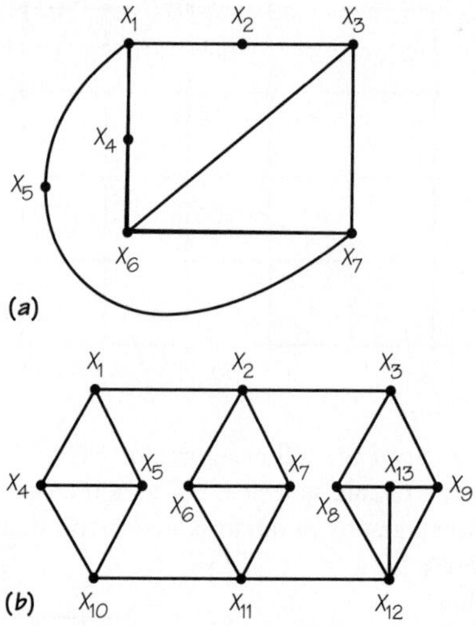

14. If an edge from X_2 to X_5 is added to each graph in Exercise 13, do the new graphs that result have a Hamiltonian circuit?

15. For each of the following graphs, determine whether there is a Hamiltonian circuit.

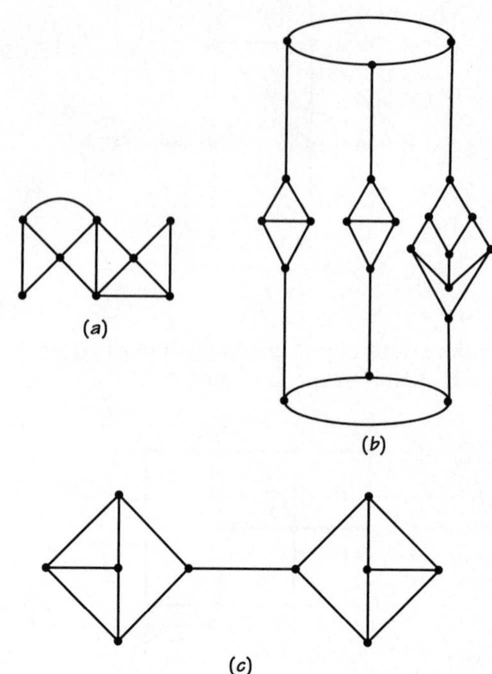

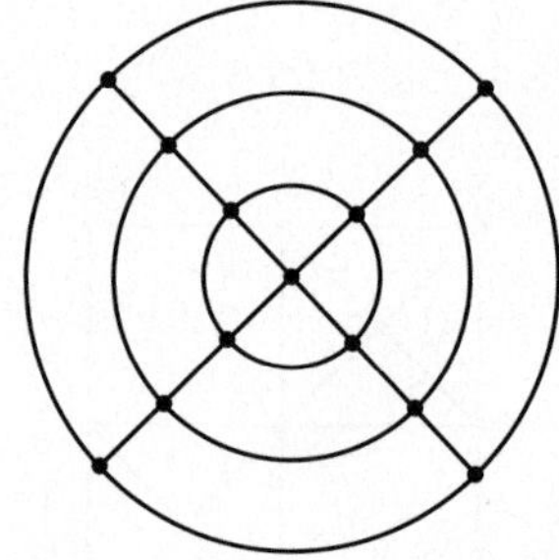

■ 16. (a) The graph below is known as a four spokes and three concentric circles graph. What conditions on m and n guarantee that an m spokes and n concentric circles graph has a Hamiltonian circuit? (Assume $m \geq 2$, $n \geq 1$.)

(b) The graph below is known as a 3×4 grid graph. What conditions on m and n guarantee that an $m \times n$ grid graph has a Hamiltonian circuit?

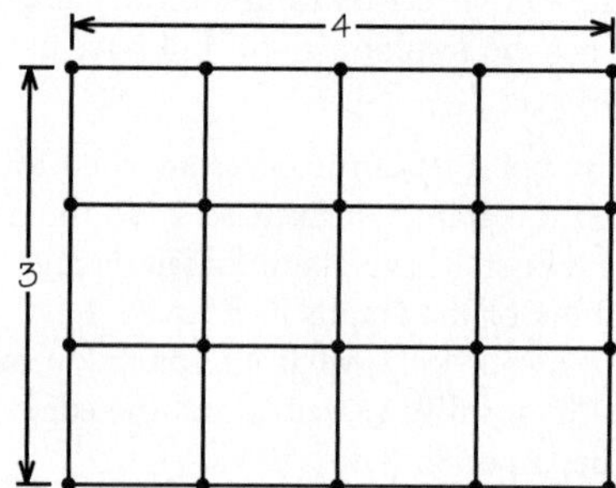

Can you think of a real-world situation in which finding a Hamiltonian circuit in an $m \times n$ grid graph would represent a solution to the problem? If an $m \times n$ grid graph has no Hamiltonian circuit, can you find a tour that repeats a minimum number of vertices and starts and ends at the same vertex?

17. A Hamiltonian path in a graph is a tour of the vertices of the graph that visits each vertex once and only once and starts and ends at different vertices.

(a) For each of the graphs shown in Exercise 13, does the graph have a Hamiltonian path?
(b) Does each of these graphs have a Hamiltonian path that starts at X_1 and ends at X_2?
(c) Describe three real-world situations where finding a Hamiltonian path in a graph would be required.

18. Using the terminology of Exercise 17, draw a graph that has

(a) a Hamiltonian path but no Hamiltonian circuit.
(b) an Euler circuit but no Hamiltonian path.
(c) a Hamiltonian path but no Euler circuit.

19. To practice your understanding of the concepts of Euler circuits and Hamiltonian circuits, determine for the following graphs (a) through (d) whether there is an Euler circuit and/or a Hamiltonian circuit. If so, write it down.

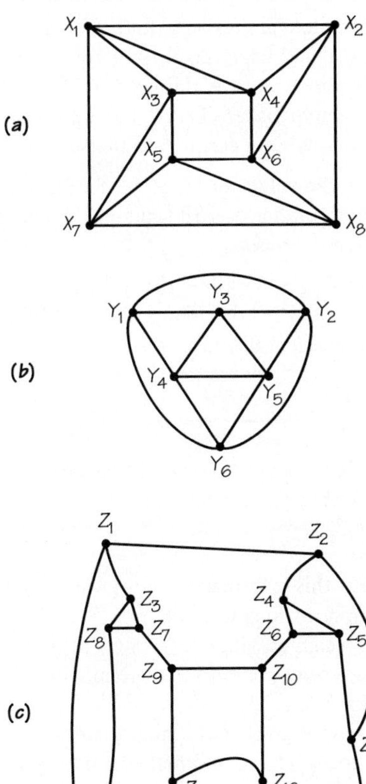

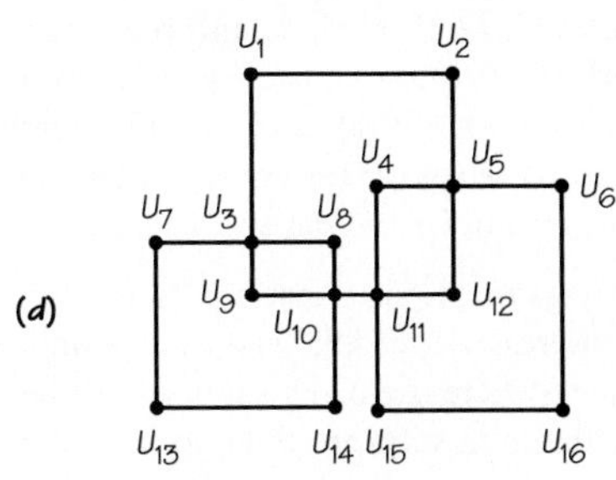

■ **20. (a)** The n-dimensional cube is obtained from two copies of an $(n-1)$-dimensional cube by joining corresponding vertices. (The process is illustrated for the 3-cube and the 4-cube in the following figure.) Can you show that every n-cube has a Hamiltonian circuit? [*Hint:* Show that if you know how to find a Hamiltonian circuit on an $(n-1)$-cube, then you can use two copies of this to build a Hamiltonian circuit on an n-cube.]

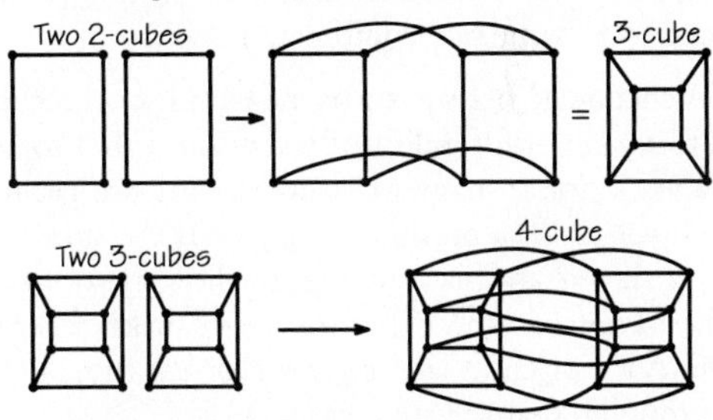

(b) Find formulas for the number of vertices and the number of edges of an n-cube.

21. If an edge is added from the vertex with subscript 4 to the vertex with subscript 5 in each graph in Exercise 19, which of the resulting graphs will have Hamiltonian circuits and which will have Euler circuits?

22. Find a family of graphs none of which have Hamiltonian circuits but for which adding a single edge to the first graph in the family creates a Hamiltonian circuit, adding two edges to the second graph in the family creates a Hamiltonian circuit, and so forth.

23. A Hamiltonian path in a graph is a tour of the vertices that visits each vertex once and only once and that starts and ends at different vertices.

(a) Draw an example of a graph that has no Hamiltonian path and where all the vertices are 3-valent.
(b) Draw a graph that has no Hamiltonian path but that does have an Euler circuit.
(c) By analogy with the Hamiltonian path, develop a definition of "Euler path."

24. (a) When going outside on a cold winter day, Jill can choose from three winter coats, five wool scarves, four pairs of boots, and three ski hats. How many outfits might her friends see her in?
(b) If Jill always insists on wearing her green wool scarf, how many outfits might her friends see her in?

25. The notes C, D, E, F, G, A, and B are to be used to form an ordered five-note musical logo. In how many ways can this be done if **(a)** no note can be repeated; **(b)** notes can be repeated; **(c)** notes can be repeated but all the notes cannot be the same?

26. A lottery game requires that a person select an upper- or lowercase letter followed by five different two-digit numbers (where the digits cannot both be zero). How many different ways are there to fill out a lottery ticket?

27. **(a)** In designing a security system for its accounts, a bank asks each customer to choose a five-digit number, all the digits to be distinct and nonzero. How many choices can a customer make?
(b) A suitcase with a liquid-crystal display allows one to unlock it with a specific combination of three capital letters that are not necessarily different. How many choices would a thief have to go through to be sure that all the possibilities had been tried? How does this compare to a "standard" combination lock?

28. To encourage her son to try new things, a mother offers to take him for a dish of ice cream with a topping once a week, for as many weeks as he does not get the same choice as on a previous occasion. If the store offers 12 flavors and six toppings, for how many weeks will she have to do this if her son never picks either of the two types of chocolate ice cream or the three types of nut topping that the store carries?

29. A large corporation has found that it has "outgrown" its current code system for routing interoffice mail. The current system places a code of three ordered, distinct nonzero digits on the mail. The new proposal calls for the use of two ordered capital letters. Does the new system have more code numbers than the old system? If so, how many more locations will the new system enable the company to encode over the current system?

30. Repeat Exercise 25a, except that exactly one of the notes in the musical logo must be a sharp and the note chosen to be sharped cannot appear elsewhere (for example, BCD#AG, where D# denotes D sharp).

◆ **31.** **(a)** In New York State, one type of license plate has three letters followed by three numbers. Suppose the digits from 0, 1, . . . , 9 can be used, except that all three digits cannot be zero, and that any letter from A to Z (repeats allowed) can be used. How many plates are possible?
(b) Investigate what schemes for license plates are used in your state and determine how many different plates are possible.

32. A restaurant offers 5 soups, 10 entrees, and 8 desserts. How many different choices for a meal can a customer make if one selection is made from each category? If 3 of the desserts are pies and the customer will never order pie, how many meals can the customer choose?

33. In the last several years, heavily populated regions that previously had only one area code have been divided into service areas with more than one area code. What is the largest number of different phone numbers that can be served using one area code? If an area code cannot begin with a zero, how many different area codes are possible?

34. **(a)** A credit-card company makes it easier for customers to memorize their PIN (personal identification number) by using a four-digit PIN that consists of three different digits selected from 0, 1, 2, . . . , 9 where one of the digits must be a zero, another is a nonzero digit that is repeated, and another is a digit different from these two. How many different PINs of this kind are there?
(b) How many PINs are possible if there are no restrictions on repeats of the 10 possible digits that can be used?

2.2 Traveling Salesman Problem

2.3 Helping Traveling Salesmen

35. Draw complete graphs with four, five, and six vertices. How many edges do these graphs have? Can you generalize to n vertices? How many TSP tours would these graphs have? (Tours yielding the same Hamiltonian circuit are considered the same.)

36. Calculate the values of 5!, 6!, 7!, 8!, 9!, and 10!. Then find the number of TSP tours in the complete graph with nine vertices.

37. The following table shows the mileage between four cities: Springfield, Ill. (S); Urbana, Ill. (U); Effingham, Ill. (E); and Indianapolis, Ind. (I).

	E	I	S	U
E	–	147	92	79
I	147	–	190	119
S	92	190	–	88
U	79	119	88	–

(a) Represent this information by drawing a weighted complete graph on four vertices.
(b) Use the weighted graph in part (a) to find the cost of the three distinct Hamiltonian circuits in the graph. (List them starting at U.)
(c) Which circuit gives the minimum cost?
(d) Would there be any different in parts (b) and (c) if the start vertex were at I?
(e) If one applies the nearest-neighbor method starting at U, what circuit would be obtained? Does the answer change if one applies the nearest-neighbor algorithm starting at S? At E? At I?

(f) If one applies the sorted-edges method, what circuit would be obtained? Does one get the optimal answer?

38. After a party at her house, Francine (F) has agreed to drive Mary (M), Rachel (R), and Constance (C) home. If the times (in minutes) to drive between her friends' homes are shown below, what route gets Francine back home the quickest?

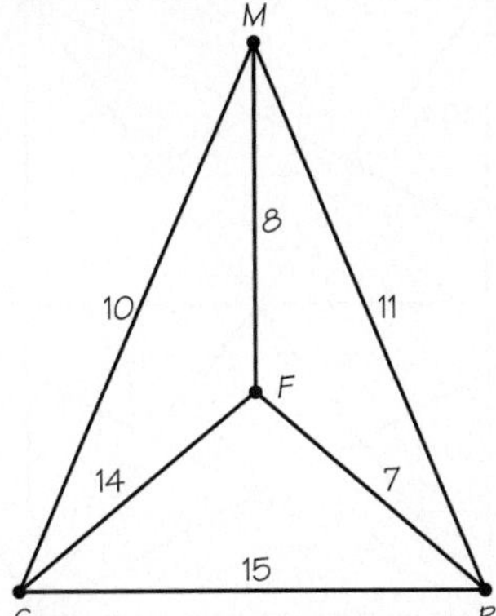

39. In Exercise 38, what route would Francine have to follow to get home as quickly as possible, assuming she promised to drive Mary home first?

40. In Exercise 38, Francine is planning to deliver her friends home and then spend the night at Rachel's house. What would her fastest route be?

41. Starting from the location where she moors her boat (M), a fisherwoman wishes to visit three areas–A, B, and C–where she has set fishing nets. If the times (in minutes) between the locales are given in the figure below, what route to visit the three sites and return to the mooring place would be optimal?

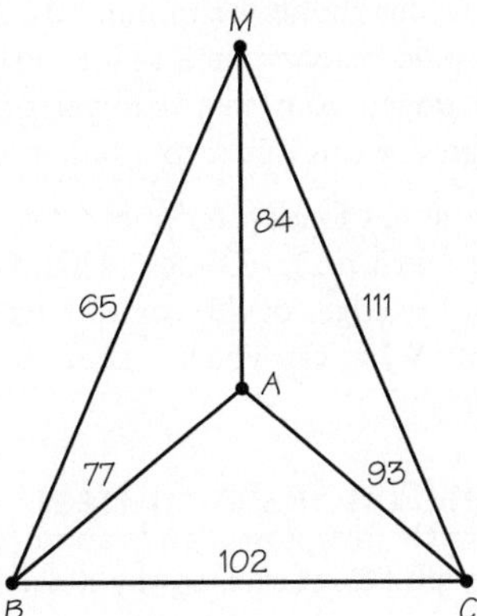

42. (a) For the two complete graphs that follow, find the costs of the nearest-neighbor tour starting at B and of the tour generated by the sorted-edges algorithm.

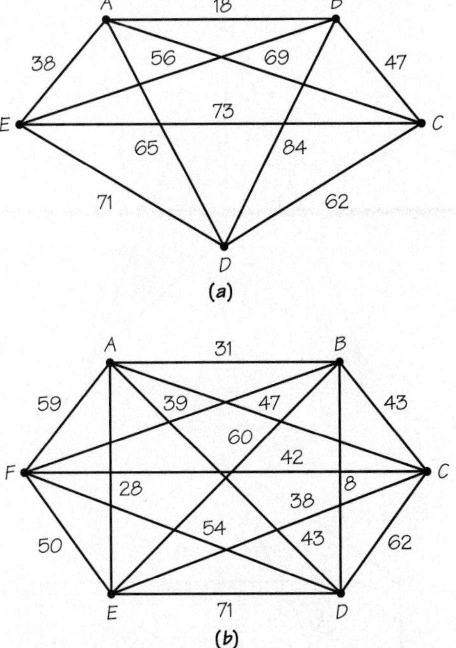

(a)

(b)

■ **(b)** How many Hamiltonian circuits would have to be examined to find a shortest route for part (a) by the brute force method?

(c) Invent an algorithm different from the sorted-edges and nearest-neighbor algorithms that is easy to apply for finding TSP solutions.

43. An airport limo must take its five passengers from the airport to different downtown hotels. Is this a traveling salesman problem, a Chinese postman problem, or an Euler circuit problem?

44. For each of the following graphs with weights, apply the nearest-neighbor method (starting at vertex A) and the sorted-edges method to find (it is hoped) a cheap tour.

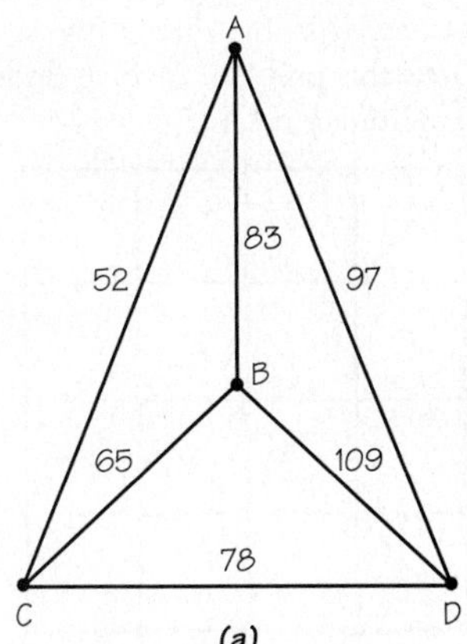

(a)

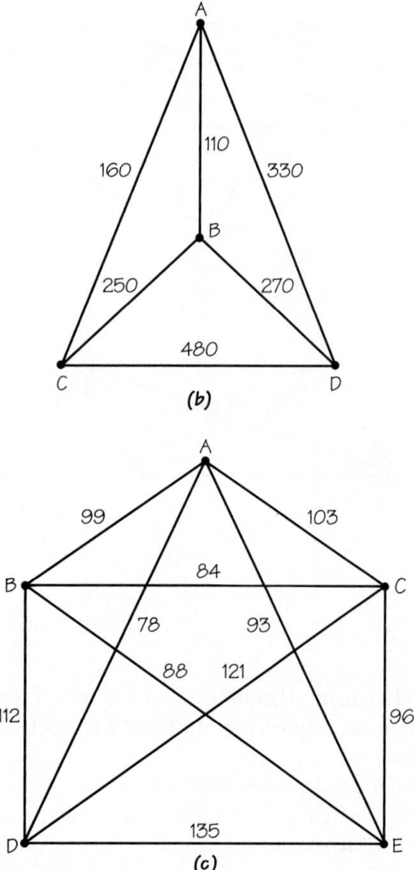

45. The following figure represents a town where there is a sewer located at each corner (where two or more streets meet). After every thunderstorm, the department of public works wishes to have a truck start at its headquarters (at vertex H) and make an inspection of sewer drains to be sure that leaves are not clogging them. Can a route start and end at H that visits each corner exactly once? (Assume that all the streets are two-way streets.) Does this problem involve finding an Euler circuit or a Hamiltonian circuit?

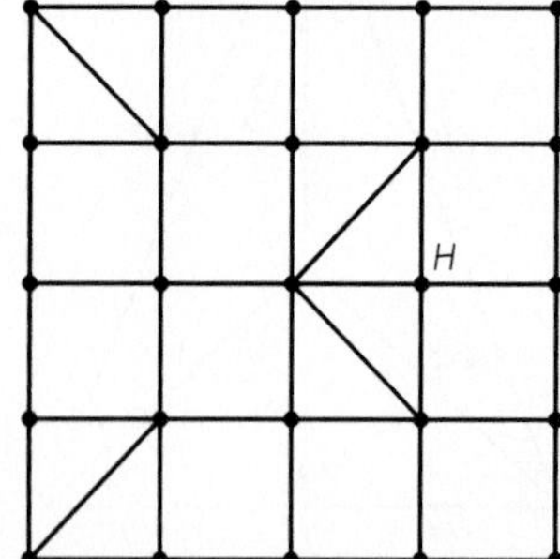

Assume that at equally spaced intervals along the blocks in this graph there are storm sewers that must be inspected after each thunderstorm to see if they are clogged. Is this a Hamiltonian circuit problem, an Euler circuit problem, or a Chinese postman problem? Find an optimal tour to do this inspection.

46. (a) Solve the six-city TSP shown in the diagram using the nearest-neighbor algorithm starting at vertex A and starting at vertex B.

(b) Apply the sorted-edges method.

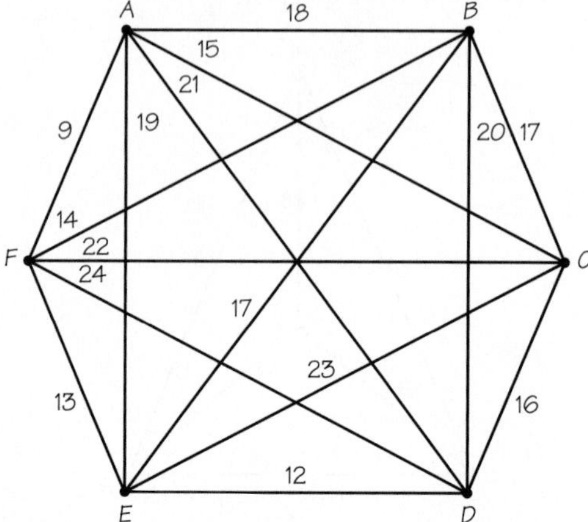

■ **47.** Construct an example of a complete graph of five vertices, with distinct weights on the edges for which the nearest-neighbor algorithm starting at a particular vertex and the sorted-edges algorithm yield different solutions for the traveling salesman problem. Can you find a five-vertex complete graph with weights on the edges in which the optimal solution, the nearest-neighbor solution, and the sorted-edges algorithm solution are all different?

■ **48.** If the brute force method of solving a 20-city TSP is employed, use a calculator to determine how many Hamiltonian circuits must be examined. How long would it take to determine the minimum-cost tour if the cost of tours could be computed at the rate of 1 billion per second? (Convert your answer to years by seeing how many years are equivalent to a billion seconds!)

49. Suppose one has found an optimal tour for a given 10-city TSP problem to have weight 4200. Now suppose the weights on the edges of the complete graph are increased by 50. What can you say about the optimal tour and its weight?

2.4 Minimum-Cost Spanning Trees

50. For each graph below, explain why it is or is not a tree.

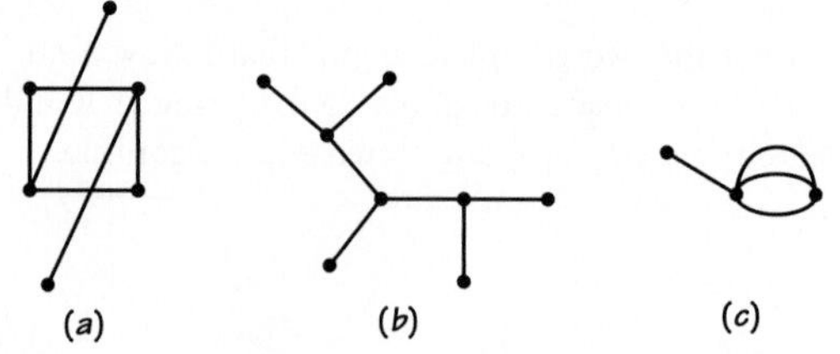

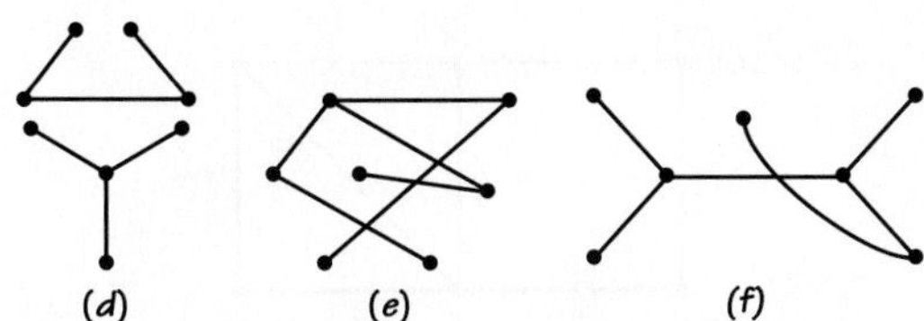

51. For each of the diagrams below, explain why the wiggly edges are not

(a) a spanning tree.
(b) a Hamiltonian circuit.

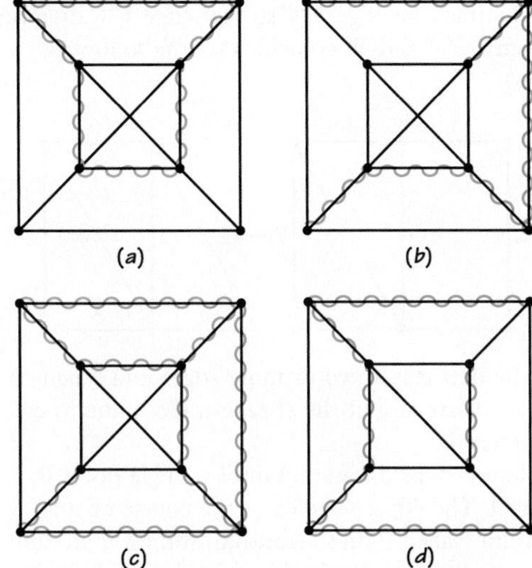

52. Find all the spanning trees in the graphs below.

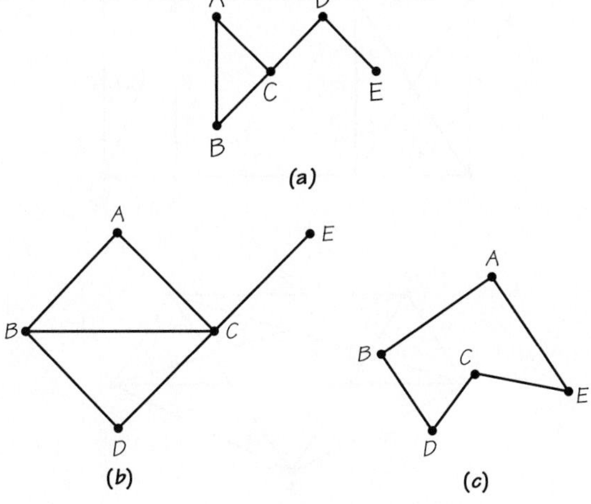

53. Use Kruskal's algorithm to find a minimum-cost spanning tree for the following graphs (a), (b), (c), and (d). In each case, what is the cost associated with the tree?

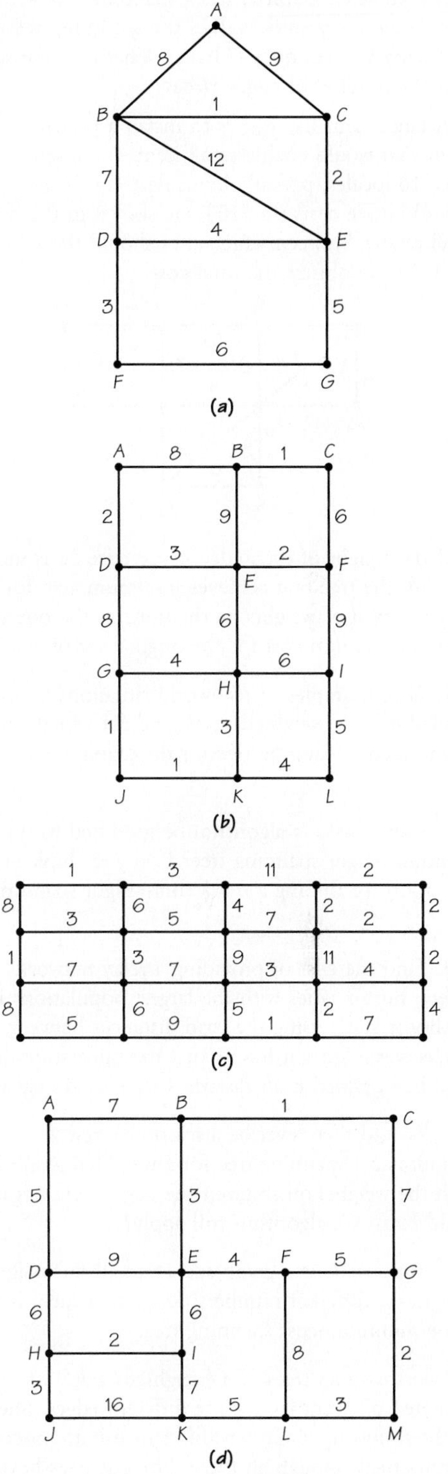

54. A connected graph G has 16 vertices. How many edges does a spanning tree of G have? How many vertices does a spanning tree of G have? What can one say about the number of edges G has?

55. A connected graph H has a spanning tree with 26 edges. How many vertices does the spanning tree have? How many vertices does H have? What can one say about the number of edges H has?

56. A large company wishes to install a pneumatic tube system that would enable small items to be sent between any of 10 locales, possibly by using relay. If the nonprohibitive costs (in \$100) are shown in the graph model below, between which sites should the tube be installed to minimize the total cost?

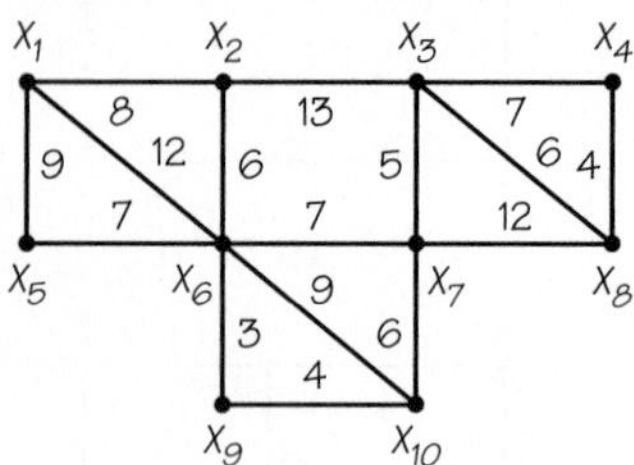

57. If the weight of each edge in Exercise 56 is increased by 3, will the tree that achieves minimum cost for the new collection of weights be the same as the one that achieves minimum cost for the original set of weights?

◆ **58.** Give examples of real-world situations that can be modeled using a weighted graph and for which finding a minimum-cost spanning tree for the graph would be of interest.

■ **59.** Can Kruskal's algorithm be modified to find a maximum-weight spanning tree? Can you think of an application for finding a maximum-weight spanning tree?

◆ **60.** Find the cost of providing a relay network between the six cities with the largest populations in your home state, using the road distances between the cities as costs. Does it follow that the same solution would be obtained if air distances were used instead?

■ **61.** Would there ever be a reason to find a minimum-cost spanning tree for a weighted graph in which the weights on some of the edges were negative? Would Kruskal's algorithm still apply?

■ **62.** Suppose G is a graph such that all the weights on its edges are different numbers. Show that there is a unique minimum-cost spanning tree.

63. Two spanning trees of a (weighted) graph are considered different if they use different edges. Show that the following graph has different minimum-cost spanning trees, though all these different trees have the same cost.

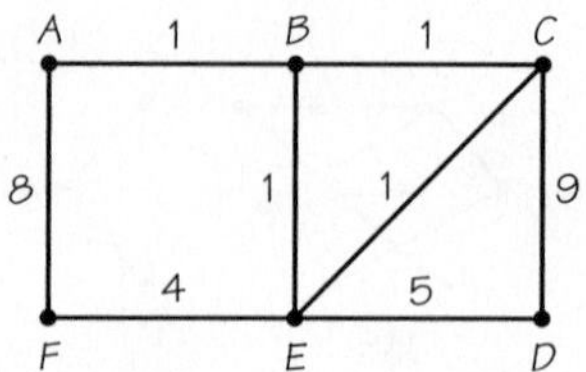

64. Let G be a graph with weights assigned to each edge. Consider the following algorithm:

(a) Pick any vertex V of G.

(b) Select an edge E with a vertex at V that has a minimum weight. Let the other endpoints of E be W.

(c) Contract the edge VW so that edge VW disappears and vertices V and W coincide (see the following figures).

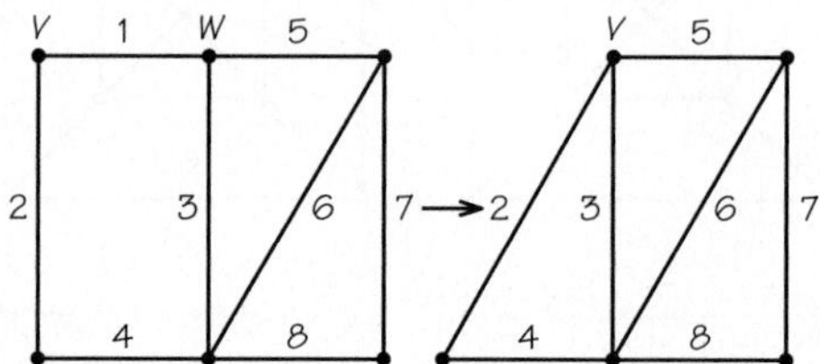

If in the new graph two or more edges join a pair of vertices, delete all but the cheapest. Continue to call the new vertex V.

(d) Repeat steps (b) and (c) until a single point is obtained. The edges selected in the course of this algorithm (called Prim's algorithm) form a minimum-cost spanning tree. Apply this algorithm to the following graphs.

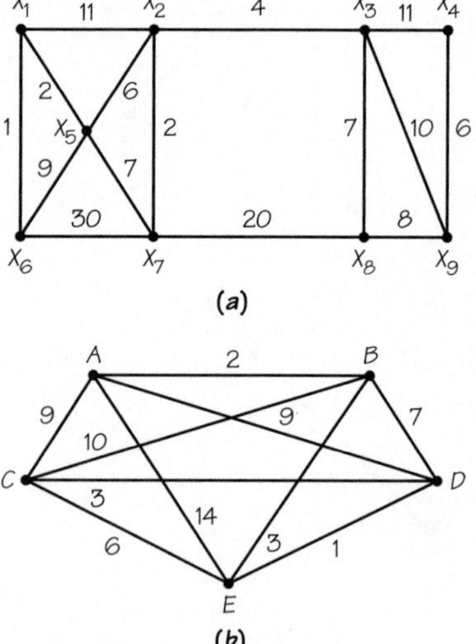

65. Determine whether each of the following statements is true or false for a minimum-cost spanning tree T for a weighted connected graph G:

(a) T contains a cheapest edge in the graph.

(b) T cannot contain a most expensive edge in the graph.

(c) T contains one fewer edge than there are vertices in G.
(d) There is some vertex in T to which all others are joined by edges.
(e) There is some vertex in T that has valence 3.

■ **66.** In the following graphs, the number in the circle for each vertex is the cost of installing equipment at the vertex if relaying must be done at the vertex, while the number on an edge indicates the cost of providing service between the endpoints of the edge.

In each case, find the minimum cost (allowing relays) for sending messages between any pair of vertices, taking vertex relay costs into account.

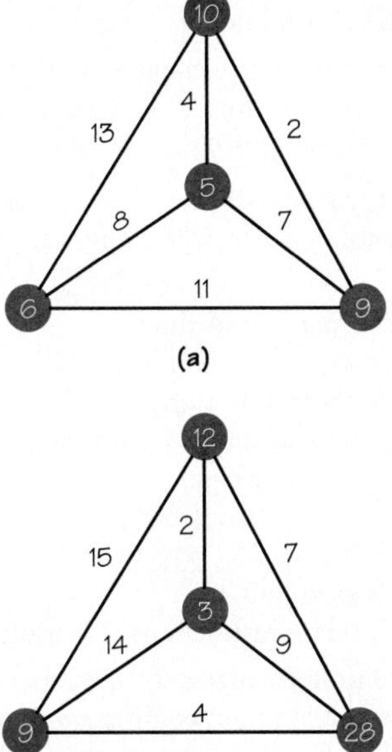

Would your answer be different if vertex relay costs were neglected? (*Warning:* Kruskal's algorithm cannot be used to answer the first question. This problem illustrates the value of having an algorithm over relying on "brute force.")

67. (a) Show that for each edge of graph J below there is a spanning tree of J that avoids that edge.
(b) For each spanning tree that you found in graph J, count the number of vertices and edges. Do you notice any pattern?
(c) For graph H below and each edge in the graph, is there a spanning tree that does not include that edge of H?

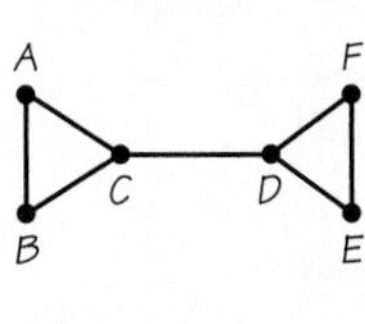

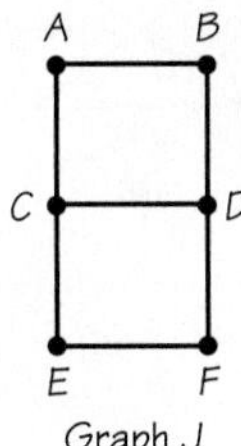

68. (a) The table shown gives the "closeness" or distance values between four objects. Construct a four-vertex tree with weights on its edges such that the distances between pairs of vertices of the tree (as measured by the sum of the weights on the path in the tree between these vertices) give rise to this table.

	A	B	C	D
A	0	3	10	14
B	3	0	7	11
C	10	7	0	4
D	14	11	4	0

(b) Produce several real-world contexts that might give rise to the situation described here.

69. The figure below represents four objects using a tree with weights on the edges. Construct a table with four rows and four columns recording how "close" pairs of vertices in the tree are to each other. To find how close a pair of objects is, add together the weights along the path that joins these two objects.

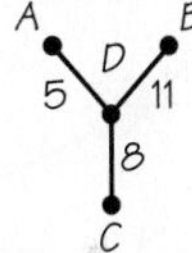

2.5 Critical-Path Analysis

70. Find the earliest completion time and critical paths for the order-requirement digraphs below.

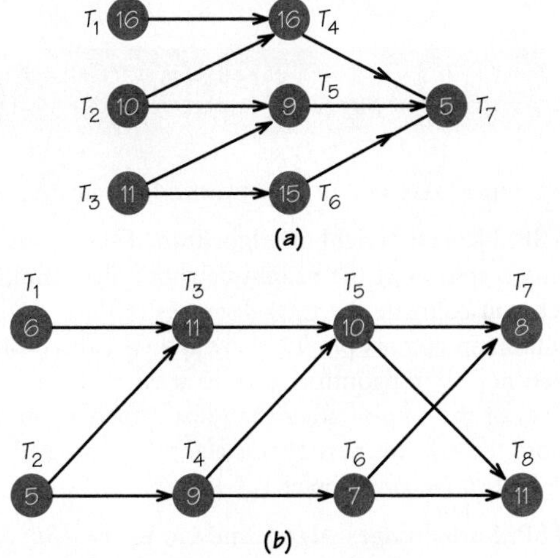

71. Find the earliest completion time and critical paths for the following order-requirement digraphs.

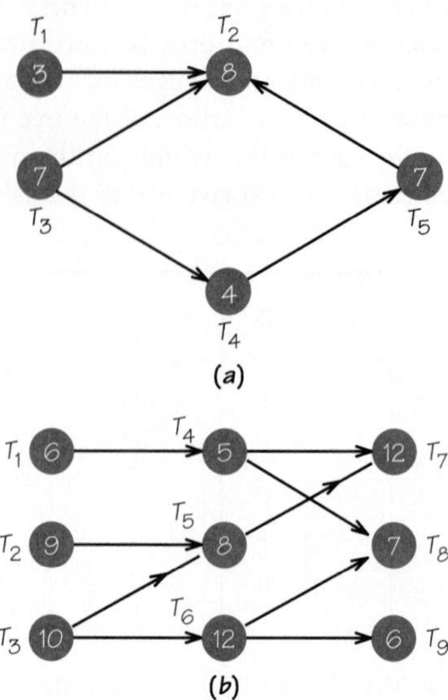

(*a*)

(*b*)

72. Construct an example of an order-requirement digraph with two different critical paths.

73. In the order-requirement digraph below, determine which tasks, if shortened, would reduce the earliest completion time and which would not. Then find the earliest completion time if task T_5 is reduced to time length 7. What is the new critical path?

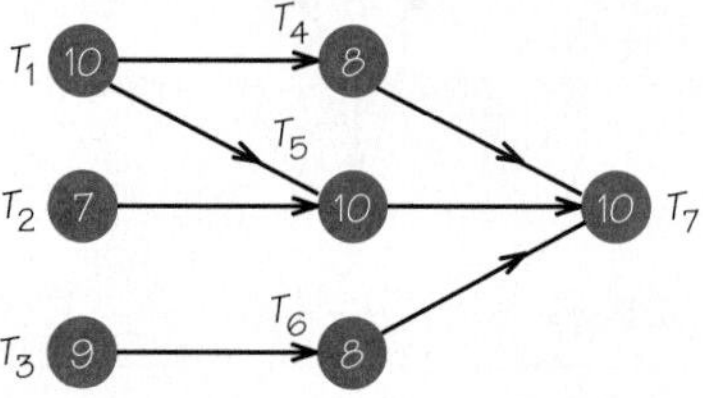

74. For the order-requirement digraph in Exercise 73, find the critical path and the task(s) in the critical path whose time, when reduced the least, creates a new critical path.

75. To build a new addition on a house, the following tasks must be completed:

(a) Lay foundation.
(b) Erect sidewalls.
(c) Erect roof.
(d) Install plumbing.
(e) Install electric wiring.
(f) Lay tile flooring.
(g) Obtain building permits.
(h) Put in door that connects new room to existing house.
(i) Install track lighting on ceiling.
(j) Install wall air-conditioner.

Construct reasonable time estimates for these tasks and a reasonable order-requirement digraph. What is the fastest time in which these tasks can be completed?

76. At a large toy store, scooters arrive unassembled in boxes. To assemble a scooter, the following tasks must be performed:

TASK 1. Remove parts from the box.
TASK 2. Attach wheels to the footboard.
TASK 3. Attach vertical housing.
TASK 4. Attach handlebars to vertical housing.
TASK 5. Put on reflector tape.
TASK 6. Attach bell to handlebars.
TASK 7. Attach decals.
TASK 8. Attach kickstand.
TASK 9. Attach safety instructions to handlebars.

Give reasonable time estimates for these tasks and construct a reasonable order-requirement digraph. What is the earliest time by which these tasks can be completed?

77. Construct an order-requirement digraph with six tasks that has three critical paths of length 26.

APPLET EXERCISES

To do these exercises, go to www.whfreeman.com/fapp8e.

1. TSP: Nearest-Neighbor Algorithm. There is an extended version of the nearest-neighbor algorithm, in which you compare the total distances of the Hamiltonian circuits produced by applying the ordinary nearest-neighbor algorithm starting at each of the vertices of the graph (rather than just a specific one). Explore the effectiveness of this algorithm using the *TSP: Nearest-Neighbor* applet.

2. TSP: Sorted-Edges Algorithm. Go to the *TSP: Sorted Edges* applet, where you can apply the sorted-edges algorithm to see if it solves the traveling salesman problem for the following graphs (and others):

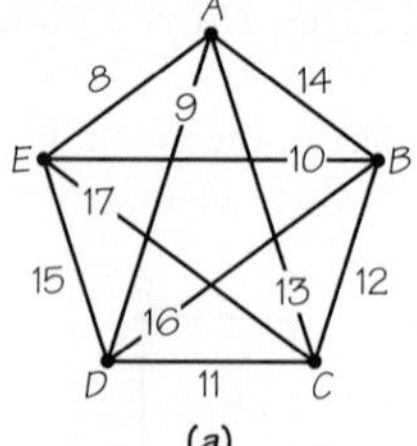

(*a*)

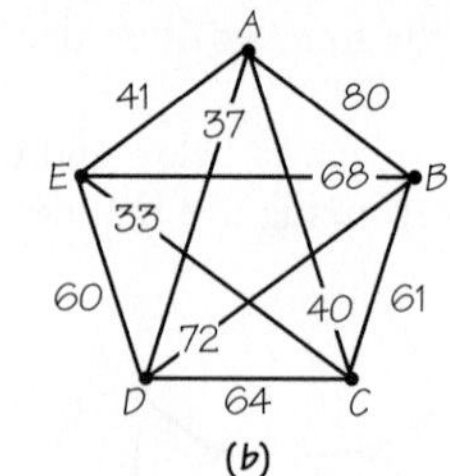

(*b*)

3. Kruskal's Algorithm. Go to the *Kruskal's Algorithm* applet, where you can apply Kruskal's algorithm to find the minimum-cost spanning trees in the following graphs (and others):

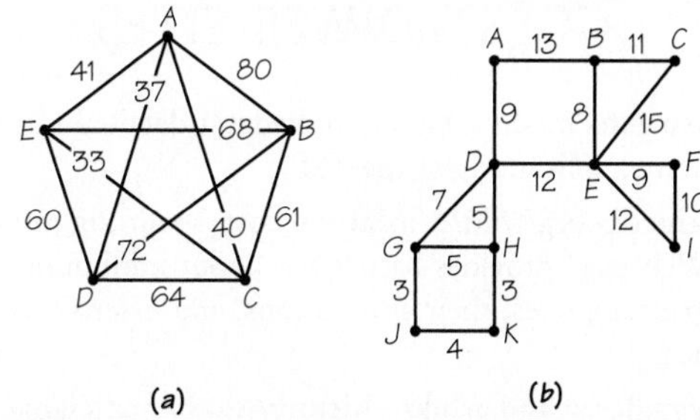

(*a*) (*b*)

WRITING PROJECTS

1. Write an essay about a variety of situations in which you are personally involved for which a solution of the TSP is (perhaps implicitly) required. Explain under what circumstances it might be valuable to carry out a formal mathematical solution to such TSPs rather than use an ad hoc solution.

2. Construct an example, of the kind suggested on page 42, that shows that in a situation where three day campers must be picked up and brought to camp, it may make a difference if the optimization criterion is minimizing distance traveled by the camp bus versus minimizing average time that the children spend on the bus.

3. Determine the six largest cities in the state in which you live. By consulting a road atlas (or by some other means) construct the graph that represents the road distances between your hometown and these six other cities. Now apply (a) the nearest-neighbor method, (b) the sorted-edges method, and (c) the nearest neighbor from each city, and pick the minimum tour method to solve the associated TSP. Do you have reason to believe that the answers you get might include an optimum solution among them?

SUGGESTED READINGS

BODIN, LAWRENCE. Twenty years of routing and scheduling, *Operations Research*, 38 (1990): 571–579. A survey of real-world situations where routing and scheduling were used, written by a pioneer in this area.

DOLAN, ALAN, and JOAN ALDUS. *Networks and Algorithms: An Introductory Approach*, Wiley, Chichester, England, 1993. An excellent introduction to graph theory algorithms.

GUSFIELD, DAN. *Algorithms on Strings, Trees, and Sequences*, Cambridge University Press, New York, 1997. Details applications of graph theory in pattern recognition and reconstruction problems.

JONES, NEIL C., and PAVEL A. PEVZNER, *An Introduction to Bioinformatics Algorithms*, MIT Press, Cambridge, Mass., 2004. This book has material on how graph theory ideas, particularly those related to Hamiltonian circuits, are being used in molecular genetics and computational biology.

LAWLER, EUGENE, J. LENSTRA, RINNOY KAN, and D. SHMOYS, eds. *The Traveling Salesman Problem*, Prentice-Hall, Englewood Cliffs, N.J., 1985. Includes survey and technical articles on all aspects of the TSP.

LUCAS, WILLIAM, FRED ROBERTS, and ROBERT THRALL, eds. *Discrete and Systems Models*, vol. 3: *Modules in Applied Mathematics*, Springer-Verlag, New York, 1983. Chapter 6, "A Model for Municipal Street Sweeping Operations," by A. Tucker and L. Bodin, describes street-sweeping and related models in detail. Other chapters detail many recent applications of mathematics.

ROBERTS, FRED S., and BARRY TESMAN, *Applied Combinatorics*, 2nd ed., Pearson Prentice Hall, Upper Saddle River, N.J., 2004. The material on network-optimization problems is excellent.

ROBERTS, FRED. *Graph Theory and Its Applications to Problems of Society*, Society for Industrial and Applied Mathematics, Philadelphia, 1978. A very readable account of how graph theory is finding a wide variety of applications.

 SUGGESTED WEB SITES

www.tsp.gatech.edu This site provides a detailed history and many applications of the TSP.

en.wikipedia.org/wiki/Minimum_cost_spanning_tree This Web page provides basic ideas about minimum-cost spanning trees, their applications, and extensions of this idea.

www-gap.dcs.st-and.ac.uk/~history/Mathematicians/Hamilton.html This site provides biographical information about William Rowan Hamilton, for whom Hamiltonian circuits are named.

www.ams.org/featurecolumn/archive/tsp.html
www.ams.org/featurecolumn/archive/trees.html These sites provide some history and information about applications of the Traveling Salesman Problem and of minimum-cost spanning trees.

CHAPTER 3

Planning and Scheduling

In a society as complex as ours, everyday problems such as providing services efficiently and on time require accurate planning of both people and machines. Take the example of a hospital in a major city. Around-the-clock scheduling of nurses and doctors must be provided to guarantee that people with particular expertise are available during each shift. The operating rooms must be scheduled in a manner flexible enough to deal with emergencies. Equipment used for X-ray, CT, or MRI scans must be scheduled for maximum efficiency.

Although many scheduling problems are often solved on an ad hoc basis, we can also use mathematical ideas to gain insight into the complications that arise in scheduling. The ideas we develop in this chapter have practical value in a relatively narrow range of applications, but they shed light on many characteristics of more realistic, and hence more complex, scheduling problems.

3.1 Scheduling Tasks

Assume that a certain number of identical **processors** (machines, humans, or robots) work on a series of tasks that make up a job. Associated with each task is a specified amount of time required to complete the task. For simplicity, we assume that any of the processors can work on any of the tasks. Our problem, known as the **machine-scheduling problem**, is to decide how the tasks should be scheduled so that the completion time for the tasks collectively is as early as possible.

Even with these simplifying assumptions, complications in scheduling will arise. Some tasks may be more important than others and perhaps should be scheduled first. When "ties" occur, they must be resolved by special rules. As an example, suppose we are scheduling patients to be seen in a hospital emergency room staffed by one doctor. If two patients arrive simultaneously, one with a bleeding foot, the other with a bleeding arm, which patient should be examined first? Suppose the doctor treats the arm patient first, and while treatment is

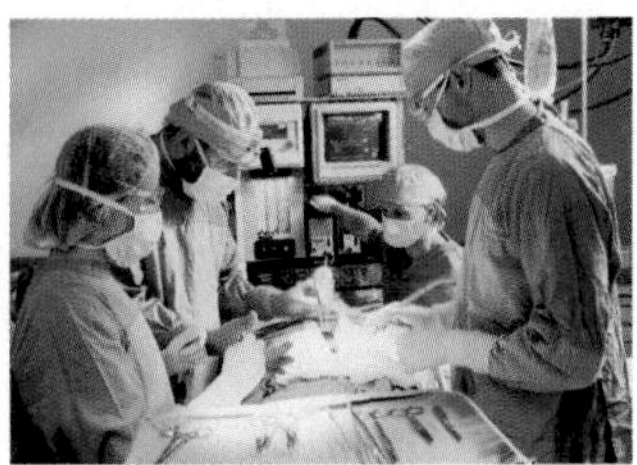

Hospitals are increasingly making use of mathematical techniques applied to scheduling problems. To make efficient use of one or more operating rooms requires the complicated assembly of a team of doctors, nurses, equipment, and support staff. Mathematical techniques for scheduling have made it possible to carry out more operations in less time. (*Eyewire/Punchstock.*)

going on, a person in cardiac arrest arrives. Scheduling rules must establish appropriate priorities for cases such as these.

Another common complication arises with jobs consisting of several tasks that cannot be done in an arbitrary order. For example, if the job of putting up a new house is treated as a scheduling problem, the task of laying the foundation must precede the task of putting up the walls, which in turn must be completed before work on the roof can begin. The plumbing system can be scheduled for installation later.

Assumptions and Goals

To simplify our analysis, we need to make clear and explicit assumptions:

1. If a processor starts work on a task, the work on that task will continue without interruption until the task is completed.
2. No processor stays voluntarily idle. In other words, if there is a processor free and a task available to be worked on, then that processor will immediately begin work on that task.
3. The requirements for ordering the tasks are given by an order-requirement digraph. (A typical example is shown in Figure 3.1, with task times highlighted within each vertex. The ordering of the tasks imposed by the order-requirement digraph often represents constraints of physical reality. For example, you cannot fly a plane until it has taken fuel on board.)
4. The tasks are arranged in a **priority list** that is independent of the order requirements. (The priority list is a ranking of the tasks according to some criterion of "importance.")

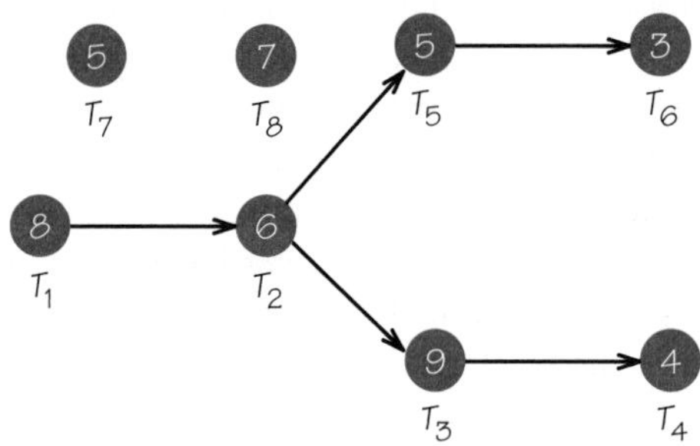

FIGURE 3.1 A typical order-requirement digraph. Tasks with no edges entering or leaving the vertices representing them (T_7, T_8) can be more flexibly scheduled than the other tasks.

EXAMPLE 1 ■ Home Construction

Let's see how these assumptions might work for an example involving a home construction project. In this case, the processors are human workers with identical skills. Assumption 1 means that once a worker begins a task, the work on this task is finished without interruption. Assumption 2 means that no worker will stay idle if there is some task for which the predecessors are finished. Assumption 3 requires that the ordering of the tasks be summarized in an order-requirement digraph. This digraph would code facts such as that the site must be cleared before the task of laying the foundation is begun. Assumption 4 requires that the tasks be ranked in a list from some perspective, perhaps a subjective view.

The task with highest priority rank is listed first in the list, followed left to right by the other tasks in priority rank. The priority list might be based on the size of the payments made to the construction company when a task is completed, even though these payments have no relation to the way the tasks must be done, as indicated in the order-requirement digraph. Alternatively, the priority list might reflect an attempt to find an algorithm to schedule the tasks needed to complete the whole job more quickly.

When considering a scheduling problem, there are various goals we might want to achieve. Among these are:

Goal 1. Minimizing the completion time of the job.

Goal 2. Minimizing the total time that processors are idle.

Goal 3. Finding the minimum number of processors necessary to finish the job by a specified time.

In the context of the construction example, goal 1 would complete the home as quickly as possible. Goal 2 would ensure that workers, who are perhaps paid by the hour, were not paid for doing nothing. One way of accomplishing this would be to hire one fewer worker even if it means the house takes longer to finish. Goal 3 might be reasonable if the family wants the house done before the first day of school, even if they have to pay a lot more workers to get the house done by this time.

For now we will concentrate on goal 1, finishing all the tasks at the earliest possible time. Note, however, that optimizing with respect to one goal may not optimize with respect to another. Our discussion here goes beyond what was discussed in Chapter 2 (see section 2.4) by dealing with how to assign tasks in a job to the processors that do the work. To build a new skyscraper involves designing a schedule for who will do what work when.

List-Processing Algorithm

The scheduling problem we have described sounds more complicated than the traveling salesman problem (TSP). Indeed, like the TSP, it is known to be NP-complete. This means that it is unlikely that anyone will ever find a computationally fast algorithm that can find an optimal solution. Thus, we will be content to seek a solution method that is computationally fast and gives only approximately optimal answers.

List-Processing Algorithm: Part I and Ready Task — PROCEDURE

The algorithm we use to schedule tasks is the **list-processing algorithm**. In describing it, we will call a task **ready** at a particular time if all its predecessors as indicated in the order-requirement digraph have been completed at that time. In Figure 3.1 at time 0, the ready tasks are T_1, T_7, and T_8, while T_2 cannot be ready until 8 time units after T_1 is started. The algorithm works as follows: At a given time, assign to the lowest-numbered free processor the first task on the priority list that is *ready* at that time and that hasn't already been assigned to a processor.

In applying this algorithm, we will need to develop skill at coordinating the use of the information in the order-requirement digraph and the priority list. It will be helpful to cross out the tasks in the priority list as they are assigned to a processor to keep track of which tasks remain to be scheduled.

EXAMPLE 2 ■ Applying the List-Processing Algorithm

Let's apply the list-processing algorithm to one possible priority list–T_8, T_7, T_6, . . . , T_1–using two processors and the order-requirement digraph in Figure 3.1. The result is the schedule shown in Figure 3.2, where idle processor time (time during which a processor is not at work on a task) is indicated by white. How does the list-processing algorithm generate this schedule?

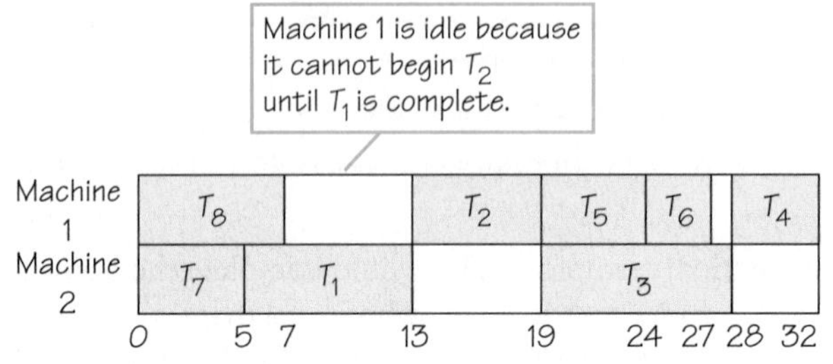

FIGURE 3.2 The schedule produced by applying the list-processing algorithm to the order-requirement digraph in Figure 3.1 using the list T_8, T_7, . . . , T_1.

T_8 (task 8) is first on the priority list and ready at time 0 since it has no predecessors. It is assigned to the lowest-numbered free processor, processor 1. Task 7, next on the priority list, is also ready at time 0 and thus is assigned to processor 2. The first processor to become free is processor 2 at time 5. Recall that by assumption 1, once a processor starts work on a task, its work cannot be interrupted until the task is complete. Task 6, the next unassigned task on the list, is not ready at time 5, as can be seen by consulting Figure 3.1. The reason task 6 is not ready at time 5 is that task 5 has not been completed by time 5. In fact, at time 5, the only ready task on the list is T_1, so that task is assigned to processor 2. At time 7, processor 1 becomes free, but no task becomes ready until time 13.

Thus, processor 1 stays idle from time 7 to time 13. At this time, because T_2 is the first ready task on the list not already scheduled, it is assigned to processor 1. Processor 2, however, stays idle because no other ready task is available at this time. The remainder of the scheduling shown in Figure 3.2 is completed in this manner. ■

We can summarize this procedure as follows:

List-Processing Algorithm: Part II PROCEDURE

As the priority list is scanned from left to right to assign a task to a processor at a particular time, we pass over tasks that are not ready to find ones that are ready. If no task can be assigned in this manner, we keep one or more processors idle until such time that, reading the priority list from the left, there is a ready task not already assigned. After a task is assigned to a processor, we resume scanning the priority list, starting over at the far left, for unassigned tasks.

When Is a Schedule Optimal?

The schedule in Figure 3.2 has a lot of idle time, so it may not be optimal. Indeed, if we apply the list-processing algorithm for two processors to another possible priority list T_1, . . . , T_8, using the digraph in Figure 3.1, the resulting schedule is that shown in Figure 3.3.

Here are the details of how we arrived at this schedule. Remember that we must coordinate the list T_1, T_2, . . . , T_8 with the information in the order-requirement digraph shown in Figure 3.3a. At time 0, task T_1 is ready, so this task is assigned to processor 1. However, at time 0, tasks $T_2, T_3, \ldots, T_6$ are not ready because their predecessors are not done. For example, T_2 is not ready at time 0 because T_1, which precedes it, is not done at time 0. The first ready task on the list, reading from left to right, that is not already assigned is T_7, so task T_7 gets assigned to processor 2. Both processors are now busy until time 5, at which point processor 2 becomes available to work on another task (Figure 3.3b).

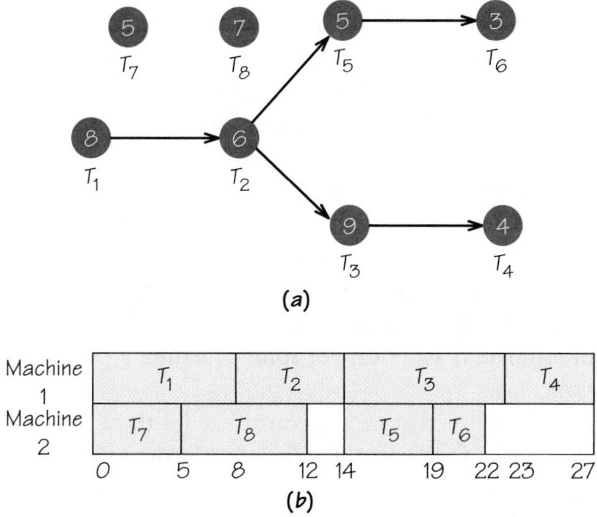

FIGURE 3.3 (a) A typical order-requirement digraph (repeat of Figure 3.1). (b) The schedule produced by applying the list-processing algorithm to the order-requirement digraph in Figure 3.3a using the list $T_1, T_2, \ldots, T_8$.

Tasks T_1 and T_7 have been assigned. Reading from left to right along the list, the first task not already assigned whose predecessors are done by time 5 is T_8, so this task is started at time 5 on processor 2; processor 2 will continue to work on this task until time 12, because the task time for this task is 7 time units. At time 8, processor 1 becomes free, and reading the list from left to right we find that T_2 is ready (because T_1 has just been completed). Thus, T_2 is assigned to processor 1, which will stay busy on this task until time 14. At time 12, processor 2 becomes free, but the tasks that have not already been assigned from the list, T_3, T_4, T_5, T_6 are not ready, because they depend on T_2 being completed before these tasks can start. Thus, processor 2 must stay idle until time 14. At this time, T_3 and T_5 become ready. Since both processors 1 and 2 are idle at time 14, the lower numbered of the two, processor 1, gets to start on T_3 because it is the first ready task left to be assigned on the list scanned from left to right. Task T_5 gets assigned to processor 2 at time 14. The remaining tasks are assigned in a similar manner.

The schedule shown in Figure 3.3b is optimal because the path T_1, T_2, T_3, T_4, with length 27, is the critical path in the order-requirement digraph. As we saw in Chapter 2, the earliest completion time for the job made up of all the tasks is the length of the longest path in the order-requirement digraph.

There is another way of relating optimal completion time to the completion time that is yielded by the list-processing algorithm. Suppose that we add all the task times given in the order-requirement digraph and divide by the number of processors. The completion time using the list-processing algorithm must be at least as large as this number. For example, the task times for the order-requirement digraph in Figure 3.3a sum to 47. Thus, if these tasks are scheduled on two processors, the completion time is at least $\frac{47}{2} = 23.5$ (in fact, 24, because the list-processing algorithm applied to integer task times must yield an integer solution), while for three processors the completion time is at least $\frac{47}{3}$ (in fact, 16).

Why is it helpful to take the total time to do all the tasks in a job and divide this number by the number of processors? Think of each task that must be scheduled as a rectangle that is 1 unit high and t units wide, where t is the time allotted for the task. Think of the scheduling diagram with m processors as a rectangle that is m units high and whose width, W, is the completion time for the tasks. The scheduling diagram is to be filled up by the rectangles that represent the tasks. How small can W be? The area of the rectangle that represents the scheduling diagram must be at least as large as the sum of all the rectangles representing tasks that are "packed" into it. The area of the scheduling diagram rectangle is mW. The combined areas of all the tasks, plus the area of rectangles corresponding to idle time, will equal mW. Width W is smallest when the idle time is zero. Thus, W must be at least as big as the sum of all the task times divided by m.

Sometimes the estimate for completion time given by the list-processing algorithm from the length of the critical path gives a more useful value than the approach based on adding task times. Sometimes the opposite is true. For the order-requirement digraph in Figure 3.1, except for a schedule involving one processor, the critical-path estimate is superior. For some scheduling problems, both these estimates may be poor.

The number of priority lists that can be constructed if there are n tasks is $n!$ and can be computed using the fundamental principle of counting. For example, for eight tasks, $T_1, \ldots, T_8$, there are $8 \times 7 \times 6 \cdots \times 1 = 40{,}320$ possible priority lists. For different choices of the priority list, the list-processing algorithm may schedule the tasks, subject to the constraints of the order-requirement digraph, in different ways. More specifically, two different lists may yield different completion times or the same completion time, but the order in which the tasks are carried out will be different. It is also possible that two different lists produce identical ordering of the assignments of the tasks to processors and completion times. A little later we will see a method that can be used to select a list that, if we are lucky, will give a schedule with a relatively good completion time. In fact, no method is known, except for very specialized cases, of how to choose a list that can be guaranteed to produce an optimal schedule when the list algorithm is applied to it.

Strange Happenings

The list-processing algorithm involves four factors that affect the final schedule. The answer we get depends on the following:

1. The times to carry out the tasks
2. Number of processors
3. Order-requirement digraph
4. Ordering of the tasks on the priority list

To see the interplay of these four factors, consider another scheduling problem, this time asociated with the order-requirement digraph shown in Figure 3.4 (the highlighted numbers are task time lengths). The schedule generated by the list-processing algorithm applied to the list $T_1, T_2, \ldots, T_9$, using three processors, is given in Figure 3.5.

Treating the list $T_1, \ldots, T_9$ as fixed, how might we make the completion time earlier? Our alternatives are to pursue one or more of these strategies:

1. Reduce task times.
2. Use more processors.
3. "Loosen" the constraints by having fewer directed edges in the order-requirement digraph.

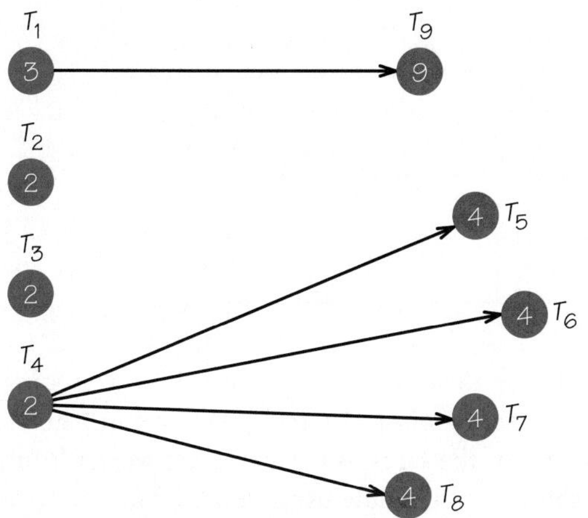

FIGURE 3.4 An order-requirement digraph designed to help illustrate some paradoxical behavior produced by the list-processing algorithm.

	0–3	3–12
Machine 1	T_1	T_9

	0–2	2–4	4–8	8–12
Machine 2	T_2	T_4	T_5	T_7
Machine 3	T_3		T_6	T_8

0 2 3 4 8 12

FIGURE 3.5 The schedule produced by applying the list-processing algorithm to the order-requirement digraph in Figure 3.4 using the list $T_1, T_2, \ldots, T_9$ with three processors.

Let's consider each alternative in turn, changing one feature of the original problem at a time, and see what happens to the resulting schedule. If we use strategy 1 and reduce the time of each task by one unit, we would expect the completion time to go down. Figure 3.6 shows the new order-requirement digraph, and Figure 3.7 shows the schedule produced for this problem, using the list-processing algorithm with three processors applied to the list $T_1, \ldots, T_9$.

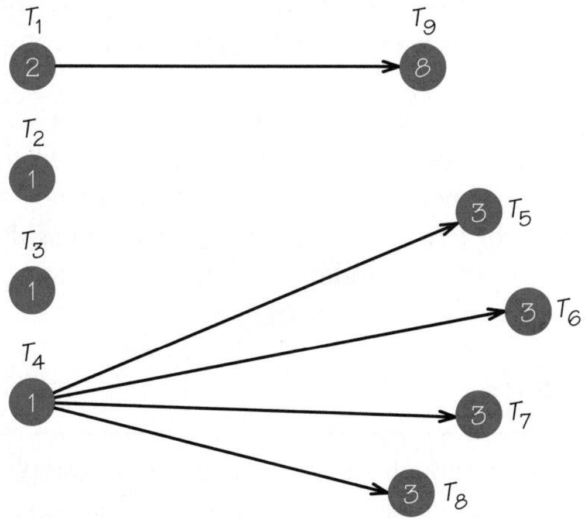

	0–2	2–5	5–8	8–13
Machine 1	T_1	T_5	T_8	

	0–1	1–2	2–5	5–13
Machine 2	T_2	T_4	T_6	T_9
Machine 3	T_3		T_7	

0 1 2 5 8 13

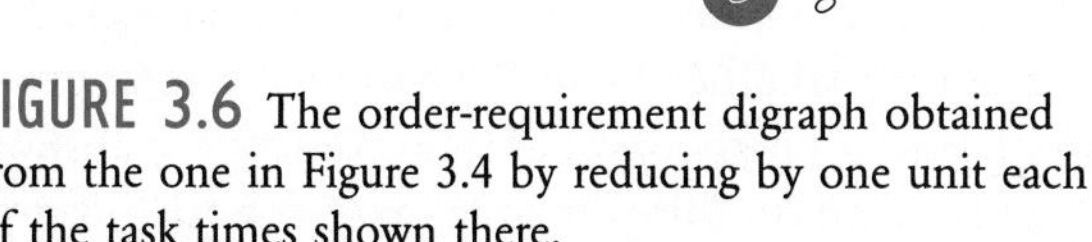

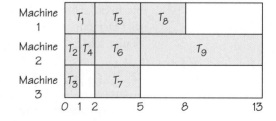

FIGURE 3.6 The order-requirement digraph obtained from the one in Figure 3.4 by reducing by one unit each of the task times shown there.

FIGURE 3.7 The schedule produced by applying the list-processing algorithm to the order-requirement digraph in Figure 3.6 using the list $T_1, T_2, \ldots, T_9$ with three processors.

The completion time is now 13–longer than the completion time of 12 for the case (Figure 3.5) with longer task times. Here is something unexpected! Let's explore further and see what happens.

Next we consider strategy 2, increasing the number of machines. Surely this should speed matters up. When we apply the list-processing algorithm to the original graph in Figure 3.4, using the list $T_1, \ldots, T_9$ and four machines, we get the schedule shown in Figure 3.8. The completion time is now 15–an even later completion time than for the previous alteration.

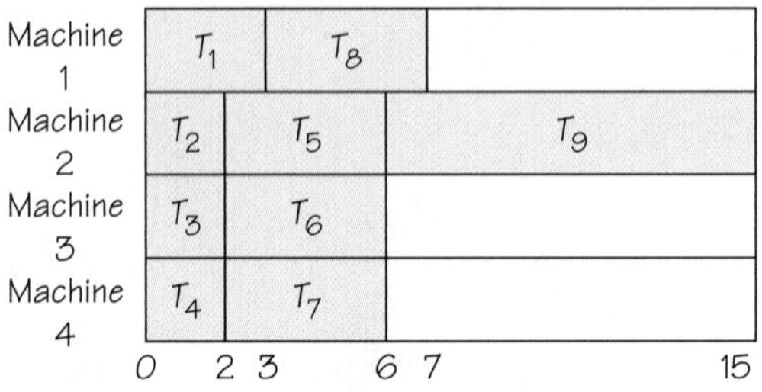

Machine 1: T_1 T_8
Machine 2: T_2 T_5 T_9
Machine 3: T_3 T_6
Machine 4: T_4 T_7
0 2 3 6 7 15

FIGURE 3.8 The schedule produced by applying the list-processing algorithm to the order-requirement digraph in Figure 3.4 using the list $T_1, T_2, \ldots, T_9$ with four processors.

Finally, we consider strategy 3, trying to shorten completion time by erasing all constraints (edges with arrows) in the order-requirement digraph shown in Figure 3.4. By increasing flexibility of the ordering of the tasks, we might guess we could finish our tasks more quickly. Figure 3.9 shows the schedule using the list $T_1, \ldots, T_9$–now it takes 16 units! This is the worst of our three strategies to reduce completion time.

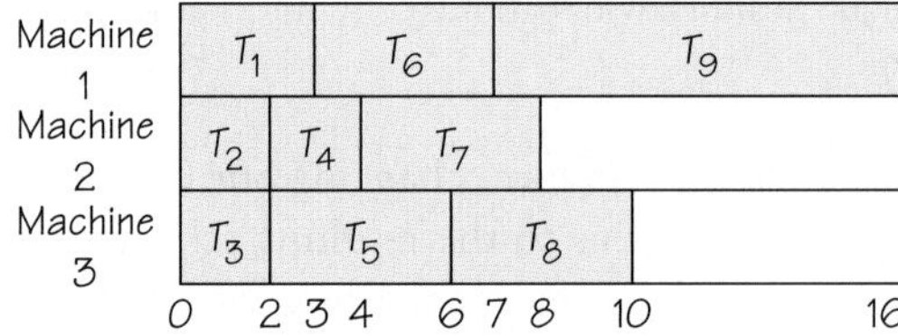

FIGURE 3.9 The schedule produced by applying the list-processing algorithm to the order-requirement digraph in Figure 3.4, modified by erasing all its directed edges, using the list $T_1, T_2, \ldots, T_9$ with three processors.

The failures we have seen here are surprising at first glance, but they are typical of what can happen when a situation is too complex to analyze with naïve intuition. The value of using mathematics rather than intuition or trial and error to study scheduling and other problems is that it points out flaws that can occur in unguarded intuitive reasoning.

It is tempting to believe that we can make an adjustment in the rules for scheduling that we adopted to avoid the paradoxical behavior that has just been illustrated. Unfortunately, operations research experts have shown that there are no "simple fixes." This means that, in practice, for large scheduling problems such as those that face our hospitals and transportation system, finding the best solution to a particular scheduling problem cannot be guaranteed (see Spotlight 3.1).

3.2 Critical-Path Schedules

In our discussion so far, we have acted as though the priority list used in applying the list-processing algorithm was given to us in advance based on external considerations. Let's now consider the question of whether there is a systematic method of *choosing* a priority list that yields optimal or nearly optimal schedules. We will show how to construct a specific priority list based on this principle, to which the list-processing algorithm can then be applied.

Recall from our discussion of critical-path analysis in Chapter 2 that no matter how a schedule is constructed, the finish time cannot be earlier than the length of the longest path in the order-requirement digraph. This suggests that we should try

SPOTLIGHT 3.1 Management Science and Disaster Recovery

The city of New York depends on a public transportation system of subways and roads to bring hundreds of thousands of people who live in the four outer boroughs (Queens, Brooklyn, the Bronx, and Staten Island) into Manhattan to work and "play." New York City also has a communication system of telephones, radio and television stations, and computer networks. These systems speed information between New York's citizens and people outside the city and around the world. The area in southern Manhattan, in the vicinity of the World Trade Center (WTC), was a center for banking, insurance, financial markets, and domestic and international commerce. The attack on the World Trade Center on September 11, 2001, disrupted these networks and markets but did not destroy them, partly because the principles of operations research and management science were used in the design and development of these systems over a long period of time.

(AP/World Wide Photo.)

The diagram below shows a very simple subway (train) system between an eastern and a western terminus.

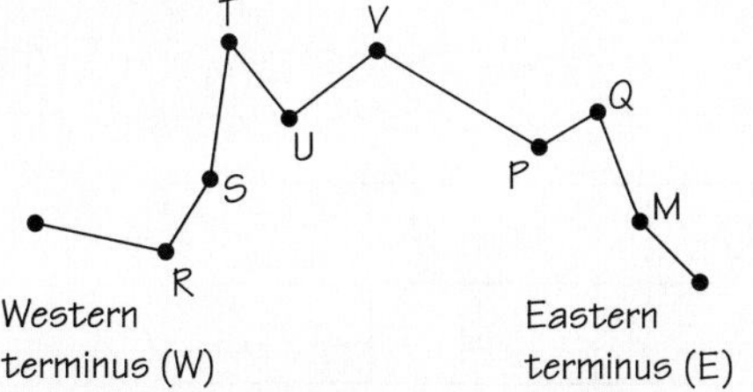

There are two tracks, each dedicated for use by westbound or eastbound trains to run between the two termini. The only place where trains can be turned around is at these termini. Simple graph theory tells us that in such a system, if a vertex is "destroyed" or out of service, or an edge is "destroyed" or out of service, the system totally breaks down. However, the simple provision that trains can be turned around at *U*, even though this is usually only one stop on the way from *W* to *E*, gives much greater flexibility to the system if there is a water main break, or a gas leak, etc. Thanks to simple principles of this kind and creating routes that use independent lines with many transfer points, New Yorkers were able to use the subway system in a flexible way after the World Trade Center disaster. In the days right after the WTC collapsed, trains were not allowed past the geographic area near the WTC for fear that the tunnels' structural foundation had been weakened and that subway vibrations could cause the collapse of damaged buildings. After it was ascertained that running the subways was safe both for partially damaged buildings and for the subways themselves, routes were altered several times to give rescue workers and people returning to their daily routines maximum support. One line's tunnels did collapse, and several stations had to be closed for extended periods, but due to the redundancy and flexibility of the design of the system, a remarkable amount of service was quickly restored.

Good planning and wise application of the principles of management science make it possible to minimize the effects of natural and manmade disasters.

to schedule first those tasks that occur early in long paths, because they might be a bottleneck for the other tasks. This idea leads to **critical-path scheduling**.

EXAMPLE 3 ■ Scheduling Two Processors

To illustrate this method, consider the order-requirement digraph in Figure 3.10a. Suppose we wish to schedule these tasks on two processors. Initially, there are two critical paths of length 64: T_1, T_2, T_3 and T_1, T_4, T_3. Thus, we place T_1 first on the priority list. With T_1 "gone," there is a new critical path of length 60 (T_5, T_6, T_4, T_3) that starts with T_5, so T_5 is placed second on the priority list. At this stage, with T_1 and T_5 removed, we have the residual order-requirement digraph shown in Figure 3.10b. In this diagram there are paths of length 50 (T_2, T_3), 56 (T_6, T_4, T_3), 36 (T_6, T_4, T_7), and 24 (T_8, T_4, T_{10}). Because T_6 heads the path that is currently longest in length, it is placed third in the priority list. Once T_6 is removed from Figure 3.10b, there is a tie for which is the longest path remaining, because both T_2, T_3 and T_4, T_3 are paths of length 50.

When there is a tie between two longest paths, we place next on the priority list in the lowest-numbered task heading a longest path. In the example shown here, this means that T_2 is placed next on the priority list, to be followed by T_4. Continuing in this fashion, we obtain the priority list T_1, T_5, T_6, T_2, T_4, T_3, T_8, T_9, T_7, T_{10}. Note that the order of T_7 and T_{10} was decided using the rule for breaking ties. The list-processing algorithm is now applied using this priority list and the order-requirement digraph in Figure 3.10a. We obtain the schedule in Figure 3.11.

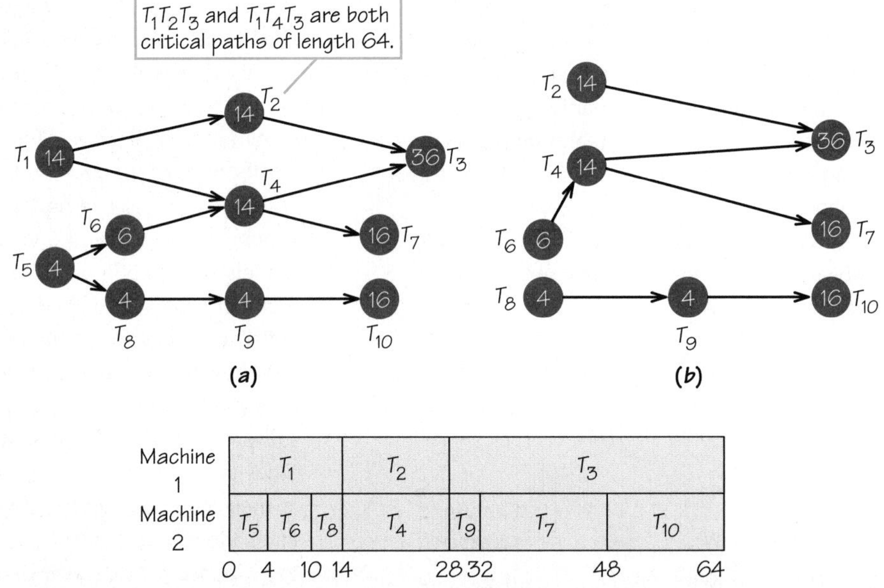

FIGURE 3.10 (a) An order-requirement digraph used to illustrate the critical-path scheduling method. (b) Residual order-requirement digraph after tasks T_1 and T_5 have been removed.

FIGURE 3.11 The optimal schedule produced by applying the critical-path scheduling method to the order-requirement digraph in Figure 3.10. The list used was T_1, T_5, T_6, T_2, T_4, T_3, T_8, T_9, T_7, T_{10}.

Critical-Path Scheduling

PROCEDURE

The **critical-path scheduling** algorithm applies the list-processing algorithm using the priority list L obtained as follows:

1. Find a task that heads a critical (longest) path in the order-requirement digraph. If there is a tie, choose the task with the lower number.
2. Place the task found in step 1 next on the list L. (The first time through the process this task will head the list.)
3. Remove the task found in step 1 and the edges attached to it from the current order-requirement digraph, obtaining a new (modified) order-requirement digraph.
4. If there are no vertices left in the new order-requirement digraph, the procedure is complete; if there are vertices left, go to step 1.

This procedure will terminate when all the tasks in the original order-requirement digraph have been placed on the list L.

The preceding example shows that critical-path scheduling can sometimes yield optimal solutions. Unfortunately, this algorithm does not always perform well. For example, the critical-path method employing four processors applied to the order-requirement digraph shown in Figure 3.12 yields the list T_1, T_8, T_9, T_{10}, T_{11}, T_5, T_6, T_7, T_{12}, T_2, T_3, T_4 and then the schedule in Figure 13.3. (Note that T_5, T_6, T_7 are thought of as heading paths of length 10.) In fact, there can be no worse schedule than this one. An optimal schedule is shown in Figure 3.14.

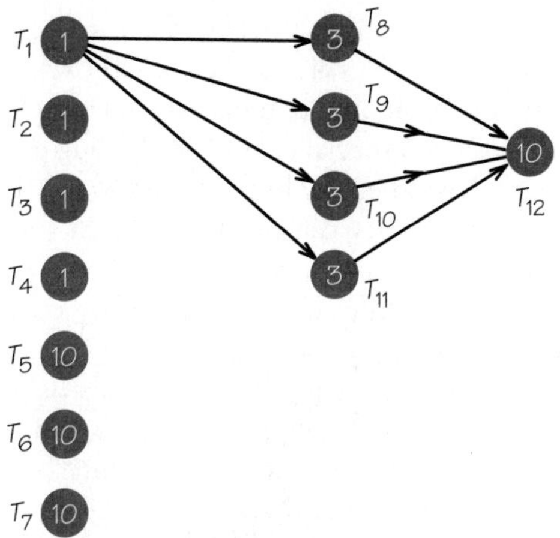

FIGURE 3.12 An order-requirement digraph used to illustrate how poorly the critical-path scheduling method can sometimes behave.

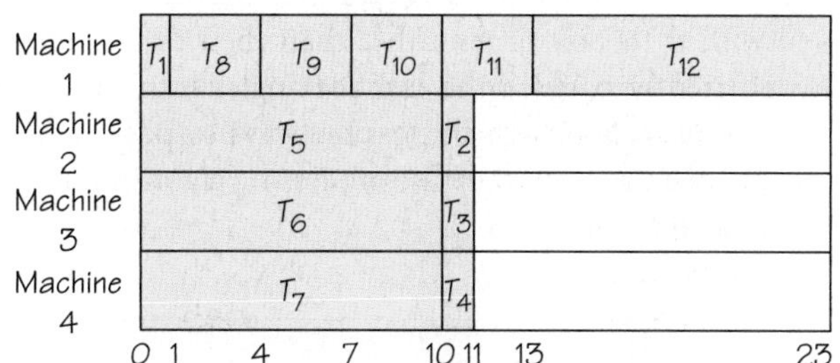

FIGURE 3.13 The schedule produced by applying the critical-path scheduling method to the order-requirement digraph in Figure 3.12 using four processors. The list used was T_1, T_8, T_9, T_{10}, T_{11}, T_5, T_6, T_7, T_{12}, T_2, T_3, T_4.

	0–1	1–4	4–14
Machine 1	T_1	T_8	T_5
Machine 2	T_2	T_9	T_6
Machine 3	T_3	T_{10}	T_7
Machine 4	T_4	T_{11}	T_{12}

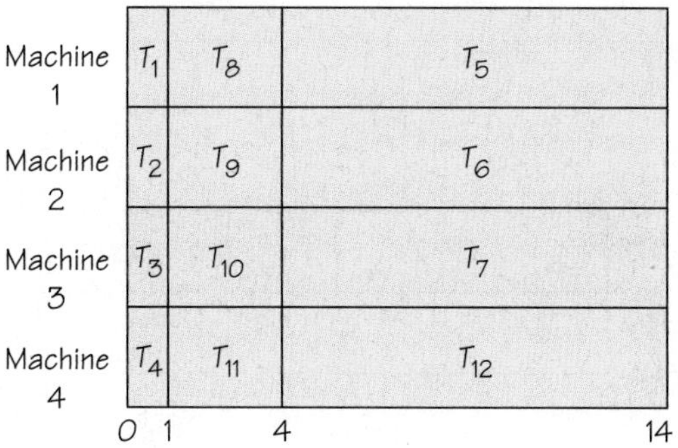

FIGURE 3.14 An optimal schedule for the order-requirement digraph in Figure 3.12 using four processors.

Many of the results we have examined so far are negative because we are dealing with a general class of problems that defy our using computationally efficient algorithms to find an optimal schedule. But we can close on a more positive note. Consider an arbitrary order-requirement digraph, but assume all the tasks take equal time. It turns out that we can always construct an optimal schedule using two processors in this situation. Ironically, we can choose among many algorithms to produce these optimal schedules. The algorithms are easy to understand (though not easy to prove optimal) and have all been discovered since 1969! Many people think that mathematics is a subject that is no longer alive, and that all its ideas and methods were discovered hundreds of years ago—but as we have just seen, this is not true. In fact, more new mathematics has been discovered and published in the last 30 years than during any previous 30-year period.

3.3 Independent Tasks

Mathematicians suspect that no computationally efficient algorithm for solving general scheduling problems optimally will ever be found. Owing to our limited success in designing algorithms for finding optimal schedules for general order-requirement digraphs, we will consider a special class of scheduling problems for which the order-requirement digraph has no edges. In this case, we say that the tasks are *independent* of one another, because they can be performed in any order. (No edges in the order-requirement digraph indicates that no tasks need to precede others; that is, the tasks can be done in any order.) In this section we consider the problem of scheduling **independent tasks**.

Geometrically, we can think of the independent tasks as rectangles of height 1 whose lengths are equal to the time length of the task. Finding an optimal schedule amounts to packing the task rectangles, with no "idle time" gaps between adjacent rectangles, into a longer rectangle whose height equals the number of machines. For example, Figure 3.15 shows two different ways to schedule tasks of length 10, 4, 5, 9, 7, 7 on two machines. (For convenience, the rectangles in the case of independent tasks are labeled with their task times rather than their task numbers.) Scheduling basically means efficiently packing the task rectangles into the machine rectangle. Finding the optimal answer among all possible ways to pack these rectangles is like looking for a needle in a haystack. The list-processing algorithm produces a packing, but it may not be a good one.

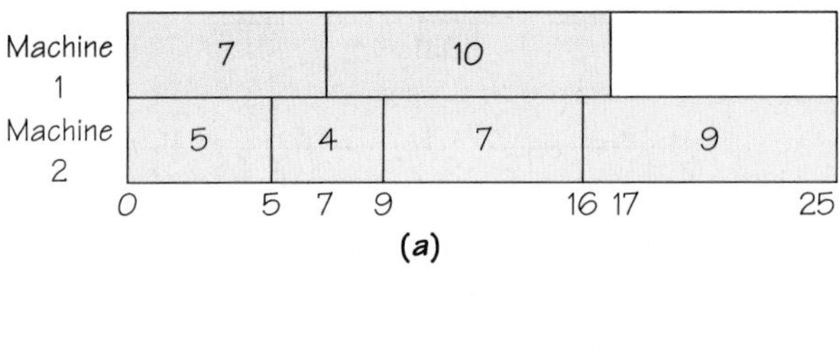

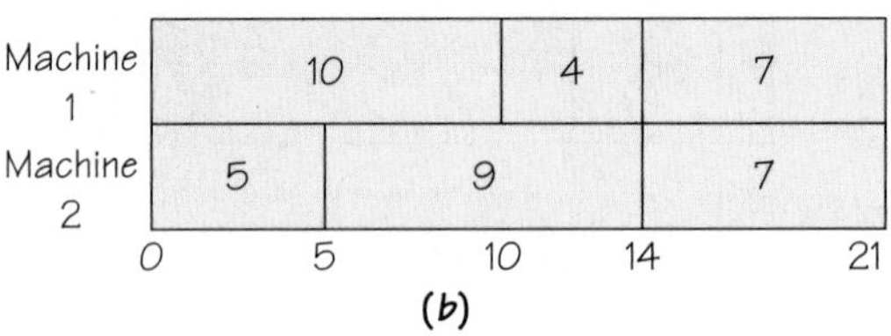

FIGURE 3.15 (a) A nonoptimal way to schedule independent tasks of time lengths 10, 4, 5, 9, 7, 7 using two processors. (b) An optimal way to schedule independent tasks of time lengths 10, 4, 5, 9, 7, 7 using two processors.

There are two approaches we can consider. To study **average-case analysis**, we might ask: *Is the average (mean) of the completion times arrived at by using the list-processing algorithm with all the possible different lists close to the optimal possible completion time?* To study **worst-case analysis**, we might ask: *How far from optimal is a schedule obtained using the list-processing algorithm with one particular priority list?* What is being contrasted with these two points of view is that an algorithm may work well most of the time (give an answer close to optimal) even though there may be a few cases in which it performs very badly. Average-case analysis is amenable to mathematical solution but requires methods of great sophistication.

Decreasing-Time Lists

Is there some way of choosing a priority list for independent tasks that consistently yields relatively good schedules? The surprising answer is yes! The idea is that when long tasks appear toward the end of the list, they often seem to "stick out" on the right end, as in Figure 3.15a. This suggests that before one tries to schedule a collection of tasks, the tasks should be placed in a list where the longest tasks are listed first.

Decreasing-Time-List Algorithm — PROCEDURE

The list-processing algorithm applied to a list of task times arranged in order of nonincreasing size is called the **decreasing-time-list algorithm**.

If we apply it to the set of tasks listed previously (10, 4, 5, 9, 7, 7), we obtain the times 10, 9, 7, 7, 5, 4 and the schedule (packing) shown in Figure 3.16. This packing is again optimal, but it is different from the optimal scheduling in Figure 3.15b. It is worth noting that the decreasing-time list and the list obtained by the critical-path method discussed earlier will coincide in the case of independent tasks. The decreasing-time list can also be constructed for the case in which the tasks are not independent. For general order-requirement digraphs, the decreasing-time list does not produce particularly good schedules.

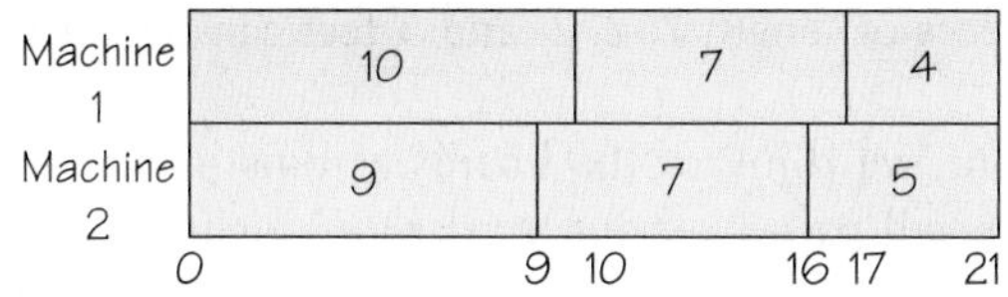

FIGURE 3.16 The optimal schedule resulting from applying the decreasing-time-list algorithm to a collection of independent tasks. The list used, written in terms of task times only, is 10, 9, 7, 7, 5, 4.

It is important to remember that the decreasing-time-list algorithm does not *guarantee* optimal solutions. This can be seen by scheduling the tasks with times 11, 10, 9, 6, 4 (Figure 3.17). The schedule has a completion time of 21. However, the rearranged list 9, 4, 6, 11, 10 yields the schedule in Figure 3.18, which finishes at time 20. This solution is obviously optimal because the machines finish at the same time and there is no idle time. Note that when tasks are independent, if there are m machines available, the completion time cannot be less than the sum of the task times divided by m.

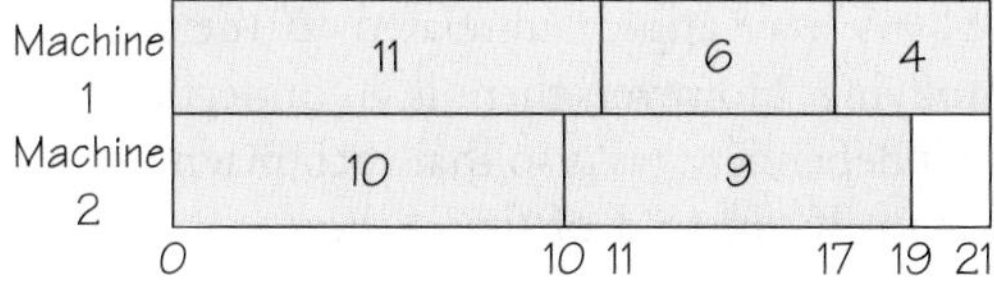

FIGURE 3.17 The nonoptimal schedule resulting from applying the decreasing-time-list algorithm to a collection of independent tasks. The list used, written in terms of task times only, is 11, 10, 9, 6, 4.

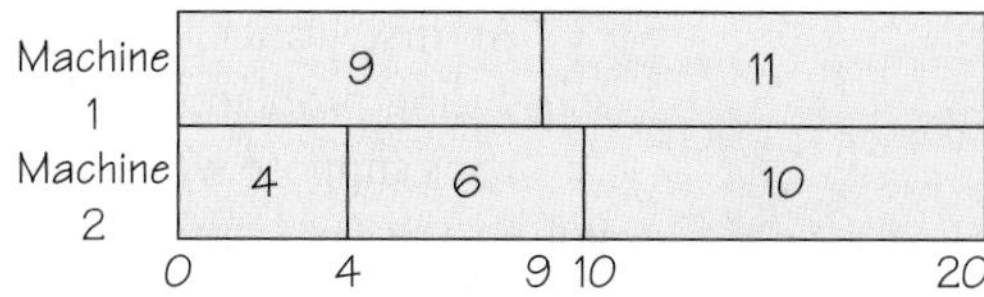

FIGURE 3.18 The optimal schedule resulting from applying the list-processing algorithm to a collection of independent tasks. The list used, written in terms of task times only, is 9, 4, 6, 11, 10.

EXAMPLE 4 ■ Photocopy Shop and Data Entry Problems

A modern copy shop provides a wide array of services ranging from copying a few sheets for a "drop in" customer, to printing elaborate reports for small businesses, to publishing monographs and advertising flyers. Using mathematical scheduling techniques can save time and cost by ensuring the many tasks are completed most efficiently. *(Christopher Robbins/ Digital Vision/Getty Images.)*

Imagine a photocopy shop with three photocopiers. Photocopying tasks that must be completed overnight are accepted until 5 P.M. The tasks are to be done in any manner that minimizes the finish time for all the work. Because this problem involves scheduling machines for independent tasks, the decreasing-time-list algorithm would be a good heuristic to apply.

For another example, consider a data entry pool at a large corporation or college, where individual entry tasks can be assigned to any data entry specialist. In this setting, however, the assumption that the data entry workers are identical in skill is less likely to be true. Hence, the tasks might have different times with different processors. This phenomenon, which occurs in real-world scheduling problems, violates one of the assumptions of our mathematical model.

3.4 Bin Packing

Suppose you plan to build a wall system for your books, CDs, DVDs, and stereo set. It requires 24 wooden shelves of various lengths: 6, 6, 5, 5, 5, 4, 4, 4, 4, 2, 2, 2, 2, 3, 3, 7, 7, 5, 5, 8, 8, 4, 4, and 5 feet. The lumberyard, however, sells wood only in boards of length 9 feet. If each board costs $8, what is the minimum cost to buy sufficient wood for this wall system?

Because all shelves required for the wall system are shorter than the boards sold at the lumberyard, the largest number of boards needed is 24, the precise number of shelves needed for the wall system. Buying 24 boards would, of course, be a waste of wood and money because several of the shelves you need could be cut from one board. For example, pieces of length 2, 2, 2, and 3 feet can be cut from one 9-foot board.

To be more efficient, we think of the boards as bins of capacity W (9 feet in this case) into which we will pack (without overlap) n weights (in this case, lengths) whose values are $w_1, \ldots, w_n$, where each $w_i \leq W$. We wish to find the minimum number of bins into which the weights can be packed. In this formulation, the problem is known as the **bin-packing problem**.

Bin-Packing Problem DEFINITION

The **bin-packing problem** involves finding the minimum number of bins of weight capacity W into which weights $w_1, w_2, \ldots, w_n$ (each less than or equal to W) can be packed without exceeding the capacity of the bins.

At first glance, bin-packing problems may appear unrelated to the machine-scheduling problems we have been studying. However, there is a connection.

Let's suppose we want to schedule independent tasks so that each machine working on the tasks finishes its work by time W. Instead of fixing the number of machines and trying to find the earliest completion time, we must find the minimum number of machines that will guarantee completion by the fixed completion time (W). Despite this similarity between the machine-scheduling problem and the bin-packing problem, the discussion that follows will use the traditional terminology of bin packing.

By now, it should come as no surprise to learn that no one knows a fast algorithm that always picks the optimal (smallest) number of bins (boards). In fact, the bin-packing problem belongs to the class of NP-complete problems (see Spotlight 2.1), which means that most experts think it unlikely that any fast optimal algorithm will ever be found. Relatively good algorithms for problems that come up in actual applications are known.

Bin-Packing Heuristics

We will think of the items to be packed, in any particular order, as constituting a list. In what follows we will use the list of 24 shelf lengths given for the wall system. We will consider various **heuristic algorithms**, namely, methods that can be carried out quickly but cannot be guaranteed to produce optimal results. Probably the easiest approach is simply to put the weights into the first bin until the next weight won't fit, and then start a new bin. (Once you open a new bin, don't use leftover space in an earlier, partially filled bin.) Continue in the same way until as many bins as necessary are used.

The resulting solution is shown in Figure 3.19. This algorithm, called **next-fit (NF),** has the advantage of not requiring knowledge of all the weights in advance. Only the remaining space in the bin currently being packed must be remembered. The disadvantage of this heuristic is that a bin packed early on may have had room for small items that come later in the list.

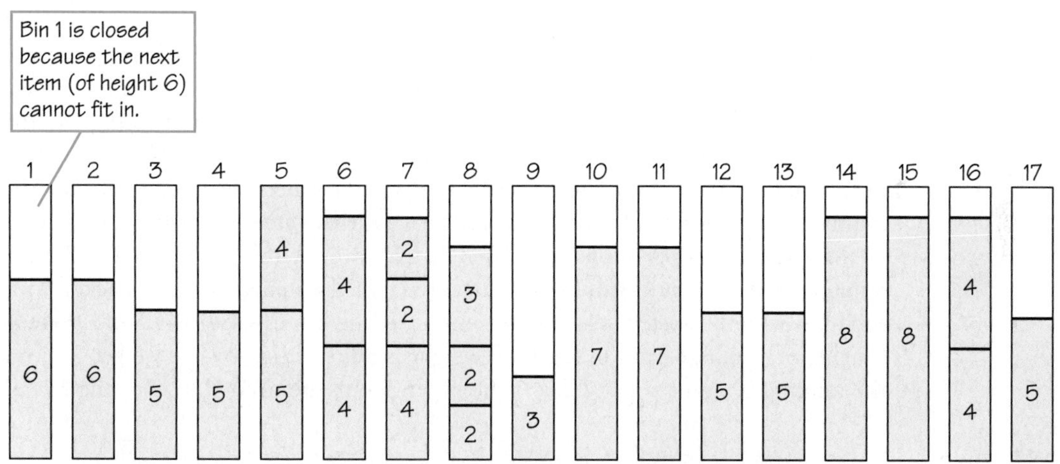

FIGURE 3.19 The list 6, 6, 5, 5, 5, 4, 4, 4, 4, 2, 2, 2, 2, 3, 3, 7, 7, 5, 5, 8, 8, 4, 4, 5 packed in bins using next fit.

Our wish to avoid permanently closing a bin too early suggests a different heuristic—**first-fit (FF)**: Put the next weight into the first bin already opened that has room for this weight. If no such bin exists, start a new bin. Note that a computer program to carry out first fit would have to keep track of how much room was left in all the previously opened bins. For the 24 wall-system shelves, the first-fit algorithm would generate a solution that uses only 14 bins (see Figure 3.20) instead of the 17 bins generated by the next-fit algorithm.

If we are keeping track of how much room remains in each partially filled bin, we can put the next item to be packed into the bin that currently has the most room available. This heuristic will be called **worst-fit (WF)**. The name *worst fit* refers to the fact that an item is packed into a bin with the most room available, that is, into which it fits "worst," rather than into a bin that will leave little room left over after it is placed in that bin ("best fit"). The solution generated by this approach looks the same as that shown in Figure 3.20. Although this heuristic also leads to 14 bins, the items are packed in a different order. For example, the first item of size 2, the tenth item in the list, is put into bin 6 in worst fit, but into bin 1 in first fit.

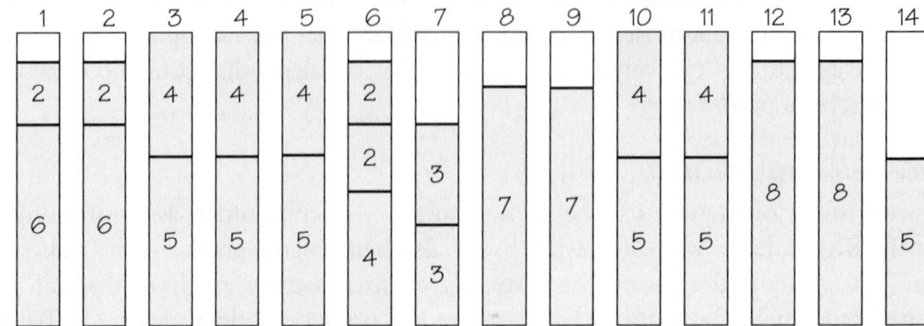

FIGURE 3.20 The list 6, 6, 5, 5, 5, 4, 4, 4, 4, 2, 2, 2, 2, 3, 3, 7, 7, 5, 5, 8, 8, 4, 4, 5 packed in bins using first fit. Worst fit would yield a packing that would look identical.

Decreasing-Time Heuristics

One difficulty with all three of these heuristics is that large weights that appear late in the list can't be packed efficiently. Therefore, we should first sort the items to be packed in order of decreasing size, assuming that all items are known in advance. We can then pack large items first and the smaller items into leftover spaces. This approach yields three new heuristics: **next-fit decreasing (NFD), first-fit decreasing (FFD)**, and **worst-fit decreasing (WFD)**. Here is the original list sorted by decreasing size: 8, 8, 7, 7, 6, 6, 5, 5, 5, 5, 5, 5, 4, 4, 4, 4, 4, 4, 3, 3, 2, 2, 2, 2. Packing using first-fit-decreasing order yields the solution in Figure 3.21. This solution uses only 13 bins.

Is there any packing that uses only 12 bins? No. In Figure 3.21, there are only 2 free units (1 unit each in bins 1 and 2) of space in the first 12 bins, but 4 occupied units (two 2's) in bin 13. We could have predicted this by dividing the total length of the shelves (110) by the capacity of each bin (board): $\frac{110}{9} = 12\frac{2}{9}$. Thus, no packing could squeeze these shelves into 12 bins–there would always be at least 2 units left over for the 13th bin. (In Figure 3.21, there are 4 units in bin 13 because of the 2 wasted empty spaces in bins 1 and 2.) Even if this division created a zero remainder, there would still be no guarantee that the items could be packed to fill each bin without wasted space. For example, if the bin capacity is 10 and there are weights of 6, 6, 6, 6, and 6, the total weight is 30; dividing by 10, we get 3 bins as the minimum requirement. Clearly, however, 5 bins are needed to pack the five 6's.

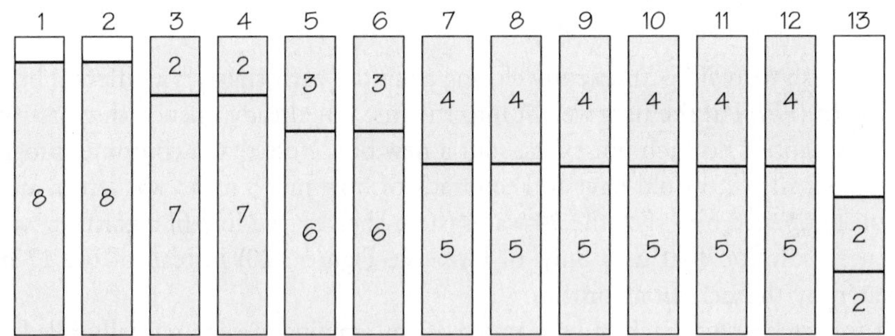

FIGURE 3.21 The bin packing resulting from applying first-fit decreasing to the wall-system numbers. The list involved, which uses the original list sorted in decreasing order, is 8, 8, 7, 7, 6, 6, 5, 5, 5, 5, 5, 5, 4, 4, 4, 4, 4, 4, 3, 3, 2, 2, 2, 2.

None of the six heuristic methods shown will necessarily find the optimal number of bins for an arbitrary problem. How can we decide which heuristic to use? One approach is to see how far from the optimal solution each method might stray.

Various formulas have been discovered to calculate the maximum discrepancy between what a bin-packing algorithm actually produces and the best possible result. For example, in situations where a large number of bins are to be packed, FF can be off by as much as 70%, but FFD is never off by more than 22%. Of course, FFD doesn't give an answer as quickly as FF, because extra time for sorting a large collection of weights may be considerable. Also, FFD requires knowing the whole list of weights in advance, whereas FF does not. It is important to emphasize that a 22% margin of error is a worst-case figure. In many cases, FFD will perform much better. Results obtained by computer simulation indicate excellent average-case performance for this algorithm.

When solving real-world problems, we always have to look at the relationship between mathematics and the real world. Thus, first-fit decreasing usually results in fewer bins than next fit, but next fit can be used even when all the weights are not known in advance. Next fit also requires much less computer storage than first fit, because once a bin is packed, it need never be looked at again.

Fine-tuning of the conditions of the actual problem often results in better practical solutions and in interesting new mathematics as well. See Spotlight 3.2 for a discussion of some of the tools mathematicians use to verify and even extend mathematical truths by raising new mathematical problems.

SPOTLIGHT 3.2 Using Mathematical Tools

The tools of a carpenter include the saw, T square, level, and hammer. A mathematician also requires tools of the trade. Some of these tools are the proof techniques that enable verification of mathematical truths. Another set of tools consists of strategies to sharpen or extend the mathematical truths already known. For example, suppose that if *A* and *B* hold, then *C* is true. What happens if only *A* holds? Will *C* still be true? Similarly, if only *B* holds, will *C* still be true?

This type of thinking is of value because such questions will result either in more general cases where *C* holds or in examples showing that *B* alone and/or *A* alone can't imply *C*. For example, we saw that if a graph *G* is connected (hypothesis *A*) and even-valent (hypothesis *B*), then *G* has a circuit which uses each edge only once (conclusion *C*). If either hypothesis is omitted, the conclusion fails to hold. The figures illustrate this point. On the left is an even-valent but nonconnected graph; on the right, a connected graph with two odd-valent vertices. Neither graph has an Euler circuit.

Here is another way that a mathematician might approach extending mathematical knowledge. If *A* and *B* imply *C*, will *A* and *B* imply both *C* and *D*, where *D* extends the conclusion of *C*? For example, not only can we prove that a connected, even-valent (hypotheses *A* and *B*) graph has an Euler circuit, but we can also show that the first edge of the Euler circuit can be chosen arbitrarily (conclusions *C* and *D*).

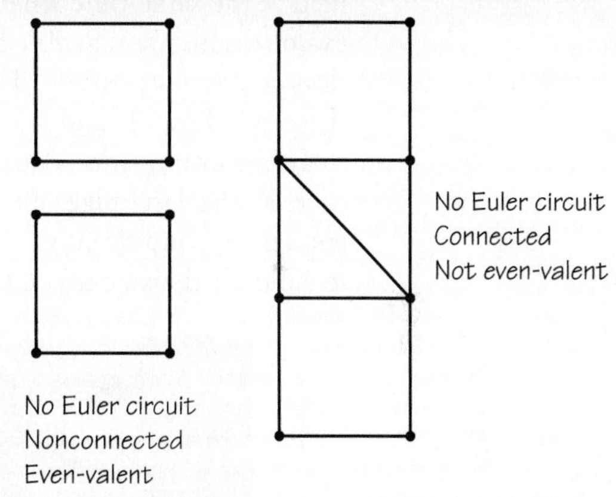

It turns out that being able to specify the first two edges of the Euler circuit may not always be possible. Mathematicians are trained to vary the hypotheses and conclusions of results they prove, in an attempt to clarify and sharpen the range of applicability of the results.

We have seen that machine scheduling and bin packing are probably computationally difficult to solve because they are NP-complete. A mathematician could then try to find the simplest version of a bin-packing problem that would still be NP-complete: What if the items to be packed can have only eight weights? What if the weights are only 1 and 2? Asking questions like these is part of the mathematician's craft. Such questions help to extend the domain of mathematics and hence the applications of mathematics.

3.5 Resolving Conflict via Coloring

In attempting to understand situations that involve scheduling, one might desire to achieve a wide variety of goals. For example, in certain types of scheduling problems, as we have seen here, one is interested in optimization issues. What is the earliest completion time for getting a collection of tasks done on two identical processors? However, in other situations a different goal may arise. For example, in sports, consider a league of baseball teams. Each team has to play some games during the day, some at night, some at home, and some away from home. In the interests of *equity*, it may be desirable for each team to play the same number of day games and night games both at home and away against each of the other teams in the league. If, for example, team *A* plays 8 games away against team *B* and 2 games at home against *B*, then if *A* wins both home games but loses 7 out of 8 away games, it may appear that *B* had an advantage due to the way its games against *A* were scheduled.

Another goal of scheduling, other than optimization and equity, may be to prevent conflicts from occurring. We can use our knowledge of graph theory to solve some interesting scheduling problems where the goal is "conflict resolution." For example, at most colleges, every semester and summer session final examinations must be scheduled. From the point of view of students and faculty both, it would be desirable to schedule these examinations so that (1) no two examinations are scheduled at the same time when a student is enrolled in both of the courses and (2) the examinations are scheduled in as "compact" a way as possible, that is, in as few time slots or days as possible. The administration of the college may share the desire for these two features and want still another property for the scheduling: (3) no more than five examinations are scheduled for any time slot. The reason for a condition such as the last might be that during the summer only five rooms with reliable enough air conditioning are available (or there might be only five rooms large enough to hold all the students taking the common final for multiple-section courses).

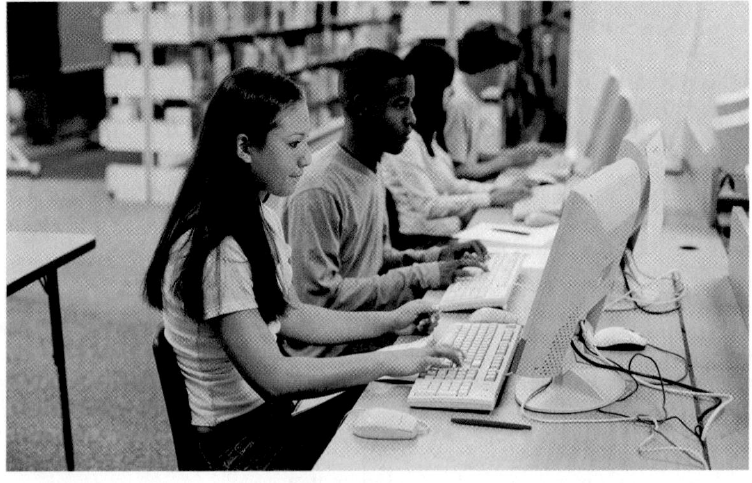

Graph theory can be used to resolve scheduling conflicts that occur in trying to provide students access to limited database or computer resources. (*Bananastock/Picturequest.*)

EXAMPLE 5 ■ Scheduling Examinations

Small State is offering eight courses during its summer session. The table shows with an X which pairs of courses have one or more students in common. Only two air-conditioned lecture halls are available for use at any one time. To design an efficient way to schedule the final examinations, we can represent the information in this table by using a graph, as shown in Figure 3.22a. In the graph, courses are represented by vertices and two courses are joined by an edge if there is any student enrolled in both courses.

	F	M	H	P	E	I	S	C
French (F)		X		X	X	X		X
Mathematics (M)	X				X	X		
History (H)						X	X	X
Philosophy (P)	X							X
English (E)	X	X				X		
Italian (I)	X	X	X		X		X	
Spanish (S)			X			X		
Chemistry (C)	X		X	X				

We are faced with the following graph theory problem: Can we assign labels to the vertices of the graph in such a way that vertices that are joined by an edge get different labels? We think of the labels as the time slots the courses are assigned for final examinations. Traditionally, in graph theory such labels are referred to as *colors.* In this language we seek to color the vertices of the graph so that vertices that are joined by an edge get different colors. Such a coloring is called a **vertex coloring**.

Vertex Coloring

DEFINITION

The **vertex coloring** problem for a graph requires assigning each vertex of the graph a color (label) such that two vertices joined by an edge are assigned different colors.

Figure 3.22b shows one way to color the vertices of the graph so that each vertex gets a different color. Note that numbers are being used to represent the different colors. This solution is not very valuable, however, because it means that each course must be given its own time slot.

To minimize the number of time slots used, we assign colors so that no two vertices that are joined by an edge get the same color. Thus, vertices *F, M, I,* and *E* must get four different colors. These four colors can then be used to color the remaining vertices, ensuring that no two connected vertices have the same color.

The coloring in Figure 3.22c is a major improvement over the one in Figure 3.22b. It uses only four colors. In fact, this is the smallest number of colors that can be used. To see this, notice that the vertices *F, M, I, E* in Figure 3.22a are all joined by edges to each other. Thus, in any coloring of this graph they would require four different colors. The improved coloring in Figure 3.22c was found by trial and error.

Chromatic Number

DEFINITION

The **chromatic number** is the minimum number of colors needed to label the vertices of a graph so that no two vertices of the graph joined by an edge get the same color.

The examination graph we have been studying has chromatic number 4; hence, we can schedule the eight examinations in four time slots without a conflict. Notice, however, that the coloring in Figure 3.22c schedules three different courses for the time slot corresponding to color 2. This means that not enough rooms with air conditioning will be available. Is there a way to recolor the graph with four colors so that each of the four colors is used only twice? Figure 3.22d shows that the answer is yes.

Thus, we are able to schedule the eight final examinations in four time slots, using only two air-conditioned rooms, and no student will have a conflict under this schedule!

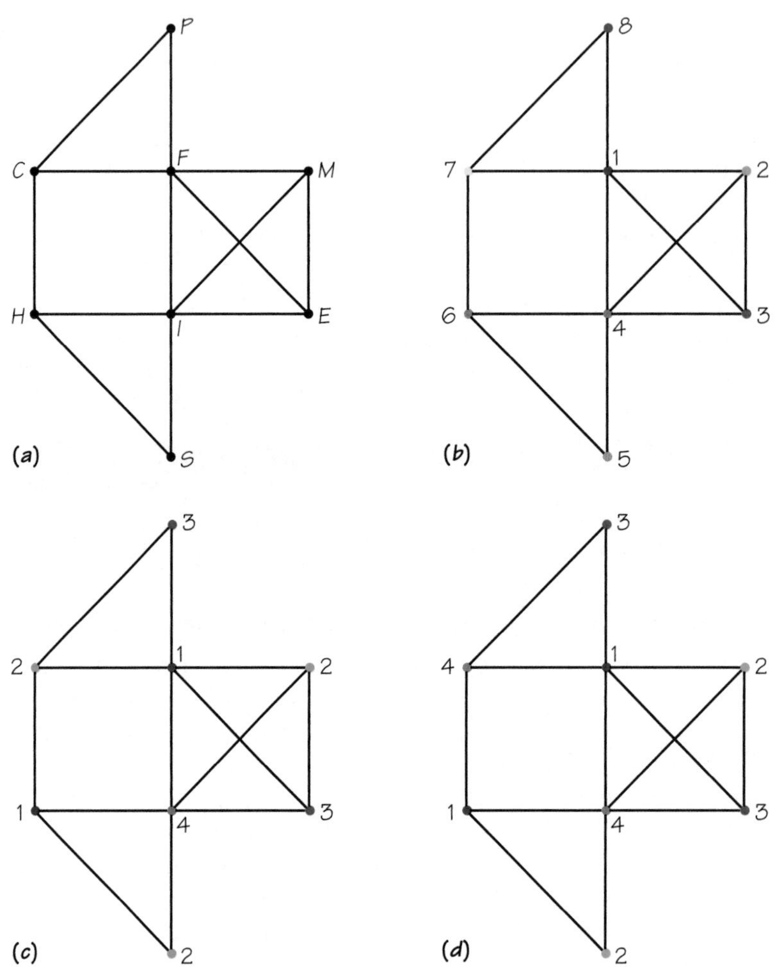

FIGURE 3.22 (a) A graph used to represent conflict information about courses. When two courses have a common student, an edge is drawn between the vertices that represent these courses. (b) A coloring of the scheduling graph with 8 colors, representing 8 time slots. Using this coloring would lead to a schedule where 8 time slots are used to schedule the examinations. This number is far from optimal. (c) A coloring of the scheduling graph with 4 colors. This translates into a way of scheduling the examinations during 4 time slots, and it is not possible to design a schedule with fewer time slots. However, this schedule calls for the use of three different rooms, because three examinations are scheduled during time slot 2. (d) A coloring of the scheduling graph with 4 colors. This means that the examinations can be scheduled in 4 time slots. However, because each color appears only twice, all the examinations can be scheduled in two air-conditioned rooms.

Realistic problems in scheduling government committees, high school and university final examinations, and job interviews (see Spotlight 3.4) are usually so large that graph coloring algorithms have to be incorporated into elaborate software packages to solve them.

Mathematicians have examined many kinds of coloring problems. Many developments about coloring graphs have been an outgrowth of work on the Four Color Problem (see Spotlight 3.3). One can study problems that involve the coloring of the edges of a graph rather than its vertices. Using techniques that have emerged from the study of coloring problems, problems involving such diverse contexts as scheduling government committees, using runways at airports efficiently, assigning frequencies for use by mobile pagers and cell phones, and designing timetables for public transportation have been solved–all these benefits from a problem that at first glance looks as if it belongs to recreational mathematics!

SPOTLIGHT 3.3 Four Color Problem

Many people perceive mathematics as complex because it often uses strange notations and algebraic symbols. Thus, it may come as a surprise that a problem that is relatively easy to state and understand without complex symbolism eluded solution for about 100 years. When it was finally solved, it set off a "firestorm" that it had not truly been solved. More importantly, many of the ideas that have been developed in the theory of graphs were expanded or developed in the course of trying to prove this "guess."

When a graph can be drawn on a flat piece of paper so that edges meet only at vertices, we can talk about not only the vertices and edges of the graph, but also about its regions or *faces*. Such graphs are known as *plane graphs*. Two examples of plane graphs are shown in the diagram below.

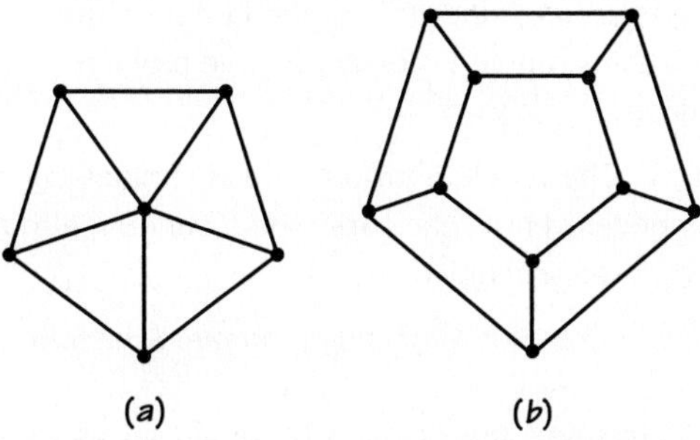

Graph (a) has 6 regions (the area "outside" of the graph is counted as one of the 6 regions), 5 of which have 3 sides and 1 of which has 5 sides, while graph (b) has 7 regions, 2 of which have 5 sides and 5 of which have 4 sides. To count the number of sides of a region, imagine you are a small ant and are following the edges around the region, starting at some vertex *w*. You count edges until you get back to *w*. Note that for each of these graphs there is one *unbounded* (goes off to "infinity") region, in addition to the other regions. When you color the regions of a plane map do not forget to assign a color to the unbounded region.

If you think of the regions of a plane graph as being distinct countries on a page that is to appear in an atlas, it would be nice if countries that share a border got different colors so that they can be distinguished. Countries that meet at a vertex, but do not share an edge representing a common border, can be colored with the same color. It is convenient to use the term map for the regions created by the drawing of a plane graph.

The following provocative question was raised in a letter (1852) from Augustus De Morgan to William Rowan Hamilton that was based on a problem posed to De Morgan by his student Fredrick Guthrie, who heard the question from his brother Francis:

Can the regions of any (plane) map always be colored with four or fewer colors?

A clever proposal as to how to prove the "Four Color Conjecture" was proposed by Alfred Kempe. Kempe's "proof" had a subtle error, which defied detection for many years, showing that proofs in mathematics really depend on the community of mathematicians to guarantee their accuracy. The British mathematician Percy Heawood discovered the error Kempe made. Heawood adapted Kempe's proof to show correctly that any map can be colored with five or fewer colors. Approximately 100 years elapsed before a proof that the Four Color Conjecture was true was found. This occurred in 1976, but there was a curious loose end: The proof found by Wolfgang Haken and Kenneth Appel required that a computer verify a large collection of "calculations," which were too numerous to be done by hand. This proof troubled some philosophers and mathematicians, but has been widely accepted by the mathematics community. In 1995, Neil Robertson, Daniel Sanders, Paul Seymour, and Robin Thomas found another proof. This proof, while simpler and shorter than the earlier Haken–Appel proof, also required computer calculations too numerous to be checked by "hand." Though it is possible that some new approach to the Four Color Conjecture will avoid the use of computers, this is not widely thought to be likely. However, human ingenuity sometimes surprises us!

SPOTLIGHT 3.4 Scheduling Job Interviews

A group of companies is coming to campus for job interviews. Different companies may want different numbers of time slots to hold their interviews. In each time slot one student can be interviewed. In the example below, all the companies have requested contiguous time slots for the interviews, but this need not be the case. Due to the fact that classes are going on at the same time, five departmental conference rooms have been made available to the companies to conduct their interviews.

The interviews will follow the school's regular hourly periods, which start at 9 A.M. and end at 4 P.M. (Companies will be scheduled for continuous interviews during lunch-hour times. Interviews cannot be scheduled beyond the end of the period that starts at 4 P.M. and ends at 5 P.M.)

	Company	Time Slot Requested
A	(Apricot Computers)	7
B	(Big Green)	1
C	(Challenge Insurance)	4, 5
D	(Daisy Printers)	7, 8
E	(Earnest Engine)	4, 5, 6
F	(Flexible Systems)	2, 3
G	(Gutter Leaders)	1, 2
H	(Halley's Combs)	6, 7
I	(Indelible Ink Corporation)	7, 8
J	(Jay's Produce)	4, 5
K	(Kelly's Detective Agency)	2, 3
L	(Large Clothes)	4, 5, 6
M	(Metropolitan TV)	1, 2
N	(Nationwide Bank)	4, 5, 6, 7

Look at the list of time blocks that the companies requested (where 1 = 9–10 A.M., . . . , 8 = 4–5 P.M.). Is it possible to accommodate all the companies that wish to do interviewing in the five rooms available while meeting their desired schedule times?

Problems of this kind seem simple enough, and you should try your hand at solving this particular one, for which a schedule does exist! However, this situation is not simple at all. The following facts are known about problems of this kind.

Fact 1. Suppose there are i interviews, p time periods, and r rooms where interviews can be scheduled. Each interviewer has specified periods during which he or she wishes to conduct interviews. Is it possible to design a schedule that meets the desired specifications? It turns out that this problem is NP-complete (see Spotlight 2.1), that is, it belongs to a large group of problems for which, among other things, the fastest known algorithms run very slowly on large-problem versions.

Fact 2. The problem just described remains NP-complete even for the case where only three rooms have to be scheduled ($p = 3$).

The moral is *surprisingly simple:* Scheduling problems are very hard to solve.

However, the situation is not always as hopeless as it might seem. If you look at the list of time requests for the corporations, you will note again that, not surprisingly, each company has requested a contiguous block of times. It turns out that when this condition holds, it is possible to determine whether there is a feasible schedule using an algorithm that works relatively quickly.

REVIEW VOCABULARY

Average-case analysis The study of the list-processing algorithm (more generally, any algorithm) from the point of view of how well it performs in all the types of problems it may be used for and seeing on average how well it does. *See also* worst-case analysis. (p. 78)

Bin-packing problem The problem of determining the minimum number of containers of capacity W into which objects of size $w_1, \ldots, w_n$ ($w_i \leq W$) can be packed. (p. 80)

Chromatic number The chromatic number of a graph G is the minimum number of colors (labels) needed in any vertex coloring of G. (p. 85)
Critical-path scheduling A heuristic algorithm for solving scheduling problems where the list-processing algorithm is applied to the priority list obtained by listing next in the priority list a task that heads a longest path in the order-requirement digraph. This task is then deleted from the order-requirement digraph, and the next task placed in the priority list is obtained by repeating the process. (p. 75)
Decreasing-time-list algorithm The heuristic algorithm that applies the list-processing algorithm to the priority list obtained by listing the tasks in decreasing order of their time length. (p. 79)
First-fit (FF) A heuristic algorithm for bin packing in which the next weight to be packed is placed in the lowest-numbered bin already opened into which it will fit. If it fits in no open bin, a new bin is opened. (p. 81)
First-fit decreasing (FFD) A heuristic algorithm for bin packing where the first-fit algorithm is applied to the list of weights sorted so that they appear in decreasing order. (p. 82)
Heuristic algorithm An algorithm that is fast to carry out but that doesn't necessarily give an optimal solution to an optimization problem. (p. 81)
Independent tasks Tasks are independent when there are no edges in the order-requirement digraph. These are tasks that can be performed in any order. (p. 78)
List-processing algorithm A heuristic algorithm for assigning tasks to processors: Assign the first ready task on the priority list that has not already been assigned to the lowest-numbered processor that is not working on a task. (p. 69)
Machine scheduling problem The problem of assigning tasks to processors so as to complete the tasks by the earliest time possible. (p. 67)
Next-fit (NF) A heuristic algorithm for bin packing in which a new bin is opened if the weight to be packed next will not fit in the bin that is currently being filled; the current bin is then closed. (p. 81)
Next-fit decreasing (NFD) A heuristic algorithm for bin packing where the next-fit algorithm is applied to the list of weights sorted so that they appear in decreasing order. (p. 82)
Priority list An ordering of the collection of tasks to be scheduled for the purpose of attaining a particular scheduling goal. One such goal is minimizing completion time when the list-processing algorithm is applied. (p. 68)
Processor A person, machine, robot, operating room, or runway with time that must be scheduled. (p. 67)
Ready task A task is called ready at a particular time if its predecessors, as given by the order-requirement digraph, have been completed by that time. (p. 69)
Vertex coloring A vertex coloring of a graph G is an assignment of labels, which can be thought of as "colors," to the vertices of G so that vertices joined by an edge get different labels (colors). (p. 85)
Worst-case analysis The study of the list-processing algorithm (more generally, any algorithm) from the point of view of how well it performs on the hardest problems it may be used on. *See also* average-case analysis. (p. 78)
Worst-fit (WF) A heuristic algorithm for bin packing in which the next weight to be packed is placed into the open bin with the largest amount of room remaining. If the weight fits in no open bin, a new bin is opened. (p. 81)
Worst-fit decreasing (WFD) A heuristic algorithm for bin packing where the worst-fit algorithm is applied to the list of weights sorted so that they appear in decreasing order. (p. 82)

✓ SKILLS CHECK

1. What is the minimum time required to complete 8 independent tasks with a total task time of 64 minutes on 4 machines?

(a) Less than 8 minutes
(b) Between 8 and 10 minutes
(c) More than 12 minutes

2. Given the order-requirement digraph below (time in minutes) and the priority list $T_1, T_2, T_3, T_4, T_5, T_6$, apply the list-processing algorithm to construct a schedule using two processors. **The completion time of the resulting schedule is ____________________.**

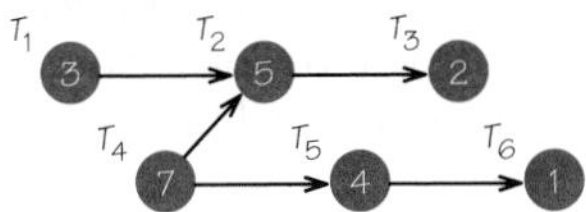

3. The following digraph cannot be an order-requirement digraph because

(a) no vertex has three edges that enter that particular vertex.
(b) it has a directed circuit.
(c) all the tasks require the same time to complete.

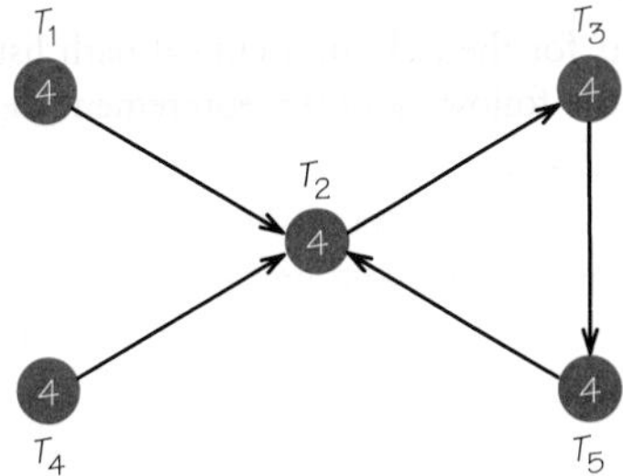

(d) Is there any list that produces a schedule where the second processor has no idle time?

22. (a) In Exercise 21, how many different lists are there that do not list T_1 first?
(b) Would it make any sense not to list T_1 first in a list?
(c) Construct a list and schedule the tasks on two processors.
(d) Can you find another list that leads to a different completion time than the schedule you found for part (c)?
(e) Find a list that leads to an optimal schedule.

23. Can you find an order-requirement digraph with five tasks for which every possible list yields exactly the same schedule?

24. Can you find an order-requirement digraph involving three tasks such that the schedule corresponding to every list is different?

25. At a large toy store, scooters arrive unassembled in boxes. To assemble a scooter, the following tasks must be performed:

Task 1. Remove parts from the box.
Task 2. Attach wheels to the footboard.
Task 3. Attach vertical housing.
Task 4. Attach handlebars to vertical housing.
Task 5. Put on reflector tape.
Task 6. Attach bell to handlebars.
Task 7. Attach decals.
Task 8. Attach kickstand.
Task 9. Attach safety instructions to handlebars.

(a) Give reasonable time estimates for these tasks and construct a reasonable order-requirement digraph. What is the earliest time by which these tasks can be completed?
(b) Schedule this job on two processors (humans) using the decreasing-time-list algorithm.

26. If two schedules for the same number of processors have the same completion time, can one schedule have more idle time than the other?

27. Could the schedule below be obtained by applying the list-scheduling algorithm to some order-requirement digraph?

	0	5	11	16	20
Machine 1	T_1	T_3		T_5	
Machine 2	T_2		T_4		

3.3 Independent Tasks

28. Could the following schedule be obtained by applying the list-scheduling algorithm to some order-requirement digraph?

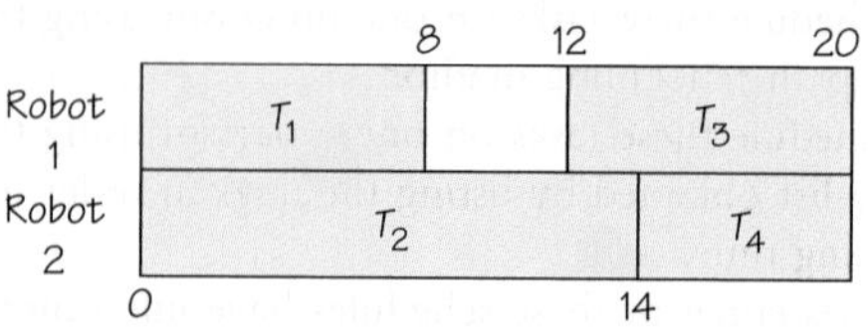

29. For the following schedules, can you produce a list so that the list-processing algorithm produces the schedule shown when the tasks are independent? What are the task times for each task?

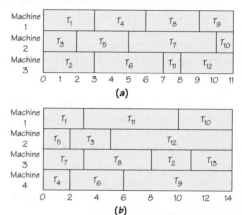

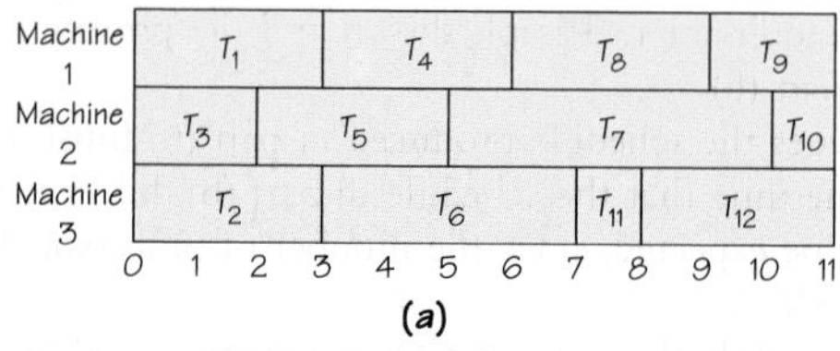

(a)

Machine 1: T_1, T_{11}, T_{10}
Machine 2: T_5, T_3, T_{12}
Machine 3: T_7, T_8, T_2, T_{13}
Machine 4: T_4, T_6, T_9
0 2 4 6 8 10 12 14

(b)

◆ **30.** Once an optimal schedule has been found for independent tasks (see diagrams in Exercise 29), usually the scheduling of the tasks can be rearranged and the same optimal time achieved.
One can, among other things, reorder the tasks done by a particular processor. Discuss criteria that might be used to implement the rearrangement process.

31. The task times of eight independent tasks T_1 to T_8 are 1, 2, 3, 4, 5, 6, 7, 8.

(a) Schedule the tasks on two processors using the lists (i) $T_1, T_2, \ldots, T_8$ and (ii) $T_8, T_7, \ldots, T_1$.
(b) Is either of the schedules you get in part (a) optimal? If not, find a list that gives an optimal schedule.

32. Repeat Exercise 31, but schedule the tasks (with the same lists) on three processors. If the schedules you get are not optimal, find a list that gives an optimal schedule.

◆ **33.** Discuss different criteria that might be used to construct a priority list for a scheduling problem.

◆ **34.** Some scheduling projects have due dates for tasks (times by which a given task should be completed) and release dates (times before which a task cannot have work begun on it). Give examples of circumstances where these situations might arise.

35. Using the lists you found in Exercise 29 and the task times you computed for those independent tasks, schedule the tasks for (a) on four processors and the tasks for (b) on five processors. Can you see why for any schedule you may produce for (a) on four processors and (b) on five processors there must be some idle time for one or more processors?

36. Given the following order-requirement digraph:

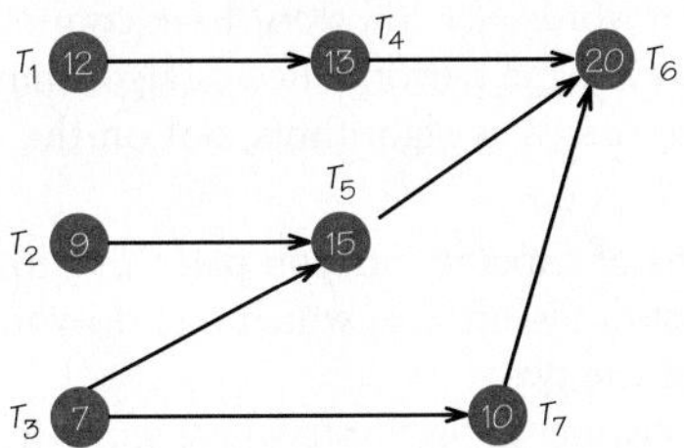

(a) Use the list-processing algorithm to schedule these seven tasks on two processors using these lists:

(i) T_1, T_3, T_7, T_2, T_4, T_5, T_6

(ii) T_1, T_3, T_2, T_4, T_5, T_6, T_7

(iii) The list obtained by listing the tasks in order of decreasing time

(b) Try to determine if any of the resulting schedules are optimal.

(c) Schedule the tasks using the critical-path scheduling method. Try to determine if this schedule is optimal.

37. Repeat the questions in Exercise 36 using the order-requirement digraph obtained by erasing all the (directed) edges shown there. How do the schedules you get compare with the ones you originally got?

38. **(a)** Find the completion time for independent tasks of length 8, 11, 17, 14, 16, 9, 2, 1, 18, 5, 3, 7, 6, 2, 1 on two processors, using the list-processing algorithm.

(b) Find the completion time for the tasks in part (a) on two processors, using the decreasing-time-list algorithm.

(c) Does either algorithm give rise to an optimal schedule?

(d) Repeat for tasks of lengths 19, 19, 20, 20, 1, 1, 2, 2, 3, 3, 5, 5, 11, 11, 17, 18, 18, 17, 2, 16, 16, 2.

39. Repeat parts (a)–(c) of Exercise 38 for independent tasks of lengths 19, 19, 20, 20, 1, 1, 2, 2, 3, 3, 5, 5, 11, 11, 17, 17, 18, 18, 17, 2, 16, 16, 2.

40. Suppose that independent tasks require a total of 36 minutes, while only one of the tasks takes as long as 12 minutes. If these tasks are scheduled on two machines, show by an example that the earliest completion time may be as long as 22 minutes.

41. A photocopy shop must schedule independent batches of documents to be copied. The times for the different sets of documents are (in minutes): 12, 23, 32, 13, 24, 45, 23, 23, 14, 21, 34, 53, 18, 63, 47, 25, 74, 23, 43, 43, 16, 16, 76.

(a) Construct a schedule using the list-processing algorithm on three machines.

(b) Construct a schedule using the list-processing algorithm on four machines.

(c) Repeat parts (a) and (b), but use the decreasing-time-list algorithm.

(d) Suppose union regulations require that an 8-minute rest period be allowed for any photocopy task over 45 minutes. Use the decreasing-time-list algorithm, with the preceding times modified to take into account the union requirement, to schedule the tasks on three human-operated machines.

42. Find a list that produces the following optimal schedule when the list-processing algorithm is applied to this list. (Assume the tasks are independent.)

	0–8	8–12	12–14	14–30
Machine 1	T_2	T_1	T_1	T_5
Machine 2	T_3	T_3	T_4	T_4

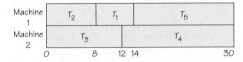

0 8 12 14 30

What completion time and schedule are obtained when the decreasing-time-list algorithm is applied to this list?

43. Can you think of situations other than those mentioned in the text where scheduling independent tasks on processors occurs?

44. Can you think of real-world scheduling situations in which all the tasks have the same time and are independent? Find an algorithm for solving this problem optimally. (If there are n independent tasks of time length k, when will all the tasks be finished?)

45. Show that when tasks to be scheduled are independent, the critical-path method and the decreasing-time-list method are identical.

3.4 Bin Packing

46. Two wooden wall systems are to be made of pieces of wood with lengths shown in the accompanying diagram. If wood is sold in 10-foot planks and can be cut with no waste, what number of boards would be purchased if one uses the first-fit-decreasing, next-fit-decreasing, and worst-fit-decreasing heuristics, respectively?

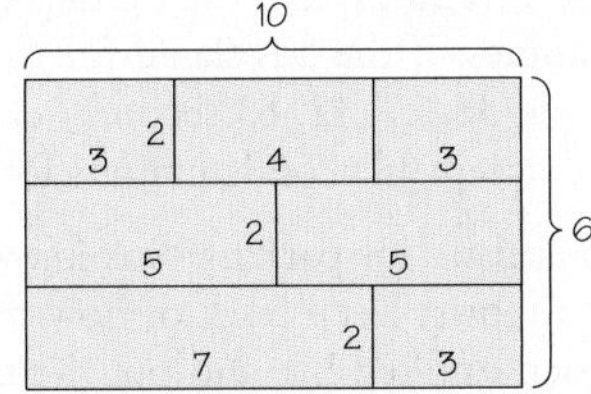

In solving this problem, does it make a difference if the 10-foot horizontal shelves and 6-foot vertical boards employ single-length pieces as compared with using pieces of boards that add up to 10- and 6-foot lengths?

47. It takes 4 seconds to photocopy one page. Manuscripts of 10, 8, 15, 24, 22, 24, 20, 14, 19, 12, 16, 30, 15, and 16 pages are to be photocopied. How many photocopy machines would be required, using the first-fit-decreasing algorithm, to guarantee that all manuscripts are photocopied in 2 minutes or less? Would the solution differ if worst-fit decreasing were used?

48. A radio station's policy allows advertising breaks of no longer than 2 minutes, 15 seconds. Using first-fit and first-fit-decreasing algorithms, determine the minimum number of breaks into which the following ads will fit (lengths given in seconds): 80, 90, 130, 50, 60, 20, 90, 30, 30, 40. Can you find the optimal solution? Do the same for these ad lengths: 60, 50, 40, 40, 60, 90, 90, 50, 20, 30, 30, 50.

49. Fiberglass insulation comes in 36-inch precut sections. A plumber must install insulation in a basement on piping that is interrupted often by joints. The distances between the joints on the stretches of pipe that must be insulated are 12, 15, 16, 12, 9, 11, 15, 17, 12, 14, 17, 18, 19, 21, 31, 7, 21, 9, 23, 24, 15, 16, 12, 9, 8, 27, 22, 18 inches. How many precut sections would he have to use to provide the insulation if he bases his decision on

(a) next-fit?
(b) next-fit decreasing?
(c) worst-fit?
(d) worst-fit decreasing?

50. The files that a company has for its employees dealing with utilities occupy 100, 120, 60, 90, 110, 45, 30, 70, 60, 50, 40, 25, 65, 25, 55, 35, 45, 60, 75, 30, 120, 100, 60, 90, 85 sectors. If, after operating systems are installed, a disk can store up to 480 sectors, determine the number of disks needed to store the utilities if each of these heuristics is used to pack the disk with files:

(a) next-fit
(b) next-fit decreasing
(c) first-fit
(d) first-fit decreasing

51. Advertisements for the TV show *Q* are permitted to last up to a total of 8 minutes, and each group of ads can last up to 2 minutes. If the ads slated for *Q* last 63, 32, 11, 19, 24, 87, 64, 36, 27, 42, 63 seconds, determine if FF and FFD yield acceptable configurations for the ads.

52. Consider the heuristic for packing bins known as *best-fit* described as follows: Keep track of how much room remains in each unfilled bin and put the next item to be packed into that bin that would leave the least room left over after the item is put into the bin. (For example, suppose that bin 4 had 6 units left, bin 7 had 5 units left, and bin 9 had 8 units left. If the next item in the list had size 5, then first-fit would place this item in bin 4, worst-fit would place the item in bin 9, while best-fit would place the item in bin 7.) If there is a tie, place the item into the bin with the lowest number. Apply this heuristic to the list 8, 7, 1, 9, 2, 5, 7, 3, 6, 4, where the bins have capacity 10.

◆ **53.** We have described two algorithms for bin packing called worst-fit and best-fit (see page 81 and Exercise 52). The words *best* and *worst* have connotations in English. However, the performance of algorithms depends on their merits as algorithms, not on the names we give them.

(a) On the basis of experiments you perform with the best-fit and worst-fit algorithms, which one do you think is the "better" of the two?
(b) Can you construct an example where worst-fit uses fewer bins than best-fit?

54. The best-fit heuristic (see Exercise 52) also has a "decreasing" version, where the list is first sorted in decreasing order. Using bins of capacity 10, apply the best-fit heuristic and its decreasing version to the following list: 6, 9, 5, 8, 3, 2, 1, 9, 2, 7, 2, 5, 4, 3, 7, 6, 2, 8, 3, 7, 1, 6, 4, 2, 5, 3, 7, 2, 5, 2, 3, 6, 2, 7, 1, 3, 5, 4, 2, 6.

■ **55.** One pianist's recording of the complete Mozart piano sonatas takes the following times (given in minutes and seconds): 13:46, 6:15, 3:29, 5:37, 7:52, 2:55, 5:00, 4:28, 4:21, 7:39, 7:55, 6:42, 4:23, 3:52, 4:21, 4:20, 5:46, 6:29, 5:34, 6:23, 6:39, 7:19, 5:54, 6:54, 2:58, 5:22, 1:42, 5:00, 1:29, 5:47, 7:30, 8:19, 4:44, 4:57, 4:09, 14:31, 3:55, 4:04, 4:01, 6:06, 6:50, 5:27, 4:28, 5:40, 2:52, 5:16, 5:34, 3:10, 7:22, 4:40, 3:08, 6:32, 4:47, 6:59, 5:38, 7:57, 3:38. If the maximum time that can be recorded on a compact disc is 70:30, can all the music be performed on four compact discs? Can all the music be performed on five compact discs?

■ **56.** In the wall-system example in the text, first-fit and worst-fit required equal numbers of bins (see Figure 3.20). Can you find an example where first-fit and worst-fit yield different numbers of bins? Can you find an example where first-fit, worst-fit, and next-fit yield answers with different numbers of bins?

◆ **57.** A common suggestion for heuristics for the bin-packing problem with bins of capacity W involves finding weights that sum to exactly W. Discuss the pros and cons of a heuristic of this type.

■ **58.** A recording company wishes to record all the Beethoven string quartets (16 quartets, each consisting of several consecutive parts called movements) on LPs. It wishes to complete the project on as few records as possible. Recording can be done on two sides as long as the movements are consecutive. Is this an example of a bin-packing problem? (Defend your answer.) If the project were to record the quartets on (standard) tape cassettes or compact discs, would your answer be different?

59. Give examples where it would be realistic to keep bins open as more items "arrive" to be packed, rather than to close a bin permanently based on some criterion.

60. Give examples where it would be unrealistic to keep bins open as more items "arrive" to be packed, rather than to close a bin permanently based on some criterion.

61. A data entry group must handle 30 (independent) tasks that will take the following amounts of time (in minutes) to type: 25, 18, 13, 19, 30, 32, 12, 36, 25, 17, 18, 26, 12, 15, 31, 18, 15, 18, 16, 19, 30, 12, 16, 15, 24, 16, 27, 18, 9, 14. Using these times as a priority list:

(a) Use the list-processing algorithm to find the completion time for scheduling tasks with four secretaries. Also, solve with five secretaries.
(b) Repeat the scheduling using the decreasing-time-list algorithm.
(c) Can you show that any of the schedules that you get are optimal?

If one needs to finish the typing in one hour:
(d) Use the FFD heuristic to find how many typists would be needed.
(e) Repeat for the NFD and WFD heuristics.
(f) Can you show that any of the solutions you get are optimal?

62. Find the minimum number of bins necessary to pack items of size 8, 5, 3, 4, 3, 7, 8, 8, 6, 5, 3, 2, 1, 2, 1, 2, 1, 3, 5, 2, 4, 2, 6, 5, 3, 4, 2, 6, 7, 7, 8, 6, 5, 4, 6, 1, 4, 7, 5, 1, 2, 4 in bins of capacity (a) through (d) using the first-fit and first-fit-decreasing algorithms. Can you determine if any of the packings you get are optimal?

(a) 9
(b) 10
(c) 11
(d) 12

■ 63. Two-dimensional bin packing refers to the problem of packing rectangles of various sizes into a minimum number of $m \times n$ rectangles, with the sides of the packed rectangles parallel to those of the containing rectangle.

(a) Suggest some possible real-world applications of this problem.
(b) Devise a heuristic algorithm for this problem.
(c) Give an argument to show that the problem is at least as hard to solve as the usual bin-packing problem.
(d) If you have $1 \times m$ rectangles with total area W to be packed into a single rectangle of area $p \times q = W$, can the packing always be accomplished?

◆ 64. In what situations would packing bins of different capacities be the appropriate model for real-world situations? Suggest some possible algorithms for this type of problem.

■ 65. Find an example of weights that, when packed into bins using first-fit, use fewer bins than the number of bins used when the first-fit algorithm is applied with the first weight on the list removed.

◆ 66. Formulate "paradoxical" situations for bin packing that are analogous to those we found for scheduling processors.

3.5 Resolving Conflict via Coloring

67. For each of the graphs below:

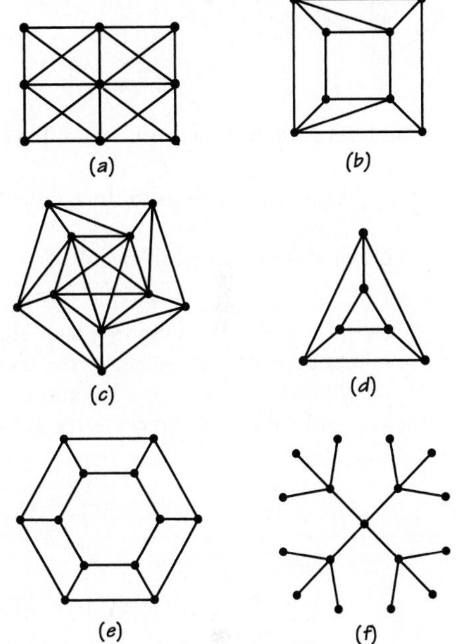

(a) Color the vertices (if possible) with three different colors.
(b) Color the vertices (if possible) with four different colors.
(c) Find the chromatic number of the graph.

68. For each of the following graphs:

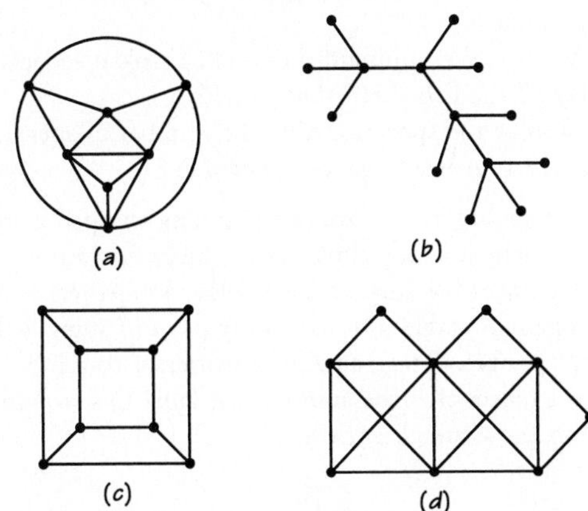

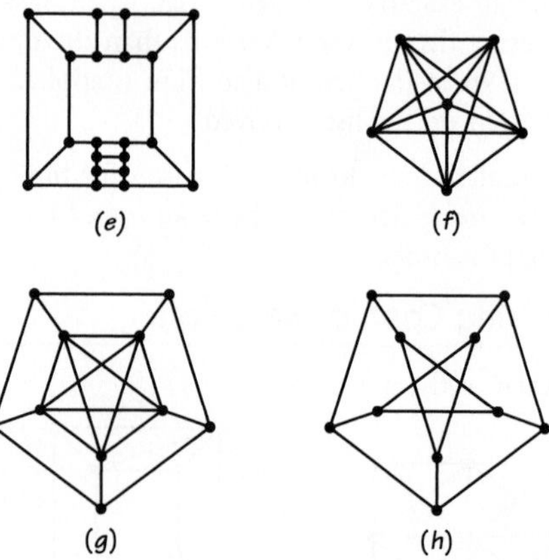

(a) Color the vertices (if possible) with two different colors.
(b) Color the vertices (if possible) with three different colors.
(c) Find the chromatic number of the graph.

69. The owner of a new pet store wishes to display tropical fish in display tanks. The following table shows the incompatibilities between the species, in the sense that an X indicates that it is unwise to allow those species in the row and column that meet at the X to be in the same tank.

	A	B	C	D	E	F	G	H	I
A						X	X		X
B			X					X	
C		X			X			X	
D					X	X		X	
E			X	X			X		
F	X			X			X		X
G	X				X	X		X	X
H		X	X	X			X		
I	X					X	X		

(a) Draw an appropriate graph to represent the information in the table.
(b) What is the minimum number of tanks needed to display all the fish she wishes to sell?
(c) Display the species so that the number of species in each tank is as nearly equal as possible.

70. The managers of a zoo are planning to open a small satellite branch. The animals are to be in enclosures in which compatible animals are displayed together. The accompanying table indicates those pairs of animals that are compatible. (Thus, an X in a particular row and column means that the animals that label this row and column *can* share an enclosure.)

	A	B	C	D	E	F	G	H	I	J
A	X	X		X	X	X	X			
B	X	X			X	X	X		X	X
C			X		X	X	X			
D	X			X	X	X	X		X	X
E	X	X	X	X	X			X	X	
F	X	X	X	X		X	X	X	X	
G	X	X	X	X		X	X	X		
H					X	X	X	X		
I		X		X	X	X			X	
J		X		X						X

(a) Draw an appropriate graph to represent the information in the table.
(b) What is the minimum number of enclosures needed to avoid housing incompatible animals in the same enclosure?
(c) Is it possible to enclose the animals in such a way that each enclosure contains the same number of animals?
(d) Why might that be desirable? Why might this approach to grouping the animals not be ideal?

71. The nine standing committees of a state legislature are designing a schedule for when the committees can meet. The matrix shown in the following table has an X in a position where the committees corresponding to the row and column have a common member and, hence, should not be scheduled to meet at the same hour. The committees involved are Agriculture (A), Commerce (C), Consumer Affairs (CA), Education (E), Forests (F), Health (H), Justice (J), Labor (L), and Rules (R).

	A	C	CA	E	F	H	J	L	R
A		X	X			X			
C	X		X	X	X				
CA	X	X					X		X
E		X			X	X			
F		X		X		X	X		
H	X			X	X			X	
J			X		X			X	X
L						X	X		X
R			X				X	X	

(a) Draw a graph that will be of value in determining the minimum number of time slots the committees can meet in without any legislator having to be in two places at one time.
(b) What is the minimum number of time slots in which the committees can be scheduled without a conflict?
(c) How many different rooms are needed at any time that a committee is scheduled to meet? (Why might this issue matter?)

72. Determine the minimum number of colors, and how often each color is used, in a vertex coloring of the graphs below.

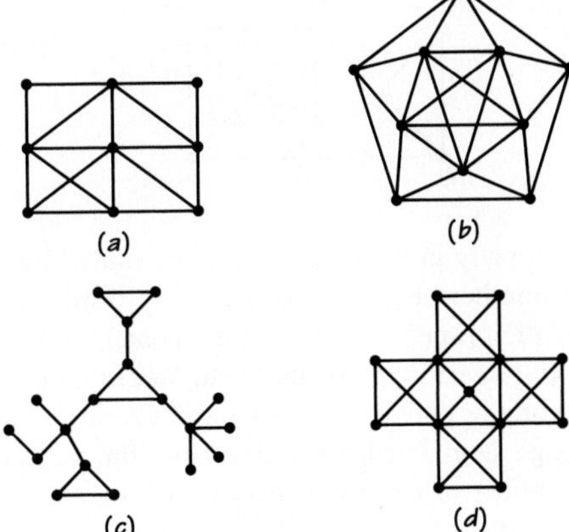

73. The faculty–student governing council at All State College has nine standing committees (such as Curriculum, Academic Standards, Campus Life) that are designed A, B, C, D, . . . , I for convenience. The following table shows which committees have no member in common.

	A	B	C	D	E	F	G	H	I
A		X		X		X	X		X
B	X				X	X		X	X
C				X		X	X	X	X
D	X		X			X		X	
E		X					X	X	X
F	X	X	X	X					
G	X		X		X			X	
H		X	X	X	X		X		X
I	X	X	X		X			X	

(a) Draw an appropriate graph to represent the information in the table.
(b) What is the minimum number of time slots in which all the committee meetings can be scheduled?
(c) How many rooms are needed during each time slot to accommodate the committees that are scheduled to meet in that time slot?

74. When two towns are within 145 miles of each other, the frequency used by a certain type of emergency response system for the towns requires that they be on different frequencies to avoid possible interference with each other. The following table shows the mileage distances between six towns.

	E	F	G	I	S	T
Evansville (E)		290	277	168	303	133
Ft. Wayne (F)	290		132	83	79	201
Gary (G)	277	132		153	58	164
Indianapolis (I)	168	83	153		140	71
South Bend (S)	303	79	50	140		196
Terre Haute (T)	113	201	164	71	196	

(a) What would be the minimum number of frequencies that are needed for each town to have its emergency broadcasts not conflict with those of any other town using this system?
(b) How many different towns would be assigned to each frequency used?

75. Show that the vertices of any tree can be colored with two colors.

76. Can you find a family of graphs H_n $(n \geq 1)$ that require n colors to color their vertices?

77. The edge-coloring number of a graph G is the minimum number of colors needed to color the edges of G so that edges that share a common vertex get different colors. Determine the edge-coloring number for each of the graphs in Exercise 67. Can you make a conjecture about the value of the minimum number of colors needed to color the edges of any graph?

78. Can you think of any applications that require determining the minimum number of colors needed to color the edges of a graph?

79. When a graph has been drawn on a piece of paper so that edges meet only at vertices, the graph divides the paper up into regions called *faces.* The faces include one called the "infinite" face, which surrounds the whole graph. The face-coloring number of a graph G (which can be drawn in this special way) is the minimum number of colors needed to color the faces of G so that two faces that share an edge receive different colors. (Note that if two faces meet only at a vertex, they can be colored the same color.)

(a) Determine the minimum number of colors needed to color the faces of the following graphs. In each case, remember to color the infinite face, which is labeled I (for "infinite").

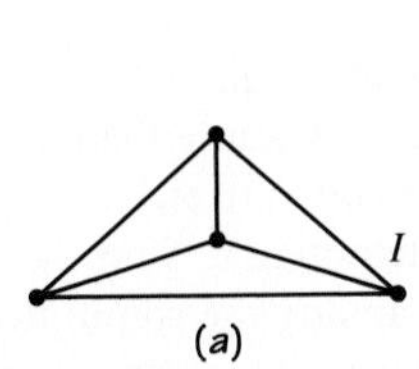

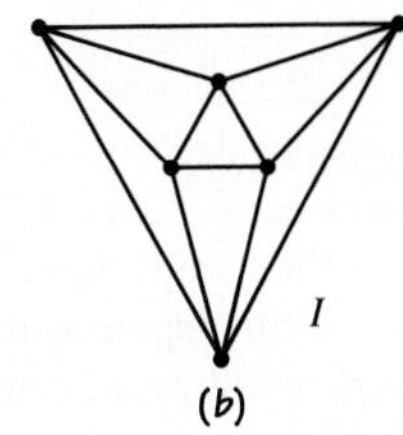

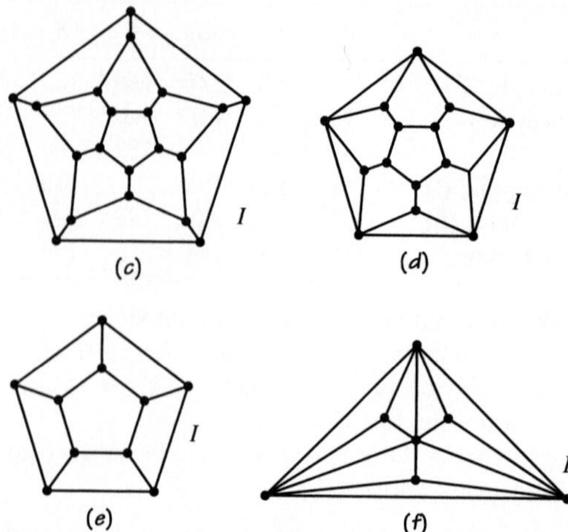

(b) Can you think of an application of the problem of coloring the faces of a graph with a minimum number of colors?

80. For each of the graphs in Exercise 68 where the graph shown has edges that meet only at vertices, verify that the Four Color Theorem holds by showing that the regions (faces) of the graph can be colored with four or fewer colors so that regions that share an edge get different colors. (Remember to assign a color to the unbounded, so-called infinite region.)

81. A company sells herbs, each of which requires a certain level of proper watering. The following graph is constructed by having one vertex for each type of herb. The vertices representing two herbs are joined by an edge if they must have different levels of watering. What is the minimum number of terrariums that the herbs can be displayed in so that herbs in the same terrarium can be watered at the same level?

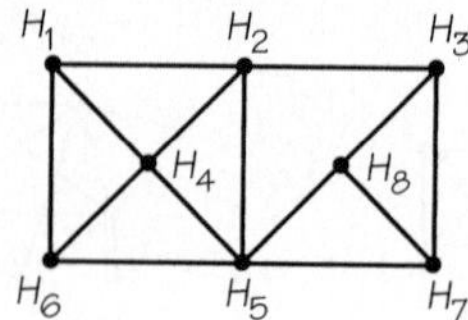

82. The company in Exercise 81 is disappointed by the minimum number of terrariums needed to display the herbs with the proper watering requirements. One company employee suggests that if the information about watering requirements is altered for a single pair of herbs (e.g., a single edge is erased from the diagram), then the number of terrariums needed will be reduced by 1. Is this true?

83. Each vertex in the graph below represents a child who attends a day care center. An edge between two children indicates these children tend to cause problems when they are in the same play group. What is the minimum number of play groups that will ensure that no conflicts arise? Can conflict-free play groups with the same number of children in each group be formed?

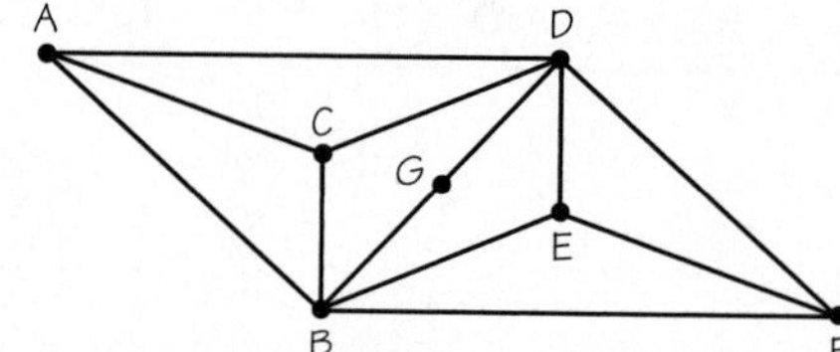

APPLET EXERCISES

To do these exercises, go to www.whfreeman.com/fapp8e.

Graph Coloring

Solving a scheduling problem such as the one below can be accomplished by constructing a related graph and then coloring it in a way that adjacent vertices have different colors. Explore the problem of graph coloring in the *Graph Coloring* applet.

Scheduling

A mathematics department has seven faculty committees–A, B, C, D, E, F and G. Because there is overlap in the composition of the committees, the chairman of the department is attempting to work out a schedule that will avoid conflicts among the committees. The following chart indicates the overlapping committee structure:

	A	B	C	D	E	F	G
A		X		X		X	
B	X		X			X	
C		X			X		X
D	X						X
E			X			X	X
F	X	X			X		
G			X	X	X		

Help the chairman arrange a schedule without conflicts in the *Scheduling* applet.

WRITING PROJECTS

1. Scheduling is important for hospitals, schools, transportation systems, police services, and fire services. Pick one of these areas and write about the different scheduling situations that come up, types of processors, and extent to which the assumptions of the list-processing model hold for the area you pick.

2. Compare and contrast the basic scheduling problem we investigated with the scheduling version of the bin-packing problem.

3. One of the oversimplifications made in our discussion of scheduling was that there were no "due dates" involved for the tasks making up a job. Develop an algorithm for solving a scheduling problem under the assumption that each task has a due date as well as a time length. You will probably want to decide on a penalty amount that will occur when a due date is exceeded.

4. Consider the problem of scheduling tasks on a single machine. Design different algorithms for achieving different goals. You will probably wish to assume that each task has a due date such that if the task is not finished by this date, some penalty payment must be made.

5. Discuss the role of graph colorings for scheduling committee meetings so as to avoid conflicts. Research whether or not these ideas are used in the legislature of your home state.

6. In choosing a location (vertex) for trains to turn around in the graph shown in Spotlight 3.1, explain why it seems to be a much better choice to use V as a place to allow the turn arounds, rather than at M or at R.

SUGGESTED READINGS

BRUCKER, P. *Scheduling Algorithms,* 4th ed., Springer-Verlag, Heidelberg, Germany, 2004. A detailed mathematical look at scheduling.

GRAHAM, RONALD. Combinatorial scheduling theory, in Lynn Steen (ed.), *Mathematics Today,* Springer-Verlag, New York, 1978, pp. 183–211. This essay on scheduling is one of many excellent accounts of recent developments in mathematics in this book.

GRAHAM, RONALD. The combinatorial mathematics of scheduling. *Scientific American,* March 1978, pp. 124–132. A very readable introduction to scheduling and bin packing.

JENSEN, T. R., and BJARNE TOFT. *Graph Coloring Problems,* Wiley, New York, 1995. A detailed summary of what is known about coloring problems and many questions that await answering.

LAWLER, E., et al. Sequencing and scheduling algorithms and complexity, in S. C. Graves et al. (eds.), *Handbooks in OR and MS,* vol. 4, Elsevier, New York, 1993, pp. 445–522. A recent survey of results about scheduling.

LEUNG, JOSEPH Y-T., *Handbook of Scheduling,* Chapman & Hall/CRC, Boca Raton, Florida, 2004. This book has an encyclopedic treatment of scheduling algorithms and the great variety of situations where mathematical analysis has assisted schedulers, ranging from sports to hospitals.

PARKER, R. GARY, *Deterministic Scheduling Theory,* Chapman & Hall, London, 1995. A wide-ranging look at scheduling methods and their applications.

SUGGESTED WEB SITES

www.ctl.ua.edu/math103/scheduling/schedmnu.htm This site provides an overview of scheduling as discussed in this chapter.

www.ams.org/featurecolumn/archive/machines1.html, www.ams.org/featurecolumn/archive/packings1.html, www.ams.org/featurecolumn/archive/bins1.html These Web pages describe mathematical aspects of machine scheduling and bin packing and give a discussion of the relationship between these two mathematical problems.

www.ie.bilkent.edu.tr/~ie672/docs/resources.html This Web page contains links to many aspects of scheduling theory, including research on the frontier.

CHAPTER 4

Linear Programming

A manager's job often calls for making very complicated decisions. One set of decisions involves planning what products the business is to make and determining what resources are needed. In the modern business world, diversification of products provides a company with stability in a climate of changing tastes and needs. So it is not surprising that companies would produce many products, some of which share resource needs. For example, any bakery uses many resources–like butter, sugar, eggs, and flour–to make its products such as cookies, cakes, pies, and breads. Similarly, car manufacturers use many kinds of metals in the different models of cars they make, and manufacturers of gasoline use different kinds of crude oils to make their product.

Resources can include more than just raw materials. A labor force with appropriate skills, farmland, time, and machinery are also resources. Typically, resources are limited: A farmer owns only so much land; there are only so many hours in a day; in a year of drought the wheat crop is very small. Resource availability is also limited by location and competition.

Because resources are limited, management faces important questions: How should the available resources be shared among the possible products? One goal of management is to maximize profit. How can that determine how much of each product should be produced? There are usually so many alternative product mixes that it is impossible to evaluate them all individually. Despite this complexity, millions of dollars may ride on management's decision.

Many business and government agencies must deal with supply-and-demand problems. The general idea is that goods or services can be provided by different providers to individuals or businesses who need these goods or services. There are varying costs to the suppliers to provide different recipients with these goods or services. The goal is to find how to meet the demands for the supplies as cheaply as possible. For example, what is the cheapest way for a company with several oil refineries to provide oil distributors, in many different geographical locations, with the oil they need?

2. It is not possible to visualize the feasible region as a part of two-dimensional space when there are more than two products. Each product is represented by an unknown, and each unknown is represented by a dimension of space. If we have 50 products, we would need 50 dimensions and couldn't visualize the feasible region.

Another type of complication can occur even in simple two-dimensional regions: Corner points can have fractional coordinates, not the integer ones we see in the specially constructed problems in this text. Making 3.75 skateboards and 5.45 dolls is not possible. Integer programming, a special type of linear programming, is used when it is not possible to use fractional answers.

The Simplex Method

Several methods are used for the typically large linear-programming problems solved in practice. The oldest method is the **simplex method**, which is still the most commonly used. Devised by the American mathematician George Dantzig (see Spotlight 4.2), this ingenious mathematical invention makes it possible to find the best corner point by evaluating only a tiny fraction of all the corners. With the use of the simplex method, a problem that might be impossible to solve if each corner point had to be checked can be solved in a few minutes or even a few seconds on a typical business computer.

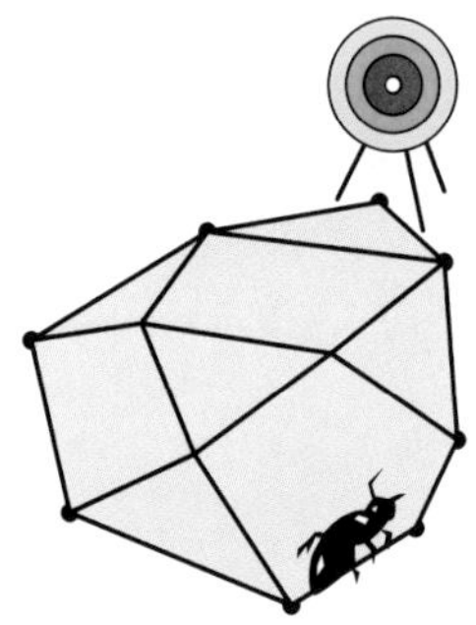

FIGURE 4.15 The simplex method can be compared to an ant crawling along the edges of a polyhedron, looking for the "target"–the optimal corner point.

The operation of the simplex method may be likened to the behavior of an ant crawling on the edges of a polyhedron (a solid with flat sides) looking for an optimal corner point–one that gives the highest profit (Figure 4.15). The ant cannot see where the optimal corner is. As a result, if it were to wander along the edges randomly, it might take a long time to reach that corner. The ant will do much better if it has a temperature clue to let it know it is getting warmer (closer to the optimal corner) or colder (farther from the optimal corner).

Think of the simplex method as a way of calculating these temperature hints. We begin at any corner. All neighboring corners are evaluated to see which ones are warmer and which are colder. A new corner is chosen from among the warmer ones, and the evaluation of neighbors is repeated–this time checking neighbors of the new corner. The process ends when we arrive at a corner all of whose neighbors are colder than it is.

Part of what the simplex method has going for it is that it works faster in practice than its worst-case behavior would lead us to believe. Although mathematicians have devised artificial cases for which the simplex method bogs down in unacceptable amounts of arithmetic, the examples arising from real applications are never like that. This may be the world's most impressive counterexample to Murphy's law, which says that if something can go wrong, it will.

Although the simplex method usually avoids visiting every corner, it may require visiting many intermediate ones as it moves from the starting corner to the optimal one. The simplex method has to search along edges on the boundary of the polyhedron. If it happens that there are a great many small edges lying between the starting corner and the optimal one, the simplex method must operate like a slow-moving bus that stops on every block.

Many computer programs are available that will use the simplex method to produce an optimal production policy if we just supply the computer with the constraint inequalities and profit formula. Simplex method programs can be found in a variety of places, among which are spreadsheets, packages of mathematics programs designed for business applications or finite mathematics courses, and large "all-purpose" mathematics packages. A graphical solution is possible only for problems limited to two products; these special exercises involve more than two products.

SPOTLIGHT 4.2 Father of Linear Programming Recalls Its Origins

George Dantzig, who died in 2005, spent most of his career as a professor of operations research and computer science at Stanford University. He is credited with inventing the linear-programming technique called the simplex method. Since its invention in the 1940s, the simplex method has provided solutions to linear-programming problems that have saved both industry and the military time and money. Here Dantzig talks about the background of his famous technique:

> Initially, all the work we did had to do with military planning. During World War II, we were planning on a very extensive scale. The civilian population and the military were all performing scheduling and planning tasks, perhaps on a larger scale than at any time in history. And this was the case up until about 1950. From 1950 on, the whole emphasis shifted from military planning to practical planning for the civilian population, and industry picked it up.
>
> The first areas of industry to use linear programming were the petroleum refineries. They used it for blending gasoline. Nowadays, all of the refineries in the world (except for one) use linear programming methods. They are one of the biggest users of it, and it's been picked up by every other industry you can think of—the forestry industry, the steel industry—you could fill up a book with all the different places it's used.
>
> The question of why linear programming wasn't invented before World War II is an interesting one. In the postwar period, various technologies just evolved that had never been there before. Computers were one example. These technologies were talked about before. You can go back in history and you'll find papers on them, but these were isolated cases that never went anywhere. . . .

George Dantzig

George Dantzig (left), sometimes referred to as the "father" of linear programming, shown with Leonid Khaciyan (right) who developed an important new approach to solving linear programming problems. (*Kees Roos.*)

> The problems we solve nowadays have thousands of equations, sometimes a million variables. One of the things that still amazes me is to see a program run on the computer—and to see the answer come out. If we think of the number of combinations of different solutions that we're trying to choose the best of, it's akin to the stars in the heavens. Yet we solve them in a matter of moments. This, to me, is staggering. Not that we can solve them—but that we can solve them so rapidly and efficiently.
>
> The simplex method has been used now for roughly 70 years. There has been steady work going on trying to use different versions of the simplex method, nonlinear methods, and interior methods. It has been recognized that certain classes of problems can be solved much more rapidly by special algorithms than by using the simplex method. If I were to say what my field of specialty is, it is in looking at these different methods and seeing which are more promising than others. There's a lot of promise in this—there's always something new to be looked at.

An Alternative to the Simplex Method

In 1984, Narendra Karmarkar (see Figure 4.16), a mathematician working at Bell Laboratories, devised an alternative method for linear programming that finds the optimal corner point in fewer steps than the simplex algorithm by making use of search routes through the interior of the feasible region. The applications of Karmarkar's algorithm are important to a lot of industries, including telephone communications and the airlines (see Spotlight 4.3). Routing millions of long-distance calls, for example, means deciding how to use the resources of long-distance landlines, repeater amplifiers, and satellite terminals to best advantage. The problem is similar to the juice company's need to find the best use of its stocks of juice to create the most profitable mix of products.

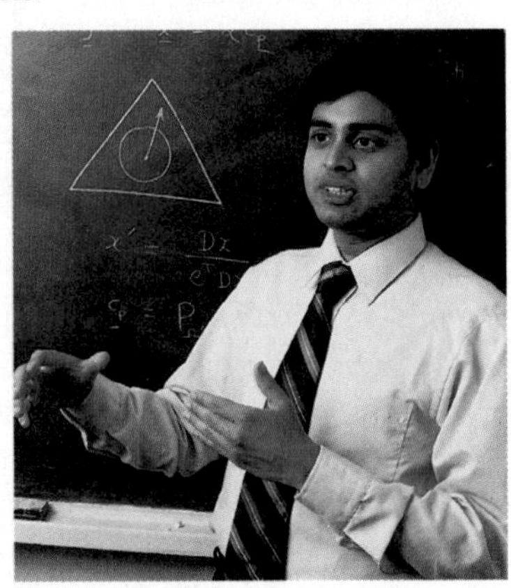

FIGURE 4.16 Narendra Karmarkar, a researcher at AT&T Bell Laboratories, invented a powerful new linear-programming algorithm that solves many complex linear-programming problems faster and more efficiently than any previous method. (*Courtesy of AT&T Labs.*)

Many airlines use software based on Karmarkar's algorithm to reduce fuel costs and deal with delays caused by storms.

In the 1980s, scientists at Bell Labs applied Karmarkar's algorithm to a problem of unprecedented complexity: deciding how to economically build telephone links between cities so that calls can get from any city to any other, possibly being relayed through intermediate cities. Figure 4.17 shows one such linking. The number of possible linkings is unimaginably large, so picking the most economical one is difficult. For any given linking, there is also the problem of deciding how to economically route calls through the network to reach their destinations.

Now, similar approaches are being used to route email packets and phone calls over the Internet.

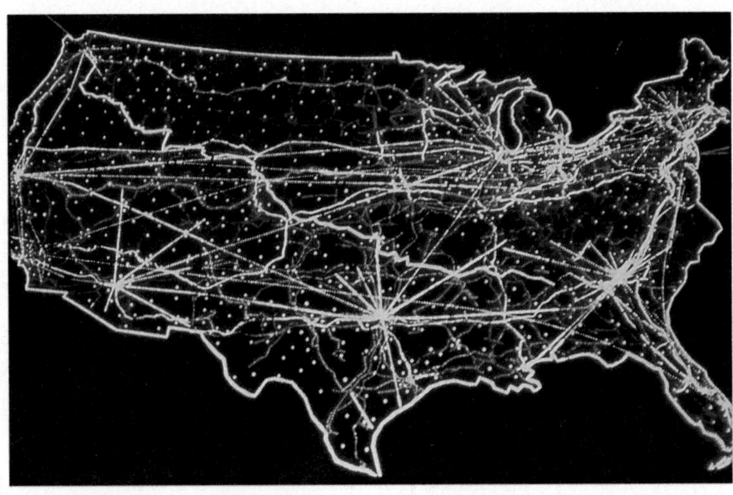

FIGURE 4.17 A map of the United States showing one conceivable network of major communication lines connecting major cities. Routing millions of calls over this immense network requires sophisticated linear-programming techniques and high-speed computers. (*Courtesy of AT&T Labs.*)

SPOTLIGHT 4.3 Finding Fast Algorithms Means Better Airline Service

Linear-programming techniques have a direct impact on the efficiency and profitability of major airlines. Thomas Cook, once director of operations research at American Airlines, made these comments concerning why optimal solutions are essential to the airline business:

> Finding an optimal solution means finding the best solution. Let's say you are trying to minimize a cost function of some kind. For example, we may want to minimize the excess costs related to scheduling crews, hotels, and other costs that are not associated with flight time. So we try to minimize that excess cost, subject to a lot of constraints, such as the amount of time a pilot can fly, how much rest time is needed, and so forth.
>
> An optimal solution, then, is either a minimum-cost solution or a maximizing solution. For example, we might want to maximize the profit associated with assigning aircrafts to the schedule; so we assign large aircraft to high-need segments and small aircraft to low-load segments.
>
> The simplex method, which was developed some 50 years ago by George Dantzig, has been very useful at American Airlines and, indeed, at a lot of large businesses. The difference between his method and Narendra Karmarkar's is speed. Finding fast solutions to linear-programming problems is also essential. With an algorithm like Karmarkar's, which is 50 to 100 times faster than the simplex method, we could do a lot of things that we couldn't do otherwise. For example, some applications could be real-time applications, as opposed to batch applications. So instead of running a job overnight and getting an answer the next morning, we could actually key in the data or access the database, generate the matrix, and come up with a solution that could be implemented a few minutes after keying in the data.
>
> A good example of this kind of application is what we call a major weather disruption. If we get a major weather disruption at one of the hubs, such as Dallas or Chicago, then a lot of flights may get canceled, which means we have a lot of crews and airplanes in the wrong places. What we need is a way to put that whole operation back together again so that the crews and airplanes are in the right places. That way, we minimize the cost of the disruption as well as passenger inconvenience.

4.5 A Transportation Problem: Delivering Perishables

A supermarket chain gets bread deliveries from a bakery chain that does its baking in different places. Each supermarket store needs a certain number of loaves each day, and the supplier bakes in total enough breads to exactly meet the demands. Figure 4.18 shows the cost to ship a loaf from a particular baking location to the store involved. How many breads should be shipped from each locale to each of the stores to stay within the demands and to minimize the cost?

Similarly, after a long holiday weekend, a car rental company will have extra cars in some cities and too few cars in other cities. It is faced with the problem of reshuffling the cars at minimal cost so that each city has the right number of cars. Problems such as these go under the general name of **transportation problems** and they form a special class of linear-programming problems that can be solved by a specialized method.

Transportation Problem DEFINITION

A group of suppliers must meet the needs of users of these supplies. There is a cost for shipping from a particular supplier to a particular user (demander). The **transportation problem** involves minimizing the total shipping cost of meeting the required demands from the supplies available.

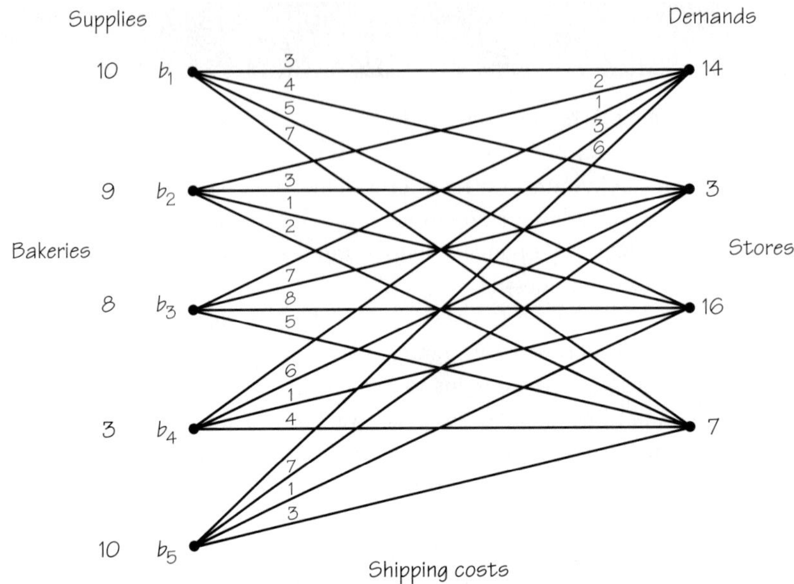

FIGURE 4.18 A graph theory representation of a supply-and-demand transportation problem that involves shipping breads from bakeries to stores.

EXAMPLE 8 ■ Delivering Bread

Imagine we have three bakeries and three stores, though the ideas we develop will also solve problems where the number of stores and bakeries need not be the same. The three stores require 3 dozen, 7 dozen, and 1 dozen loaves of bread, respectively, while the three bakeries can supply 8 dozen, 1 dozen, and 2 dozen loaves, respectively. The information given so far can be displayed in Figure 4.19, where the "suppliers" are represented by the rows of the table (labeled with Roman numerals) and the "demanders" are represented by the columns (labeled with Hindu-Arabic numerals).

The numbers of breads available and the numbers being required are shown on the right side and bottom of the table and will be referred to as **rim conditions**. Each entry of the table shown in Figure 4.19 is known as a cell. It is convenient to have a name for each of these cells. For example, the cell in the third row and second column will be denoted (III, 2). The first number always corresponds to a row, the second to a column. Thus cell (I, 2) refers to bakery I and store 2.

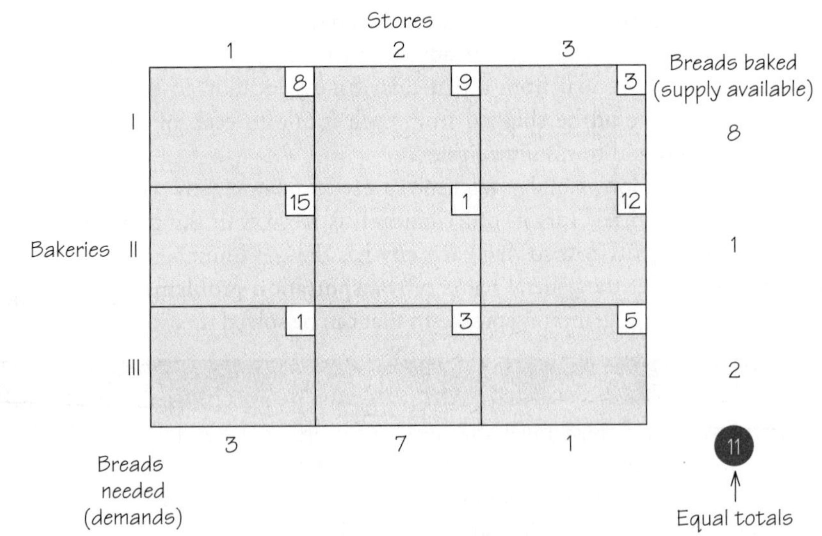

FIGURE 4.19 A representation of a specific problem involving meeting the demands of three stores for breads from the supplies available at three bakeries. Shipping costs between bakeries and stores are also shown.

In deciding which bakeries should ship to which stores, it seems natural to take into account the costs of shipping a dozen breads from a particular bakery to a particular store. If bakery I is farther from store 2 than is bakery II, it seems reasonable that the shipping cost for I will be higher than for II when shipping to that particular store.

However, the costs of shipping may also involve time considerations. (The distance to a store may be shorter, but it may be that this route is a very slow one.) Also, it may take extra time for a truck coming from I to park when making the next delivery.

The numbers we use in our diagrams are "aggregate" costs. The nice thing about what we are doing is that the solution method works independently of the way the costs are computed or arrived at. These costs (see Figure 4.19) are shown in the upper-right-hand corner of a cell. Thus the 9 shown in the cell (I, 2) means that it costs nine units to ship a bread from bakery I to store 2. Our goal will be to supply the stores with the breads they require from the supplies available at the bakeries so that the total cost of providing the breads to the stores is as small as possible (a minimum).

The tools for solving transportation problems like these were developed during World War II in conjunction with getting supplies from different ports in the United States to different ports in Europe (mostly the United Kingdom) in as efficient a manner as possible. (The U.S. ports were like the bakeries, and the British ports were like the stores that needed the breads.)

We can think of finding a solution to a problem like this as a special kind of linear-programming problem, because we can express the objective of minimizing the cost using a linear relationship. The constraints that express that the rim conditions are met can also be expressed using linear equations. However, it turns out there are algorithms that make it possible to solve problems of this kind that are rather larger than general linear-programming problems that can be solved by hand. These algorithms are intuitively appealing.

We can divide the problem of finding a solution to a transportation problem into two phases, as we did for general linear-programming problems. First, find a solution that is feasible (that is, a solution that does not violate any of the constraints of the problem). Second, if the current solution is not optimal, we move to a better one. Thus, we will first find a solution that meets the constraints and then try to find an improved solution. If there is no better solution than the one we have, under suitable circumstances we show that there is never a better one. Thus, the solution that we have found is an optimal solution. We will work our way through a simple example that is typical of what is required in general transportation problems.

Let's turn to the table shown in Figure 4.20, where certain numbers have been inserted with circles around them.

Tableau DEFINITION

A table showing costs and rim conditions for a transportation problem is known as a **tableau**.

When we see a circled number such as the 6 in row I, column 2, this means that we plan to ship six breads to store 2 from bakery I. Similarly, the circled number 1 in row III and column 3 means we plan to ship one bread from bakery III to store 3. The cells that have no circled numbers are thought of as having zero entries; no breads are being shipped between these stores and these bakeries. Note, for example, that the row sum of the circled numbers in the first row is 8. This means that all the breads available at bakery I are being shipped to some store.

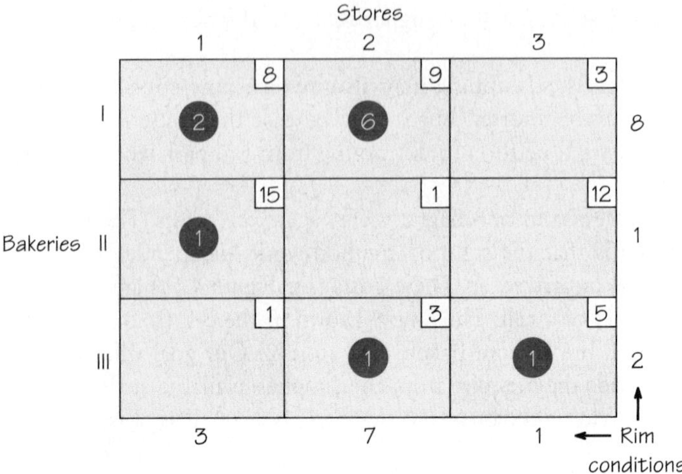

FIGURE 4.20 A possible solution to meeting the needs of three stores for bread from supplies available at three bakeries. The circled numbers show the amounts shipped from the bakeries to the stores.

Similarly, the fact that circled entries in column 2 add up to 7 means that all the breads needed by store 2 are being supplied to it. You can verify that all the row sums and column sums add to exactly the numbers that we want to ship from each bakery to each store. Note that 11 breads have been shipped by the bakeries and received by the stores. When this happens, the circled numbers are said to be a *feasible* solution to the problem.

How much will it cost to ship these amounts of breads (see Figure 4.20) to the stores? The number can be computed by multiplying the circled numbers by the cost shown in the associated cell. For example, to ship two breads from bakery I to store 1 costs 2(8) = 16 because the cost associated with the cell in which the 2 appears is 8. The cost of shipping six breads from bakery I to store 2 is 6(9) = 54. To get the total cost of this "shipment plan," we sum all the shipped amounts by the associated costs to get

$$2(8) + 6(9) + 1(15) + 1(3) + 1(5) = 16 + 54 + 15 + 3 + 5 = 93$$

However, at this point we do not know if there is a cheaper way to ship the breads to the stores. Notice that the number of cells with circled numbers is exactly equal to the number of rows m plus the number of columns n minus 1. This is the general pattern with transportation problems. Cells that are used for shipping are circled. On occasion, we ship a zero amount because the procedure works only when $m + n - 1$ cells are circled.

If we look at the pattern of circled numbers in the tableau in Figure 4.21, we see that there is a difficulty even though 11 breads are involved (the sum of all the circled numbers).

The numbers in the first row add to 6, which means that there will be breads left over at bakery I that have not been shipped. In row 2 the sum of the circled numbers is 2, but this means that something is wrong. How can bakery II, which has a supply of only one bread, ship two breads? Furthermore, column 3 sums to 3, which means that three breads have been shipped to store 1 despite the fact that it only requested one bread! These facts add up to the realization that this assignment of numbers to the cells violates the rules we are requiring. This proposed shipment plan also violates our rule that we are not allowed to circle more than five cells.

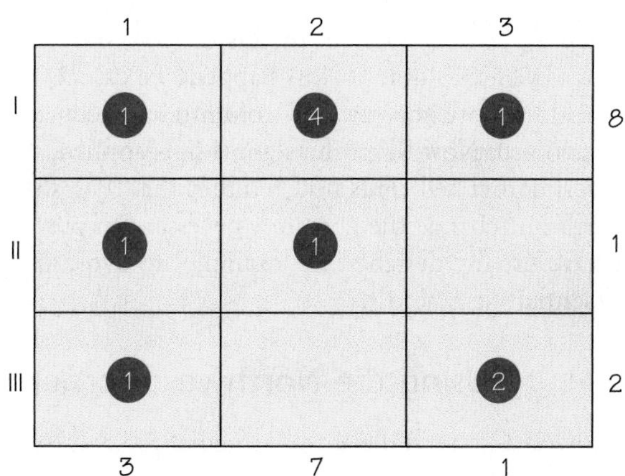

FIGURE 4.21 The circled numbers are not a possible solution to this transportation problem because the rim conditions are not satisfied.

How can we find a solution that meets the constraints of the problem (the rim conditions)? We will show two ways to do this. The first is "fast and dirty" but typically does not find a very good solution with which to start. The second (developed in the exercises) usually gives a better "initial" solution but is a little harder to carry out.

This pair of approaches displays a common tension in problem solving: the ease of getting started but requiring more work later on, or more work at the start, which often proves to be a good investment of extra effort because less work is needed to find an optimal solution. If we know in advance the method being chosen to solve a problem, we can often find an example where this particular method does poorly. Mathematicians work hard to find methods that work well on the kinds of problems that come up in genuine applications.

Northwest Corner Rule

The easier approach involves what is called the **Northwest Corner Rule (NCR)**. This rule is simple because it is based on the geometry of the table that is involved and does not even look at the costs associated with the cells in the table, which in the long run cannot be a good idea, because these costs come into play when trying to get an optimal solution.

How does the Northwest Corner Rule work? The algorithm carries out the following procedure until exactly one cell remains in the "altered tableau."

Northwest Corner Rule — PROCEDURE

1. Locate that cell of the current tableau that is as far to the top and to the left as possible (that is, in the northwest corner). Ship via this cell the smaller of the two rim values (call the value s) associated with the row and column of this cell. (Indicate that this cell is being used by putting a circle around the entry in the tableau.)
2. Cross out the row or column that had rim value s and reduce the other rim value for this cell by s.
3. When a single cell remains, there will be a tie for the rim conditions of both the row and column involved, and this amount is entered into the cell and circled.

Note that it is possible (when there is more than one cell at the start) for there to be a tie when step 1 above is applied. In this case we simultaneously fulfill the rim conditions for a row and column. If this happens we can always choose to cross out, say, the column (not both the row and column) and reduce to 0 the rim condition for the row involved. Now when the algorithm is applied, one has a rim value of 0 for the northwest corner cell. This now requires that 0 be shipped via that cell. Even though this will not change the cost, it is necessary to put a 0 in this cell and circle it. Here we have usually designed the examples to avoid ties so as to make it easier to get the essential ideas across.

EXAMPLE 9 ■ Using the Northwest Corner Rule

Applying the Northwest Corner Rule to our original tableau (see Figure 4.19), we get the sequence of tableaux in Figure 4.22 as we cross out the rows or columns, where for clarity the costs associated with the cells are suppressed. The last diagram in the sequence shows the results on the original tableau, with the cost restored. Note that, at the steps in between, the costs played no role. It is a good idea to check that the circled numbers in each row and column really add up to the rim value for that row and column and that exactly $m + n - 1$ cells are filled.

We can now compute the cost of the associated solution that we have found (feasible solution), which obeys the rim conditions. As we did previously, we add up the cost multiplied by the amount shipped for each cell with a circled entry. We get the following calculation:

$$3(8) + 5(9) + 1(1) + 1(3) + 1(5) = 78$$

This shows a cost that is smaller than the solution we found earlier. That solution involved a cost of 93. But is this solution the cheapest one? Since finding this feasible solution did not make use of the costs on the cells, it suggests that it is not very likely. ■

Improving the Feasible Solution

The next phase of the transportation problem algorithm attempts to answer the question of how to tell if the feasible solution found by using the Northwest Corner Rule is the best. If this solution is not the best, we should be able to find a way to improve it.

Suppose we decided to ship an additional bread from bakery II to store 3. Now, this would violate the fact that we had shipped exactly the right numbers of breads before this new additional shipment, so we have shipped one bread too many from bakery II. We can compensate for this by reducing from 1 to 0 the bread shipped from bakery II to store 2. But this now means that store 2 has not gotten all the breads it needs. We can take care of this by shipping one more bread from bakery III to store 2, but again we now have one extra bread shipped from bakery III. We can compensate for this by reducing the number of breads shipped via cell (III, 3)–that is, from bakery III to store 3. This step will ensure that the rim conditions will hold for the circled numbers. This is because we have located a circuit–(II, 3), (II, 2), (III, 2), (III, 3), (II, 3)–where, if we increase and decrease the breads alternately going around that circuit, we maintain the rim conditions (see Figure 4.23).

Check for yourself that the tableau on the right in Figure 4.23 with the circled entries meets the rim conditions. To the left we show the circuit of cells with plus and minus signs (+ and −) where we have increased the amounts in the cells with +

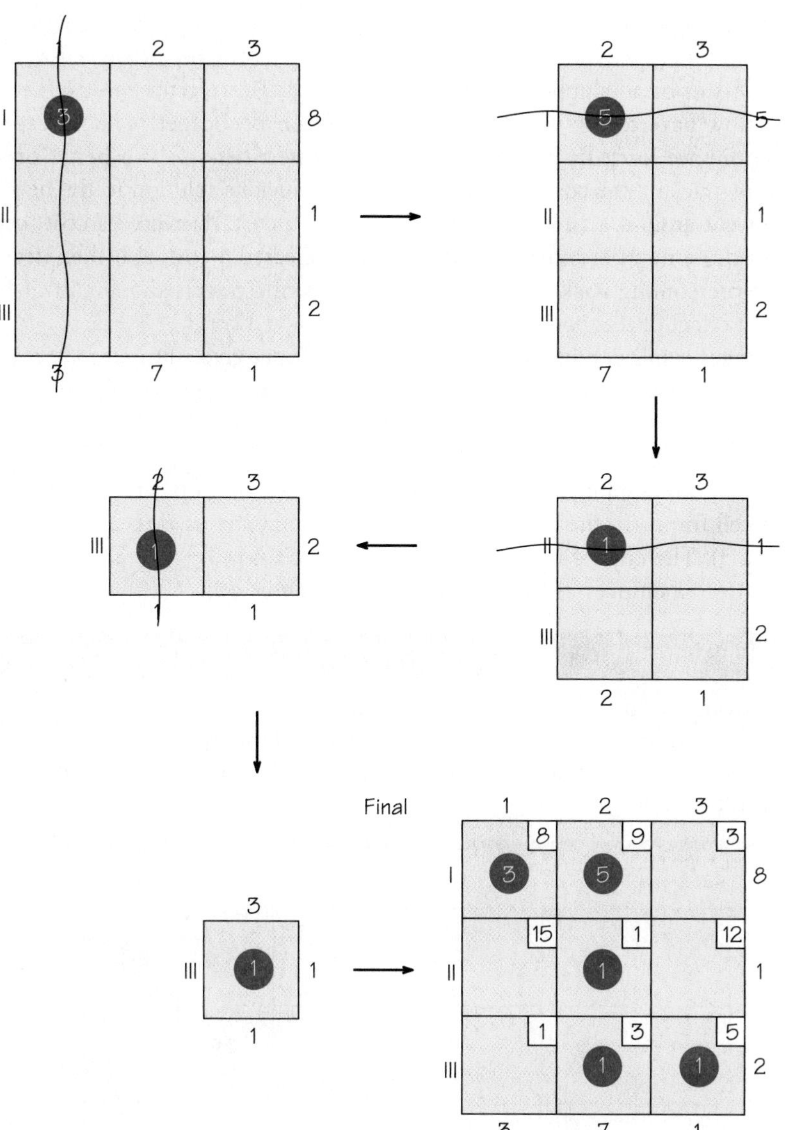

FIGURE 4.22 The construction of an initial solution to a transportation problem using the Northwest Corner Rule.

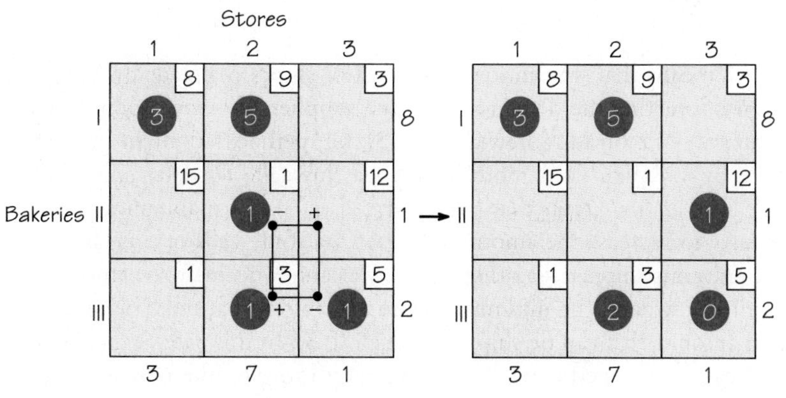

FIGURE 4.23 An illustration of how to take a current solution to a transportation problem and try to get an improved cheaper solution that still meets the rim conditions.

and decreased the amounts in the cells with − by 1 unit. (Note that to keep a total of five cells circled, we have set one of the circled cells to 0, because when we reduce the amounts of bread shipped in cells (II, 2) and (III, 3), we get a tie of value 0.)

We now have to ask whether this new solution is cheaper or more expensive than the one we started with. We can figure out whether this is a better or worse solution by tracking the costs of moving from the previous solution to the new one.

We went around a circuit where we increased a cost, decreased a cost (because we reduced the number of breads in that cell), increased a cost, and then decreased a cost before coming back to where we started, having traversed a circuit of length 4. The net effect of this collection of cost changes is $+12 - 1 + 3 - 5 = +9$. Thus, these changes, while producing a new feasible solution, give a more costly solution!

Perhaps increasing the amount shipped via a different circuit of cells would be better. Suppose we try the same process for cell (I, 3) (Figure 4.24)–that is, increase the shipping of breads from bakery I to store 3. To see if this is worthwhile, check the circuit formed by shipping more via cell (I, 3). It takes a bit of practice to find the circuit this cell forms. In the case of cell (I, 3), the circuit we get is (I, 3), (I, 2), (III, 2), (III, 3), (I, 3). The cost of moving around this circuit is $+3 - 9 + 3 - 5 = -8$. We will refer to this number as the **indicator value** for this cell.

Indicator Value

DEFINITION

The **indicator value of a cell** *C* (not currently a circled cell) is the cost change associated with increasing or decreasing the amounts shipped in a circuit of cells starting at *C*. It is computed by summing with alternating signs the costs of the cells in the circuit.

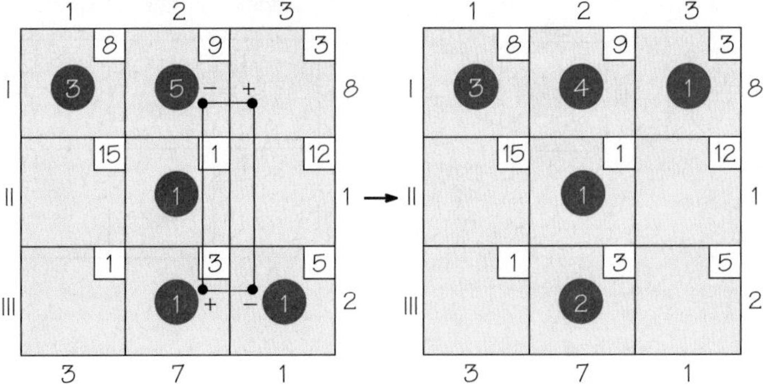

FIGURE 4.24 Using a cell with a negative indicator value, we can find a cheaper way of meeting the demands from the supplies.

The −8 means that we can lower the cost of shipping breads by using a different pattern of meeting the demands from the supplies. We have computed the saving for shipping one bread more via cell (I, 3), but perhaps we might be able to save even more by shipping even more breads via this cell. To determine whether we could, we look at the circuit that begins at cell (I, 3). To maintain a feasible solution, we have to increase the amounts shipped via some cells of this circuit and decrease the amounts shipped via others. Since we cannot decrease the amount shipped via any cell below zero, the minimum value of any cell that must be reduced is the maximum amount that can be shipped via cell (I, 3). In this case, it means that only one bread can be shipped via cell (I, 3), thereby lowering the cost from the previous solution by 8.

When we looked to improve the solution shown in Figure 4.22, we have now seen that by shipping via cell (I, 3) we can get a better solution. However, there might be several cells in the solution shown in Figure 4.22 that would lead to improvement. Which one should we choose? The answer is that we should adopt a greedy point of view. If there are several cells with a negative indicator value, pick the one that is "most negative" to improve the solution.

Given a current feasible solution (one that satisfies the rim condition), we check each cell that does not have a circled number for improvement if we ship via that cell. If a cell leads to a positive indicator value with the circuit associated with it, no improvement is possible. If a cell has a negative indicator value associated with the circuit for that cell, we can get an improvement. We select as the cell to increase that cell with the largest negative indicator value. We now have a new feasible solution that is cheaper than the one we started with and can repeat our procedure just described starting from this new feasible solution.

It turns out that there was no better cell than (I, 3) (using this greedy approach) to get an improved solution. We will take the current best solution and see if we can improve it further. It turns out that for the current tableau (Figure 4.24), all the cells have a positive indicator except for cell (III, 1):

$$\text{Indicator for cell (III, 1):} \qquad +1 - 3 + 9 - 8 = -1$$

Since the minimum of the circled numbers in the cell with a negative label is 2 in cell (III, 2), we can increase by 2 the amount shipped in cell (III, 1) and get a new solution as shown in Figure 4.25.

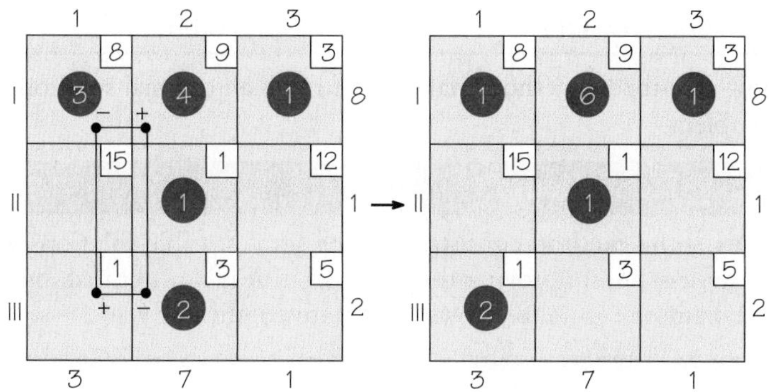

FIGURE 4.25 We can find an even cheaper way of meeting the demands from the supplies available using a cell with a negative indicator value.

Now, for this tableau, all the empty cells have positive indicator values.

$$\begin{aligned}
&\text{Indicator (II, 1):} && +15 - 1 + 9 - 8 = 15\\
&\text{Indicator (II, 3):} && +12 - 3 + 9 - 1 = 17\\
&\text{Indicator (III, 2):} && +3 - 9 + 8 - 1 = 1\\
&\text{Indicator (III, 3):} && +5 - 3 + 8 - 1 = 9
\end{aligned}$$

This means that the current solution is optimal. The cost of this solution is

$$1(8) + 6(9) + 1(3) + 1(1) + 2(1) = 8 + 54 + 3 + 1 + 2 = 68$$

It turns out that if all the cells associated with a feasible solution have positive indicator values, then the solution one has reached is optimal. (Cells with zero indicator value show that there are other solutions that achieve the same optimal value.)

How to Recognize an Optimal Solution THEOREM

We are given a transportation problem with m suppliers and n demanders where the amount of the supplies equals the amount of demands. A collection of $m + n - 1$ circled cells is optimal (that is, the circled cells determine a minimum cost solution) if the indicator value associated with the empty cells is positive. If some indicator cells are positive and some are zero, there are multiple solutions for an optimal value.

This theorem is the analog of the result for linear programming that states that if a corner point is feasible, and if no neighbor of the corner point has a better value of the objective function, then the corner point we are at is already an optimal one. Note that there may be other optimal solutions that use a different number of cells than $m + n - 1$, but we can never do any better in terms of the cheapness of a solution than what we have described above.

For those interested in the exciting fact that one piece of mathematics is often useful for other mathematics, we see an example of that here. The reason an empty cell gives rise to a unique circuit with which we can try to improve the current solution of a transportation problem results from the fact that when an edge not in a tree is added to a tree, it creates a unique circuit (see Chapter 3). Since we have m rows and n columns, a tree associated with a graph on $m + n$ vertices has $m + n - 1$ edges, exactly the number of cells we need to fill in a transportation problem!

4.6 Improving on the Current Solution

We have now described a method guaranteed to find an optimal solution to a transportation problem.

The Stepping Stone Method DEFINITION

The **stepping stone method** consists of taking some feasible solution of a transportation problem and improving this solution, if it is not optimal, by shipping an additional amount using a cell with a negative indicator value.

EXAMPLE 10 ■ Applying the Stepping Stone Method

We will work out another small example to illustrate the technique of applying the Northwest Corner Rule to get an initial solution, and then improving this solution if it is not optimal. Again, we do so by computing the indicator values of the cells and improving the current solution by shipping using a cell with a negative indicator value.

We start with an initial tableau where there are two mines that can supply ore to three companies that extract ore. There are 10 units of ore being mined and the extractors need 10 units to run at full capacity. The initial tableau for the problem is displayed in Figure 4.26.

Using the Northwest Corner Rule we find an initial feasible solution as shown in Figure 4.27. When applying the Northwest Corner Rule we eliminate a row or column as follows: first column 1, then column 2, then row I, and we are now left with a single cell. The cost of the feasible solution shown is

$$2(7) + 4(1) + 1(3) + 3(12) = 14 + 4 + 3 + 36 = 57$$

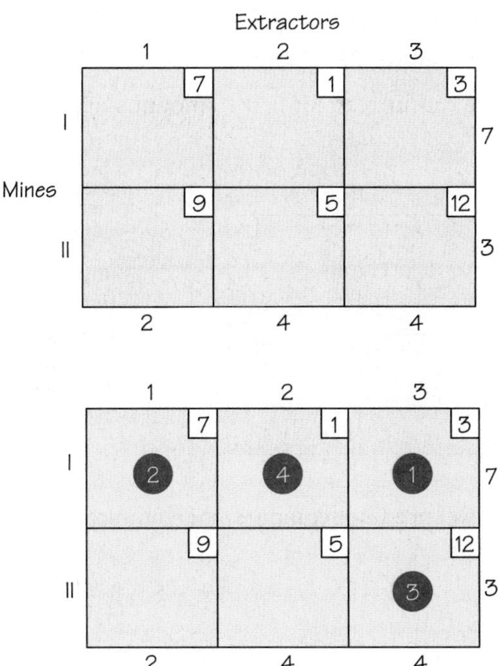

FIGURE 4.26 A transportation problem where two mines supply ore to three companies that extract metal from the ore. The shipping costs are indicated.

FIGURE 4.27 The Northwest Corner Rule has been used to find a possible way to meet the demands from the supplies for the tableau in Figure 4.26.

The two empty cells we have are (II, 1) and (II, 2). We compute the indicator value for each of these cells:

$$\text{Indicator for cell (II, 1):} \qquad +9 - 12 + 3 - 7 = -7$$

$$\text{Indicator for cell (II, 2):} \qquad +5 - 12 + 3 - 1 = -5$$

Since cell (II, 1) has a more negative indicator value, we can reduce the cost more by using that cell. Increasing by 2 (since this is the minimum of circled numbers with negative signs in the computation of the indicator) the amount of metal shipped via cell (II, 1) and cell (I, 3) and reducing by 2 the amount in cells (I, 1) and (II, 3), we obtain the new tableau in Figure 4.28. This has cost

$$4(1) + 3(3) + 2(9) + 1(12) = 4 + 9 + 18 + 12 = 43$$

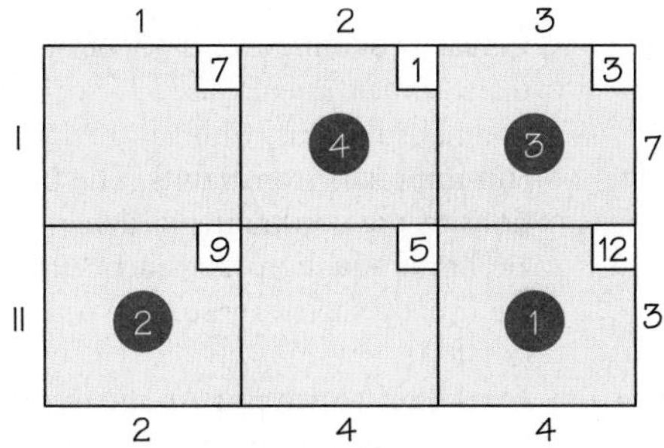

FIGURE 4.28 An improved solution based on the negative indicator value for cell (II, 1) in Figure 4.27.

Note that as a partial check on our work, if we multiply the indicator (-7) by 2, this is -14 and $57 - 43 = 14$, so we reduced the cost of our first solution by 14, as expected.

We now repeat this procedure for this new tableau. We must compute the indicator value of cells (I, 1) and (II, 2).

Indicator for cell (I, 1): $+7 - 9 + 12 - 3 = +7$

Indicator for cell (II, 2): $+5 - 12 + 3 - 1 = -5$

Thus, it turns out that we can increase by 1 the amount shipped by (II, 2), and get the tableau in Figure 4.29.

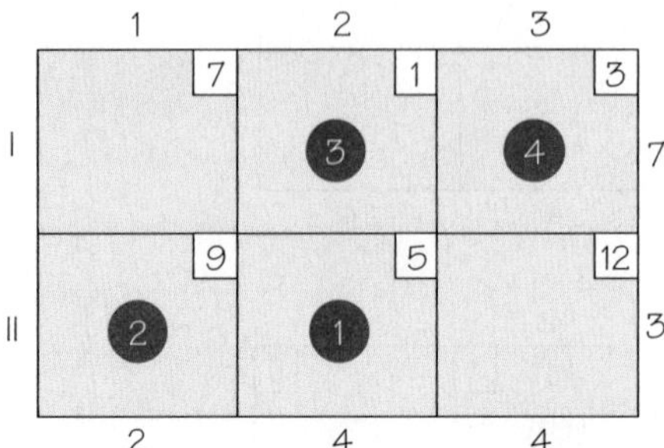

FIGURE 4.29 An improved solution, which turns out to be optimal, based on the negative indicator value for cell (II, 2) in Figure 4.28.

From this tableau, we need to compute the indicator values for the cells (I, 1) and (II, 3). We obtain

Indicator for cell (I, 1): $+7 - 9 + 5 - 1 = +2$

Indicator for cell (II, 3): $+12 - 3 + 1 - 5 = +5$

Not surprisingly, the cell (II, 2) has a positive indicator value because in the previous tableau, that cell was the one that, when we shipped less via it, enabled us to reduce the cost. The fact that both of these indicator values are positive means that the current shipping schedule is an optimal one; that is, using a shipping schedule that ships via only four cells, we cannot find any other solution with the same value. ■

Transportation problems arise in a very large range of situations including shipping milk from dairies to supermarkets, vegetables to health food stores, and vitamins to your local drug store. The next time you sit down to breakfast, think about how many mathematics problems were solved for you to have a healthy breakfast!

REVIEW VOCABULARY

Corner point principle The principle states that there is a corner point of the feasible region that yields the optimal solution. (p. 112)

Feasible points A possible solution (but not necessarily the best) to a linear-programming problem. With just two products, we can think of a feasible point as a point on the plane. (p. 136)

Feasible region The set of all **feasible points**, that is, possible solutions to a linear-programming problem. For problems with just two products, the feasible region is a part of the plane. Also called **feasible set.** (p. 109)

Indicator value of a cell The change in cost due to shipping an increased or decreased amount, using the cells in a transportation tableau that form a circuit consisting of circled cells together with a selected cell that is not circled. When an indicator value is negative, a cheaper solution can be found by shipping using this cell. (p. 132)

Linear programming A set of organized methods of management science used to solve problems of finding optimal solutions, while at the same time respecting certain important constraints. The mathematical formulations of the constraints in linear-programming problems are linear equations and inequalities. Mixture problems are usually solved by some type of linear programming. (p. 104)

Minimum constraint An inequality in a mixture problem that gives a minimum quantity of a product. Negative quantities can never be produced. (p. 108)

Mixture chart A table displaying the relevant data in a linear-programming mixture problem. The table has a row for each product and a column for each resource, for any nonzero minimums, and for the profit. (p. 107)

Mixture problem A problem in which a variety of resources available in limited quantities can be combined in different ways to make different products. It is usually desired to find the way of combining the resources that produces the most profit. (p. 105)

Northwest Corner Rule (NCR) A method for finding an initial but rarely optimal solution to a transportation problem starting from a tableau with rim conditions. The amounts to be shipped between the suppliers and demanders are indicated by circling numbers in the cells in the tableau. The number of cells circled after applying the method will equal the number of rows plus the number of columns minus 1. The method depends on locating at each stage the "northwest corner" of the original tableau or a part of it. (p. 129)

Optimal production policy A corner point of the feasible region where the profit formula has a maximum value. (p. 106)

Profit line In a two-dimensional, two-product, linear-programming problem, the set of all feasible points that yield the same profit. (p. 118)

Resource constraint An inequality in a mixture problem that reflects the fact that no more of a resource can be used than what is available. (p. 108)

Rim conditions The supplies available (listed in a column at the right of a transportation tableau) and demands required (listed in a row at the bottom of a transportation tableau) in a transportation problem. The supplies available are usually taken to exactly meet the demands required. (p. 126)

Simplex method One of a number of algorithms for solving linear-programming problems. (p. 122)

Stepping stone method A method for solving a transportation problem that improves the current solution, when it is not optimal, by increasing the amount shipped using a cell with a negative indicator value. (p. 134)

Tableau A table for a transportation problem indicating the supplies available and demands required, as well as the cost of shipping from a supplier to a demander. The amounts to be shipped from different suppliers to different users are indicated by circled cells in the tableau. The number of such circled cells is always the number of rows plus the number of columns diminished by 1 for the tableau. (p. 127)

Transportation problem A special type of linear-programming problem where we have sources of supplies and users of, or demand for, these supplies. There is a cost to ship an item from a supplier to a demander. The goal is to minimize the total shipping cost to meet the demands from the supplies. (p. 125)

✓ SKILLS CHECK

1. Where do the lines $6x + 2y = 26$ and $2x + 3y = 18$ intersect?

(a) At the point (3, 4)
(b) At the point (6, 2)
(c) At the point (3, 2)

2. The lines $x + 3y = 12$ and $y = 2$ intersect at the point with x-coordinate _____ and y-coordinate _____ .

3. Which of these points lie in the region $4x + 3y \geq 24$, $x \geq 0, y \geq 0$?

(a) Points (5, 2) and (3, 4)
(b) Points (2, 5) and (3, 4)
(c) Points (5, 2) and (2, 5)

4. Producing a bench (x) requires 2 boards, and producing a table (y) requires 5 boards. There are 25 boards available. The resource constraint associated with this situation is _____ x + _____ $y \leq 25$.

5. A tart requires 3 oz of fruit and 2 oz of dough; a pie requires 13 oz of fruit and 7 oz of dough. There are 140 oz of fruit and 90 oz of dough available. Each tart earns 6 cents profit; each pie earns 25 cents profit. What are the resource inequalities of this situation?

(a) $3x + 2y \leq 140$
$13x + 7y \leq 90$
$x \geq 0, y \geq 0$

(b) $3x + 13y \leq 140$
$2x + 7y \leq 90$
$x \geq 0, y \geq 0$

(c) $3x + 2y \leq 6$
$13x + 7y \leq 25$
$x \geq 0, y \geq 0$

6. A tart requires 3 oz of fruit and 2 oz of dough; a pie requires 13 oz of fruit and 7 oz of dough. There are 140 oz of fruit and 90 oz of dough available. Each tart earns 6 cents profit; each pie earns 25 cents profit. The profit formula for this situation, if x represents the numbers of pies produced and y represents the number of tarts produced, is given by P (in cents) = _____ x + _____ y.

7. Graph the feasible region identified by the following inequalities:

$2x + 4y \leq 20$
$4x + 2y \leq 16$
$x \geq 0, y \geq 0$

Which of these points is *not* in the feasible region of the graph drawn?

(a) (2, 4)
(b) (1, 1)
(c) (10, 0)

8. Suppose the feasible region has four corners, at points (0, 0), (4, 0), (0, 3), and (3, 2). If the profit formula is $3x - $2y, the maximum value for the profit is ______________________.

9. Suppose the feasible region has four corners, at points (0, 0), (4, 0), (0, 3), and (3, 2). For which of these profit formulas is the profit maximized by producing a mix of products?

(a) $2x − $2y
(b) $x + $2y
(c) $2x − $y

10. The corner point method cannot be applied to find the optimal answer for the value of the profit $P = 3x + 7y$, where the feasible region is shown in the diagram, because the feasible region is not ______________.

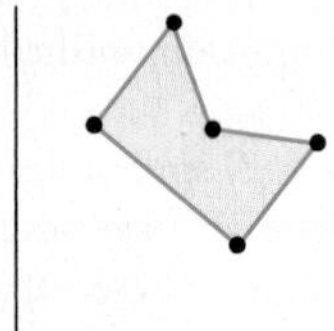

11. The shaded region in the accompanying diagram is an example of a region

(a) whose area is not bounded.
(b) that is not convex.
(c) that is not bounded by straight-line segments.

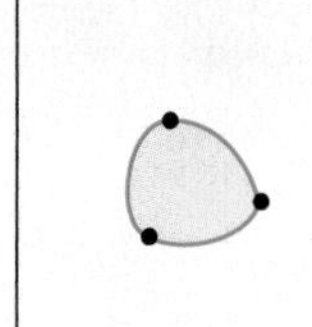

12. Suppose the feasible region has five corners, at points (1, 1), (2, 1), (3, 2), (2, 4), (1, 5). If the profit formula is $5x − $3y, the corner point which maximizes the profit has x-coordinate ____ and y-coordinate ____.

13. Suppose the feasible region has five corners, at points (1, 1), (2, 1), (3, 2), (2, 4), (1, 5). Which of these points is *not* in the feasible region?

(a) (1, 3)
(b) (2, 2)
(c) (0, 0)

14. Consider the feasible region identified by the inequalities $x \geq 0$, $y \geq 0$, $3x + y \leq 10$, $x + 2y \leq 6$. The corner point of this region, which is not (0, 0), that has x-coordinate 0 has y-coordinate ______________.

15. How does the line representing the maximum feasible profit intersect the feasible region?

(a) No points of intersection
(b) Only one point of intersection
(c) At least one point of intersection, and sometimes more than one point of intersection

16. Consider the feasible region for a linear programming problem involving the inequalities $x \geq 0$, $y \geq 0$, $3x + y \leq 10$, $x + 2y \leq 5$. The corner point for this feasible region that has no zero coordinates has x-coordinate ______ and y-coordinate ______.

17. When the Northwest Corner Rule is applied to the accompanying transportation problem tableau, the cells that remain empty are

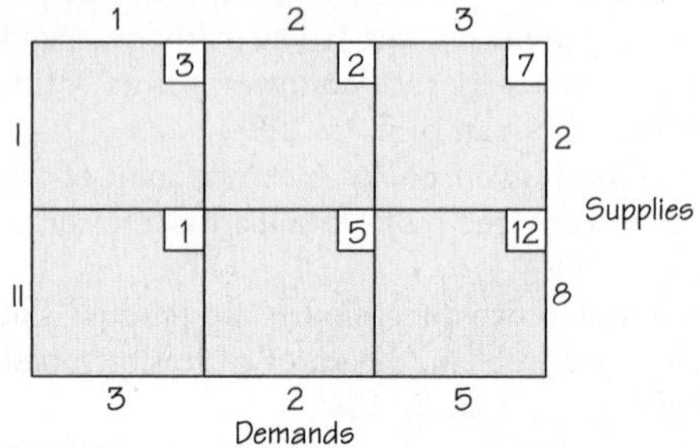

(a) cell (II, 2) and cell (I, 2).
(b) cell (I, 2) and cell (II, 3).
(c) cell (I, 2) and cell (I, 3).

18. The circled cells in the accompanying tableau give a solution that satisfies the rim conditions. The cost associated with this solution is ______________.

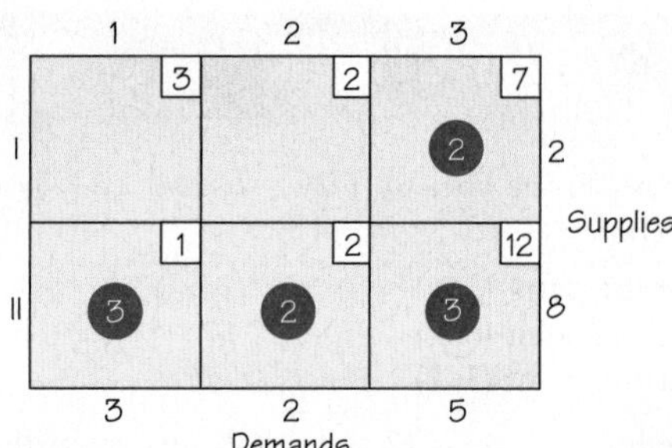

19. The circled cells in the accompanying tableau satisfy the rim conditions. When the indicator value for cell (I, 2) is computed,

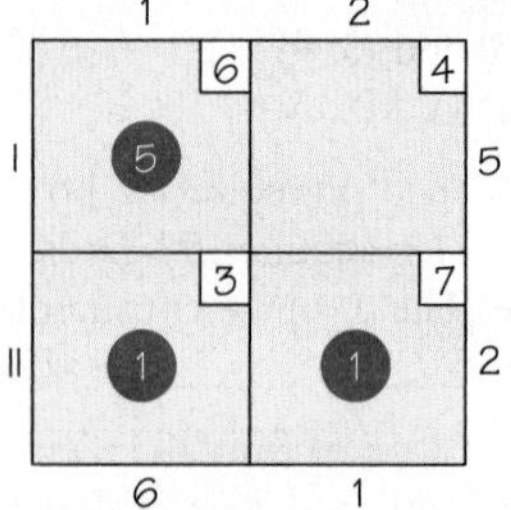

(a) the result being positive, the current solution is optimal.
(b) the result being positive, this tableau has no minimal cost solution.
(c) the result being negative means this tableau does not give a minimal cost solution.

20. The indicator value associated with cell (I, 2) of the accompanying tableau is ________________ .

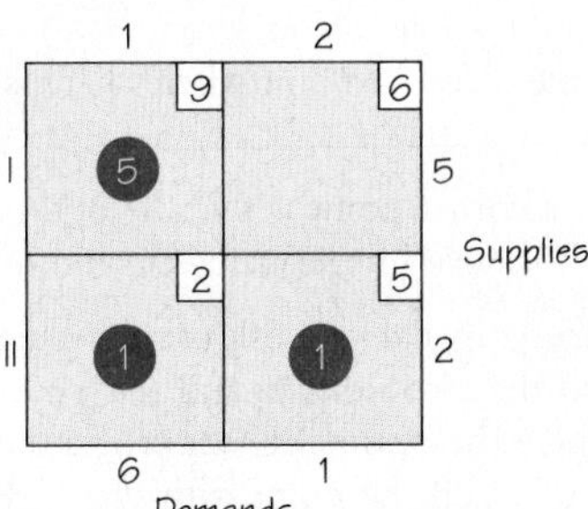

CHAPTER 4 EXERCISES

■ Challenge ◆ Discussion

4.1 Mixture Problems: Combining Resources to Maximize Profit

1. Using intercepts, the points where the lines cross the axes, graph each line.

(a) $2x + 3y = 12$
(b) $3x + 5y = 30$
(c) $4x + 3y = 24$

2. Using intercepts, the points where the lines cross the axes, graph each line.

(a) $7x + 4y = 42$
(b) $x = -3$
(c) $y = 6$

3. Graph both lines on the same axes. Put a dot where the lines intersect. Use algebra to find the x- and y-coordinates of the point of intersection.

(a) $4x + 3y = 18$ and $x = 0$
(b) $5x + 3y = 45$ and $y = -5$
(c) $5x + 3y = 45$ and $x = 3$

4. Graph both lines on the same axes. Put a dot where the lines intersect. Use algebra to find the x- and y-coordinates of the point of intersection.

(a) $x = 3$ and $y = -4$
(b) $3x + 5y = 45$ and $x = -5$
(c) $5x + 3y = 45$ and $x = -3$

5. Graph both lines on the same axes. Put a dot where the lines intersect. Use algebra to find the x- and y-coordinates of the point of intersection.

(a) $x + y = 10$ and $x + 2y = 14$
(b) $y - 2x = 0$ and $x = 2$

6. Graph the line and half-plane corresponding to the inequality, a typical constraint from a mixture problem.

(a) $x \geq 7$ (c) $5x + 3y \leq 15$
(b) $y \geq 4$ (d) $4x + 5y \leq 30$

7. Graph the line and half-plane corresponding to the inequality, a typical constraint from a mixture problem.

(a) $x \geq 3$ (c) $3x + 2y \leq 18$
(b) $y \geq 8$ (d) $7x + 2y \leq 42$

In Exercises 8–10, for each description, write one or more appropriate resource-constraint inequalities. The unknown to use for each product is given in parentheses.

8. **(a)** One bridesmaid's bouquet (x) requires 2 roses, and one corsage (y) requires 4 roses. There are 28 roses available.
(b) Maintaining a large tree (x) takes 2 hours of pruning time and 30 minutes of shredder time; maintaining a small tree (y) takes 30 minutes of pruning time and 15 minutes of shredder time. There are 40 hours of pruning time and 2 hours of shredder time available.

9. **(a)** Manufacturing one package of hot dogs (x) requires 6 oz of beef, and manufacturing one package of bologna (y) requires 4 oz of beef. There are 300 oz of beef available.
(b) It takes 30 ft of 12-in. board to make one bookcase (x); it takes 72 ft of 12-in. board to make one table (y). There are 420 ft of 12-in. board available.

10. Manufacturing one salami (x) requires 12 oz of beef and 4 oz of pork. Manufacturing one bologna (y) requires 10 oz of beef and 3 oz of pork. There are 40 lb of beef and 480 oz of pork available.

In Exercises 11–16, graph the feasible region, label each line segment bounding it with the appropriate equation, and give the coordinates of every corner point.

11. $x \geq 0; y \geq 0; 2x + y \leq 10$

12. $x \geq 0; y \geq 0; x + 2y \leq 12$

13. $x \geq 0; y \geq 0; 2x + 5y \leq 60$

14. $x \geq 10; y \geq 0; 3x + 5y \leq 120$

15. $x \geq 0; y \geq 4; x + y \leq 20$

16. $x \geq 2; y \geq 6; 3x + 2y \leq 30$

In Exercises 17–18, determine whether the points (2, 4) and/or (10, 6) are points of the given feasible regions of:

17. Exercises 11, 13, and 15.

18. Exercises 12, 14, and 16.

19. In the toy problem, x represents the number of skateboards and y the number of dolls. Using the version of that problem whose feasible region is presented in Figure 4.3b, with the profit formula $\$2.30x + \$3.70y$, write a sentence giving the maximum profit and describing the production policy that gives that profit.

20. In the toy problem, x represents the number of skateboards and y the number of dolls. Using the version of that problem whose feasible region is presented in Figure 4.3b, with the profit formula $\$5.50x + \$1.80y$, write a sentence giving the maximum profit and describing the production policy that gives that profit.

21. Graph both lines on the same axes. Put a dot where the lines intersect. Use algebra to find the x- and y-coordinates of the point of intersection.

(a) $5x + 4y = 22$ and $5x + 10y = 40$
(b) $x + y = 7$ and $3x + 4y = 24$

In Exercises 22–25, graph the feasible region, label each line segment bounding it with the appropriate equation, and give the coordinates of every corner point.

22. $x \geq 0; y \geq 0; 3x + y \leq 9; x + y \leq 7$

23. $x \geq 0; y \geq 0; 2x + y \leq 4; 4x + 4y \leq 12$

24. $x \geq 0; y \geq 2; 5x + y \leq 14; x + 2y \leq 10$

25. $x \geq 4; y \geq 0; 5x + 4y \leq 60; x + y \leq 13$

26. Determine whether the points (4, 2) and/or (1, 3) are points of the given feasible regions of Exercises 23 and 25.

4.2 Finding the Optimal Production Policy

4.3 Why the Corner Point Principle Works

4.4 Linear Programming: Life Is Complicated

27. Find the maximum value of P where $P = 3x + 2y$ subject to the constraints $x \geq 3, y \geq 2, x + y \leq 10, 2x + 3y \leq 24$.

28. Find the maximum value of P where $P = 3x - 2y$ subject to the constraints $x \geq 2, y \geq 3, 3x + y \leq 18, 6x + 4y \leq 48$.

29. Find the maximum value of P where $P = 5x + 2y$ subject to the constraints $x \geq 2, y \geq 4, x + y \leq 10$.

30. Given profit $P = 21x + 11y$ subject to the constraints $x \geq 0, y \geq 0, 7x + 4y \leq 13$:

(a) Graph the feasible region.
(b) Determine a corner point where there is an optimal solution.

(*Warning:* The corner point where the optimal solution occurs may not have integer values for both x and y.)

31. (a) Referring to Exercise 30, use the usual rounding rule to round the x-coordinate and the y-coordinate of the point where the optimal linear-programming solution occurs. Call the point with these coordinates Q.
(b) Determine if Q's coordinates define a feasible point by checking them against the constraints.
(c) Evaluate the profit value P at point Q. How does the profit value compare with the point where the optimal value occurred in Exercise 30?
(d) Let R be the point with coordinates (0, 3). Is R in the feasible region? Evaluate P at point R and compare the result with the answer at Q and where the optimum linear-programming value occurred.
(e) Explain the significance of the situation here for solving maximization problems where $P = ax + by$ (a and b are known in advance) is subject to linear constraints but where the variables must be nonnegative integers rather than arbitrary nonnegative decimal numbers.

Exercises 32–43 each have several steps leading to a complete solution to a mixture problem. Practice a specific step of the solution algorithm by working out just that step for several problems. The steps are:

(a) Make a mixture chart for the problem.
(b) Using the mixture chart, write the profit formula and the resource- and minimum-constraint inequalities.
(c) Draw the feasible region for those constraints and find the coordinates of the corner points.
(d) Evaluate the profit information at the corner points to determine the production policy that best answers the question.
(e) (Requires technology) Compare your answer with the one you get from running the same problem on a simplex algorithm computer program.

32. A clothing manufacturer has 600 yd of cloth available to make shirts and decorated vests. Each shirt requires 3 yd of material and provides a profit of \$5. Each vest requires 2 yd of material and provides a profit of \$2. The manufacturer wants to guarantee that under all circumstances there are minimums of 100 shirts and 30 vests produced. How many of each garment should be made to maximize profit? If there are no minimum quantities, how, if at all, does the optimal production policy change?

33. A car maintenance shop must decide how many oil changes and how many tune-ups can be scheduled in a typical week. The oil change takes 20 min, and the tune-up

requires 100 min. The maintenance shop makes a profit of $15 on an oil change and $65 on a tune-up. What mix of services should the shop schedule if the typical week has available 8000 min for these two types of services? How, if at all, do the maximum profit and optimal production policy change if the shop is required to schedule at least 50 oil changes and 20 tune-ups?

34. A clerk in a bookstore has 90 min at the end of each workday to process orders received by mail or on voice mail. The store has found that a typical mail order brings in a profit of $30 and a typical voice-mail order brings in a profit of $40. Each mail order takes 10 min to process and each voice-mail order takes 15 min. How many of each type of order should the clerk process? How, if at all, do the maximum profit and optimal processing policy change if the clerk must process at least three mail orders and two voice-mail orders?

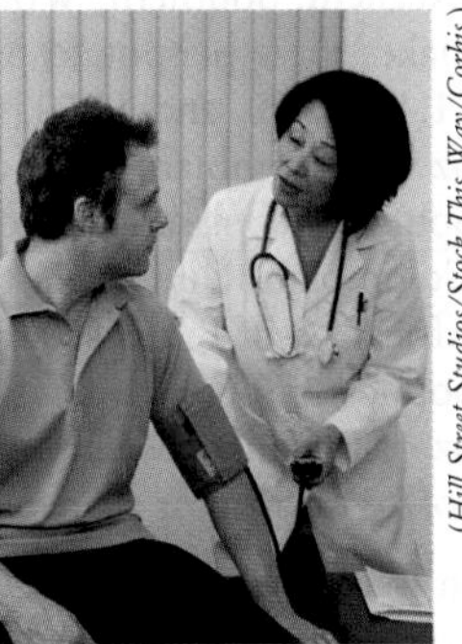
(Hill Street Studios/Stock This Way/Corbis.)

35. In a certain medical office, a routine office visit requires 5 min of doctors' time and a comprehensive office visit requires 25 min of doctors' time. In a typical week, there are 1800 min of doctors' time available. If the medical office clears $30 from a routine visit and $50 from a comprehensive visit, how many of each should be scheduled per week? How, if at all, do the maximum profit and optimal production policy change if the office is required to schedule at least 20 routine visits and 30 comprehensive ones?

36. A bakery makes 600 specialty breads–multigrain or herb–each week. Standing orders from restaurants are for 100 multigrain breads and 200 herb breads. The profit on each multigrain bread is $8 and on herb bread, $10. How many breads of each type should the bakery make in order to maximize profit? How, if at all, do the maximum profit and optimal production policy change if the bakery has no standing orders?

37. A student has decided that passing a mathematics course will, in the long run, be twice as valuable as passing any other kind of course. The student estimates that to pass a typical math course will require 12 hr a week to study and do homework. The student estimates that any other course will require only 8 hr a week. The student has available 48 hr for study per week. How many of each kind of course should the student take? (*Hint:* The profit could be viewed as 2 "value points" for passing a math course and 1 "value point" for passing any other course.) How, if at all, do the maximum value and optimal course mix change if the student decides to take at least two math courses and two other courses?

Exercises 38–43 require finding the point of intersection of two lines, each corresponding to a resource constraint.

38. The firm WebsAreUs creates and maintains Web sites for client companies. There are two types of Web sites: "Hot" sites change their layout frequently but keep their content for long times; "cool" sites keep their layout for a while but frequently change their content. To maintain a hot site requires 1.5 hr of layout time and 1 hr for content changes. To maintain a cool site requires 1 hr of layout time and 2 hr for content changes. Every day, WebsAreUs has available 12 hr for layout changes and 16 hr for content changes. Net profit is $50 for a set of changes on a hot site and $250 for a set of changes on a cool site. In order to maximize profit, how many of each type of site should WebsAreUs maintain daily? How, if at all, do the maximum profit and optimal policy change if the company must maintain at least two hot and three cool sites daily?

39. A paper recycling company uses scrap cloth and scrap paper to make two different grades of recycled paper. A single batch of grade A recycled paper is made from 25 lb of scrap cloth and 10 lb of scrap paper, whereas one batch of grade B recycle paper is made from 10 lb of scrap cloth and 20 lb of scrap paper. The company has 100 lb of scrap cloth and 120 lb of scrap paper on hand. A batch of grade A paper brings a profit of $500, whereas a batch of grade B paper brings a profit of $250. What amounts of each grade should be made? How, if at all, do the maximum profit and optimal production policy change if the company is required to produce at least one batch of each type?

40. Jerry Wolfe has a 100-acre farm that he is dividing into one-acre plots, on each of which he builds a house. He then sells the house and land. It costs him $20,000 to build a modest house and $40,000 to build a deluxe house. He has $2,600,000 to cover these costs. The profits are $25,000 for a modest house and $60,000 for a deluxe house. How many of each type of house should he build to maximize profit? How, if at all, do the maximum profit and optimal production policy change if Wolfe is required to build at least 20 of each type of house?

41. The maximum production of a soft-drink bottling company is 5000 cartons per day. The company produces regular and diet drinks, and must make at least 600 cartons of regular and 1000 cartons of diet per day. Production costs are $1.00 per carton of regular and $1.20 per carton of diet. The daily operating budget is $5400. How many cartons of each type of drink should be produced if the profit is $0.10 per regular and $0.11 per diet? How, if at all, do the maximum profit and optimal bottling policy change if the company has no minimum required production?

(c) Is the current solution optimal? If not, find a cheaper solution.

57. The accompanying tableau arose by applying the Northwest Corner Rule.

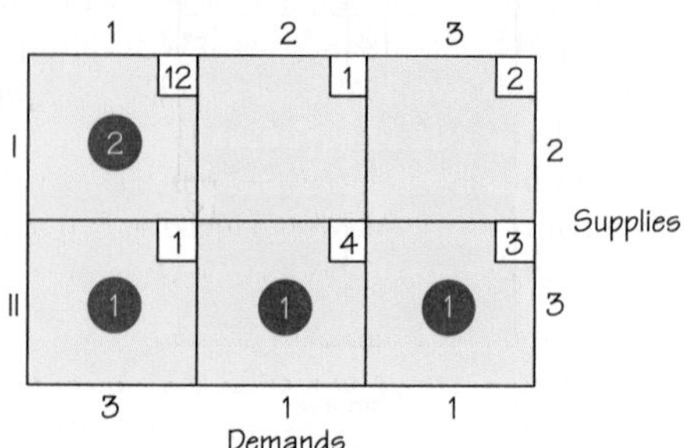

The accompanying graph was constructed so that there is one edge for each circled vertex in the tableau above.

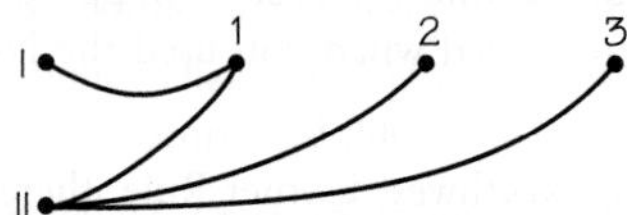

(a) Verify that the graph is a tree.
(b) Show that for each empty cell in the tableau, adding to the graph the unique edge that corresponds to the empty cell creates one circuit.
(c) Show that this circuit corresponds to the one used to find the indicator value of the empty cell.

58. (a) For each row of the following tableau, compute the minimum cost for that row. Now select the row R that among all the rows has the smallest row minimum. In a way similar to the Northwest Corner Rule, use the cheapest cell in row R and ship as much as possible via that cell, crossing out a row or a column, and adjust the rim conditions and repeat the process. This is known as the *minimum row entry method.* Use the minimum row entry method to find an initial solution to the following transportation problem, which shows the costs of returning rental cars from cities that have more cars than necessary to cities that have too few cars.

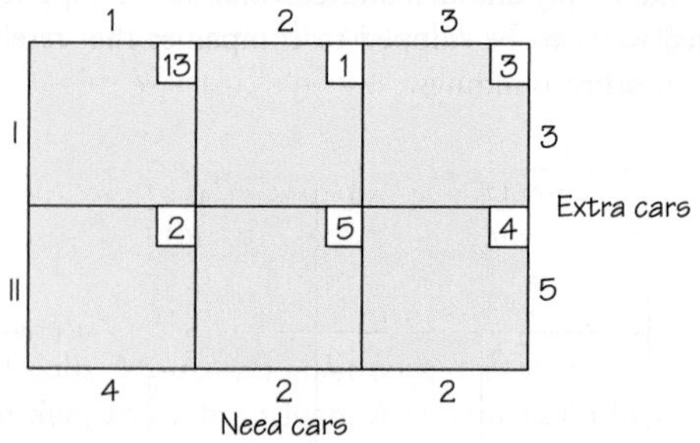

(b) Compute the cost of the solution you find using the minimum row entry method.
(c) Compare the cost found in part (b) with the cost of the initial solution obtained using the Northwest Corner Rule.

59. (a) For each of the following tableaux, find an initial solution using the Northwest Corner Rule.

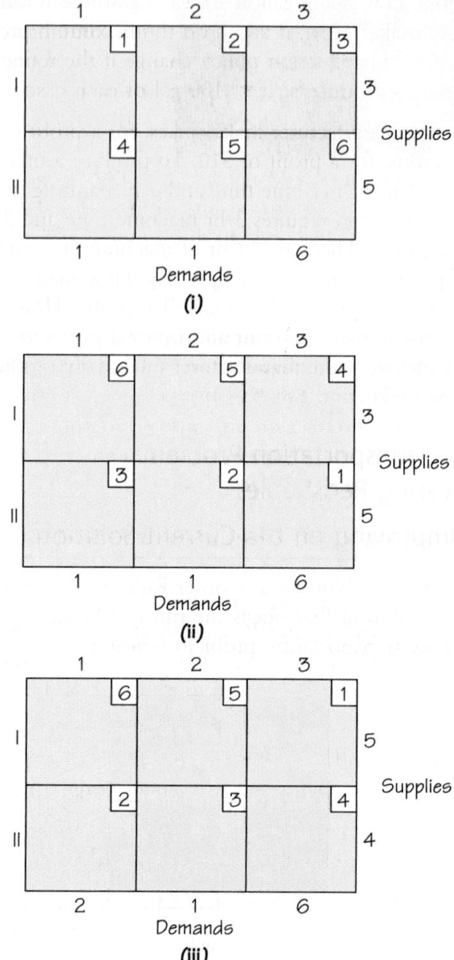

(b) If the solution you find using the Northwest Corner Rule is not optimal, then apply the stepping stone algorithm to find an optimal solution.

60. (a) Apply the Northwest Corner Rule to the following tableau.

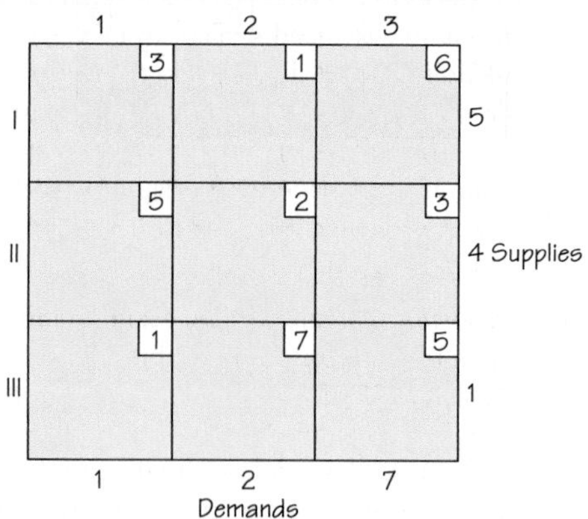

(b) Determine the cost associated with the solution you found.
(c) Compute the indicator value for each noncircled cell.
(d) Does the Northwest Corner Rule give rise to an optimal solution?

61. For each tableau at right, a solution for the associated transportation problem has been proposed using the circled cells that obey the rim conditions.
(a) Determine the cost associated with the indicated feasible solution.
(b) Is the solution shown optimal?
(c) If it is not an optimal solution, apply the stepping stone method to obtain an optimal solution.

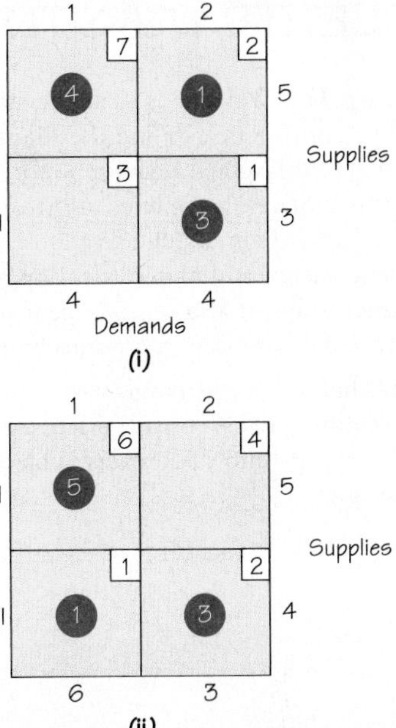

 WRITING PROJECTS

1. Interview a local businessperson who is in charge of deciding the product mix for a business. Must this business take into consideration situations other than minimum and resource constraints? If so, what are these considerations? Find out what methods the person uses to make production policy decisions. Is linear programming used? Are other methods used? If so, what are they? Write a report of your findings, and add some of your own conclusions about the usefulness of linear programming for this business.

2. In economics, it is often useful to distinguish between a firm that has a monopoly (for example, is the only supplier of a product) and firms that supply only a small share of the market. How would the presence of a monopoly affect the relation between production and price? Would the presence of a monopoly tend to ensure the fixed-profit assumption of linear programming, or would it make it more likely that the interplay of supply and demand would have to be considered in order to have a truly realistic model? Write an essay addressing these issues.

 SUGGESTED READINGS

ANDERSON, DAVID R., DENNIS J. SWEENEY, and THOMAS A. WILLIAMS. *An Introduction to Management Science: Quantitative Approaches to Decision Making,* West, St. Paul, Minn., 1985. A business management text with seven chapters on linear programming.

DOLAN, ALAN, and JOAN ALDUS, *Networks and Algorithms: An Introductory Approach,* Wiley, New York, 1993. A graph theoretical approach to network optimization problems, including the transportation problem.

GASS, SAUL I. *Decision Making, Models, and Algorithms,* Krieger, Melbourne, Fla., 1991. This book demonstrates how to use linear programming and related ideas to solve a variety of industrial and governmental problems.

HARDWICK, I., *Decision and Discrete Mathematics,* Albion Publishing, Chichester, England, 1996. A survey of situations that can be modeled using graphs in the area of operations research. It treats both the simplex method for solving linear-programming problems and the transportation problem.

Note: Simplex software can be found in *Maple* (keyword is *simplex*), *Mathematica* (keyword is *Linear Programming*), in both *Lotus 1-2-3* and *MSExcel* via *Solver,* and in other software packages, especially those intended for quantitative mathematics courses focusing on business applications.

 SUGGESTED WEB SITES

www.informs.org This Web site is maintained by the Institute for Operations Research and the Management Sciences, the main professional organization in these fields in the United States. It contains information on (and/or links to) news items about operations research and management science and employment opportunities and summer internships; it also has a student newsletter. Much of the material is written in a nontechnical style.

www.hsor.org/what_is_or.cfm?name=linear_programming This Web page discusses how linear programming fits into the broader subject of operations research.

www-gap.dcs.st-and.ac.uk/~history/Mathematicians/Dantzig_George.html This site contains biographical information about George Dantzig, who, by developing the simplex method, greatly expanded the use and applicability of linear programming.

www-unix.mcs.anl.gov/otc/Guide/faq/linear-programming-faq.html This site is the "frequently asked questions" section of an online newsgroup for people interested in linear programming.

en.wikipedia.org/wiki/Linear_programming This Web page outlines the theory of linear programming.

Statistics: The Science of Data PART II

What CDs are big sellers this week? When you buy a CD, the checkout scanner probably reports your choice to a company that tallies sales and reports the winners. Are there genetic differences between two related types of cancer? To find out, biologists use "microarrays" to report the activity of thousands of genes at once. Checkout scanners and microarrays produce immense amounts of *data*, numerical facts. So do opinion polls, medical studies, and even the sports pages. *Statistics* is the science of collecting, organizing, and interpreting data.

Chapters 5 and 6 concern *data analysis*, the art of seeing what data say. We learn from data by making graphs and doing calculations, guided by principles that help us decide what graphs to make, what to look for in our graphs, and what calculations are helpful based on what we see. Sometimes we want to know more: An opinion poll or a medical study looks at only some people, but we want conclusions that apply to all voters or all patients. This is called *statistical inference*, because we infer conclusions about a large group from data on a small part of the group. Chapter 7 discusses inference from beginning to end: from how to produce data when we have inference in mind to how to say just how much confidence we can have in our conclusions. Confidence, uncertainty, risk, chance–the mathematics that describes all these ideas is *probability theory*, the topic of Chapter 8. Probability is the mathematics behind statistical inference, but that's just a small part of its usefulness. ■

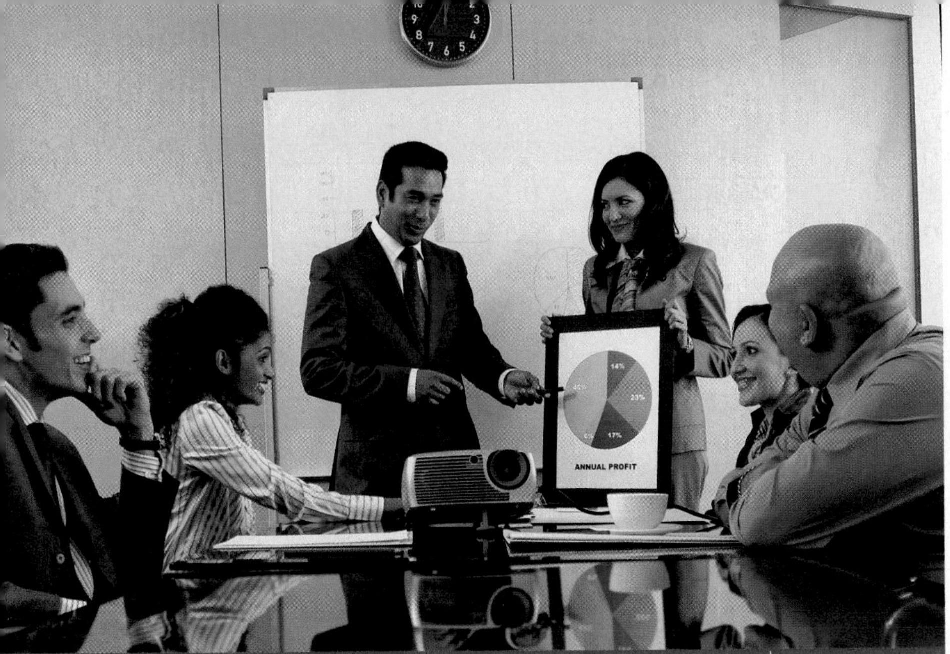

CHAPTER 5

Exploring Data: Distributions

A flood of data is a prominent feature of modern society. Data are essential for making decisions in almost every area of life and work. Like other great floods, the flood of numbers threatens to overwhelm us. We must control the flood by careful organization and interpretation. A corporate database, for example, contains an immense volume of data–on employees, sales, inventories, customer accounts, equipment, taxes, and other topics. These data are useful only if we can organize them and present them so that their meaning is clear. The penalties for ignoring data can be severe–several banks have suffered billion-dollar losses from unauthorized trades in financial markets by their employees, trades that were hidden in a mass of data that the banks' management did not examine carefully.

This chapter and the next show you how to use graphs and numerical summaries to work with data. Always remember that although knowing which tools to use and how to use them is important, interpreting your work is even more important. Data come from the real world, and in the end, your goal is to use data to learn something about the real world.

Any set of data contains information about some group of **individuals**. The information is organized in **variables**.

Individuals DEFINITION

Individuals are the objects described by a set of data. Individuals may be people, but they may also be animals or things.

Variable DEFINITION

A **variable** is any characteristic of an individual. A variable can take different values for different individuals.

FIGURE 5.1 Part of a data set as displayed by the Excel spreadsheet program.

	A	B	C	D	E	F
1	SEX	HAND	HEIGHT	STUDY	COINS	
2	F	L	65	200	50	
3	M	L	72	30	35	
4	M	R	62	95	35	
5	F	L	64	120	0	
6	M	R	63	220	0	
7	F	R	58	60	76	
8	F	R	67	150	215	
9						

Sheet1 Sheet2 Sheet3

EXAMPLE 1 ■ Data from a Student Questionnaire

Figure 5.1 is a small part of a data set that describes the students in a large statistics class. The data come from anonymous responses to a class questionnaire. Each row records data on one *individual,* that is, one student. Each column contains the values of one *variable* for all the individuals. There are five variables. Sex (female or male) and handedness (left-handed or right-handed) are variables that are usually described as categorical or qualitative because they categorize individuals by traits and do not take numerical values. You have probably already had much experience with the usual ways to summarize categorical data (for example, proportions, pie charts, and bar graphs), so we will not take time to repeat that here.

The remaining three variables are described as measurement or quantitative variables because they do take numerical values. They are: height (inches), time spent studying (in minutes per weeknight), and "How many cents in coins (not bills) are you carrying?" Our main focus will be on variables involving numerical data.

Most data tables follow this format–each row is an individual, and each column is a variable. This data set appears in a *spreadsheet* program that has rows and columns ready for your use. Spreadsheets are commonly used to enter and transmit data, and spreadsheet programs also have functions for basic statistics.

Knowing the context of the data–that these are student responses to a class questionnaire–helps us make sense of them. For example, one student claimed to study 30,000 minutes on a typical night. We know that this is impossible! ■

Statistical tools and ideas help us examine data in order to describe their main features. This examination is called **exploratory data analysis**. Like an explorer crossing unknown lands, we want first to describe simply what we see. In this chapter and the next, we use both numbers and graphs to explore data. Here are two principles that provide the tactics for exploratory analysis of data.

Exploring Data PROCEDURE

1. Begin by examining each variable by itself. Then move on to study the relationships among the variables.
2. Begin with a graph or graphs. Then add numerical summaries of specific aspects of the data.

These principles also organize the material in Chapters 5 and 6. In this chapter, we look at data on a single variable. Chapter 6 moves on to relations among several variables. In each chapter, we first display data in graphs, then add numerical summaries.

5.1 Displaying Distributions: Histograms

Data analysis begins with graphical displays of the values of a single variable. For example, you may want to compare the study times claimed by female and male students. Because individual study times vary so much, we are interested in the **distribution** of study time for female and male students.

Distribution of a Variable DEFINITION

The **distribution** of a variable tells us what values the variable takes and how often it takes these values.

Numerical variables often take many values. A graph of the distribution is clearer if nearby values are grouped together. The most common graph of the distribution of one numerical variable is a **histogram**.

Histogram DEFINITION

A **histogram** is a graph of the distribution of outcomes (often divided into classes) for a single numerical variable. The height of each bar is the number of observations in the class of outcomes covered by the base of the bar. All classes should have the same width and each observation must fall into exactly one class.

EXAMPLE 2 ■ Population Distribution

Every 10 years, the Census Bureau (www.census.gov) tries to contact every household in the United States. One of the most striking findings of the 2000 Census was the growth of the Hispanic population of the United States. Table 5.1 presents the percent of adult (age 18 and over) residents in each of the 50 states who identified themselves in the 2000 Census as "Spanish/Hispanic/Latino." The *individuals* in this data set are the 50 states. The *variable* is the percent of Hispanics in a state's population. To make a histogram of the distribution of this variable, proceed as follows:

Making a Histogram

Step 1. Choose the classes. Divide the range of the data into some reasonable number of classes of equal width. The data in Table 5.1 range from 0.7 to 42.1, so here's one way to choose classes:

$$0.0 \leq \text{percent Hispanic} < 5.0$$
$$5.0 \leq \text{percent Hispanic} < 10.0$$
$$\vdots$$
$$40.0 \leq \text{percent Hispanic} < 45.0$$

Be sure to specify the classes precisely so that each individual falls into exactly one class. A state with 4.9% Hispanic residents would fall into the first class, but a state with 5.0% falls into the second.

TABLE 5.1 Percent of Adult Population of Hispanic Origin, by State (2000 Census)

State	Percent	State	Percent	State	Percent
Alabama	1.5	Louisiana	2.4	Ohio	1.9
Alaska	4.1	Maine	0.7	Oklahoma	5.2
Arizona	25.3	Maryland	4.3	Oregon	8.0
Arkansas	2.8	Massachusetts	6.8	Pennsylvania	3.2
California	32.4	Michigan	3.3	Rhode Island	8.7
Colorado	17.1	Minnesota	2.9	South Carolina	2.4
Connecticut	9.4	Mississippi	1.3	South Dakota	1.4
Delaware	4.8	Missouri	2.1	Tennessee	2.0
Florida	16.8	Montana	2.0	Texas	32.0
Georgia	5.3	Nebraska	5.5	Utah	9.0
Hawaii	7.2	Nevada	19.7	Vermont	0.9
Idaho	7.9	New Hampshire	1.7	Virginia	4.7
Illinois	10.7	New Jersey	13.3	Washington	7.2
Indiana	3.5	New Mexico	42.1	West Virginia	0.7
Iowa	2.8	New York	15.1	Wisconsin	3.6
Kansas	7.0	North Carolina	4.7	Wyoming	6.4
Kentucky	1.5	North Dakota	1.2		

Step 2. Count the individuals in each class. Here are the counts (sometimes called frequencies, since they tell how frequently values fall in a class):

Class	Count	Class	Count	Class	Count
0.0 to 4.9	27	15.0 to 19.9	4	30.0 to 34.9	2
5.0 to 9.9	13	20.0 to 24.9	0	35.0 to 39.9	0
10.0 to 14.9	2	25.0 to 29.9	1	40.0 to 44.9	1

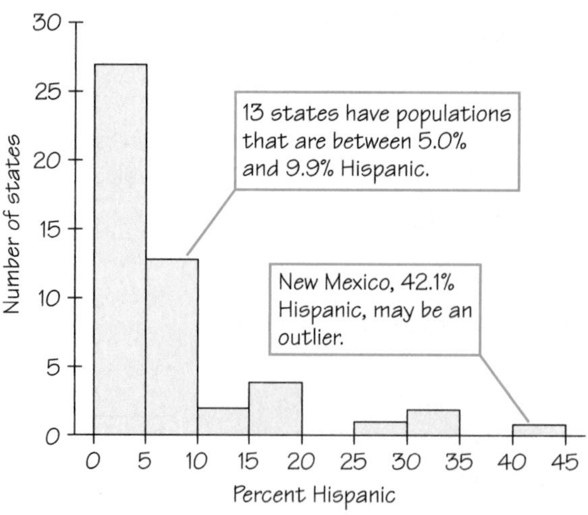

FIGURE 5.2 Histogram of the percent of Hispanics among the adult residents of the states.

Step 3. Draw the histogram. First mark the scale for the variable whose distribution you are displaying on the horizontal axis. That's the percent of a state's population who are Hispanic. The scale runs from 0 to 45 because that is the span of the classes we chose. The vertical axis contains the scale of counts. Each bar represents a class. The base of the bar covers the class, and the bar height is the class count. There is no horizontal space between the bars unless a class is empty, so that its bar has height zero. Figure 5.2 is our histogram.

The bars of a histogram should cover the entire range of values of a variable. When the possible values of a variable have gaps between them, extend the bases of the bars to meet halfway between two adjacent possible values. For example, in a histogram of the ages in years of university faculty, the bars representing 25 to 29 years and 30 to 34 years would meet at 29.5.

Our eyes respond to the *area* of the bars in a histogram. Because the classes are all the same width, area is determined by height and all classes are fairly represented. There is no one right choice of the classes in a histogram. Too few classes will give a "skyscraper" graph, with all values in a few classes with tall bars. Too many will produce a "pancake" graph, with most classes having one or no observations. Neither choice will give a good picture of the shape of the distribution. You must use your judgment in choosing classes to display the shape. Statistics software will choose the classes for you. The computer's choice is usually a good one, but you can change it if you want.

5.2 Interpreting Histograms

Making a statistical graph is not an end in itself. The purpose of the graph is to help us understand the data. After you make a graph, always ask, "What do I see?" Once you have displayed a distribution, you can see its important features as follows.

Outlier — DEFINITION

In any graph of data, look for the overall pattern and for striking deviations from that pattern. You can describe the overall pattern of a distribution by its shape, center, and spread. An important kind of deviation is an **outlier**, an individual value that falls outside the overall pattern.

We will soon learn how to describe center and spread numerically. For now, you can describe the center of a distribution by its middle value–with roughly half the observations taking smaller values and half taking larger values. You can give a rough description of the spread of a distribution by giving the *smallest and largest values.*

EXAMPLE 3 ■ Describing a Distribution

Look again at the histogram in Figure 5.2. **Shape:** The distribution has a *single peak*, which represents states in which less than 5% of adults are Hispanic. The distribution is *skewed to the right.* Most states have no more than 10% Hispanics, but some

(*Corbis/Punchstock.*)

states have much higher percentages, so that the graph trails off to the right. **Center:** Table 5.1 shows that about half the states have less than 4.7% Hispanics among their adult residents and half have more. So the midpoint of the distribution is close to 4.7%. **Spread:** The spread is from about 0% to 42%, but only four states fall above 20%. **Outliers:** Arizona, California, New Mexico, and Texas stand out. Whether these are outliers or just part of the long right tail of the distribution is a matter of judgment. There is no universal rule for calling an observation an outlier. Once you have spotted possible outliers, look for an explanation. Some outliers are due to mistakes, such as typing 4.2 as 42. Other outliers point to the special nature of some observations. These four states are heavily Hispanic by history and location.

When you describe a distribution, concentrate on the main features. Look for major peaks, not for minor ups and downs in the bars of the histogram. Look for clear outliers, not just for the smallest and largest observations. Look for rough **symmetry** or clear **skewness**.

Some variables have distributions with predictable shapes. For example, incomes, house prices and other money amounts usually have right-skewed distributions because there are always CEOs and celebrities well to the right of the rest of us! A simple test designed to measure basic achievement may yield a left-skewed distribution because most students will cluster together with high scores, but there are usually still a few people who perform low (due to lack of attendance or effort) and give the distribution a tail stretching out to the left.

Skewed Distribution — DEFINITION

A distribution is **skewed to the right** if the longer tail of the histogram is on the right side. (Because positive numbers are on the right side of a number line, such a distribution is also referred to as positively-skewed.) For example, see Figure 5.2 or Figure 5.7.

Similarly, a distribution is **skewed to the left** (or negatively-skewed) if the longer tail is on the left side.

Other distributions, however, are *symmetric* and may have little or no skewness. For example, the distribution of heights (or handspans) in an adult population may look like two hills of equal size (visualize a two-humped camel) if males cluster around one value and females cluster around another. A more common and more important symmetric shape is the bell-shaped histogram (Figure 5.3) yielded by many standardized tests as well as by many biological measurements (such as, height, length of thigh bone, and so on) on specimens from the same species and sex.

Symmetric Distribution — DEFINITION

A distribution is **symmetric** if the right and left sides of the histogram are approximately mirror images of each other. (That is, if you "folded" the histogram in half, the left half would fall roughly onto the right half.) For example, see Figure 5.3.

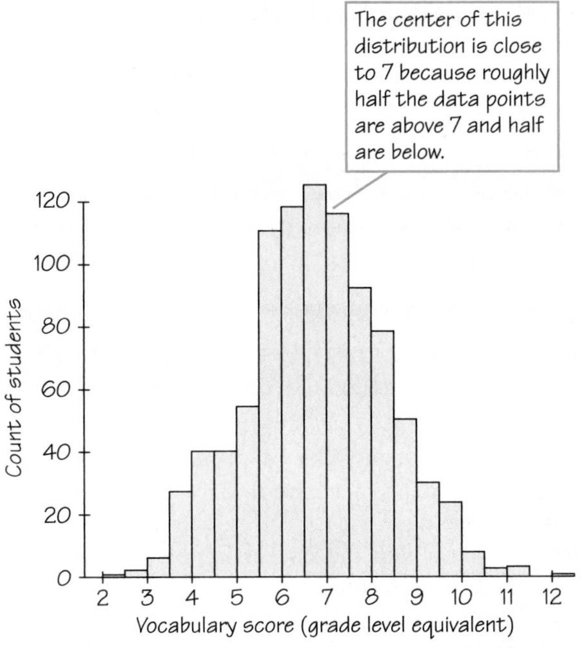

FIGURE 5.3 Histogram of the Iowa Test of Basic Skills vocabulary scores for 947 seventh-grade students.

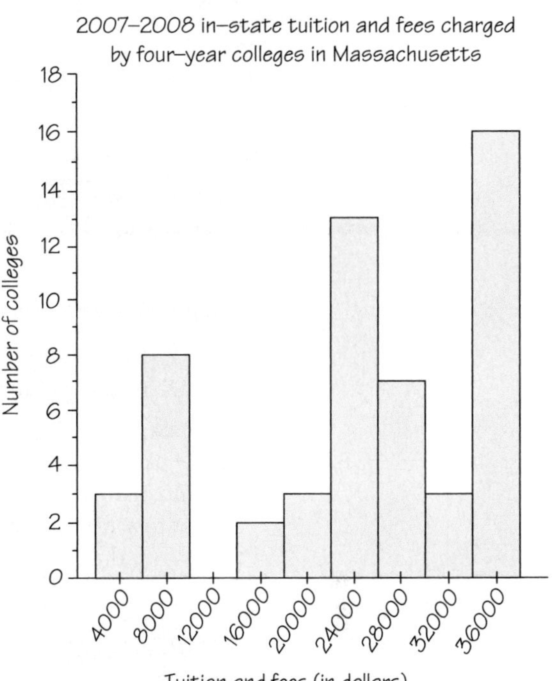

FIGURE 5.4 Histogram of the tuition and fees charged by four-year colleges in Massachusetts.

EXAMPLE 4 ■ Iowa Test Scores

Figure 5.3 displays the scores of all 947 seventh-grade students in the public schools of Gary, Indiana, on the vocabulary part of the Iowa Test of Basic Skills. The distribution is *single-peaked* and *symmetric*. In mathematics, the two sides of symmetric patterns are exact mirror images, but real-life data are almost never exactly symmetric. We are content to describe Figure 5.3 as symmetric. The center (half above, half below) is close to 7. This is a seventh-grade reading level. The scores range from 2.0 (second-grade level) to 12.1 (twelfth-grade level). ■

EXAMPLE 5 ■ College Tuition

Jeanna plans to attend college in her home state of Massachusetts. She looks up the tuition and fees for the 2007–2008 academic year for all 55 four-year colleges in Massachusetts (omitting art schools and other special colleges). Figure 5.4 is a histogram of the data. For example, the tallest bar tells us there are 16 colleges charging between \$34,000 and \$38,000. As is often the case, we can't call this irregular distribution either symmetric or skewed. It does show two separate *clusters* of colleges, 11 with tuition less than \$10,000 and the remaining 44 costing more than \$16,000. Clusters suggest that two types of individuals are mixed in the data set. In fact, the histogram distinguishes the 11 state colleges in Massachusetts from the 44 private colleges, which charge much more. ■

5.3 Displaying Distributions: Stemplots

Histograms are not the only way to graphically display distributions. For small data sets, a **stemplot** is quicker to make and presents more detailed information.

Stemplot DEFINITION

A **stemplot** is a display of the distribution of a variable that attaches the final digits of the observations as leaves on stems made up of all but the final digit.

Making a Stemplot PROCEDURE

To make a **stemplot**:

1. Separate each observation into a *stem* consisting of all but the final (rightmost) digit and a *leaf*, the final digit. Stems may have as many digits as needed, but each leaf contains only a single digit.
2. Write the stems in a vertical column with the smallest at the top, and draw a vertical line at the right of this column. Include all stems, even if they are not used.
3. Write each leaf in the row to the right of its stem, in increasing order out from the stem.

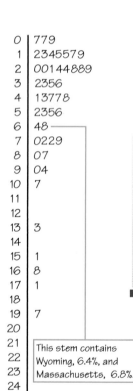

```
 0 | 779
 1 | 2345579
 2 | 00144889
 3 | 2356
 4 | 13778
 5 | 2356
 6 | 48
 7 | 0229
 8 | 07
 9 | 04
10 | 7
11 |
12 |
13 | 3
14 |
15 | 1
16 | 8
17 | 1
18 |
19 | 7
20 |
21 |
22 |
23 |
24 |
25 | 3
```

This stem contains Wyoming, 6.4%, and Massachusetts, 6.8%.

High 32.0 32.4 42.1

FIGURE 5.5 Stemplot of the percent of Hispanics among the adult residents of the states.

EXAMPLE 6 ■ Making a Stemplot

For the "percent Hispanic" percents in Table 5.1, take the whole-number part of the percent as the stem and the final digit (tenths) as the leaf. The Massachusetts entry, 6.8%, has stem 6 and leaf 8. Wyoming, at 6.4%, places leaf 4 on the same stem. These are the only observations on this stem. We then arrange the leaves in order, as 48, so that 6|48 is one row in the stemplot. Figure 5.5 is the complete stemplot for the data in Table 5.1. To save space, we left out California, Texas, and New Mexico, which have stems 32 and 42. These observations are listed as "High" below the stemplot. ■

If we rotate Figure 5.5 a quarter-turn counterclockwise, the stemplot would look like a histogram (of a distribution skewed to the right). Comparing the stemplot in Figure 5.5 with the histogram in Figure 5.2 reveals the strengths and weaknesses of stemplots. The stemplot, unlike the histogram, preserves the actual value of each observation. But you can choose the classes in a histogram, whereas the classes (the stems) of a stemplot are forced on you. Whether the large number of classes in Figure 5.5 is an improvement over Figure 5.2 is a matter of taste. Stemplots do not work well for large data sets like the 947 Iowa Test scores in Figure 5.3, because each stem must hold a large number of leaves.

When the observed values have many digits, it is often best to *round* the numbers to just a few digits before making a stemplot. For example, a stemplot of data like

3.468 2.567 2.981 1.095 . . .

would have very many stems and no leaves or just one leaf on most stems. You can round these data to

3.5 2.6 3.0 1.1 . . .

before making a stemplot.

Graphical summaries are good for analyzing the shape of distribution of values. To answer precise questions about features of a dataset, such as its center, however, it helps to have numerical summaries as well. We explore this next.

5.4 Describing Center: Mean and Median

What kind of gas mileage do you get with the new cars in the Environmental Protection Agency's "midsized cars" category? Table 5.2 gives the city and highway gas mileage for a representative sample of Model Year 2008 midsized cars.

TABLE 5.2 Fuel Economy (Miles per Gallon) for Model Year 2008 Vehicles

Model	City Mileage	Highway Mileage
Acura RL	16	24
BMW 550i	15	23
Chevrolet Malibu	22	30
Dodge Avenger	21	30
Hyundai Elantra	24	33
Lexus ES 350	19	27
Mercury Milan	20	29
Mitsubishi Galant	20	27
Nissan Sentra	21	29
Nissan Versa	27	33
Pontiac Grand Prix	18	28
Toyota Camry	21	31
Toyota Prius	48	45

We start with graphs. Figure 5.6 is a *dotplot* of the city mileages of the 13 cars in the sample of midsized cars. As is often the case when there only a few observations, the shape of the distribution is irregular. The most striking feature is a high outlier on the right end of the dotplot. Upon closer examination, to see if the outlier value may be a typographical error, we see that the Toyota Prius is the only hybrid gas-electric car in the sample.

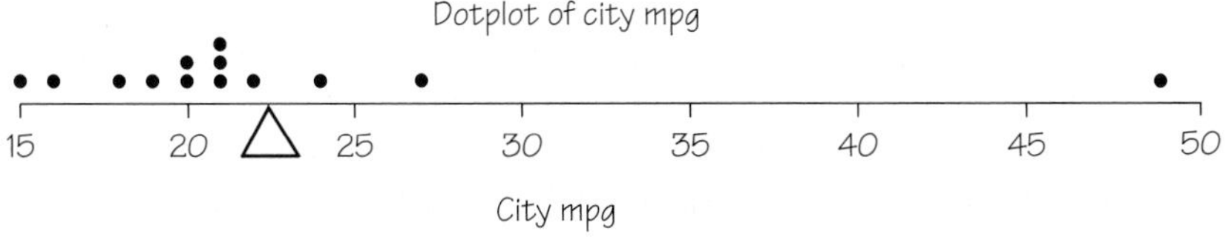

FIGURE 5.6 Dotplot of the city gas mileages of the sample of midsized cars. The Toyota Prius is an outlier.

Numerical summaries make the comparison we want more specific. Numerical description of a distribution begins with a measure of its center. The two most common measures of center are the **mean** and the **median**. Basically, the mean is the arithmetic "average value" and the median is the "middle value." We need to explore the precise procedures for calculating these measures and observe how they behave differently.

Finding the Mean $\bar{x}$ — PROCEDURE

To find the **mean** of a set of observations, add their values and divide by the number of observations. If the n observations are $x_1, x_2, \ldots, x_n$, their mean is

$$\bar{x} = \frac{x_1 + x_2 + \cdots + x_n}{n}$$

The bar over the x indicates the mean of all the x-values. Pronounce the mean $\bar{x}$ as "x-bar." This notation is very common. When writers who are discussing data use $\bar{x}$ or $\bar{y}$, they are talking about a mean.

One way to visualize the value of the mean of a dataset is to imagine where the fulcrum would have to be placed for its dotplot to "balance." This metaphor tells us the mean is always between the largest and smallest values, and by visual inspection we can further estimate that the balance point appears to be somewhere between 20 and 25. Let's see what exact value the formula yields.

EXAMPLE 7 ■ Calculating the Mean

The mean city mileage for the 13 midsized cars in Table 5.2 is

$$\begin{aligned}\bar{x} &= \frac{x_1 + x_2 + \cdots + x_n}{n}\\ &= \frac{16 + 15 + 22 + 21 + 24 + 19 + 20 + 20 + 21 + 27 + 18 + 21 + 48}{13}\\ &= \frac{292}{13} = 22.5 \text{ miles per gallon (mpg)}\end{aligned}$$

We said that the Toyota Prius may not belong with the other cars. If we exclude the Prius, the mean city mileage drops to $\frac{244}{12} = 20.3$ mpg. The single outlier adds more than 2 mpg to the mean city mileage. This illustrates an important weakness of the mean as a measure of center: *The mean is sensitive to the influence of a few extreme observations.* These may be outliers, but a skewed distribution that has no outliers will also pull the mean toward its long tail.

We have used the middle of a distribution as an informal measure of center. The *median* is the formal version of the middle, with a specific rule for calculation. The **median** M is the midpoint of a distribution, the number such that half the observations are smaller and the other half are larger.

Be sure to write down each individual observation in the data set, even if several observations repeat the same value. And be sure to arrange the observations in order of size before locating the median. Note that the recipe $(n + 1)/2$ gives the *position* of the median in the ordered list of observations, *not* the median itself.

Finding the Median M PROCEDURE

To find the median of a distribution:

1. Arrange all observations in increasing order (from smallest to largest).
2. If the number of observations is *odd*, the median M is the center observation in the ordered list.
3. If the number of observations is *even*, the median M is the average of the two center observations in the ordered list.

EXAMPLE 8 ■ Calculating the Median

Since we're exploring the gas mileage cars get on the road, you might notice the connection that just as a median divides a road into two halves (with opposite directions of travel), a median divides a dataset into two halves! To find the median city mileage for 2008 midsized cars, arrange the data in increasing order:

15 16 18 19 20 20 **21** 21 21 22 24 27 48

The median is the bold **21**, which you can find by eye–there are 6 observations to the left and 6 to the right. Or visualize (or construct) a long paper strip divided into 13 equal-sized squares, where the sorted values are written into the squares. When the strip is folded in half, the fold line falls on the median!

What happens if we drop the Toyota Prius? The remaining 12 cars have city mileages:

15 16 18 19 20 **20** **21** 21 21 22 24 27

Because the number of observations $n = 12$ is even, there is no single center observation. There is a center *pair* of observations (20 and 21) that has 5 observations to its left and 5 to its right. The median M is the mean of the center pair, which is $(20 + 21)/2 = 20.5$. ■

You see that the median resists the influence of extreme observations better than the mean does. A very high value like the Toyota Prius is simply one observation to the right of center and removing it hardly changed the median at all. In fact, removing an extreme outlier can leave the median completely unchanged while significantly changing the mean. The *Mean and Median* applet (at www.whfreeman.com/fapp8e) is an excellent way to compare the resistance of M and $\bar{x}$. See Applet Exercises 1 and 2.

(Dan Forer/Beateworks/Corbis.)

The median and mean are the most common measures of the center of a distribution. The mean and median of a symmetric distribution are close together. If the distribution is exactly symmetric, the mean and median are exactly the same. In a skewed distribution, the mean is generally farther out in the long tail than is the median (see Figure 5.7). For example, the distribution of house prices is skewed to the right. There are many moderately priced houses and a few very expensive mansions. The mansions pull the mean up but do not affect the median. The mean price of existing single-family houses in the U.S. sold in September 2007 was \$257,800, but the median price for these same houses was only \$211,700.

Another common numerical summary of a distribution is the **mode**–the most frequently occurring value. Like the mean and median, it is a measure of location

SPOTLIGHT 5.1 Think!

Computer software will do all of your statistical graphs and calculations for you. But you can think and the computer can't. Here are some examples of the importance of thinking as well as calculating.

What's not on the plot matters Abraham Wald (1902–1950), like many statisticians, worked on war problems during World War II. Wald invented some statistical methods that were military secrets until the war ended. Here is one of his simpler contributions. Asked where extra armor should be added to airplanes, Wald studied the location of enemy bullet holes in planes returning from combat. He plotted the locations on an outline of the plane. As data accumulated, most of the outline filled up. Put the armor in the few spots with no bullet holes, said Wald. That's where bullets hit the planes that didn't make it back.

What are your units? Not paying attention to units of measurement can get you into trouble. In 1999, the *Mars Climate Orbiter* spacecraft burned up in the Martian atmosphere. It was supposed to be 93 miles (150 kilometers) above the planet, but was in fact only 35 miles (57 kilometers) up. It seems that Lockheed Martin, which built the *Orbiter*, specified important measurements in English units (for example, pounds and miles). The National Aeronautics and Space Administration team who directed the spacecraft thought the numbers were in metric system units (for example, kilograms and kilometers). There went $125 million!

What are we counting? A news report says that 22% of American high school students smoke. That sounds serious. But what counts as "smoking"? It turns out that this is the percent who smoked at least once in the past month. If we say that a smoker is someone who smoked on at least 20 of the past 30 days, only 9.7% of high school students smoke. The number you get depends on how you define your variable—and you can make a problem sound more or less serious by changing the definition. Interest groups do this all the time, so pay attention.

for a distribution. However, the mode is not necessarily a good measure of center. Consider a skewed histogram with its maximum height at one end and a long tail on the other end, or a symmetric histogram with a tall peak at each end and a valley in between!

The Mode — DEFINITION

The **mode** is the most frequently occurring value in a set of numerical observations.

The mode of the city mileage numbers in Table 5.2 would be 21, because it occurs 3 times and no other value occurs more than twice. Some datasets, however, may have two or more values that "tie" for the honor of being a mode, such as the values 27, 29, 30, and 33 in the Table 5.2 highway mileage numbers. We can also identify the mode(s) of a histogram. For example, Figure 5.4 does not give the raw data of individual values, but we can say that the modal class is $34,000–$38,000.

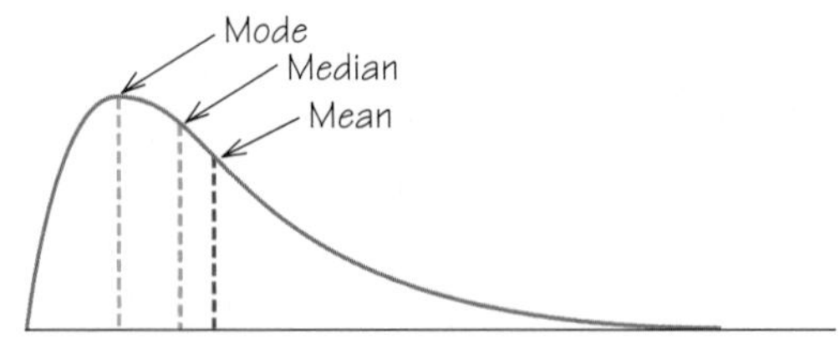

FIGURE 5.7 This smooth curve describes a right-skewed distribution with one mode. The mean is pulled toward the long tail more than the median is.

5.5 Describing Spread: The Quartiles

The mean and median provide two different measures of the center of a distribution. But a measure of center alone can be misleading. Two neighborhoods with median house price $193,000 are very different if one has both mansions and modest homes and the other has little variation among houses. We are interested in the spread or variability of house prices as well as their centers. *The simplest useful numerical description of a distribution consists of both a measure of center and a measure of spread.*

The simplest way to measure spread is with the **range**, which is the difference between the smallest and largest observations. For example, the percents of Hispanics in the states are as low as 0.7% (Maine and West Virginia) and as high as 42.1% (New Mexico), so the range would be 42.1% − 0.7% = 41.4%. Also, the range of the city mileage numbers in Table 5.2 is 48 − 15 = 33 mpg. The range tells us the full span of the data, but it may be greatly affected by one or more outliers. Without the Toyota Prius, the preceding answer becomes 27 − 15 = 12 mpg.

The Range — DEFINITION

The **range** is a measure of spread of a set of observations. It is obtained by subtracting the smallest observation from the largest observation.

We can improve our description of spread by also looking at the spread of the middle half of the data. The first and third *quartiles* separate out the middle half. At the end of the first quarter of a football game, one quarter of the game is complete. Similarly, the first quartile of a distribution or dataset is the point that exceeds one-quarter (or 25%) of the values. The third quartile is the point that exceeds three quarters (or 75%) of the values. (The second quartile exceeds two quarters (or 50%) of the values, and so is equivalent to the median.) To make the idea of quartiles more exact, we need a rule to find them:

Calculating the Quartiles Q_1 and Q_3 — PROCEDURE

To calculate the **quartiles**:

1. Use the median to split the data set into two halves–an upper half and a lower half.
2. The **first quartile** Q_1 is the median of the lower half.
3. The **third quartile** Q_3 is the median of the upper half.

EXAMPLE 9 ■ Calculating Quartiles

The city mileages of the 12 gasoline-powered midsized cars, after sorting:

15 16 18 19 20 20 21 21 21 22 24 27

The first quartile is the median of the 6 observations in the lower half, so $Q_1 = 18.5$. Similarly, the third quartile is the median of the upper half: $Q_3 = 21.5$.

For an example with an odd number of observations, try the city mileages of all 13 midsized cars in Table 5.2. Below are the mileages in increasing order with **bold** used to denote the median which will be excluded to form two equal-sized groups:

15 16 18 19 20 20 **21** 21 21 22 24 27 48

Ignoring the bold **21**, we find the quartiles by finding the median of each half of the dataset: $Q_1 = 18.5$ and $Q_3 = 23$.

Some software packages or calculators use a slightly different rule to find the quartiles, so computer results may be a bit different from your own work. Don't worry about this. The differences will always be too small to be important.

5.6 The Five-Number Summary and Boxplots

We started by using the smallest and largest observations to indicate the spread of a distribution. These single observations tell us little about the distribution as a whole, but they give information about the tails of the distribution that is missing if we know only Q_1, M, and Q_3. To get a quick summary of both center and spread, combine all five numbers.

The Five-Number Summary — DEFINITION

The **five-number summary** of a distribution consists of the smallest observation, the first quartile, the median, the third quartile, and the largest observation, written in order from smallest to largest. In symbols, the five-number summary is

$$\text{Minimum} \quad Q_1 \quad M \quad Q_3 \quad \text{Maximum}$$

These five numbers offer a reasonably complete description of center and spread. For the 12 gasoline-powered midsized cars, you can verify that the five-number summary for city gas mileage is

15 18.5 20.5 21.5 27

and is

23 27 29 30.5 33

for highway gas mileage.

The five-number summary breaks the dataset into four equal groups with equal numbers of observations. A **boxplot** (sometimes called "box and whisker plot") can visually represent the spread of the data across these groups from the five-number summary. Figure 5.8 shows boxplots for both city and highway gas mileages for midsized cars.

Boxplot — DEFINITION

A **boxplot** is a graph of the five-number summary.

- A central box spans the quartiles Q_1 and Q_3.
- A line somewhere in the middle of the box marks the median M of the data set.
- Lines extend from the box out to the smallest and largest observations.

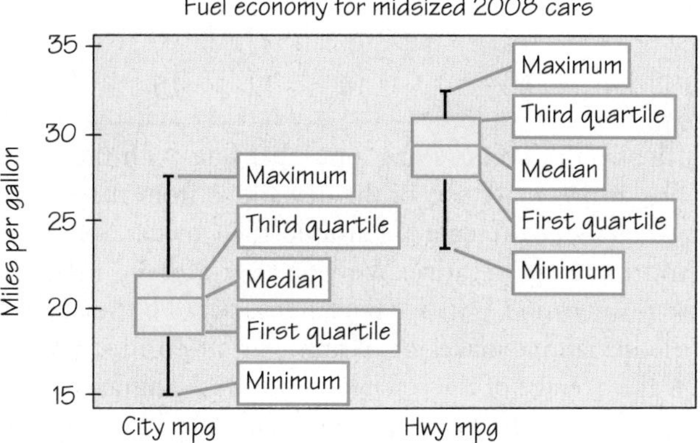

FIGURE 5.8 Boxplots of the highway and city gas mileages for cars classified as midsized by the Environmental Protection Agency. These boxplots are drawn vertically, but it is equally correct to draw them horizontally.

Because boxplots show less detail than histograms or stemplots, they are best used for side-by-side comparison of more than one distribution, as in Figure 5.8. When you look at a boxplot, first locate the median, which marks the center of the distribution. Then look at the spread. The quartiles show the spread of the middle half of the data, and the extremes (the smallest and largest observations) show the spread of the entire data set. So for non-hybrid cars, is there really much of a difference in gas mileages between city and highway? From the boxplots, we see at once that highway mileages are noticeably higher than city mileages: The maximum city mileage reaches only the first quartile of highway mileages! We also see that the spread of highway mileages is roughly the same as the spread of city mileages. Boxplots can also be an indicator of a distribution's skewness, so we have gotten a lot of mileage from this vehicle for exploratory data analysis.

Be aware that some calculators and software packages offer an alternative option for boxplots in which the lines go to the furthest values within 1.5 box lengths of the quartiles, but do not automatically go out to the minimum and maximum values. The advantage of this is that any values beyond these can be individually marked as outliers.

5.7 Describing Spread: The Standard Deviation

Although the five-number summary is the most generally useful numerical description of a distribution, it is not the most common. That distinction belongs to the combination of the mean with the **standard deviation**. The mean, like the median, is a measure of center. The standard deviation, like the quartiles and extremes in the five-number summary, measures spread. The standard deviation and its close relative, the *variance*, measure spread by looking at how far the observations are from their mean.

EXAMPLE 10 ■ Understanding the Standard Deviation

Starting October 10, 2007, the English rock band Radiohead allowed people to choose how much they wanted to pay to download a digital copy of the group's new album *In Rainbows*. According to a survey reported by Associated Press, about 60% of Americans worldwide who downloaded the album chose to pay nothing, but the remaining 40% voluntarily paid an average payment of \$8.05. A sample

(*AP Photo/Robert E. Klein.*)

(sorted in increasing order) of the dollar amounts Americans in the "paying group" paid is:

3 4 5 7 10 12 15

Figure 5.9 displays the data as points along a number line, with their mean marked by an asterisk (*). The arrows mark two of the deviations from the mean. These deviations show how spread out the data are around their mean. Some of the deviations are positive and some are negative. We won't get a useful measure of spread by totaling up these positive and negative deviations, because they will always sum to zero! Squaring the deviations makes these numbers all positive and a reasonable measure of spread is the average of the squared deviations. This average is called the *variance.* The variance is large if the observations are widely spread around their mean. It is small if the observations are all close to the mean.

But the variance does not have meaningful units. With the Radiohead sales data measured in dollars, the variance of the purchase prices has units of "squared dollars." Taking the square root of the variance yields the standard deviation, which gets us back to dollars. The standard deviation has other uses as well, as discussed in Section 5.9.

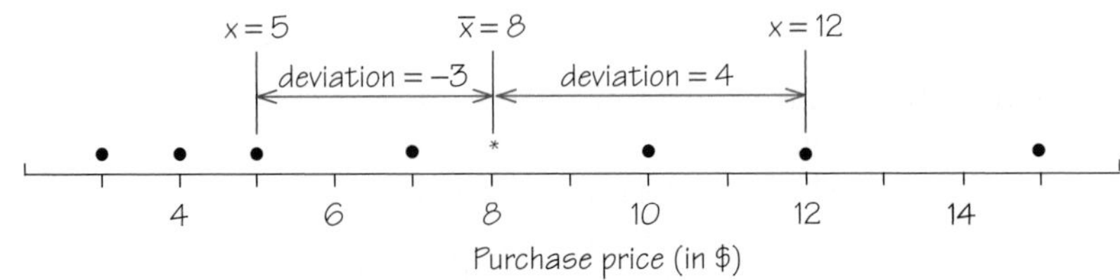

FIGURE 5.9 The variance and standard deviation measure spread by looking at the deviations of observations from their mean.

The Standard Deviation s

DEFINITION

The **standard deviation** is a kind of "standard" or average amount that observed data values deviate from their mean. More precisely, it is the square root of the mean of the squared deviations, except that the mean involves dividing by $n-1$ instead of n. (It turns out $n-1$ makes the formula more accurate, but the justification is beyond the scope of this book.) In symbols, the standard deviation s of n observations $x_1, x_2, \cdots, x_n$ is

$$s = \sqrt{\frac{(x_1 - \bar{x})^2 + (x_2 - \bar{x})^2 + \cdots + (x_n - \bar{x})^2}{n - 1}}$$

In practice (especially with large data sets), you will use your calculator or software to obtain the standard deviation with a single command. However, going through all the steps of an example first will give you understanding about how the standard deviation works.

EXAMPLE 11 ▪ Calculating the Standard Deviation

To find the standard deviation of the 7 purchase prices, first find the mean:

$$\bar{x} = \frac{3 + 4 + 5 + 7 + 10 + 12 + 15}{7} = \frac{56}{7} = 8 \text{ dollars}$$

The deviations shown in Figure 5.9 are the starting point for calculating the standard deviation.

Observations x_i	**Deviations (of observation from mean)** $x_i - \bar{x}$	**Squared Deviations** $(x_i - \bar{x})^2$
3	$3 - 8 = -5$	$(-5)^2 = 25$
4	$4 - 8 = -4$	$(-4)^2 = 16$
5	$5 - 8 = -3$	$(-3)^2 = 9$
7	$7 - 8 = -1$	$(-1)^2 = 1$
10	$10 - 8 = 2$	$2^2 = 4$
12	$12 - 8 = 4$	$4^2 = 16$
15	$15 - 8 = 7$	$7^2 = 49$
		sum = 120

FIGURE 5.10 This table shows a step-by-step approach to form the building blocks for the calculation of standard deviation, whether done by hand, calculator, or spreadsheet.

The variance is the sum of the squared deviations divided by 1 less than the number of observations, so it would be $\frac{120}{7-1} = 20$. The standard deviation is the square root of the variance, and so we obtain $s = \sqrt{20} = 4.47$ dollars. This value can be considered large for this context and range, which suggests that people vary quite a bit in what they consider a fair price for music when they have the power to choose.

If the 7 observations still had a mean of \$8, but were spread out further, their deviations from 8 would be larger and the standard deviation would then be even larger. To explore this relationship, redo the calculation after decreasing \$5 to \$2 and increasing \$15 to \$18. The mean remains the same, but the resulting standard deviation will be larger since the numbers are more spread out.

SPOTLIGHT 5.2 Calculating Standard Deviation

While the formula in the Definition box for standard deviation has conceptual clarity and a straightforward implementation (as in Figure 5.10), it can be tedious to apply to large datasets. Even with the most basic calculator, you'll get the same answer faster using this more computationally-oriented formula:

$$\sqrt{\frac{(x_1^2 + x_2^2 + \cdots + x_n^2) - n(\bar{x})^2}{n-1}}.$$

If you have a *scientific calculator*, put it into a "STAT MODE" if required, clear out any old data, then enter your data one number at a time (after each number, press your calculator's data-entry button—it may say DATA or have a symbol such as [Σ+] or [M+]). Once the data is entered, you can find the standard deviation by hitting the key labeled something like [σn−1] or [σxn−1] or [s].

If you have a *graphing calculator* in the TI-83/84+ family (and you already used (STAT)→ EDIT to enter one variable of quantitative data in a column, say, L1), then hit this sequence of buttons: [2ND] (CATALOG) (it's above the "0" key)→stdDev (ENTER) [2ND] (L1) (it's above the "1" key)(ENTER). If you use the alternative command sequence (STAT)→CALC→1-Var Stats [2ND] (L1)(ENTER), you will obtain not only the standard deviation s_x but also other descriptive statistics, including the five-number summary. Keystrokes for other specific models can be found online, for example, at http://www.geocities.com/calculatorhelp/.

More important than the details of hand calculation are the properties that determine the usefulness of the standard deviation:

- s measures spread about the mean $\bar{x}$. Use s to describe the spread of a distribution only when you use $\bar{x}$ to describe the center.
- $s = 0$ only when there is *no spread.* This happens only when all observations have the same value. Otherwise $s > 0$. As the observations become more spread out about their mean, s gets larger.
- s has the same units of measurement as the original observations. For example, if you measure metabolic rates in calories, both the mean $\bar{x}$ and the standard deviation s are also in calories.
- The use of squared deviations makes s even more sensitive than $\bar{x}$ to a few extreme observations. For example, dropping the Toyota Prius from our list of midsized cars cuts the standard deviation of city mileages more than half, from 8.3 mpg with the Prius to 3.3 mpg without it. Distributions with outliers and strongly skewed distributions have large standard deviations. The number s does not give much helpful information about such distributions.

We now have a choice between two descriptions of the center and spread of a distribution: the five-number summary, or $\bar{x}$ and s. Because $\bar{x}$ and s are sensitive to extreme observations, they can be misleading when a distribution is strongly skewed or has outliers. In fact, because the two sides of a skewed distribution have different spreads, no single number such as s describes the spread well. The five-number summary, with its two quartiles and two extremes, does a better job.

Choosing a Summary RULE

The five-number summary is usually better than the mean and standard deviation for describing a skewed distribution or a distribution with outliers. Use $\bar{x}$ and s only for reasonably symmetric distributions that are free of outliers.

Although the standard deviation is widely used, it is not a natural or convenient measure of the spread of a distribution. The real reason for the popularity of the standard deviation is that it is the natural measure of spread for **normal distributions**, an important class of distributions that we will meet next.

Do remember that a graph gives the best overall picture of a distribution. Numerical measures of center and spread report specific facts about a distribution, but they do not describe its entire shape. Numerical summaries do not disclose the presence of clusters, for example. *Always start with a graph of your data.*

5.8 Normal Distributions

We now have a kit of graphical and numerical tools for describing distributions. What is more, we have a clear strategy for exploring data on a single numerical variable:

1. Always plot your data: make a graph, usually a histogram, dotplot, or a stemplot.

2. Look for the overall pattern (shape, center, spread) and for striking deviations such as outliers.
3. Calculate a numerical summary to give some description of center and spread.

Here is one more step to add to this strategy:

4. Sometimes the overall pattern of a large number of observations is so regular that we can describe it by a smooth curve.

Figure 5.3 is a histogram of the Iowa Test vocabulary scores of 947 seventh-grade students. Like most histograms from national standardized tests, the histogram is symmetric, is single-peaked, and has a distinctive bell shape. In Figure 5.11, we draw a smooth curve through the tops of the histogram bars to describe the shape. The curve is an idealized description of the distribution. It gives a compact picture of the overall pattern of the data but ignores minor irregularities as well as any outliers. The curve in Figure 5.11 is a *normal curve.* A distribution whose shape is described by a normal curve is a *normal distribution.*

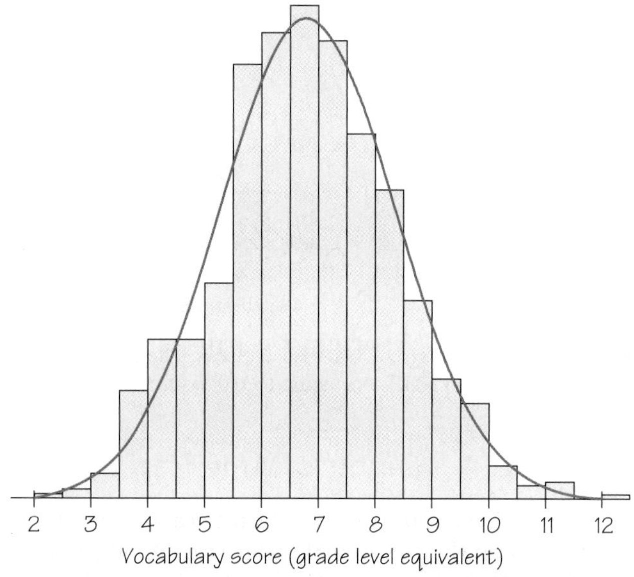

FIGURE 5.11 Histogram of the vocabulary scores of 947 seventh-grade students in Gary, Indiana. The smooth curve shows the overall shape of the distribution.

Normal Distribution DEFINITION

The *distribution* of a variable tells us what values the variable takes and how often it takes these values. A **normal distribution** is described by a *normal* curve. The area under the curve above any interval of values tells us what proportion of all values of the variable lie in that interval. The total area under the curve is exactly 1.

EXAMPLE 12 ■ From Histogram to Normal Curve

You can think of a normal curve as a smoothed-out histogram when there is symmetry and one mode. Our eyes respond to the *areas* of the bars in a histogram. The bar areas represent proportions of the observations. Figure 5.12a is a copy of Figure 5.11

with the leftmost bars shaded. The area of the shaded bars in Figure 5.12a represents the students with vocabulary scores 6.0 or lower. There are 287 such students, who make up the proportion 287/947 = 0.303 of all Gary seventh graders.

Now look at the curve drawn through the bars. In Figure 5.12b, the area under the curve to the left of 6.0 is shaded. We know that the areas of histogram bars represent proportions of all the observations, but we don't worry about the actual total area. Note that all the bars together represent 100% of the students and so we treat the total area under the normal curve as 1 = 100%. Now areas under the curve actually *are* proportions of the observations. This curve is a normal curve. The shaded area under the normal curve in Figure 5.12b is the proportion of students with score 6.0 or lower. This area turns out to be 0.293, only 0.010 away from the histogram result. You see that areas under the normal curve give quite good approximations of areas given by the histogram.

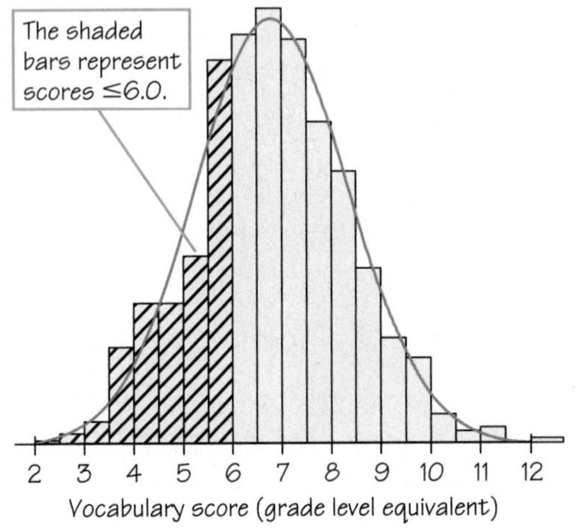

FIGURE 5.12a The proportion of scores less than or equal to 6.0 from the histogram is 0.303.

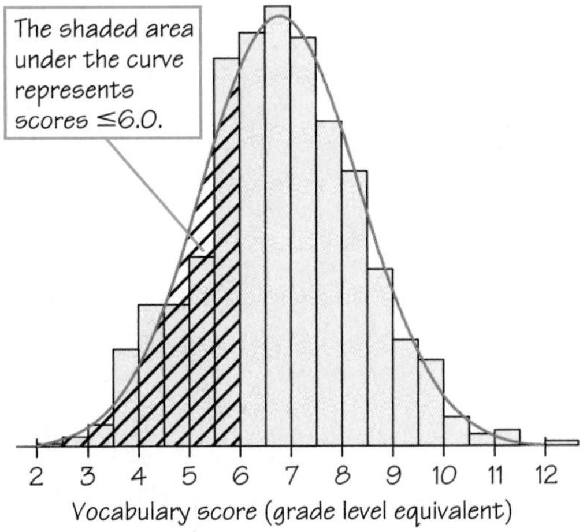

FIGURE 5.12b The proportion of scores less than or equal to 6.0 from the normal curve is 0.293.

EXAMPLE 13 ■ Heights of American Women

The normal curve is a good approximation of the real-life distribution for a variety of biological measures (height, weight, heart rate, blood pressure, and so on), when examined for a particular species and gender. Figure 5.13 shows the heights of American women between the ages 18–24. The proportion of young women who are between 60 inches (5 feet) and 65 inches tall is given by the area under the curve between 60 and 65. This area is about 0.54, so approximately 54% of these women are between 60 and 65 inches tall.

Normal distributions play a large role in statistics, but they are rather special and not at all "normal" in the everyday sense of being typical or natural. Normal curves can be specified exactly by an equation, but we will be content with pictures. Figure 5.14 shows two normal curves. All normal curves have the same overall ***shape.*** They are symmetric and bell-shaped, with tails that fall off rapidly from a central peak. The *center* of the normal curve is the center of the distribution in several senses. It is the mean of the distribution. It is also the median since half the observations (half the area under the curve) lie on each side of the center.

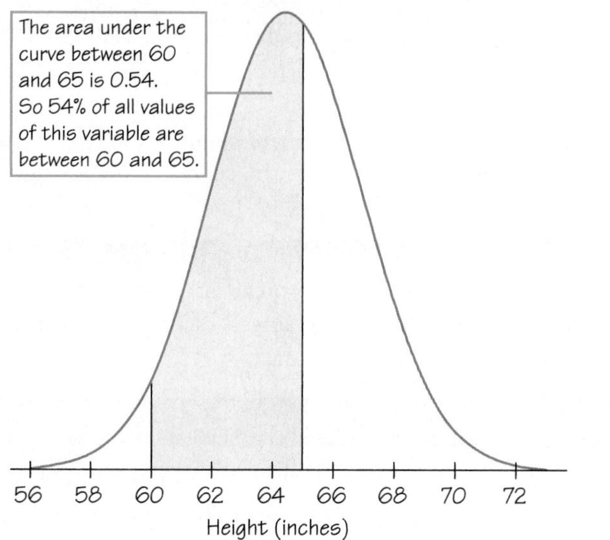
The area under the curve between 60 and 65 is 0.54. So 54% of all values of this variable are between 60 and 65.
56
58
60
62
64
66
68
70
72
Height (inches)

FIGURE 5.13 Areas under a normal curve describe a normal distribution. This normal curve describes the distribution of heights of American women.

FIGURE 5.14 Two normal curves with the same mean but different standard deviations. The standard deviation for each curve is the distance from the center (the mean) to the change-of-curvature point on one side of the center.

What about the *spread* of a normal curve? Normal curves have the special property that their spread is completely determined by a single number, the standard deviation. We have learned how to calculate the standard deviation from a set of observations. For normal distributions, the standard deviation, like the mean, can be found directly from the curve. Here's how. Imagine that you are skiing down a mountain that has the shape of a normal curve. At first, you descend at an ever-steeper angle as you go out from the peak:

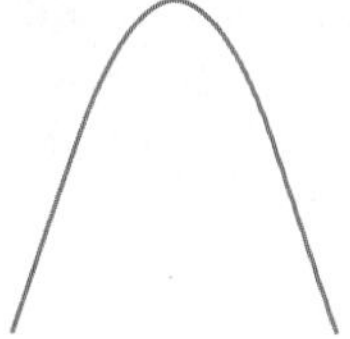

Fortunately, before you find yourself going straight down, the slope now begins to grow flatter rather than steeper as you continue downhill:

The points at which this change of curvature takes place are located one standard deviation from the mean on either side. You can feel the change as you run your finger along a normal curve, and so find the standard deviation. Try it with the two normal curves in Figure 5.14. Normal curves with the same standard deviation have exactly the same shape. Changing the mean just moves the center of the curve to a new location. Changing the standard deviation changes the spread of the curve, as Figure 5.14 shows.

Mean and Standard Deviation of Normal Distributions — DEFINITION

The shape of a normal distribution is completely determined by two numbers, the mean and the standard deviation.

The *mean* of a normal distribution is at the center of symmetry of the normal curve. The *standard deviation* is the distance from the center to the change-of-curvature points on either side.

We have often used the quartiles to indicate the spread of a distribution. Because the standard deviation completely describes the spread of any normal distribution, it tells us where the quartiles are. Here are the facts.

Quartiles of Normal Distributions — DEFINITION

The *first quartile* of any normal distribution is located about 0.67 (a bit more than 2/3) standard deviation below the mean; by symmetry, the *third quartile* is located 0.67 standard deviation above the mean.

EXAMPLE 14 ■ Heights of American Women

The distribution of heights of American women (aged 18–24) is approximately normal with mean 64.5 inches and standard deviation 2.5 inches. Figure 5.15 shows this normal curve. The quartiles are 0.67 standard deviation, or

$$(0.67)(2.5 \text{ inches}) = 1.7 \text{ inches}$$

away from the mean. The first quartile is 64.5 − 1.7, or 62.8 inches. The third quartile is 64.5 + 1.7, or 66.2 inches. The middle 50% of women's heights lie approximately between 62.8 inches and 66.2 inches. These numbers are exact for the normal distribution with mean 64.5 inches and standard deviation 2.5 inches, but only approximately true for the actual heights of the women because real-life distributions of biological measurements such as heights are only approximately normal. ■

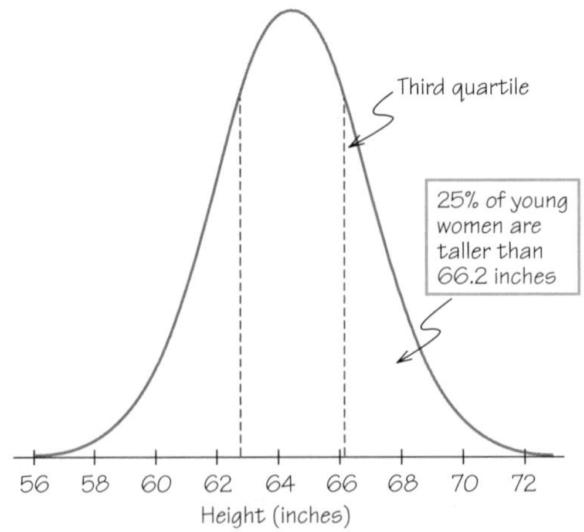

FIGURE 5.15 The quartiles of a normal distribution are located 0.67 standard deviation on either side of the mean. For this normal curve, the mean is 64.5 inches and the standard deviation is 2.5 inches.

SPOTLIGHT 5.3 Density Estimation

Smooth curves that describe the overall pattern of distributions of data are called *density curves.* Normal curves are one type of density curve. There are many other types used for different purposes. However, you don't have to call for a specific type such as the normal curves. Clever software for "density estimation" will calculate a density curve to describe any set of observations you give it.

The figure shows a strongly skewed distribution, the survival times of 72 guinea pigs in a medical experiment. Two graphs of the distribution are overlaid: a histogram and a density curve produced by software from the data. The histogram and density curve agree on the overall shape and on the "bumps" in the long right tail. The density curve shows a higher single peak as a main feature of the distribution. The histogram divides the observations near the peak between two bars, thus reducing the height of the peak. Because density estimators don't depend on dividing the data into classes, as histograms do, many statisticians prefer them when they need a picture of a distribution.

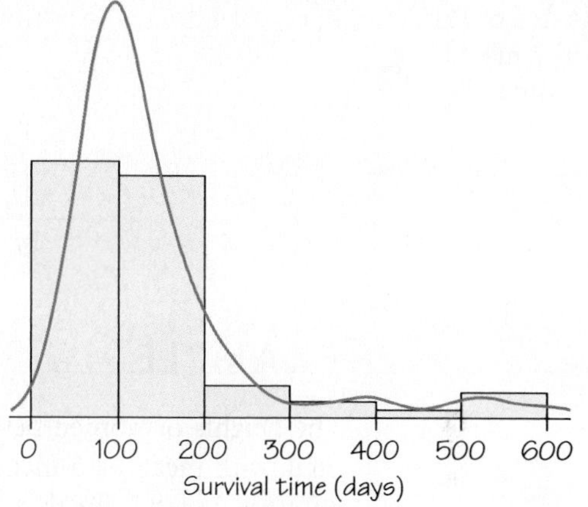

Density estimation software fits this smooth curve to data on the survival time of 72 guinea pigs.

Why are the normal distributions important in statistics? Here are two reasons. First, normal distributions are good models or approximations for some distributions of *real data.* Distributions that are often close to normal include scores on tests taken by many people (such as SAT exams and many psychological tests), repeated careful measurements of the same quantity, and characteristics of biological populations (such as heights of young women and yields of corn). Second, normal distributions are good approximations to the results of many kinds of *chance outcomes,* such as tossing a coin many times. We will return to normal curves in Chapter 8 when we study probability, the mathematics of chance. However, many sets of data do not follow a normal distribution. Most income distributions, for example, are skewed to the right and so are not normal.

5.9 The 68–95–99.7 Rule

Because any particular normal distribution is completely determined by its mean and standard deviation, it is not surprising that all normal distributions are the same in terms of what proportion of observations are any given number of standard deviations from the mean. Here is an important rule based on this fact.

The 68–95–99.7 Rule for Normal Distributions RULE

According to the **68–95–99.7 rule**, in any normal distribution:

- about 68% of the observations fall within 1 standard deviation of the mean.
- about 95% of the observations fall within 2 standard deviations of the mean.
- about 99.7% of the observations fall within 3 standard deviations of the mean.

Figure 5.16 illustrates the 68–95–99.7 rule. By remembering these three numbers, you can think about normal distributions without making detailed calculations. For more detailed information, you can use tables or software that give areas under normal curves, but the 68–95–99.7 rule is adequate for our purposes.

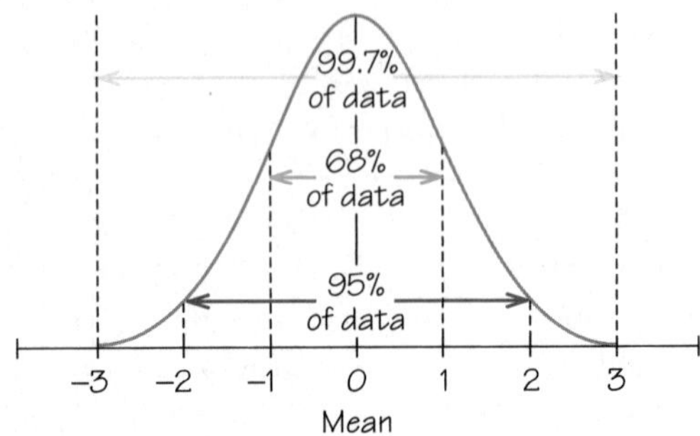

FIGURE 5.16 The 68–95–99.7 rule for normal distributions.

EXAMPLE 15 ■ Heights of American Women

The heights of women between the ages of 18 and 24 are roughly normally distributed, with mean 64.5 inches and standard deviation 2.5 inches. Two standard deviations is 5 inches for this distribution. The 95 part of the 68–95–99.7 rule says that the middle 95% of young women are between 64.5 − 5 and 64.5 + 5 inches tall, that is, between 59.5 inches and 69.5 inches. This fact is exactly true for an exactly normal distribution. It is approximately true for the heights of young women because the distribution of heights is approximately normal.

The other 5% of American women have heights outside the range from 59.5 to 69.5 inches. Because the normal distributions are symmetric, half of these women are on the tall side and half on the short side. So the tallest 2.5% of young women are taller than 69.5 inches. ■

EXAMPLE 16 ■ SAT Reasoning Test Scores

The distribution of scores on tests such as the SAT college entrance examination is close to normal. Scores on each of the three sections (math, critical reading, writing) of the SAT are adjusted so that the mean score is about $\mu = 500$ and the standard deviation is about $\sigma = 100$. This information allows us to answer many questions about SAT scores.

► *How high must a student score to fall in the top 25%?*

The third quartile is (0.67)(100) = 67 points above the mean. So scores above 567 are in the top 25%.

► *What percent of scores fall between 200 and 800?*

Scores of 200 and 800 are 3 standard deviations on either side of the mean. The 99.7 part of the 68–95–99.7 rule says that 99.7% of all scores lie in this range. (In fact, 200 and 800 are the lowest and highest scores that are reported on the SAT. The few scores higher than 800 are reported as 800.)

► *What percent of scores are above 700?*

A score of 700 is 2 standard deviations above the mean. By the 95 part of the 68–95–99.7 rule, 95% of all scores fall between 300 and 700 and 5% fall below 300 or above 700. Because normal curves are symmetric, half of this 5% are above 700. So a score above 700 places a student in the top 2.5% of test-takers.

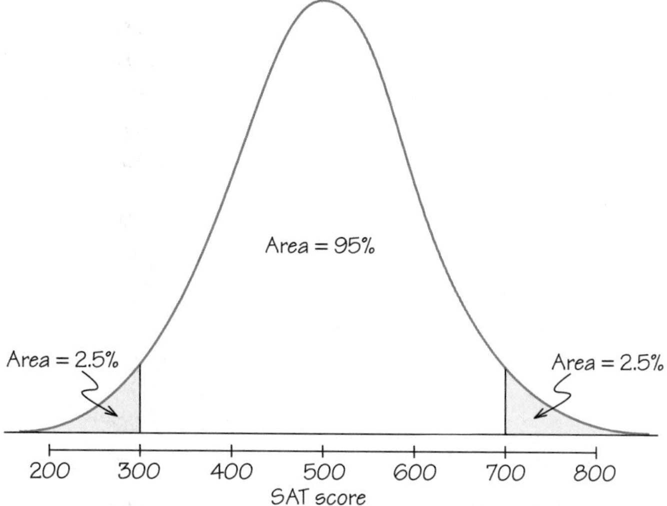

FIGURE 5.17 Using the 68–95–99.7 rule to find the percent of SAT scores that are above 700. This normal curve has mean 500 and standard deviation 100.

Sketching a normal curve with the points 1, 2, and 3 standard deviations from the mean marked can help you use the 68–95–99.7 rule. Figure 5.17 shows the distribution of SAT scores with the areas needed to find the percent of scores above 700. Note that the tails of Figure 5.17, like any bell curve, technically stretch out forever in both directions (even as the amount of far-away area becomes vanishingly small). This is another reminder that the bell-curve is a very good, but not perfect, model of reality. We know that real-life SAT subtest scores are scaled so that they do not go beyond 200 or 800.

The 68–95–99.7 rule allows you to find selected areas under a normal curve—areas for outcomes bounded by 1, 2, or 3 standard deviations away from the mean. You can use software, a graphing calculator, or the *Normal Curve* applet to find any area under a normal curve. See Applet Exercises 3 to 5 for use of the applet.

REVIEW VOCABULARY

Boxplot A graph of the five-number summary. A box spans the quartiles, with an interior line marking the median. Lines extend out from this box to the extreme high and low observations. (p. 162)

Distribution The pattern of outcomes of a variable. The distribution describes what values the variable takes and how often each value occurs. (p. 151)

Exploratory data analysis The practice of using graphs and numbers to examine data for overall patterns and special features, without necessarily seeking answers to specific questions. (p. 150)

Five-number summary A summary of a distribution that gives the smallest observation, first quartile, median, third quartile, and largest observation, in that order. (p. 162)

Histogram A graph of the distribution of outcomes (often divided into classes) for a single numerical variable. The height of each bar is the number of observations in the class of outcomes covered by the base of the bar. All classes should have the same width and each observation must fall into exactly one class. (p. 151)

Individuals The people, animals, or things described by a data set. (p. 149)

Mean The ordinary arithmetic average of a set of observations. To find the mean, add all the observations and divide the sum by the number of observations summed. (p. 158)

Median The middle of a set of ordered observations. Half the observations fall below the median and half fall above. (p. 158)
Mode The most frequently occurring value in a set of numerical observations. (p. 159)
Normal distributions A family of distributions that describe how often a variable takes its values by areas under a curve. The normal curves are symmetric and bell-shaped. A specific normal curve is completely described by giving its mean and its standard deviation. (p. 167)
Outlier A data point that falls clearly outside the overall pattern of a set of data. (p. 153)
Quartiles The first quartile (Q_1) of a distribution is the point with one quarter of the observations falling below it; the third quartile (Q_3) is the point with three quarters below it. Q_1 is the median of the lower half of the observations; Q_3 is the median of the upper half. (p. 161)
Range Measure of spread obtained by subtracting the smallest observation from the largest observation. (p. 161)
68–95–99.7 rule In any normal distribution, 68% of the observations lie within 1 standard deviation on either side of the mean, 95% lie within 2 standard deviations of the mean, and 99.7% lie within 3 standard deviations of the mean. (p. 171)
Skewed distribution A distribution in which observations on one side of the median extend notably farther from the median than do observations on the other side. In a right-skewed distribution, the larger observations extend farther to the right of the median than the smaller observations extend to the left. (p. 154)
Standard deviation A measure of the spread of a distribution about its mean as center. It is the square root of the average squared deviation of the observations from their mean. (p. 163)
Standard deviation of a normal curve The standard deviation of a normal curve is the distance from the mean to the change-of-curvature points on either side. (p. 170)
Stemplot A display of the distribution of a variable that attaches the final digits of the observations as leaves on stems made up of all but the final digit. (p. 156)
Symmetric distribution A distribution with a histogram or stemplot in which the part to the left of the median is roughly a mirror image of the part to the right of the median. (p. 154)
Variable Any characteristic of an individual. (p. 149)

✓ SKILLS CHECK

1. Here are the first lines of a professor's data set at the end of a mathematics course:

Name	Major	Points	Grade
ADVANI, SURA	COMM	397	B
BARTON, DAVID	HIST	323	C
BOAZ, JUDAH	BIOL	446	A
CHIU, SUN	PSYC	405	B
DAVIS, LAUREN	PSYC	461	A

The individuals in these data are

(a) the students.
(b) the total points.
(c) the course grades.

Figure 5.4 is a histogram of the tuition and fee charges for the 2007–2008 academic year for 55 four-year colleges in Massachusetts. Exercises 2 and 3 are based on this histogram.

2. The number of colleges with tuition and fee charges covered by the leftmost bar in the histogram is ______ .

3. The leftmost bar in the histogram covers tuition and fee charges ranging from about

(a) \$2000 to \$6000.
(b) \$3000 to \$5000.
(c) \$4000 to \$8000.

4. The distribution in Figure 5.2 is best described as ________-skewed.

5. You look at real estate ads for houses in Sarasota, Florida. There are many houses ranging from \$200,000 to \$400,000 in price. The few houses on the water, however, have prices up to \$15 million. The distribution of house prices will be

(a) skewed to the left.
(b) roughly symmetric.
(c) skewed to the right.

6. Here are the systolic blood pressures of 10 randomly chosen adults:

147	141	120	124	127
132	98	112	120	128

In a stemplot of these scores, the largest stem is _____ .

7. For Figure 5.5, interpret the meaning of 10|7.

(a) 10 states have 7% Hispanic population.
(b) 7 states have 10% Hispanic population.
(c) 1 state has 10.7% Hispanic population.

8. The mean blood pressure of the 10 adults in Exercise 6 is _________ .

9. The median of the blood pressures in Exercise 6 is

(a) 127.
(b) 125.5.
(c) 124.9.

10. If a single-peaked distribution is skewed to the right, the median is generally to the __________ of the mean.

11. The mode of the 10 blood pressures in Exercise 6 is
(a) 147.
(b) 120.
(c) 2.

12. Between the first quartile and the third quartile lie ________ percent of the observations in a distribution.

13. Which of these is *not* in a five-number summary?
(a) Median
(b) Minimum
(c) Mean

14. The five-number summary of the 10 blood pressures in Exercise 6 is _____. (Remember to list the 5 numbers in increasing order).

15. The standard deviation of the 10 blood pressures in Exercise 6 (use your calculator) is
(a) 13.23.
(b) 13.95.
(c) 194.6.

16. You have data on the weights (measured in grams) of 5 crackers. The correct units for the standard deviation of these weights are: ________________ .

17. What are all the values that a standard deviation s can possibly take?
(a) $0 \leq s$
(b) $0 \leq s \leq 1$
(c) $-1 \leq s \leq 1$

18. To completely specify the shape of a normal distribution, you must give its mean and its ________ .

19. The scale of scores on an IQ test is approximately normal with mean 100 and standard deviation 15. The organization MENSA, which calls itself "the high IQ society," requires an IQ score of 130 or higher for membership. What percent of adults would qualify for membership?
(a) 95%
(b) 5%
(c) 2.5%

20. The length of human pregnancies from conception to birth varies according to a distribution that is approximately normal with mean 266 days and standard deviation 16 days. We can expect that about ________ percent of all completed pregnancies are between 234 and 298 days.

CHAPTER 5 EXERCISES

■ Challenge ◆ Discussion

Make and Model	Vehicle Type	Transmission Type	Number of Cylinders	City mpg	Highway mpg
Mazda MX-5	Two-seater	Manual	4	22	27
Toyota Yaris	Subcompact	Automatic	4	29	35
Bentley Azure	Compact	Automatic	12	9	15
Audi S4 Avant	Small Station Wagon	Manual	8	13	20

Some exercises require use of a calculator (or software or Internet applet) that will find mean and standard deviation from keyed-in data.

1. Above is a small part of a data set that describes the fuel economy (in miles per gallon) of year 2008 model motor vehicles:
(a) What are the individuals in this data set?
(b) For each individual, what variables are given? For which of these variables would a histogram be helpful? (That is, which variables do not yield categorical data)?

5.1 Displaying Distributions: Histograms

5.2 Interpreting Histograms

◆ 2. Figure 5.18 is a histogram of the lengths of words used in Shakespeare's plays. Because there are so many words in the plays, the vertical axis of the graph is the percent that are of each length, rather than the count. What is the overall shape of this distribution? What does this shape say about word lengths in Shakespeare? Do you expect other authors to have word-length distributions of the same general shape? Why?

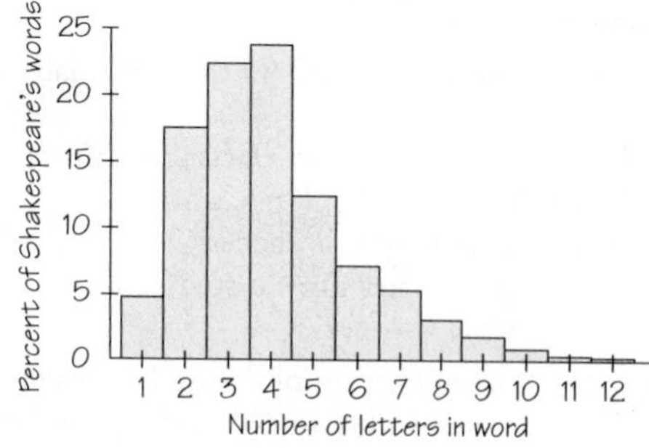

FIGURE 5.18 Histogram of the lengths of words used in Shakespeare's plays, for Exercise 2.

TABLE 5.3 Carbon Dioxide Emissions, Metric Tons per Person

Country	CO_2	Country	CO_2	Country	CO_2	Country	CO_2
Algeria	2.3	Germany	10.0	Mexico	3.7	South Africa	8.1
Argentina	3.9	Ghana	0.2	Morocco	1.0	Spain	6.8
Australia	17.0	India	0.9	Myanmar	0.2	Sudan	0.2
Bangladesh	0.2	Indonesia	1.2	Nepal	0.1	Tanzania	0.1
Brazil	1.8	Iran	3.8	Nigeria	0.3	Thailand	2.5
Canada	16.0	Iraq	3.6	Pakistan	0.7	Turkey	2.8
China	2.5	Italy	7.3	Peru	0.8	Ukraine	7.6
Colombia	1.4	Japan	9.1	Philippines	0.9	United Kingdom	9.0
Congo	0.0	Kenya	0.3	Poland	8.0	United States	19.9
Egypt	1.7	Korea, North	9.7	Romania	3.9	Uzbekistan	4.8
Ethiopia	0.0	Korea, South	8.8	Russia	10.2	Venezuela	5.1
France	6.1	Malaysia	4.6	Saudi Arabia	11.0	Vietnam	0.5

◆ **3.** Suppose that you and your friends emptied your pockets of coins and recorded the year marked on each coin. Would you expect the histogram for the distribution of dates to be skewed to the left or right? Explain your answer and make a sketch of this histogram.

4. Make a histogram of the city gas mileages of the midsized cars in Table 5.2 on page 157. Use classes with width 5 mpg. Do you prefer the histogram or the representation in Figure 5.6 of the same data? Why?

5. Burning fuels in power plants or motor vehicles emits carbon dioxide (CO_2), which contributes to global warming. Table 5.3 displays CO_2 emissions per person from countries with population at least 20 million.

(a) Why do you think we choose to measure emissions per person rather than total CO_2 emissions for each country?

(b) Display the data of Table 5.3 in a histogram. Describe the shape, center, and spread of the distribution. Which countries appear to be outliers?

■ **6.** A survey of a large college class asked the following questions:

1. Are you female or male? (In the data, male = 0, female = 1.)
2. Are you right-handed or left-handed? (In the data, right = 0, left = 1.)
3. What is your height, in inches?
4. How many minutes do you study on a typical weeknight?

Figure 5.19 shows histograms of the student responses, in scrambled order and without scale markings. Which histogram goes with each variable? Explain your reasoning.

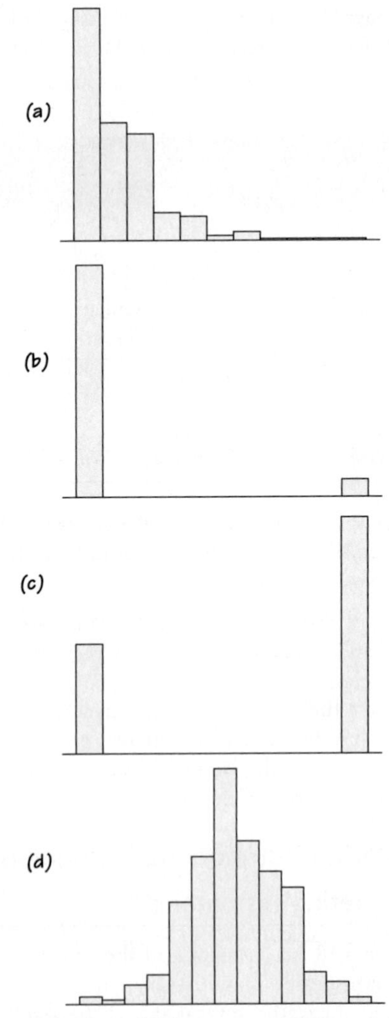

FIGURE 5.19 Match each histogram with its variable, for Exercise 6.

5.3 Displaying Distributions: Stemplots

7. The population of the United States is aging, though less rapidly than in other developed countries. Figure 5.20 is a stemplot of the percents of residents aged 65 and over in the 50 states, according to the 2000 census. The stems are whole percents and the leaves are tenths of a percent.

```
 5 | 7
 6 |
 7 |
 8 | 5
 9 | 679
10 | 6
11 | 02233677
12 | 0011113445789
13 | 00012233345568
14 | 034579
15 | 36
16 |
17 | 6
```

FIGURE 5.20 Stemplot of the percentages of residents aged 65 and over in the 50 states, for Exercise 7.

(a) There are two outliers: Alaska has the lowest percent of older residents, and Florida has the highest. What are the percents for these two states?
(b) Ignoring Alaska and Florida, describe the shape, center, and spread of this distribution.

8. People with diabetes must monitor and control their blood glucose level. The goal is to maintain "fasting plasma glucose" between about 90 and 130 milligrams per deciliter (mg/dl). Here are the fasting plasma glucose levels for 18 diabetics enrolled in a diabetes control class, five months after the end of the class:

78	103	141	148	172	255
95	112	145	153	172	271
96	134	147	158	200	359

Round these values to the nearest 10 and then drop the zero. For example, 141 rounds to 14 and 158 rounds to 16. Make a stemplot of the rounded data. Describe the main features of the distribution. Are there outliers? How well is the group as a whole achieving the goal for controlling glucose levels?

9. The Survey of Study Habits and Attitudes (SSHA) is a psychological test that evaluates college students' motivation, study habits, and attitudes toward school. A private college gives the SSHA to 18 of its incoming first-year women students. Their scores are (sorted):

101	115	129	140	154	165
103	126	137	148	154	178
109	126	137	152	165	200

Make a stemplot of these data. The overall shape of the distribution is irregular, as often happens when only a few observations are available. Are there any outliers? About where is the center of the distribution (the score with half the scores above it and half below)? What is the spread of the scores (ignoring any outliers)?

10. In 1798 the English scientist Henry Cavendish measured the density of the earth in a careful experiment with a torsion balance. In sorted order here are his 29 measurements of the same quantity (the density of the earth relative to that of water) made with the same instrument. [S. M. Stigler, Do robust estimators work with real data? *Annals of Statistics*, 5 (1977): 1055–1078.]

4.88	5.29	5.36	5.47	5.58	5.68
5.07	5.29	5.39	5.50	5.61	5.75
5.10	5.30	5.42	5.53	5.62	5.79
5.26	5.34	5.44	5.55	5.63	5.85
5.27	5.34	5.46	5.57	5.65	

Make a stemplot of the data. Describe the distribution: Is it approximately symmetric or distinctly skewed? Are there gaps or outliers?

11. Here is a stemplot for percentage of live births to unmarried mothers for each state in the United States in 2006. (*Source*: Centers for Disease Control Web site)

```
1 | 9
2 | 4
2 | 89
3 | 222223344444
3 | 5556677778888999
4 | 000111111224
4 | 5669
5 | 13
```

(a) Explain how and why there are repeated stems.
(b) Describe the shape of the distribution.

5.4 Describing Center: Mean and Median

■ **12.** In Malay, the expression for the *mean* is *sama rata*, which roughly translates as "same level." To understand this cultural and conceptual connection, take some poker chips (or other equal-sized, stackable objects) and make stacks with 3, 7 and 8 chips. Redistribute chips among the stacks until they are at the same level and explain how this relates to the mean. (*Optional extension*: How might such redistribution enter into discussions of social justice?)

13. Refer to the data and the stemplot from Exercise 9:

101	115	129	140	154	165
103	126	137	148	154	178
109	126	137	152	165	200

(a) Find the mean score from the formula for the mean. That is, add the 18 scores, record the sum, and divide by 18.
(b) Your stemplot of the scores suggests that the score 200 is an outlier. Use your calculator to find the mean for the 17 observations that remain when you drop the outlier. How does the outlier change the mean?

14. The Major League Baseball career and single-season home run records are held by Barry Bonds of the San Francisco Giants. Here are Bonds's home run totals from 1986 (his first year) through 2007:

16	25	24	19	33	25	34	46
37	33	42	40	37	34	49	73
46	45	45	5	26	28		

(a) Make a stemplot of the data. Are there any outliers?
(b) Find his career mean and median number of home runs. How do these change when you drop 73? What general fact about the mean and median does your result illustrate?

(Lucy Nicholson/Reuters/Corbis.)

◆ **15.** The distribution of income in the United States is skewed to the right. According to a Census Bureau report, the mean and median incomes of American households were \$48,201 and \$66,570 in 2006. Which of these numbers is the mean and which is the median? Explain your reasoning.

◆ **16.** Which team is #1? In addition to polls of coaches and journalists, rankings from six computer programs, which have various ways to value factors such as the quality of the opponent played, determine the Bowl Championship Series (BCS) Standings in major college football.
(a) At the end of the 2007 regular season, Hawaii (the only undefeated team) received these computer rankings: 12th, 8th, 14th, 10th, 8th, 13th. The BCS formula throws out the high and low of the six computer rankings and uses the mean of the remaining four ranks. Find this mean.
(b) Why do you think the high and low values are excluded from the mean? Is your reason connected to why the median is sometimes preferred to the mean?

17. Make up an example of a small set of data for which the mean lies in the top 25% of the observations.

■ **18.** According to the 1.5 × *IQR* rule explained in Exercise 29, which countries in Table 5.3 are suspected outliers? Based on your histogram (Exercise 5), do you agree with the rule's suggestions about which countries are and are not outliers?

5.5 Describing Spread: The Quartiles
5.6 The Five-Number Summary and Boxplots

19. The stemplot in Figure 5.20 (p. 177) displays the distribution of the percents of residents aged 65 and over in the 50 states. Stemplots help you find the five-number summary because they arrange the observations in increasing order. Give the five-number summary of this distribution.

20. In chronological order, here are the percents of the popular vote won by each successful candidate in the last 15 presidential elections, starting with 1948:

49.6	55.1	57.4	49.7	61.1
43.4	60.7	50.1	50.7	58.8
53.9	43.2	49.2	47.9	51.2

(a) Make a stemplot of the winners' percents.
(b) What is the median percent of the vote won by the successful candidate in presidential elections?
(c) Call an election a landslide if the winner's percent falls at or above the third quartile. Find the third quartile. Which elections were landslides?
(d) Find the range.

21. Figure 5.4 is a histogram of the tuition and fees charged by the 55 four-year colleges in the state of Massachusetts. Here are those charges (in dollars), arranged in increasing order. [Data for 2007–2008, from the College Board Web site, www.collegeboard.com.]

5799	5864	5992	6034	6124
6168	6210	8595	8732	8840
9924	16080	17750	20000	21330
21850	22073	22500	22950	23600
23755	24075	24250	24617	25748
25755	25850	25942	25990	26080
26250	27485	27497	28302	28440
29810	31899	32865	32896	34186
34830	34986	34994	34998	35142
35418	35670	35674	35702	35940
36232	36550	36645	36690	36700

Find the five-number summary and make a boxplot. What distinctive feature of the histogram do these summaries miss? Remember that numerical summaries are not a substitute for looking at the data.

22. Find the five-number summary of Cavendish's measurements of the density of the earth in Exercise 10. How is the symmetry of the distribution reflected in the five-number summary?

23. Table 5.3 gives carbon dioxide (CO_2) emissions per person for countries with population at least 20 million. The distribution is strongly skewed to the right. The United States and several other countries appear to be high outliers. Give the five-number summary. Explain why this summary suggests that the distribution is right-skewed.

24. Find the five-number summary of the data from Exercise 8.

◆ **25.** Figure 5.21 shows boxplots of the incomes of a large sample of people who have a high school diploma but no further education and another large group of people with a bachelor's degree but no higher degree. The data come from a Census Bureau survey, so that they represent all people aged 25 to 64 in the United States. Because there are a few extremely high incomes, the boxplot leaves out the highest 5% in each group. Based on the plot, compare the distributions of income for these two levels of education. Comment on both center and spread.

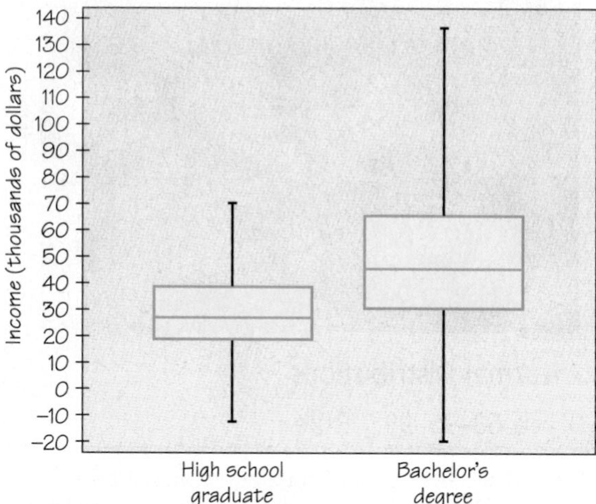

FIGURE 5.21 Boxplots comparing the incomes (in thousands of dollars) of people aged 25 to 64 years who worked full time for two levels of education. Because the highest incomes in any large group are very high indeed, the plot omits the top 5% of incomes in each group.

26. The data for Figure 5.21 include the incomes of 14,959 people whose highest level of education is a bachelor's degree.

(a) What is the position of the median in the ordered list of incomes (1 to 14,959)? From the boxplot, about what is the median income of people with a bachelor's degree?

(b) What is the position of the first and third quartiles in the ordered list of incomes for these people? About what are the numerical values of Q_1 and Q_3?

■ **27.** How much oil the wells in a given field will ultimately produce is key information in deciding whether to drill more wells. Here are the estimated total amounts of oil recovered from 64 wells in the Devonian Richmond Dolomite area of the Michigan basin, in thousands of barrels. [J. Marcus Jobe and Hutch Jobe, A statistical approach for additional infill development, *Energy Exploration and Exploitation*, 18 (2000): 89–103.]

2.0	18.5	34.6	47.6	69.5
2.5	20.1	34.6	49.4	69.8
3.0	21.3	35.1	50.4	79.5
7.1	21.7	36.6	51.9	81.1
10.1	24.9	37.0	53.2	82.2
10.3	26.9	37.7	54.2	92.2
12.0	28.3	37.9	56.4	97.7
12.1	29.1	38.6	57.4	103.1
12.9	30.5	42.7	58.8	118.2
14.7	31.4	43.4	61.4	156.5
14.8	32.5	44.5	63.1	196.0
17.6	32.9	44.9	64.9	204.9
18.0	33.7	46.4	65.6	

(a) Make a histogram and describe its main features.

(b) Find the mean and median of the amounts recovered. Explain how the relationship between the mean and the median reflects the shape of the distribution.

(c) Give the five-number summary and explain briefly how it reflects the shape of the distribution.

■ **28.** Look at the histogram of lengths of words in Shakespeare's plays, Figure 5.18. The heights of the bars tell us what percent of words have each length. (Analysis of writing tendencies can help determine authorship of a newly-discovered manuscript.) The median length is the middle, the length with half of all words shorter and half longer. What is the median length of words used by Shakespeare? Similarly, what are the quartiles? Give the five-number summary for Shakespeare's word lengths.

■ **29.** A common criterion for identifying an outlier in a set of data is if an observation falls more than 1.5 × *IQR* above the third quartile or below the first quartile. (IQR stands for the interquartile range, which is the difference between the quartiles: $Q_3 - Q_1$.)

So which states are suspected outliers in the distribution of percent of Hispanics among adult residents, Table 5.1?

5.7 Describing Spread: The Standard Deviation

30. Do you think the standard deviation of the tuition and fees of the public colleges in Massachusetts is likely to be bigger or smaller than the standard deviation for the private colleges? Why?

31. Many standard statistical methods are intended for use with distributions that are symmetric and have no outliers. These methods start with the mean and standard deviation, $\bar{x}$ and s. An example of scientific data for which standard methods should work well are Cavendish's measurements of the density of the earth in Exercise 10.

(a) Summarize this data set by giving $\bar{x}$ and s.

(b) Find the median. Is the median quite close to the mean, as we expect it to be for symmetric distributions?

32. The level of various substances in the blood influences our health. Here are measurements of the level of phosphate in the blood of a patient, in milligrams of phosphate per deciliter of blood, made on six consecutive visits to a clinic.

5.6   5.2   4.6   4.9   5.7   6.4

(a) Find the mean.

(b) Find the standard deviation.

33. The mean $\bar{x}$ and standard deviation s measure center and spread but are not a complete description of a distribution. Data sets with different shapes can have the same mean and standard deviation. To demonstrate this fact, use your calculator to find $\bar{x}$ and s for these two small data sets. Then make a stemplot of each and comment on the shape of each distribution.

Data A:	9.14	8.14	8.74	8.77
	9.26	8.10	6.13	3.10
	9.13	7.26	4.74	
Data B:	7.46	6.77	12.74	7.11
	7.81	8.84	6.08	5.39
	8.15	6.42	5.73	

34. Your data consist of observations on the age of several subjects (measured in years) and the reaction times of these subjects (measured in seconds). In what units are each of the following descriptive statistics measured?

(a) The mean age of the subjects
(b) The standard deviation of the subjects' reaction times
(c) The variance of the subjects' reaction times
(d) The median age of the subjects

■ **35.** This is a standard deviation contest! You must choose four numbers from the whole numbers 0 to 10, with repeats allowed.

(a) Choose four numbers that have the smallest possible standard deviation.
(b) Choose four numbers that have the largest possible standard deviation.
(c) Is more than one choice possible in part (a)? Explain.
(d) Is more than one choice possible in part (b)? Explain.

■ **36.** "Conservationists have despaired over destruction of tropical rainforest by logging, clearing, and burning." These words begin a report on a statistical study of the effects of logging in Borneo. [C. H. Cannon, D. R. Peart, and M. Leighton, Tree species diversity in commercially logged Bornean rainforest, *Science,* 281 (1998): 1366–1367.] Researchers compared forest plots that had never been logged (Group 1) with similar plots nearby that had been logged 1 year earlier (Group 2) and 8 years earlier (Group 3). All plots were 0.1 hectare in area. Here are the counts of trees for plots in each group, courtesy of Charles Cannon:

Group 1:	27	22	29	21	19	33
	16	20	24	27	28	19
Group 2:	12	12	15	9	20	18
	17	14	14	2	17	19
Group 3:	18	4	22	15	18	
	19	22	12	12		

Give a complete comparison of the three distributions, using both graphs and numerical summaries. To what extent has logging affected the count of trees? The researchers used an analysis based on $\bar{x}$ and s. Explain why this is reasonably well justified.

(Edward Parker/Alamy.)

5.8 Normal Distributions

5.9 The 68–95–99.7 Rule

37. Some teachers graded "on a curve" based on the belief that classroom test scores are normally distributed. One way of doing this is to assign a "C" to all scores within 1 standard deviation of the mean. Then, the teacher would assign a "B" to all scores between 1 and 2 standard deviations above the mean, an "A" to all scores more than 2 standard deviations above the mean, and use symmetry to define the regions for "D" and "F" on the left side of the normal curve. If 200 students take an exam, determine the number of students who would receive a B.

38. The length of human pregnancies from conception to birth varies according to a distribution that is approximately normal, with mean 266 days and standard deviation 16 days. Draw a normal curve for this distribution on which the mean and standard deviation are correctly located. (*Hint:* First draw the curve, then mark the axis.)

39. Figure 5.22 shows a smooth curve used to describe a distribution that is not symmetric. The mean and median do not coincide. Which of the points marked is the mean of the distribution, and which is the median? Explain your answer.

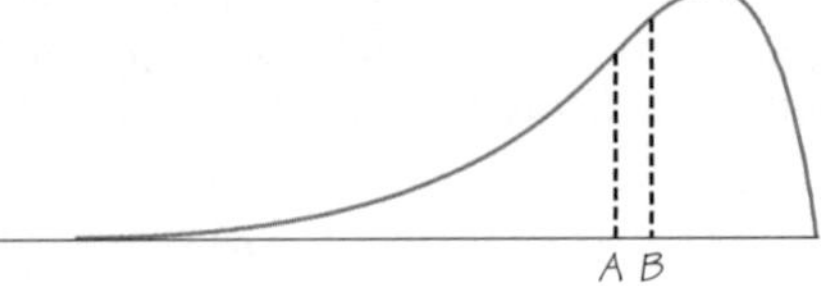

FIGURE 5.22 A curve describing a skewed distribution, for Exercise 39.

40. Sketch a smooth curve that describes a distribution that is symmetric but has two peaks (that is, two strong clusters of observations).

41. Bigger animals tend to carry their young longer before birth. The length of horse pregnancies from conception to birth varies according to a roughly normal distribution, with mean 336 days and standard deviation 3 days. Use the 68–95–99.7 rule to answer the following questions.

(a) Almost all (99.7%) horse pregnancies fall in what range of lengths?
(b) What percent of horse pregnancies are longer than 339 days?

42. Scores on the three-section SAT Reasoning college entrance test for the class of 2007 were roughly normal, with mean 1511 and standard deviation 194.

(a) What was the range of the middle 68% of SAT scores?
(b) How high must a student score to be in the top 2.5% of SAT scores?

43. What are the quartiles of SAT Reasoning scores, according to the distribution in Exercise 42?

44. The Wechsler Adult Intelligence Scale (WAIS) is the most common "IQ test." The scale of scores is set separately for each age group and is approximately normal, with mean 100 and standard deviation 15. People with WAIS scores below 70 are considered mentally retarded for purposes of applying for Social Security disability benefits. By this criterion, what percent of adults are retarded?

45. The yearly rate of return on the Standard & Poor's 500 (an index of 500 large-cap corporations) is approximately normal. From January 1956 through September 2007, the S&P 500 had a mean yearly return of 10.51%, with a standard deviation of about 15.51%. Take this normal distribution to be the distribution of yearly returns over a long period.

(a) In what range do the middle 95% of all yearly returns lie?
(b) Stocks can go down as well as up. What are the worst 2.5% of annual returns?

46. What is the range of the middle 50% of annual returns on stocks, according to the distribution given in the previous exercise? (*Hint:* What two numbers mark off the middle 50% of any distribution?)

47. The concentration of the active ingredient in capsules of a prescription painkiller varies according to a normal distribution with $\mu = 10\%$ and $\sigma = 0.2\%$.

(a) What is the median concentration? Explain your answer.
(b) What range of concentrations covers the middle 95% of all the capsules?
(c) What range covers the middle half of all capsules?

48. Answer the following questions for the painkiller in Exercise 47.

(a) What percent of all capsules have a concentration of active ingredient higher than 10.4%?
(b) What percent have a concentration higher than 10.6%?

49. One reason that normal distributions are important is that they describe how the results of an opinion poll would vary if the poll were repeated many times. About 40% of adult Americans say they are afraid to go out at night because of crime. Take many randomly chosen samples of 1050 people. The proportions of people in these samples who stay home for fear of crime will follow the normal distribution with mean 0.4 and standard deviation 0.015. Use this fact and the 68–95–99.7 rule to answer these questions.

(a) In many samples, what percent of samples give results above 0.4? Above 0.43?
(b) In a large number of samples, what range contains the central 95% of proportions of people who stay home because of crime?

■ **50.** You can compare observations from different normal distributions if you measure in standard deviations away from the mean. Scores expressed in standard deviation units are called *standard scores* (or *z-scores*).

(a) Scores on the ACT college entrance exam in a recent year were roughly normal, with mean 21.2 and standard deviation 4.8. Jermaine scores 27 on the ACT. Express his score in standard deviation units by calculating

$$\text{standard score} = \frac{\text{score} - \text{mean}}{\text{standard deviation}}$$

(b) Scores on the SAT Reasoning college entrance exam in the same year were roughly normal, with mean 1511 and standard deviation 194. Tonya scores 1718 on the SAT. What is her standard score?
(c) Assuming that the ACT and the SAT measure the same thing, did Jermaine or Tonya have the higher score?

Chapter Review

Different varieties of the tropical flower *Heliconia* are fertilized by different species of hummingbirds. Over time, the lengths of the flowers and the form of the hummingbirds' beaks have evolved to match each other. Here are data on the lengths in millimeters of two varieties of these flowers on the island of Dominica.

***H. caribaea* Red**

37.40	38.07	38.87	40.66	41.93
37.78	38.10	39.16	41.47	42.01
37.87	38.20	39.63	41.69	42.18
37.97	38.23	39.78	41.90	43.09
38.01	38.79	40.57		

H. caribaea Yellow

34.57	35.45	36.03	36.66	37.02
34.63	35.68	36.11	36.78	37.10
35.17	36.03	36.52	36.82	38.13

SOURCE: Thanks to Ethan J. Temeles of Amherst College for providing the data. His work is described in Ethan J. Temeles and W. John Kress, Adaptation in a plant-hummingbird association, *Science*, 300 (2003): 630–633.

Exercises 51 to 55 use these data.

51. Make stemplots of the lengths of each of the two varieties (red and yellow). Briefly describe the overall shape of the two distributions.

52. Find the five-number summaries of the two distributions of flower lengths. Make side-by-side boxplots to give a quick picture that compares the two distributions.

53. The biologists who collected the flower length data compared the two *Heliconia* varieties using statistical methods based on the mean and standard deviation. Find $\bar{x}$ and s for each variety. Based on your stemplots in Exercise 51, which distribution is more suitable for use of $\bar{x}$ and s as summaries? Why?

54. Your stemplot in Exercise 51 suggests that the distribution of lengths of yellow *Heliconia* flowers is roughly normal. Suppose that the distribution is exactly normal. Use the mean and standard deviation you found in Exercise 53 as the μ and σ of the distribution.

(a) What range of lengths covers the middle 50% of yellow flowers?

(b) What range of lengths covers the middle 95% of yellow flowers?

■ **55.** Continue to work with the normal distribution of lengths of yellow flowers from the previous exercise. The shortest red flower was 37.4 millimeters long. Using the 68–95–99.7 rule and the location of the quartiles in normal distributions, what can you say about what percent of yellow flowers that are longer than 37.4 millimeters?

56. By hand, find the standard deviation of these five numbers: 0, 1, 3, 4, 12. Use the approach in the standard deviation definition box on page 164 and Figure 5.9.

57. If every number in a data set is increased by 10, which of these will increase: range, standard deviation, mode, mean, median?

APPLET EXERCISES

To do these exercises, go to www.whfreeman.com/fapp8e.

1. The *Mean and Median* applet allows you to place observations on a line and see their mean and median visually. Place two observations on the line, by clicking below it. Why does only one arrow appear?

2. In the *Mean and Median* applet, place three observations on the line by clicking below it, two close together near the center of the line and one somewhat to the right of these two. Pull the single rightmost observation out to the right. (Place the cursor on the point, hold down a mouse button, and drag the point.) How does the mean behave? How does the median behave? Explain briefly why each measure acts as it does.

3. In Example 16 we used the fact that SAT scores are close to normal and are adjusted so that the mean is close to 500 and the standard deviation is close to 100. (Actual scores in a particular year have slightly different mean and standard deviation.) Use the *Normal Curve* applet with $\mu = 500$ and $\sigma = 100$ to answer these questions:

(a) What proportion of SAT scores are above 640?

(b) What proportion of SAT scores are between 420 and 640? (If you drag one flag across the other, the applet shows the area between the flags.)

4. Because Web browsers have limited resolution, the *Normal Curve* applet can't always get exactly the values you want. Use the applet to come close to exact answers to these questions:

(a) How high must an SAT score be to fall in the top 10% of all scores?

(b) How high must an SAT score be to fall in the top 1% of all scores?

5. The 68–95–99.7 rule for normal distributions is a useful approximation. You can use the *Normal Curve* applet to see how accurate the rule is. Drag one flag across the other so that the applet shows the area under the curve between the two flags.

(a) Place the flags 1 standard deviation on either side of the mean. What is the area between these two values? What does the 68–95–99.7 rule say this area is?

(b) Repeat for locations 2 and 3 standard deviations on either side of the mean. Again compare the 68–95–99.7 rule with the area given by the applet.

WRITING PROJECTS

1. Many social issues involve data and interpreting data. For example, income inequality (roughly speaking, the gap in income between people toward the top of the income scale and people toward the bottom) has increased in the past few decades. A good place to find data is on the Web site of the Census Bureau, www.census.gov. Click on "Income" and look for the latest report on income in the United States. Select a few facts from this detailed collection of income data to describe the extent of income inequality. Write a few paragraphs based on these facts.

2. Let's produce some data and describe them in order to gain insight into chance behavior. The mathematics of chance is the topic of Chapter 8, but for now we will concentrate on data rather than math. You need two things: a standard six-sided die (raid your Monopoly game) and a thumbtack with a rounded back (like a satellite dish). Toss the thumbtack 100 times (to speed things up, you could do 10 tosses of 10 tacks each) and record each outcome (pointing straight up or angled down). Also, toss the die 180 times and record each outcome (1, 2, 3, 4, 5, or 6). Use graphs and numbers to describe each set of results. Is the die roughly balanced, so that all six outcomes come up about equally? What about the thumbtack: Is point up or point down much more common?

SUGGESTED READINGS

CLEVELAND, WILLIAM S. *The Elements of Graphing Data,* rev. ed., Hobart Press, Summit, N.J., 1994. A careful study of the most effective elementary ways to present data graphically, with much sound advice on improving simple graphs.

LESSER, LAWRENCE M. Critical values and transforming data: Teaching statistics with social justice, *Journal of Statistics Education* (2007): www.amstat.org/publications/jse/v15n1/lesser.html. Article filled with resources for finding social justice data to expand upon Writing Project 1.

MOORE, DAVID S. *The Basic Practice of Statistics,* 3rd ed., W. H. Freeman, New York, 2004. This text is a natural next step for more detail on all the material in Part II at about the same mathematical level. The first three chapters provide a more extensive treatment of the material of Chapter 5.

ROSSMAN, ALLAN J., and BETH L. CHANCE. *Workshop Statistics: Discovery with Data,* 3rd ed., Springer, New York, 2008. A different approach to basic data analysis, using hands-on activities. There are several versions, keyed to graphing calculators and to several different software packages.

SUGGESTED WEB SITES

The Web site of the U.S. Census Bureau, www.census.gov, is a prime source of information on many topics. The latest estimates for the populations of the United States and the world are on the home page, updated regularly. Data appear under many headings, not only as numbers but as maps. Try clicking on "American Fact Finder," then on "Maps and Geography," then on "Reference Maps." Enter your zip code, then zoom in on your street. The 1000 tables in the *Statistical Abstract of the United States* are also available. If you need data for a report, this is the place to start. Canadians can find similar help at the Web site of Statistics Canada: www.statcan.ca.

Interested in data about schools, colleges, and students? The National Center for Education Statistics, nces.ed.gov, is the place to look. Go to the "What's New" section. There are useful statistics applets at www.shodor.org/interactivate/activities/.

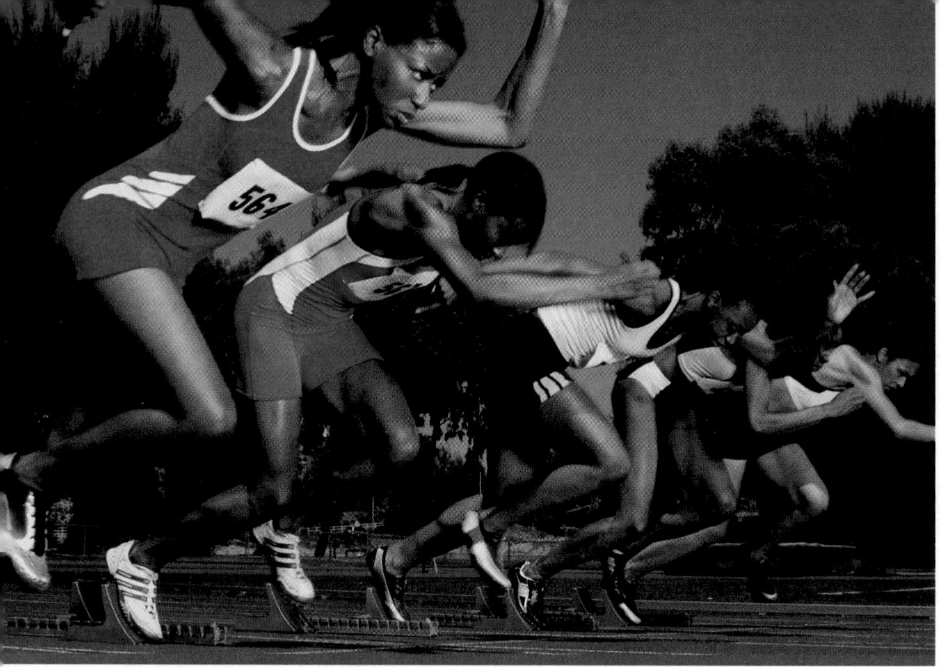

CHAPTER 6

Exploring Data: Relationships

A medical study finds that short women are more likely to have heart attacks than women of average height, while tall women have fewer heart attacks. An insurance group reports that heavier cars have fewer accident deaths per 100,000 vehicles registered than do lighter cars. These and many other statistical studies look at the *relationship between two variables*. To understand such a relationship, we must often examine other variables as well. To conclude that shorter women have higher risk from heart attacks, for example, the researchers had to eliminate the effect of other variables such as weight, diet and exercise habits. Our topic in this chapter is relationships between variables.

To study the relationship between two variables, we measure both variables on the *same individuals*. If we measure both the height and the weight of each of a large group of people, we know which height goes with each weight. These data allow us to study the connection between height and weight. A list of heights and a separate list of weights, two sets of single-variable data, do not show the connection between the two variables.

Height and weight are connected: Taller people also tend to be heavier. Neither height nor weight explains or causes the other. They go together in describing bigger or smaller people. Smoking and life expectancy are also connected, and in this case we think that smoking does explain or influence life expectancy: People who smoke more cigarettes per day tend not to live as long as those who smoke fewer. So we call smoking an **explanatory variable** and life expectancy a **response variable**.

Response Variable DEFINITION

A **response variable** measures an outcome or result of a study.

Explanatory Variable DEFINITION

An **explanatory variable** is a variable that we think explains or causes changes in the response variables.

6.1 Displaying Relationships: Scatterplot

The most useful graph for displaying the relationship (whether it fits a trend perfectly or not) between two numerical variables is a **scatterplot**.

Scatterplot DEFINITION

A **scatterplot** shows the relationship between two numerical variables measured on the same individuals. The values of one variable appear on the horizontal axis, and the values of the other variable appear on the vertical axis. Each individual in the data appears as the point in the plot fixed by the values of both variables for that individual.

Always plot the explanatory variable, if there is one, on the horizontal axis (the x-axis) of a scatterplot. As a reminder, we usually call the explanatory variable x and the response variable y. If the variables don't naturally fall into "explanatory" and "response," either variable can go on the horizontal axis.

EXAMPLE 1 ■ Beer and Blood Alcohol

How well does the number of beers a student drinks predict his or her blood alcohol content? In a study at The Ohio State University, 16 student volunteers drank a randomly assigned number of cans of beer. Thirty minutes later, a police officer measured their blood alcohol content (BAC) in grams of alcohol per deciliter of blood. In all states of the U.S., the legal BAC limit is 0.08. Here are the data:

Student	**1**	**2**	**3**	**4**	**5**	**6**	**7**	**8**
Beers	5	2	9	8	3	7	3	5
BAC	0.10	0.03	0.19	0.12	0.04	0.095	0.07	0.06

Student	**9**	**10**	**11**	**12**	**13**	**14**	**15**	**16**
Beers	3	5	4	6	5	7	1	4
BAC	0.02	0.05	0.07	0.10	0.085	0.09	0.01	0.05

The students were equally divided between men and women and differed in weight and usual drinking habits. Because of this variation, many students don't believe that number of drinks predicts blood alcohol well. What do the data say?

Figure 6.1 is a scatterplot of these data. Because we think that number of beers helps explain BAC, it is the explanatory variable. We plot number of beers on the horizontal axis. One student drank 2 beers and had BAC 0.03. This student's point on the scatterplot is (2, 0.03), above $x = 2$ and to the right of $y = 0.03$. We have marked this point in Figure 6.1. ■

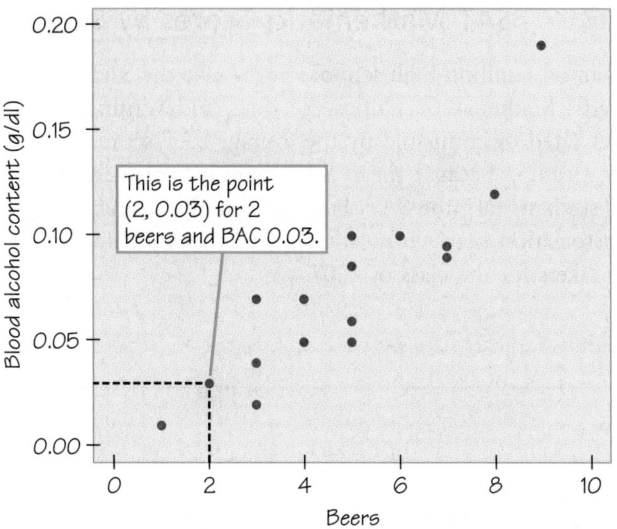

FIGURE 6.1 Scatterplot of blood alcohol content (response variable) against the number of beers a student drinks (explanatory variable).

To interpret a scatterplot, apply the usual strategies of data analysis.

Examining a Scatterplot — PROCEDURE

In any graph of data, look for the *overall pattern* and for striking *deviations* from that pattern.

You can describe the overall pattern of a scatterplot by the *form*, *direction*, and *strength* of the relationship.

An important kind of deviation is an **outlier**, an individual value that falls outside the overall pattern of the relationship.

The *form* of the relationship in Figure 6.1 is roughly a straight-line pattern. If you look ahead a bit, Figure 6.3 draws a line through the plot to describe the overall pattern. The *direction* of the relationship is clear: As number of beers increases, BAC also increases. We call this a **positive association** between the two variables.

Positive Association — DEFINITION

Two variables are **positively associated** if an increase in one variable tends to accompany an *increase* in the other variable.

Negative Association — DEFINITION

Two variables are **negatively associated** if an increase in one variable tends to accompany a *decrease* in the other variable.

The *strength* of a relationship describes how closely the points in a scatterplot follow a simple form such as a straight line. Figure 6.1 shows only a small amount of scatter about the straight line, so the relationship is moderately strong. We will soon learn a numerical measure of the strength of a straight-line relationship.

EXAMPLE 2 ▪ SAT Mathematics Scores by State

(*Cleve Bryant/PhotoEdit Inc.*)

Each year, more than one million high school seniors take the SAT Reasoning Test, which has three parts: Mathematics, Critical Reading, and Writing. We sometimes see individual states rated or compared by the average SAT scores of their seniors. However, this is misleading because mean SAT score is explained largely by what percent of a state's students take the SAT. For example, the scatterplot in Figure 6.2 shows a negative association between mean score on the Mathematics section and the percent of test takers for the class of 2007.

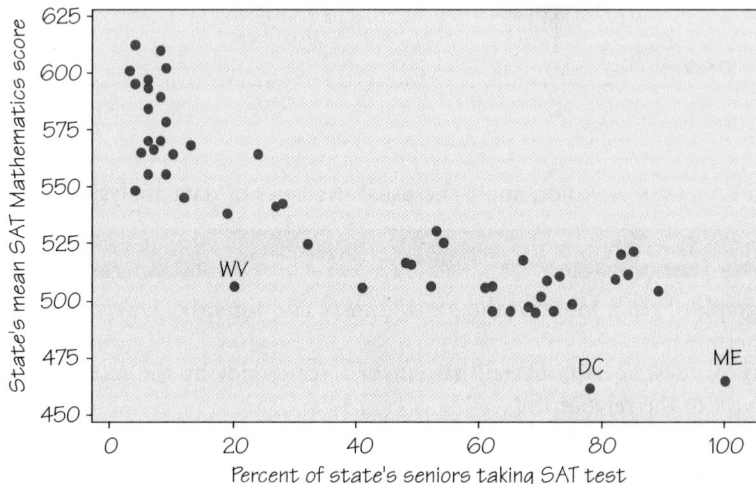

FIGURE 6.2 Scatterplot of the mean SAT Mathematics scores (response variable) against the percent of high school seniors who take the SAT (explanatory variable).

The *form* of Figure 6.2 is a bit irregular, but there are two distinct *clusters* of states. In one cluster, more than half of high school seniors take the SAT, and the mean scores are low. Fewer than 40% of seniors in states in the other cluster take the SAT–fewer than 20% in most of these states–and these states have higher mean scores. Clusters in a graph suggest that the data describe several distinct kinds of individuals. The two clusters in Figure 6.2 do in fact describe two distinct sets of states. There are two common college entrance examinations, the SAT and the ACT. Each state tends to prefer one or the other. In ACT states (the left cluster in Figure 6.2) most students who take the SAT are applying to selective out-of-state colleges. This select group performs well. In SAT states (the right cluster), many seniors take the SAT, and this broader group has a lower mean score.

The relationship in Figure 6.2 also has a clear *direction:* States in which a higher percent of students take the SAT tend to have lower mean scores. This is true both between the clusters and within each cluster. That is, there is a **negative association** between the two variables.

There are no clear *outliers* in Figure 6.2, but each cluster does include a state whose mean SAT Mathematics score is lower than we would expect from the percent of its students who take the SAT. In the cluster of ACT states, this occurs with West Virginia (WV). In the cluster of SAT states, this occurs with the District of Columbia (DC)–a city, not a state–and Maine (ME).

6.2 Making Predictions: Regression Line

If a scatterplot shows a straight-line relationship, we would like to summarize this overall pattern by drawing a line on the scatterplot. A **regression line** summarizes the relationship between two variables, but only in a specific setting: when one of the variables helps explain or predict the other. That is, regression describes a relationship between an explanatory variable and a response variable.

Regression Line DEFINITION

A **regression line** is a straight line that describes how a response variable y changes as an explanatory variable x changes. We often use a regression line to *predict* the value of y for a given value of x.

EXAMPLE 3 ▪ Predicting Blood Alcohol

Figure 6.1 shows a straight-line relationship between how many beers a student drinks and his or her blood alcohol content (BAC) 30 minutes later. Figure 6.3 repeats this scatterplot, and adds a regression line that we can use to predict BAC for a student based on the number of beers consumed.

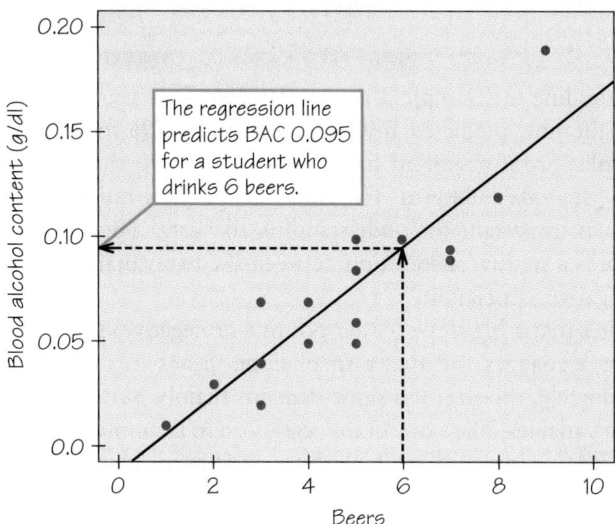

FIGURE 6.3 Regression line for predicting blood alcohol content from the number of beers a student drinks.

Figure 6.3 shows the prediction in graphical form for a student who drinks 6 beers. Start at $x = 6$, go up to the line, then head left to the y-axis. We hit the y-axis at BAC 0.095. This is the BAC that corresponds to 6 beers, according to the regression line. (Recall that the legal limit for driving is 0.08.) The line represents only the overall pattern of the data, so the BAC of a randomly chosen student after 6 beers will probably not be exactly 0.095. But because the points for the 16 students in the Ohio State study are not far from the line, we expect the prediction to be reasonably accurate.

It is easier to use the *equation of the line* for prediction. Applying formulas that will be given in Section 6.4, the equation of the line in Figure 6.3 is

$$\text{predicted BAC} = -0.0127 + 0.01796 \times \text{beers}$$

For a student who drinks 6 beers, we have

$$\text{predicted BAC} = -0.0127 + (0.01796)(6) = 0.095$$

You can plot a line from its equation by substituting two values of x, such as $x = 2$ and $x = 8$. Find the corresponding values of y, plot the two points, and draw the line through them.

Statistical software and many calculators will give you the equation of a regression line from keyed-in data. You should know how to use a regression line even if you don't look into the details needed to calculate the line from data. First, recall some basic facts about the (slope and intercept) coefficients in the equation of a line.

Equation of a Regression Line

DEFINITION

Suppose that y is a response variable (plotted on the vertical axis) and x is an explanatory variable (plotted on the horizontal axis). If we call $\hat{y}$ the predicted value of y, then the resulting regression line for predicting y from x has an equation of the form[1]:

$$\hat{y} = mx + b$$

In this equation, m is the **slope**, the amount by which y changes when x increases by 1 unit. The number b is the ***y*-intercept**, the value of y when $x = 0$.

The slope of the line in Example 3 is $m = 0.01796$. This says that as we move to the right along the line, predicted BAC goes up by 0.01796 for each additional beer a student drinks. So, if a student has 3 additional beers, the BAC would increase by $3 \times 0.01796 = 0.05388$ g/dl. The slope tells us how quickly y changes as we change x, so it is important for understanding the data. The slope is positive ($m > 0$) when there is a positive association between the two variables. It is negative when there is a negative association.

You might think that a big slope (either positive or negative) says that there are big changes in y as x changes and that a small slope means that x has little influence on y. Unfortunately, the size of a slope depends mainly on the units in which we measure the two variables. The slope of the regression in Example 3 is $m = 0.01796$ when we measure BAC y in grams of alcohol per deciliter of blood. That is, when beers consumed increases by 1, BAC increases by 0.01796 grams. There are 1000 milligrams in a gram, so if we measured BAC in milligrams of alcohol, the slope would be 1000 times as large, $m = 17.96$. When beers consumed increases by 1, BAC increases by 17.96 milligrams. *You can't say how important a relationship is by looking at how big the slope is.*

The intercept of the regression line in Example 3 is $b = -0.0127$. This is the predicted value of y when $x = 0$. Although we need the value of the intercept to draw the line, it is statistically meaningful only when x can actually take values close to zero. Even then, you should think of the intercept as describing the line rather than taking it seriously as a prediction. If a student drinks no beers, his or her blood alcohol should be exactly zero. The intercept of the regression line in Example 3 is close to zero, but it is not exactly zero.

[1] The letters m and b are from the slope-intercept form from algebra class, but be aware that some books and technologies use different letters, such as b and a. To be safe, check that the letter used for slope corresponds to the number multiplied by the explanatory variable.

6.3 Correlation

A scatterplot displays the form, direction, and strength of the relationship ("co-relation") between two numerical variables. Straight-line relations are particularly important because a straight line is a simple pattern that is quite common. We say a straight-line association is strong if the points lie close to a line, and weak if they are widely scattered about a line. This is vague. We need to follow our strategy for data analysis by using a numerical measure along with the graph. **Correlation** is the measure we use. Correlation is usually written as r, thanks to nineteenth-century statistician Sir Francis Galton, who was studying related ideas of ***r***egression and ***r***eversion.

Correlation DEFINITION

The **correlation** measures the direction and strength of the straight-line relationship between two numerical variables.

A correlation r is always a number between -1 and 1, inclusive. It has the same sign as the slope of a regression line: $r > 0$ for positive association and $r < 0$ for negative association.

Perfect correlation $r = 1$ or $r = -1$ occurs only when all points lie exactly on a straight line. The correlation moves away from 1 or -1 as the straight-line relationship gets weaker. Correlation $r = 0$ indicates no straight-line relationship.

SPOTLIGHT 6.1 Scatterplot Smoothers: Crash Test Dummies

Our eyes are good at seeing the overall pattern of a scatterplot. Sometimes the pattern has a simple form, such as a straight line, that we can draw on the plot to summarize the pattern. Clever software for "smoothing" a scatterplot can pick out much more complex overall patterns.

Crash a motorcycle into a wall. The rider, fortunately, is a dummy with an instrument to measure acceleration (change of velocity) mounted in its head. The figure is a scatterplot of the acceleration of the dummy's head against time. Acceleration is measured in g's, or multiples of the acceleration due to gravity at the earth's surface. The motorcycle approaches the wall at a constant speed (acceleration near 0). As it hits, the dummy's head snaps forward and decelerates violently (negative acceleration reaching more than 100 g's), then snaps back again (up to 75 g's) and wobbles a bit before coming to rest. We see that the plot has a clear, but complicated, overall pattern. A scatterplot smoother picks out this pattern and draws a curve on the plot to display it.

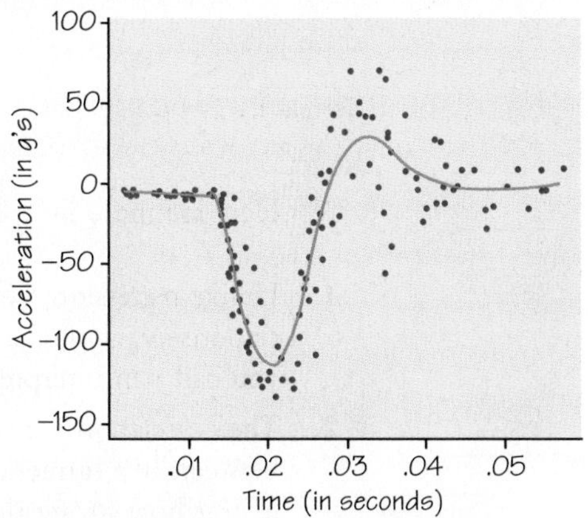

Smoothing a scatterplot. Software draws the smooth curve to describe the overall pattern of the relationship.

EXAMPLE 4 ■ Scatterplots and Correlation

The scatterplots in Figure 6.4 illustrate how values of r closer to 1 or -1 correspond to stronger straight-line relationships. To make the meaning of r clearer, the standard deviations of both variables in these plots are equal and the horizontal and vertical scales are the same. In general, it is not so easy to guess the value of r from the appearance of a scatterplot. Changing the plotting scales in a scatterplot can alter the appearance of the graph, but it does not change the correlation.

We said that Figure 6.1 shows a moderately strong positive straight-line relationship between how many beers a student drinks and his or her blood alcohol content. The correlation between these variables is $r = 0.894$. Figure 6.2, despite the clusters, also shows a quite strong straight-line relationship between the percent of a state's high school seniors who take the SAT exam and their mean SAT score. The association is negative: Higher percents taking the SAT go with lower mean scores. The correlation is $r = -0.877$.

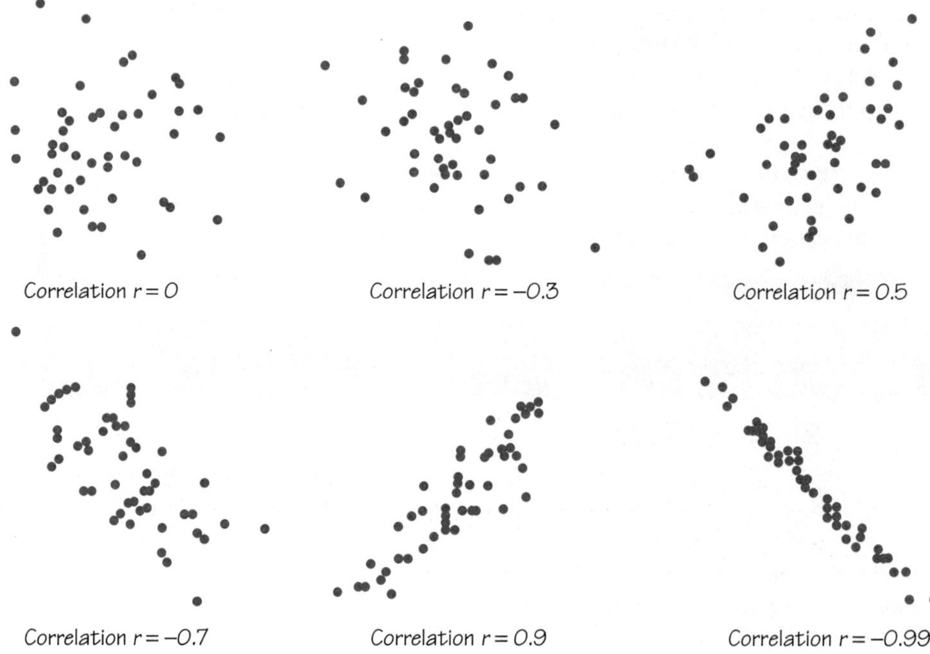

FIGURE 6.4 How the correlation r measures the direction and strength of straight-line association.

Here are more facts about the correlation r:

1. Unlike regression, correlation makes no distinction between explanatory and response variables. It makes no difference which variable you call x and which you call y in interpreting a correlation.
2. The correlation r measures the strength of only a straight-line relationship between two numerical variables. It does not describe curved relationships, no matter how strong they are. Correlation makes no sense for non-numerical variables such as ethnicity and occupation.
3. The correlation r does not change when we change the units of measurement of x, y, or both. Measuring height in inches rather than centimeters and weight in pounds rather than kilograms does not change the correlation between height and weight. The correlation r itself has no unit of measurement; it is just a number. So correlation is easier to interpret than the slope of a regression line, which changes when the units of x and y change.

4. Like the mean and standard deviation, the correlation is strongly affected by a few outlying observations. Use r with caution when outliers appear in the scatterplot.

In practice, you will use a calculator or software to find the correlation from keyed-in data. That's fortunate, because using the formula for correlation is quite a bit of work. Nonetheless, all the properties of r come from the formula that defines it.

Formula for Correlation

PROCEDURE

Suppose that we have data on variables x and y for n individuals. The means and standard deviations of the two variables are $\bar{x}$ and s_x for the x-values, and $\bar{y}$ and s_y for the y-values. The correlation r between x and y is

$$r = \frac{1}{n-1}\left[\left(\frac{x_1-\bar{x}}{s_x}\right)\left(\frac{y_1-\bar{y}}{s_y}\right) + \left(\frac{x_2-\bar{x}}{s_x}\right)\left(\frac{y_2-\bar{y}}{s_y}\right) + \cdots + \left(\frac{x_n-\bar{x}}{s_x}\right)\left(\frac{y_n-\bar{y}}{s_y}\right)\right]$$

This formula starts by *standardizing* each observation value (as you did in Exercise 50 in Chapter 5). That is, subtract the mean for that variable from the observation and then divide by the standard deviation. Standardizing turns each original data value into "number of standard deviations above the mean." A value of, say, −2 indicates 2 standard deviations below the mean. This removes the original units and explains why r has no units and doesn't change when we change the units of x or y. The formula says that the correlation is an average of the products of the standardized x and y values for n individuals. Exercise 26 asks you to calculate a correlation step by step from the formula to solidify your understanding.

SPOTLIGHT 6.2 Calculating Correlation

While the preceding formula for correlation has conceptual clarity, it can be tedious to apply to large datasets. Even with the most basic calculator, you'll get the same answer faster using this more computationally-oriented formula:

$$\frac{(x_1y_1 + x_2y_2 + \cdots + x_ny_n) - n\bar{x}\bar{y}}{(n-1)s_xs_y}.$$

If you have a scientific calculator, select the calculator mode to be able to do two-variable regression statistics, clear out any old data, and then enter your new x and y values. Once the data is entered, you can find the correlation (or regression line slope and y-intercept, for that matter) by hitting the appropriate key (probably with some kind of SHIFT or 2nd button). On the Internet, you can find several Web sites such as www.geocities.com/calculatorhelp that can help with keystrokes for specific models.

If you have a graphing calculator in the TI-83/84+ family (and you have the two variables entered in two columns using (STAT)→EDIT), then hit this sequence of buttons: (STAT)→TESTS→ LinRegTTest (ENTER). Enter what data columns are your independent(Xlist) and dependent(Ylist) variables (note that the labels L1 through L6 are above the keys (1) through (6), respectively). Then select Calculate, hit (ENTER), and scroll to the end of the output to see the correlation r. (Note that the output also contains the slope and y-intercept coefficients for the least-squares regression line.)

6.4 Least-Squares Regression

In Example 3, we used the straight line given by the equation

$$\text{predicted BAC} = -0.0127 + 0.01796 \times \text{beers}$$

to predict blood alcohol content from the number of beers consumed. Where did this equation come from? We will now see that the equation is the result of saying what we mean by the *best* line for predicting BAC from beers consumed. Once we say exactly what we mean by the best line, finding that line becomes a mathematical problem. The line in Example 3 is the solution to this problem for the beer and BAC data.

Different people might draw different lines by eye on a scatterplot. This is especially true when the points are widely scattered. We need a way to draw a regression line that doesn't depend on our guess as to where the line should go. No line will pass exactly through all the points, but we want one that is as close as possible. We will use the line to predict y from x, so we want a line that is as close as possible to the points in the *vertical* direction. That's because the prediction errors we make are errors in y, which is the vertical direction in the scatterplot.

The table in Example 1 shows that student #12 drank 6 beers and was observed to have a BAC of 0.10. However, the regression line equation in Example 3 showed that the predicted BAC for a student who drinks 6 beers is 0.095. These values are close, but are not the same. Indeed, from Figure 6.3, we can see the observed data point (6, 0.10) lies a bit off the line, and the vertical deviation of this gap is the prediction error.

$$\text{prediction error} = \text{observed BAC} - \text{predicted BAC} = 0.10 - .095 = .005$$

SPOTLIGHT 6.3 Regression in Action

No other statistical method is used as much as regression. Here are some more applications:

Did the vote counters cheat? Republican Bruce Marks was ahead of Democrat William Stinson when the voting machines were tallied in their Pennsylvania election. But Stinson was ahead after absentee ballots were counted by the Democrats who controlled the election board. A court fight followed. The court called in a statistician, who used regression with data from past elections to predict the counts of absentee ballots based on the results from the voting machines. Marks's lead of 564 votes from the machines predicted that he would get 133 more absentee votes than Stinson. In fact, Stinson got 1025 more absentee votes than Marks. This looks suspicious.

Is regression garbage? No—but garbage can be the setting for regression. The Census Bureau once asked if weighing a neighborhood's garbage would help count its people. So 63 households had their garbage sorted and weighed. It turned out that pounds of plastic in the trash gave the best garbage prediction of the number of people in a neighborhood. Alas, the prediction wasn't good enough to help the Census Bureau.

Can college success be predicted? Colleges with more applicants than spaces want to admit students who are most likely to succeed. To predict this, admissions officers consider variables such as high school GPA, scores on standardized tests (ACT or SAT), number of advanced (AP) classes taken, and so on. Multiple regression extends regression to allow more than one explanatory variable to contribute to explaining a response variable. By tracking first-year college GPA or graduation rates of admitted students, colleges can continue to assess and fine-tune their equation.

When the observed response lies above the line (as we saw when the number of beers is 6), the error is positive. And when the response lies below the line (as it does when the number of beers is 8), the error is negative. The most common way to make the collection of prediction errors for the entire dataset as small as possible is **least-squares regression**. The line in Figure 6.3 is the least-squares regression line.

Least-Squares Regression Line — DEFINITION

The **least-squares regression line** is the line that makes the sum of the *squares* of the vertical distances of the data points from the line the *least* value possible.

The least-squares idea says what we mean by the best-fitting line. How can we find this line from data? Starting with n observations on variables x and y, finding the line that makes the sum of the squares of the vertical errors as small as possible is a mathematical problem. Here is the solution to this problem.

Finding the Least-Squares Regression Line — PROCEDURE

We have data on an explanatory variable x and a response variable y for n individuals. From the data, calculate the means $\bar{x}$ and $\bar{y}$ and the standard deviations s_x and s_y of the two variables, as well as their correlation r (recall Spotlights 5.2 and 6.2). If we call $\hat{y}$ the predicted value of y, then the **least-squares regression line** is the line

$$\hat{y} = mx + b$$

with **slope** m given by

$$m = r\frac{s_y}{s_x}$$

and **y-intercept** b given by

$$b = \bar{y} - m\bar{x}$$

This equation gives insight into the behavior of least-squares regression by showing that it is related to the means and standard deviations of the x and y observations and to the correlation between x and y. For example, it is clear that the slope m always has the same sign as the correlation r. In practice, you don't need to calculate the means, standard deviations, and correlation first. Statistical software or your calculator will give the slope m, intercept b, and equation of the least-squares line from keyed-in values of the variables x and y. (Recall the footnote on notation in Section 6.2.) Notice that if you confuse whether y is your explanatory or your response variable, you will get a different slope value!

EXAMPLE 5

Least-Squares Regression of BAC on Number of Beers

Go back to the data in Example 1. Use your calculator to verify that the mean and standard deviation of x, number of beers consumed, are

$$\bar{x} = 4.8125 \quad \text{and} \quad s_x = 2.1975$$

The mean and standard deviation of y, blood alcohol content, are

$$\bar{y} = 0.07375 \qquad \text{and} \qquad s_y = 0.04414$$

The correlation between number of beers and BAC is $r = 0.8943$. The least-squares regression line of BAC y on number of beers x has slope

$$\begin{aligned} m = r\frac{s_y}{s_x} &= 0.8943 \times \frac{0.04414}{2.1975} \\ &= 0.01796 \end{aligned}$$

and intercept

$$\begin{aligned} b = \bar{y} - m\bar{x} &= 0.07375 - (0.01796)(4.8125) \\ &= -0.0127 \end{aligned}$$

The equation of the least-squares line is therefore

$$\hat{y} = -0.0127 + 0.01796x$$

just as we claimed earlier.

When doing calculations like this by hand, you may need to carry extra decimal places in the preliminary calculations to get accurate values of the slope and intercept. Using software or a calculator with a regression function eliminates this worry. ■

You now see that correlation and least-squares regression are closely connected. The expression $m = rs_y/s_x$ for the slope says that along the regression line, a change of one standard deviation in x corresponds to a change of r standard deviations in y. When the variables are perfectly correlated ($r = 1$ or $r = -1$), the change in the predicted response is the same (in standard deviation units) as the change in x. Otherwise, because $-1 \leq r \leq 1$, the change in the predicted y is less than the change in x. As the correlation grows less strong, the prediction moves less in response to changes in x.

6.5 Interpreting Correlation and Regression

Correlation and regression are among the most-used statistical methods. Here are a few cautions to keep in mind when you use these methods or see others use them.

Both the correlation r and the least-squares regression line can be strongly influenced by a few outlying points. Always make a scatterplot before doing any calculations. Here is an artificial example that illustrates what can happen.

EXAMPLE 6 ■ Beware the Outlier!

Figure 6.5 shows a scatterplot of data that have a strong positive straight-line relationship. In fact, the correlation is $r = 0.987$, close to the $r = 1$ of a perfect straight line. The line on the plot is the least-squares regression line for predicting y from x. One point is an extreme outlier in both the x and y directions. Let's examine the influence of this outlier.

First, suppose we drop the outlier. The correlation for the 5 remaining points (the cluster at the lower left) is $r = 0.523$. The outlier extends the straight-line pattern and greatly increases the correlation.

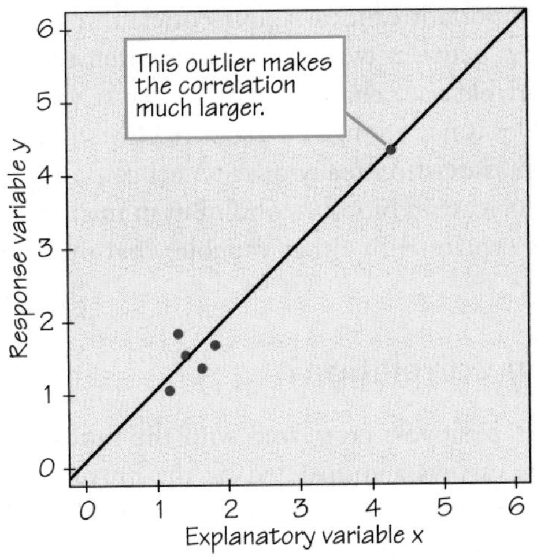

FIGURE 6.5 The outlier increases the correlation and fixes the location of the least-squares line.

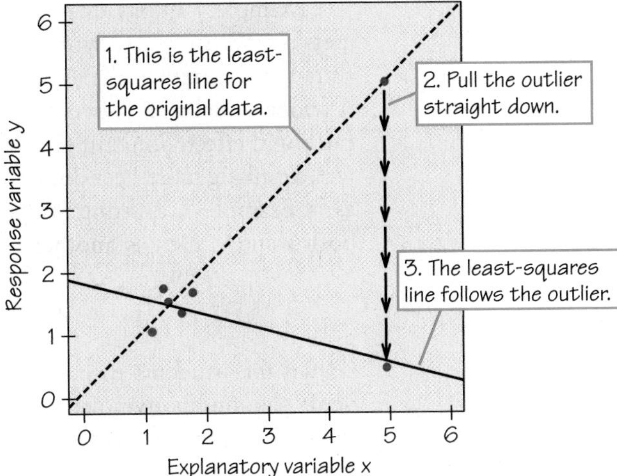

FIGURE 6.6 Moving the outlier unduly changes the correlation and moves the least-squares line.

Next, grab the outlier and pull it straight down, as in Figure 6.6. The least-squares line chases the outlier down, pivoting until it has a negative slope. This is the least-squares idea at work: The line stays close to all 6 points. However, its location is determined almost entirely by the one outlier. Of course, the correlation is now also negative, $r = -0.796$. Never trust a correlation or a regression line if you have not plotted the data.

Even if the correlation is strong and there are no outliers in the data we used to find our regression line, we also must not be quick to extrapolate and make predictions well beyond the data collected: Just because the data fits a particular linear trend over a window, there is no guarantee that that trend will continue into the future. For example, the rate of growth of a newborn may fit a line with a steep slope for the first several months, but then the slope (while still positive) starts to decrease.

A good way to see how outlying points can influence the correlation and the regression line is to use the *Correlation and Regression* applet. Applet Exercise 1 asks you to animate Example 6 above, watching r change and the regression line move as you pull the outlier down.

Correlation and regression *describe* relationships. *Interpreting* relationships requires more thought. *Often the relationship between two variables is strongly influenced by other variables.* You should always think about the possible effect of other variables before you draw conclusions based on correlation or regression.

EXAMPLE 7 ■ Money Helps SAT?

The College Board, which administers the SAT Reasoning Test, offers information on its Web site about college-bound seniors who take the test. This information shows a strong positive association between test score and a test taker's family's income. But there's no direct mechanism–wealthy families are not sending secret bribes to the College Board! It may simply be that children of wealthy parents are more likely to have advantages such as: well-educated role models, high expectations, access to extra tutoring or test preparation, and schools with experienced, qualified teachers and smaller class sizes.

Example 7 brings us to the most important caution about correlation and regression. When we study the relationship between two variables, we often hope to show that changes in the explanatory variable *cause* changes in the response variable. A strong association between two variables is not enough to draw conclusions about cause and effect. Sometimes an observed association really does reflect cause and effect. Drinking more beer does cause an increase in blood alcohol. But in many cases, as in Example 7, a strong association is explained by other variables that influence both x and y. Here is another example.

EXAMPLE 8 ■ Evaluation Correlation?

Grades that students earn in courses are positively correlated with the ratings students give on anonymous end-of-course surveys administered by the university. A simple interpretation is that instructors give easy tests with "low standards" which in turn causes students to express appreciation through high instructor ratings. But perhaps there is a third variable that drives the other two variables: A professor who is a skillful teacher and motivator may be more likely both to be rated well and to inspire high performance. Or perhaps a course that includes group projects (rather than only in-class, timed tests) as a significant component of the grade naturally results in higher levels of both performance and satisfaction. Or perhaps courses that have higher grade distributions are more likely to be upper level courses for majors in that subject, and such students would be more favorably inclined towards the course. ■

EXAMPLE 9 ■ Does Running Lead to Winning in Football?

(*Getty Images.*)

A football broadcaster discussed how often a team wins when it runs the ball at least 30 times in a game. For the most recent NFL regular season, the correlation between wins and number of running plays was indeed strongly positive ($r = 0.643$). Could this mean that running leads to winning–that all any team has to do to improve is to run the ball more? No. To take it to the extreme, if a team chose to run on every single play, the other team would simply adjust its entire defense to focus on and stop the run. Basically, once teams get a good lead in a game (regardless of their mix of running, passing and special team plays), they tend to start running the ball more often as a way to minimize the risk of losing the ball (pass plays are riskier) and to use up clock time faster (an incomplete pass stops the clock). And when teams get far behind late in the game, they begin passing more often as a last chance to get back into the game before time runs out. ■

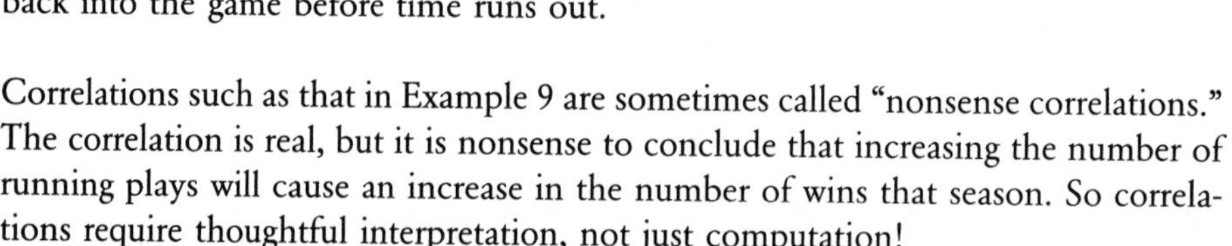

Correlations such as that in Example 9 are sometimes called "nonsense correlations." The correlation is real, but it is nonsense to conclude that increasing the number of running plays will cause an increase in the number of wins that season. So correlations require thoughtful interpretation, not just computation!

Association Does Not Imply Causation RULE

An association between an explanatory variable x and a response variable y, even if it is very strong, is not by itself good evidence that changes in x actually cause changes in y.

Here is a final example in which we use a scatterplot, correlation, and a regression line to understand data.

EXAMPLE 10
What Does Growth Hormone Do in Adults?

In most species, adults stop growing, but still release growth hormone from the pituitary gland to regulate metabolism. Physiologists subjected groups of adult rats to various conditions that activated muscle tissue that was either fast-twitch (like sprinters use) or slow-twitch (like distance runners use). They then measured levels of a bioassayable form of growth hormone (BGH) in the blood and in pituitary tissue. Units are 100s of nanograms per milliliter of blood and micrograms per milligram of tissue, respectively.

Here are the data, courtesy of neurobiologist Kristin Gosselink:

blood	15.8	20.0	26.7	25.0	23.0	23.8	24.7	16.3	0.8	0.8
tissue	38.0	36.7	27.8	28.3	34.9	34.1	33.2	32.7	38.1	39.1
blood	0.6	10.8	37.6	41.3	39	57.5	84.8	82.8	28.8	16.5
tissue	43.9	42.8	19.3	13.7	11.2	14.2	9.7	9.5	31.7	32.8

SOURCE: G.E. McCall et al., Muscle afferent-pituitary axis: a novel pathway for modulating the secretion of a pituitary growth factor, *Exercise and Sport Science Reviews* 29 (2001): 164-169.

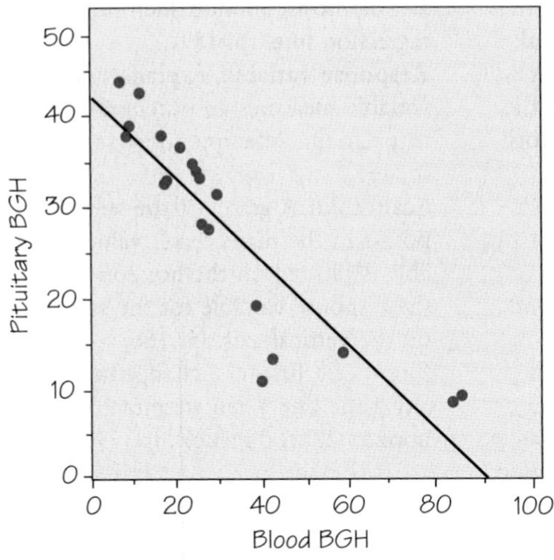

FIGURE 6.7 This scatterplot of BGH level in pituitary tissue versus BGH level in the blood shows a strong negative association.

Figure 6.7 is a scatterplot of these data. The plot shows a strong negative straight-line association with correlation $r = -0.90$. When there is a higher BGH level in the blood, we can assume that means BGH must have been recently secreted by the pituitary gland so that less BGH is now remaining in pituitary tissue. The least-squares regression line is

$$\hat{y} = 41.081 - 0.43343x, \text{ or}$$

$$\text{predicted pituitary BGH} = 41.081 + ((-0.43343) \times \text{blood BGH}).$$

The slope $m = -0.43343$ is negative, which is consistent with how blood and pituitary tissue levels of BGH move in opposite directions. The y-intercept $b = 41.08$ is the estimated amount of BGH the pituitary gland has if it does not release any into the blood.

Furthermore, the two uppermost points of the scatterplot represent groups of rats whose slow-twitch muscles were activated, the five points on the lower right of the scatterplot involved activation of fast-twitch muscles, and all but one of the remaining points represent groups that were untreated. These data come from an *experiment* that assigned rats randomly to treatment (or no treatment) conditions to make us reasonably confident that slow-twitch muscle activation *causes* a decrease in BGH and that fast-twitch muscle activation *causes* an increase. We will discuss experiments in detail in Chapter 7.

REVIEW VOCABULARY

Correlation A measure of the direction and strength of the straight-line relationship between two numerical variables. Correlations take values between 0 (no straight-line relationship) and ±1 (perfect straight-line relationship). (p. 191)

Intercept of a line The vertical (y) coordinate of the point on the line above 0 on the horizontal (x) axis. (p. 190)

Least-squares regression line A line drawn on a scatterplot that makes the sum of the squares of the vertical distances of the data points from the line as small as possible. The regression line can be used to predict the response variable y for a given value of the explanatory variable x. (p. 195)

Negative association Two variables are negatively associated if an increase in one variable tends to accompany a *decrease* in the other variable. The scatterplot has a northwest-to-southeast pattern, and the correlation and regression slope are both negative. (p. 187)

Outlier An outlier in a scatterplot is a point that lies outside the overall pattern of the other points. Outliers sometimes strongly influence the value of the correlation and the position of the least-squares regression line. (p. 187)

Positive association Two variables are positively associated if an increase in one variable tends to accompany a *increase* in the other variable. The scatterplot has a southwest-to-northeast pattern, and the correlation and regression slope are both positive. (p. 187)

Regression line Any line that describes how a response variable y changes as we change an explanatory variable x. The most common such line is the least-squares regression line. (p. 189)

Response variable, explanatory variable A response variable measures an outcome of a study. An explanatory variable attempts to explain the observed outcomes. (p. 185)

Scatterplot A graph of the values of two variables as points in the plane. Each value of the explanatory variable is plotted on the horizontal axis, and the value of the response variable for the same individual is plotted on the vertical axis. (p. 186)

Slope of a line The change in the vertical (y) direction along the line when we move 1 unit to the right in the horizontal (x) direction. (p. 190)

✓ SKILLS CHECK

1. You have data for many families on the parents' income and the years of education their eldest child completes. When you make a scatterplot, the explanatory variable on the x-axis

(a) is parents' income.
(b) is years of education.
(c) doesn't matter.

2. You expect to see a __________ association between parents' income and the years of education their eldest child completes.

3. Figure 6.8 is a scatterplot of reading test scores against IQ test scores for 15 fifth-grade children. There is one low outlier in the plot. The IQ and reading scores for this child are

(a) IQ = 10, Reading = 124.
(b) IQ = 124, Reading = 72.
(c) IQ = 124, Reading = 10.

4. The line in Figure 6.8 is a regression line for predicting reading score from IQ score. If another child in this class has IQ score 125, then ____ is the multiple

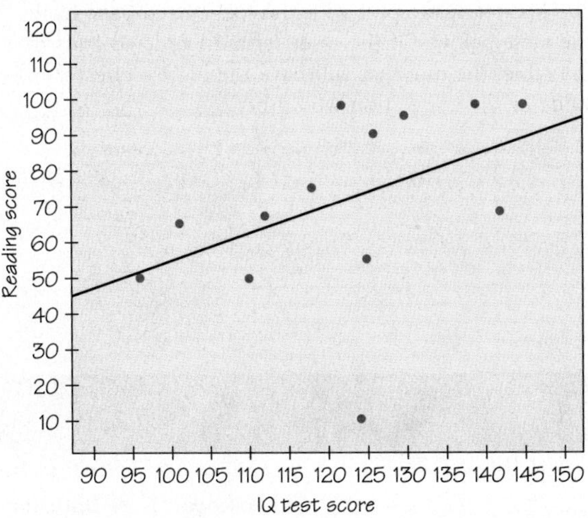

FIGURE 6.8 Scatterplot of the reading test scores of fifth-grade children (response variable) against the children's IQ scores (explanatory variable).

of 10 to which that predicted reading score would be closest.

5. The slope of the line in Figure 6.8 is closest to

(a) −1.
(b) 0.
(c) 1.

6. The points on a scatterplot lie close to the line whose equation is $y = 2 - 5x$. The slope of this line is ______ .

7. Starting with a fresh bar of soap, you weigh the bar each day after you take a shower. Then you find the regression line for predicting weight from number of days elapsed. The slope of this line will be

(a) positive.
(b) negative.
(c) can't tell without seeing the data.

8. Fred keeps his savings in his mattress. He began with \$500 from his mother and adds \$100 each year. In the form $y = mx + b$, the equation for his total savings y after x years would be $y =$ ____________________ .

9. The amount of water discharged by the Mississippi River has changed over time in roughly a straight-line pattern. A regression line for predicting water discharged (in cubic kilometers) from year is

$$\text{predicted discharge} = -7792 + (4.226 \times \text{year})$$

How much (on the average) does the volume of water increase with each passing year?

(a) −7792 cubic kilometers
(b) 4.226 cubic kilometers
(c) 7792 cubic kilometers

10. According to the regression line in the previous exercise, the predicted Mississippi River discharge in the year 2010 is ________ cubic kilometers.

11. You have data on the body weight x and brain weight y for many species of mammals. Body weight is given in kilograms and brain weight is given in grams. There are 1000 grams in a kilogram. The slope of the regression line for predicting y from x is $m = 1.4$. If brain weight were given in kilograms, the slope would

(a) still be 1.4.
(b) change to 0.0014.
(c) change to 1400.

12. The correlation between brain weight and body weight in Exercise 11 is $r = 0.86$. If brain weight had been measured in kilograms rather than grams, the correlation would have a value of ______________ .

13. Given the following set of five ordered pairs, the correlation r equals

x	0	1	2	3	4
y	2	3	5	6	14

(a) 0.3
(b) 0.6
(c) 0.9

14. Suppose $y = 2x + 3$, where x and y are measured in meters. If x is reexpressed in centimeters instead, the equation becomes $y =$ ____________________ .

15. The points on a scatterplot lie very close to the line whose equation is $y = 5 - 3x$. The correlation between x and y is close to

(a) −3.
(b) −1.
(c) 1.

16. High coffee prices give farmers in Indonesia an incentive to cut forest in order to plant more coffee. Here are data on coffee price x (cents per pound) and percent y of deforestation in a national park for five years:

x	29	40	54	55	72
y	0.49	1.59	1.69	1.82	2.98

Using a calculator, we can determine that the correlation between x and y has a value of ________ , to the nearer hundredths place.

17. Using the Table in Example 1 and the prediction equation in Example 3, the prediction error for Student #4 is what?

(a) −.01
(b) .01
(c) .13

18. Look again at the coffee data in Exercise 16. Using your calculator, you can find that the equation (in $y = mx + b$ form, with m and b to the nearest hundredths place) of the least-squares regression line for predicting y from x is: $\hat{y} =$ ____________________ .

19. There is a strong positive correlation between the number of firefighters at a fire and the amount of damage the fire does. The reason for this is that

(a) more firefighters cause more damage at the fire scene.
(b) bigger fires require more firefighters and also do more damage.
(c) more damage requires more firefighters to clean it up.

20. Make a scatterplot with the six ordered pairs from the table below. Of the three leftmost ordered pairs in the table, the one that will have the biggest effect on the value of the correlation would be ________________ .

x	0	8	2	0	1	1
y	0	3	4	1	1	0

CHAPTER 6 EXERCISES

■ Challenge ◆ Discussion

Some exercises require use of a calculator (or software or Internet applet) that will find correlation and the slope and intercept of the least-squares regression line from keyed-in data.

1. In each of the following situations, is it more reasonable simply to explore the relationship between the two variables or to view one of the variables as an explanatory variable and the other as a response variable? In the latter case, which is the explanatory variable?

(a) The amount of time spent studying for a statistics exam and the grade on the exam.
(b) The weight in kilograms and height in centimeters of a person.
(c) Inches of rain in the growing season and the yield of corn in bushels per acre.
(d) A student's scores on the SAT math exam and the SAT verbal exam.

6.1 Displaying Relationships: Scatterplot

2. Figure 6.9 shows the calories and salt content (milligrams of sodium) in 17 brands of meat hot dogs. Describe the overall pattern (form, direction, and strength) of these data. In what way is the point marked A unusual?

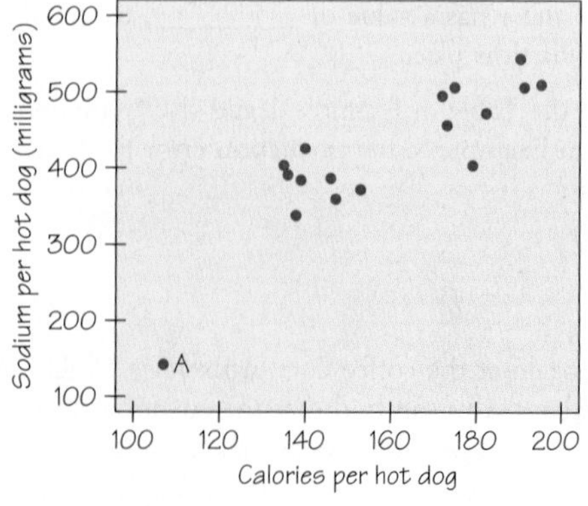

◆ 3. Figure 6.10 is a scatterplot of data from the World Bank. The individuals are all the world's nations for which data are available. The explanatory variable is a measure of how rich a country is, the gross domestic product (GDP) per person. GDP is the total value of the goods and services produced in a country, converted into dollars. The response variable is life expectancy at birth. Three African nations are outliers, with lower life expectancy than usual for their GDP. A full study would ask what special circumstances explain these outliers.

(a) Describe the direction and form of the relationship. Aside from the outliers, it is moderately strong.
(b) Explain why the direction and form of this relationship make sense.

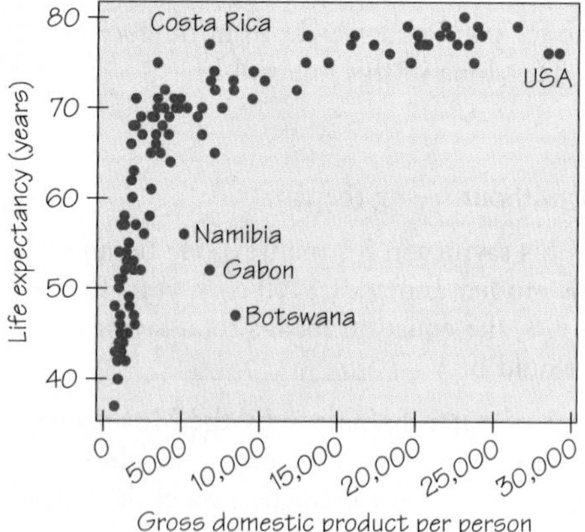

FIGURE 6.10 Scatterplot of the life expectancy of people in many nations against each nation's gross domestic product per person, for Exercise 3.

FIGURE 6.9 Scatterplot of sodium content versus calories in 17 brands of meat hot dogs, for Exercise 2.

4. Global warming may be due to increased concentrations of greenhouse gases such as carbon dioxide (CO_2). Here are data from the National Oceanic and Atmospheric Administration Web site (www.noaa.gov), where CO_2 is measured in parts per million by volume:

CO_2	316.91	325.68	338.70	354.16	369.41
year	1960	1970	1980	1990	2000

(a) Which is the explanatory variable?
(b) Make a scatterplot. Is the association between these variables positive or negative? Explain why you expect the relationship to have this direction.
(c) Describe the form and strength of the relationship.

5. Table 5.2 gives the city and highway gas mileages for 13 midsized cars. Omit the hybrid car (Toyota Prius) and make a scatterplot, taking city mileage as the explanatory variable. Describe in words the form, direction, and strength of the relationship between highway mileage and city mileage.

6. How fast do icicles grow? Here are data for one set of conditions: no wind, temperature −11°C, and water flowing over the icicle at 12 milligrams per second.

Time (minutes)	10	20	30	40	50
Length (centimeters)	0.6	1.8	2.9	4.0	5.0
Time (minutes)	60	70	80	90	100
Length (centimeters)	6.1	7.9	10.1	10.9	12.7
Time (minutes)	110	120	130	140	150
Length (centimeters)	14.4	16.6	18.1	19.9	21.0

SOURCE: N. Maeno et al., Growth rates of icicles, *Journal of Glaciology*, 40 (1994): 319–326.

Which is the explanatory variable? Make a scatterplot. Describe in words the direction, form, and strength of the relationship.

◆ **7.** How does the fuel consumption of a car change as its speed increases? Here are data for a British Ford Escort. Fuel consumption is measured in liters of gasoline used per 100 kilometers traveled.

Speed	10	20	30	40	50
Fuel	21.00	13.00	10.00	8.00	7.00
Speed	60	70	80	90	100
Fuel	5.90	6.30	6.95	7.57	8.27
Speed	110	120	130	140	150
Fuel	9.03	9.87	10.79	11.77	12.83

SOURCE: T. N. Lam, Estimating fuel consumption from engine size, *Journal of Transportation Engineering*, 111 (1985): 339–357.

(a) Which is the explanatory variable?
(b) Make a scatterplot. Describe the form of the relationship. Explain why the form of the relationship makes sense.
(c) How would you describe the direction of this relationship?
(d) Is the relationship reasonably strong or quite weak? Explain your answer.

◆ **8.** Give an example of two variables from everyday life that have a positive association. Give an example of two variables that have a negative association.

6.2 Making Predictions: Regression Line

9. Figure 6.9 shows the calories and salt content (milligrams of sodium) in 17 brands of meat hot dogs. If we ignore the outlying point marked A, a regression line for predicting sodium from calories passes close to these two observations:

calories = 139, sodium = 386 mg
calories = 191, sodium = 506 mg

Use this fact to estimate the slope of this regression line. (*Hint:* Remember that the slope of a line is the "rise" (vertical change) divided by the "run" (horizontal change) for any two points on the line.)

10. Exercise 4 gives data on carbon dioxide concentration (in parts per million) over time. A regression line for predicting carbon dioxide from time is

predicted carbon dioxide concentration = 314.276 + 1.3348 × (years elapsed since 1960)

(a) What is the slope of this line? What does the slope say about how carbon dioxide is changing over time?
(b) Predict the carbon dioxide concentration for 2006. In fact, the observed value was 381.85. How accurate is your prediction?

11. Researchers studying acid rain measured the acidity of precipitation in a Colorado wilderness area for 150 consecutive weeks. Acidity is measured by pH. Lower pH values show higher acidity. The acid rain researchers observed a straight-line pattern over time. They reported that the regression line

predicted pH = 5.43 − (0.0053 × weeks)

fit the data well. [W. M. Lewis and M. C. Grant, Acid precipitation in the western United States, *Science*, 207 (1980): 176–177.]

(a) Draw a graph of this line. Explain what the line says about how pH was changing over time.
(b) According to the regression line, what was the pH at the beginning of the study (weeks = 1)? At the end (weeks = 150)?
(c) What is the slope of the regression line? Explain what this slope says about the rate of change in pH.

12. A University of Massachusetts Amherst study published in the May 2007 *Journal of Marriage and Family* found that married women do about one less hour of housework a week for every $7500 they earn as full-time workers outside the home, regardless of the husband's income.

(a) What would be the numerical value of the slope coefficient in the regression model that predicts women's housework from their income? What does the sign of the slope (positive or negative) tell us about the relationship between these variables?
(b) Suppose Lynette's salary is $30,000 greater than Gabrielle's. What would you predict to be the difference in hours of housework they each do?

13. If heterosexual women always married men who were two years older than they are, what would be the slope of the regression line for predicting husband's age from wife's age? (*Hint:* Draw a scatterplot for several ages.)

14. Suppose that the slope of the regression line of weight on height for a group of young men is $m = 1.1$ when we measure height x in centimeters and weight y in kilograms. That is, when height increases by 1 centimeter, weight increases by 1.1 kilograms. There are 1000 grams in a kilogram. If we measured weight in grams, what would be the slope?

6.3 Correlation

15. Find the correlation between the city and highway gas mileages for the 12 non-hybrid midsized cars in Table 5.2. (That is, omit the Toyota Prius.) Explain why the value of r matches the scatterplot that you made in Exercise 5.

16. Exercise 4 gives data on carbon dioxide concentration (in parts per million) over time.

(a) Use a calculator to find the correlation r. Explain from looking at the scatterplot why this value of r is reasonable.
(b) Suppose that the concentration had been recorded in parts per billion instead of parts per million. For example, the value 354.16 would become 354160. How would the value of r change?

17. Find the correlation between city and highway mileage for all 13 midsized cars in Table 5.2, including the Toyota Prius. Compare your r with the value you found in Exercise 15. Explain why adding the Prius changes r in this direction.

18. Find the correlation between time and icicle length for the data in Exercise 6. Explain why the value of r matches the scatterplot that you made in Exercise 6.

19. Exercise 7 gives data on gas used versus speed for a small car. Make a scatterplot, if you did not do so in Exercise 7. Calculate the correlation. Explain why r is close to 0 despite a strong relationship between speed and gas use.

20. The length of the icicle in Exercise 6 is measured in centimeters. There are 2.54 centimeters in an inch. If length were measured in inches, how would the correlation you found in Exercise 18 change?

21. If heterosexual women always dated men who are three years older than they are, what would be the correlation between the ages of the man and the woman? (*Hint:* Draw a scatterplot for several ages.)

22. We want to find the correlation:

(a) between the heights of fathers and the heights of their adult sons.
(b) between the heights of husbands and the heights of their wives.
(c) between the heights of women at age 4 and their heights at age 18.

The answers (in scrambled order) are $r = 0.2$, $r = 0.5$, and $r = 0.8$. Match the answers to the variable pairings and explain your choice.

23. For each of the following pairs of variables, would you expect a substantial negative correlation, a substantial positive correlation, or a small correlation?

(a) The age of used cars and their prices.
(b) The weight of new cars and their gas mileages in miles per gallon.
(c) The heights and the weights of adult men.
(d) The heights and the IQ scores of adult men.

◆ **24.** Each of the following statements contains a mistake. Explain what is wrong in each case.

(a) "There is a high correlation ($r = 0.89$) between the hair color of American workers and their income."
(b) "We found a high correlation ($r = 1.09$) between students' ratings of faculty teaching and ratings made by other faculty members."
(c) "The correlation between age and income was found to be $r = 0.53$ years."

■ **25.** Mutual-fund reports often give correlations to describe how the prices of different investments are related. You look at the correlations between three Fidelity funds and the Standard & Poor's 500 Stock Index, which describes stocks of large U.S. companies. The three funds are Dividend Growth (stocks of large U.S. companies), Small Cap Stock (stocks of small U.S. companies), and Emerging Markets (stocks in developing countries). For a recent year, the three correlations are $r = 0.35$, $r = 0.81$, and $r = 0.98$.

(a) Which correlation goes with each fund? Explain your answer.
(b) The correlations of the three funds with the index are all positive. Does this tell you that stocks went up that year? Explain your answer.

■ **26.** *Archaeopteryx* is an extinct beast having feathers like a bird but teeth and a long bony tail like a reptile. Only six fossil specimens are known. If the specimens belong to the same species and differ in size because some are younger than others, there should be a straight-line relationship between the lengths of a pair of bones from all individuals. An outlier from this relationship would suggest a different species. Here are data on the lengths in centimeters of the femur (a leg bone) and the humerus (a bone in the upper arm) for the five specimens that preserve both bones.

Femur length x	38	56	59	64	74
Humerus length y	41	63	70	72	84

SOURCE: M. A. Houck et al., Allometric scaling in the earliest fossil bird, *Archaeopteryx lithographica*, *Science*, 247 (1990): 195–198.

(a) Make a scatterplot. Do you think that all five specimens come from the same species?
(b) Find the correlation r step by step. *Step 1*: Find the mean $\bar{x}$ and standard deviation s_x of the five femur lengths. Find the mean $\bar{y}$ and the standard deviation s_y of the five humerus lengths. (Use your calculator.) *Step 2*: Find the standardized values $(x - \bar{x})/s_x$ of each of the five femur lengths, and do the same for the five humerus lengths. *Step 3*: Substitute your numbers from Steps 1 and 2 into the formula for r.
(c) Now use one of the faster methods in Spotlight 6.2 to find r and check that you get the same result as in part (b).

6.4 Least-Squares Regression

27. In Exercise 5 you made a scatterplot of city and highway gas mileage for the 12 non-hybrid midsized cars (omitting the Prius) in Table 5.2.

(a) What is the least-squares regression line for predicting highway mileage from city mileage?

(David R. Frazier Photolibrary, Photo Researchers.)

(b) If a midsized car gets 17 mpg in the city, predict its highway mileage.
(c) Based on the scatterplot you made in Exercise 5, do you expect the prediction in part (b) to be quite accurate? Why?

28. In Exercise 6 you made a scatterplot of the length of an icicle and the number of minutes water has been flowing over the icicle.

(a) What is the equation of the least-squares regression line for predicting icicle length from time?
(b) Use your regression line to predict the length of the icicle after 75 minutes.

29. Redo your scatterplot of highway mileage against city mileage from Exercise 5. Add your regression line from Exercise 27 to the plot. Be sure to show how you were able to plot the line starting with its equation. Finally, use the "up-and-across" method illustrated in Figure 6.3 to show the predicted highway mileage of a car that gets 18 mpg in the city.

30. Redo your scatterplot of icicle length against time from Exercise 6. Add your regression line from Exercise 28 to the plot. Be sure to show how you were able to plot the line starting with its equation. Finally, use the "up-and-across" method illustrated in Figure 6.3 to show the predicted length of the icicle after 75 minutes.

31. Exercise 7 gives data on gas used versus speed for a small car. Make a scatterplot, if you did not do so in Exercise 7. The least-squares regression line for these data is

$$\text{predicted fuel} = 11.058 - 0.0147 \times \text{speed}$$

Draw this line on your scatterplot. What are the predicted and observed fuel consumption values for speeds of 10, 70, and 150 kilometers per hour (km/h)? *You can fit a regression line to any set of two-variable data. The line is of little use if the plot does not show a straight-line pattern.*

32. The length of the icicle in Exercise 6 is measured in centimeters. There are 2.54 centimeters in an inch. If length were measured in inches, how would the slope of the regression line you found in Exercise 28 change?

33. The mean height of American women in their early twenties is about 64.5 inches and the standard deviation is about 2.5 inches. The mean height of men the same age is about 68.5 inches, with standard deviation about 2.7 inches. If the correlation between the heights of husbands and wives is about $r = 0.5$, what is the equation of the regression line of the husband's height on the wife's height in young couples? Predict the height of the husband of a woman who is 67 inches tall.

34. This data, from the National Oceanic and Atmospheric Administration Web site (www.noaa.gov), is the mean annual number of named Atlantic storms (hurricanes, tropical storms, tropical depressions), during five-year windows ending with the year shown in the table.

2007	2002	1997	1992	1987	1982	1977
16.2	13.6	11	10.4	8.2	10	8.8
1972	1967	1962	1957	1952	1947	1942
11.2	9.2	8.8	10.6	10.4	9.4	7.4

(a) What is the slope of the least-squares regression line of named storms on year? What is the intercept?
(b) Use the regression line to predict the number of named storms for the five-year window 2008–2012.

■ 35. Use the equation for the least-squares regression line to show that this line always passes through the point $(\bar{x}, \bar{y})$. That is, set $x = \bar{x}$ and show that the line predicts that $y = \bar{y}$.

■ **36.** Exercise 6 gives data on the growth of an icicle.
(a) Find the mean and standard deviation of the times and icicle lengths. Find the correlation between the two variables. Use these five numbers to find the equation of the regression line for predicting length from time. Verify that your result agrees with that in Exercise 28.
(b) Use the same five numbers to find the equation of the regression line for predicting from an icicle's length the time it has been growing. Use your line to predict the time that an icicle 15 centimeters long has been growing. *There is just one correlation between two variables, but there are two different least-squares lines, depending on which you choose as the response variable.*

■ **37.** Fidelity Investments, like other large mutual fund companies, offers many "sector funds" that concentrate their investments in narrow segments of the stock market. These funds often rise or fall by much more than the market as a whole. Here are the percent returns for 23 Fidelity "Select Portfolios" funds for the years 2002 (when stocks fell) and 2003 (when stocks went up).

2002 return	2003 return	2002 return	2003 return	2002 return	2003 return
−17.1	23.9	−0.7	36.9	−37.8	59.4
−6.7	14.1	−5.6	27.5	−11.5	22.9
−21.1	41.8	−26.9	26.1	−0.7	36.9
−12.8	43.9	−42.0	62.7	64.3	32.1
−18.9	31.1	−47.8	68.1	−9.6	28.7
−7.7	32.3	−50.5	71.9	−11.7	29.5
−17.2	36.5	−49.5	57.0	−2.3	19.1
−11.4	30.6	−23.4	35.0		

Do a careful statistical analysis of these data using both graphs and whatever numerical measures you think are appropriate. Make a side-by-side comparison of the distributions of returns in 2002 and 2003 and also describe the relationship between the returns of the same funds in these two years. What are your most important findings? (The outlier is Fidelity Gold Fund.)

6.5 Interpreting Correlation and Regression

38. Make a scatterplot of the following data:

x	1	2	3	4	10	10
y	1	3	3	5	1	11

Use your calculator to show that the correlation is about 0.5. What feature of the data is responsible for reducing the correlation to this value despite a strong straight-line association between x and y in most of the observations?

39. Table 6.1 on the next page has four data sets prepared by statistician Frank Anscombe to show dangers of calculating without first plotting the data.
(a) Without making scatterplots, find the correlation and the least-squares regression line for all four data sets. What do you notice? Use the regression line to predict y for $x = 10$.
(b) Make a scatterplot for each of the data sets and add the regression line to each plot.
(c) In which of the four cases would you be willing to use the regression line to describe the dependence of y on x? Explain your answer in each case.

◆ **40.** Children who watch many hours of television get lower grades in school on the average than those who watch less TV. Explain clearly why this fact does not show that watching TV *causes* poor grades. In particular, suggest some other characteristics of households where children watch lots of TV that may contribute to poor grades.

◆ **41.** People who use artificial sweeteners in place of sugar tend to be heavier than people who use sugar. Does this mean that artificial sweeteners cause weight gain? Give a more plausible explanation for this association.

◆ **42.** "Based on an examination of twenty-two companies that announced large layoffs during 1994, Downs found a strong (.31) correlation between the size of the layoffs and the compensation of the CEOs." [K. Phillips, *Wealth and Democracy,* Broadway Books, New York, 2002, p. 151.] Discuss why this correlation is probably explained by a third variable, the size of the company as measured by its number of employees.

■ **43.** "The positive correlation between health and income per capita is one of the best-known relations in international development. This correlation is commonly thought to reflect a causal link running from income to health. . . . Recently, however, another intriguing possibility has emerged: that the health-income correlation is partly explained by a causal link running the other way–from health to income." [D. E. Bloom and D. Canning, The health and wealth of nations, *Science,* 287 (2000): 1207–1208.] Explain how higher income in a nation can cause better health. Then explain how better health can cause higher national income. There is no simple way to determine the direction of the link.

■ **44.** The effect of an outside variable can be surprising when individuals are divided into groups. In recent years, the mean SAT score of all high school seniors has increased. But the mean SAT score has decreased for students at each level of high school grades (A, B, C, and so on). Explain how grade inflation in high school

- Government economists inquire about average household income.

In all these cases, we want to gather information about a large group of individuals. Time, cost, and inconvenience preclude contacting every individual. So we gather information about only part of the group in order to draw conclusions about the whole. Also, when an observation is destructive, it is necessary to use only a sample. For example, testing a shipment of fuses to see if they are defective would wipe out the whole shipment if every single fuse were tested. And if your doctor's appointment includes a blood test, you want only *some* of your blood removed!

Population

DEFINITION

The **population** in a statistical study is the entire group of individuals about which we want information.

Sample

DEFINITION

A **sample** is a part of the population from which we actually collect information used to draw conclusions about the whole. Sampling refers to the process of choosing a sample from the population.

We often draw conclusions about a whole on the basis of a sample. Everyone has sipped a spoonful of soup and judged the entire bowl on the basis of that taste. But a bowl of soup is homogeneous, so that the taste of a single spoonful represents the whole. On the other hand, a spoonful of salad dressing may be misleading since its elements may separate if the bottle has not been shaken recently. Choosing a representative sample from a large and varied population is not so easy. The first step in a proper *sample survey* is to say carefully just what population we want to describe. The second step is to say exactly what we want to measure. These preliminary steps can be complicated, as the following example illustrates.

EXAMPLE 1

How Can a Survey Measure Unemployment?

The monthly unemployment rate comes from the government's Current Population Survey (CPS; www.census.gov/cps/), a sample of about 60,000 households each month conducted by the Census Bureau (see Figure 7.1). To measure unemployment, we must first specify the population we want to describe. Which age groups will we include? Will we include illegal aliens or people in prisons? The CPS defines its population as all U.S. residents (whether citizens or not) 16 years of age and over who are civilians and are not in an institution such as a prison. The civilian unemployment rate announced in the news refers to this specific population.

The second question is harder: What does it mean to be "unemployed"? Someone who is not looking for work–for example, a full-time student–should not be called unemployed just because she is not working for pay. If you are chosen for the CPS sample, the interviewer first asks whether you are available to work and whether you actually looked for work in the past four weeks. If not, you are neither employed nor unemployed–you are not in the labor force.

TABLE 6.1 Four Data Sets for Exploring Correlation and Regression

Data Set A

x	10	8	13	9	11	14	6	4	12	7	5
y	8.04	6.95	7.58	8.81	8.33	9.96	7.24	4.26	10.84	4.82	5.68

Data Set B

x	10	8	13	9	11	14	6	4	12	7	5
y	9.14	8.14	8.74	8.77	9.26	8.10	6.13	3.10	9.13	7.26	4.74

Data Set C

x	10	8	13	9	11	14	6	4	12	7	5
y	7.46	6.77	12.74	7.11	7.81	8.84	6.08	5.39	8.15	6.42	5.73

Data Set D

x	8	8	8	8	8	8	8	8	8	8	19
y	6.58	5.76	7.71	8.84	8.47	7.04	5.25	5.56	7.91	6.89	12.50

SOURCE: Frank J. Anscombe, Graphs in statistical analysis, *The American Statistician*, 27 (1973): 17–21.

can account for this pattern. *A relationship that holds for each group within a population need not hold for the population as a whole. In fact, the relationship can even change direction.*

Chapter Review

45. Recent major recalls of toys with lead paint refocused people on the dangers of lead exposure. Below is data from research exploring the association with student achievement for blood lead levels below the "danger threshold" of 10 micrograms/deciliter set by the Centers for Disease Control [M.L. Miranda et al., The relationship between early childhood blood lead levels and performance on end-of-grade tests, *Environmental Health Perspectives*, 115 (2007): 1242-1247].

Blood lead level	1	2	3	4	5
Mean 4th grade reading score	255.9	253.8	252.6	251.0	250.4

Blood lead level	6	7	8	9
Mean 4th grade reading score	249.5	248.5	247.8	249.3

What are the explanatory and response variables? Do you expect a positive or negative association between these variables? Why? Does the scatterplot support this?

46. A study of reading ability in schoolchildren chose 60 fifth-grade children at random from a school. The researchers had the children's scores on an IQ test and on a test of reading ability. Figure 6.11 plots reading test score (response) against IQ score (explanatory).

(a) Explain why we should expect a positive association between IQ and reading score for children in the same grade. Does the scatterplot show a positive association?
(b) A group of four points appear to be outliers. In what way do these children's IQ and reading scores deviate from the overall pattern?
(c) Ignoring the outliers, is the form of the association between IQ and reading scores roughly a straight line? Is it very strong? Explain your answers.

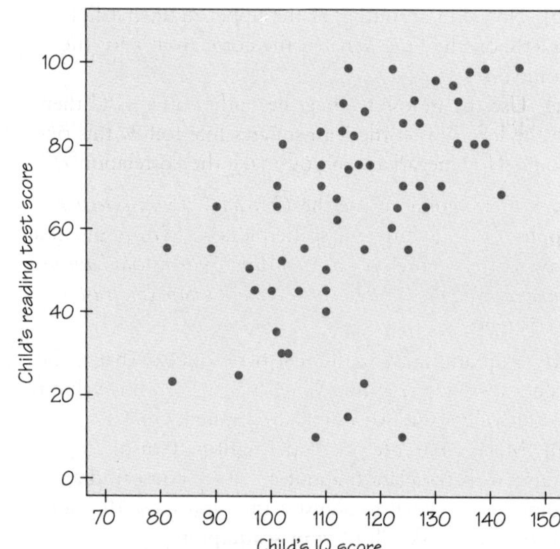

FIGURE 6.11 IQ and reading test scores for 60 fifth-grade children, for Exercise 46.

If you are in the labor force, the interviewer goes on to ask about employment. Any work for pay or in your own business the week of the survey counts you as employed. So does at least 15 hours of unpaid work in a family business. You are also employed if you have a job but didn't work because of vacation, being on strike, or for some other good reason. An unemployment rate of 4.7% means that 4.7% of the sample was unemployed, using the exact CPS definitions of both "labor force" and "unemployed."

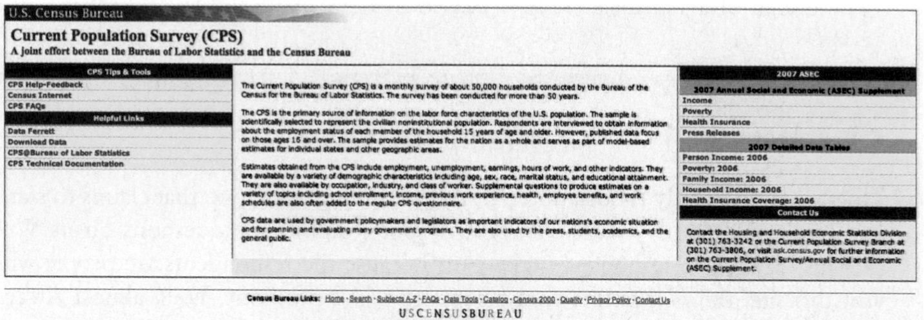

FIGURE 7.1 The Web page of the Current Population Survey.

7.2 Bad Sampling Methods

How can we choose a sample that is truly representative of the population? The easiest–but not the best–way to select a sample is to choose individuals close at hand. If we are interested in finding out how many people have jobs, for example, we might go to a shopping mall and ask people passing by if they are employed.

Convenience Sample — DEFINITION

A **convenience sample** is a sample of individuals who are selected because they are members of a population who are the most convenient to reach, such as people passing by in the street. Usually, such a sample cannot be trusted to be representative of the population.

EXAMPLE 2

The Inconvenient Truth About Convenience Samples

Taking a sample of shoppers at a mall seems like a fast and inexpensive way of finding out Americans' opinions. But people at malls tend to be more prosperous than typical Americans. They are also more likely to be teenagers or retired. Also, when we decide which people to question, we may tend (even unconsciously) to avoid poorly dressed or tough-looking individuals. In short, our shopping mall interviews will result in a sample that is not representative of the entire population because we underrepresent those people we may avoid.

Closer to home, your professor may try to "sample" the understanding the class has about a topic by calling on the next two students who raise their hands or by simply asking the nearest two students on the front row. If students who sit near the front and/or raise their hands have higher levels of preparation, interest and engagement, the professor may overestimate how well the class understands the material.

In both cases, the inaccuracies obtained cannot simply be explained as a sample's "bad luck," but are likely to happen every time with the same pattern because unscientific sampling methods have **bias**. In this context, bias refers matter-of-factly to the built-in systematic error of the procedure itself and not to the kind of political or personal bias an individual human being may have.

Bias DEFINITION

The design of a statistical study is **biased** if it systematically favors certain outcomes.

EXAMPLE 3 ■ Are Online Polls in Line?

The American Family Association (AFA) is a conservative group that claims to stand for "traditional family values." It regularly posts online poll questions on its Web site–just click on a response to take part. Because the respondents are people who visit this site, the poll results always support AFA's positions. Well, almost always. Recently, AFA's online poll asked about the heated issue of allowing same-sex marriage. Soon, email lists and social-network sites favored mostly by young liberals pointed to the AFA poll. Almost 850,000 people responded, and 60% of them favored legalization of same-sex marriage. AFA claimed that homosexual rights groups had skewed its poll. ■

Online polls are now everywhere–some sites will even provide help in conducting your own online poll. As the AFA poll illustrates, you can't trust the results. People who take the trouble to respond to an open invitation are not representative of the entire adult population. That's true of regular visitors to AFA's site, of the activists who made a special effort to vote in the marriage poll, and of the people who bother to respond to write-in, call-in, or online polls in general. Polls like these are examples of **voluntary response sampling.**

Voluntary Response Sample DEFINITION

A **voluntary response sample** consists of people who choose themselves by responding to a general appeal. Voluntary response samples are biased because people with strong opinions are most likely to respond.

7.3 Simple Random Samples

In a voluntary response sample, people choose whether to respond. In a convenience sample, the interviewer makes the choice. In both cases, personal choice produces bias. The statistician's remedy is to allow impersonal chance to choose the sample. A sample chosen by chance allows neither favoritism by the sampler nor self-selection by respondents. Choosing a sample by chance avoids bias by giving all individuals an equal chance to be chosen. Any individual has the same chance to be in the sample, whether rich or poor, young or old, black or white, and so on.

The simplest way to use chance to select a sample is to place names (the population) in a hat and draw out a handful (the sample). This is the idea of **simple random sampling.**

Simple Random Sample DEFINITION

A **simple random sample (SRS)** of size n consists of n individuals from the population chosen in such a way that every set of n individuals has an equal chance to be the sample actually selected.

Picturing drawing names from a hat helps us understand what an SRS is. The same picture helps us see that an SRS is a better method of choosing samples than convenience or voluntary response sampling because it doesn't favor any part of the population. But writing names on slips of paper and drawing them from a hat is slow and inconvenient. That's especially true if, as in the Current Population Survey, we must draw a sample of size 60,000. We can speed up the process by using a **table of random digits**. In practice, samplers use computers to do the work, but we can do it by hand for small samples.

Table of Random Digits DEFINITION

A **table of random digits** is a list of the digits 0, 1, 2, 3, 4, 5, 6, 7, 8, 9 with these two properties:

1. Each entry in the table is equally likely to be any of the 10 digits 0 through 9.
2. The entries are independent of one another. That is, knowledge of one part of the table gives no information about any other part.

Table 7.1 is a table of random digits. The digits in the table appear in groups of five to make the table easier to read, and the rows are numbered so we can refer to them, but the groups and row numbers are just for convenience. The entire table is one long string of randomly chosen digits. There are two steps in using the random-digit table to choose a simple random sample:

Step 1. Label Give each member of the population a numerical label of the *same length*. Up to 100 items can be labeled with two digits: 01, 02, . . . , 99, 00. Up to 1000 items can be labeled with three digits, and so on.

Step 2. Table To choose a simple random sample, read from Table 7.1 successive groups of digits of the length you used as labels. Your sample contains the individuals whose labels you find in the table. This gives all individuals the same chance because all labels of the same length have the same chance of being found in the table. For example, any pair of digits in the table is equally likely to be any of the 100 possible labels 01, 02, . . . , 99, 00. Ignore any group of digits that was not used as a label or that duplicates a label already in the sample.

EXAMPLE 4 ■ Sampling Songs

Professor Lesser has all 27 songs from *The Beatles One* CD stored on a digital media player and wants to play 4 randomly chosen songs to accompany his morning commute.

Step 1. Label Give each song a numerical label. Because two digits are needed to label the 27 songs, all labels will have two digits. In the table below, we have listed the 27 songs with labels from 01 to 27. Always specify how you label the members of the population. If the player had 500 songs, we would label them 001, 002, …, 499, 500.

(*Steve Prezant/Corbis.*)

01 Love Me Do	10 Help!	19 Hello, Goodbye
02 From Me to You	11 Yesterday	20 Lady Madonna
03 She Loves You	12 Daytripper	21 Hey Jude
04 Ticket to Ride	13 We Can Work it Out	22 Get Back
05 Can't Buy Me Love	14 Paperback Writer	23 All You Need is Love
06 A Hard Day's Night	15 Yellow Submarine	24 Something
07 I Feel Fine	16 Eleanor Rigby	25 Come Together
08 Eight Days a Week	17 Penny Lane	26 Let it Be
09 I Want to Hold Your Hand	18 The Ballad of John and Yoko	27 The Long and Winding Road

Step 2. Table Go to Table 7.1 and pick any row (we picked line 125). Now read across that row, left to right, two digits at a time, until you have chosen 4 songs. Remember to skip any two-digit groups that repeat numbers already chosen (like 18) and also skip any groups representing numbers beyond the size of our population (27). Here's how you view line 125 with the 4 selected songs in **bold**.

96 74 61 **21** 49 37 82 37 **18** 68 18 44 **23** 51 **19** 62 10 33 92 44

So our media player will play "Hey Jude," "The Ballad of John and Yoko," "All You Need is Love," and "Hello, Goodbye." Instead of using Table 7.1, you can select an SRS using a spreadsheet (for example, the Excel command RANDBETWEEN), statistical software, the *Simple Random Sample* applet (see Applet Exercise 1), or most types of calculators. For example, you can use: (MATH) → PRB → randInt(1, 27, 4) → (ENTER) on the TI-84 calculator. And, of course, many digital media players have a "shuffle" option! ■

Online polls and mall interviews produce samples. We can't trust results from these samples because they are chosen in ways that invite bias. We have more confidence in results from an SRS because it uses impersonal chance to avoid bias. The first question to ask about any sample is whether it was chosen at random. Opinion polls and other sample surveys carried out by people who know what they are doing use random sampling. Most national sample surveys use sampling schemes more complex than an SRS. They may, for example, dial the last four digits of a telephone number at random separately within each exchange (the area code and first three digits). The national sample is pieced together from many smaller samples. The big idea remains the deliberate use of *chance* to choose the sample. Because simple random sampling is the essential principle behind all random sampling and because it is also the main building block for more complex samples, we will focus our study on simple random sampling.

(John Kelly/The Image Bank/ Getty Images.)

EXAMPLE 5 ■ Exercising Judgment

A Gallup poll released in 2008 shows that Americans' self-reported rates of physical exercise have changed little since 2001. When asked how often each week they participated in "vigorous sports or physical activities for at least 20 minutes that cause large increases in breathing or heart rate," 45% of Americans answered "not at all." Can we trust that 45%? Ask first how Gallup selected its sample. Later in the press release we learn that the results are based on telephone interviews with a randomly selected national sample of 1014 adults, aged 18 and older, conducted November 11–14, 2007.

It is a good start toward gaining our confidence in the poll to know the intended population, the sample size, the tight window of time (so that there is minimal influence from changes in current events), and—most importantly—random selection. In the next section, we address a few other important considerations.

TABLE 7.1 Random Digits

101	19223	95034	05756	28713	96409	12531	42544	82853
102	73676	47150	99400	01927	27754	42648	82425	36290
103	45467	71709	77558	00095	32863	29485	82226	90056
104	52711	38889	93074	60227	40011	85848	48767	52573
105	95592	94007	69971	91481	60779	53791	17297	59335
106	68417	35013	15529	72765	85089	57067	50211	47487
107	82739	57890	20807	47511	81676	55300	94383	14893
108	60940	72024	17868	24943	61790	90656	87964	18883
109	36009	19365	15412	39638	85453	46816	83485	41979
110	38448	48789	18338	24697	39364	42006	76688	08708
111	81486	69487	60513	09297	00412	71238	27649	39950
112	59636	88804	04634	71197	19352	73089	84898	45785
113	62568	70206	40325	03699	71080	22553	11486	11776
114	45149	32992	75730	66280	03819	56202	02938	70915
115	61041	77684	94322	24709	73698	14526	31893	32592
116	14459	26056	31424	80371	65103	62253	50490	61181
117	38167	98532	62183	70632	23417	26185	41448	75532
118	73190	32533	04470	29669	84407	90785	65956	86382
119	95857	07118	87664	92099	58806	66979	98624	84826
120	35476	55972	39421	65850	04266	35435	43742	11937
121	71487	09984	29077	14863	61683	47052	62224	51025
122	13873	81598	95052	90908	73592	75186	87136	95761
123	54580	81507	27102	56027	55892	33063	41842	81868
124	71035	09001	43367	49497	72719	96758	27611	91596
125	96746	12149	37823	71868	18442	35119	62103	39244

SPOTLIGHT 7.1 Is It Really Random?

Are the random digits in Table 7.1 really random? Not a chance. They were produced by a computer program. A computer program implements an algorithm that does exactly what you tell it to do.

Give the program the same input and it will produce exactly the same "random" digits. You can get quite respectable random digits by calculating $\pi = 3.14159265358979\ldots$ to more and more decimal places. Go to http://oldweb.cecm.sfu.ca/pi/pi.html and you will see these digits stream by. You get the same digits on every visit, of course. Clever people have devised algorithms that produce output that *looks* like random digits. These are called "pseudo-random numbers," and that's what Table 7.1 contains. Pseudo-random numbers work fine for statistical randomizing, but they have hidden nonrandom patterns that can mess up more refined uses. *(continued on page 218)*

Is It Really Random? *(continued)*

For purists, the RAND Corporation long ago published a book titled *One Million Random Digits*. The book lists 1 million digits that were produced by a very elaborate physical randomization and really are random. An employee of RAND once commented that this is not the most boring book that RAND has ever published.

Cryptologists and computer scientists would like an endless supply of really random digits. Really random digits must come from nature, not from a computer program. Radioactive decay is really random, and so is the "thermal noise" in an amplifier, which you can hear as a soft whoosh if you turn up the volume with no music playing. Alas, extracting random digits from these really random sources requires various human devices, and these often impose subtle patterns. As of now, in fact, pseudo-random numbers from the best algorithms actually look more random than numbers refined from the randomness in nature by some human apparatus. "Easy to say, hard to do" applies to making random digits as well as to many other human aspirations.

7.4 Cautions About Sample Surveys

Random sampling eliminates bias in the choice of a sample from a list of the population. Sample surveys of large human populations, however, require more than a good sampling design.

To begin, we need an accurate and complete list of the population. Because such a list is rarely available, most samples suffer from some degree of **undercoverage**. A sample survey of households, for example, will miss not only homeless people but also prison inmates and students in dormitories. An opinion poll conducted by telephone will miss the 6% of American households without residential phones. Also, about 3% of the population (often younger adults) have cell phones only, and most random digit dialing does not select cell phones. The results of national sample surveys therefore have some bias if the people not covered–who most often are young or poor people–differ from the rest of the population.

Undercoverage — DEFINITION

Undercoverage occurs when some groups in the population are left out of the process of choosing the sample.

A more serious source of bias in most sample surveys is **nonresponse**, which occurs when a selected individual cannot be contacted or refuses to cooperate. Nonresponse to sample surveys often reaches 50% or more, even with careful planning and several callbacks. Because nonresponse is higher in urban areas, most sample surveys substitute other people in the same area to avoid favoring rural areas in the final sample. If the people contacted differ from those who are rarely at home or who refuse to answer questions, some bias remains.

Nonresponse — DEFINITION

Nonresponse occurs when an individual chosen for the sample can't be contacted or refuses to participate.

EXAMPLE 6 ■ How Bad Is Nonresponse?

The Current Population Survey (CPS) has the lowest nonresponse rate of any poll we know: Only about 4% of the households in the CPS sample refuse to take part and another 3% or 4% can't be contacted. People are more likely to respond to a government survey such as the CPS, and the CPS contacts its sample in person before doing later interviews by phone.

What about polls done by the media and by market research and opinion polling firms? We don't know their rates of nonresponse because they won't say. That nondisclosure is a bad sign. The Pew Research Center imitated a careful telephone survey and published the results: Out of 2879 households called, 1658 were never at home, refused, or would not finish the interview. That's a nonresponse rate of 58%.

When people do respond, we can't rely on them to always tell the truth. People know that they should take the trouble to vote, for example, so many who didn't vote in the last election will tell a pollster that they did.

EXAMPLE 7 ■ Encouraging Honesty

The Centers for Disease Control and Prevention previously used face-to-face interviews to ask Americans about their sexual activity and illegal drug use. In a new version of the survey, released in 2007, data were gathered using computer-assisted self-interviews in which each participant was alone in a room, heard questions through a headset, and touched a computer screen with responses.

Another tool for encouraging honesty with sensitive topics is "randomized response," invented by sociologist S. L. Warner in 1965. By introducing randomness in a structured way, researchers use their knowledge of probability distributions to get reasonably accurate information about the overall group, while allowing each potentially embarrassing answer to be "camouflaged."

Finally, the *wording of questions* strongly influences the answers given to a sample survey. Confusing or leading questions can introduce strong bias, and even minor changes in wording or order can change a survey's outcome. Here are some examples.

EXAMPLE 8 ■ Watch That Wording

How do Americans feel about government help for the poor? Only 13% think we are spending too much on "assistance to the poor," but 44% think we are spending too much on "welfare." How do the Scots feel about the movement to become independent from England? Well, 51% would vote for "independence for Scotland," but only 34% support "an independent Scotland separate from the United Kingdom." It seems that "assistance to the poor" and "independence" are nice, hopeful words. "Welfare" and "separate" are negative words. Other topics that have produced survey results that vary greatly with wording include abortion, gay rights, and affirmative action.

The statistical design of sample surveys is a science, but this science is only part of the art of sampling. Because of nonresponse, false responses, and the difficulty of posing clear and neutral questions, you should analyze critically before fully trusting reports about complicated issues based on surveys of large human populations.

Insist on knowing the exact questions asked, the rate of nonresponse, and the date and method of the survey before you trust a poll result.

7.5 Experiments

Sample surveys gather information on part of the population in order to draw conclusions about the whole. When the goal is to describe a population, statistical sampling is the right tool to use.

Suppose, however, that we want to study the response to a stimulus, to see how one variable affects another when we change existing conditions. For example:

- Will a new mathematics curriculum improve the scores of sixth-graders on a standard test of mathematics achievement?
- Will taking small amounts of aspirin daily reduce the risk of suffering a heart attack?
- Does a mother's smoking during pregnancy reduce the IQ of her child?

Studies that simply *observe and describe* are ineffective tools for answering these questions. **Experiments** give us clearer answers.

Experiment DEFINITION

An **experiment** deliberately imposes a *treatment* on individuals in order to observe their responses. The purpose of an experiment is to study whether the treatment *causes* a change in the response.

Experiments are the preferred method for examining the effect of one variable on another. By imposing the specific treatment of interest and controlling other influences, we can pin down cause and effect. A sample survey may show that two variables are related, but it cannot demonstrate that one causes the other. Statistics has something to say about how to arrange experiments, just as it suggests methods for sampling.

EXAMPLE 9 ■ An Uncontrolled Experiment

A college regularly offers a review course to prepare candidates for the Graduate Management Admission Test (GMAT) required by most graduate business schools. This year, it offered only an online version of the course. The average GMAT score of students in the online course was 10% higher than the long-time average for those who took the classroom review course. Is the online course more effective?

This experiment has a very simple design. A group of subjects (the students) were exposed to a treatment (the online course), and the outcome (GMAT scores) was observed. Here is the design:

$$\text{Online course} \rightarrow \text{Observe GMAT scores}$$

or, in general form

$$\text{Treatment} \rightarrow \text{Observe response}$$

Most laboratory experiments use a design like that in the example: Apply a treatment and measure the response. In the controlled environment of the laboratory, simple designs often work well. But field experiments and experiments with human subjects are exposed to more variable conditions and deal with more variable subjects. It isn't possible to control outside factors that can influence the outcome. With greater variability comes a greater need for statistical design.

A closer look at the GMAT review course showed that the students in the online review course were quite different from the students who in past years took the classroom course. In particular, they were older and more likely to be employed. An online course appeals to these mature people, but we can't compare their performance with that of the undergraduates who previously dominated the course. The online course might even be less effective than the classroom version. The effect of online versus in-class instruction is hopelessly mixed up with influences lurking in the background. Figure 7.2 shows the mixed-up influences in picture form. We say that student age and background is **confounded** with whatever effect the change to online instruction may have. In everyday usage, someone who is confounded is confused or mixed up. In statistics, confounded variables have their effects mixed together so that it's hard to tell what effect is due to each variable separately.

Confounding

DEFINITION

Variables, whether intentionally part of a study or not, are said to be **confounded** when their effects on the outcome cannot be distinguished from each other.

Online vs. classroom

CAUSE?

GMAT score after course

Students' ages and backgrounds

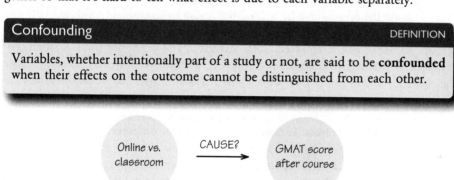

FIGURE 7.2
Confounding. We can't distinguish the effects of the treatment from those of other influences.

The remedy for confounding is to do a *comparative experiment* in which some students are taught in the classroom and other similar students take the course online. The first group is called a **control group**. Most well-designed experiments compare two or more treatments. Of course, comparison alone isn't enough to produce results we can trust. If the treatments are given to groups that differ markedly when the experiment begins, bias will result. For example, if we allow students to elect online or classroom instruction, older employed students are likely to sign up for the online course. Personal choice will bias our results in the same way that volunteers bias the results of call-in opinion polls. The solution to the problem of bias is the same for experiments and for samples: Use impersonal chance to select the groups.

EXAMPLE 10
A Randomized Comparative Experiment

The college decides to compare the progress of 25 on-campus students taught in the classroom with that of 25 students taught the same material online. Select which students will be taught online by taking a simple random sample of size 25 from the 50 available students. The remaining 25 students form the control group. They will receive classroom instruction. The result is a **randomized comparative experiment** with two groups. Figure 7.3 outlines the design in graphical form.

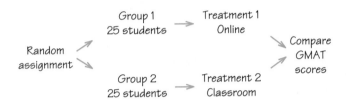

	Group 1 25 students	→	Treatment 1 Online	
Random assignment ↗↘				Compare GMAT scores
	Group 2 25 students	→	Treatment 2 Classroom	

FIGURE 7.3 The design of a randomized comparative experiment to compare online and classroom instruction.

The selection procedure is exactly the same as it is for sampling: label and table. Step 1: *Label* the 50 students 01 to 50. Step 2: Go to the *table* of random digits and read successive two-digit groups. The first 25 labels encountered select the online group. As usual, ignore repeated labels and groups of digits not used as labels. For example, if you begin at line 125 in Table 7.1, the first five students chosen are those labeled 21, 49, 37, 18, and 44. The *Simple Random Sample* applet makes it particularly easy to choose treatment groups at random.

The GMAT experiment is *comparative* because it compares two treatments (the two instructional settings). The experiment could have even had a third treatment if there had been a "hybrid" (part classroom, part online) course option. It is *randomized* because the subjects are assigned to the treatments by chance. Randomization creates groups that are similar to each other before we start the experiment. Possible confounding variables act on both groups at once so their effects tend to balance out and do not greatly impact the results of the study. The *only* difference between the groups is the online versus in-class setting. So if we see a difference in performance, it must be due to the different setting. That is the basic logic of randomized comparative experiments. This logic shows why experiments can give good evidence that the different treatments really *caused* different outcomes.

There is a fine point: The performance of the two groups will differ even if the treatments are identical, just because the individuals assigned at random to the groups differ. It is only differences *larger than would plausibly occur just by chance* that show the effects of the treatments. The laws of chance allow statisticians to say how big an effect is **statistically significant**. You have intuition for this concept because you would not think it was unusual if the first two children in an extended family were girls, but you would if the first ten were!

Statistical Significance — DEFINITION

An observed effect so large that it would rarely occur by chance is called **statistically significant**. ("Rarely" usually means $< 5\%$ of the time.)

EXAMPLE 11 ■ Cervical Cancer Screening

On October 18, 2007, *The New England Journal of Medicine* reported on a randomized comparative experiment in which 10,154 Canadian women (aged 30–69) were given two different cervical cancer screening tests (the standard Pap test and a test for the DNA of the HPV virus) in a randomly assigned sequence. Although it might have seemed "fair" to randomly assign each woman to one treatment or the other, it was deemed better to give each woman the benefit of both medical tests for ethical reasons. Because of the randomization in the sequence, however, the first test was able to be analyzed as if it had been done alone. The DNA-based test proved to be much more powerful than the Pap test in detecting cancer.

7.6 Experiments Versus Observational Studies

Randomized comparative experiments are common tools of industrial and academic research. They are also widely used in medical research. For example, federal regulations require that the safety and effectiveness of new drugs be demonstrated by randomized comparative experiments. Let's look at a medical experiment.

EXAMPLE 12 ■ St. John's Wort for Depression?

Although prescription drugs must pass the test of randomized comparative experiments before being sold, herbs and other "natural remedies" are exempt. Because these treatments are so popular, some are now being studied more carefully. Fans of natural remedies often use extracts of the herb St. John's wort to treat depression. Is the herb safe? Does it work? The *Journal of the American Medical Association* reported a "randomized, double-blind, placebo-controlled clinical trial" in which 200 patients with major depression were assigned at random to take either herb extract or a dummy pill that looked and tasted the same. Results: The herb is safe, but "[i]n this study, St. John's wort was not effective for treatment of major depression."

If you read accounts of medical studies, you will often meet language like "randomized, double-blind, placebo-controlled clinical trial." A clinical trial is a medical experiment with actual patients as subjects. "Randomized" and "controlled" tell us that this was a randomized comparative experiment (that's good). A "placebo" is a fake treatment, the dummy pill in this study. Here we meet a new idea, the importance of the **placebo effect**, a special kind of confounding. The placebo effect is the tendency of patients to respond favorably to any treatment, even a placebo. If depressed patients given St. John's wort are compared with patients who receive no treatment, the first group gets the benefit of both the herb and the placebo effect. Any beneficial effect that St. John's wort may have is confounded with the placebo effect. To prevent confounding, it is important that some treatment be given to all subjects in any medical experiment.

The depression study was a **double-blind experiment**: Neither the subjects nor the experimenters who worked with them knew which treatment any subject received. Subjects might react differently if they knew they were getting "only a placebo." Knowing that a particular subject was getting "only a placebo" could also influence the health workers who interviewed and examined the subjects. So both subjects and workers were kept "blind." Only the study's statistician knew which treatment each subject received.

The difference between the St. John's wort and placebo groups was *not statistically significant*—that is, it was no larger than would be expected when we divide 200 depressed patients at random into two groups and do nothing else. Larger numbers of subjects would give more precise results. It's unlikely that there is exactly *no* difference between St. John's wort and a placebo. If the clinical trial had used 2000 patients rather than 200, it might have picked up a small effect (in either direction). The researchers thought that 200 patients was enough to pick up any effect large enough to be medically important.

The logic of experimentation, the statistical design of experiments, and the laws that govern chance behavior combine to give compelling evidence of cause and effect. Only experimentation can produce the most convincing evidence of causation.

EXAMPLE 13 ■ Smoking and Health

By way of contrast, consider the statistical evidence linking cigarette smoking to lung cancer. We can't ethically assign groups of people to smoke or not, so a direct experiment isn't possible. The most careful studies have selected samples of smokers and nonsmokers, then followed them for many years, eventually recording the cause of death. These are called **prospective (observational) studies** because they follow the subjects forward in time. (A **retrospective study** looks backward in time.) Prospective studies are comparative, but they are not experiments because the subjects themselves choose whether or not to smoke. A large prospective study of British doctors found that the death rate from lung cancer among cigarette smokers was 20 times that among nonsmokers. Another study of American men aged 40 to 79 found that the lung cancer death rate was 11 times higher among smokers than among nonsmokers. These and many other **observational studies** show a strong connection between smoking and lung cancer.

Observational Study — DEFINITION

An **observational study** does not try to manipulate the environment (such as by assigning treatments to people), but simply observes the measurements of variables of interest that result from people's free choices. This kind of study is generally done when a treatment is unethical (for example, smoking while pregnant) and/or impossible (such as ethnicity) to assign to a person.

The connection between smoking and lung cancer is statistically significant. That is, it is far stronger than would occur by chance. We can be confident that something other than chance links smoking to cancer. But observation of samples cannot tell us *what* factors other than chance are at work. Perhaps there is something in the genetic makeup of some people that predisposes them both to nicotine addiction and to lung cancer. In that case, we would observe a strong link even if smoking itself had no effect on the lungs.

The statistical evidence that points to cigarette smoking as a cause of lung cancer is about as strong as nonexperimental evidence can be. First, the connection has been observed in many studies in many countries. This eliminates factors peculiar to one group of people or to one specific study design. Second, there is a *dose-response relationship*: People who smoke more are more likely to get lung cancer than those who smoke less, and quitting cigarettes reduces the cancer risk. Third, specific ways in which smoking could cause cancer have been identified–cigarette smoke contains tars that have been shown by experiment to cause tumors in animals. Finally, no plausible alternative explanation is available. For example, the genetic hypothesis cannot explain the increase in lung cancer among women that occurred as more and more women became smokers. Lung cancer, which has long been the leading cause of cancer deaths in men, has now passed breast cancer as the most fatal cancer for women. This evidence is convincing, but it is not quite as strong as the conclusive statistical evidence we get from randomized comparative experiments.

Despite their attractions, experiments can have weaknesses. The most common weakness is a *lack of realism* that makes it hard to say exactly how far the results of an experiment apply beyond controlled or contrived laboratory settings.

EXAMPLE 14 ■ Is the Experiment Realistic?

Clinical trials give medical treatments to actual patients with the condition that the treatments are supposed to help. Many experiments are less realistic. A psychologist studying the effects of stress on teamwork observes teams of students carrying out tasks in a psychology laboratory under different conditions. The students know it's "just an experiment" and that the stress will only last an hour. Do the conclusions of such experiments apply to real-life stress? An engineer uses a small pilot production process in a laboratory to find the choices of pressure and temperature that maximize yield from a complex chemical reaction. Do the results apply to a full-scale manufacturing plant?

These are not statistical questions. The psychologist and the engineer must use their understanding of psychology and engineering to judge how far their results apply. The statistical design enables us to trust the results for the students and the pilot process but not to generalize the conclusions to other settings. ■

7.7 Inference: From Sample to Population

A market research firm interviews a random sample of 2500 adults. Result: 66% find shopping for clothes frustrating and time consuming. This applies to the 2500 people in the sample. What is the truth about the 230 million American adults who make up the population? Because the sample was chosen at random, it's reasonable to think that these 2500 people represent the entire population fairly well. So the market researchers turn the *fact* that 66% of the *sample* find shopping frustrating into an *estimate* that about 66% of *all adults* feel this way. That's a fundamental operation in statistics: Use a fact about a sample to estimate the truth about the whole population. We call this *statistical inference.*

Statistical Inference — DEFINITION

Statistical inference refers to methods for drawing conclusions about an entire population on the basis of data from a sample. A **confidence interval** is one type of inference method.

If the selected individuals were chosen at random, we think that they fairly represent the population and inference makes sense. If we have data from only a convenience sample or a voluntary response sample, the data do not represent the population and we can't use them for inference. *Statistical inference works only if the data come from a random sample or randomized comparative experiment.* That's why this chapter starts with producing reliable data before moving on to inference from the data to a larger population.

To think about inference, we must keep straight whether a number describes a sample or a population (recall Section 7.1). Here is the vocabulary we use.

Parameter DEFINITION

A **parameter** is a fixed (usually unknown) number that describes a population.

Statistic DEFINITION

A **statistic** is a number that describes a sample. The value of a statistic is known when we have taken a sample, but it can change from sample to sample. We often use a statistic to estimate an unknown parameter.

EXAMPLE 15 ■ Do You Find Shopping Frustrating?

(*Carol Kohen/Getty Images.*)

Sample surveys show that fewer people enjoy shopping than in the past. A survey by the market research firm Yankelovich Clancy Shulman asked a nationwide random sample of 2500 adults if they agreed or disagreed that "I like buying new clothes, but shopping is often frustrating and time consuming." Of the respondents, 1650 said they agreed. The proportion of the sample who agree is

$$\hat{p} = \frac{1650}{2500} = 0.66 = 66\%$$

The symbol $\hat{p}$ is read "p-hat." The ^ symbol here tells us a quantity has been estimated, just as the use of $\hat{y}$ in Chapter 6 told us a value was estimated by using a regression line model. The number $\hat{p} = 0.66$ is a *statistic.* The corresponding *parameter* is the proportion (call it p) of all adult U.S. residents who would have said "Agree" if asked the same question. We don't know the value of the parameter p, so we use the statistic $\hat{p}$ to estimate it. ■

If Yankelovich took a second random sample of 2500 adults, the new sample would have different people in it. It is almost certain that there would not be exactly 1650 positive responses. That is, the value of the statistic $\hat{p}$ will vary from sample to sample. If the variation when we take repeat samples from the same population is too great, we can't trust the results of any one sample. We are saved by the second great advantage of random samples. The first advantage is that choosing at random eliminates favoritism. That is, random sampling avoids bias. The second advantage is that if we take lots of random samples of the same size from the same population, the variation from sample to sample will follow a predictable pattern.

All of statistical inference is based on one idea: to see how trustworthy a procedure is, ask what would happen if we repeated it many times. So we must ask, "What would happen if we took many samples?" Here's how to answer that question:

- Take a large number of random samples from the same population.
- Calculate the sample proportion $\hat{p}$ for each sample.

- Make a histogram of the values of $\hat{p}$.
- Examine the distribution displayed in the histogram for shape, center, and spread, as well as outliers or other deviations.

In practice it is too expensive to take many samples from a large population such as all adult U.S. residents. But we can use a computer to imitate drawing many samples at random from a population that we specify. This is called *simulation.* Here's what happens when we do this.

EXAMPLE 16 ■ What Happens in Many Samples?

Figure 7.4 illustrates a result of choosing many samples and finding the sample proportion $\hat{p}$ for each one. The histogram shows the distribution of the values of $\hat{p}$ from 1000 separate SRSs of size 100 drawn from a population that we suppose has a parameter value $p = 0.6$.

Of course, Yankelovich interviewed 2500 people, not just 100. Figure 7.5 is parallel to Figure 7.4. It shows a result from choosing 1000 SRSs, each of size 2500, from a population in which the true proportion is $p = 0.6$. The 1000 values of $\hat{p}$ from these samples form the histogram. Figures 7.4 and 7.5 are drawn on the same scale. Comparing them shows what happens when we increase the size of our samples from 100 to 2500. These histograms display the **sampling distribution** of the statistic $\hat{p}$ for two sample sizes. For intuition, consider a sample size of only 2. There would be only 3 possible $\hat{p}$ values from a sample of size 2: 0, 0.5, or 1. For a sample of size 100, there would be 101 possible $\hat{p}$ values: 0, 0.01, 0.02, ..., 1.00. Of course not all values are equally likely–the ones near 0.6 are most common. ■

Sampling Distribution — DEFINITION

The **sampling distribution** of a statistic is the distribution of values taken on by the statistic in all possible samples of the same size from the same population.

Strictly speaking, the sampling distribution is the ideal pattern that would emerge if we looked at all possible samples of the same size from our population. A distribution obtained from a fixed number of trials, like the 1000 trials in these figures, is only an approximation to the sampling distribution. Probability theory, the mathematics of chance behavior, can sometimes describe sampling distributions exactly. Chapter 8 will introduce you to basic probability theory. The interpretation of a sampling distribution is the same, however, whether we obtain it by simulation or by the mathematics of probability.

We can use the tools of data analysis from Chapter 5 to describe any distribution. Let's apply those tools to Figures 7.4 and 7.5.

- **Shape:** The histograms look normal. The normal curves drawn through the histograms describe the overall shape quite well.
- **Center:** In both cases, the values of the sample proportion $\hat{p}$ vary from sample to sample, but the values are centered at 0.6. Recall that we are assuming $p = 0.6$ is the true population parameter. Some samples have a $\hat{p}$ less than 0.6 and some greater, but there is no tendency to be always low or always high. That is, $\hat{p}$ has *no bias* as an estimator of p. This is true for both

large and small samples. (Want the details? The mean of the 1000 values of $\hat{p}$ is 0.598 for samples of size 100 and 0.6002 for samples of size 2500. The median value of $\hat{p}$ is exactly 0.6 for samples of both sizes.)

- **Spread:** The values of $\hat{p}$ from samples of size 2500 are much less spread out than the values from samples of size 100. In fact, the standard deviations are 0.051 for Figure 7.4 and 0.0097, or about 0.01, for Figure 7.5.

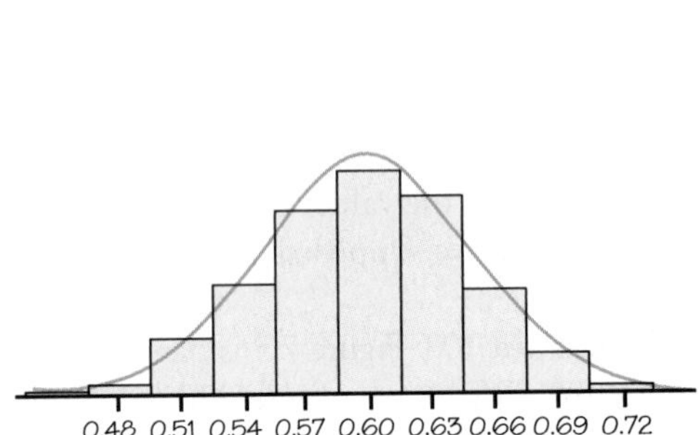

FIGURE 7.4 Draw 1000 SRSs of size 100 from a population with proportion $p = 0.60$ of successes. The histogram shows the distribution of the 1000 sample proportions $\hat{p}$.

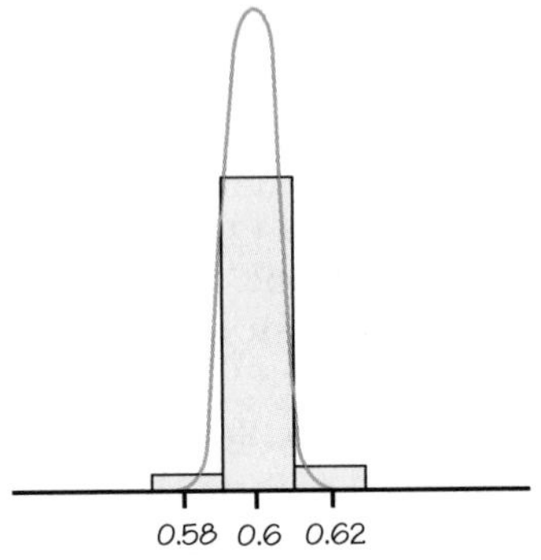

FIGURE 7.5 Draw 1000 SRSs of size 2500 from the same population as in Figure 7.4. The histogram shows the distribution of the 1000 sample proportions $\hat{p}$, using the same scale as Figure 7.6. The statistic from the larger sample is less variable.

Although these results describe just two sets of simulations, they reflect facts that are true whenever we use random sampling. We now turn to probability theory to learn the mathematical facts that lie behind the simulations. We'll use the word "success" for whatever we are counting, such as "Agree" responses in the shopping survey. Note that "success" does not necessarily have the positive (or negative) association it does in real life, but is simply a convenient way to identify an outcome.

Sampling Distribution of a Sample Proportion — THEOREM

Choose an SRS of size n from a large population that contains population proportion p of successes. Let $\hat{p}$ be the **sample proportion** of successes,

$$\hat{p} = \frac{\text{count of successes in the sample}}{n}$$

Then:

- **Shape:** For large ($n \geq 30$) sample sizes, the sampling distribution of $\hat{p}$ is *approximately normal.*
- **Center:** The *mean* of the sampling distribution of $\hat{p}$ is p.
- **Spread:** The *standard deviation* of the sampling distribution of $\hat{p}$ is

$$\sqrt{\frac{p(1-p)}{n}}$$

(This will be confirmed in Section 8.6.)

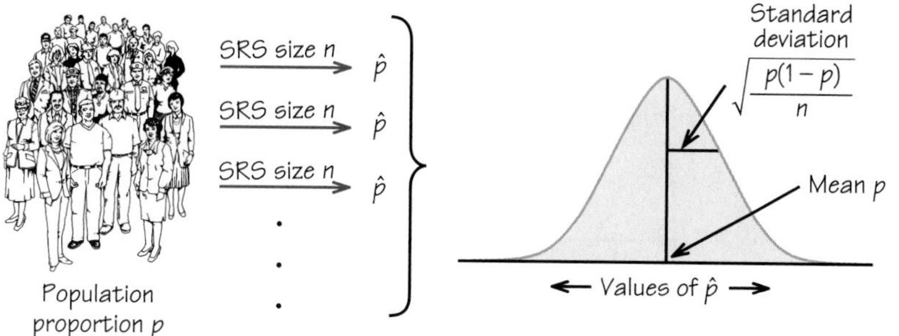

FIGURE 7.6 Repeat many times the process of selecting an SRS of size n from a population of which the proportion p are successes. The values of the sample proportion of successes $\hat{p}$ have this normal sampling distribution.

Figure 7.6 summarizes these facts in a form that reminds us that a sampling distribution describes the results of lots of samples from the same population.

EXAMPLE 17 ■ Do You Find Shopping Frustrating?

Suppose that 60% of all adults find shopping for clothes frustrating and time consuming. The population proportion is $p = 0.6$. Take a simple random sample of 2500 adults. That's exactly the setting of the second simulation in Example 16. Now we can apply mathematics to learn how the sample proportion $\hat{p}$ would behave if we took many samples. The distribution of $\hat{p}$ in many samples

- is close to normal;
- has mean 0.6;
- has standard deviation

$$\sqrt{\frac{p(1-p)}{n}} = \sqrt{\frac{(0.6)(0.4)}{2500}} = 0.0098$$

The mean 0.6 and standard deviation 0.0098 from the mathematics are very close to the mean 0.6002 and standard deviation 0.0097 we observed in our simulation. If the simulation used more than 1000 trials, the results would be yet closer to the mathematical truth. ■

7.8 Confidence Intervals

The sampling distribution shows why we can trust the results of a large random sample: Almost all such samples give results that are close to the truth about the population.

EXAMPLE 18 ■ The 68–95–99.7 Rule Again

In Example 17, the population parameter, the proportion of adults who find shopping frustrating, is $p = 0.6$. If we take SRSs of size 2500, the sample proportions $\hat{p}$ follow the normal distribution with mean 0.6 and standard deviation about 0.01. The 95 part of the 68–95–99.7 rule from Section 5.9 says that 95% of all samples give a $\hat{p}$ within 2 standard deviations of the truth about the population. So in this example, 95% of all samples have $\hat{p}$ within 2×0.01 of 0.6, that is, between 0.58 and 0.62. Figure 7.7 illustrates this use of the 68–95–99.7 rule. ■

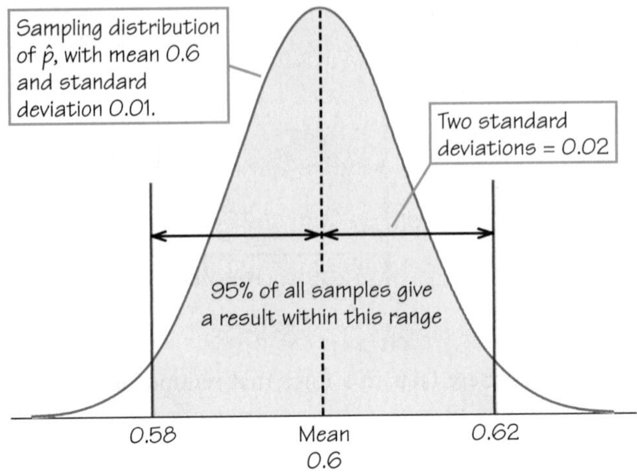

FIGURE 7.7 The sampling distribution of $\hat{p}$ for Example 17. By the 68–95–99.7 rule, 95% of all samples have a sample proportion $\hat{p}$ within ± 0.02 of the true population proportion $p = 0.6$.

We can repeat this reasoning for any value of the parameter p and the sample size n. It is always true that 95% of all samples give a sample proportion $\hat{p}$ within 2 standard deviations of the population proportion p. That is, 95% of all samples catch p in the interval extending 2 standard deviations on either side of $\hat{p}$. That's the interval

$$\hat{p} \pm 2\sqrt{\frac{p(1-p)}{n}}$$

This formula tells us how close the unknown parameter p lies to the observed statistic $\hat{p}$ in 95% of all samples. There is one catch: We can't calculate the interval from the data because the standard deviation involves the population proportion p, and in practice we don't know p. In Examples 17 and 18, we applied the formula for $p = 0.6$, but this may not be the true p for the actual population of all American adults.

What to do? The standard deviation of the statistic $\hat{p}$ does depend on the parameter p, but it doesn't change a lot when p changes. Go back to Example 17 and redo the calculation for other values of p. Here's the result:

Value of p	0.4	0.5	0.6	0.7	0.8
Standard deviation	0.0098	0.01	0.0098	0.0092	0.008

The standard deviations are all 0.01 when rounded to two places. You see that if we guess a value of p reasonably close to the true value, the standard deviation found from the guessed value will be about right. We know that when we take a large random sample, the statistic $\hat{p}$ is almost always close to the parameter p. So we will use $\hat{p}$ as the guessed value of the unknown p. Now we have an interval that we can calculate from the sample data. We call it a *confidence interval.*

Confidence Interval DEFINITION

A **95% confidence interval** is an interval obtained from the sample data by a method in which 95% of all samples will produce an interval containing the true population parameter.

Choose an SRS of size n from a large population that contains an unknown proportion p of successes. A 95% confidence interval for p is approximately

$$\hat{p} \pm 2\sqrt{\frac{\hat{p}(1-\hat{p})}{n}}$$

The ± sign is read "plus or minus," so 0.5 ± 0.2 yields two numbers, 0.3 and 0.7, which can be written as an interval: (0.3, 0.7).

This formula is only approximately correct but is quite accurate when the sample size n is large (≥ 030). Here $\hat{p}$ is the proportion of successes in the sample and $2\sqrt{\hat{p}(1-\hat{p})/n}$ is the **margin of error.**

Margin of Error DEFINITION

The **margin of error** is the number to the right of the ± sign in a 95% confidence interval and is equal to half of the width of the full interval. It equals about 2 standard deviations of the sampling distribution of the estimated parameter. If you conducted a very large number of polls, about 95% of the time the difference between a particular poll's result and the true value of the population parameter would be within the margin of error.

This interval is only approximately correct for two reasons. The sampling distribution of the sample proportion $\hat{p}$ isn't exactly normal. And we don't get the standard deviation of $\hat{p}$ exactly right because we used $\hat{p}$ in place of the unknown p. Both of these difficulties go away as the sample size n gets larger. Our method works well enough for many practical uses. More important, it shows how we get a confidence interval from the sampling distribution of a statistic. That's the reasoning behind any confidence interval.

EXAMPLE 19 ■ Risky Behavior in the Age of AIDS

How common is behavior that puts heterosexuals at risk for AIDS? In 1990–1991, the National AIDS Behavioral Survey interviewed a random sample of 2673 adult heterosexuals. Of these people, 170 had had more than one sexual partner in the past year. The sample proportion who admit to multiple partners is

$$\hat{p} = \frac{170}{2673} = 0.0636$$

A 95% confidence interval for the proportion p of all adult heterosexuals with multiple partners is therefore

$$\begin{aligned}\hat{p} \pm 2\sqrt{\frac{\hat{p}(1-\hat{p})}{n}} &= 0.0636 \pm 2\sqrt{\frac{(0.0636)(0.9364)}{2673}} \\ &= 0.0636 \pm 0.0094, \text{ or } 6.36\% \pm 0.94\% \\ &= 0.0542 \text{ to } 0.0730, \text{ or } 5.42\% \text{ to } 7.30\%\end{aligned}$$

A report of these calculations might say, "The study found that 6.36% of heterosexuals had more than one sexual partner. The margin of error for this result is 0.94%." ■

We got the interval in Example 19 by using a formula that catches the true unknown population proportion in 95% of all samples. The shorthand for this is: We are **95% confident** that the true proportion of heterosexuals with multiple partners lies between 5.42% and 7.30%. The margin of error refers to the spread needed to capture the true p in 95% of all samples. The truth lies outside the interval (5.42%, 7.30%) in 5% of all samples.

Figure 7.8 lays out the meaning of "95% confidence." The vertical line is the true value of the population proportion p. The normal curve at the top of the figure is the sampling distribution of the sample statistic $\hat{p}$, which is centered at the true p. The 95% confidence intervals from 25 SRSs appear below, one after the other. The central dots are the values of $\hat{p}$, the centers of the intervals. The arrows on either side span the confidence interval. In the long run, 95% of the intervals will cover the true p and 5% will miss. Of the 25 intervals in Figure 7.8, 24 hit and 1 misses. (Remember that the sampling distribution describes what happens in a very large number of samples–we don't expect exactly 95% of 25 intervals to capture the true parameter.) The *Confidence Interval* applet animates Figure 7.8. You can use the applet to watch confidence intervals from one sample after another capture or fail to capture the true parameter.

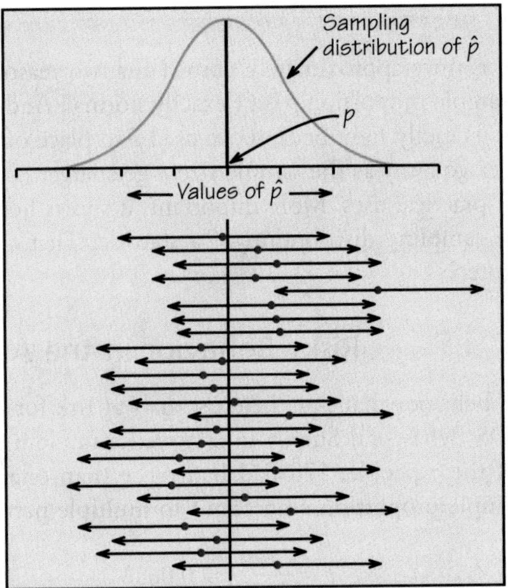

FIGURE 7.8 Twenty-five samples from the same population give these 95% confidence intervals. In the long run, 95% of all such intervals cover the true population proportion, marked by the vertical line.

The length of a confidence interval depends on the size n of the sample. Larger samples give shorter intervals because of the $\sqrt{n}$ in the denominator of the margin of error. But the interval does *not* depend on the size of the population. This is true as long as the population is much larger than the sample. The confidence interval in Example 19 works for a sample of 2673 from a city with 100,000 adults as well as for a sample of 2673 from a nation of 220 million. What matters is how many people we interview, not what percent of the population we contact.

The length of a confidence interval also depends on how confident we want to be that the interval does capture the true parameter value. It is common to use 95% confidence, but you can ask for higher or lower confidence if you want. Our 95% confidence interval was based on the middle 95% of a normal distribution. A 99% confidence interval requires the middle 99% of the distribution and so is wider (has a larger margin of error).

EXAMPLE 20 ■ Understanding the News

Here's what the TV news announcer says: "A new Gallup poll on American exercise habits finds that 45% of adults are not engaging in vigorous sports or physical activities. The margin of error for the poll was 3 percentage points." Plus or minus 3% starting at 45% is 42% to 48%. People with minimal statistics knowledge may think that the truth about the entire population must be in that interval, but now we know better!

This is the full background Gallup actually gives: "For results based on this sample, one can say with 95% confidence that the maximum error attributable to sampling and other random effects is 3 percentage points. In addition to sampling error, question wording and practical difficulties in conducting surveys can introduce error or bias into the findings of public opinion polls." That is, Gallup tells us that the margin of error only works for 95% of all its samples. "95% confidence" is shorthand for that. The news report left out the "95% confidence." In fact, *almost all margins of error in the news are for 95% confidence.* If you don't see the confidence level in a scientific poll, it's usually safe to assume 95%. ■

Gallup's mention of "question wording and practical difficulties" takes us back to our cautions about sample surveys. *The margin of error does not address nonresponse and other practical difficulties.* The margin of error in a confidence interval comes from the sampling distribution of the statistic. The sampling distribution describes the variation of the statistic due to chance in repeated random samples. This random variation is the *only* source of error covered by the margin of error. Real-life samples also suffer from undercoverage and nonresponse. Errors from these practical difficulties are usually more serious and harder to quantify than random sampling error. The actual error in sample surveys may be much larger than the announced margin of error. Worse, we can't say how much larger. Statistical conclusions are approximations to a complicated truth, not mathematical results that are simply true.

EXAMPLE 21 ■ Measuring Risky Behavior

What about the National AIDS Behavioral Surveys? The interviews were carried out by telephone. This is acceptable for surveys of the general population, because about 94% of American households have telephones. However, some groups at high risk for AIDS, such as intravenous drug users, often don't live in settled households and are underrepresented in the sample. About 30% of the people reached refused to cooperate. A nonresponse rate of 30% is not unusual in large sample surveys, but it may cause some bias if those who refuse differ systematically from those who cooperate. The survey used statistical methods that adjust for unequal response rates in different groups. Finally, some respondents may not have told the truth when asked about their sexual behavior. The survey team tried hard to make respondents feel comfortable. For example, Hispanic women were interviewed by Hispanic women, and Spanish speakers were interviewed by Spanish speakers with the same regional accent (Cuban, Mexican, or Puerto Rican). Nonetheless, the survey report says that some bias is probably present:

> *It is more likely that the present figures are underestimates; some respondents may underreport their numbers of sexual partners and intravenous drug use because of embarrassment and fear of reprisal, or they may forget or not know details of their own*

or of their partner's HIV risk and their antibody testing history. [Joseph H. Catania et al., Prevalence of AIDS-related risk factors and condom use in the United States, *Science*, 258 (1992): 1104.]

Reading the report of a large study like the National AIDS Behavioral Surveys reminds us that statistics in practice involves much more than formulas for confidence intervals. ■

SPOTLIGHT 7.2 Truth in Polling

Responsible polling organizations tell the public something about both the precision and limitations of their poll results. For example, here is the statement from the Harris Poll Web site that accompanied results of a poll (of about 1000 randomly selected people):

"In theory, with a probability sample of this size, one can say with 95 percent certainty that the results have a statistical precision of plus or minus 3 percentage points of what they would be if the entire adult population had been polled with complete accuracy. Unfortunately, there are several other possible sources of error in all polls or surveys that are probably more serious than theoretical calculations of sampling error. They include refusals to be interviewed (nonresponse), question wording and question order, interviewer bias, weighting by demographic control data and screening (e.g., for likely voters). It is difficult or impossible to quantify the errors that may result from these factors."

Your college student newspaper may not have the resources to conduct polls using random sampling, but it is refreshing when the polls it publishes from voluntary response samples are accompanied by a disclaimer such as this one from The University of Texas at El Paso's *The Prospector*:

"This poll is not scientific and reflects the opinions of only those Internet users who have chosen to participate. The results cannot be assumed to represent the opinions of Internet users in general, nor the public as a whole."

Because of this limitation, *The Prospector* simply reports the breakdown of responses given but without any margin of error, since sampling error cannot be quantified from a (voluntary response) sample that is not probability-based.

REVIEW VOCABULARY

Bias A systematic error that tends to cause the observations to deviate in the same direction from the truth about the population whenever a sample or experiment is repeated. (p. 214)

Confidence interval An interval of values used to estimate a population parameter with a specific level of confidence. A **95% confidence interval** is an interval computed from a sample by a method that surrounds the unknown parameter 95% of the time, so when we calculate the interval for a single sample, we are 95% confident that the interval contains the unknown parameter. (pp. 225, 230)

Confounding Two variables are confounded when their effects on the outcome of a study cannot be distinguished from each other. (p. 221)

Control group A group of experimental subjects that is given a standard treatment or no treatment (such as a placebo). (p. 221)

Convenience sample A sample that consists of the individuals who are most easily available, such as people passing by in the street. A convenience sample is usually biased. (p. 213)

Double-blind experiment An experiment in which neither the experimental subjects nor the persons who interact with them know which treatment each subject received. (p. 223)

Experiment A study in which treatments are applied to people, animals, or things in order to observe the effect of the treatments. (p. 220)

Margin of error The number to the right of the ± sign in a 95% confidence interval and is equal to half of the width of the full interval. It equals about 2 standard deviations of the sampling distribution of the estimated parameter. If you conducted a very large number of polls, about 95% of the time the difference between a particular poll's result and the true value of the population parameter would be within the margin of error. (p. 231)
Nonresponse Some individuals chosen for a sample cannot be contacted or refuse to participate. (p. 218)
Observational study A study (such as a sample survey) that observes individuals and measures variables of interest but does not attempt to influence the responses. (p. 224)
Parameter A number that describes the population. In statistical inference, the goal is often to estimate an unknown parameter or make a decision about its value. (p. 226)
Placebo effect The effect of a dummy treatment (such as an inert pill in a medical experiment) on the response of subjects. (p. 223)
Population The entire group of people or things about which we want information. (p. 212)
Prospective study An observational study that follows two or more groups of subjects forward in time. (p. 224)
Randomized comparative experiment An experiment to compare two or more treatments in which people, animals, or things are assigned to treatments by chance. (p. 222)
Retrospective study An observational study that uses interviews or records to collect information about past behaviors of subjects in two or more groups. (p. 224)
Sample A part of the population that is actually observed and used to draw conclusions, or inferences, about the entire population. (p. 212)
Sample proportion The proportion $\hat{p}$ of the members of a sample having some characteristic (such as agreeing with an opinion poll question). The sample proportion from a simple random sample is used to estimate the corresponding proportion p in the population from which the sample was drawn. (p. 228)
Sampling distribution The distribution of values taken by a statistic when all possible random samples of the same size are drawn from the same population. The sampling distributions of sample proportions are approximately normal. (p. 227)
Simple random sample (SRS) A sample chosen by chance, so that every possible sample of the same size has an equal chance to be the one selected. (p. 214)
Statistic A number that describes a sample. A statistic can be calculated from the sample data alone; it does not involve any unknown parameters of the population. (p. 226)
Statistical inference Methods for drawing conclusions about an entire population on the basis of data from a sample. Confidence intervals are one type of inference method. (p. 211)
Statistical significance An observed effect is statistically significant if it is so large that it is unlikely to occur just by chance in the absence of a real effect in the population from which the data were drawn. (p. 222)
Table of random digits A table whose entries are the digits 0, 1, 2, 3, 4, 5, 6, 7, 8, 9 in a completely random order. That is, each entry is equally likely to be any of the 10 digits and no entry gives information about any other entry. (p. 215)
Undercoverage The process of choosing a sample may systematically leave out some groups in the population, such as households without telephones. (p. 218)
Voluntary response sample A sample of people who choose themselves by responding to a general invitation to give their opinions. Such a sample is usually strongly biased. (p. 214)

✓ SKILLS CHECK

1. An opinion poll contacts 1021 adults and asks them, "Which political party do you think has better ideas for leading the country in the twenty-first century?" In all, 723 of the 1021 say, "The Democrats." The sample in this setting is

(a) all 220 million adults in the United States.
(b) the 1021 people interviewed.
(c) the 723 people who chose the Democrats.

2. A committee on community relations in a college town plans to survey local businesses about the importance of students as customers. From the 10,000 businesses listed in the telephone book, the committee chooses 150 businesses at random. Of these, 72 return the questionnaire mailed by the committee. The nonresponse rate is ________________ percent.

3. The sample in the setting of the previous exercise is

(a) all 10,000 businesses in the college town.
(b) the 150 businesses chosen.
(c) the 72 businesses that returned the questionnaire.

4. A call-in poll asks who people are planning to vote for in the next Presidential election. People who think major change is needed are likely to be represented in this poll ________ than they should be, if the goal is to get results that are representative of all voters.

5. On January 2, 2008, the American Idol Web site (www.americanidol.com) had an online poll that asked who you think would win among six former contestants. To become part of the sample, you simply clicked on a response. Of the 941,434 responses to this poll, 55% went to Clay Aiken. We can conclude that:

(a) most Americans prefer Clay Aiken out of those former contestants.
(b) the sample is too small a fraction of the millions of people who watched the TV show to draw any conclusion.
(c) the poll uses voluntary response, so the results tell us little about the population of all adults.

6. You are using the table of random digits to choose a simple random sample of 6 students from a class of 30 students. You label the students 01 to 30 in alphabetical order. Go to line 113 of Table 7.1. Of the labels corresponding to the six students selected for your sample, the label that is largest is ______________ .

7. You must choose an SRS of 10 of the 420 retail outlets in New York that sell your company's products. How would you label this population in order to use Table 7.1?

(a) 001, 002, 003, ..., 419, 420
(b) 000, 001, 002, ..., 419, 420
(c) 1, 2, 3, ...,419, 420

8. From an alphabetical list of the 7200 salaried employees of a corporation, you label the employees 0001 to 7200. Using line 111 of Table 7.1, choose an SRS of 5 of the 7200 salaried employees of a corporation. Of the five employees selected for your sample, the label that is the largest is ___________ .

9. A sample of households in a community is selected at random from the telephone directory. In this community, 4% of households have no telephone and another 35% have unlisted telephone numbers. The sample will certainly suffer from

(a) nonresponse.
(b) undercoverage.
(c) false responses.

10. To learn about the population of a county containing 100,000 people, a sample of 3000 households was selected to be interviewed. For 1200 of the 3000 households researchers attempted to contact, there was no one home willing to participate. For this survey, the nonresponse rate was ____________ percent.

11. A clinical trial compares an antidepression medicine with a placebo for relief of chronic headaches. There are 36 headache patients available to serve as subjects. To choose 18 patients to receive the medicine, you would

(a) assign labels 01 to 36 and use Table 7.1 to choose 18.
(b) assign labels 01 to 18 because only 18 need be chosen.
(c) assign the first 18 who signed up to get the medicine.

12. A study of cell phones and the risk of brain cancer looked at a group of 519 people who have brain cancer. The investigators matched each cancer patient with a person of the same sex, age, and race who did not have brain cancer, then asked about use of cell phones. This kind of study is known as ______________________ .

13. Studies that follow subjects forward in time are called

(a) retrospective.
(b) prospective.
(c) double-blind.

14. A treatment consisting of a "dummy pill" that looks like (but isn't) real medicine is known as a __________ .

15. A study of religious practices among college students interviewed a sample of 125 students; 105 of the students said that they prayed at least once in a while. The sample proportion who said they pray is what?

(a) 107
(b) 84
(c) 0.84

16. An opinion poll asks a simple random sample of 1000 adults how they view the state of the economy. Suppose that 35% of all adults would say "good" if they were asked. In repeated samples, the sample proportion $\hat{p}$ who say "good" would follow a normal distribution with mean having a value of ____________________ .

17. The standard deviation of the distribution of the sample proportion in the previous exercise is about

(a) 0.00023.
(b) 0.015.
(c) 0.03.

18. To the nearer half of a percentage point, the margin of error is ______ when we use the result of Exercise 15 to estimate what percent of all college students pray.

19. The sample survey in Exercise 15 actually called 150 students, but 23 of the students refused to say whether they pray. This nonresponse could cause the survey result to be in error. The error due to nonresponse

(a) is in addition to the margin of error found in Exercise 18.
(b) is included in the margin of error found in Exercise 18.
(c) can be ignored because it isn't random.

20. A survey of folk music fans yields this 95% confidence interval estimate of the proportion of fans who love the music of David Wilcox: 0.74 to 0.86. To the nearer percentage point, the margin of error for this survey is ______________________________ .

CHAPTER 7 EXERCISES

■ **Challenge** ◆ **Discussion**

7.1 Sampling

1. A Gallup poll asked, "In general, are you satisfied or dissatisfied with the way things are going in your personal life at this time?" 84% of Americans answered "satisfied." Interestingly, only 27% of Americans in the same survey said they were satisfied with the way things are going in the United States at this time. Gallup's report said, "Results are based on telephone interviews with 1027 national adults, aged 18 and older, conducted Dec. 6–9, 2007."

(a) What is the population for this sample survey?
(b) What is the sample size?

2. The American Community Survey (ACS) will replace the census "long form" starting with the 2010 census. The main part of the ACS contacts households at 250,000 addresses by mail each month, with follow-up by phone and in person if there is no response. Each household answers questions about its housing, economic, and social status. What is the population for the ACS?

7.2 Bad Sampling Methods

◆ **3.** You see a woman student standing in front of the student center, now and then stopping other students to ask them questions. She says that she is collecting student opinions for a class assignment. Explain why this sampling method is almost certainly biased.

◆ **4.** A member of Congress is interested in whether her constituents favor a proposed gun-control bill. Her staff reports that letters on the bill have been received from 361 constituents and that 323 of these oppose the bill. What is the population of interest? What is the sample? Is this sample likely to represent the population well? Explain your answer.

◆ **5.** Highway planners made a main street in a college town one-way. Local businesses were against the change. The local newspaper invited readers to call a telephone number to record their comments. The next day, the paper reported:

> *Readers overwhelmingly prefer two-way traffic flow to one-way streets. By nearly a 7-1 margin, callers to the newspaper's Express Yourself opinion line on Wednesday complained about the one-way streets that have been in place since May. Of the 98 comments received, all but 14 said no to one-way.*

(a) What population do you think the newspaper wants information about?
(b) Is the proportion of this population who favor one-way streets almost certainly larger or smaller than the proportion 14/98 in the sample? Why?

◆ **6.** Your college wants to gather student opinion about a proposed student fee increase. It isn't practical to contact all students.

(a) Give an example of a way to choose a sample of students that is poor practice because it depends on voluntary response.
(b) Give another example of a bad way to choose a sample that doesn't use voluntary response.

7.3 Simple Random Samples

7. You have just been blessed with triplets (all girls). You decide to select their names using an SRS of three names from the following list of the most popular names given to American girls born in this decade. To do this, use Table 7.1, starting at line 117.

1) Emily	2) Madison	3) Hannah	4) Emma
5) Ashley	6) Alexis	7) Samantha	8) Sarah
9) Abigail	10) Olivia	11) Elizabeth	12) Alyssa
13) Jessica	14) Grace	15) Lauren	16) Taylor
17) Kayla	18) Brianna	19) Isabella	20) Anna

8. (a) Would pulling out and lining up several dollar bills to use the 8-digit serial numbers be a reasonable substitute for Table 7.1? Explain.
(b) How about using the telephone numbers on a page of the phone book? Explain.

9. There are approximately 371 active telephone area codes covering Canada, the United States, and some Caribbean areas. (More are created regularly.) You want to choose an SRS of 25 of these area codes for a study of available telephone numbers.

(a) How would you label the area codes in order to use Table 7.1?
(b) Use Table 7.1, starting at line 125, to choose the first 3 members of this sample.

10. Each March, the Current Population Survey is expanded to gather a wider variety of information. On the Bureau of Labor Statistics Web site, you can find data from this survey on 14,959 people aged 25 to 64 whose highest level of education is a bachelor's degree. Think of these people as a population.

(a) In order to select an SRS of these people, how would you assign labels?
(b) Use Table 7.1, starting at line 107, to choose the first three members of the SRS.

◆ **11.** In using Table 7.1 repeatedly to choose samples, you should not always begin at the same place, such as line 101. Why not?

■ **12.** Which of the following statements are true of a table of random digits and which are false? Explain your answers.

(a) There are exactly four 0's in each row of 40 digits.
(b) Each pair of digits has chance 1/100 of being 00.
(c) The digits 0000 can never appear as a group because this pattern is not random.

■ **13.** The last stage of the Current Population Survey uses a *systematic sample*. An example will illustrate the idea of a systematic sample. Suppose that we must choose 4 rooms out of the 100 rooms in a dormitory. Because 100/4 = 25, we can think of the list of 100 rooms as 4 lists of 25 rooms each. Choose 1 of the first 25 rooms at random, using Table 7.1. The sample will contain this room and the rooms 25, 50, and 75 places down the list from it. If 13 is chosen, for example, then the systematic random sample consists of the rooms numbered 13, 38, 63, and 88.

(a) Use Table 7.1 to choose a systematic random sample of 5 rooms from a list of 200. Enter the table at line 120.
(b) Your sample gives every room the same chance to be chosen. Explain why. Yet this systematic sample is not a simple random sample. Explain why.

■ **14.** At a party there are 30 students over age 21 and 20 students under age 21. You choose at random 3 of those over 21 and separately choose at random 2 of those under 21 to interview about attitudes toward alcohol. You have given every student at the party the same chance to be interviewed: What is that chance? Why is your sample not an SRS?

7.4 Cautions About Sample Surveys

◆ **15.** An opinion poll calls 1334 randomly chosen residential telephone numbers, then the interviewer asks to speak with an adult member of the household to ask, "How many movies have you watched in a movie theater in the past 12 months?"

(a) What population do you think the poll has in mind?
(b) In all, 931 people respond. What is the rate (percent) of nonresponse?
(c) Many responses to this question are likely to be inaccurate. Why?

◆ **16.** Randomized Response: Suppose 30 students in a class participate in a survey in which they each flip a coin and do not tell the result. If the result was "heads," the student is supposed to say "yes." If the result was "tails," the student is supposed to give an honest answer to the question "Have you ever used a fake ID?" Suppose the results in the class are 18 "yes" answers and 12 "no" answers.

(a) If students follow the procedure correctly, is it true that all students who answered "no" have not used a fake ID?
(b) If students follow the procedure correctly, is it true that all students who have not used a fake ID answered "no"?
(c) On average, about half of the students who have not used a fake ID flipped "tails," so what is your best estimate of the true number of students who have not used a fake ID?
(d) Based on the answer to part (c), what is your estimate of the true number and proportion of students who have used a fake ID?
(e) Do we have any way to know which of the 18 "yes" answers are truthful?

◆ **17.** The wording of questions can strongly influence the results of a sample survey. You are writing an opinion poll question about a proposed amendment to the Constitution. You can ask if people are in favor of "changing the Constitution" or "adding to the Constitution" by approving the amendment. One of these choices of wording will likely produce a much higher percent in favor. Which one? Why?

7.5 Experiments

◆ **18.** As reported in College Teaching in 2006, R. L. Garner randomly assigned 117 undergraduates to "review lecture videos" on statistics research methods; the videos either did or did not have short bits of humor inserted. Students who viewed the humor-added version of the video gave significantly higher ratings in their opinion of the lesson, how well the lesson communicated information, and quality of the instructor. Even more importantly, that same group of students also recalled and retained significantly more information on the topic. What are the explanatory and response variables? Why is this an experiment? Why were students not initially told that the true purpose of the study was to assess the use of humor? Why do you think the study was done using a fixed video format rather than through live teaching?

◆ **19.** Could the magnetic fields from power lines cause leukemia in children? Investigators spent five years and $5 million comparing 638 children who had leukemia and 620 who did not. They went into the homes and actually measured the magnetic fields in the children's bedrooms, in other rooms, and at the front door. They recorded facts about nearby power lines for the family home and also for the mother's residence when she was pregnant. Result: No evidence of more than a chance connection between magnetic fields and childhood leukemia. Explain carefully why this study is *not* an experiment and what kind of study it is.

◆ **20.** A typical hour of prime-time television shows three to five violent acts. Linking family interviews and police records shows a clear association between time spent watching TV as a child and later aggressive behavior.

(a) Explain why this is an observational study rather than an experiment.
(b) Suggest several variables describing a child's home life that may be confounded with how much TV he or she watches. Explain why confounding makes it difficult to conclude that more TV *causes* more aggressive behavior.

◆ **21.** The Nurses' Health Study has interviewed a sample of more than 100,000 female registered nurses every two years since 1976. Beginning in 1980, the study asked questions about diet, including alcohol consumption. The researchers concluded that "light-to-moderate drinkers had a significantly lower risk of death" than either nondrinkers or heavy drinkers.

(a) Is the Nurses' Health Study an observational study or an experiment? Why?
(b) What does "significant" mean in a statistical report?
(c) Suggest some confounding variables that might explain why moderate drinkers have lower death rates than nondrinkers. (The study adjusted for these variables.)

22. You can use your computer to make telephone calls over the Internet. How would cost affect the behavior of users of this service? You will offer the service to all 200 rooms in a college dormitory. Some rooms will pay a low flat rate. Others will pay higher rates at peak periods and very low rates off-peak. You are interested in the amount and time of use and in the effect on the congestion of the network. Outline the design of an experiment to study the effect of rate structure.

23. Will classroom programs explaining the health advantages of drinking water rather than sugary sodas reduce obesity among children aged 7 to 11 years? Because children are already in school classrooms, we must randomize classes rather than individual children. An experiment assigned 15 classes to receive the program and another 14 to form a control group. After 12 months, obesity had increased in the control group and remained steady in the treatment group. Outline the design of the experiment, label the available classes, and use Table 7.1, beginning at line 103, to carry out the random assignment.

◆ **24.** A college allows students to choose either classroom or self-paced instruction in a basic mathematics course. The college wants to compare the effectiveness of self-paced and regular instruction. Someone proposes giving the same final exam to all students in both versions of the course and comparing the average score of those who took the self-paced option with the average score of students in regular sections.

(a) Explain why confounding makes the results of that study worthless.
(b) Given 30 students who are willing to use either regular or self-paced instruction, outline an experimental design to compare the two methods of instruction. Then use Table 7.1, starting at line 108, to carry out the randomization.

25. Will people spend less on health care if their health insurance requires them to pay some part of the cost themselves? An experiment on this issue asked if the percent of medical costs that are paid by health insurance has an effect either on the amount of medical care that people use or on their health. The treatments were four insurance plans. Each plan paid all medical costs above a ceiling. Below the ceiling, the plans paid 100%, 75%, 50%, or 0% of costs incurred. Outline the design of a randomized comparative experiment suitable for this study.

26. Track down a print or online copy of the Bible. The opening chapter of the book of Daniel (especially verses 12–16) appears to have the first clinical trial in recorded history. Outline the design of the experiment. Discuss how you know whether it is an uncontrolled experiment, a comparative experiment, or a randomized comparative experiment.

27. Stores advertise price reductions to attract customers. What type of price cut is most attractive? Market researchers prepared ads for athletic shoes announcing different levels of discounts (20%, 40%, or 60%). The student subjects who read the ads were also given "inside information" about the fraction of shoes on sale (50% or 100%). Each subject then rated the attractiveness of the sale on a scale of 1 to 7.

(a) Each treatment in this experiment combines values of two explanatory variables, discount level and fraction on sale. List the treatments. How many treatments are there?
(b) Outline a randomized comparative experiment using 60 student subjects. Use Table 7.1 at line 123 to choose the subjects for the first treatment.

28. Healthcare providers are giving more attention to relieving the pain of cancer patients. An article in the journal *Cancer* surveyed a number of studies and concluded that controlled-release (CR) morphine tablets, which release the painkiller gradually over time, are more effective than giving standard morphine when the patient needs it. The "methods" section of the article begins: "Only those published studies that were controlled (i.e., randomized, double-blind, and comparative), repeated-dose studies with CR morphine tablets in cancer pain patients were considered for this review." Explain the terms in parentheses to someone who knows nothing about medical trials.

29. Eye cataracts are responsible for over 40% of blindness around the world. Can drinking tea regularly slow the growth of cataracts? We can't experiment on people, so we use rats as subjects. Researchers injected 14 young rats with a substance that causes cataracts. Half the rats also received tea extract; the other half got a placebo. The response variable was the growth of cataracts over the next six weeks. Yes, the tea extract did slow cataract growth.

(a) Outline the design of this experiment.
(b) Use Table 7.1, starting at line 108, to assign rats to treatments.

■ **30.** The rats in the previous exercise were labeled 01 to 14 in order to use the table of random digits. Unknown to the researchers, the 5 rats labeled 01 to 05 have a genetic defect that favors cataracts. If we simply put rats 01 to 07 in the tea group, the experiment would be biased against tea. We can observe how random selection works to reduce bias by keeping track of how many of these 5 rats get assigned to the tea group. Carry out the random assignment of 7 rats to the tea group 20 times, keeping track of how many of rats 01 to 05 are in the tea group each time. Make a histogram of the count of rats 01 to 05 assigned to tea. What is the average number in your 20 tries?

7.6 Experiments Versus Observational Studies

◆ **31.** People who eat lots of fruits and vegetables have lower rates of colon cancer than those who eat little of these foods. Fruits and vegetables are rich in antioxidants such as vitamins A, C, and E. Will taking antioxidants help prevent colon cancer? A clinical trial studied this question with 864 people who were at risk for colon cancer. The subjects were divided into four groups: daily beta-carotene, daily vitamins C and E, all three vitamins every day, and daily placebo. After four years, the researchers were surprised to find no significant difference in colon cancer among the groups.

(a) Outline the design of the experiment. Use your judgment in choosing the group sizes.
(b) Assign labels to the 864 subjects and use Table 7.1, starting at line 118, to choose the first five subjects for the beta-carotene group.
(c) The study was double-blind. What does this mean?
(d) What does "no significant difference" mean in describing the outcome of the study?
(e) Suggest some characteristics of the kind of people who eat lots of fruits and vegetables that might explain lower rates of colon cancer. The experiment suggests that these variables, rather than the antioxidants, may be responsible for the observed benefits of fruits and vegetables.

◆ **32.** The financial aid office of a university asks a sample of students about their employment and earnings. The report says that "for academic year earnings, a statistically significant difference was found between the sexes, with men earning more on the average. No significant difference was found between the earnings of black and white students." Explain both of these conclusions, for the effects of sex and of race on average earnings, in language understandable to someone who knows no statistics.

◆ **33.** Do those high center brake lights, required on all cars sold in the United States since 1986, really reduce rear-end collisions? Randomized comparative experiments with fleets of rental and business cars, done before the lights were required, showed that the third brake light reduced rear-end collisions by as much as 50%. Alas, requiring the third light in all cars led to only a 5% drop. Explain why the experiment did not realistically imitate conditions after the lights were required.

◆ **34.** A psychologist studies how much people disclose about themselves to other people met at a party. He arranges for student subjects to be introduced to new people. The subjects are both female and male and both black and white. The results show that "there were no significant race effects, but self-disclosure was significantly higher among females than among males." Explain what this means in language understandable to someone who knows no statistics. Do not use the word *significance* in your answer.

◆ **35.** In the July 15, 2007 issue of *Cancer*, a study reported on 533,715 women at least 40 years old who were diagnosed with invasive breast cancer and reported to the National Cancer Data Base. The study found strong evidence that patients without health insurance were more likely to have a more advanced stage (i.e., III or IV) of cancer. Is this an experiment or observational study and how do you know?

7.7 Inference: From Sample to Population

36. An opinion poll uses random digit dialing equipment to dial 2000 randomly chosen residential telephone numbers. Of these, 631 are unlisted numbers. This isn't surprising, because 35% of all residential numbers are unlisted. For each underlined number, state whether it is a parameter or a statistic.

37. The Tennessee STAR experiment randomly assigned children to regular or small classes during their first four years of school. When these children reached high school, 40.2% of blacks from small classes took the ACT or SAT college entrance exams. Only 31.7% of blacks from regular classes took one of these exams. For each underlined number, state whether it is a parameter or a statistic.

38. The College Alcohol Study interviewed an SRS of 14,941 college students about their drinking habits. Suppose that half of all college students "drink to get drunk" at least once in a while. That is, $p = 0.5$.

(a) What are the mean and standard deviation of the proportion $\hat{p}$ of the sample who drink to get drunk?
(b) In what range of values do the proportions $\hat{p}$ from 95% of all samples fall?
(c) In what range of values do the proportions $\hat{p}$ from 99.7% of all samples fall?

39. Harley-Davidson motorcycles make up 14% of all the motorcycles registered in the United States. You plan to interview an SRS of 500 motorcycle owners.

(Peter Turnley/Corbis.)

(a) What is the approximate distribution of the proportion of your sample who own Harleys?
(b) In 95% of all samples like this one, the proportion of the sample who own Harleys will fall between ____________ and ____________. What are the missing numbers?

40. Exercise 38 asks what values the sample proportion $\hat{p}$ is likely to take when the population proportion is $p = 0.5$ and the sample size is $n = 14{,}941$. What range covers the middle 95% of values of $\hat{p}$ when $p = 0.5$ and $n = 1000$? When $n = 4000$? When $n = 16{,}000$? What general fact about the behavior of $\hat{p}$ do your results illustrate?

■ **41.** You can use a table of random digits to *simulate* sampling from a population. Suppose that 60% of the population bought a lottery ticket in the last 12 months. We will simulate the behavior of random samples of size 40 from this population.

(a) Let each digit in the table stand for one person in this population. Digits 0 to 5 stand for people who bought a lottery ticket, and 6 to 9 stand for people who did not. Why does looking at one digit from Table 7.1 simulate drawing one person at random from a population with 60% "yes"?
(b) Each row in Table 7.1 contains 40 digits. So the first 10 rows represent the results of 10 samples. How many digits between 0 and 5 does the top row contain? What is the percent of "yes" responses in this sample? How many of your 10 samples overestimated the population truth 60%? How many underestimated it? You could program a computer to continue this process, say 1000 times, to produce a pattern like that in Figure 7.4.

7.8 Confidence Intervals

42. In a random sample of students who took the SAT Reasoning college entrance examination twice, it was found that 427 of the respondents had paid for coaching courses and that the remaining 2733 had not. Give a 95% confidence interval for the proportion of coaching among students who retake the SAT.

43. A Gallup poll asked each of 1785 randomly selected adults whether she happened to attend a house of worship in the previous seven days. Of the respondents, 750 said "yes." Give a 95% confidence interval for the proportion of all adults who claim that they attended a house of worship during the week preceding the poll. (The proportion who actually attended may be lower–some people say "yes" if they often attend, even if they didn't attend that particular week.)

44. *The New York Times* and CBS News conducted a nationwide survey of 1048 randomly selected 13- to 17-year-olds. Of these teenagers, 692 had a television in their room.

(a) Give a 95% confidence interval for the proportion of all teens who have a TV set in their room.
(b) The news article says, "In theory, in 19 cases out of 20, the survey results will differ by no more than three percentage points in either direction from what would have been obtained by seeking out all American teenagers." Explain how your results agree with this statement.

◆ **45.** A telephone survey of 880 randomly selected drivers asked, "Recalling the last 10 traffic lights you drove through, how many of them were red when you entered the intersections?" Of the 880 respondents, 171 admitted that at least one light had been red.

(a) Give a 95% confidence interval for the proportion of all drivers who ran one or more of the last 10 red lights they met.
(b) A practical problem with this survey is that people may not give truthful answers. What is the likely direction of the bias: Do you think more or fewer than 171 of the 880 respondents really ran a red light? Why?

◆ **46.** The Harris poll asked a sample of 1009 adults which causes of death they thought would become more common in the future. Topping the list was gun violence: 70% of the sample thought deaths from guns would increase.

(a) How many of the 1009 people interviewed thought deaths from gun violence would increase?
(b) Harris says that the margin of error for this poll is plus or minus 3 percentage points. Explain to someone

who knows no statistics what "margin of error plus or minus 3 percentage points" means.
(c) Give a 95% confidence interval for this survey. Does your margin of error agree with the 3 percentage points announced by Harris?

◆ **47.** Consider the margin of error formula

$$2\sqrt{\hat{p}(1-\hat{p})/n}\,.$$

(a) For a fixed value of n, what value of $\hat{p}$ causes this formula to be the largest value it can be?
(b) Using the answer to part (a), what is a simplified (and slightly more conservative) formula for the margin of error?

◆ **48.** A news article reports that in a recent Gallup poll, 78% of the sample of 1108 adults said they believe there is a heaven. Only 60% said they believe there is a hell. The news article ends, "The poll's margin of sampling error was plus or minus four percentage points." Can we be certain that between 56% and 64% of all adults believe there is a hell? Explain your answer.

◆ **49.** A survey of Internet users found that males outnumbered females by nearly 2 to 1. This was a surprise, because earlier surveys had put the ratio of men to women closer to 9 to 1. Later in the article we find that surveys were sent to 13,000 organizations and that 1468 of these responded. The survey report claims that "the margin of error is 2.8 percent, with 95 percent confidence."

(a) What was the *response rate* for this survey? (The response rate is the percent of the planned sample that responded.)
(b) Do you think that the small margin of error is a good measure of the accuracy of the survey's results? Explain your answer.

50. A recent Gallup poll found that 68% of adult Americans favor teaching creationism along with evolution in public schools. The Gallup press release says:

> *For results based on samples of this size, one can say with 95 percent confidence that the maximum error attributable to sampling and other random effects is plus or minus 3 percentage points.*

Give one example of a source of error in the poll result that is *not* included in this margin of error.

■ **51.** The Internal Revenue Service plans to examine an SRS of individual income tax returns from each state that were filed electronically. One variable of interest is the proportion of returns that were filed by a tax practitioner rather than by an individual taxpayer. The total number of e-filed tax returns in a state varies from 4.9 million in California to 97,000 in Vermont.

(a) Will the margin of error for estimating the proportion change from state to state if an SRS of 1000 e-filed returns is selected in each state? Explain your answer.
(b) Will the margin of error change from state to state if an SRS of 1% of all e-filed returns is selected in each state? Explain your answer.

■ **52.** Exercise 46 describes a Harris poll that interviewed 1009 people. Suppose you want a margin of error half as large as the one you found in that exercise. How many people must you plan to interview?

■ **53.** Though opinion polls usually make 95% confidence statements, some sample surveys use other confidence levels. The monthly unemployment rate, for example, is based on the Current Population Survey of about 60,000 households. The margin of error in the unemployment rate is announced as about ±0.15% with 90% confidence. Is the margin of error for 90% confidence larger or smaller than the margin of error for 95% confidence? Why? (*Hint:* Look at Figure 7.7 again.)

Chapter Review

54. The proportion of one's body that is fat is a key indicator of fitness. The many ways to estimate this have different margins of error (given in percentage points):

method	calipers pinch	bioelectrical impedance	body mass index calculator	hydrostatic weighing (dunk test)
margin of error	±3	±4	±10	±1

(a) Which of these tests is the least accurate?
(b) If the pinch test says you have 21% body fat, what is the 95% confidence interval for this estimate?

55. Many medical trials randomly assign patients to either an active treatment or a placebo. These trials are always double-blind. Sometimes the patients can tell whether or not they are getting the active treatment. This defeats the purpose of blinding. Reports of medical research usually ignore this problem. Investigators looked at a random sample of 97 articles reporting on placebo-controlled randomized trials in the top five general medical journals. Only 7 of the 97 discussed the success of blinding. Give a 95% confidence interval for the proportion of all such articles that discuss the success of blinding. [Dean Fergusson et al., Turning a blind eye: The success of blinding reported in a random sample of randomised, placebo-controlled trials, *British Medical Journal,* 328 (2004): 432–436.]

56. Tomeka wants to ask a sample of students at her college, "Do you think that Social Security will still be paying benefits when you retire?" She obtains the college email addresses of the 2654 students.

(a) How would you label the addresses in order to choose a simple random sample of 100 students?
(b) Use Table 7.1, starting at line 103, to choose the first three labels in the sample.
(c) Tomeka sends her question by email to the 100 addresses in her sample. Although she has chosen an

SRS, a serious practical difficulty may make it hard to draw clear conclusions from her sample. What practical difficulty do you expect Tomeka to encounter?

■ **57.** Suppose that exactly 10% of all articles in major medical journals that describe placebo-controlled randomized trials discuss the success of blinding. That is, the proportion of "successes" in the population is $p = 0.1$. What is the approximate probability that fewer than 7% of an SRS of 97 articles from this population discuss the success of blinding?

58. Ability to grow in shade may help pines found in the dry forests of Arizona resist drought. How well do these pines grow in shade? Investigators planted pine seedlings in a greenhouse in either full light or light reduced to 5% of normal by shade cloth. At the end of the study, they dried the young trees and weighed them.

(a) Explain why this study is an experiment.
(b) What are the individuals, the treatments, and the response variable in this experiment?
(c) You have 200 pine seedlings available. Outline the design you would use for this experiment.

APPLET EXERCISES

To do these exercises, go to www.whfreeman.com/fapp8e.

1. Use the *Simple Random Sample* applet to choose the sample of songs in Example 4. Assign labels 01 to 27 by entering 27 in the "Population 1 to" box and clicking "Reset." Then enter 4 in the "Select a sample of size" box and click "Sample." Which songs from the list in Example 4 make up your sample? Click "Reset" and choose another sample. Which songs did you choose this time? You see that random sampling gives different samples each time–what matters is that all songs have the same chance to be chosen.

2. The *Simple Random Sample* applet is handy when the population or sample is large. (The applet will handle population sizes up to 500.) Skills Check 5 asks you to choose an SRS of 10 out of 440 retail outlets. Use the applet to do this and report your result.

3. You can use the *Simple Random Sample* applet to choose treatment groups at random for a randomized comparative experiment. Exercise 29 asks you to choose the subjects to get the first treatment in an experiment that compares two treatments.

(a) Use the applet to choose an SRS of 7 out of 14 to receive the first treatment. Which subjects make up this group?
(b) The applet allows you to randomly assign subjects to more than two groups. Suppose you had a total of 36 rats and you wanted to assign a different treatment to each of four 9-rat groups. After you choose the first group, the "Population hopper" contains the 27 subjects that were not chosen, in scrambled order. Click "Sample" again to choose 9 of these remaining subjects to receive the second treatment. Do this once more to choose the third group. The 9 subjects that remain in the "Population hopper" form the fourth group. Which of the 36 subjects will receive each of the four treatments?

4. You can use the *Probability* applet to speed up and improve Exercise 41. You have a population in which 60% of the individuals approve of legal gambling. You want to take many samples from this population to observe how the sample proportion that approves of gambling varies from sample to sample. Set the "Probability of heads" in the applet to 0.6 and the number of tosses to 40. This simulates an SRS of size 40 from a large population. Each head in the sample is a person who approves of legal gambling, and each tail is a person who disapproves. By alternating between "Toss" and "Reset" you can take many samples quickly.

(a) Take 50 samples, recording the proportion that approves of gambling in each sample. (The applet gives this proportion at the top left of its display.) Make a histogram of the 50 sample proportions.
(b) Another population contains only 20% who approve of legal gambling. Take 50 samples of size 40 from this population, record the number in each sample that approves, and make a histogram of the 50 sample proportions. How do the centers of your two histograms reflect the differing truths about the two populations?

5. The idea of an 80% confidence interval is that the interval captures the true parameter value in 80% of all samples. That's not high enough confidence for practical use, but 80% hits and 20% misses make it easy to see how a confidence interval behaves in repeated samples from the same population. Go to the *Confidence Interval* applet.

(a) Set the confidence level to 80%. Click "Sample" to choose an SRS and calculate the confidence interval. Do this 10 times to simulate 10 SRSs with their 10 confidence intervals. How many of the 10 intervals captured the true mean μ? How many missed?
(b) You see that we can't predict whether the next sample will hit or miss. The confidence level, however, tells us what percent will hit in the long run. Reset the applet and click "Sample 50" to get the confidence intervals from 50 SRSs. How many hit? Keep clicking "Sample 50" and record the percent of hits among 100, 200, 300, 400, and 500 SRSs. Even 500 samples is not truly "the long run," but we expect the percent of hits in 500 samples to be fairly close to the confidence level, 80%.

WRITING PROJECTS

1. Go to the Web site of the Gallup Organization (www.gallup.com) and click on "Gallup Poll." You should be able to find a press release you can access and read without charge. Newspapers publish short articles based on press releases. Write a news article about two paragraphs long based on the press release.

2. Recall how Example 8 shows how wording can affect survey results. You can explore this by doing an experiment disguised as a survey! Choose a topic, then design two questions with a key difference in wording. Use randomization to choose which version of the question you give each person. Don't reveal the design of the experiment to participants until after they have provided their answers. After you have roughly 20 or more responses to each version, compare and interpret your results. For an example of such an experiment, see John Rubin's article "Weighing Anchors" in the June 1990 issue of *Omni*.

3. Do one of the following:

(a) How would you design a double-blind experiment in which participants test which of two brands of tissue are preferred? Conduct this experiment and write up the results. How do the results compare with any claims made in advertisements for the products?

(b) Choose an issue of current interest to students at your school. Prepare a short questionnaire (no more than five questions) to determine opinions on this issue. Choose a sample of about 25 students, administer your questionnaire, and write a brief description of your findings. Also write a short discussion of your experiences in designing and carrying out the survey. Although 25 students are too few for you to be statistically confident of your results, this project centers on the practical work of a survey. You must first identify a population; if it is not possible to reach a wider student population, use students enrolled in this course. Did the subjects find your questions clear? Did you write the questions so that it was easy to tabulate the responses? At the end, did you wish you had asked different questions?

SUGGESTED READINGS

ANDERSON-COOK, C. M. and SUNDAR DORAI-RAJ. An active learning in-class demonstration of good experimental design, *Journal of Statistics Education*, 9(1) (2001): http://www.amstat.org/publications/jse/v9n1/anderson-cook.html. A nice example of issues that arise when designing a randomized experiment. This article includes an applet students can use to experience the experiment.

BOCK, DAVID E., PAUL F. VELLEMAN and RICHARD D. DE VEAUX. *Stats*, Addison Wesley, Boston, 2007. Another text at the mathematical level just above ours. Like this chapter, it uses estimating a population proportion to introduce confidence intervals (Chapter 19).

LESSER, LAWRENCE M. and ERIK NORDENHAUG. Ethical statistics and statistical ethics: making an interdisciplinary module, *Journal of Statistical Education*, 12(3) (2004): http://www.amstat.org/publications/jse/v12n3/lesser.html. This article's discussion includes ethical issues associated with surveys, experiments and observational studies.

MOORE, DAVID S. *The Basic Practice of Statistics*, 3rd ed., Freeman, New York, 2004. Chapters 7 and 8 discuss samples and experiments. Chapter 13 presents the reasoning of confidence intervals, and Chapter 18 presents confidence intervals for a population proportion.

SUGGESTED WEB SITES

The National Council on Public Polls, www.ncpp.org, has a statement on "20 Questions a Journalist Should Ask About a Poll" that makes interesting reading. The explanations expand our cautions about sample surveys in practice. You can find similar information on the Web site of the American Association for Public Opinion Research, www.aapor.org. Click the "For Journalists" box and take a look at the "Guide to Best Practices" and "Survey Practices that AAPOR Condemns" listings. Also, read the American Statistical Association publication "What is a Survey (2nd ed)" at www.whatisasurvey.info.

The single most important sample survey in the United States is probably the government's monthly

Current Population Survey (CPS), carried out by the Census Bureau on behalf of the Bureau of Labor Statistics (BLS). The CPS Web site, www.bls.census.gov/cps/, contains an abundant wealth of information. You might click on "Publications," then on "Chronological List" and look at the latest monthly statement of the BLS Commissioner to Congress about employment and unemployment.

CHAPTER 8

Probability: The Mathematics of Chance

Have you ever wondered how gambling, which is a recreation or an addiction for individuals, can be a business for the casino? A business requires predictable revenue from the service it offers, even when the service is a game of chance. Individual gamblers may win or lose. They can never say whether a day at the casino will turn a profit or a loss. But the casino isn't gambling. Casinos are consistently profitable, and state governments make money both from running lotteries and from selling licenses for other forms of gambling.

It is a remarkable fact that the aggregate result of many thousands of chance outcomes can be known with near certainty. The casino need not load the dice, mark the cards, or alter the roulette wheel. It knows that in the long run each dollar bet will yield its five cents or so of revenue. It is therefore good business to concentrate on free floor shows or inexpensive bus fares to increase the flow of dollars bet. The flow of profit will follow.

Gambling houses are not alone in profiting from the fact that a chance outcome many times repeated is firmly predictable. For example, although a life insurance company does not know *which* of its policyholders will die next year, it can predict quite accurately *how many* will die. It sets its premiums according to this knowledge, just as the casino sets its jackpots. Statisticians also rely on the regular behavior of chance: A 95% confidence interval works 95% of the time because, in the long run, chance behavior is predictable.

Random DEFINITION

A phenomenon or trial is said to be **random** if individual outcomes are uncertain but the long-term pattern of many individual outcomes is predictable.

To a statistician, "random" does not mean "haphazard." Randomness is a kind of order, an order that emerges only in the long run, over many repetitions. Many phenomena, both natural and of human design, are random. The hair colors of children, the spread of epidemics, and the decay of radioactive substances are examples of natural randomness. Indeed, quantum mechanics asserts that at the subatomic level the natural world is inherently random.

Games of chance are examples of randomness deliberately produced by human effort. Casino dice are carefully machined, and their drilled holes are filled with material equal in density to the plastic body. This guarantees that the side with six spots has the same weight as the opposite side, which has only one spot. Thus, each side is equally likely to land upward. All the odds and payoffs of dice games rest on this carefully planned randomness. Random sampling and randomized comparative experiments are also examples of planned randomness, although they use tables of random digits rather than dice and cards. The reasoning of statistical inference rests on asking, "How often would this method give a correct answer if I used it very many times?" Probability theory, the mathematical description of randomness, is the basis for gambling, insurance, much of modern science, and statistical inference. **Probability** is the topic of this chapter.

8.1 Probability Models and Rules

Toss a coin, or choose a simple random sample (SRS). The result can't be predicted in advance, because the result will vary when you toss the coin or choose the sample repeatedly. But there is nonetheless a regular pattern in the results, a pattern that emerges clearly only after many repetitions. This remarkable fact is the basis for the idea of probability.

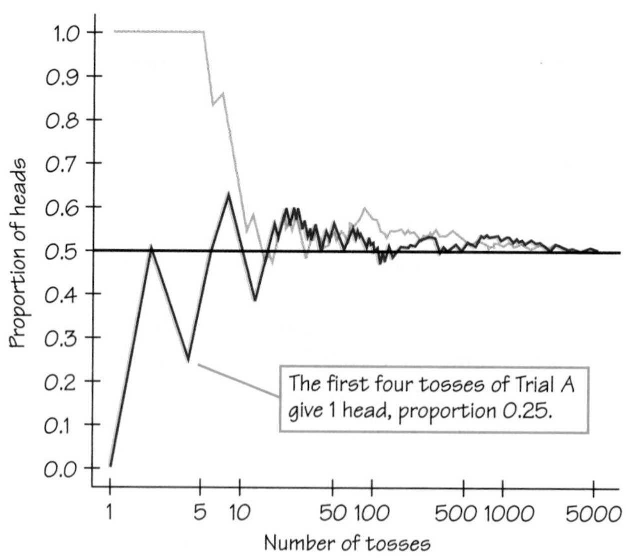

FIGURE 8.1 The proportion of tosses of a coin that give a head varies as we make more tosses. Eventually, however, the proportion approaches 0.5, the probability of a head. This figure shows the results of two trials of 5000 tosses each. The horizontal scale is transformed using logarithms to show both short-term and long-term behavior.

EXAMPLE 1 ■ Tossing a Coin

When you toss a coin, there are only two possible outcomes, heads or tails. Figure 8.1 shows the results of tossing a coin 5000 times twice. For each number of tosses from 1 to 5000, we have plotted the proportion of those tosses that gave a head.

Trial *A* (red line) begins tail, head, tail, tail. You can see that the proportion of heads for Trial *A* starts at 0 on the first toss, rises to 0.5 when the second toss gives a head, then falls to 0.33 and 0.25 as we get two more tails. Trial *B* (blue line), on the other hand, starts with five straight heads, so the proportion of heads is 1 until the sixth toss.

The proportion of tosses that produce heads is quite variable at first. Trial *A* starts low and Trial *B* starts high. As we make more and more tosses, however, the proportions of heads for both trials get close to 0.5 and stay there. If we made yet a third trial at tossing the coin a great many times, the proportion of heads would again settle down to 0.5 in the long run. We say that 0.5 is the *probability* of a head. The probability 0.5 appears as a horizontal line on the graph.

Probability

DEFINITION

The **probability** of any outcome of a random phenomenon is the proportion of times the outcome would occur in a very long series of repetitions. We will soon see a concrete expression of this in the procedure box for "Equally Likely Outcomes."

The *Probability* applet (see Applet Exercise 1) animates Figure 8.1. It allows you to choose the probability of a head and simulate any number of tosses of a coin with that probability. Try it. You will see that the proportion of heads gradually settles down close to the probability. Equally important, you will also see that the proportion in a small or moderate number of tosses can be far from the probability. *Probability describes only what happens in the long run.* Random phenomena are irregular and unpredictable in the short run.

We might suspect that a coin has probability 0.5 of coming up heads just because the coin has two sides. As Exercise 1 illustrates, such suspicions are not always correct. The idea of probability is *empirical.* That is, it is based on observation rather than theorizing. Probability describes what happens in very many trials, and we must actually observe many trials to pin down a probability.

Gamblers have known for centuries that the fall of coins, cards, and dice displays clear patterns in the long run. In fact, a question about a gambling game launched probability as a formal branch of mathematics. The idea of probability rests on the observed fact that the average result of many thousands of chance outcomes can be known with near certainty. But a definition of probability as "long-run proportion" is vague. Who can say what "the long run" is? We can always toss the coin another 1000 times. Instead, we give a mathematical description of *how probabilities behave,* based on our understanding of long-run proportions. To see how to proceed, think first about a very simple random phenomenon, tossing a coin once. When we toss a coin, we cannot know the outcome in advance. What do we know? We are willing to say that the outcome will be either heads or tails. We believe that each of these outcomes has probability 1/2. This description of coin tossing has two parts:

- A list of possible outcomes
- A probability for each outcome

This description is the basis for all probability models. Here is the vocabulary we use.

Sample Space DEFINITION

The **sample space** *S* of a random phenomenon is the set of all possible outcomes that cannot be broken down further into simpler components.

Event DEFINITION

An **event** is any outcome or any set of outcomes of a random phenomenon. That is, an event is a subset of the sample space.

Probability Model DEFINITION

A **probability model** is a mathematical description of a random phenomenon consisting of two parts: a sample space *S* and a way of assigning probabilities to events.

The sample space *S* can be very simple or very complex. When we toss a coin once, there are only two outcomes, heads and tails. So the sample space is $S = \{H, T\}$. If we draw a random sample of 1000 U.S. residents age 18 and over, as opinion polls often do, the sample space contains all possible choices of 1000 of the more than 230 million adults in the country. This *S* is extremely large: 1.3×10^{5794}. Each member of *S* is a possible opinion poll sample, which explains the term **sample space**.

EXAMPLE 2 ▪ Tossing Two Coins

Probabilities can be hard to determine without detailing or diagramming the sample space. For example, E. P. Northrop notes that even the great eighteenth-century French mathematician Jean le Rond d'Alembert tripped on the question: "In two coin tosses, what is the probability that heads will appear at least once?" Because the number of heads could be 0, 1 or 2, d'Alembert reasoned (incorrectly) that each of those possibilities would have an equal probability of 1/3, and so he reached the (wrong) answer of 2/3. What went wrong? Well, {0, 1, 2} could not be the fully-detailed sample space because "1 head" can happen in more than one way. For example, if you flip a dime and a penny once each, you could display the sample space with a *table*:

Another way is with a *tree diagram*, in which all possible left-to-right pathways through the branches generate outcomes.

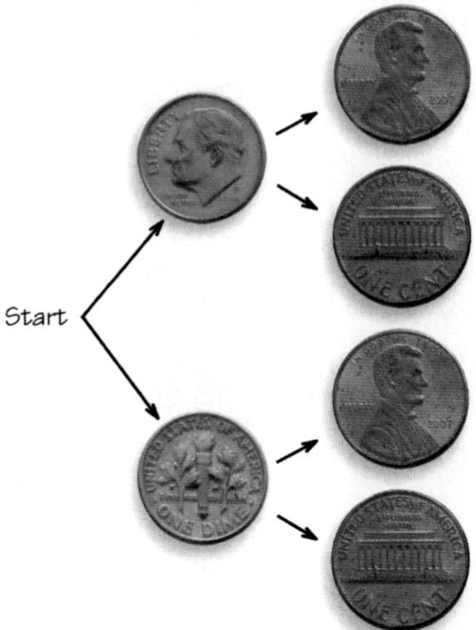

Either way, we can see that the sample space has 4, not 3, equally likely outcomes: {HH, HT, TH, TT}. With the table or tree diagram in front of us, you may already see that the correct probability of at least 1 head is not 2/3, but 3/4.

EXAMPLE 3

Pair-a-Dice: Outcomes for Rolling Two Dice

Rolling two dice is a common way to lose money in casinos. There are 36 possible outcomes when we roll two dice and record the up faces in order (first die, second die). Figure 8.2 displays these outcomes. They make up the sample space S.

If the dice are carefully made, experience shows that each of the 36 outcomes in Figure 8.2 comes up equally often. So a reasonable probability model assigns probability 1/36 to each outcome.

In craps and most other games, all that matters is the *sum* of the spots on the up faces. Let's change the random outcomes we are interested in: Roll two dice and count the spots on the up faces. Now there are only 11 possible outcomes, from a sum of 2 (for rolling a double 1) to a sum of 12 (for rolling a double 6). The sample space is now

(George Diebold/Stone/ Getty Images.)

$$S = \{2, 3, 4, 5, 6, 7, 8, 9, 10, 11, 12\}$$

Comparing this S with Figure 8.2 reminds us that we can change S by changing the detailed description of the random phenomenon we are describing. The outcomes in this new sample space are *not* equally likely, because there are six ways to roll a 7 and only one way to roll a 12. The probability aspect of this example is developed further in Example 4.

There are many ways to assign probabilities, so it is convenient to start with some general rules that any assignment of probabilities to outcomes must obey. These facts follow from the idea of probability as "the long-run proportion of

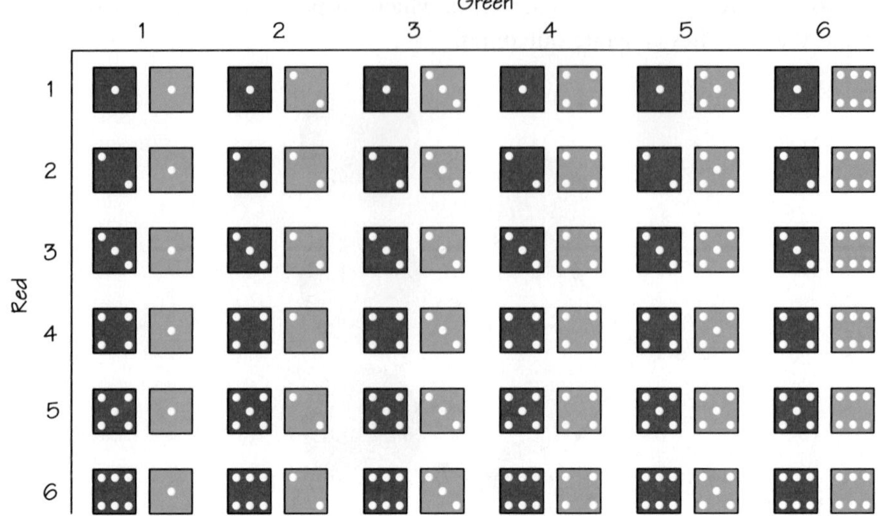

FIGURE 8.2 The 36 possible outcomes for rolling two dice, for Example 3.

repetitions on which an event occurs." Some rules apply only to special kinds of events, which we define here:

Complement of an Event DEFINITION

The **complement of an event** A is the event that A does *not* occur, written as A^C.

Disjoint Events DEFINITION

Two events are **disjoint events** if they have no outcomes in common. Disjoint events are also called *mutually exclusive events*.

Independent Events DEFINITION

Two events are **independent events** if the occurrence of one event has no effect on the probability of the occurrence of the other event.

1. **Any probability is a number between 0 and 1 inclusive**. Any proportion is a number between 0 and 1 inclusive, so any probability is also a number between 0 and 1 inclusive. An event with probability 0 never occurs, and an event with probability 1 always occurs, and an event with probability 0.5 occurs in half the trials in the long run.
2. **All possible outcomes together must have probability 1**. Because some outcome must occur on every trial, the sum of the probabilities for all possible (simplest) outcomes must be exactly 1.
3. **The probability that an event does not occur is 1 minus the probability that the event does occur**. If an event occurs in (say) 70% of all trials, it fails to occur in the other 30%. The probability that an event occurs and the probability that it does not occur always add to 100%, or 1 (see Figure 8.3).

4. **If two events are *independent*, then the probability that one event and the other both occur is the product of their individual probabilities.** Consider event A is "red die is a 1 or 2" and event B is "green die is 6." The red die and green die logically have no influence over each other's outcomes, but we can also look at Figure 8.2 and see that the chance of being in the top two rows does not affect and is not affected by the chance of being in the sixth column. And so Rule 4 for independent events applies and the probability that A and B both happen is the product $(1/3)(1/6) = 1/18$. Note that we can also see from Figure 8.2 that the *intersection* or "overlap" of events A and B happens in 2 of the 36 outcomes and $2/36 = 1/18$. Also, since A and B overlap, they are *not* disjoint, even though the everyday use of the word "independent" might (incorrectly) suggest that kind of separateness.
5. **The probability that one event or the other occurs is the sum of their individual probabilities minus the probability of their intersection.** This general addition rule makes sense if we look at Rule 5 in Figure 8.3. Simply adding the probabilities of the two events would overshoot the answer because we would be incorrectly "double-counting" the overlap. The way to adjust for this is to subtract the overlap so that it is counted only once. Note that the mathematical "or" is inclusive, which means that the event "A or B" happens as long as at least one of the two events happens. In the set theory, it is the *union* of A and B, which includes A's and B's "separate property" as well as their "community property." Consider event A is "red die is a perfect square," which has probability of 2/6. Consider event B is "red die is an odd number" (that is, 1, 3, or 5), which has probability of 3/6. The intersection of events A and B corresponds to rolling a "1," which has a probability of 1/6. So the probability that A or B occurs is $2/6 + 3/6 - 1/6 = 4/6 = 2/3$. Notice that if events A and B had been disjoint, there would be no intersection to worry about double counting and this rule would simply turn into this next one:
6. **If two events are *disjoint*, the probability that one or the other occurs is the sum of their individual probabilities.** If one event occurs in 40% of all trials, a different event occurs in 25% of all trials, and the two can never occur together, then one or the other occurs on 65% of all trials because 40% + 25% = 65%.

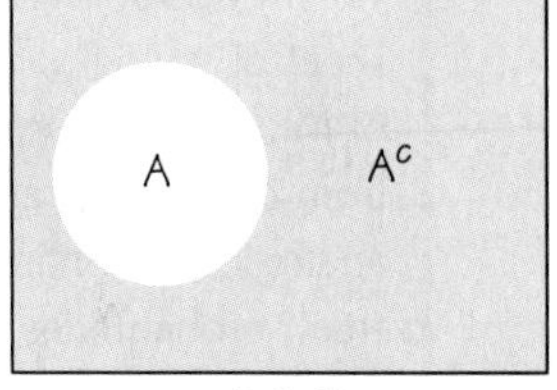

Rule 3

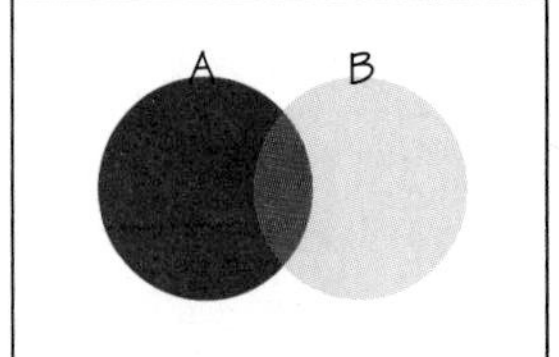

Rule 5

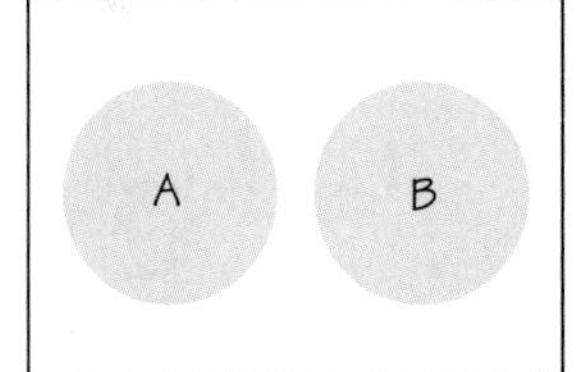

Rule 6

FIGURE 8.3 Each rectangle represents the whole sample space in these illustrations of Rules 3, 5 and 6.

We can use mathematical notation to state Rules 1 to 6 more concisely. We use capital letters near the beginning of the alphabet to denote events. If A is any event, we write its probability as $P(A)$. Here are our probability facts in formal language. As you apply these rules, remember that they are just another form of intuitively true facts about long-run proportions.

Probability Rules

RULE

Rule 1. The probability $P(A)$ of any event A satisfies $0 \leq P(A) \leq 1$.
Rule 2. If S is the sample space in a probability model, then $P(S) = 1$.
Rule 3. The **complement rule**: $P(A^C) = 1 - P(A)$
Rule 4. The **multiplication rule** for *independent* events: $P(A \text{ and } B) = P(A) \times P(B)$
Rule 5. The *general* **addition rule**: $P(A \text{ or } B) = P(A) + P(B) - P(A \text{ and } B)$
Rule 6. The *addition rule* for *disjoint* events: $P(A \text{ or } B) = P(A) + P(B)$

SPOTLIGHT 8.1 Probability and Psychology

Our judgment of probability can be affected by psychological factors. Our desire to become instantly rich may lead us to overestimate the tiny probability of winning the lottery. Our feeling that we are "in control" when we are driving may make us underestimate the probability of an accident. (This may be why some people prefer driving to flying even though flying has a lower probability of death per mile traveled.)

The probability of winning (a share of) the twelve-state Mega Millions jackpot is 1 in 175,711,536. This is like guessing a particular sheet of typing paper from a stack twice the height of Mt. Everest. Or guessing a particular second from a period of about 5.5 years. Without concrete analogies, it is hard to grasp the meaning of very small probabilities and some players may greatly overestimate their chances of winning even if they buy lots of tickets. For example, suppose someone buys 20 $1 Mega Millions tickets every week for 50 years. She would have spent over $50,000 and yet her probability of winning at least one jackpot in that whole time would still be only 1 in 3368. For comparison, the probability of dying in a car accident during a lifetime of driving is about 50 times greater than this!

Andrew Gelman reports that most people say they would not switch to a situation in which they had a small probability *p* of dying and a large probability 1-*p* of gaining $1000. And yet, people will not necessarily spend that much for air bags for their cars. Becoming more aware of our inconsistencies and biases can help us make better use of probability when deciding what risks to take.

EXAMPLE 4 ■ Probabilities for Rolling Two Dice

Figure 8.2 displays the 36 possible outcomes of rolling two dice. For casino dice, it is reasonable to assign the same probability to each of the 36 outcomes in Figure 8.2. Because all 36 outcomes together must have probability 1 (Rule 2), each outcome must have probability 1/36.

What is the probability of rolling a sum of 5? Because the event "roll a sum of 5" contains the four outcomes displayed in Figure 8.2, the addition rule for disjoint events (Rule 6) says that its probability is

$$P(\text{roll a sum of 5}) = P(⚀⚃) + P(⚁⚂) + P(⚂⚁) + P(⚃⚀)$$

$$= \frac{1}{36} + \frac{1}{36} + \frac{1}{36} + \frac{1}{36}$$

$$= \frac{4}{36} = 0.111$$

Continue using Figure 8.2 in this way to get the full probability model (sample space and assignment of probabilities) for rolling two dice and summing the spots on the up faces. Here it is:

Outcome	2	3	4	5	6	7	8	9	10	11	12
Probability	$\frac{1}{36}$	$\frac{2}{36}$	$\frac{3}{36}$	$\frac{4}{36}$	$\frac{5}{36}$	$\frac{6}{36}$	$\frac{5}{36}$	$\frac{4}{36}$	$\frac{3}{36}$	$\frac{2}{36}$	$\frac{1}{36}$

This model assigns probabilities to individual outcomes. Note that Rule 2 is satisfied because all the probabilities add up to 1. To find the probability of an event, just add the probabilities of the outcomes that make up the event. For example:

$$\begin{aligned} P(\text{outcome is odd}) &= P(3) + P(5) + P(7) + P(9) + P(11) \\ &= \frac{2}{36} + \frac{4}{36} + \frac{6}{36} + \frac{4}{36} + \frac{2}{36} \\ &= \frac{18}{36} = \frac{1}{2} \end{aligned}$$

What is the probability of rolling any sum other than a 5? The "long way" to find this would be

$$P(2) + P(3) + P(4) + P(6) + P(7) + P(8) + P(9) + P(10) + P(11) + P(12).$$

A much better way would be to use the complement rule (Rule 3):

$$\begin{aligned} P(\text{roll sum that is } \textit{not } 5) &= 1 - P(\text{roll sum of } 5) \\ &= 1 - \frac{4}{36} = \frac{32}{36} = 0.889 \end{aligned}$$

Another good time to use the complement rule would be to find the probability of getting a sum greater than 3. Compare the calculation of $P(\text{sum} > 3)$ with $1 - P(\text{sum} \leq 3)$.

For an example of Rule 5, let event A be "sum is odd" and event B be "sum is a multiple of 3." We previously calculated $P(A) = 1/2$. You can verify that $P(B) = 1/3$ and $P(A \text{ and } B)$ is 1/6. And so, $P(A \text{ or } B) = 1/2 + 1/3 - 1/6 = 2/3$.

When the outcomes for a probability model are numbers, we can use a histogram to display the assignment of probabilities to the outcomes. Figure 8.4 is a **probability histogram** of the probability model in Example 4. The height of each bar shows the probability of the outcome at its base. Because the heights are probabilities, they add to 1. Think of Figure 8.4 as an idealized picture of the results of very many rolls of a die. As an idealized picture, it is perfectly symmetric.

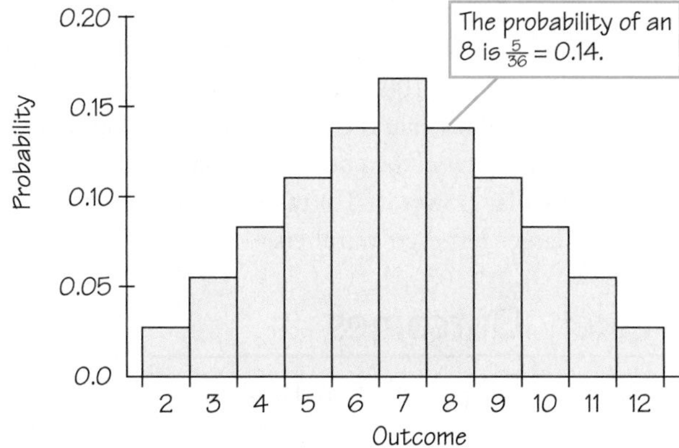

FIGURE 8.4 Probability histogram showing the probability model for rolling two balanced dice and counting the spots on the up faces.

Example 4 illustrates one way to assign probabilities to events: Assign a probability to every individual outcome, then add these probabilities to find the probability of any event. This idea works well when there are only a finite (fixed and limited) number of outcomes.

8.2 Discrete Probability Models

We will work with two kinds of probability models. The first kind is illustrated by Example 4 and is called a **discrete probability model**. (The second kind is in Section 8.4.)

Discrete Probability Model DEFINITION

A probability model is called discrete if its sample space has a countable number of outcomes. To assign probabilities in a discrete model, list the probability of all the individual outcomes. By Rules 1 and 2, these probabilities must be numbers between 0 and 1 inclusive and must have sum 1.
The probability of any event is the sum of the probabilities of the outcomes making up the event.

EXAMPLE 5 ■ Benford's Law

Faked numbers in tax returns, invoices, or expense account claims often display patterns that aren't present in legitimate records. Some patterns, like too many round numbers, are obvious and easily avoided by a clever crook. Others are more subtle. It is a striking fact that the first (leftmost) digits of numbers in legitimate records often follow a model known as Benford's law. Here it is (note that a first digit can't be 0):

First digit	1	2	3	4	5	6	7	8	9
Probability	0.301	0.176	0.125	0.097	0.079	0.067	0.058	0.051	0.046

Check that the probabilities of the outcomes sum exactly to 1. This is therefore a legitimate discrete probability model. Investigators can detect fraud by comparing the first digits in records such as invoices paid by a business with these probabilities. For example, consider the events A = "first digit is 1" and B = "first digit is 2." Applying Rule 6 to the table of probabilities yields $P(A \text{ or } B) = 0.301 + 0.176$, which is 0.477 (almost 50%). Crooks trying to "make up" the numbers probably would not make up numbers starting with 1 or 2 this often.

Let us use some intuition about why first digits behave this way. Note that the increase from 1 to 2 is an increase of 100%, but from 2 to 3 is only 50%, from 3 to 4 is only 33%, and so on. So data values that increase at an approximately constant percentage (which a lot of financial data does, for example) will naturally "spend more time" (within any particular power of 10) taking on values whose left digit is 1, and successively less for larger left-digit numbers. ■

8.3 Equally Likely Outcomes

A simple random sample gives all possible samples an equal chance to be chosen. Rolling two casino dice gives all 36 outcomes the same probability. When randomness is the product of human design, it is often the case that the outcomes in the sample space are all equally likely. Rules 1 and 2 force the assignment of probabilities in this case.

Finding Probabilities of Equally Likely Outcomes PROCEDURE

If a random phenomenon has equally likely outcomes, then the probability of event A is

$$P(A) = \frac{\text{count of outcomes in event } A}{\text{count of outcomes in sample space } S}$$

As an aside, a less common way of expressing likelihood that you may encounter in some gambling contexts is *odds*. The *odds* of an event A happening can be expressed as:

$$\frac{\text{count of outcomes in which } A \text{ happens}}{\text{count of outcomes where } A \text{ does not happen.}}$$

The *odds* against an event A happening can be expressed as:

(count of outcomes A does not happen) / (count of outcomes A happens).

For example, let event A be "a die is rolled and lands on a 4 or 5." Since there are twice as many ways A does not happen as there are that A happens, the odds of event A happening are 1:2 and the odds against event A are 2:1. Notice that these numbers are different from the respective probability values $P(A) = 1/3$ and $P(A^C) = 2/3$, but we can express odds in terms of probabilities as follows:

$$\text{odds of } A \text{ happening} = \frac{P(A)}{P(A^C)}, \text{ and odds against } A \text{ happening} = \frac{P(A^C)}{P(A)}.$$

We have included this aside so you will recognize what odds mean in the rare occasions you encounter them, but be aware that odds values do *not* follow the six rules of probability.

EXAMPLE 6 ■ Are First Digits Equally Likely?

You might think that first (leftmost) digits are distributed "at random" among the digits 1 to 9. Under such a "discrete uniform distribution," the 9 possible outcomes would then be equally likely. The sample space is $S = \{1, 2, 3, 4, 5, 6, 7, 8, 9\}$, and the probability model is:

First digit	1	2	3	4	5	6	7	8	9
Probability	$\frac{1}{9}$	$\frac{1}{9}$	$\frac{1}{9}$	$\frac{1}{9}$	$\frac{1}{9}$	$\frac{1}{9}$	$\frac{1}{9}$	$\frac{1}{9}$	$\frac{1}{9}$

The probability of the event that a randomly chosen first digit is a 1 or 2 is

$$\begin{aligned} P(1 \text{ or } 2) &= P(1) + P(2) \\ &= \frac{1}{9} + \frac{1}{9} = \frac{2}{9} = 0.222 \end{aligned}$$

This answer of 0.222 is less than half of what we found for $P(1 \text{ or } 2)$ using the Benford's Law probability model in Example 5–a huge difference that illustrates one way an auditor could easily detect data that was faked–the crook would have too few 1's and 2's. Figure 8.5 displays probability histograms that compare the probability model for random digits with the model given by Benford's law. ■

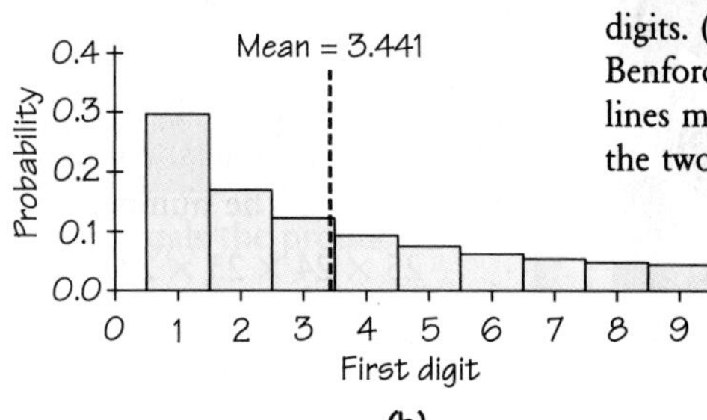

FIGURE 8.5 Probability histograms of two models for first digits in numerical records. (a) Equally likely digits. (b) Digits follow Benford's law. The vertical lines mark the means of the two models.

These facts can be stated more precisely and then proved mathematically. The law of large numbers brings the idea of probability to a natural completion. We first observed that some phenomena are random in the sense of showing long-run regularity. Then we used the idea of long-run proportions to motivate the basic laws of probability. Those laws are mathematical idealizations that can be used without interpreting probability as proportion in many trials. Now the law of large numbers tells us that in many trials the proportion of trials on which an outcome occurs will always approach its probability.

The law of large numbers also explains why gambling can be a business. The winnings (or losses) of a gambler on a few plays are uncertain–that's why gambling is exciting. It is only *in the long run* that the mean outcome is predictable. The house plays many tens of thousands of times. So the house, unlike individual gamblers, can count on the long-run regularity described by the law of large numbers. The average winnings of the house on tens of thousands of plays will be very close to the mean of the distribution of winnings. Needless to say, gambling games have mean outcomes that guarantee the house a profit.

We know that the simplest description of a distribution of data requires both a measure of center and a measure of spread. The same is true for probability models. The *mean* is the average value for both a set of data and a discrete probability model. All the observations are weighted equally in finding the mean $\bar{x}$ for data, but the values are weighted by their probabilities in finding the mean μ of a probability model. The measure of spread that goes with the mean is the **standard deviation.** For data, the standard deviation s is the square root of the average squared deviation of the observations from their mean. We apply exactly the same idea to probability models, using probabilities as weights in the average. Here is the definition.

Standard Deviation of a Discrete Probability Model — DEFINITION

Suppose that the possible outcomes $x_1, x_2, \ldots, x_k$ in a sample space S are numbers, and that p_j is the probability of outcome x_j. The **standard deviation σ of a discrete probability model** with mean μ is

$$\sigma = \sqrt{(x_1 - \mu)^2 p_1 + (x_2 - \mu)^2 p_2 + \cdots + (x_k - \mu)^2 p_k}$$

EXAMPLE 14 ■ First Digits

If the first digits in a set of records obey Benford's law, the discrete probability model is

First digit	1	2	3	4	5	6	7	8	9
Probability	0.301	0.176	0.125	0.097	0.079	0.067	0.058	0.051	0.046

We saw in Example 13 that the mean is $\mu = 3.441$. To find the standard deviation,

$$\begin{aligned}
\sigma &= \sqrt{(x_1 - \mu)^2 p_1 + (x_2 - \mu)^2 p_2 + \cdots + (x_k - \mu)^2 p_k} \\
&= \sqrt{(1 - 3.441)^2(0.301) + (2 - 3.441)^2(0.176) + \cdots + (9 - 3.441)^2(0.046)} \\
&= \sqrt{1.7935 + 0.3655 + \cdots + 1.4215} \\
&= \sqrt{6.061} = 2.46
\end{aligned}$$

You can follow the same pattern to find the standard deviation of the equally likely model and show that the Benford's law model is less spread out than the equally likely model.

Finding the standard deviation of a continuous probability model usually requires advanced mathematics (calculus). Chapter 5 told us the answer in one important case: The standard deviation of a normal curve is the distance from the center (the mean) to the change-of-curvature points on either side.

SPOTLIGHT 8.3 Birthday Coincidences

If 366 people are gathered, you can see why there's a 100% chance at least two people share the same birthday (ignoring leap days). Now, if only 23 people are gathered, what do you think is the probability of any birthday matches? Guess before reading further.

Now imagine these 23 people enter a room one at a time, adding their birthday to a list in the order they enter. Using $n = 365$ and $k = 23$, Rule A gives us the total number of lists of 23 birthdays, and Rule B gives us how many of those lists have birthdays that are all different. Using the rule for Equally Likely Outcomes (each day of the year is equally likely to be a randomly chosen person's birthday), we conclude that the probability of all birthdays being different is the result from Rule B divided by the result from Rule A:

$$\frac{{}_{365}P_{23}}{365^{23}} = \frac{365 \times 364 \times \ldots \times 343}{365^{23}}$$

Alternatively, we could assume independence of birthdays and use Probability Rule 4. The second person that walks in has a 364/365 chance of not matching person #1. The third person that walks in has a 363/365 chance of not matching persons #1 or #2, and so on. Verify that you get the same product by multiplying this string of fractions:

$$\frac{364}{365} \times \frac{363}{365} \times \ldots \times \frac{343}{365}$$

Either way, our final step to find the probability of getting at least one match is to subtract that answer from 1 (using Probability Rule 3), and we obtain the surprisingly high value of 51%! Maybe it is not so surprising if we consider that the combinations formula tells us that there are 253 ways to choose 2 people from 23 to ask each other if they have the same birthday. Because we underestimate the huge number of potential opportunities for "coincidences," we are surprised that they happen as often as they do. As Jessica Utts points out, if something has a 1 in a million chance of happening to any person on a given day, this rare event will happen to roughly 300 people in the United States each day!

(Michael Rosenfeld/Photographer's Choice/Getty Images.)

8.6 The Central Limit Theorem

The key to finding a confidence interval that estimates a population proportion (Chapter 7) was the fact that the sampling distribution of a population proportion is close to normal when the sample is large. This fact is an application of one of the most important results of probability theory, the **Central Limit Theorem**. This theorem says that the distribution of any random phenomenon tends to be normal if we average it over a large number of independent repetitions. The Central Limit Theorem allows us to analyze and predict the results of chance phenomena when we average over many observations.

The word "limit" in Central Limit Theorem reflects that the normal curve is the limit or target shape to which the sampling distribution gets closer and closer as the sample size increases. The theorem also tells us the mean of the sampling distribution, and the mean is a measure of "central" tendency.

Central Limit Theorem THEOREM

Draw an SRS of size n from any large population with mean μ and finite standard deviation σ. Then

- The mean of the sampling distribution of $\bar{x}$ is μ.
- The standard deviation of the sampling distribution of $\bar{x}$ is $\sigma/\sqrt{n}$.
- The **Central Limit Theorem** says that the sampling distribution of $\bar{x}$ is approximately normal when the sample size n is large ($n > 30$).

The first two parts of this statement can be proved from the definitions of the mean and the standard deviation. They are true for any sample size n. The Central Limit Theorem is a much deeper result. Pay attention to the fact that the standard deviation of a mean decreases as the number of observations n increases. Together with the Central Limit Theorem, this makes exact two general statements that help us understand a wide variety of random phenomena:

Averages are less variable than individual observations.

Averages are more normal than individual observations.

The *Central Limit Theorem* applet allows you to watch the Central Limit Theorem in action: It starts with a distribution that is strongly skewed, not at all normal. As you increase the size of the sample, the distribution of the mean $\bar{x}$ gets closer and closer to the normal shape.

Consider dice. Rolls of a single die would have a uniformly flat probability histogram, with each of the six possible values having the probability 1/6. Now consider the mean of rolling a pair of dice. The probability model for the mean of two dice simply divides by 2 the outcome sum in Example 4. (So, the probability that the mean of two dice equals 4.5 must be the same as the probability that their sum equals 9.) And the histogram in Figure 8.4 is certainly less variable and closer to looking "normal" than is the flat histogram for rolling a single die.

EXAMPLE 15 ▪ Heights of Young Women

The distribution of heights of young adult women is approximately normal, with mean 64.5 inches and standard deviation 2.5 inches. This normal distribution describes the population of young women. It is also the probability model for choosing one woman at random from this population and measuring her height. For example, the 68–95–99.7 rule says that the probability is 0.95 that a randomly chosen woman is between 59.5 and 69.5 inches tall.

Now choose an SRS of 25 young women at random and take the mean $\bar{x}$ of their heights. The mean $\bar{x}$ varies in repeated samples–the pattern of variation is the sampling distribution of $\bar{x}$. The sampling distribution has the same center $\mu = 64.5$ inches as the population of young women. In statistical terms, the sample mean $\bar{x}$ has *no bias* as an estimator of the population mean μ. If we take many samples, $\bar{x}$ will sometimes be smaller than μ and sometimes larger, but it has no systematic tendency to be too small or too large.

The standard deviation of the sampling distribution of $\bar{x}$ is

$$\frac{\sigma}{\sqrt{n}} = \frac{2.5}{\sqrt{25}} = \frac{2.5}{5} = 0.5 \text{ inch}$$

The standard deviation σ describes the variation when we measure many individual women. The standard deviation $\sigma/\sqrt{n}$ of the distribution of $\bar{x}$ describes the variation in the average heights of samples of women when we take many samples. The average height is less variable than individual heights.

Figure 8.10 compares the two distributions: Both are normal and both have the same mean, but the average height of 25 randomly chosen women is much less spread out. For example, the 68–95–99.7 rule says that 95% of all averages $\bar{x}$ lie between 63.5 and 65.5 inches because 2 standard deviations of $\bar{x}$ make 1 inch. This 2-inch span is just one-fifth as wide as the 10-inch span that catches the middle 95% of heights for individual women.

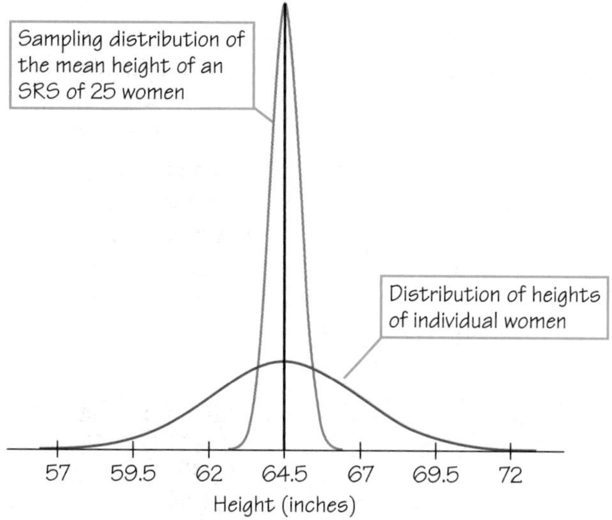

FIGURE 8.10 The sampling distribution of the average height of an SRS of 25 women has the same center as the distribution of individual heights but is much less spread out.

The Central Limit Theorem says that in large samples the sample mean $\bar{x}$ is approximately normal. In Figure 8.10, we show a normal curve for $\bar{x}$ even though sample size 25 is not very large. Is that acceptable? How large a sample is needed for the Central Limit Theorem to work depends on how far from a normal curve the model we start with is. The closer to normality we start, the quicker the distribution of the sample mean becomes normal. In fact, if individual observations follow a normal curve, the sampling distribution of $\bar{x}$ is exactly normal for any sample size. So Figure 8.10 is accurate. The Central Limit Theorem is a striking result because as n gets large it works for *any* model we may start with, no matter how far from normal. Here is an example that starts very far from normal.

EXAMPLE 16 ■ Red or Black in Roulette

FIGURE 8.11 A gambler may win or lose at roulette, but in the long run the casino always wins. (*Ingram Publishing/PictureQuest.*)

An American roulette wheel has 38 slots, of which 18 are black, 18 are red, and 2 are green. The dealer spins the wheel and whirls a small ball in the opposite direction within the wheel. Gamblers bet on where the ball will come to rest (see Figure 8.11). One of the simplest wagers chooses red (or black). A bet of $1 on red pays off an additional $1 if the ball lands in a red slot. Otherwise, the player loses his $1. The two green slots always belong to the house.

Lou bets on red. He wins if the ball stops in one of the 18 red slots. He loses if it lands in one of the 20 slots that are black or green. Because casino roulette wheels are carefully balanced so that all slots are equally likely, the probability model is

	Net Outcome for Gambler	
	Win \$1	Lose \$1
Probability	18/38 = .474	20/38 = .526

The mean outcome of a single \$1 bet on red is

$$\mu = (\$1)\left(\frac{18}{38}\right) + (-\$1)\left(\frac{20}{38}\right)$$
$$= -\$\frac{2}{38} = -\ \$0.053 \text{(a loss of 5.3 cents)}$$

The law of large numbers says that the mean μ is the average outcome of a very large number of individual bets. In the long run, gamblers will lose (and the casino will win) an average of 5.3 cents per bet. We can similarly find the standard deviation for a single \$1 bet on red:

$$\sigma = \sqrt{(1-(-0.053))^2\frac{18}{38} + (-1-(-0.053))^2\frac{20}{38}}$$
$$= \sqrt{(1.053)^2\frac{18}{38} + (-0.947)^2\frac{20}{38}}$$
$$= \sqrt{0.9972} = 0.9986$$

Lou certainly starts far from any normal curve. The probability model for each bet is discrete, with just two possible outcomes. Yet the Central Limit Theorem says that the average outcome of many bets follows a normal curve. Lou is a habitual gambler who places fifty \$1 bets on red almost every night. Because we know the probability model for a bet on red, we can simulate Lou's experience over many nights at the roulette wheel. The histogram in Figure 8.12 shows Lou's average winnings for 1000 nights. As the Central Limit Theorem says, the distribution looks normal.

EXAMPLE 17 ■ Lou Gets Entertainment

The normal curve in Figure 8.12 comes from the Central Limit Theorem and the values of the mean μ and standard deviation σ in Example 16. It has

$$\text{mean} = \mu = -0.053$$
$$\text{standard deviation} = \frac{\sigma}{\sqrt{n}} = \frac{0.9986}{\sqrt{50}} = 0.141$$

Apply the 99.7 part of the 68–95–99.7 rule: Almost all average nightly winnings will fall within 3 standard deviations of the mean, that is, between

$$-0.053 - (3)(0.141) = -0.476$$

and

$$-0.053 + (3)(0.141) = 0.370$$

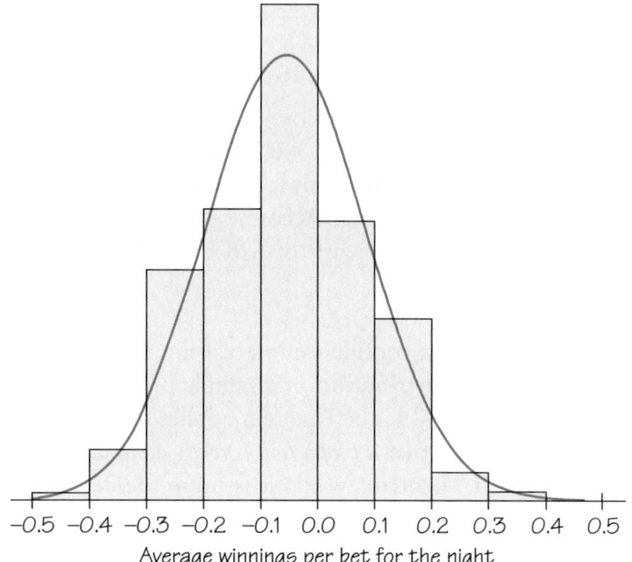

FIGURE 8.12 A gambler's winnings in a night of 50 bets on red or black in roulette vary from night to night. Here is the distribution for 1000 nights. It is approximately normal.

Lou's total winnings after 50 bets of $1 each will then almost surely fall between

$$(50)(-0.476) = -23.80$$

and

$$(50)(0.370) = 18.50$$

Lou may win as much as $18.50 or lose as much as $23.80. Some find gambling exciting because the outcome, even after an evening of bets, is uncertain. It is possible to walk away a winner. It's all a matter of luck.

The casino, however, is in a different position. It doesn't want excitement, just a steady income.

EXAMPLE 18 ■ The Casino Gets Rich

The casino bets with all its customers–perhaps 100,000 individual bets on black or red in a week. The Central Limit Theorem guarantees that the distribution of average customer winnings on 100,000 bets is very close to normal. The mean is still the mean outcome for one bet, −0.053, a loss of 5.3 cents per dollar bet. The standard deviation is much smaller when we average over 100,000 bets. It is

$$\frac{\sigma}{\sqrt{n}} = \frac{0.9986}{\sqrt{100{,}000}} = 0.003$$

Here is what the spread in the average result looks like after 100,000 bets:

$$\begin{aligned} \text{Spread} &= \text{mean} \pm 3 \text{ standard deviations} \\ &= -0.053 \pm (3)(0.003) \\ &= -0.053 \pm 0.009 \\ &= -0.062 \text{ to } -0.044 \end{aligned}$$

Because the casino covers so many bets, the standard deviation of the average winnings per bet becomes very small. And because the mean is negative, almost all outcomes will be negative. The gamblers' losses and the casino's winnings are almost certain to average between 4.4 and 6.2 cents for every dollar bet.

The gamblers who collectively place those 100,000 bets will lose money. The probable range of their losses is

$$(100{,}000)(-0.062) = -6200 \quad \text{to} \quad (100{,}000)(-0.044) = -4400$$

The gamblers are almost certain to lose–and the casino is almost certain to take in–between \$4400 and \$6200 on those 100,000 bets. What's more, the range of average outcomes continues to narrow as more bets are made. That is how a casino can make a business out of gambling. According to *Forbes* magazine, the third richest American (with an estimated worth of \$28 billion) in 2007 was casino mogul Sheldon Adelson.

In Chapter 7, we based a confidence interval for a population proportion p on the fact that the **sampling distribution** of a sample proportion $\hat{p}$ is close to normal for large samples. The Central Limit Theorem applies to means. How can we apply it to proportions? By seeing that *a proportion is really a mean.* This is our final example of the Central Limit Theorem. While it is more theoretical than our other examples, it gives us an important foundation.

EXAMPLE 19

The Sampling Distribution of a Proportion

If we can express the sample proportion of successes as a sample mean, we can apply tools we have learned to derive the formula (in Section 7.7) for the standard deviation of the sample proportion.

Consider an SRS of size n from a population that contains proportion p of "having a particular trait." For each of the n individuals, we can define a simple numerical variable x_i to equal 1 for a success and 0 for a failure. For example, if the third individual has the trait of interest, then $x_3 = 1$. So the sum of all n of the x_i values is the total number of "successes" (that is, people that had the trait of interest). So the proportion $\hat{p}$ of successes is given by

$$\hat{p} = \frac{\textit{number of successes}}{n} = \frac{x_1 + x_2 + \cdots + x_n}{n} = \bar{x}$$

So $\hat{p}$ is really a mean, and so its sampling distribution (by the Central Limit Theorem) is close to normal when the sample size n is large ($n > 30$).

Because $\hat{p}$ is the mean of the x_i, we can find the mean and standard deviation of $\hat{p}$ from the mean and standard deviation of one observation x_i. Each observation has probability p of being a success, so the probability model for one observation is

Outcome	Success, $x_i = 1$	Failure, $x_i = 0$
Probability	p	$1 - p$

Using the tools of Section 8.5, the mean of x_i is therefore

$$\mu = (1)(p) + (0)(1 - p) = p$$

In the same way, after a bit more algebra, the tools of Section 8.5 show that the standard deviation of one observation x_i is

$$\sigma = \sqrt{(1 - p)^2 p + (0 - p)^2 (1 - p)} = \sqrt{p(1 - p)}$$

From the Central Limit Theorem (Section 8.6), the standard deviation of the mean of n observations is $\frac{\sigma}{\sqrt{n}}$, so we simply substitute in our expression for σ and obtain:

$$\frac{\sigma}{\sqrt{n}} = \frac{\sqrt{p(1-p)}}{\sqrt{n}} = \sqrt{\frac{p(1-p)}{n}}.$$

This last expression is precisely the fact we used in Section 7.7.

Examples 15 to 19 illustrate the importance of the Central Limit Theorem and the reason for the importance of normal distributions. We can often replace tricky calculations about a probability model by simpler calculations for a normal distribution, courtesy of the Central Limit Theorem.

REVIEW VOCABULARY

Addition rule The probability that one event or the other occurs is the sum of their individual probabilities minus the probability of any overlap they have. (p. 253)

Central Limit Theorem The average of many independent random outcomes is approximately normally distributed. When we average n independent repetitions of the same random phenomenon, the resulting distribution of outcomes has mean equal to the mean outcome of a single trial and standard deviation proportional to $1/\sqrt{n}$. (p. 267)

Combination An unordered collection of k items chosen (without allowing repetition) from a set of n distinct items. (p. 260)

Combinatorics The branch of mathematics that counts arrangements of objects. (p. 258)

Complement of an event The complement of an event A is the event "A does not occur," which is denoted A^C. (p. 252)

Complement rule $P(A^C) = 1 - P(A)$. (p. 253)

Continuous probability model A probability model that assigns probabilities to events as areas under a density curve. (p. 262)

Density curve A curve that is always on or above the horizontal axis and has area exactly 1 underneath it. A density curve describes a continuous probability model. (p. 261)

Discrete probability model A probability model that assigns probabilities to each of a finite number of possible outcomes. (p. 255)

Disjoint events Events that have no outcomes in common. (Also called *mutually exclusive events*.) (p. 252)

Event A collection of possible outcomes of a random phenomenon. A subset of the sample space. (p. 250)

Factorial The product of the first n positive integers, denoted "$n!$" (p. 259)

Fundamental principle of counting A multiplicative method for counting outcomes of multistage processes. (p. 258)

Independent events Events that do not affect each other's probability of occurring. (p. 252)

Law of large numbers As a random phenomenon is repeated many times, the mean $\bar{x}$ of the observed outcomes approaches the mean μ of the probability model. (p. 265)

Mean of a discrete probability model The average outcome of a random phenomenon with numerical values. When possible values $x_1, x_2, \dots, x_k$ have probabilities $p_1, p_2, \dots, p_k$, the mean is the average of the outcomes weighted by their probabilities, $\mu = x_1p_1 + x_2p_2 + \dots + x_kp_k$. (Also called *expected value*.) (p. 264)

Multiplication rule $P(A \text{ and } B) = P(A) \times P(B)$, when A and B are independent events. (p. 253)

Permutation An ordered arrangement of k items chosen (without allowing repetition) from a set of n distinct items. (p. 258)

Probability A number between 0 and 1 that gives the long-run proportion of repetitions of a random phenomenon on which an event will occur. (p. 248)

Probability histogram A histogram that displays a discrete probability model when the outcomes are numerical. The height of each bar is the probability of the event at the base of the bar. (p. 255)

Probability model A sample space S together with an assignment of probabilities to events. The two main types of probability models are *discrete* and *continuous*. (p. 250)

Random A phenomenon or trial is random if it is uncertain what the next outcome will be but each outcome nonetheless tends to occur in a fixed proportion of a very long sequence of repetitions. These long-run proportions are the probabilities of the outcomes. (p. 247)

Sample space A list of all possible (simplest) outcomes of a random phenomenon. (p. 250)

Sampling distribution The distribution of values taken by a statistic when many random samples are drawn under the same circumstances. A sampling distribution consists of an assignment of probabilities to the possible values of a statistic. (p. 272)

Standard deviation of a discrete probability model A measure of the variability of a probability model. When the possible values $x_1, x_2, \dots, x_k$ have probabilities $p_1, p_2, \dots, p_k$, the standard deviation is the square root of the average (weighted by probabilities) of the squared deviations from the mean:

$$\sigma = \sqrt{(x_1 - \mu)^2 p_1 + (x_2 - \mu)^2 p_2 + \cdots + (x_k - \mu)^2 p_k}.$$

(p. 266)

SKILLS CHECK

1. You read in a book on poker that the probability of being dealt three of a kind in a five-card poker hand is 1/50. What does this mean?

(a) If you deal thousands of poker hands, the fraction of them that contain three of a kind will be very close to 1/50.
(b) If you deal 50 poker hands, exactly one of them will contain three of a kind.
(c) If you deal 10,000 poker hands, exactly 200 of them will contain three of a kind.

2. If two coins are flipped and then a die is rolled, the sample space would have ________ different outcomes.

Exercises 3 to 5 use this probability model for the blood type of a randomly chosen person in the United States:

Blood type	O	A	B	AB
Probability	0.45	0.40	0.11	?

3. The probability that a randomly chosen American has type AB blood is

(a) 0.044.
(b) 0.04.
(c) 0.4.

4. Maria has type A blood. She can safely receive blood transfusions from people with blood types O and A. The probability that a randomly chosen American can donate blood to Maria is ________________.

5. What is the probability that a randomly chosen American does not have type O blood?

(a) 0.55
(b) 0.45
(c) 0.04

6. Figure 8.2 shows the 36 possible outcomes for rolling two dice. These outcomes are equally likely. A "soft 4" is a roll of 1 on one die and 3 on the other. The probability of rolling a soft 4 is ________________.

7. In a table of random digits such as Table 7.1, each digit is equally likely to be any of 0, 1, 2, 3, 4, 5, 6, 7, 8, or 9. What is the probability that a digit in the table is a 0?

(a) 1/9
(b) 1/10
(c) 9/10

8. In a table of random digits such as Table 7.1, each digit is equally likely to be any of 0, 1, 2, 3, 4, 5, 6, 7, 8, or 9. The probability that a digit in the table is 7 or greater is ________________.

9. Toward the end of a game of Scrabble, you hold the letters J, U, D, A, and H. In how many orders can you arrange these 5 letters?

(a) 5
(b) (5)(4)(3)(2)(1) = 120
(c) (5)(5)(5)(5)(5) = 3125

10. Toward the end of a game of Scrabble, you hold the letters D, O, G, and Q. You can choose 3 of these 4 letters and arrange them in order in ____________ different ways.

11. A 52-card deck contains 13 cards from each of the four suits: clubs ♣, diamonds ♦, hearts ♥, and spades ♠. You deal 4 cards without replacement from a well-shuffled deck, so that you are equally likely to deal any 4 cards. What is the probability that all 4 cards are clubs?

(a) 1/4, because 1/4 of the cards are clubs.
(b) (13)(12)(11)(10)/(52)(51)(50)(49) = 0.0026
(c) (13)(12)(11)(10)/(52)(52)(52)(52) = 0.0023

12. You deal 4 cards as in the previous exercise. The probability that you deal no clubs is ____________.

13. Figure 5.3 (page 155) shows that the normal distribution with mean $\mu = 6.8$ and standard deviation $\sigma = 1.6$ is a good description of the Iowa Test vocabulary scores of seventh-grade students in Gary, Indiana. The probability that a randomly chosen student has a score higher than 8.4 is

(a) 0.68.
(b) 0.32.
(c) 0.16.

14. Figure 8.7 shows the density curve of a continuous probability model for choosing a number at random between 0 and 1 inclusive. The probability that the number chosen is less than or equal to 0.4 is ____________.

15. Annual returns on the more than 5000 common stocks available to investors vary a lot. In a recent year, the mean return was 8.3% and the standard deviation of returns was 28.5%. The law of large numbers says:

(a) you can get an average return higher than the mean 8.3% by investing in a large number of stocks.

(b) as you invest in more and more stocks chosen at random, your average return on these stocks gets ever closer to 8.3%.
(c) if you invest in a large number of stocks chosen at random, your average return will have approximately a normal distribution.

16. Suppose you are trying to decide between buying many shares of a promising individual stock and spending that same amount of money on a mutual fund consisting of a variety of different stocks. Choosing the mutual fund would result in an investment that is _____ variable than the individual stock.

17. Figure 8.7 shows the density curve of a continuous probability model for choosing a number at random between 0 and 1 inclusive. The mean of this model is

(a) 0.5 because the curve is symmetric.
(b) 1 because there is area 1 under the curve.
(c) can't tell–this requires advanced mathematics.

18. Scores on the SAT Reasoning college entrance test in a recent year were roughly normal, with mean 1511 and standard deviation 194. You take an SRS of 100 students and average their SAT scores. If you do this many times, the mean of the average scores you get from all those samples would be ________________ .

19. The number of hours a light bulb burns before failing varies from bulb to bulb. The distribution of burnout times is strongly skewed to the right. The Central Limit Theorem says that

(a) as we look at more and more bulbs, their average burnout time gets ever closer to the mean μ for all bulbs of this type.
(b) the average burnout time of a large number of bulbs has a distribution of the same shape (strongly skewed) as the distribution for individual bulbs.
(c) the average burnout time of a large number of bulbs has a distribution that is close to normal.

20. Referring to Question #18, the standard deviation of the average scores you get from all those samples would be ________________________________ .

CHAPTER 8 EXERCISES

■ Challenge ◆ Discussion

8.1 Probability Models and Rules

1. Estimating probabilities empirically:

(a) Hold a penny upright on its edge under your forefinger on a hard surface, then snap it with your other forefinger so that it spins for some time before falling. Based on 30 spins, estimate the probability of heads.
(b) Toss a thumbtack (with a gently curved back) on a hard surface 100 times. (To speed it up, toss 10 at a time.) How many times did it land with the point up? What is the approximate probability of landing point up?

2. Some situations refer not to probabilities, but to odds. The odds against an event E are equal to $P(E^C)/P(E)$. If there are 3:2 odds against a particular horse winning a race, what is the probability that the horse wins?

3. The table of random digits (Table 7.1) was produced by a random mechanism that gives each digit probability 0.1 of being a 0. What proportion of the first five lines in the table are 0's? This proportion is an estimate of the true probability, which in this case is known to be 0.1.

4. Probability is a measure of how likely an event is to occur. Match one of the probabilities that follow with each statement about an event. (The probability is usually a much more exact measure of likelihood than is the verbal statement.)

0, 0.01, 0.3, 0.6, 0.99, 1

(a) This event is impossible. It can never occur.
(b) This event is certain. It will occur on every trial of the random phenomenon.
(c) This event is very unlikely, but it will occur once in a while in a long sequence of trials.
(d) This event will occur more often than not.

In each of Exercises 5 to 7, describe a reasonable sample space S for the random phenomena mentioned. In some cases, you must use judgment to choose a reasonable S.

5. Toss a coin 10 times.

(a) Count the number of heads observed.
(b) Calculate the percent of heads among the outcomes.
(c) Record whether or not at least five heads occurred.

6. A randomly chosen subject arrives for a study of exercise and fitness.

(a) The subject is either female or male.
(b) After 10 minutes on an exercise bicycle, you ask the subject to rate his or her effort on the Rate of Perceived Exertion (RPE) scale. RPE ranges in whole-number steps from 6 (no exertion at all) to 20 (maximal exertion).
(c) You also measure the subject's maximum heart rate (beats per minute).

7. A basketball player shoots four free throws.

(a) You record the sequence of hits and misses.
(b) You record the number of shots she makes.

8. The Punnett square is a diagram biologists use to determine the probability of offspring having certain genetic makeup. Suppose "B" represents the gene for brown eyes and "b" represents the gene for blue eyes. In genetics, capital letters refer to dominant traits, so a person receiving both "B" and "b" generally has brown eyes. This diagram shows the possibilities for the child of two Bb parents. Each parent gives the child one of its 2 genes with equal probability. What is the probability that this child will receive the genetic makeup for brown eyes? Discuss how this relates to Example 2.

	Mother gives B	Mother gives b
Father gives B	BB	Bb
Father gives b	Bb	bb

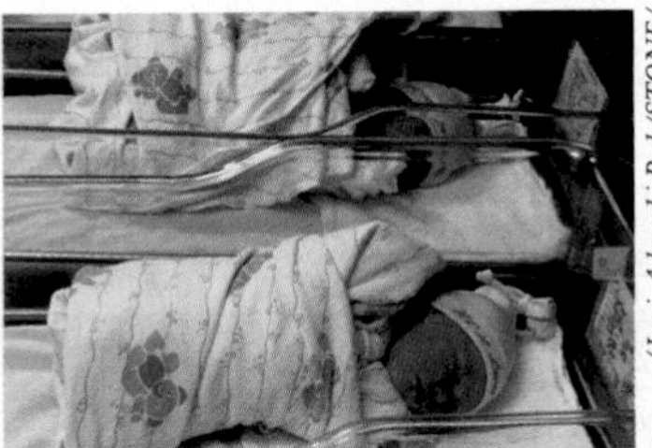
(Lori Adamski Peek/STONE/Getty Images.)

9. Many email messages are "spam." Choose a spam email message at random. Here is the probability model for the topic of a randomly chosen spam email message:

Topic	Adult	Financial	Health
Probability	0.145	0.162	0.073
Topic	Leisure	Products	Scams
Probability	0.078	0.210	0.142

(a) What is the probability that a spam email does not concern one of these topics?
(b) Corinne is particularly annoyed by spam offering "adult" content (that is, pornography) and scams. What is the probability that a randomly chosen spam email falls into one of these categories?

10. Choose a young adult (age 25 to 34 years) at random. The probability is 0.12 that the person chosen did not complete high school, 0.31 that the person has a high school diploma but no further education, and 0.29 that the person has at least a bachelor's degree.

(a) What must be the probability that a randomly chosen young adult has some education beyond high school but does not have a bachelor's degree?
(b) What is the probability that a randomly chosen young adult has at least a high school education?

■ **11.** What is the probability that Laurie rolls doubles (both dice match) each of her first three rolls in the game of Monopoly? (This matters because rolling three consecutive doubles sends you right to jail!)

8.2 Discrete Probability Models

12. Choose a new car or light truck at random and note its color. Here are the probabilities of the most popular colors:

Color	Silver	White	Black
Probability	0.201	0.184	0.116
Color	Gray	Dark blue	Light brown
Probability	0.115	0.088	0.085

(a) What is the probability that the car you choose has any color other than the six listed?
(b) What is the probability that a randomly chosen car is either silver or white?

13. North Carolina State University posts the grade distributions for its courses online. Students in Statistics 101 in a recent semester earned 21% A's, 43% B's, 30% C's, 5% D's, and 1% F's. Here is the probability model for the grade of a randomly chosen Statistics 101 student.

Grade	0 (= F)	1 (= D)	2 (= C)	3 (= B)	4 (= A)
Probability	0.01	0.05	0.30	0.43	0.21

(a) Make a probability histogram for this model.
(b) What is the probability that the student got a grade of B or better?

14. How do rented housing units differ from units occupied by their owners? Here are probability models for the number of rooms for owner-occupied units and renter-occupied units, according to the Census Bureau:

# of Rooms	1	2	3	4	5
Owned	0.000	0.001	0.014	0.099	0.238
Rented	0.011	0.027	0.229	0.348	0.224
# of Rooms	6	7	8	9	10
Owned	0.266	0.178	0.107	0.050	0.047
Rented	0.105	0.035	0.012	0.004	0.005

Make probability histograms of these two models, using the same scale. What are the most important differences between the models for owner-occupied and rented housing units?

15. In each of the following situations, state whether or not the given assignment of probabilities to individual outcomes is legitimate, that is, satisfies the rules of probability. If not, give specific reasons for your answer.

(a) Choose a college student at random and record gender and enrollment status: P(female full-time) = 0.56, P(female part-time) = 0.24, P(male full-time) = 0.44, P(male part-time) = 0.17.

(b) Choose a college student at random from a class and record the season of her birth: P(spring) = 0.39, P(summer) = 0.28, P(fall) = 0, P(winter) = 0.33.

16. What is the probability that a housing unit has five or more rooms? Use the models in Exercise 14 to answer this question for both owner-occupied and rented units.

■ **17.** Balanced six-sided dice with altered labels can produce interesting distributions of outcomes. Construct the probability model (sample space and assignment of probabilities for each sum) for rolling two "weird dice" from a *Math Horizons* article by Joseph Gallian. Instead of using the regular values {1,2,3,4,5,6}, one die has the labels 1,2,2,3,3,4 and the other die has the labels 1,3,4,5,6,8. How does this model compare to the model for regular dice?

■ **18.** Role-playing games like Dungeons & Dragons use many different types of dice. Suppose that a balanced four-sided die has faces marked 1, 2, 3, 4. The intelligence of a character is determined by rolling this die twice and adding 1 to the sum of the spots at the bottom of the die, since a triangular pyramidal die has no up "side," just a point! Give a probability model for the character's intelligence. (Start with a display like Figure 8.2 for the outcomes of the two rolls of the die. These outcomes are equally likely.) What is the probability that the character has intelligence 7 or higher?

8.3 Equally Likely Outcomes

19. A party host gives a door prize to one guest chosen at random. There are 42 men and 48 women at the party. What is the probability that the prize goes to a woman?

20. Abby, Boaz, Carmen, Dani, and Eduardo work in a firm's public relations office. Their employer must choose two of them to attend a conference in Paris. To avoid unfairness, the choice will be made by drawing two names from a hat. (This is an SRS of size 2.)

(a) Write down all possible choices of two of the five names. This is the sample space.
(b) The random drawing makes all choices equally likely. What is the probability of each choice?
(c) What is the probability that Abby is chosen?
(d) What is the probability that neither of the two men (Boaz and Eduardo) is chosen?

21. You toss a balanced coin 10 times and write down the resulting sequence of heads and tails, such as HTTTHHTHHH.

(a) How many possible outcomes are there for the 10 tosses?
(b) What is the probability that your 10-toss sequence is either all heads or all tails?

22. In the Texas Hold 'Em style of poker, play begins with each player being dealt two cards face down. From a standard 52-card deck, how many possible 2-card hands could be dealt to you?

23. A computer assigns three-character log-in IDs that may contain the digits 0 to 9 as well as the letters a to z, with repeats allowed.

(a) What is the probability that your ID contains no x?
(b) What is the probability that your ID contains no digits?

24. Consider a typical "combination lock" on a locker or briefcase.

(a) If you ask for the three numbers in the combination needed to open the lock, and they are given to you in numerical order as 3–5–8, why is this not enough information to open the lock?
(b) What would be the probability that you could open the lock with one try?
(c) Is such a lock accurately named or is it really a "permutation lock"?

25. You may have heard that a monkey hitting keys at random on a typewriter keyboard for an infinite amount of time could eventually type a particular chosen text, such as the complete works of Shakespeare. Let's focus on a monkey who just types the letters a, p, and s in random order.

(a) How many possible three-letter "words" can the monkey type using only these letters?
(b) Which of these are words in an English dictionary?
(c) What is the probability that the word the monkey typed is in an English dictionary?

■ **26.** Mozart composed a 16-bar Viennese minuet ("Musical Dice Game") in which bars #1–7 each have 11 choices, bar #8 has 2, bars #9–15 each have 11 and bar #16 has 1. How many possible versions of this minuet are there?

■ **27.** In poker, a royal flush is a 5-card hand containing (in any order) an Ace, King, Queen, Jack and 10 all of the same suit.

(a) How many royal flush hands are possible?
(b) What is the number of 5-card hands possible from a 52-card deck?
(c) What is the probability that 5 cards drawn at random from a 52-card deck yield a royal flush?

■ **28.** Biblical Permutations: The King James Version Old Testament has its 39 books canonized in a different order than the Hebrew Bible does. What mathematical expression would yield the number of possible orders of these 39 books? Is this number larger than you expected?

8.4 Continuous Probability Models

29. Generate two random numbers between 0 and 1 and take their sum. The sum can take any value between 0 and 2. The density curve is the shaded triangle shown in Figure 8.13.

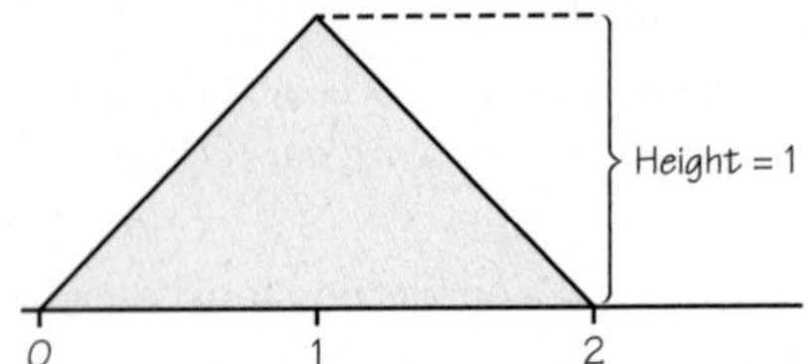

FIGURE 8.13 The density curve for the sum of two random numbers, for Exercise 29.

(a) Verify by geometry that the area under this curve is 1.
(b) What is the probability that the sum is less than 1? (Sketch the density curve, shade the area that represents the probability, then find that area. Do this for part (c) also.)
(c) What is the probability that the sum is less than 0.5?

30. Many random-number generators allow users to specify the range of the random numbers to be produced. Suppose that you specify that the range is to be all numbers between 0 and 2. The density curve for the outcome has constant height between 0 and 2, and height 0 elsewhere.

(a) What is the height of the density curve between 0 and 2? Draw a graph of the density curve.
(b) Use your graph from part (a) and the fact that probability is area under the curve to find the probability of getting an outcome less than or equal to 1.
(c) Find the probability of an outcome between 0.5 and 1.3.

31. On the TV show *The Price is Right*, there is a pricing game called "Range Game" in which a prize has a price somewhere on a scale with a range of \$600. The contestant must try to guess where to position a window spanning \$150 of the range so that the price falls within the window. If the actual price is equally likely to be anywhere, what is the probability the contestant will be successful with a random guess?

8.5 The Mean and Standard Deviation of a Probability Model

32. Linda is a sales associate at a large auto dealership. She expects to earn \$400 for each vehicle she sells. Linda motivates herself by using probability estimates of her sales. For a sunny Saturday in April, she estimates her sales as follows:

Vehicles sold	0	1	2	3
Probability	0.3	0.4	0.2	0.1

Give the probability model for Linda's earnings. What are her mean earnings?

33. Exercise 13 gives a probability model for the grade of a randomly chosen student in Statistics 101 at North Carolina State University, using the 4-point scale. What is the mean grade in this course? What is the standard deviation of the grades?

34. In Exercise 18, you gave a probability model for the intelligence of a character in a role-playing game. What is the mean intelligence for these characters?

35. Exercise 14 gives probability models for the number of rooms in owner-occupied and rented housing units. Find the mean number of rooms for each type of housing. Make probability histograms for the two models and mark the mean on each histogram. You see that the means describe an important difference between the two models: Owner-occupied units tend to have more rooms.

36. Typographical and spelling errors can be either "nonword errors" or "word errors." A nonword error is not a real word, as when "the" is typed as "teh." A word error is a real word, but not the right word, as when "lose" is typed as "loose." When undergraduates are asked to write a 250-word essay (without spell-checking), the number of nonword errors has this probability model:

Errors	0	1	2	3	4
Probability	0.1	0.2	0.3	0.3	0.1

The number of word errors has this model:

Errors	0	1	2	3
Probability	0.4	0.3	0.2	0.1

What are the mean numbers of nonword errors and word errors in an essay? How does the difference between the means describe the difference between the two models?

37. Find (and explain how you found) the mean for:

(a) the continuous probability model in Exercise 29.
(b) the probability model in Exercise 30.

■ **38.** The idea of insurance is that we all face risks that are unlikely but carry high cost. Think of a fire destroying your home. Insurance spreads the risk: We all pay a small amount, and the insurance policy pays a large amount to those few of us whose homes burn down. An insurance company looks at the records for millions of homeowners and sees that the mean loss from fire in a year is μ = \$250 per person. (The great majority of us have no loss, but a few lose their homes. The \$250 is the average loss.) The company plans to sell fire insurance for \$250 plus enough to cover its costs and profit. Explain clearly why it would be unwise to sell only 12 policies. Then explain why selling thousands of such policies is a safe business.

◆ **39.** Should you buy the extended warranty on a new washing machine? Suppose there are two outcomes–a 85% probability of needing no repairs, and a 15% probability of needing a $200 repair during the warranty period. Based on the mean outcome for this model, what would be a "break-even" price to you for the extended warranty? (The company, of course, will charge more than this in order to make a profit.)

40. An American roulette wheel has 38 slots numbered 0, 00, and 1 to 36. The ball is equally likely to come to rest in any of these slots when the wheel is spun. The slot numbers are laid out on a board on which gamblers place their bets. One column of numbers on the board contains multiples of 3–that is, 3, 6, 9, . . . , 36. Joe places a $1 "column bet" that pays out $3 if any of these numbers comes up.

(a) What is the probability model for the outcome of one bet, taking into account the $1 cost of a bet?
(b) What are the mean and standard deviation for this model?
(c) Joe plays roulette every day for years. What does the law of large numbers tell us about his results?

41. This table shows the prizes and respective probabilities for a lottery:

Net prize	$1,000,000	$1000	$100	$4
Probability	1/10,000,000	1/10,000	1/1,000	3/100

On average, how much money from a $1 ticket comes back to you in prizes?

■ **42.** A state lottery "Pick 3" game offers a choice of several bets. You choose a three-digit number and bet $1. The lottery commission announces the winning three-digit number, chosen at random, at the end of each day. The "box" pays $82.33 if the number you chose has the same digits as the winning number, in any order. Otherwise, you lose your dollar. Find the mean winnings for a bet on the box, taking into account that you paid $1 to play. (Assume that you chose a number having three distinct digits.)

■ **43.** Suppose a test is designed in which each question has 5 possible answer choices (ABCDE).

(a) If you get +1 point for every correct answer, what would the "penalty" for a wrong answer need to be if you wanted guessing to neither help nor hurt the score on average? (*Hint:* Consider a test with 5 questions on it.)
(b) If you are able to eliminate some but not all of the wrong answers, does it help on average to guess among the remaining choices?

■ **44.** Here is a simple way to create a probability model that has specified mean μ and standard deviation σ: There are only two outcomes, $\mu - \sigma$ and $\mu + \sigma$, each with probability 0.5. Use the definition of the mean and variance for probability models to show that this model does have mean μ and standard deviation σ.

8.6 The Central Limit Theorem

45. Newly manufactured automobile radiators may have small leaks. Most have no leaks, but some have one, two, or more. The number of leaks in radiators made by one supplier has mean 0.15 and standard deviation 0.4. The distribution of the number of leaks cannot be normal because only whole-number counts are possible. The supplier ships 400 radiators per day to an auto assembly plant. Take $\bar{x}$ to be the mean number of leaks in these 400 radiators. Over several years of daily shipments, what range of values will contain the middle 95% of the many $\bar{x}$'s?

46. The scores of eighth-grade students on the National Assessment of Educational Progress (NAEP) year 2007 mathematics test have a distribution that is approximately normal, with mean $\mu = 281$ and standard deviation $\sigma = 35$.

(a) Choose one eighth-grader at random. What is the probability that her score is higher than 281? Higher than 316?
(b) Now choose an SRS of four eighth-graders. What is the probability that their mean score is higher than 281? Higher than 316?

◆ **47.** Antonio measures the alcohol content of whiskey for his Chemistry 101 lab. He actually measures the mass of 5 milliliters of whiskey–a chemical calculation then finds the percent alcohol from the mass. The standard deviation of students' measurements of mass is $\sigma = 10$ milligrams (mg). Antonio repeats the measurement three times and records the mean $\bar{x}$ of his three measurements.

(a) What is the standard deviation of Antonio's mean result?
(b) How many times must Antonio repeat the measurement to reduce the standard deviation of $\bar{x}$ to 5 mg? Explain to someone who knows no statistics the advantage of reporting the average of several measurements rather than the result of a single measurement.

48. In Exercise 40 you found the mean and standard deviation of the outcome of a column bet in roulette. The Central Limit Theorem says that the average outcome of a large number of bets has a distribution that is close to normal.

(a) What is the spread (mean ± 3 standard deviations) of a gambler's average winnings after 100 bets?
(b) What is the spread of a gambler's average winnings after 1000 bets?

49. Averages of several measurements are less variable than individual measurements. The true mass of the whiskey sample in Exercise 47 is 4.6 grams, or 4600 milligrams (mg). Antonio's measurements have the normal distribution with mean 4600 mg and standard deviation 10 mg. In this case, the mean of his three measurements also has a normal distribution.

(a) Sketch on the same graph the two normal curves, for individual measurements and for means of three measurements. Figure 8.10 is an example of this kind of graph.
(b) What spread of values covers the middle 95% of Antonio's measurements?
(c) What spread of values covers the middle 95% of averages of three measurements?

50. Exercise 18 gives the probability model for the intelligence assigned by chance to a character in a role-playing game. You found the mean intelligence of such characters in Exercise 34. Jermaine plays this character often. What range covers (approximately) the middle 68% of average intelligence scores for 100 of Jermaine's games?

51. The scores of high school seniors on the ACT college entrance examination in 2003 were roughly normal with mean $\mu = 20.8$ and standard deviation $\sigma = 4.8$.

(a) What is the approximate probability that a single student randomly chosen from all those taking the test scores 25.6 or higher?
(b) Now take an SRS of nine students who took the test. What are the mean and standard deviation of the sample mean score $\bar{x}$ of these nine students?
(c) What is the approximate probability that the mean score $\bar{x}$ of these nine students is 25.6 or higher?

■ **52.** Although cities encourage carpooling to reduce traffic congestion, most vehicles carry only one person. For example, 70% of vehicles on the roads in the Minneapolis–St. Paul metropolitan area are occupied by just the driver. You choose 84 vehicles at random.

(a) What are the mean and standard deviation of the proportion of vehicles in your sample that carry only one person?
(b) What is the probability that more than 60% of the vehicles in your sample carry only one person? (Use the Central Limit Theorem.)

Chapter Review

53. License plates in Florida have the form A12BCD, that is, a letter followed by two digits followed by three more letters.

(a) How many possible different license plates are there?
(b) Jerry would like a plate that ends in AAA. How many such plates are there?
(c) If license plates are issued at random from all possible plates, what is the probability that Jerry will get a plate that ends in AAA?

54. After you tell Jerry the probability you calculated in the previous exercise, he realizes that he's unlikely to get a plate ending in AAA. So he asks you, "What's the probability I will get a plate in which all four letters are from my name?" These letters are J, E, R, and Y.

(a) Suppose Jerry insists that the letters appear in order, so that his plate reads JnmERY, where "n" and "m" stand for any number. What is the probability?
(b) Suppose Jerry allows his letters to appear in the plate in any order and also allows repeats. What is now the answer to Jerry's question?

55. Choose a person age 19 to 25 years at random and ask, "In the past four days, how many days did you do physical exercise or work out?"
Based on a large sample survey, here is a discrete probability model for the answer you will get:

# of Days	0	1	2	3	4
Probability	0.61	0.17	0.10	0.08	0.04

(a) What is the probability that the person you choose worked out either two or three days in the past four?
(b) What is the probability that the person you choose worked out at least one day in the past four?

56. What is the mean number of days that randomly chosen 19- to 25-year-olds worked out in the past four days? (Use the information in the previous exercise.) If you interview many people in this age group, what does the law of large numbers say about the average number of days these people work out?

■ **57.** Use the information in Exercise 55 and your result from Exercise 56 to answer these questions.

(a) What is the standard deviation of the number of days in the past four that a randomly chosen 19- to 25-year-old has worked out?
(b) You interview 100 randomly chosen 19- to 25-year olds. You ask each how many days in the past four they have worked out and you calculate the average number of days. According to the Central Limit Theorem, there is probability 0.95 that your average will fall between what two values?

58. In Example 16, we saw that a \$1 bet on red has a mean outcome of −\$2/38. It turns out not all American roulette \$1 bets have the same mean outcome. The "5-number bet" {0,00,1,2,3} pays an additional \$6 if one of those 5 numbers comes up–otherwise the player loses his \$1. Find the expected value for this 5-number bet. Is this 5-number bet a better or worse bet than a bet on red?

59. Suppose you select 10 people at random. Find the probability of each event below:

(a) At least one match in the day of the week they were born.
(b) At least one match in the day of the month they were born (assume 31 days per month).
(c) At least one match in the day of the year they were born.

60. Combination Connections:
(a) Give an intuitive or algebraic argument to explain why ${}_nC_k = {}_nC_{n-k}$.
(b) Generate several small values of ${}_nC_k$ and explain how they relate to the numbers in Pascal's Triangle (see for example, Chapter 11).

APPLET EXERCISES

To do these exercises, go to www.whfreeman.com/fapp8e.

1. When we toss a coin, experience shows that the probability (long-term proportion) of a head is close to 1/2. Suppose now that we toss the coin repeatedly until we get a head. What is the probability that the first head comes up in an odd number of tosses (1, 3, 5, and so on)? Use the *Probability* applet to estimate this probability. Set the probability of heads to 0.5. Toss coins one at a time until the first head appears. Do this 50 times (click "Reset" after each trial). What is your estimate of the probability that the first head appears on an odd toss?

2. The table of random digits (Table 7.1) was produced by a random mechanism that gives each digit probability 0.1 of being a 0.
(a) What proportion of the digits in the first row of Table 7.1 are 0's? This proportion is an estimate, based on 40 repetitions, of the true probability, which in this case is known to be 0.1.
(b) The *Probability* applet can imitate random digits. Set the probability of heads in the applet to 0.1. Check "Show true probability" to show this value on the graph. A head stands for a 0 in the random digit table and a tail stands for any other digit. Simulate 200 digits (40 at a time–don't click "Reset"). If you kept going forever, presumably you would get 10% heads. What was the percent of heads in your 200 tosses?

3. The basketball player Shaquille O'Neal makes about half of his free throws over an entire season. Use the *Probability* applet to simulate 100 free throws shot independently by a player who has probability 0.5 of making each shot. (Toss 40, 40, and 20 without clicking "Reset.")
(a) What percent of the 100 shots did he hit?
(b) Examine the sequence of hits and misses after each click on "Toss" and keep track of the longest run of shots made and the longest run of shots missed. How long were the longest runs in the 100 shots taken? (Sequences of random outcomes often show longer runs than our intuition expects.)

WRITING PROJECTS

1. Psychologists have shown that our intuitive understanding of chance behavior is rather poor. Amos Tversky (1937–1996) was a leader in the study of how we make decisions in the face of uncertainty. In its obituary of Tversky, the *New York Times* cited the following example:

> *Tversky asked subjects to choose between two public health programs that affect 600 people. One has probability 1/2 of saving all 600 and probability 1/2 that all 600 will die. The other is guaranteed to save exactly 400 of the 600 people. Most people chose the second program. He then offered a different choice. One program has probability 1/2 of saving all 600 and probability 1/2 of losing all 600, while the other will definitely lose exactly 200 lives. Most people chose the first program.*

Discuss this example. What is the difference between the two choices offered? What is the mean number of people saved by the two options in each choice? What do the reactions of most subjects to these choices show about how people make decisions?

2. There are about 1×10^{44} air molecules in the atmosphere and about 2×10^{22} molecules of air in a single breath taken at rest. What is the probability that the breath you took just now contained at least one molecule of air that was exhaled by Pythagoras in his last breath? What probability rules did you use to calculate this? What assumptions did you make and why do you think they were reasonable?

3. Double or Nothing: Gambler's Ruin. We have seen that by betting on "red" in roulette, you have a 18/38 chance of winning, and therefore doubling the money you bet. Suppose you have $5 and you want to bet until either you reach (and stop with) $10 or you go broke. Is placing individual $1 bets on red more, less, or equally likely to reach this goal than just placing a single $5

bet? First, try to give an answer based on intuition, taking into account the casino's advantage.

You could also explore the following formula that gives the probability of going from h dollars to N dollars by making $1 bets on red in American roulette:

$$\frac{1 - (20/18)^h}{1 - (20/18)^N}$$

Discuss how the strategy for maximizing the chance of reaching a financial target compares to the strategy for maximizing the length of time your money lasts (for entertainment value).

SUGGESTED READINGS

COMAP. *Principles and Practices of Mathematics*, Springer, New York, 1997. Chapter 4 of this team-authored text presents combinatorics, and Chapter 8 discusses probability. This is a good choice for going beyond *For All Practical Purposes* in these areas.

LESSER, LAWRENCE M. Take a chance by exploring the statistics in lotteries. *Statistics Teacher Network*, 65 (2004): 6–7. This article shows how lotteries can illustrate all major topics of an introductory statistics course, using a graphing calculator. Available online at www.amstat.org/education/stn/pdfs/STN65.pdf.

MOSTELLER, FREDERICK, ROBERT E. K. ROURKE, and GEORGE B. THOMAS. *Probability with Statistical Applications*, Addison-Wesley, Reading, Mass., 1970. A rich treatment of basic probability that requires only high school algebra but is somewhat sophisticated. Although out of print, this book is a classic that nonetheless deserves mention.

SUGGESTED WEB SITES

"Buffon's needle" is a probability problem first stated in 1777 by Count Buffon: If you drop a needle on a sheet of lined paper, what is the probability that the needle crosses one of the lines? In the simplest case, the length of the needle is the same as the distance between the lines. Some fairly advanced math shows that the answer is $2/\pi$, or about 0.637. A number of Web sites simulate dropping a needle many times in order to estimate this probability. One good simulation is at the Web site of Charles Stanton, www.math.csusb.edu/faculty/stanton/probstat/buffon.html. (If this site has disappeared, type "Buffon's needle" into Google. Be sure to choose a site that actually pictures the needle's position.)

You may be interested in the debate over legalized gambling. For the case against, visit the National Coalition Against Legalized Gambling at www.ncalg.org. For the defense by the casino industry, visit the American Gaming Association at www.americangaming.org.

Voting and Social Choice PART III

The application of mathematics to the study of human beings–their behavior, values, interactions, conflicts, and methods of making decisions–is generally considered to be a recent revolution. Yet the study of voting and social choice, which is very much the root of this revolution, goes back several centuries.

We begin in Chapter 9 with the question of how a group of individuals, each with his or her own set of values, selects one outcome from a list of possibilities. While majority rule is a good system for deciding an election with just two candidates, it turns out that there is no perfect way of deciding an election in which there are three or more candidates.

Group decision-making is often a strategic encounter, and citizens need to be aware of the difficulties that can arise when some participants have an incentive to manipulate the outcome. It is this issue that we turn to in Chapter 10.

In Chapter 11 we consider decision-making bodies in which the individual voters or parties do not have equal power. In particular, we look at weighted voting systems in which a voter's power need not be proportional to the number of votes that he or she is entitled to cast.

In Chapter 12 we analyze not only how the Electoral College influences resource-allocation in a campaign but also how polls and the positioning of candidates on a left–right continuum affect the strategies of candidates and the choices of voters. ■

CHAPTER 9

Social Choice: The Impossible Dream

The basic question of *social choice,* of how groups can best arrive at decisions, has occupied social philosophers and political scientists for centuries. One primary example of a social-choice problem is the selection of a "good" voting system. Indeed, voting is a subject that lies at the very heart of representative government and participatory democracy.

Social-choice theory attempts to address the problem of finding good procedures that will turn individual preferences for different candidates into a single choice by the whole group. An example of such a choice would be the selection of a *winner* of an election. The goal is to find such procedures that will result in an outcome that "reflects the will of the people."

This search for good voting systems, as we shall see, is plagued by a variety of counterintuitive results and disturbing outcomes. In fact, it turns out that one can prove (mathematically) that no one will ever find a completely satisfactory voting system for three or more candidates.

The elections with which we are most familiar often involve only two candidates, and we will begin our discussion of voting systems with this two-candidate case.

There are, however, real-world situations in which elections must be held to choose a single winner from among three or more candidates, as in the presidential election of 2004 in which George W. Bush, John Kerry, and Ralph Nader were the candidates.

There are several methods that can be used to elect a single candidate from a choice of three or more, and we will investigate some in this chapter. Most of these methods use a ballot in which a voter provides a rank ordering of the candidates (without ties) that indicates the order in which he or she prefers them.

In the 2004 election, President George W. Bush was challenged by Massachusetts senator John Kerry. President Bush received a slim majority of the vote. The election results have been dogged by charges of irregularities in the important swing state of Ohio.
(Left: Reuters/Corbis; right: John Gress/Reuters/Corbis.)

Preference List Ballot DEFINITION

A ballot consisting of such a rank ordering of candidates (which we often picture as a vertical list with the most preferred candidate on top and the least preferred on the bottom) is called a **preference list ballot** because it is a statement of the preferences of the individual who is voting.

Preference list ballots allow voters to make a much clearer statement of their preferences than do ballots allowing a single vote. Preference list ballots are already used in a wide range of applications, such as rating football teams and scoring track meets.

Although we do not allow ties in a preference list ballot, most voting rules of interest will, in some elections, result in a tie for the win among two or more of the candidates. In the real world, the number of voters is often so large that ties seldom occur. Nevertheless, to avoid excessive annoyances in the theory we develop, and to simplify what we do in this chapter, we make the following assumption throughout:

The Number of Voters Assumption RULE

Throughout this chapter, we assume that the number of voters is odd.

9.1 Majority Rule and Condorcet's Method

When a choice is being made between two candidates, the first type of voting system to suggest itself is **majority rule**: Each voter indicates a preference for one of the two candidates and the one with the most votes wins. With two candidates, there is no real distinction between a ballot that indicates a voter's choice for one of the two candidates and what we have called a preference list ballot. The point is that we can, for example, identify a choice for A (however indicated) with the list that has A over B and a choice B with the list that has B over A.

Majority rule has at least three desirable properties:

1. All voters are treated equally. That is, if any two voters were to exchange (marked) ballots before submitting them, the outcome of the election would be the same.
2. Both candidates are treated equally. That is, if a new election were held and every voter were to reverse his or her vote, then the outcome of the previous election would be reversed as well.
3. It is **monotone**. If a new election were held and a single voter were to change his or her ballot from being a vote for the loser of the previous election to being a vote for the winner of the previous election, and everyone else voted exactly as before, then the outcome of the new election would be the same as the outcome of the previous election.

It is easy to devise voting systems for two candidates in which these fail, but each such voting system quickly reveals its undesirability. For example, condition 1 is not satisfied by a *dictatorship* (in which all ballots except that of the dictator are ignored); condition 2 is not satisfied by *imposed rule* (in which candidate X wins regardless of who votes for whom); and condition 3 is not satisfied by *minority rule* (in which the candidate with the fewest votes wins).

But maybe there are voting systems in the two-candidate case that are superior to majority rule in the sense of satisfying the three properties just listed *and* some other properties that we might also wish to have satisfied. This, however, turns out not to be the case. In 1952, Kenneth May proved the following:

May's Theorem THEOREM

Among all two-candidate voting systems that never result in a tie, majority rule is the *only* one that treats all voters equally, treats both candidates equally, and is monotone.

This is an important and elegant result. Thus, mathematical reasoning spares us the trouble of searching for a better voting system for two candidates.

But what if there are three or more candidates? Perhaps we can design a voting system for this situation that, in some way, builds on the success of majority rule in the two-candidate case. In point of fact, there does exist a voting system that arises from precisely this hope, and it is known today as **Condorcet's method**.

Our description of Condorcet's method begins with the observation that if we have a sequence of preference list ballots, then–for each pair of candidates–we can determine who the winner would have been had the election involved only these two in a one-on-one contest using majority rule.

To illustrate this notion of a one-on-one contest, consider the following preference list ballots:

Rank	**Number of Voters (3)**		
First	*A*	*B*	*C*
Second	*B*	*C*	*A*
Third	*C*	*A*	*B*

In this election, candidate *A* would defeat candidate *B* in a one-on-one contest (two votes to one), while *B* would, in turn, defeat *C* in a one-on-one contest, again by a score of 2 to 1. We'll return to this example in a moment, but we now have at hand all we need to describe Condorcet's voting system for three or more candidates.

Description of Condorcet's Method — PROCEDURE

With the voting system known as Condorcet's method, a candidate is a winner precisely when he or she would, on the basis of the ballots cast, defeat every other candidate in a one-on-one contest using majority rule.

Historically, the voting system we are calling Condorcet's method dates back at least to Ramon Llull in the thirteenth century (see Spotlight 9.1). It was rediscovered and popularized in the eighteenth century by the Marquis de Condorcet (1743–1794).

EXAMPLE 1 ■ Condorcet's Method

Suppose we have four candidates (*GB, AG, RN,* and *PB,* with these initials chosen for a soon-to-be-revealed reason) and the following sequence of preference list ballots, where the heading of "6" indicates that 6 of the 15 voters hold the ballot with *GB* over *AG* over *PB* over *RN,* the heading of "5" indicates that 5 of the 15 voters hold the ballot with *AG* over *RN* over *GB* over *PB,* and so on.

	Number of Voters (15)			
Rank	**6**	**5**	**3**	**1**
First	*GB*	*AG*	*RN*	*PB*
Second	*AG*	*RN*	*AG*	*GB*
Third	*PB*	*GB*	*GB*	*AG*
Fourth	*RN*	*PB*	*PB*	*RN*

We claim that *AG* is the winner in this election if we use Condorcet's method. Let's check the one-on-one scores for each possible pair of opponents:

AG versus *GB: AG* is over *GB* on 5 + 3 = 8 of the ballots, while the reverse is true on 6 + 1 = 7 of the ballots. Thus, *AG* defeats *GB* by a score of 8 to 7.

AG versus *RN: AG* is over *RN* on 6 + 5 + 1 = 12 of the ballots, while the reverse is true on 3 of the ballots. Thus, *AG* defeats *RN* by a score of 12 to 3.

AG versus *PB: AG* is over *PB* on 6 + 5 + 3 = 14 of the ballots, while the reverse is true on 1 of the ballots. Thus, *AG* defeats *PB* by a score of 14 to 1.

This shows that *AG* is the winner using Condorcet's method. ■

Like majority rule, Condorcet's method satisfies some very desirable properties, as we'll see later in this section. But it also has a tragic flaw, and this flaw is called **Condorcet's voting paradox.**

Condorcet's Voting Paradox THEOREM

With three or more candidates, there are elections in which Condorcet's method yields *no* winners. In particular, the following ballots (often called the "Condorcet voting paradox ballots") constitute an election in which Condorcet's method yields no winner.

Rank	**Number of Voters (3)**		
First	A	B	C
Second	B	C	A
Third	C	A	B

The Condorcet voting paradox ballots given above are the same ones we used earlier in illustrating the notion of a one-on-one contest. We pointed out then that A defeats B one on one and B defeats C one on one. The additional observation needed is that we also have C defeating A one on one. Thus A cannot be a winner using Condorcet's method (it loses to C), B cannot be a winner (it loses to A), and C cannot be a winner (it loses to B). We will revisit Condorcet's voting paradox in section 9.3.

Notice that, because of our assumption that the number of voters is odd, Condorcet's method yields either no winner or a unique winner (see Exercise 5).

It is tempting at this point to suggest modifying Condorcet's method as we have presented it by declaring all the candidates to be tied for the win if there is no candidate who defeats each of the others one on one. The drawback to this modification is that a number of the upcoming desirable properties possessed by Condorcet's method then evaporate. We'll explore this in the exercises.

9.2 Other Voting Systems for Three or More Candidates

With three or more candidates, we find no shortage of additional procedures that suggest themselves and that seem to represent perfectly reasonable ways to choose a winner. Closer inspection, however, reveals shortcomings with all of these. We illustrate this with a consideration of several well-known procedures. Additional procedures (and additional shortcomings) can be found in the exercises.

Plurality Voting and the Condorcet Winner Criterion

In **plurality voting**, only first-place votes are considered. Thus, while we will consider plurality voting in the context of preference list ballots, a ballot here might just as well be a single vote for a single candidate. The candidate with the most votes wins, even though this may be considerably fewer than one-half the total votes cast. This is perhaps the most common system in use today. It is how the voters in Florida chose George W. Bush over Al Gore, Ralph Nader, and Patrick Buchanan in the presidential election of 2000.

SPOTLIGHT 9.1 The Historical Record

The following letter was written by Friedrich Pukelsheim of the University of Augsburg, Germany. He is imagining what Ramon Llull (1232–1316) might say if he were alive today.

Dear Editors:

It is my distinct pleasure to respond "from the beyond" to your kind invitation to set the historical record straight. I was born in 1232 on the Island of Mallorca in the Mediterranean Sea, which in your times is known as a popular tourist place. In my days it was a strong political center of that part of the world, with a population that was a mix of Christians, Jews, and Muslims. It was my dream to persuade people of the virtues of Christian belief by relying, not on force, but on reason.

Unfortunately, people did not find it easy to follow my arguments, so I was more than pleased to discover some down-to-earth applications, including an election system. My idea was to oppose every pair of candidates, one on one, and ask the electors whom of the two they would prefer—very much like a medieval jousting tournament. But how to combine the results from all the duels into a winner of the election? I first proposed electing the candidate who won the most duels, then later suggested a system of successive eliminations.

I wrote three papers on the topic, the second of which I "smuggled" into my novel Blanquerna in 1283. More than a century after my death, in 1428, the young German scholar Nicolaus Cusanus (1401–1464) journeyed to Paris to read my works in libraries there. He even copied out the third of my electoral writings, which I had completed on 1 July 1299 in Paris, and his manuscript is the only copy handed down to your days. Reading my papers, Cusanus was inspired to invent his own electoral system. Did he not understand mine, or just find it inadequate? Who knows?

While I had been concerned with electing church officials, Cusanus sought a system to elect the Holy Roman Emperor. In his system, each elector assigns each candidate a rank score, with the lowest candidate getting a score of 1, the second lowest a score of 2, and the best candidate the highest score possible, that is, 10 when there are 10 candidates. The scores are totaled for each candidate and the candidate with the highest score wins. If you are a soccer player or a hockey player, you will have a good sense for one difference between our systems: Whereas I count victories, Cusanus adds up goals. Cusanus applauded himself for having invented an absolutely ingenious and novel electoral system.

Also, I advocated open voting, whereas Cusanus favored a secret ballot. He was concerned that voters might sell their votes, or that the candidates might pressure the voters. Well, that certainly happened all of the time in elections for worldly authorities! But for election to clerical office, I thought it good enough if electors take an oath to vote for the most worthy candidate and submit themselves to the social control that comes with an open election.

Cusanus was famous in his times, as I was in mine, but fame indeed is transitory. Sure enough, my electoral system was reinvented by the Marquis de Condorcet (1743–1794), and Cusanus's system was proposed afresh by the Chevalier de Borda (1733–1799)—neither of whom, I am sure, wasted a thought on the possibility that "their" systems might already be on record. But, as my works had fallen into oblivion as had those of Cusanus, neither Condorcet nor Borda should be blamed for failing to acknowledge our priority.

My first electoral paper—actually the one that is longest and most detailed, written around 1280—was rediscovered only in 2000, filed away in the Vatican Library. How would you feel if your work attracts fresh attention after more than 700 years? Actually, I am utterly pleased that mine has resurfaced at last! The text was excavated by a mathematician interested in voting systems, Friedrich Pukelsheim of the University of Augsburg, Germany. Since the text is handwritten in Latin, handling it became an interdisciplinary project that brought together experts on medieval manuscripts, Church Latin and theology, and even computer scientists. As a result, my electoral writings are

The Historical Record *(continued)*

now on the Internet (in the original and in translations into English and German) at www.uni-augsburg.de/llull/.

Looking back on my lack of success in preaching peace among Christians, Jews, and Muslims, and all the writing and copying by hand of my works, I hope you can appreciate how highly I value the printed book (such as this one) and, even more, instant communication worldwide over the Internet. May that ease of communication help facilitate the religious peace that I so dearly sought.

Yours truly,
Ramon Llull (1232–1316)
Left Choir Chapel
San Francisco Cathedral
Palma de Mallorca

EXAMPLE 2

Plurality Voting and the 2000 Presidential Election

On the evening of December 12, 2000, Al Gore conceded the presidential election of 2000 to George W. Bush, thus bringing to a close one of the most remarkable elections in modern times. The outcome, ultimately decided in the electoral college, came down to which of Bush or Gore would carry Florida. With more than six million votes cast in Florida, the ultimate margin of victory for George W. Bush was only a few hundred votes.

There is little doubt that if the 2000 presidential election had pitted Al Gore solely against any one of the other three candidates, then Gore would have won both the election in Florida and the presidency. The point is that while most of the Buchanan supporters would have voted for Bush, the far more numerous Nader supporters would have gone largely for Gore. In fact, the illustration of Condorcet's method that we gave in Example 1 is a simplified version of this Florida election (with *GB* standing for George Bush, *AG* for Al Gore, *PB* for Patrick Buchanan, and *RN* for Ralph Nader).

Thus, although plurality voting led to Bush's winning the 2000 election in Florida (and hence the presidency), Gore was, in this example, what is called a **Condorcet winner**: He would have won the election if Condorcet's method had been used.

Governor George W. Bush and Vice President Al Gore debate the issues before the 2000 election, possibly the most controversial election in U.S. history. Gore was the Condorcet winner of the election, but Bush eked out a victory that relied on the rules of the Electoral College. Many voters were suddenly put on notice that the U.S. Constitution makes the election of the president indirect–and not a pure expression of the majority's choice. *(Reuters/Corbis.)*

Condorcet Winner Criterion (CWC) DEFINITION

A voting system is said to satisfy the **Condorcet winner criterion** (CWC) provided that, for every possible sequence of preference list ballots, either (1) there is no Condorcet winner (as is often the case) or (2) the voting system produces exactly the same winner for this election as does Condorcet's method.

The CWC is certainly a property that one would like to see satisfied. We record plurality voting's failure in this respect with the following.

The Failure of the CWC with Plurality Voting THEOREM

The Florida vote in the 2000 presidential election shows that plurality voting fails to satisfy the Condorcet winner criterion.

Perhaps a more fundamental drawback of plurality voting is the extent to which the ballots provide no opportunity for a voter to express any preferences except for naming his or her top choice. No use is made, for example, of the fact that a candidate may be no one's first choice but everyone's close second choice.

Finally, there is yet another shortcoming of plurality voting: There are elections in which it is to a voter's advantage to submit a ballot that misrepresents his or her true preferences.

Manipulability DEFINITION

A voting system is subject to **manipulabity** (or is **manipulable**) if there are elections in which it is to a voter's advantage to submit a ballot that misrepresents his or her true preferences.

For example, in the presidential election of 2000, many voters who ranked Ralph Nader or Patrick Buchanan over George W. Bush and Al Gore chose to vote for Bush or Gore rather than to "throw away" their vote on a candidate they felt had no chance. Condorcet's method, it turns out, is not manipulable, and this is one of its most desirable properties. We'll explore this further in the next chapter.

The Borda Count and Independence of Irrelevant Alternatives

In many elections that use preference list ballots, the goal is to arrive at a final group rank ordering of all the contestants that best expresses the desires of the electorate. The purpose is not only to determine the winner–say, the class valedictorian–but also to arrive at who finished second, third, and so on, as in the case of one's rank in the senior class. In other applications, such as an election to a hall of fame, the first few finishers each receive the award, while the remaining nominees are also-rans.

One common mechanism for achieving this objective is to assign points to each voter's rankings and then to sum these for all voters to obtain the total points for each candidate. If there are 10 candidates, for example, then we could assign 10 points to each first-place vote for a given candidate, 9 points for each second-place vote, 8 for each third-place vote, and so forth. The candidate with the highest total number of points is the winner. Subsequent positions are assigned to those with the next-highest tallies.

Description of Rank Methods and the Borda Count — PROCEDURE

A *rank method* of voting assigns points in a nonincreasing manner to the ordered candidates on each voter's preference list ballot and then sums these points to arrive at a group's final ranking. The special case in which there are n candidates with each first-place vote worth $n - 1$ points, each second-place vote worth $n - 2$ points, and so on down to each last-place vote worth zero points is known as the **Borda count.** The actual point totals are referred to as a candidate's **Borda score.**

The Borda count is named after Jean-Charles de Borda (1733–1799). He was a contemporary of Condorcet.

Rank methods other than the Borda count are common. For example, a track meet can be thought of as an "election" in which each event is a "voter" and each of the schools competing is a "candidate." If the order of finish in the 100-meter dash is school A, school B, school C, school D, then points are often awarded to each school as follows: 5 points for first place, 3 for second place, 2 for third place, and 1 for fourth place.

Sports polls often use point assignments that qualify as rank methods according to our definition. The following example provides an illustration of this.

EXAMPLE 3 ■ Rank Methods and a Basketball Poll

In November of 2007, the Associated Press issued the early-season ranking of the top 25 teams in women's college basketball, shown to the right.

An interesting question is whether or not this is a ranking system. If it is, who are the candidates and how many are there? In fact, this can be regarded as a ranking system, but the number of candidates is not 25. That is, although 25 teams appeared on each ballot, at least one ballot included each of the teams listed at the bottom in the category "Others receiving votes."

To regard this as a ranking system, the set of candidates must include the entire set of eligible collegiate women's basketball teams. We must also infer that each ballot lists all teams other than that voter's top 25 *below* that voter's top 25, perhaps, in alphabetical order. The point assignments are then like those in the newspaper clipping, except that we also assign 0 points for a 26th-place vote, 0 points for a 27th-place vote, and so on. This is why our definition states that a rank method "assigns points in a *nonincreasing* manner" instead of "assigns points in a *decreasing* manner."

We can use this poll to illustrate how total points are arrived at with a ranking method. With the top-ranked team, Tennessee, it's quite easy. Each first-place vote is worth 25 and Tennessee received all 50 first-place votes. This accounts for its total of $25 \times 50 = 1250$ points. But the calculation is more interesting for the second-ranked team, Connecticut, and requires some speculation on our part, since we

WOMEN'S TOP 25

The top 25 teams in The Associated Press' women's college basketball poll, with first-place votes in parentheses, records through Sunday, total point based on 25 points for a first-place vote through one point for a 25th-place vote and previous ranking:

	Record	Pts	Pvs
1. Tennessee (50)	1-0	1250	1
2. Connecticut	1-0	1187	2
3. Maryland	2-0	1139	4
4. LSU	2-0	1072	5
5. Stanford	2-0	1040	7
6. Rutgers	0-1	973	3
7. North Carolina	2-0	969	8
8. Georgia	2-0	881	9
9. Oklahoma	0-1	833	6
10. Duke	1-0	818	10
11. Texas A&M	1-0	753	11
12. California	1-0	631	13
13. Baylor	3-0	576	15
14. G. Washington	1-0	560	14
15. Arizona St.	0-1	554	12
16. Ohio St.	1-0	506	16
17. Michigan St.	2-0	390	17
18. Florida St.	2-0	365	19
19. West Virginia	1-0	345	18
20. Vanderbilt	2-0	254	23
21. Texas	1-0	247	22
22. Louisville	1-0	212	21
23. Notre Dame	1-0	112	24
24. DePaul	1-0	97	25
25. Wisconsin	1-0	77	–

Others receiving votes: N.C. State 76, Pittsburgh 76, Purdue 68, Penn St. 53, Auburn 46, Wyoming 21, Oklahoma St. 13, Marquette 9, Middle Tennessee 9, Georgia Tech 8, Old Dominion 8, Illinois 7, Xavier 4, Marist 3, Kentucky 2, New Mexico 2, S. Dakota St. 2, Florida 1, Iowa St. 1.

(*AP Photo/Bill Kostroun.*)

don't actually have the ballots to examine. We know that there were 50 ballots (because there were exactly 50 first-place votes all together), and we know Connecticut had no first-place votes. It stands to reason that Connecticut's 1187 points must have come from a vast majority of the second-place votes, together with a few lower rankings. (We know it didn't receive *all* the second place-votes, otherwise its point total would have been $24 \times 50 = 1200$.)

One possibility is that Connecticut received:

0 first-place votes (at 25 points each)
40 second-place votes (at 24 points each)
8 third-place votes (at 23 points each)
1 fourth-place votes (at 22 points each)
1 fifth-place vote (at 21 points)

The total would then be:

$$0 + 960 + 184 + 22 + 21 = 1187$$

There is an easy way to calculate the Borda score of a candidate. You can count the number of occurrences of other candidate names that are below this candidate's name. For example, consider the following ballots:

Rank	Number of Voters (5)					Points
First	*A*	*A*	*A*	*B*	*B*	2
Second	*B*	*B*	*B*	*C*	*C*	1
Third	*C*	*C*	*C*	*A*	*A*	0

Because there are three candidates, each first-place vote is $n - 1$, or $3 - 1 = 2$, each second-place vote is $n - 2 = 1$, and each third-place vote is $n - 3 = 0$. If we were to calculate the Borda score of candidate B algebraically, we would say that B has two first-place votes, worth 2 points each (a total of 4 points), and three second-place votes, worth 1 point each (a total of 3 more points). Thus, the Borda score of candidate B is $4 + 3 = 7$.

But instead of calculating this Borda score algebraically, we can mentally replace each occurrence of a letter below B by a box, □, and simply count the boxes.

Rank	Number of Voters (5)				
First	*A*	*A*	*A*	*B*	*B*
Second	*B*	*B*	*B*	□	□
Third	□	□	□	□	□

Notice that there are seven boxes, giving us the correct value of 7 as the Borda score for candidate B. Of course, you don't actually have to draw any boxes. We are just emphasizing the fact that, in the counting process, it is "spaces" that we are counting, without regard to which letter occurs in the space. A quick glance at the original ballots (without the boxes) reveals that the Borda score of candidate A is 6 and the Borda score of candidate C is 2. When calculating Borda scores this way, be sure that each individual ballot is listed separately, as opposed to using a single list to represent the ballots of several voters (as we often do).

The Borda count certainly seems to be a reasonable way to choose a winner from among several candidates (or to arrive at a group ranking of the candidates). It also has its shortcomings, however, one of which is the failure of a property known as **independence of irrelevant alternatives**.

Independence of Irrevelant Alternatives (IIA) DEFINITION

A voting system is said to satisfy independence of **irrelevant alternatives (IIA)** if it is impossible for a candidate *B* to move from nonwinner status to winner status unless at least one voter reverses the order in which he or she had *B* and the winning candidate ranked.

To describe this property, suppose that an election yields one candidate (call it *A*) as a winner and another candidate (call it *B*) as a nonwinner. Suppose that a new election is now held and that, although some of the voters may have changed their preference list ballots, no one who had previously ranked *A* and *B* changed his or her ballot to rank *B* over *A* now.

If this new election were to yield *B* as a winner, the new outcome would seem strange, especially because none of the relative individual preferences for *A* over *B* had changed in *B*'s favor. The ballot changes responsible for the new outcome involve candidates *other than A* or *B*. One could argue that these other candidates ought to be irrelevant to the question of whether *A* is more desirable than *B* or *B* is more desirable than *A*.

Condorcet's method satisfies IIA. That is, if we have a sequence of preference list ballots that yield *A* as a Condorcet winner and *B* as a nonwinner, then *A* defeats every other candidate, and *B* in particular, in a one-on-one contest according to these ballots. If no voter reverses the order in which he or she ranked *A* and *B*, then *A* will still defeat *B* one on one, and thus *B* remains a nonwinner.

The following illustration shows that the Borda count, unlike Condorcet's method, fails to satisfy independence of irrelevant alternatives. Suppose the initial five ballots are as follows:

Rank	Number of Voters (5)				
First	*A*	*A*	*A*	*C*	*C*
Second	*B*	*B*	*B*	*B*	*B*
Third	*C*	*C*	*C*	*A*	*A*

Our counting procedure shows that the Borda scores are as follows:

Borda score of *A* is 6
Borda score of *B* is 5
Borda score of *C* is 4

The winner is A (with 6 points), and B is a nonwinner (with 5 points). But now suppose that the two voters on the right change their ballots by moving C down between A and B. The ballots then become

Rank	Number of Voters (5)				
First	A	A	A	B	B
Second	B	B	B	C	C
Third	C	C	C	A	A

Our counting procedure shows that the Borda scores are as follows:

Borda score of A is 6

Borda score of B is 7

Borda score of C is 2

The Borda count therefore now yields B as the winner (with 7 points). Thus, B has gone from being a nonwinner to being a winner, even though no one changed his or her mind about whether B is preferred to A, or vice versa.

The above discussion establishes the following:

The Failure of IIA with the Borda Count THEOREM

The Borda count fails to satisfy independence of irrelevant alternatives.

Sequential Pairwise Voting and the Pareto Condition

In our voting-theoretic context, an **agenda** will be understood to be a listing (in some order) of the candidates. This listing is not to be confused with any of the preference list ballots, and, to avoid confusion, we will present agendas as horizontal lists and continue to present preference list ballots vertically.

Description of Sequential Pairwise Voting PROCEDURE

Sequential pairwise voting starts with an agenda and pits the first candidate against the second in a one-on-one contest. The winner then moves on to confront the third candidate in the list, one on one. Losers are deleted. This process continues throughout the entire agenda, and the one remaining at the end wins.

For a given sequence of individual preference list ballots, the particular agenda chosen can greatly affect the outcome of the election, as we'll show in the next chapter. Nevertheless, we will see later in this chapter that sequential pairwise voting arises naturally in the legislative process. Notice also that because of our assumption that the number of voters is odd, there is always a unique winner with sequential pairwise voting.

EXAMPLE 4 ■ Sequential Pairwise Voting

Assume we have four candidates and that the agenda is *A, B, C, D*. Consider the following sequence of three preference list ballots:

Rank	**Number of Voters (3)**		
First	*A*	*C*	*B*
Second	*B*	*A*	*D*
Third	*D*	*B*	*C*
Fourth	*C*	*D*	*A*

The first one-on-one pits *A* against *B*, and *A* wins by a score of 2 to 1 (meaning that two of the voters–the two on the left–prefer *A* to *B*, and one of the voters prefers *B* to *A*). Thus, *B* is eliminated and *A* moves on to confront *C*. Because *C* wins this one on one (by a score of 2 to 1), *A* is eliminated. Finally, *C* takes on *D*, and *D* wins by a score of 2 to 1. Thus, *D* is the winner. ■

There is something very troubling about the outcome of the preceding example, especially if you are candidate *B*. *Everyone* prefers *B* to *D*!

The Failure of the Pareto Condition with Sequential Pairwise Voting — THEOREM

Sequential pairwise voting fails to satisfy what is called the **Pareto condition**, which says that if everyone prefers one candidate (in this case, *B*) to another candidate (*D*), then this latter candidate (*D*) should not be among the winners of the election.

The Pareto condition is named after Vilfredo Pareto (1848–1923), an Italian economist.

Runoff Systems and Monotonicity

The voting system known as the **Hare system**, which was introduced by Thomas Hare in 1861, is also known by names such as the "single transferable vote system." In 1862, John Stuart Mill described the Hare system as being "among the greatest improvements yet made in the theory and practice of government." Today, the system is used to elect public officials in Australia, Malta, the Republic of Ireland, and Northern Ireland.

Description of the Hare System — PROCEDURE

The Hare system proceeds to arrive at a winner by repeatedly deleting candidates that are "least preferred" in the sense of being at the top of the fewest ballots. If a single candidate remains after all others have been eliminated, it alone is the winner. If two or more candidates remain and all of these remaining candidates would be eliminated in the next round (because they all have the same number of first-place votes), then these candidates are declared to be tied for the win.

EXAMPLE 5 ■ The Hare System

Suppose we have the following sequence of preference list ballots, where, as before, the heading of "5" indicates that 5 of the 13 voters hold the ballot with A over B over C, the heading of "4" indicates that 4 of the 13 voters hold the ballot with C over B over A, and so forth.

	Number of Voters (13)			
Rank	**5**	**4**	**3**	**1**
First	A	C	B	B
Second	B	B	C	A
Third	C	A	A	C

Candidates B and C have only 4 first-place votes (while A has 5). Thus, B and C are eliminated in the first round, and A wins the election.

In the preceding example, suppose that the voter in the last column moves candidate A up on his list. Let's look at the new election. Notice that, even though A won the last election, the only change we are making in ballots for the new election is one that is favorable to A. The ballots for the new election are as follows:

	Number of Voters (13)			
Rank	**5**	**4**	**3**	**1**
First	A	C	B	A
Second	B	B	C	B
Third	C	A	A	C

If we apply the Hare system again, only B is eliminated in round one, as it has 3 first-place votes to 4 for C and 6 for A. Thus, after this round, the ballots are as follows:

	Number of Voters (13)			
Rank	**5**	**4**	**3**	**1**
First	A	C	C	A
Second	C	A	A	C

We now have A on top of 6 lists and C on top of 7 lists. Thus, at stage two, A (our previous winner!) is eliminated and C is the winner of this new election.

Clearly, this is once again quite counterintuitive. Alternative A won the original election, the only change in ballots made was one favorable to A (and no one else), and then A lost the next election.

Failure of Monotonicity with the Hare System THEOREM

Example 5 shows that the Hare system does not satisfy **monotonicity**, which, with three or more candidates, says that if a candidate is a winner, and a new election is held in which the only ballot change made is for some voter to move the former winning candidate higher on his or her ballot, then the original winner should remain a winner.

The fact that the Hare system does not satisfy monotonicity is considered by many–and with good reason–to be a glaring defect. A 17-voter example in which only a single candidate is eliminated in the first round can also be used to show that the Hare system does not satisfy monotonicity–see Exercise 28. For an even more glaring version of this defect, one in which alternative *A* goes from winning to losing because voters move *A* from last place on their ballots to first place on their ballots, see Exercise 29.

In spite of these drawbacks, the Hare system is used in important ways today. For example, it is essentially the method that was used to choose Sydney, Australia, as the site of the 2000 Summer Olympics. Beijing would have been the plurality winner, but after the elimination of Istanbul, Berlin, and Manchester (in that order), Sydney defeated Beijing by a vote of 45 to 43. Four years later, Beijing finally did prevail; it was the site of the 2008 Summer Olympics.

There are other runoff systems, some more frequently used than the Hare system. One such example is the following.

Description of the Plurality Runoff Method PROCEDURE

Plurality runoff is the voting system in which there is a runoff (that is, a new election using the same ballots) between the two candidates receiving the most first-place votes. If there are ties, then the runoff is among either those tied for the most first-place votes, or the lone candidate with the most first-place votes along with those tied for the second-most first-place votes (and plurality voting is used).

Alas, the plurality runoff method also is not monotone. Exercise 25 asks you to verify this by making use of the following ballots:

	Number of Voters (13)				
Rank	4	3	3	2	1
First	*A*	*B*	*C*	*D*	*E*
Second	*B*	*A*	*A*	*B*	*D*
Third	*C*	*C*	*B*	*C*	*C*
Fourth	*D*	*D*	*D*	*A*	*B*
Fifth	*E*	*E*	*E*	*E*	*A*

EXAMPLE 6 ■ Plurality Runoff

The plurality runoff method is somewhat similar in spirit to the Hare system. In fact, you might wonder if they aren't just two different descriptions of the same voting system. That is, you might ask if the plurality runoff method and the Hare system always yield the same winner.

The answer is no, however, as we now demonstrate. Consider the following sequence of preference list ballots:

	Number of Voters (13)			
Rank	**4**	**4**	**3**	**2**
First	*A*	*B*	*C*	*D*
Second	*B*	*A*	*D*	*C*
Third	*C*	*C*	*A*	*A*
Fourth	*D*	*D*	*B*	*B*

With the plurality runoff method, *A* and *B* initially tie with 4 first-place votes each, with 3 for *C* and 2 for *D*. In the runoff between *A* and *B*, the ballots are as follows:

	Number of Voters (13)			
Rank	**4**	**4**	**3**	**2**
First	*A*	*B*	*A*	*A*
Second	*B*	*A*	*B*	*B*

With the plurality runoff method, *A* is the winner after defeating *B* in the runoff by a score of 9 to 5.

On the other hand, with the Hare system we find that the only alternative deleted in the first round is *D*, with only 2 first-place votes. With this deletion of *D*, the ballots are as follows:

	Number of Voters (13)			
Rank	**4**	**4**	**3**	**2**
First	*A*	*B*	*C*	*C*
Second	*B*	*A*	*A*	*A*
Third	*C*	*C*	*B*	*B*

A and *B* now have only 4 first-place votes compared to the 5 first-place votes that *C* has. Hence, *A* and *B* are now deleted, leaving *C* as the winner with the Hare system.

■

9.3 Insurmountable Difficulties: Arrow's Impossibility Theorem

All of the voting systems for three or more candidates that we have discussed turn out to be flawed in one way or another. You may well ask at this point why we don't simply present *one* voting method for the three-candidate case that has all the desirable properties we want to have satisfied. That is, after all, exactly what we did for the two-candidate case (with majority rule filling the bill, and being the only one to do so by May's theorem).

The answer to this question is extremely important. The difficulties in the three-candidate case are not in any way tied to a few particular systems that we present in a text such as this (or that we choose to use in the real world). The fact is, there are difficulties that will be present *regardless* of what voting system is used, and this applies even to voting systems not yet discovered.

Nothing in the remarkable body of work produced by Nobel laureate Kenneth J. Arrow of Stanford University is as well known or widely acclaimed as the result known as **Arrow's impossibility theorem** (see Spotlight 9.2).

Arrow's Impossibility Theorem THEOREM

With three or more candidates and any number of voters, there does not exist–and there never will exist–a voting system that always produces a winner, satisfies the Pareto condition and independence of irrelevant alternatives, and is not a dictatorship.

Arrow's theorem isn't obvious, and we won't be saying anything about the proof. But we can state and prove a much weaker result of some interest in its own right. This version is taken from the 1995 text *Mathematics and Politics,* cited in the Suggested Readings, and replaces Arrow's assumption of the Pareto condition and non-dictatorship by the Condorcet winner criterion.

A Weak Version of Arrow's Impossibility Theorem THEOREM

With three or more candidates and an odd number of voters, there does not exist–and there never will exist–a voting system that satisfies both the Condorcet winner criterion and independence of irrelevant alternatives and that always produces at least one winner in every election.

To see why this is true, we'll handle only the case of exactly three voters. Our plan will be to assume that we have some kind of hypothetical voting system that satisfies both the Condorcet winner criterion and independence of irrelevant alternatives and to show that, when confronted by the Condorcet voting paradox ballots, it produces *no* winner.

The argument really comes in three separate, but extremely similar, pieces–one for each of the three candidates. Piece 1 argues that A can't be among the winners, piece 2 that B can't be among the winners, and piece 3 that C can't be among the

Kenneth J. Arrow *(continued)*

My first interest was in the theory of corporations. In a firm with many owners, how do the owners agree when they have different opinions, for example, about the prospects of the company? I was thinking of stockholders. In the course of this, I realized that there was a paradox involved—that majority voting can lead to cycles. I then dropped that discussion because I was frustrated by it.

I happened to be working with The RAND Corporation one summer about a year or two later. They were very interested in applying concepts of rationality, particularly of game theory, to military and diplomatic affairs. That summer, I felt not like an economist but instead like a general social scientist or a mathematically-oriented social scientist. There was tremendous interest in game theory, which was then new.

Someone there asked me, "What does it mean in terms of national interest?" I said, "That's a very simple matter." He then asked me to write a memorandum on the subject. That memorandum led to a sharper formulation of the social-choice question, and I realized that I had been thinking of it earlier in that other context.

Society must choose among a number of alternative policies. These policies may be thought of as quite comprehensive, covering a number of aspects: foreign policy, budgetary policy, or whatever. Each individual member of the society has a preference, or a set of preferences, over these alternatives. I guess you can say one alternative is better than another. These individual preferences have a property I call rationality or consistency, or more specifically, what is technically known as transitivity: If I prefer a to b, and b to c, then I prefer a to c.

Imagine that society has to make these choices among a set. Each individual has a preference ordering, a ranking of these alternatives. But we really want society, in some sense, to give a ranking of these alternatives. You can always produce a ranking, but you would like it to have some properties. One is that, of course, it be responsive in some sense to the individual rankings. Another is that when you finish, you end up with a real ranking, that is, something that satisfies these consistency, or transitivity, properties. And a third condition is that when choosing between a number of alternatives, all I should take into account are the preferences of the individuals among those alternatives. If certain things are possible and some are impossible, I shouldn't ask individuals whether they care about the impossible alternatives, only the possible ones.

It turns out that if you impose the conditions I just stated, there is no method of putting together the individual preferences that satisfies all of them.

The whole idea of the axiomatic method was very much in the air among anybody who studied mathematics, particularly among those who studied the foundations of mathematics. The idea is that if you want to find out something, to find the properties, you say, "What would I like it to be?" [You do this] instead of trying to investigate special cases. I was really accustomed to this approach. Of course, the actual process did involve trial and error.

But I went in with the idea that there was some method of handling this problem. I started out with some examples. I had already discovered that these led to some problems. The next thing that was reasonable was to write down a condition that I could outlaw. I constructed another example, another method that seemed to meet that problem, and something else didn't seem very right about it. Then I had to postulate that we have some other property. I found I was having difficulty satisfying all of these properties that I thought were desirable, and it occurred to me that they couldn't be satisfied.

After having formulated three or four conditions of this kind, I kept on experimenting. Lo and behold, no matter what I did, there was nothing that would satisfy these axioms. So after a few days of this, I began to get the idea that maybe there was another kind of theorem here, namely, that there was no voting method that would satisfy all the conditions that I regarded as rational and reasonable. It was at this point that I set out to prove it. It turned out to be a matter of only a few days' work.

It should be made clear that my impossibility theorem is really a theorem [showing that] the contradictions are possible, not that they are necessary. What I claim is that given any voting procedure, there will be some possible set of preference orders for individuals that will lead to a contradiction of one of these axioms.

But you say, "Well, okay, since we can't get perfection, let's at least try to find a method that works well most of the time." Then when you do have a problem, you don't notice it as much. So my theorem is not a completely destructive or negative feature any more than the second law of thermodynamics means that people don't work on improving the efficiency of engines. We're told you'll never get 100% efficient engines. That's a fact—and a law. It doesn't mean you wouldn't like to go from 40% to 50%.

9.4 A Better Approach? Approval Voting

Elections in which there are only two candidates present no problem. Majority rule is, as we have seen, an eminently successful voting system in both theory and practice. If there are three or more candidates, however, the situation changes quite dramatically. While several voting systems suggest themselves (plurality, the Borda count, sequential pairwise voting, and the Hare system), each fails to satisfy one or more desired properties (the Condorcet winner criterion, independence of irrelevant alternatives, the Pareto condition, and monotonicity). Manipulability is an ever-present problem, as we'll see in the next chapter. Moreover, when all is said and done, Arrow's impossibility theorem says that any search for an ideal voting system of the kind we have discussed is doomed to failure.

Where does this leave us? More than intellectual issues are at stake here: More than 550,000 elected officials serve in approximately 80,000 governments in the United States. Whether it is a small academic department voting on the best senior thesis or a democratic country electing a new leader, multicandidate elections will be contested in one way or another. If there is no perfect voting system–and perhaps not even a best voting system (whatever that may mean; that is, best in what way?)–what can we do?

Perhaps the answer is that different situations lend themselves to different voting systems, and what is required is a judicious blend of common sense with an awareness of what the mathematical theory has to say. For example, while both the Hare system and the Borda count are subject to manipulability, it seems easier to manipulate the latter. Thus, people may tend to vote more sincerely, rather than strategically, if the Hare system is used instead of the Borda count. This may be a consideration when choosing a voting system for a faculty governance system, for example.

For national political elections, there are also practical considerations. The kind of ballot we are considering (a preference list ballot) is certainly more complicated than the ballots we now employ, and preference list ballots cannot be used with existing voting machines. There is, however, a voting system that avoids the practical difficulties caused by the type of ballot being used that has much else to commend it. It is called **approval voting**.

Description of Approval Voting — PROCEDURE

Under approval voting, each voter is allowed to give one vote to as many of the candidates as he or she finds acceptable. No limit is set on the number of candidates for whom an individual can vote. Voters show disapproval of other candidates simply by not voting for them. The winner under approval voting is the candidate who receives the largest number of approval votes. This approach is also appropriate in situations where more than one candidate can win, for example, in electing new members to an exclusive society such as the National Academy of Sciences or the Baseball Hall of Fame.

Approval voting was proposed independently by several analysts in the 1970s. Probably the best-known official elected by approval voting today is the secretary general of the United Nations. In the 1980s, several academic and professional societies initiated the use of approval voting. Examples include the Institute of Electrical and Electronics Engineers (IEEE), with about 400,000 members, and the National Academy of Sciences. In Eastern Europe and some former Soviet republics, approval voting has been used in the form wherein one disapproves of (instead of approving of) as many candidates as one wishes.

Is approval voting the perfect voting system? Certainly not. For example, the type of ballot used limits the extent to which voter preferences can be expressed. However, it is certainly a voting system with much potential, and the reader wishing to explore it in more detail can start with Brams and Fishburn's 1983 monograph, listed in the Suggested Readings.

REVIEW VOCABULARY

Agenda An ordering of the candidates to be considered. Often used in sequential pairwise voting. (p. 296)

Approval voting A method of electing one or more candidates from a field of several in which each voter submits a ballot that indicates which candidates he or she approves of. Winning is determined by the total number of approvals a candidate obtains. (p. 305)

Arrow's impossibility theorem Kenneth J. Arrow's discovery that any voting system can give undesirable outcomes. (p. 301)

Borda count A voting system for elections with several candidates in which points are assigned to voters' preferences and these points are summed for each candidate to determine a winner. The actual point totals are referred to as a candidate's **Borda score.** (p. 293)

Condorcet's method A voting system for elections with several candidates in which a candidate is a winner precisely when he or she would, on the basis of the ballots cast, defeat every other candidate in a one-on-one contest. (p. 287)

Condorcet winner A Condorcet winner in an election is a candidate who, based on the ballots, would have defeated every other candidate in a one-on-one contest. (p. 291)

Condorcet winner criterion (CWC) A voting system satisfies the Condorcet winner criterion if, for every election in which there is a Condorcet winner, it wins the election when that voting system is used. (p. 292)

Condorcet's voting paradox The observation that there are elections in which Condorcet's method yields no winner. (p. 288)

Hare system A voting system for elections with several candidates in which candidates are successively eliminated in an order based on the number of first-place votes. (p. 297)

Independence of irrelevant alternatives (IIA) A voting system satisfies independence of irrelevant alternatives if the only way a candidate (call him A) can go from losing one election to being among the winners of a new election (with the same set of candidates and voters) is for at least one voter to reverse his or her ranking of A and the previous winner. (p. 295)

Manipulability A voting system is subject to manipulability (or is manipulable) if there are elections in which it is to a voter's advantage to submit a ballot that misrepresents his or her true preferences. (p. 292)

Majority rule A voting system for elections with two candidates (and an odd number of voters) in which the candidate preferred by more than half the voters is the winner. (p. 286)

May's theorem Kenneth May's discovery that, for two alternatives and an odd number of voters, majority rule is the only voting system satisfying three natural properties. (p. 287)

Monotonicity A voting system satisfies monotonicity provided that ballot changes favorable to one candidate (and not favorable to any other candidate) can never hurt that candidate. (p. 287)

Pareto condition A voting system satisfies the Pareto condition provided that every voter's ranking of one candidate higher than another precludes the possibility of this latter candidate winning. (p. 297)

Plurality runoff A voting system for elections with several candidates in which, assuming there are no ties, there is a runoff between the two candidates receiving the most first-place votes. (p. 299)

Plurality voting A voting system for elections with several candidates in which the candidate with the most first-place votes wins. (p. 289)

Preference list ballot A ballot that ranks the candidates from most preferred to least preferred, with no ties. (p. 286)

Sequential pairwise voting A voting system for elections with several candidates in which one starts with an agenda and pits the candidates against each other in one-on-one contests (based on preference list ballots), with losers being eliminated as one moves along the agenda. (p. 296)

✓ SKILLS CHECK

1. A preference list ballot

(a) indicates only a voter's top choice.
(b) is a rank ordering of the candidates, with no ties.
(c) will often have ties.

2. To say that a voting system treats all voters equally means that ______________________________ .

3. To say that a voting system for two candidates treats both candidates equally means that

(a) each wins if he or she receives all the votes.
(b) if all voters reverse their ballots, the election outcome changes.
(c) if any two voters exchange ballots, the election outcome is unchanged.

4. A two-candidate voting system is monotone if ______________________________ .

5. May's theorem says that, with an odd number of voters, among all two-candidate voting systems that never result in a tie, majority rule is the only one that

(a) treats both candidates equally.
(b) treats both candidates equally and all voters equally.
(c) treats both candidates equally and all voters equally and is monotone.

6. The winner with Condorcet's method is the candidate who ______________________________ .

7. Which of the following does not satisfy exactly two of the conditions in May's theorem?

(a) A dictatorship
(b) Imposed rule
(c) Minority rule
(d) None of the above

8. The flaw in Condorcet's method is that it ______________________________ .

9. Condorcet's voting paradox refers to the fact that

(a) people vote even though an individual vote virtually never affects the outcome of an election.
(b) the statement "This statement is false" can be neither true nor false.
(c) there are elections in which there is no winner using Condorcet's method.

10. With plurality voting, the winner is the candidate who ______________________________ .

11. George W. Bush's defeat of Al Gore in the state of Florida in the 2000 presidential election shows that

(a) plurality voting does not satisfy the Condorcet winner criterion.
(b) majority rule is not monotone.
(c) the Borda count does not satisfy independence of irrelevant alternatives.

12. With the Borda count, the election winner is the candidate who ______________________________ .

13. Instead of assigning points and doing arithmetic, the Borda score of a candidate can be found by

(a) scanning the ballots and counting the number of occurrences of other candidates below that one.
(b) counting the number of first-place votes and multiplying by 4.
(c) counting the number of candidates that it defeats one on one.

14. Independence of irrelevant alternatives says that a nonwinner can never switch to being a winner unless at least one voter changes his or her ballot in a way that ______________________________ .

15. The Borda count fails to satisfy

(a) monotonicity.
(b) the Pareto condition.
(c) independence of irrelevant alternatives.

16. Sequential pairwise voting is the voting system in which ______________________________ .

17. Sequential pairwise voting fails to satisfy

(a) monotonicity.
(b) the Pareto condition.
(c) the Condorcet winner criterion.

18. Both the Hare system and the plurality runoff method are defective in that ______________________________ .

19. Arrow's theorem says that with three or more candidates and any number of voters, there is no voting system that

(a) is not a dictatorship.
(b) satisfies independence of irrelevant alternatives and is not a dictatorship.
(c) satisfies the Pareto condition and independence of irrelevant alternatives, and is not a dictatorship.
(d) always produces a winner, satisfies the Pareto condition and independence of irrelevant alternatives, and is not a dictatorship.

20. The weak version of Arrow's theorem asserts that, with three or more candidates and an odd number of voters, there is no voting system that ______________________________ .

CHAPTER 9 EXERCISES

■ Challenge ◆ Discussion

9.1 Majority Rule and Condorcet's Method

1. In a few sentences, explain why minority rule (the voting procedure for two alternatives that is described on page 287) satisfies conditions (1) and (2) on page 287, but not (3).

2. In a few sentences, explain why imposed rule (the voting procedure for two alternatives that is described on page 287) satisfies conditions (1) and (3) on page 287, but not (2).

3. In a few sentences, explain why a dictatorship (the voting procedure for two alternatives that is described on page 287) satisfies conditions (2) and (3) on page 287, but not (1).

4. Find (or invent) a voting rule for two alternatives that satisfies

(a) condition (1) on page 287, but neither (2) nor (3).
(b) condition (2) on page 287, but neither (1) nor (3).
(c) condition (3) on page 287, but neither (1) nor (2).

5. In a sentence or two, explain why it's impossible, with an odd number of voters, to have two distinct candidates win the same election using Condorcet's method.

6. Construct a real-world example (perhaps involving yourself and two friends) where the individual preference lists for three alternatives are as in the voting paradox of Condorcet.

7. Condorcet's voting paradox shows that with three voters (or three equal-size groups of voters) and the three alternatives *A*, *B*, and *C*, it is possible to have two-thirds prefer *A* to *B*, two-thirds prefer *B* to *C*, and two-thirds prefer *C* to *A*. Find four preference lists that show that with four voters and the four alternatives *A*, *B*, *C*, and *D*, it is possible to have three-fourths prefer *A* to *B*, three-fourths prefer *B* to *C*, three-fourths prefer *C* to *D*, and three-fourths prefer *D* to *A*.

8. Generalize the result in Exercise 7 from four alternatives to *n* alternatives: $A_1, \ldots, A_n$.

9.2 Other Voting Systems for Three or More Candidates

9. Plurality voting is illustrated by the 1980 U.S. Senate race in New York among Alfonse D'Amato (*D*, a conservative), Elizabeth Holtzman (*H*, a liberal), and Jacob Javits (*J*, also a liberal). Reasonable estimates (based largely on exit polls) suggest that voters ranked the candidates according to the following table:

22%	23%	15%	29%	7%	4%
D	*D*	*H*	*H*	*J*	*J*
H	*J*	*D*	*J*	*H*	*D*
J	*H*	*J*	*D*	*D*	*H*

(a) Is there a Condorcet winner?
(b) Who won using plurality voting?

10. (Everyone wins.) Consider the following set of preference lists:

	Number of Voters (9)						
Rank	**3**	**1**	**1**	**1**	**1**	**1**	**1**
First	*A*	*A*	*B*	*B*	*C*	*C*	*D*
Second	*D*	*B*	*C*	*C*	*B*	*D*	*C*
Third	*B*	*C*	*D*	*A*	*D*	*B*	*B*
Fourth	*C*	*D*	*A*	*D*	*A*	*A*	*A*

Note that the first list is held by three voters, not just one. Calculate the winner using

(a) plurality voting.
(b) the Borda count.
(c) the Hare system.
(d) sequential pairwise voting with the agenda *A*, *B*, *C*, *D*.

11. Consider the following set of preference lists:

	Number of Voters (7)				
Rank	**2**	**2**	**1**	**1**	**1**
First	*C*	*D*	*C*	*B*	*A*
Second	*A*	*A*	*D*	*D*	*D*
Third	*B*	*C*	*A*	*A*	*B*
Fourth	*D*	*B*	*B*	*C*	*C*

Calculate the winner using

(a) plurality voting
(b) the Borda count.
(c) the Hare system.
(d) sequential pairwise voting with the agenda *B*, *D*, *C*, *A*.

12. Consider the following set of preference lists:

	Number of Voters (8)					
Rank	**2**	**2**	**1**	**1**	**1**	**1**
First	*A*	*E*	*A*	*B*	*C*	*D*
Second	*B*	*B*	*D*	*E*	*E*	*E*
Third	*C*	*D*	*C*	*C*	*D*	*A*
Fourth	*D*	*C*	*B*	*D*	*A*	*B*
Fifth	*E*	*A*	*E*	*A*	*B*	*C*

Calculate the winner using

(a) plurality voting.
(b) the Borda count.
(c) the Hare system.
(d) sequential pairwise voting with the agenda *B, D, C, A, E*.

13. Consider the following set of preference lists:

	Number of Voters (5)				
Rank	**1**	**1**	**1**	**1**	**1**
First	*A*	*B*	*C*	*D*	*E*
Second	*B*	*C*	*B*	*C*	*D*
Third	*E*	*A*	*E*	*A*	*C*
Fourth	*D*	*D*	*D*	*E*	*A*
Fifth	*C*	*E*	*A*	*B*	*B*

Calculate the winner using

(a) plurality voting.
(b) the Borda count.
(c) the Hare system.
(d) sequential pairwise voting with the agenda *A, B, C, D, E*.

14. Consider the following set of preference lists:

	Number of Voters (7)				
Rank	**2**	**2**	**1**	**1**	**1**
First	*A*	*B*	*A*	*C*	*D*
Second	*D*	*D*	*B*	*B*	*B*
Third	*C*	*A*	*D*	*D*	*A*
Fourth	*B*	*C*	*C*	*A*	*C*

Calculate the winner using

(a) plurality voting.
(b) the Borda count.
(c) the Hare system.
(d) sequential pairwise voting with the agenda *B, D, C, A*.

15. Consider the following set of preference lists:

	Number of Voters (7)				
Rank	**2**	**2**	**1**	**1**	**1**
First	*C*	*E*	*C*	*D*	*A*
Second	*E*	*B*	*A*	*E*	*E*
Third	*D*	*D*	*D*	*A*	*C*
Fourth	*A*	*C*	*E*	*C*	*D*
Fifth	*B*	*A*	*B*	*B*	*B*

Calculate the winner using

(a) plurality voting.
(b) the Borda count.
(c) the Hare system.
(d) sequential pairwise voting with the agenda *A, B, C, D, E*.

16. Consider the following set of preference lists:

	Number of Voters (7)						
Rank	**1**	**1**	**1**	**1**	**1**	**1**	**1**
First	*C*	*D*	*C*	*B*	*E*	*D*	*C*
Second	*A*	*A*	*E*	*D*	*D*	*E*	*A*
Third	*E*	*E*	*D*	*A*	*A*	*A*	*E*
Fourth	*B*	*C*	*A*	*E*	*C*	*B*	*B*
Fifth	*D*	*B*	*B*	*C*	*B*	*C*	*D*

Calculate the winner using

(a) plurality voting.
(b) the Borda count.
(c) sequential pairwise voting with the agenda *A, B, C, D, E*.
(d) the Hare system.

17. An interesting variant of the Hare system was proposed by the psychologist Clyde Coombs. It operates exactly as does the Hare system, but instead of deleting alternatives with the fewest first-place votes, it deletes those with the most last-place votes.

(a) Use the Coombs procedure to find the winner if the ballots are as in Exercise 16.
(b) Show that for two voters and three alternatives, it is possible to have ballots that result in one candidate winning if the Coombs procedure is used and a tie between the other two if the Hare system is used.

◆ **18.** In a few sentences, explain why Condorcet's rule satisfies

(a) the Pareto condition.
(b) monotonicity.

◆ **19.** In a few sentences, explain why plurality voting satisfies

(a) the Pareto condition.
(b) monotonicity.

◆ **20.** In a few sentences, explain why the Borda count satisfies

(a) the Pareto condition.
(b) monotonicity.

◆ **21.** In a few sentences, explain why sequential pairwise voting satisfies

(a) the Condorcet winner criterion.
(b) monotonicity.

◆ **22.** In a few sentences, explain why the Hare system satisfies the Pareto condition.

◆ **23.** In a few sentences, explain why the plurality runoff method satisfies the Pareto condition.

■ **24.** Use the following ballots to show that the plurality runoff method does not satisfy the Condorcet winner criterion:

	Number of Voters (5)		
Rank	2	2	1
First	*A*	*B*	*C*
Second	*C*	*C*	*B*
Third	*B*	*A*	*A*

■ **25.** Use the following ballots to show that the plurality runoff method does not satisfy monotonicity:

	Number of Voters (13)				
Rank	4	3	3	2	1
First	*A*	*B*	*C*	*D*	*E*
Second	*B*	*A*	*A*	*B*	*D*
Third	*C*	*C*	*B*	*C*	*C*
Fourth	*D*	*D*	*D*	*A*	*B*
Fifth	*E*	*E*	*E*	*E*	*A*

26. Consider the following two elections among candidates *A*, *B*, and *C*:

	Number of Voters (4)			
Rank	1	1	1	1
First	*A*	*A*	*B*	*C*
Second	*B*	*B*	*C*	*B*
Third	*C*	*C*	*A*	*A*

	Number of Voters (4)			
Rank	1	1	1	1
First	*A*	*A*	*B*	*B*
Second	*B*	*B*	*C*	*C*
Third	*C*	*C*	*A*	*A*

(a) Use these two elections to show that plurality voting does not satisfy independence of irrelevant alternatives.
(b) Use these two elections to show that the Hare system does not satisfy independence of irrelevant alternative.

■ **27.** Construct ballots for the alternatives *A*, *B*, and *C* to show that the Borda count does not satisfy the Condorcet winner criterion.

28. Show that the nonmonotonicity of the Hare system can also be demonstrated by the following 17-voter, 4-alternative election. (In a number of recent books, this example is used to show the nonmonotonicity of the Hare system. The 13-voter, 3-alternative example given in the text was pointed out to us by Matt Gendron, an undergraduate at Union College.)

	Number of Voters (17)			
Rank	7	5	4	1
First	*A*	*C*	*B*	*D*
Second	*D*	*A*	*C*	*B*
Third	*B*	*B*	*D*	*A*
Fourth	*C*	*D*	*A*	*C*

29. The following example illustrates how badly the Hare system can fail to satisfy monotonicity. Consider the following sequence of preference lists:

	Number of Voters (21)			
Rank	7	6	5	3
First	*A*	*B*	*C*	*D*
Second	*B*	*A*	*B*	*C*
Third	*C*	*C*	*A*	*B*
Fourth	*D*	*D*	*D*	*A*

(a) Show that *A* is the unique winner if the Hare system is used.
(b) Find the winner using the Hare system in the new election wherein the three voters on the right all move *A* from last place on their preference lists to first place on their preference lists.

◆ **30.** In a few sentences, explain why, with an odd number of voters,

(a) sequential pairwise voting always yields a unique winner.
(b) we can never have exactly two winners with the Hare system.

◆ **31.** In a few sentences, explain why the plurality runoff method can never elect a candidate ranked last on a majority of ballots, assuming there are no ties for first or second place in the voting.

32. Produce ballots showing that plurality voting can, in fact, elect a candidate ranked last on a majority of the ballots.

33. Suppose there are three voters and three alternatives: *A*, *B*, and *C*.

(a) If each alternative has exactly one first-place vote, what is the election outcome if the Hare procedure is used? What if plurality runoff is used?
(b) If an alternative has two or more first-place votes, what is the election outcome if the Hare procedure is used? What if plurality runoff is used?
(c) Can the Hare procedure and plurality runoff yield different election outcomes when there are three voters and three alternatives? Explain your answer in one sentence.

9.3 Insurmountable Difficulties: Arrow's Impossibilty Theorem

■ **34.** Complete the proof of the version of Arrow's theorem from the text by showing that neither *B* nor *C* can be a winner in the situation described. (Your argument will be almost word for word the same as the proofs in the text.)

9.4 A Better Approach? Approval Voting

35. Ten board members vote by approval voting on eight candidates for new positions on their board as indicated in the following table. An X indicates an approval vote. For example, Voter 1, in the first column, approves of candidates *A*, *D*, *E*, *F*, and *G*, and disapproves of *B*, *C*, and *H*.

	Voters									
Candidate	**1**	**2**	**3**	**4**	**5**	**6**	**7**	**8**	**9**	**10**
A	X	X	X			X	X	X		X
B		X	X	X	X	X	X	X	X	
C			X					X		
D	X	X	X	X	X		X	X	X	X
E	X		X		X		X		X	
F	X		X	X	X	X	X	X		X
G	X	X	X	X	X			X		
H		X		X		X		X		X

(a) Which candidate is chosen for the board if just one of them is to be elected?
(b) Which candidates are chosen if the top four are selected?
(c) Which candidates are elected if 80% approval is necessary and at most four are elected?
(d) Which candidates are elected if 60% approval is necessary and at most four are elected?

36. The 45 members of a school's football team vote on three nominees, *A*, *B*, and *C*, by approval voting for the award of "most improved player" as indicated in the following table. An X indicates an approval vote.

	Number of Voters (45)							
Nominee	**7**	**8**	**9**	**9**	**6**	**3**	**1**	**2**
A	X			X	X		X	
B		X		X		X	X	
C			X		X	X	X	

(a) Which nominee is selected for the award?
(b) Which nominee gets announced as runner-up for the award?
(c) Note that two of the players "abstained," that is, approved of none of the nominees. Note also that one person approved of all three of the nominees. What would be the difference in the outcome if one were to "abstain" or "approve of everyone"?

WRITING PROJECTS

1. In the 2000 presidential election in Florida, the final results were as follows:

Candidates	Number of Votes	Percentage of Votes
Bush	2,911,872	49
Gore	2,910,942	49
Nader	97,419	2
Buchanan	17,472	0

Making reasonable assumptions about voters' preference schedules, give a one-page discussion of how the election might have turned out under the different voting methods discussed in this chapter.

2. Frequently in presidential campaigns, the winner of the first few primaries is given front-runner status that can lead to the nomination of his or her party. Moreover, there are often several candidates running in early primaries such as New Hampshire. In one page, consider a recent election and discuss how the nominating process might have proceeded through the campaign if approval voting had been used to decide primary winners.

SUGGESTED READINGS

BLACK, DUNCAN. *The Theory of Committees and Elections,* Kluwer, Dordrecht, The Netherlands, 1986. The historical highlights and development of voting methods in the nineteenth and twentieth centuries are traced in this economist's volume.

BRAMS, STEVEN J., and PETER C. FISHBURN. *Approval Voting,* Birkhäuser, Boston, 1983. This volume is a research-level work on development in the recently popular (but rediscovered) method now called approval voting. The first chapter, however, is an elementary exposition of this voting method and its uses.

NURMI, HANNU. *Comparing Voting Systems,* Reidel, Dordrecht, The Netherlands, 1987. This monograph provides an excellent treatment, at a somewhat more technical level, of the topics dealt with in this chapter.

SAARI, DONALD G. *Chaotic Elections! A Mathematician Looks at Voting,* American Mathematical Society, Providence, R.I., 2001. This expository book begins with the 2000 presidential election and discusses a number of paradoxical results in voting.

TAYLOR, ALAN D. *Mathematics and Politics: Strategy, Voting, Power, and Proof,* Springer-Verlag, New York, 1995. Chapters 5 and 10 give an expanded treatment of the topics considered here, with proofs included. This book is also intended for nonmajors.

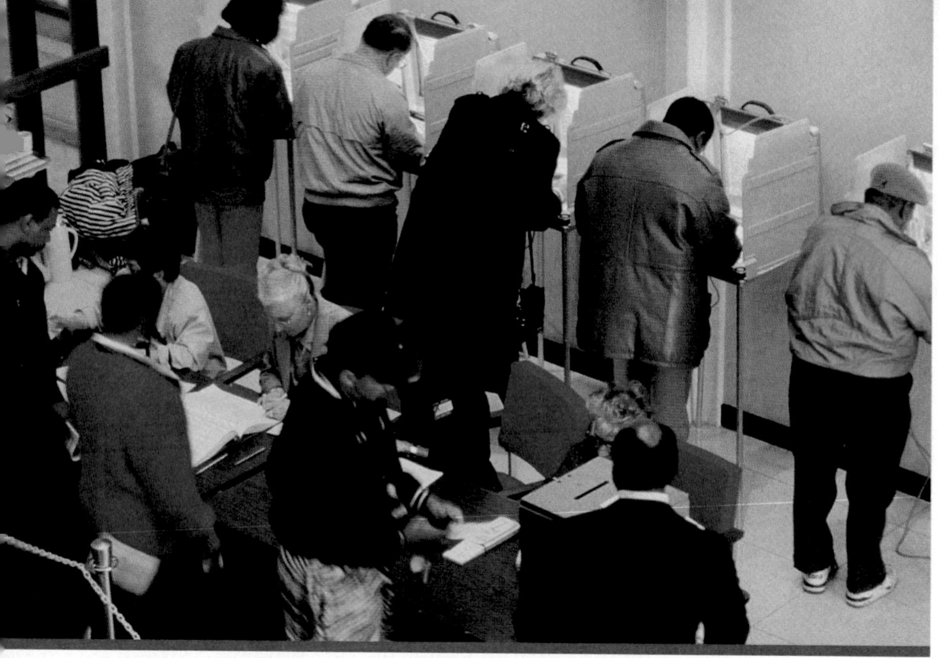

CHAPTER 10

The Manipulability of Voting Systems

People know almost by instinct that, sometimes, you can achieve the election result you prefer by submitting a ballot that misrepresents your actual preferences. This type of strategic voting is called **manipulation**, and a ballot that misrepresents a voter's true preferences is referred to as an **insincere** or **disingenuous ballot**.

All three of these terms–manipulation, insincere, disingenuous–are widely used in the social-choice literature, but in daily life we use these terms pejoratively–they aren't exactly warm praise. In point of fact, your choice to manipulate a voting system is typically no more inherently evil than your submission of a sealed bid for a lamp at an auction at a price considerably below its actual worth. "Strategy-proof"–a term with considerably less negative content–is sometimes used in place of "nonmanipulable," but "nonmanipulable" is more common so we'll stick with it here.

Historical references to the manipulability of voting systems include a comment by the nineteenth-century mathematician C. L. Dodgson (1832–1898), better known by the pseudonym Lewis Carroll, under which he wrote *Alice's Adventures in Wonderland* (1865). Dogson commented that voters have a tendency to "adopt a principle of voting which makes it more of a game of skill than a true test of the wishes of the electors," and that it would be "better for elections to be decided according to the wishes of the majority than of those who have the most skill at the game."

But the most famous manipulability quote in the history of social choice is Jean Charles de Borda's reply to a colleague who had pointed out to him how easily the Borda count can be manipulated. "My scheme," Borda replied, "is only intended for honest men!"

Let's look at an example to illustrate how the Borda count can be manipulated.

Charles L. Dodgson was a mathematical lecturer at Oxford University. Dodgson, who used the pen name Lewis Carroll, wrote on mathematical topics and even manipulability. But he achieved greater fame for his satirical works. In the *Alice* books, he refers to the mathematical operations as Ambition, Distraction, Uglification, and Derision, and his characters play nonsensical, easily manipulated games. (*Bettmann/Corbis.*)

EXAMPLE 1 ▪ Manipulating the Borda Count with Four Candidates and Two Voters

Suppose there are two voters and four candidates, and suppose the true preferences of the voters are reflected in the following ballots:

Voter 1	Voter 2
A	B
B	C
C	A
D	D

Using the Borda count with point values 3, 2, 1, 0 (or by counting the number of occurrences of other candidates below the one in question, as described in Section 9.2), we see that the Borda scores of the four candidates are as follows:

Borda score of A is 4
Borda score of B is 5
Borda score of C is 3
Borda score of D is 0

Thus, Candidate B wins this election. Voter 1, however, would have preferred to see Candidate A–his top choice, according to his true preferences–win this election rather than Candidate B, his second choice.

Assume that Voter 1 had known that Voter 2 planned to submit the ballot that he cast above. Could Voter 1 have secured a victory for Candidate A by submitting a disingenuous ballot?

The answer here, as we'll show, turns out to be yes. The intuition is fairly transparent: Voter 1 wants to pretend that B is not her second choice, but her last choice. Let's see if this is enough to bring about the desired switch in winner from B to A. The new ballots and Borda scores are as follows:

Voter 1	Voter 2
A	B
C	C
D	A
B	D

Borda score of A is 4
Borda score of B is 3
Borda score of C is 4
Borda score of D is 1

Close, but not quite what we wanted: Candidates A and C now tie for the win, and we wanted the winner to be just Candidate A. But a moment's inspection reveals that Voter 1 can achieve this if, in addition to plunging Candidate B to the bottom of her ballot, she also flip-flops C and D. That is, the desired ballots (and Borda scores) that yield Candidate A as the sole winner are as follows:

Voter 1	Voter 2
A	B
D	C
C	A
B	D

Borda score of A is 4

Borda score of B is 3

Borda score of C is 3

Borda score of D is 2

In presenting an example of a voting system's susceptibility to manipulation, we will typically present two elections–the original one ("Election 1") in which we assume all ballots are sincere, and the one that contains a disingenuous ballot from a voter ("Election 2"). For example, if we collect the pieces of what we just did, this instance of manipulation of the Borda count could be succinctly presented as follows.

Election 1		
Rank	**Number of Voters (2)**	
First	A	B
Second	B	C
Third	C	A
Fourth	D	D

Election 2		
Rank	**Number of Voters (2)**	
First	A	B
Second	D	C
Third	C	A
Fourth	B	D

There are two aspects of manipulation taking place in this example that deserve comment.

First, there is only one voter (the voter on the left, in this example) changing his or her ballot–we call this a **unilateral change** in ballot. An example involving a unilateral change of ballot is sometimes referred to as an instance of "single-voter manipulation" to distinguish it from a situation wherein a group of voters, acting in concert, can change their ballots so that all of them prefer the new winner to the original winner. We'll see examples of group manipulation in Section 10.2.

Second, the original election produced a single winner, as did the new election held after we finished constructing Voter 1's disingenuous ballot. Thus, because we know each voter's sincere preference ranking for the candidates, we also know exactly which of the two election outcomes each voter will prefer. Ties, on the other hand, present a problem. For example, if a voter has sincere preferences that rank A over B over C over D, then it's not at all obvious whether this voter will prefer an election outcome that ties A and D to an election outcome that ties B and C or vice-versa.

A voting system is **manipulable** if there is at least one scenario in which some voter can achieve a more preferred election outcome by unilaterally changing his or her ballot. The precise definition follows.

Manipulability DEFINITION

A voting system is said to be **manipulable** if there exist two sequences of preference list ballots and a voter (call the voter Jane) such that

1. Neither election results in a tie.
2. The only ballot change is by Jane.
3. Jane prefers–assuming that her ballot in the first election represents her true preferences–the outcome of the second election to that of the first election.

In this chapter, as in Chapter 9, we begin with majority rule in the two-candidate case and Condorcet's method in the case of three or more candidates. Condorcet's method again shines, leaving the voting paradox as its only blemish. We then move on to revisit the other voting systems introduced in Chapter 9 that apply to elections with three or more candidates, and we show that each succumbs to some form of manipulation. In fact, there is a striking impossibility result that arises here known as the **Gibbard–Satterthwaite manipulability theorem**. It is related to–indeed, some would say equivalent to–Arrow's impossibility theorem. We conclude with a treatment of a striking first cousin of manipulability known as the **chair's paradox**.

10.1 Majority Rule and Condorcet's Method

Throughout this section, we assume that the number of voters is odd. In Section 9.1 we pointed out that with two candidates, majority rule has three very desirable properties: It treats all voters equally, it treats both candidates equally, and it is *monotone*, meaning that a single voter's change in ballot from a vote for the loser to a vote for the winner has no effect on the election outcome. More strikingly, May's theorem told us that among all voting systems in the two-candidate case that never result in a tie, majority rule is the *only* one satisfying these three properties.

But let's consider for a moment what monotonicity is saying in this two-candidate case for voting systems that never yield ties. It says that if you rank A over B on your ballot, and the election winner is B, then the election winner will remain B if you switch to a ballot with B over A. But there are only two possible choices for a ballot in this two-candidate case: B over A and A over B. Monotonicity is thus saying that if you rank A over B, then no unilateral change in your ballot can make the outcome A. This is simply the assertion that you can't manipulate the voting system!

Thus, in the two-candidate case, nonmanipulability and monotonicity are exactly the same thing. This allows us to restate **May's theorem** from Section 9.1 with "monotonicity" replaced by "nonmanipulability."

May's Theorem for Manipulability — THEOREM

Among all two-candidate voting systems that never result in a tie, majority rule is the only one that treats all voters equally, treats both candidates equally, and is nonmanipulable.

There are examples of two-candidate voting systems that are manipulable, even though they treat all voters equally and both candidates equally. For example, the voting system that declares the winner to be the alternative with the fewest first-place votes is manipulable, as is the one that declares the winner to be whichever alternative has an odd number of first-place votes, (even if that's fewer than half). Exercises 1 and 2 ask you to provide an example of voter manipulation for each of these systems.

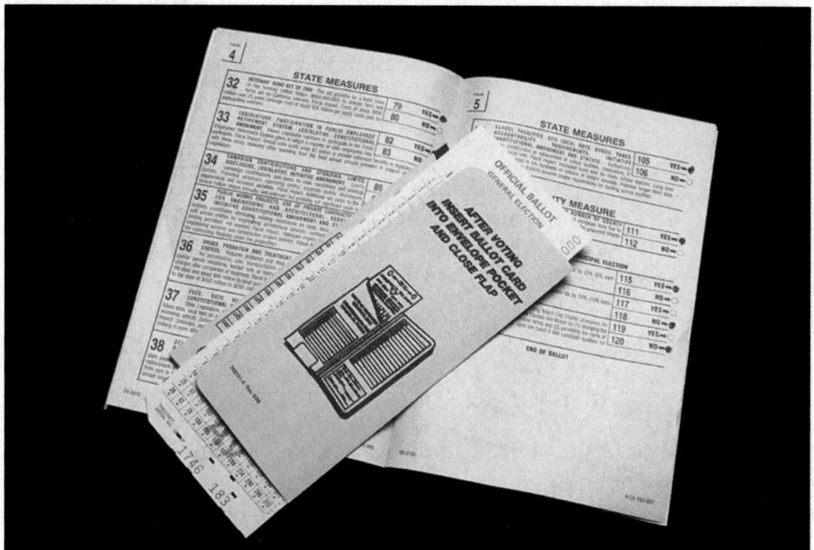

Paper ballots are still used in elections in many states. A lingering controversy from the 2004 elections is over the use of electronic ballots, which do not leave physical evidence and thus may not be possible to recount in disputes about the plurality or majority. (*Jonathan Nourok/ PhotoEdit.*)

Turning to the case of three or more candidates, we begin with Condorcet's method, as we did in Chapter 9. Condorcet's method is based on majority rule, and, as we've just seen, majority rule is nonmanipulable. So the following result, as pleasing as it is, comes as no surprise.

The Nonmanipulability of Condorcet's Method — THEOREM

Condorcet's method is nonmanipulable in the sense that a voter can never unilaterally change an election result from one candidate to another candidate that he or she prefers.

Let's see why Condorcet's method is nonmanipulable, regardless of the number of voters. Suppose that we have an election in which you, as one of the voters, prefer Candidate A to Candidate B, but B wins using Condorcet's method. We'll show that any attempt that you might make to manipulate the election so that A becomes the winner is doomed to failure, even if there are more than these two candidates in the election.

Because Candidate B is the winner using Condorcet's method, we know that B defeats every other candidate in a one-on-one contest based on the ballots cast. In particular, B defeats A in a one-on-one contest, even with your original ballot that has A over B. This means that more than half of the other voters ranked B over A, so, regardless of how you change your ballot, B will *still* defeat A in a one-on-one contest. While this need not ensure that B remains a winner with Condorcet's method, it certainly guarantees that A isn't. Hence, you cannot unilaterally cause A to be a winner using Condorcet's method, and so your attempt at manipulation will have failed.

EXAMPLE 2 ■ Exploiting the Condorcet Voting Paradox

We had to be careful in stating the above theorem because, as we've seen, there are elections in which there is no winner using Condorcet's method. With three voters and three candidates, it is possible for a voter (the one on the left in this example) to unilaterally change an election from one that yields his or her second choice as the sole winner (Candidate C in the example), to one in which there is no winner at all, as this example shows:

Election 1

Rank	Number of Voters (3)		
First	A	B	C
Second	C	C	A
Third	B	A	B

Election 2

Rank	Number of Voters (3)		
First	A	B	C
Second	B	C	A
Third	C	A	B

A voter's ability to unilaterally bring about this kind of change in an election, however, is not something that falls within the scope of our formal definition of manipulation. Nevertheless, one could argue that there are situations in which you might well prefer having an election with no outcome at all to having an election in which a candidate other than your top choice emerges as the sole winner. ■

We now move on to voting systems with three or more candidates that, unlike Condorcet's method, always produce at least one winner. As you might expect from the results in Chapter 9, these voting systems are not, in terms of nonmanipulability, as perfect as one might hope for.

10.2 The Manipulability of Other Voting Systems for Three or More Candidates

Manipulability and the Borda Count

Example 1 showed how a single voter can manipulate an election in which the Borda count is being used. But Example 1 involved four candidates. Is there a simpler example involving only three candidates?

The answer turns out to be no, provided that we continue to interpret the notion of a "more preferred election outcome" to be a switch from a single winner to another single winner (as opposed to a switch creating or breaking a tie). This negative answer is formalized in the following theorem.

The Nonmanipulability of the Borda Count with Exactly Three Candidates — THEOREM

With exactly three candidates, the Borda count cannot be manipulated in the sense of a voter unilaterally changing an election outcome from one single winner to another single winner that he or she prefers according to that voter's ballot in the first election, which we take to be sincere preferences.

Let's see why this is true. Suppose the candidates are *A*, *B*, and *C*, and that you prefer *A* to *B*, but *B* is the election winner using the Borda count. We'll show that any attempt you make to manipulate the election by changing your ballot so that *A* emerges as the winner (using the Borda count) is doomed to failure.

Because you prefer *A* to *B*, your sincere ballot can be one of only three possibilities, corresponding to whether *C* is ranked first, second, or third. We'll consider each case in turn.

Case 1. Your sincere ballot is *A* over *B* over *C*. No ballot change on your part can increase *A*'s Borda score, and you can only decrease *B*'s Borda score by at most 1. Thus, at best, you can make a unilateral change that results in *A* and *B* having the same Borda score, whereas successful manipulation on your part requires that *A* have a strictly higher Borda score than *B* after your ballot change.

Case 2. Your sincere ballot is *C* over *A* over *B*. No ballot change on your part can decrease *B*'s Borda score, and you can only increase *A*'s Borda score by at most 1. Thus, at best, you can make a unilateral change that results in *A* and *B* having the same Borda score, whereas successful manipulation on your part requires that *A* have a strictly higher Borda score than *B* after your ballot change.

Case 3. Your sincere ballot is *A* over *C* over *B*. No ballot change on your part can increase *A*'s Borda score or decrease *B*'s Borda score. Thus, after your ballot change *B* will still have a higher Borda score than *A*, so your attempt at manipulation has failed in this case also.

So with three candidates, the Borda count is nonmanipulable. With more than three candidates, the Borda count does not fare as well, regardless of how many voters there are.

The Manipulability of the Borda Count with Four or More Candidates — THEOREM

With four or more candidates (and two or more voters), the Borda count can be manipulated in the sense that there exists an election in which a voter can unilaterally change the election outcome from one single winner to another single winner that he or she prefers according to that voter's ballot in the first election, which we take to be sincere preferences.

We've already established part of this theorem, as we've seen an example of manipulation of the Borda count in the case of four candidates and two voters. This is really half the battle, as we can readily modify that example to serve in any case in which the number of voters is even as follows.

1. Any candidates in addition to A, B, C, and D can be placed below those four on every ballot.
2. The rest of the voters can be paired off with the members of each pair holding ballots that rank the candidates in exactly opposite orders (thus "canceling each other out" in terms of the Borda scores).

The following example illustrates this method of generalizing our earlier instance of manipulation of the Borda count to the case of five candidates and six voters.

EXAMPLE 3 ■

Manipulating the Borda Count with Five Candidates and Six Voters

Consider the following two elections:

Election 1						Election 2					
A	B	A	E	A	E	A	B	A	E	A	E
B	C	B	D	B	D	D	C	B	D	B	D
C	A	C	C	C	C	C	A	C	C	C	C
D	D	D	B	D	B	B	D	D	B	D	B
E	E	E	A	E	A	E	E	E	A	E	A

The ballots of the first two voters (in both elections) are the same as in Example 1 (the manipulation of the Borda count with four candidates and two voters), with the new candidate E placed at the bottom of both ballots. The last four voters contribute exactly 8 to the Borda score of each candidate, and so, taken together, they have no effect on who is the winner of the election. This is what we mean by "canceling each other out."

In the first election, as in Example 1, Candidate B wins. But if we take these ballots to represent true preferences, the voter on the far left prefers A to B. Moreover, that voter can achieve this better outcome–Candidate A–by submitting the disingenuous ballot that he or she cast in Election 2.

■

To handle the case where the number of voters is odd, we need to start with a four-candidate, three-voter example of manipulation of the Borda count. Exercise 9 provides this. We can then modify this example to work for any odd number of voters by again adding pairs of ballots that cancel each other out exactly as we did before. Exercises 10 and 11 fill in some of the details needed for this part of the argument, and ask you to provide the necessary explanations and calculations.

Manipulability of Runoff Systems

EXAMPLE 4 ■ Manipulability of Runoff Systems

Both the plurality runoff rule and the Hare system are manipulable. But rather than give the whole story away, we'll just present the sequences of sincere ballots in each case. Exercises 16 and 17 ask you to figure out how the left-most voter in each case can secure a more preferred outcome by a unilateral change of ballot.

Election 1 for the Hare System				
A	*B*	*C*	*C*	*D*
B	*A*	*B*	*B*	*B*
C	*C*	*A*	*A*	*C*
D	*D*	*D*	*D*	*A*

Election 1 for the Plurality Runoff Rule				
A	*A*	*C*	*C*	*B*
B	*B*	*A*	*A*	*C*
C	*C*	*B*	*B*	*A*

EXAMPLE 5 ■ Manipulating Sequential Pairwise Voting

Sequential pairwise voting can also be manipulated by a single voter, even in the case of three voters and three candidates. For example, consider the following two elections with the agenda *ABC*:

Election 1

Rank	Number of Voters (3)		
First	*A*	*B*	*C*
Second	*B*	*C*	*A*
Third	*C*	*A*	*B*

Election 2

Rank	Number of Voters (3)		
First	*B*	*B*	*C*
Second	*A*	*C*	*A*
Third	*C*	*A*	*B*

In the Election 1, *A* defeats *B* by a score of 2-to-1, so *A* moves on to meet *C*. But *C* defeats *A* by a score of 2-to-1, so *C* is the winner in Election 1. Election 2 is the result of Voter 1 (on the left) submitting a disingenuous ballot in which he has elevated *B* (his actual second choice) to first place on his ballot. It is now clear that *B* first defeats *A* by a score of 2-to-1 and then moves on to defeat *C* by this same score. Hence, *B* is the winner in Election 2. This is an instance of manipulation in which Voter 1 has secured a more preferred outcome by submitting an insincere ballot, because Voter 1 actually prefers *B* to *C* (assuming that his ballot in Election 1 represents his true preferences). This shows that sequential pairwise voting is manipulable.

Sequential Pairwise Voting and Agenda Manipulability

Thus, sequential pairwise voting can also be manipulated by a single voter, even in the case of three voters and three candidates. But there is another aspect of manipulability that arises with this particular voting system that is of even more interest, and this is something called **agenda manipulation**.

Agenda Manipulation DEFINITION

Agenda manipulation refers to the ability to control who wins an election with sequential pairwise voting by a choice of the agenda.

William H. Riker, in his book *The Art of Political Manipulation*, spoke of the possibility that "those in control of procedures can manipulate the agenda by, for example, restricting alternatives [candidates] or by arranging the order in which they are brought up." The following example provides a striking illustration of this with sequential pairwise voting.

EXAMPLE 6

Agenda Manipulation of Sequential Pairwise Voting

Suppose we have four candidates and three voters who we know will be submitting the following preference list ballots:

Rank	Number of Voters (3)		
First	*A*	*C*	*B*
Second	*B*	*A*	*D*
Third	*D*	*B*	*C*
Fourth	*C*	*D*	*A*

Now suppose that we have agenda-setting power in the sense that we get to choose the order in which the one-on-one contests will take place. Remarkably, we can arrange for the winner to be whichever of the four candidates we want!

The intuition behind finding an agenda that will yield a certain candidate as the winner arises from the observation that candidates who appear later in the agenda are favored over candidates who appear early in the agenda. For example, if we want *A* to win, we place *A* last and look for which candidates would, in fact, defeat *A* one on one. Here, only *C* defeats *A*, and so we want to arrange for *C* to be eliminated along the way. But *B* defeats *C* one on one, so if we choose the agenda *BCDA*, we have that *C* is eliminated by *B* in the first round, then *D* is eliminated by *B* in the second round, and finally *B* is eliminated by *A* in the third round, leaving *A* as the winner. Exercise 19 asks you to find the three other agendas that will, in turn, yield *B*, *C*, and *D* as the winner.

Plurality Voting and Group Manipulability

In the real world, all other voting systems pale in comparison to plurality voting in terms of the significance of the role played by disingenuous voting. "Throwing away your vote"–as some accuse Nader voters in Florida of doing in the 2000 presidential election–represents a choice, conscious or otherwise, to forgo obtaining a more desired outcome through strategic considerations.

Ironically, plurality voting, like Condorcet's method, is nonmanipulable according to the formal definition given on page 346. However, a *group* of voters, acting together, can change an election outcome into something they *all* prefer. We'll record this observation in the following, and then explain why it's true.

The Group Manipulability of Plurality Voting THEOREM

Plurality voting cannot be manipulated by a single individual. However, it is **group manipulable** in the sense that there are elections in which a group of voters can change their ballots so that the new winner is preferred to the old winner by everyone in the group, assuming that the original ballots represent the true preferences of each voter in the group.

First of all, let's see why no individual can manipulate plurality voting. Suppose that you prefer *A* to *B*, but *B* is the winner with plurality voting. Then *B* has at least one more first-place vote than A. Now, because you prefer A to B, we know that B is not on top of your sincere ballot, so no ballot change that you make can subtract from B's number of first-place votes. Moreover, by moving A to the top of your ballot, you only increase A's number of first-place votes by 1. Thus, the best you can do with a unilateral change in ballots is to move A into a tie with B.

To see that plurality voting is group manipulable we only have to look at any real-world election in which a third-party candidate acted as the "spoiler." As we've said, Ralph Nader was exactly this in the state of Florida in the 2000 presidential election. Another example occurs in Exercise 20.

At this point, we've seen that several of our familiar voting systems for three or more candidates–the Borda count, runoff systems, sequential pairwise voting–can be manipulated. Can't we do better than this in attempting to improve on Condorcet's method? We turn to this question next.

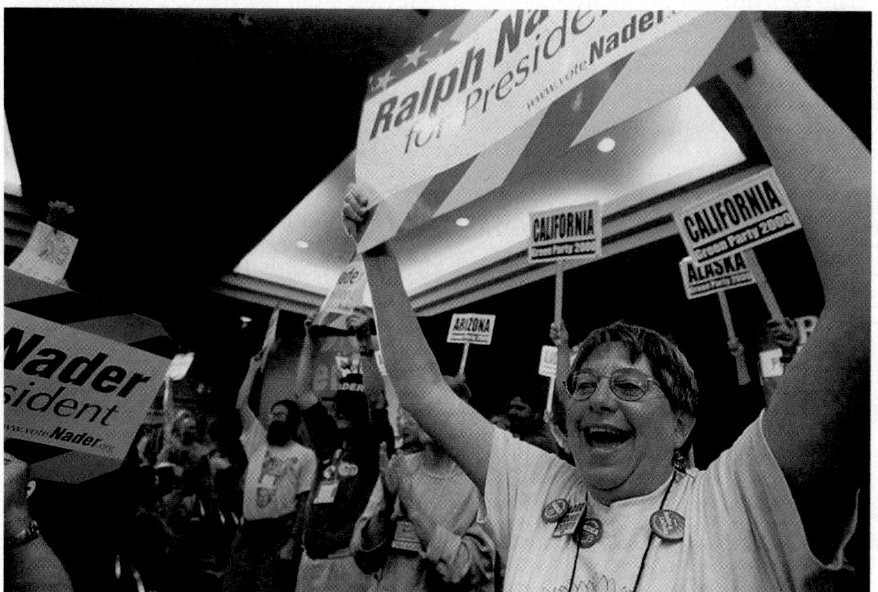

The Green Party holds its convention. Ralph Nader ran for the presidency as a Green in the 2000 election. By doing so, he brought up many questions of social choice–some would say deliberately. Was Nader a spoiler candidate? Were Nader supporters casting sincere votes for him? Were other voters who liked his positions hedging their bets and voting insincerely if they chose another candidate? (*Mark Leffingwell/AFP/Getty.*)

10.3 Impossibility

Condorcet's method, as we've seen, has a number of very desirable properties, including the following four:

1. Elections never result in ties (with an odd number of voters).
2. It satisfies the Pareto condition.
3. It is nonmanipulable.
4. It is not a dictatorship.

Unfortunately, Condorcet's voting paradox on page 289 shows that there are elections in which Condorcet's method produces no winner at all.

Can we find a voting system that satisfies all four of these properties and that, unlike Condorcet's method, always yields a winner? Several possibilities suggest themselves. For example, to avoid ties, we could modify any of our usual methods by agreeing to use a fixed ordering of the candidates to break any ties that occur. Or we could extend Condorcet's method by making the winner be the candidate with the best "win-loss record" in one-on-one contests (a method called *Copeland's rule*).

Alas, any such attempt is doomed. In the early 1970s, Allan Gibbard and Mark Satterthwaite independently proved the following remarkable result.

The Gibbard-Satterthwaite Theorem THEOREM

With three or more candidates and any number of voters, there does not exist–and there never will exist–a voting system that always produces a winner, never has ties, satisfies the Pareto condition, is nonmanipulable, and is not a dictatorship.

The Gibbard–Satterthwaite theorem (often called the GS theorem, for short) is a deep result that is related in important ways to Arrow's impossibility theorem. In particular, you shouldn't find it at all obvious, and we won't be saying anything about the proof. But we can state and prove a much weaker result that is of some interest in its own right.

A Weak Version of the GS Theorem THEOREM

Any voting system for three candidates that agrees with Condorcet's method whenever there is a Condorcet winner–and that additionally produces a unique winner when confronted by the ballots in the Condorcet voting paradox–is manipulable.

Let's see why this is true. With the Condorcet voting paradox, the winner is either A or B or C. For the moment, we'll assume it is C (and leave the other two cases to you–see Exercise 25). Consider the following two elections:

Election 1

Rank	Number of Voters (3)		
First	*A*	*B*	*C*
Second	*B*	*C*	*A*
Third	*C*	*A*	*B*

Election 2

Rank	Number of Voters (3)		
First	*B*	*B*	*C*
Second	*A*	*C*	*A*
Third	*C*	*A*	*B*

In Election 1, the winner is C (our assumption in this case) and in Election 2, the winner is B (because we are assuming that our voting system agrees with Condorcet's method when there is a Condorcet winner, as B is here). Notice that the voter on the left, by a unilateral change in ballot, has improved the election outcome from his or her third choice to being his or her second choice. This is what that voter set out to do and is the desired instance of manipulation.

But the nonintuitive nature of voting and manipulation does not end here. It also turns out that sometimes "more is less" when it comes to "voting power." We illustrate this with the so-called chair's paradox.

10.4 The Chair's Paradox

We conclude this chapter with an aspect of manipulability that is so counterintuitive that it is referred to as the chair's paradox. The situation is as follows. Suppose we have three candidates–*A*, *B*, and *C*–and three voters whom we'll call, for simplicity, the chair, you, and me.

Now the chair prefers *A* to *B* to *C*. You prefer *B* to *C* to *A*. I prefer *C* to *A* to *B*. Thus, if we were to cast sincere preference list ballots, they'd be precisely the Condorcet voting paradox ballots. But that's not what we're going to do.

We're going to assume that each of the three of us gets to vote for one of the candidates. Votes are tallied as follows: If any candidate gets at least two of the three votes, he or she wins. But if each gets one vote, then whichever candidate the chair voted for wins. Thus, the chair has what might be called **tie-breaking power**. In particular, the chair clearly has more power than you or I.

The goal now is to analyze the situation and to determine how each of us will vote if we're rational in the sense of being willing to vote strategically (that is, to manipulate the system) if it's in our own best interest. This is really a game-theoretic analysis, and it's useful to borrow a couple of pieces of game-theoretic terminology.

A choice of which candidate to vote for is called a **strategy**. So each of us has three strategies at our disposal: Vote for *A*, vote for *B*, and vote for *C*. It will be useful to have our preferences displayed as if they were ballots. But remember, these are just our preferences, not our ballots.

Chair	You	Me
A	*B*	*C*
B	*C*	*A*
C	*A*	*B*

Our first observation is that if we're all rational and acting in our own self-interest, none of us will vote for our least-preferred candidate. The point is that voting for either a first or second choice **weakly dominates** the strategies of voting for a third choice in the sense that the former choices always yield outcomes that are either the same as, or better than, the latter.

But now the chair's strategy of voting for *A* weakly dominates his strategy of voting for *B*. That is, if we both vote for *C*, the outcome is *C* regardless of how he votes, but otherwise he does strictly better by voting for *A* rather than *B*. Hence, assuming the chair is rational, we know the chair will, in fact, vote for his top choice, Candidate *A*.

Now, given that we know what the chair will do, the claim is that my strategy of voting for *C* weakly dominates my strategy of voting for *A*. That is, if you vote for *B*, the outcome is *A* regardless of whether I vote for *C* or *A*. On the other hand, if you vote for *C*, then I can secure my best outcome *C* by voting for *C*. Assuming I'm rational, we know that I will, like the chair, vote for my top choice, which is Candidate *C*.

But let's see where these decisions leave you. You know that the chair is voting for *A* and I'm voting for *C*. So if you vote for *B*, then the outcome is *A*–your last choice. However, if you vote for *C* along with me, then the outcome is *C*, your sec-

A woman casts her ballot on election day, the most important day in the American civic ritual of political campaigns and elections. Although she is acting as a responsible citizen, she may also contribute to some remarkable and contradictory results: Condorcet's voting paradox and the Gibbard–Satterthwaite theorem warn us that some elections produce strange results! (*David Paul Morris/Getty Images.*)

ond choice. There is no way that you can secure your top choice B as the winner. So if you're rational, then you'll vote for C also, and Candidate C will win the election.

So why is this paradoxical? Well, the chair had the most power, but the eventual winner of the election was his least-preferred candidate! He would have been better off handing over the tie-breaking power to either you or me.

The Chair's Paradox — THEOREM

With three voters and three candidates, the voter with tie-breaking power can, if all three voters act rationally in their own self-interest, end up with his or her least-preferred candidate as the election winner.

The chair's paradox represents only one of manipulability's first cousins, some of which involve not only the fields of mathematics and political science but psychology as well. I recall a third-grade penmanship contest in which each of us had a writing sample taped to the blackboard, and the teacher, Mrs. Levy, announced that we'd get to vote for the one we thought best, with the proviso that the voter couldn't vote for his or her own paper. She also announced that if two or more were tied, we'd have a runoff among those.

I remember being torn as to which of three particular ones to vote for, all of which I thought were very good and considerably better than the rest, including my own. When the votes were counted, these three were, in fact, tied for the win, with my writing sample alone in fourth, and only one vote out of the tie.

After announcing the results, Mrs. Levy went on to say that the runoff would involve not three of us, but four, as she had decided also to vote, and she was voting for me! I don't remember the final tally, or what Mrs. Levy then said to the class, or what my three classmates, all plenty smart enough to realize what had just happened, later said to me. But I do remember sitting back and smiling–absolutely sure of the outcome–as soon as she had announced her intention to vote for me.

REVIEW VOCABULARY

Agenda manipulation The ability to control who wins an election with sequential pairwise voting by a choice of the agenda–that is, a choice of the order in which the one-on-one contests will be held. (p. 321)

Chair's paradox The fact that with three voters and three candidates, the voter with tie-breaking power (the "chair") can, if all three voters act rationally in their own self-interest, end up with her or his least-preferred candidate as the election winner. (p. 316)

Disingenuous ballot Any ballot that does not represent a voter's true preferences. Also called an **insincere ballot**. (p. 313)

Gibbard–Satterthwaite (GS) manipulability theorem Alan Gibbard and Mark Satterthwaite's independent discovery that every voting system for three or more alternatives and any number of voters that satisfies the Pareto condition, always produces a unique winner, and is not a dictatorship can be manipulated. (p. 316)

Group manipulability A voting system is group manipulable if there exists at least one election in which a group of voters can change their ballots (with the ballots of voters not in the group left unchanged) in such a way that they all prefer the winner of the new election to the winner of the old election, assuming that the original ballots represent the true preferences of these voters. (p. 322)

Manipulation A voting system is manipulable if there exists at least one election in which a voter can change his or her ballot (with the ballots of all other voters left unchanged) in such a way that he or she prefers the winner of the new election to the winner of the old election, assuming that the original ballots represent the true preferences of the voters. (p. 313)

May's theorem for manipulability Kenneth May's discovery that for two candidates and an odd number of voters, majority rule is the only voting system that treats both candidates equally, treats all voters equally, and is nonmanipulable. (p. 317)

Strategy In the chair's paradox, a choice of which candidate to vote for is called a strategy. This is a special case of the use of the term in general game-theoretic situations. (p. 325)

Tie-breaking power That aspect of the voting rule used in the chair's paradox that says the winner will be whichever candidate the chair votes for if there is a tie (which only happens if each candidate gets exactly one vote). (p. 325)

Unilateral change A change (in ballot) by a voter while every other voter keeps her or his ballot exactly as it was. (p. 315)

Weak-dominance One strategy (for example, a choice of whom to vote for) weakly dominates another if it yields an outcome that is at least as good, and sometimes better, than the other. (p. 325)

SKILLS CHECK

1. A "unilateral change in ballot" refers to the fact that

(a) only one candidate's position is being altered.

(b) no communication is taking place.

(c) only one voter is changing his or her ballot.

2. The quote "My scheme is intended only for honest men!" is from ____________________ .

3. If a voter has sincere preferences of A over B over C over D, then

(a) she will prefer a tie between A and D to a tie between B and C.

(b) she will prefer a tie between B and C to a tie between A and D.

(c) it's not at all clear which tie–AD or BC–she will prefer.

4. A ballot that misrepresents a voter's true preferences is referred to as ____________________ .

5. Suppose that two elections show that a voting system is manipulable. Then

(a) neither election results in a tie.

(b) the winners are the same in both elections.

(c) every voter has changed his or her ballot.

6. In the two-candidate case, manipulation is equivalent to ____________________ .

7. Condorcet's method

(a) can be manipulated but always produces a winner.

(b) is nonmanipulable but sometimes produces no winner.

(c) sometimes results in a tie, so manipulability is hard to assess.

8. May's theorem for manipulability says that, with an odd number of voters, among all voting systems for two candidates that never result in a tie, majority rule is the only one that is nonmanipulable and __________ .

9. With the Borda count, two ballots "cancel each other out" if

(a) they are identical.
(b) each is arrived at by turning the other one upside down.
(c) other voters also hold these same ballots.

10. The Borda count is nonmanipulable in the special case in which ____________________ .

11. A six-voter example of manipulation with the Borda count can be modified to yield a ten-voter example by

(a) adding four ballots that are identical to each other.
(b) adding four ballots that are identical to Voter 1's ballot.
(c) adding two pairs of ballots, with the ballots in each pair canceling each other out.

12. With any voting system that satisfies the Pareto condition, an n-voter example of manipulation with k candidates can be modified to yield an n-voter example with $k + j$ candidates by ____________________ .

13. Of the Hare system and the plurality runoff method,

(a) only the Hare system is manipulable.
(b) only plurality runoff is manipulable.
(c) both are manipulable.

14. Sequential pairwise voting is susceptible to a kind of manipulation called ____________________ .

15. Plurality voting

(a) cannot be manipulated by a single voter.
(b) can be manipulated by a single voter.
(c) is subject to agenda manipulation.

16. Plurality voting is susceptible to a kind of manipulation called ____________________ .

17. The Gibbard–Satterthwaite theorem says that with three or more candidates and any number of voters, there is no voting system that

(a) is not a dictatorship.
(b) is nonmanipulable and is not a dictatorship.
(c) satisfies the Pareto condition, is nonmanipulable, and is not a dictatorship.
(d) always yields a unique winner, satisfies the Pareto condition, is nonmanipulable, and is not a dictatorship.

18. The weak version of the Gibbard–Satterthwaite theorem asserts that if we have a voting system that agrees with Condorcet's method whenever there is a Condorcet winner, and that additionally produces a unique winner when confronted by the ballots in the Condorcet voting paradox, then the system is ____________________ .

19. The voters' preferences in the paradox of the chair are

(a) precisely the Condorcet voting paradox ballots.
(b) all the same.
(c) dictated by the chair.

20. The chair's paradox is paradoxical because ____________________ .

CHAPTER 10 EXERCISES

■ Challenge ◆ Discussion

10.1 Majority Rule and Condorcet's Method

1. Consider the voting system for two candidates (A and B) and three voters in which the candidate with the *fewest* first-place votes wins. Produce two elections that show this voting system is manipulable.

2. Consider the voting system for two candidates (A and B) and three voters in which the candidate receiving an odd number of first-place votes wins. Produce two elections that show this voting system is manipulable.

3. Consider the voting system for two candidates (A and B) and three voters in which the candidate receiving an even number of first-place votes wins. Produce two elections that show this voting system is manipulable.

4. There are at least two voting systems for two candidates (A and B) and three voters that are nonmanipulable and that treat all voters the same (meaning that if two voters were to exchange ballots, then the election outcome would be unchanged).

(a) What does May's theorem tell us about such a voting system?
(b) In one sentence, give an example of such a voting system (that is, produce the rule that determines which of the two candidates, A or B, wins an election).
(c) In one sentence, give another example that is different from the example you gave in part (b) in that it produces a different winner for at least one election.

5. There are at least three voting systems for two candidates (A and B) and three voters that are nonmanipulable and that treat both candidates the same (meaning that if all three voters change their ballots, then the election outcome also changes).

(a) What does May's theorem tell us about such a voting system?
(b) In one sentence, give an example of such a voting system (that is, produce the rule that determines which of the two candidates wins an election).
(c) In one sentence, give two other examples that are different from the example you gave in part (b) in that they produce a different winner for at least one election.

6. Alfonse D'Amato (D) won the 1980 U.S. Senate race in New York by defeating Elizabeth Holtzman (H) and Jacob Javits (J). Reasonable estimates (based largely on exit polls) suggest that voters ranked the candidates according to the following table:

22%	23%	15%	29%	7%	4%
D	*D*	*H*	*H*	*J*	*J*
H	*J*	*D*	*J*	*H*	*D*
J	*H*	*J*	*D*	*D*	*H*

Who would have won if Condorcet's method (instead of plurality voting) had been used?

10.2 The Manipulability of Other Voting Systems for Three or More Candidates

7. Consider the following election with four candidates and two voters:

B	*A*
C	*D*
A	*C*
D	*B*

Show that if the Borda count is being used, the voter on the left can manipulate the outcome (assuming the above ballot represents his true preferences).

8. Example 2 showed that the Borda count is manipulable if there are five candidates and six voters. Mimic what was done there in order to construct an example with seven candidates and eight voters.

9. Use the following election to illustrate the manipulability of the Borda count with three voters and four candidates:

A	*B*	*B*
B	*A*	*A*
C	*C*	*C*
D	*D*	*D*

10. Show that the Borda count is manipulable if there are four candidates and five voters. (*Hint:* Start with the ballots in the previous exercise, and then add two ballots that cancel each other out.)

11. Building on the idea in the previous exercise, show that the Borda count is manipulable if there are six candidates and nine voters.

12. Assume the following ballots give the true preferences of the voters and that the Borda count is being used. Show that at least one of the voters can improve the election outcome from her point of view by a unilateral change in her ballot.

B	*D*	*C*	*B*
C	*C*	*A*	*A*
D	*A*	*B*	*C*
A	*B*	*D*	*D*

13. There is a modified version of Condorcet's method called the *weak Condorcet rule:* A candidate is among the winners precisely if he would defeat or tie every other candidate in a one-on-one contest. Notice that with an odd number of voters, the weak Condorcet rule is identical to Condorcet's method. Use the following ballots to show that the weak Condorcet rule is manipulable:

A	*C*	*B*	*D*
B	*A*	*D*	*C*
C	*B*	*C*	*A*
D	*D*	*A*	*B*

14. *Copeland's rule* is a voting system that, like Condorcet's method, looks at one-on-one contests. Copeland's rule, however, takes as the election winner the candidate with the best "win-loss record." Use the following ballots to show that Copeland's rule is manipulable:

A	*C*	*A*	*D*
B	*E*	*E*	*B*
C	*D*	*D*	*E*
D	*B*	*C*	*C*
E	*A*	*B*	*A*

15. *Coombs's rule* is the voting system that operates like the Hare system, except that instead of deleting candidates with the *fewest* first-place votes one after another, it deletes candidates with the ***most*** last-place votes one after another. Use the following ballots to show that Coombs's rule is manipulable:

A	*B*	*B*	*A*	*A*
B	*C*	*C*	*C*	*C*
C	*A*	*A*	*B*	*B*

16. Use the following election to show that the Hare system is manipulable:

A	*B*	*C*	*C*	*D*
B	*A*	*B*	*B*	*B*
C	*C*	*A*	*A*	*C*
D	*D*	*D*	*D*	*A*

17. Use the following election to show that the plurality runoff rule is manipulable:

A	*A*	*C*	*C*	*B*
B	*B*	*A*	*A*	*C*
C	*C*	*B*	*B*	*A*

18. Use the following election to show that sequential pairwise voting is manipulable. (Assume the agenda is *ABC*.)

A	*B*	*C*
B	*C*	*A*
C	*A*	*B*

19. Given the following ballots:

A	*C*	*B*
B	*A*	*D*
D	*B*	*C*
C	*D*	*A*

mimic what was done in Example 6 to find

(a) an agenda for which *B* is the winner using sequential pairwise voting.

(b) an agenda for which *C* is the winner using sequential pairwise voting.

(c) an agenda for which *D* is the winner using sequential pairwise voting.

◆ **20.** Suppose that we have a voting system that satisfies unanimity: If every voter ranks the same candidate first, then that candidate is the unique winner. In a few sentences, explain why it is that, if the system fails to satisfy the Pareto condition, it can be manipulated by some group.

21. Use the ballots in Exercise 6 to show that the plurality rule is group manipulable.

22. Consider the voting rule in which an alternative is among the winners if it receives at least one first-place vote. In one sentence, explain why this voting system is *not* manipulable.

23. Consider the voting rule in which an alternative is among the winners if it has at least two first-place votes.

(a) In one sentence, explain why this voting system is *not* manipulable.

(b) Explain why the following two elections don't contradict part (a).

Election 1

Rank	Number of Voters (4)			
First	*B*	*A*	*A*	*C*
Second	*C*	*B*	*B*	*B*
Third	*A*	*C*	*C*	*A*

Election 2

Rank	Number of Voters (4)			
First	*C*	*A*	*A*	*C*
Second	*B*	*B*	*B*	*B*
Third	*A*	*C*	*C*	*A*

(c) Intuitively, does it seem to you that Voter 1, on the left in part (b), has secured a better outcome by submitting a disingenuous ballot?

24. Consider the voting system in which the winner is determined by the total number of first- and second-place votes, with ties broken (when possible) according to the number of first-place votes. Thus, a candidate with no first-place votes and three second-place votes would defeat a candidate with two first-place votes and no second-place votes, but a candidate with two first-place votes and three second-place votes would defeat a candidate with one first-place vote and four second-place votes. Given Election 1 below, find a change in Voter 1's ballot that shows that this voting system is manipulable.

Election 1

Rank	Number of Voters (3)		
First	*A*	*C*	*E*
Second	*B*	*D*	*D*
Third	*C*	*A*	*A*
Fourth	*D*	*B*	*B*
Fifth	*E*	*E*	*C*

10.3 Impossibility

■ **25.** Complete the proof of the weak version of the Gibbard–Satterthwaite theorem by handling the case where

(a) the winner with the voting paradox ballots is *A*.

(b) the winner with the voting paradox ballots is *B*.

The Gibbard–Satterthwaite theorem says that the following four properties of voting systems can never be simultaneously satisfied:

(1) Elections always have unique winners.

(2) It satisfies the Pareto condition.

(3) It is nonmanipulable.

(4) It is not a dictatorship.

26. Which of the four properties are satisfied by a dictatorship?

27. Which of the four properties are satisfied by an "antidictatorship," where the election winner is whichever candidate Voter 1 ranks *last* on his or her ballot?

28. Which of the four properties are satisfied if we use the plurality rule with Voter 1's ballot used to break any ties that occur?

10.4 The Chair's Paradox

Consider the ballots from the chair's paradox:

Chair	You	Me
A	B	C
B	C	A
C	A	B

Assume that we know that the chair will vote for A, but that we don't know anything about how I will vote.

◆ **29.** In a sentence or two, explain why your strategy to vote for B does not weakly dominate your strategy of voting for C.

◆ **30.** In a sentence or two, explain why your strategy to vote for C does not weakly dominate your strategy of voting for B.

WRITING PROJECT

In a paragraph or two, explain why Condorcet's method is not group manipulable.

SUGGESTED READINGS

MOULIN, HERVÉ. *The Strategy of Social Choice*, North Holland, New York, 1983. Manipulability from an economist's point of view.

RIKER, WILLIAM. *The Art of Political Manipulation*, Yale University Press, New Haven and London, 1986. Manipulability from a political scientist's point of view.

TAYLOR, ALAN. *Social Choice and the Mathematics of Manipulation*, Cambridge University Press, Cambridge, U.K., 2005. Manipulability from a mathematician's point of view.

[illegible]

CHAPTER 11

Weighted Voting Systems

Voting is often used to decide yes or no questions. Legislatures vote on bills, stockholders vote on resolutions presented by the board of directors of a corporation, and juries vote to acquit or convict a defendant. In this chapter, we shall concentrate on situations where there are just two alternatives, such as "yes" or "no." The theorem of Kenneth May quoted in Chapter 9 says that majority rule is the only system with the following properties:

1. All voters are treated equally.
2. Both alternatives are treated equally.
3. If you vote "no," and "yes" wins, then "yes" would still win if you switched your vote to "yes," provided that no other voters switched their votes.
4. A tie cannot occur unless there is an even number of voters.

There are many situations in which one or more of these properties are not valid. For example, in a criminal trial the jury is required to reach a unanimous decision on a motion to convict (or on a motion to acquit); thus, if there is one "no" vote, the motion is not adopted. In this case, the alternatives are not treated equally. Here's another familiar example. Stockholders are allowed one vote per share that they own. If shareholder A owns 10,000 shares and shareholder B owns 100, then this voting system does not treat A and B equally.

Some systems where the voters appear to be unequal in power actually have all of the properties required by May's theorem. Any student of politics will attest that not all legislators are equally powerful (think of the speaker of the U.S. House of Representatives versus a freshman member, or the prime minister versus a backbencher in Parliament). Nevertheless, the voting system actually treats the legislators equally: Each has one vote. Our interest is in the voting system itself and not in the influence that some voters might acquire as a result of experience or accomplishment.

To justify the formula, suppose that we are listing all of the permutations. There are n voters who could be first; when the first voter is selected, there are $n - 1$ remaining voters who could be in second position, then $n - 2$ who could be third, and so on. When it is time to select for the last position, there is one voter left. By the fundamental principle of counting (see Chapter 2), the number of permutations is the product of the numbers of choices that we have had at each stage.

EXAMPLE 6 ▪ Calculating n!

Here are the first four, starting with 1!.

$$1! = 1$$
$$2! = 2 \times 1 = 2$$
$$3! = 3 \times 2 \times 1 = 6$$
$$4! = 4 \times 3 \times 2 \times 1 = 24$$

To continue this list, observe that $n! = n \times (n - 1)!$ for $n \geq 1$. Thus

$$5! = 5 \times 4! = 5 \times 24 = 120$$
$$6! = 6 \times 5! = 6 \times 120 = 720$$
$$7! = 7 \times 6! = 7 \times 720 = 5040$$
$$8! = 8 \times 7! = 8 \times 5040 = 40{,}320$$

and so on. You can imagine that $n!$ increases dramatically as n increases–an instance of the combinatorial explosion. You probably don't want to calculate 100!. It is a 158-digit number.

The Shapley-Shubik Power Index — DEFINITION

The **Shapley–Shubik power index** of each voter is computed by counting the number of permutations in which he or she is pivotal, then dividing it by $n!$, where n is the number of voters.

If we say that each voter "owns" the permutations in which he or she is pivotal, then each voter's Shapley-Shubik index is his or her share of the permutations. The Shapley-Shubik index can also be viewed as a probability. A voter's index is equal to the probability that when a permutation is selected at random, he or she will be the pivotal voter.

EXAMPLE 7 ▪ The Shapley–Shubik Power Index of a Three-Voter System

Let us calculate the Shapley–Shubik power index of the voting system $[6:5,3,1]$. We will name the participants A, B, and C, and consider their $3! = 6$ permutations. Table 11.2 displays all six permutations. Next to each permutation, the total weights of the first voter, the first two voters, and all three voters are shown in sequence. The first number in the sequence that equals or exceeds the quota (6) is underlined, and the corresponding pivotal voter's symbol is circled. We see that A is pivotal in four permutations, while B and C are each pivotal in one. Hence the Shapley–Shubik index of A is $\frac{4}{6}$, and B and C each have Shapley–Shubik indices of $\frac{1}{6}$.

TABLE 11.2 Permutations and Pivotal Voters for the Three-Person Committee

Permutations			Weights		
A	Ⓑ	C	2	3	4
A	Ⓒ	B	2	3	4
B	Ⓐ	C	1	3	4
B	C	Ⓐ	1	2	4
C	Ⓐ	B	1	3	4
C	B	Ⓐ	1	2	4

TABLE 11.3 Permutations and Pivotal Voters for the Four-Shareholder Corporation

Permutations				Weights				Pivot
A	Ⓑ	C	D	40	70	90	100	B
A	Ⓑ	D	C	40	70	80	100	B
A	Ⓒ	B	D	40	60	90	100	C
A	Ⓒ	D	B	40	60	70	100	C
A	D	Ⓑ	C	40	50	80	100	B
A	D	Ⓒ	B	40	50	70	100	C
B	Ⓐ	C	D	30	70	90	100	A
B	Ⓐ	D	C	30	70	80	100	A
B	C	Ⓐ	D	30	50	90	100	A
B	C	Ⓓ	A	30	50	60	100	D
B	D	Ⓐ	C	30	40	80	100	A
B	D	Ⓒ	A	30	40	60	100	C
C	Ⓐ	B	D	20	60	90	100	A
C	Ⓐ	D	B	20	60	70	100	A
C	B	Ⓐ	D	20	50	90	100	A
C	B	Ⓓ	A	20	50	60	100	D
C	D	Ⓐ	B	20	30	70	100	A
C	D	Ⓑ	A	20	50	60	100	B
D	A	Ⓑ	C	10	50	80	100	B
D	A	Ⓒ	B	10	50	70	100	C
D	B	Ⓐ	C	10	40	80	100	A
D	B	Ⓒ	A	10	40	60	100	C
D	C	Ⓐ	B	10	30	70	100	A
D	C	Ⓑ	A	10	30	60	100	B

EXAMPLE 8 ■ The Corporation with Four Shareholders

A corporation has four shareholders, *A*, *B*, *C*, and *D*, with 40, 30, 20, and 10 shares, respectively. The corporation uses the weighted voting system

$$[51 : 40, 30, 20, 10]$$

11.3 The Banzhaf Power Index

In contrast to the Shapley-Shubik power index, which is based on counting permutations, the Banzhaf power index is based on counting combinations.

Voting Combination DEFINITION

A **voting combination** is a list of voters indicating how each voted on an issue.

EXAMPLE 11 ■ Voting Combinations in the 2004 Presidential Election

The voting combination for the 2004 Electoral College can be determined from Table 11.1. All voters that had Bush's margin greater than 1.000 voted for the Bush–Cheney ticket; those with Bush's margin less than 1.000 voted for the Kerry–Edwards ticket.

■

In any voting combination there may be one or more voters who have the power to change the outcome by switching their votes.

Critical Voter DEFINITION

A voter in a given voting combination is a **critical voter** if the outcome would be different if that voter, and no other voter, changed his or her vote.

Although each voting permutation has exactly one pivotal voter, a voting combination may have no critical voters, or it may have many.

EXAMPLE 12 ■ A Criminal Trial

When the jury is unanimous in favor of a motion to convict (or a motion to acquit), then each juror is a critical voter. On the other hand, if all but one juror is in favor of a motion, then the motion fails, and the lone holdout is a critical voter. Voting combinations in which more than one juror opposes a motion have no critical voters.

■

EXAMPLE 13 ■ The. U.S. Presidential Elections

Table 11.1 shows that the Bush–Cheney ticket received 286 electoral votes in 2004. The quota for the Electoral College is 270, so the ticket had 16 extra votes. A state that voted for the Bush–Cheney ticket was a critical voter if and only if its voting weight was more than 16: Thus Florida, Ohio, and Texas were critical voters. If one of these states had switched to the Kerry–Edwards ticket, then Kerry–Edwards would have won. States that voted for the Kerry–Edwards ticket were not critical voters, because if they switched to the Bush–Cheney ticket the outcome of the election would not have changed.

In the closer election of 2000, in which the Bush–Cheney ticket won by only two electoral votes, every state that voted for the Bush–Cheney ticket was a critical voter[1]. In the landslide election of 1984 in which the Reagan–Bush ticket beat the

[1]Even Nebraska was critical: See Exercise 22.

Mondale–Ferraro ticket by receiving the largest number of electoral votes in history, 155 votes more than the quota–there were no critical voters at all, because no state has more than 155 electoral votes! Nevertheless, the 1984 voting permutation had a pivot–the state that brought the Reagan–Bush ticket over the 270 vote quota.

Banzhaf Power Index DEFINITION

A voter's **Banzhaf power index** is the number of voting combinations in which he or she casts a critical vote.

We have seen that a juror in a criminal trial casts a critical vote in two voting combinations: One in which the jury is unanimously in favor of a motion, and one in which the juror is the lone holdout, voting against a motion that all other jurors support. Thus, each juror has a Banzhaf index of 2.

SPOTLIGHT 11.4 A Mathematical Quagmire

A county legislature in the United States is usually called a Board of Supervisors. Unlike state legislators, who represent districts that are carefully drawn to be equal in population, supervisors in some counties represent towns within the county. Because the towns differ in population, some countries use weighted voting to compensate for the resulting inequity.

If each supervisor's voting weight is proportional to the population of the town he or she represents, there will be situations in which one or more supervisors on a board are dummy voters, even if no supervisor is dictator. In a 1965 law review article, John F. Banzhaf III showed that three of the six supervisors of Nassau County, New York, were dummies. The article inspired legal action against several elected bodies that employ weighted voting systems.

The first legal challenge to weighted voting was to invalidate the voting system of the Board of Supervisors of Washington County, New York. In its decision, the New York State Court of Appeals provided a way to fix a weighted voting system: Each supervisor's Banzhaf power index, rather than his or her voting weight, should be proportional to the population of the district that he or she represents. The court predicted that its remedy would lead to a "mathematical quagmire."

Five lawsuits, filed over a period of 25 years, challenged weighted voting in the Nassau County Board of Supervisors. These cases proved to be the mathematical quagmire that the appeals court had feared. The courts attempted to force Nassau County to comply with the Washington County decision. Although the county made a sincere attempt to do so, every voting system that it devised faced a new legal challenge. With conflicting expert testimony, the U.S. District Court finally ruled in 1993 that weighted voting was inherently unfair.

Banzhaf's law review article, which initially drew attention to weighted voting in Nassau County, was aptly titled "Weighted Voting Doesn't Work."

Nevertheless, tradition is hard to change. Many boards of supervisors of counties, particularly in the State of New York, still use weighted voting, and legal challenges to the practice, even after the Nassau County decision, have not always been successful.

In more complicated examples, it is helpful to view voting combinations in terms of **coalitions**.

Winning and Blocking Coalitions DEFINITION

If a voting combination results in the approval of a motion, the set of voters who support the the motion are said to form a **winning coalition**; the opposing voters are in a **losing coalition.** On the other hand, if the result of the voting combination is to defeat the motion, the voters who oppose are said to form a **blocking coalition,** and the voters in favor are a losing coalition.

If the result of a voting combination is to approve a motion, the critical voters, if there are any, belong to the winning coalition. In a combination that defeats a motion, the critical voters will belong to the blocking coalition.

EXAMPLE 14 ■ A Three-Member Committee

Consider a committee of three members, *A, B,* and *C*. The chairperson of the committee, *A*, has two votes, while *B* and *C* each have one. The quota is three, and this voting system is

[3: 2, 1, 1]

If the committee votes unanimously in favor of a motion, the winning coalition is {*A*,*B*,*C*}. Let's identify the critical voters in this coalition. Suppose *A* switches her vote:

A	*B*	*C*	Votes	Outcome
Yes	Yes	Yes	4	Pass
↓				
No	Yes	Yes	2	Fail

By changing her vote, *A* has changed the outcome. In this coalition, *A* is a critical voter.

Now let's see what happens if *B* changes his vote:

A	*B*	*C*	Votes	Outcome
Yes	Yes	Yes	4	Pass
	↓			
Yes	No	Yes	3	Pass

This time, the outcome doesn't change, so *B* is not a critical voter in this coalition. Since *C* has the same power as *B*, he is also not a critical voter in the combination.

Now consider the voting combination in which *A* and *B* vote "yes" and *C* votes "no." Then "yes" wins, with 3 votes, so {*A*,*B*} is a winning coalition. There are no extra votes, so both *A* and *B* are critical voters in this coalition. There is a third voting winning coalition {*A*,*C*}. Again there are no "yes" votes to spare, so both *A* and *C* are critical voters.

There are five voting combinations in which "no" wins. The corresponding blocking coalitions are {*A*,*B*,*C*}, {*A*,*B*}, {*A*,*C*}, {*A*}, and {*B*,*C*}. It takes just 2 votes to defeat a motion, so there are no critical voters in {*A*,*B*,*C*}; if any voter defects, there will still be enough votes to block. In each of the weight-3 blocking coalitions, $\{A,B\}$ and $\{A,C\}$, A is the only critical voter. In the weight-2 blocking coalitions, $\{A\}$ and $\{B,C\}$, all voters are critical.

To determine the Banzhaf index, we count the critical votes in each of the winning or blocking coalitions: A has three critical votes in winning coalitions, and another 3 in blocking coalitions; B and C each have one critical vote in a winning coalition and one critical vote in a blocking coalition. We will say that the Banzhaf index of this system is (6,2,2).

The Banzhaf index provides a comparison of the voting power of the participants in a voting system. Thus, A, with a Banzhaf index of 6, is three times as powerful as B or C. To determine the way voting power is distributed, we can add the numbers of critical voters for all three voters together to get $6 + 2 + 2 = 10$ critical votes in all. Thus, A has 60% of the voting power, while B and C each have 20%. The Shapley–Shubik model gives $\frac{2}{3}$ of the power to A, while B and C each have $\frac{1}{6}$, so the models are in close agreement in this case.

Counting Combinations

If there are three voters, A, B, and C, and A and C voted "yes" while B voted "no," we might record the voting combination as "Yes,No,Yes." A briefer notation is to visualize voting combinations as **binary numbers**. A whole number N is represented in binary form as a sequence of binary digits, or **bits**, which can be 0 or 1. This sequence expresses the way that N can be expressed as a sum of powers of 2. For example,

$$5 = 2^2 + 2^0$$

can be represented by a binary number where bits 2 and 0 are equal to 1, and the remaining bit 1, is 0. We would say that $5_2 = 101$: This binary number could stand for the voting combination "Yes,No,Yes." The largest number that can be represented with 3 bits is 7, because $7_2 = 111$: Because the smallest number that can be represented in 3 bits is 0 ($0_2 = 000$), there is a total of 8 3-bit binary numbers; thus 8 voting combinations when there are 3 voters.

Number of Voting Combinations THEOREM

The number of voting combinations with n voters is 2^n.

We have seen that the number of voting combinations with 3 voters is equal to the number of 3-bit binary numbers. By the same reasoning, the number of voting combinations with n voters is equal to the number of n-bit binary numbers. The largest n-bit binary number is the sequence of n ones, which represents $2^n - 1$: Since we start counting with 0, there are 2^n n-bit binary numbers.

We have interpreted the Shapley-Shubik index of a voter as the probability that he or she will be pivotal in a randomly selected permutation. We can also interpret the Banzhaf index in terms of probability. A voter's Banzhaf index is simply the number of voting combinations in which he or she is a critical voter. Thus, if we divide the voter's Banzhaf index by the number of possible voting combinations, we will get the probability that the voter will cast a critical vote in a randomly selected voting combination.

There is one difference to be aware of: Because each voting permutation has exactly one pivotal voter, the sum of the Shapley-Shubik indices of all voters is 1. However, not all voting combinations have a unique critical voter, so the sum of the probabilities of the voters casting critical votes is usually not equal to 1.

The easiest way to select a voting combination randomly is for each voter to decide how to vote by tossing a coin. Thus, a voter's Banzhaf index is equal to the number of possible voting combinations times the probability that he or she will be a critical voter, provided each voter decides his or her vote by a coin toss.

A 12-member jury has $2^{12} = 4096$ voting combinations. Each juror casts a critical vote in just two combinations; thus, his or her probability of casting a critical vote if the combination is randomly selected is $\frac{2}{4096}$. This is important because we don't want any juror to vote randomly! If he or she does that, we'd like to be sure that there is very little chance that it will make a difference.

How to Calculate the Banzhaf Power Index

To determine the Banzhaf power index of a voter A, we must count all possible winning and blocking coalitions of which A is a member and casts a critical vote. The weight of a winning coalition must be q or more, where q is the quota. A blocking coalition must be large enough to deny the "yes" voters the q votes they need to win. If the total weight of all the voters is n, then the weight of the blocking coalition has to be more than $n - q$. Assuming that all weights are integers, this means that the weight of a blocking coalition must be at least $n - q + 1$.

To identify the critical voters in a given winning or blocking coalition, the following principle is useful.

Extra Votes Principle — THEOREM

A winning coalition with total weight w has $w - q$ **extra votes**. A blocking coalition with total weight w has $w - (n - q + 1)$ extra votes. The critical voters are those whose weight is more than the coalition's extra votes.

We can readily identify the critical voters in any winning or blocking coalition by comparing each voter's weight with the number of extra votes that the coalition has.

Calculating the Banzhaf Power Index — PROCEDURE

To calculate the Banzhaf power index of a given voting system:

1. Make a list of the winning and blocking coalitions.
2. Use the **extra-votes principle** to identify the critical voters in each coalition.

A voter's Banzhaf power index is then the number of coalitions in which he or she appears as a critical voter.

In the examples that we have discussed so far, each participant has been a critical voter in exactly as many winning coalitions as blocking coalitions. This is not a coincidence, as we will now see.

Winning-Blocking Duality THEOREM

The number of winning coalitions in which a given voter is critical is equal to the number of blocking coalitions in which the same voter is critical.

Consider a voter A. For every winning coalition C in which A is a critical voter, let C^* be the coalition that would vote against the motion if A switched his vote from"yes" to "no." Because A was a critical voter in C, the switch in vote will change the original voting combination, where the result was "yes" to "no." In other words, C^* becomes a blocking coalition, and A is a critical voter in C^*: This correspondence,

$$C \longleftrightarrow C^*$$

shows the number of winning coalitions in which A is a critical voter is equal to the number of blocking coalitions in which A is critical. By the winning/blocking duality principle, we can determine a voter's Banzhaf power index by doubling the number of winning coalitions in which he or she is a critical voter–there is no need to count blocking coalitions.

TABLE 11.4 Winning Coalitions in the Four-Shareholder Corporation

			Critical Voters			
Coalition	**Weight**	**Extra Votes**	A	B	C	D
$\{A, B, C, D\}$	100	49				
$\{A, B, C\}$	90	39	X			
$\{A, B, D\}$	80	29	X	X		
$\{A, C, D\}$	70	19	X		X	
$\{A, B\}$	70	19	X	X		
$\{B, C, D\}$	60	9		X	X	X
$\{A, C\}$	60	9	X		X	
Critical votes			5	3	3	1

EXAMPLE 15 ■ The Corporation with Four Shareholders

The corporation with four shareholders (see Example 8) uses the weighted voting system

$$[51 : 40, 30, 20, 10]$$

Table 11.4 displays a list of all the winning coalitions of shareholders and the number of extra votes that each has. The four columns at the right are marked to indicate the critical voters in each coalition. By doubling the critical votes shown in the table, we arrive at the Banzhaf index of the corporation: $(10, 6, 6, 2)$. In this model, A has

$$\frac{10}{24} \text{ or approximately } 42\%$$

of the voting power, while B and C each have 25% (even though B has more shares than C). Shareholder D has the remaining 8% of the voting power, according to the Banzhaf model. In this case, power is distributed exactly as it was by the Shapley–Shubik model.

We have seen that each voting combination for a set of n voters corresponds to a binary number with n bits. Thus, if there are n voters, there will be 2^n voting combinations. Obviously there is exactly one voting combination where everyone votes "yes," and one voting combination where everyone votes "no." These correspond to the n-bit binary numbers with all bits equal to 1, and all bits equal to 0, respectively.

The number of voting combinations with n voters and exactly k "yes" votes is denoted ${}_nC_k$ (when speaking, ${}_nC_k$ is pronounced "n choose k"). Thus, the statement that there is exactly one combination of n voters where everyone votes "yes" would be ${}_nC_n = 1$. Similarly, we have ${}_nC_0 = 1$. There are n combinations with exactly one "yes" vote:

$$100\cdots0,\ 010\cdots0,\ \ldots,\ 000\cdots1,$$

where each combination has one 1 and $n - 1$ zeros. Thus, ${}_nC_1 = n$.

Duality Formula for Combinations — THEOREM

If each voter in a combination with k "yes" votes and $n - k$ "no" votes were to switch his or her vote to the opposite side, there would be $n - k$ "yes" votes and k "no" votes. Thus, the number of combinations of n voters with k "yes" votes is equal to the number of combinations of n voters with $n - k$ "yes" votes. This proves the following theorem:

$${}_nC_k = {}_nC_{n-k}$$

Addition Formula for Combinations — THEOREM

Now suppose that there are $n + 1$ voters, one of whom is Zoë. We would like to determine ${}_{n+1}C_k$: The number of combinations with k "yes" votes for a set of $n + 1$ voters. We will divide this set into two parts, depending on how Zoë votes. If she votes "no," there are ${}_nC_k$ voting combinations in which k of the other voters say "yes." If Zoë votes "yes" then a voting combination with k "yes" votes can be assembled by combining Zoë's vote with a combination of the other n voters with $k - 1$ "yes" votes; there are ${}_nC_{k-1}$ of these. Adding these, we obtain a valuable formula:

$${}_{n+1}C_k = {}_nC_k + {}_nC_{k-1}$$

The **addition formula** enables us to calculate the numbers ${}_nC_k$. Starting with ${}_0C_0 = {}_1C_0 = {}_1C_1 = 1$, we obtain ${}_2C_1 = {}_1C_1 + {}_1C_0 = 2$. Continuing, it is convenient to display the results in triangular form:

Pascal's Triangle THEOREM

The numbers ${}_nC_k$ can be arranged in the triangle shown below. The number ${}_nC_k$ is located on the nth row (rows are numbered downward; the 1 at the summit is the 0th row) and then counting to the kth entry from the left (again, the 1 at the left end of the row is the 0th entry).

$$
\begin{array}{c}
1 \\
1\ 1 \\
1\ 2\ 1 \\
1\ 3\ 3\ 1 \\
1\ 4\ 6\ 4\ 1 \\
1\ 5\ 10\ 10\ 5\ 1 \\
1\ 6\ 15\ 20\ 15\ 6\ 1
\end{array}
$$

Each entry in **Pascal's triangle** is determined by adding the two entries above its location on the previous row. For example, ${}_6C_3 = 20$ on the last row in the triangle above is obtained by adding ${}_5C_3 + {}_5C_2 = 10 + 10$ on the previous row. Thus, Pascal's triangle is constructed in accordance with the addition formula. The French mathematician and philosopher, Blaise Pascal (1623–1662) is credited with the discovery of his eponymous triangle.

Pascal's triangle is an intriguing pattern, but it is only useful to calculate ${}_nC_k$ when n is relatively small. The following expression gives a way to calculate ${}_nC_k$ in more general situations:

Combination Formula THEOREM

$$ {}_nC_k = \frac{n!}{k!(n-k)!} $$

To use the combination formula, cancel before multiplying.

EXAMPLE 16 ▪ Calculate ${}_{40}C_4$

From the combination formula, ${}_{40}C_4 = \dfrac{40!}{4!\,36!}$. Notice that $40! = 40 \times 39 \times 38 \times 37 \times 36!$. Thus we can cancel 36! and obtain

$$ {}_{40}C_4 = \frac{40 \times 39 \times 38 \times 37}{4 \times 3 \times 2 \times 1} = 91{,}390 $$

To verify the combination formula, let ${}_nD_k = \dfrac{n!}{k!(n-k)!}$. It's our job to show that ${}_nC_k = {}_nD_k$. Recalling that $0! = 1$, we have ${}_nD_0 = \frac{n!}{n!0!} = 1$ and ${}_nD_n = \frac{n!}{0!n!} = 1$. Also, we will see that the numbers ${}_nD_k$ obey the addition formula:

$$ {}_{n+1}D_k = {}_nD_k + {}_nD_{k-1} $$

or

$$ \frac{(n+1)!}{k!(n+1-k)!} = \frac{n!}{k!(n-k)!} + \frac{n!}{((k-1)!(n-k+1)!)} $$

Mode and Unimodal Distribution DEFINITION

A distribution that has one peak or highest point, called the **mode,** is **unimodal.**

For simplicity, we picture the distribution as continuous, although in fact, because the number of voters is finite, there cannot be voters at all points along the continuum.

More important than the mode, from the viewpoint of the candidates, is the **median M** of a distribution.

Median DEFINITION

The **median *M*** of a voter distribution is the point on the horizontal axis where half the voters have attitudes that lie to the left and half to the right.

The notion of the median of a voter distribution is the same as the notion of a median of a data sample given in Chapter 5.

The Figure 12.1a distribution is *symmetric*–the curve to the left of M is a mirror image of the curve to the right. Thus, the same numbers of voters have attitudes that are equal distances to the left and to the right of M.

Although the *attitudes* of voters are a fixed quantity in the calculations of the candidates, the *decisions* of voters will depend on the positions that the candidates take. Assuming the candidates know the distribution of voter attitudes, what positions are optimal for them?

Assume candidates A (red) and B (blue) take the positions along the left–right axis shown in Figure 12.1a, where candidate A is to the left of M and candidate B is to the right. Assume that all voters vote for the candidate whose position is closer to their own, and that all voters vote (we will consider modifications of this assumption later in the exercises). Then A will certainly attract all the voters to the left of his position, and B all the voters to the right of her position. If both candidates are an equal distance from M, as shown in Figure 12.1a, they will split any votes in the middle, with those to the left of M going to A and those to the right going to B.

Can either candidate do better by changing his or her position? If B's position remains fixed to the right of M, A could move alongside B, just to her left, and capture all the votes to B's left, as illustrated in Figure 12.1b. Because A would have moved to the right of M, he would, by changing his position in this manner, receive a majority of the votes and thereby win the election.

By analogous reasoning, there is no reason for B to stick to her original position to the right of M. By approaching A's original position to the left of M, B can capture all the votes to A's right (Figure 12.1c). In other words, both candidates, acting rationally, should approach each other and M.

If A were to move rightward past M, but B moved leftward only as far as M, their positions would be as shown in Figure 12.1d. Now B would receive not only the 50% of the votes to the left of M but also some of the votes that lie between B's position at M and A's position (now to the right of M).

Clearly, A loses by crossing M from the left. Hence, there is an incentive for both candidates to move toward M but not overstep it. In fact, taking a position at M maximizes the minimum number of votes a candidate can guarantee for himself or herself.

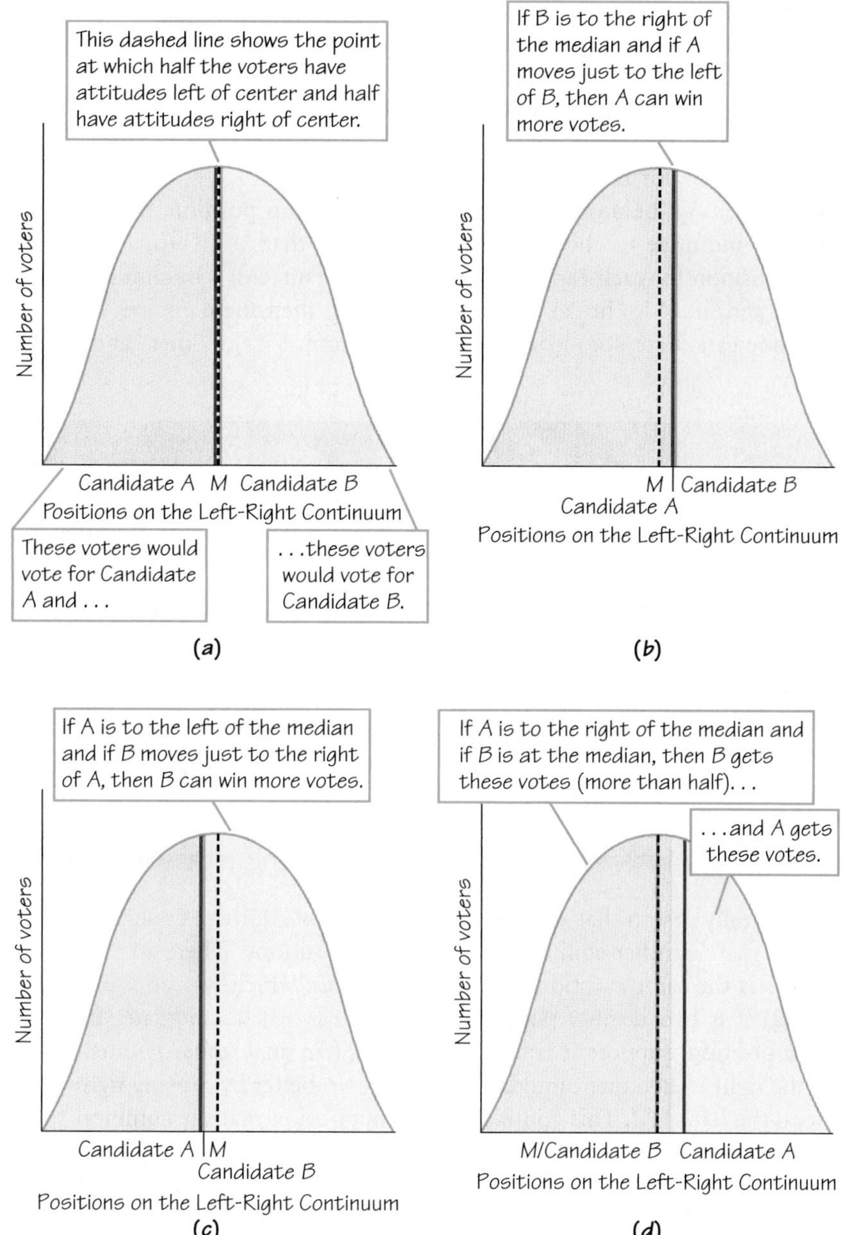

FIGURE 12.1 Unimodal distribution. The median M is the point on the horizontal axis that divides the area under the distribution curve–measuring the number of voters–exactly in half.

Maximin DEFINITION

A position is **maximin** for a candidate if there is no other position that can *guarantee* a better outcome–more votes for that candidate–whatever position the other candidate adopts.

If both candidates choose M, voters will be indifferent to the choice between them on the basis of their positions alone and would presumably make their choice on other grounds.

Taking a position at M, however, guarantees A at least 50% of the total vote *no matter what B does*. Moreover, there is no other position that can guarantee A more votes, and likewise for B.

M is also *stable*, because if one candidate adopts this position, the other candidate has no incentive to choose any position other than M. Thus, M is both the maximin position for each candidate (it offers a guarantee of a minimum of 50% of the votes) and, if M is chosen by both candidates, then these choices are in equilibrium (one candidate does worse by departing from it if the other candidate stays at it).

Equilibrium DEFINITION

A pair of positions is in **equilibrium** if, once chosen by both candidates, neither candidate has an incentive to depart from it unilaterally (that is, by himself or herself).

More formally, we have the median-voter theorem.

Median Voter Theorem THEOREM

Median-voter theorem: In a two-candidate election with an odd number of voters, M is the unique equilibrium position.

We have already shown that if both candidates choose M, these choices are in equilibrium. Is there another equilibrium position or positions? There are two possibilities: (1) It is the same position for both candidates, which we call a *common position*, or (2) it is two distinct positions, one taken by each candidate. If it were a common position, suppose it is to the left of M. (An analogous argument works if it is to the right.) Then one candidate can always do better by moving rightward but staying to the left of M. This contradicts the supposition that the common position is in equilibrium. Now suppose the equilibrium were two distinct positions. Then one candidate can always do better by moving alongside the other candidate but staying closer to M. This contradicts the supposition that these two positions are in equilibrium. Thus, in both cases, one candidate would have an incentive to depart from his or her position–holding the position of the other candidate fixed–so a nonmedian position of one or both candidates cannot be in equilibrium. Therefore, M is the only equilibrium position.

Bimodal Distribution: Median and Mean Different

The median-voter theorem is applicable *whatever* the distribution of the electorate's attitudes. Consider the distribution in Figure 12.2a, which is **bimodal** (two peaks) and is not symmetric. Applying the logic of the previous analysis, M is once again the maximin and equilibrium position of two candidates, even though the bulk of voters are concentrated at the two modes.

We next compare the M with the mean, which may be quite different:

Mean of a Voter Distribution

DEFINITION

The **Mean** $\bar{l}$ of a voter distribution is

$$\bar{l} = \frac{1}{n}\sum_{i=1}^{k} n_i l_i$$

where

k = number of different positions i that voters take on the continuum

n_i = number of voters at position i

l_i = location of position i on the continuum

$$n = \sum_{i=1}^{k} n_i = n_1 + n_2 + \cdots + n_k = \text{total number of voters}$$

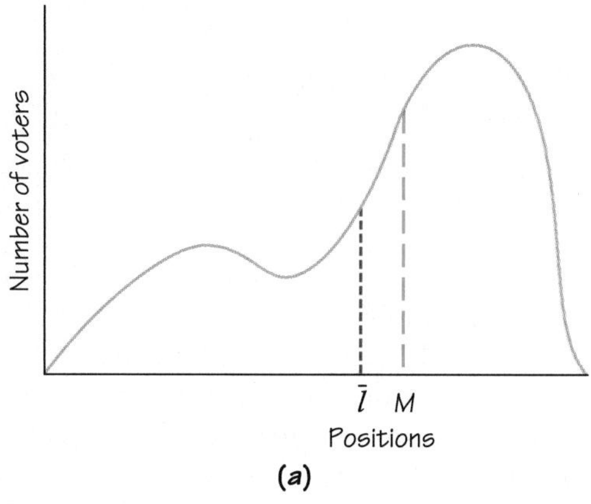

(*a*)

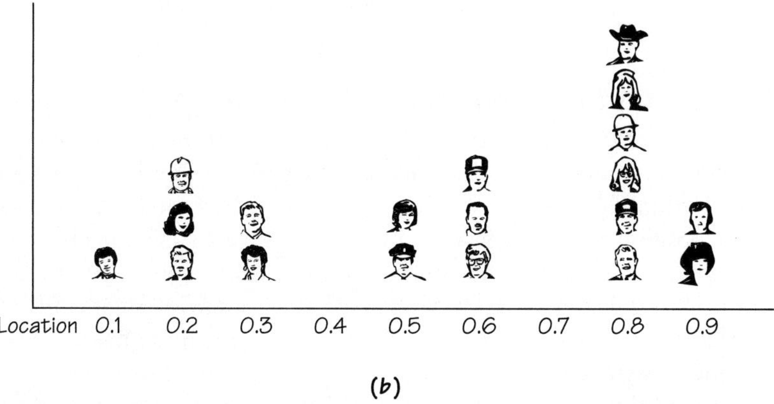

(*b*)

FIGURE 12.2 Unimodal distribution in which the median and mean are different. The distribution is skewed to the left; the median M is to the right of the mean $\bar{l}$.

The symbol Σ (sigma) is the *summation sign*. It signifies that all subscripted terms to its right (e.g., in the definition of n), beginning with the subscript 1 and continuing to the subscript k, are summed and a total obtained.

The notion of a mean is the same as that for data samples in Chapter 5, except that here we are calculating a *weighted* average: The location l_i of each position is weighted by the number of voters, n_i, at that position. Thus, the mean can be thought of as the position of a typical voter–that is, the expected position of a voter drawn randomly from the set of all voters.

The mean $\bar{l}$ need not coincide with the median M. As an illustration of this point, consider the following **discrete distribution** of $n = 19$ voters at $k = 7$ different positions over the interval between 0 and 1, or [0, 1], which is illustrated in Figure 12.2b.

Discrete Distribution of Voters DEFINITION

A **discrete distribution of voters** is one in which voters are located at only certain positions–not all points–along the left–right continuum.

Position, i	1	2	3	4	5	6	7
Location (l_i) of position i	0.1	0.2	0.3	0.5	0.6	0.8	0.9
Number of voters (n_i) at position i	1	3	2	2	3	6	2

Whereas M is 0.6 because 8 voters lie to the left and 8 voters lie to the right,

$$\bar{l} = \left(\frac{1}{19}\right) [1(0.1) + 3(0.2) + 2(0.3) + 2(0.5) + 3(0.6) + 6(0.8) + 2(0.9)] = 0.56$$

Taking a position at 0.56 against an opponent who takes a position at 0.6, a candidate would lose the election by 11 to 8 votes.

The distributions of Figures 12.2a and 12.2b are **skewed** to the left because the area under the curve, or the number of voters, is less concentrated to the left of M than to the right. These more spread-out voters put the mean $\bar{l}$ to the left of the median M. The lesson we derived from these figures is that it may *not* be rational for a candidate to take a position at $\bar{l}$ if the distribution is skewed, either to the right or to the left.

Unimodal Distribution: Median and Mean Same

A sufficient condition for M and $\bar{l}$ to coincide is that the distribution be **symmetric**, but this condition is not necessary: M and $\bar{l}$ can coincide if a distribution is asymmetric, as illustrated in Figure 12.3a. When M and $\bar{l}$ coincide, a candidate need not take a different position to ensure victory–or at least prevent defeat if his or her opponent adopts the same position. However, as Figure 12.3 demonstrates, the noncoincidence of M and $\bar{l}$ is not necessarily related to the lack of symmetry in a distribution: Half the voters may still lie to the left, and half to the right, of $M/\bar{l}$ if the distribution is asymmetric.

What can we say about equilibrium positions if there is an even number of voters? For example, consider the following discrete distribution of $n = 26$ voters at $k = 8$ different positions over the interval [0, 1], which is illustrated in Figure 12.3b.

Position, i	1	2	3	4	5	6	7	8
Location (l_i) of position i	0	0.2	0.3	0.4	0.5	0.7	0.8	0.9
Number of voters (n_i) at position i	2	3	4	4	2	3	7	1

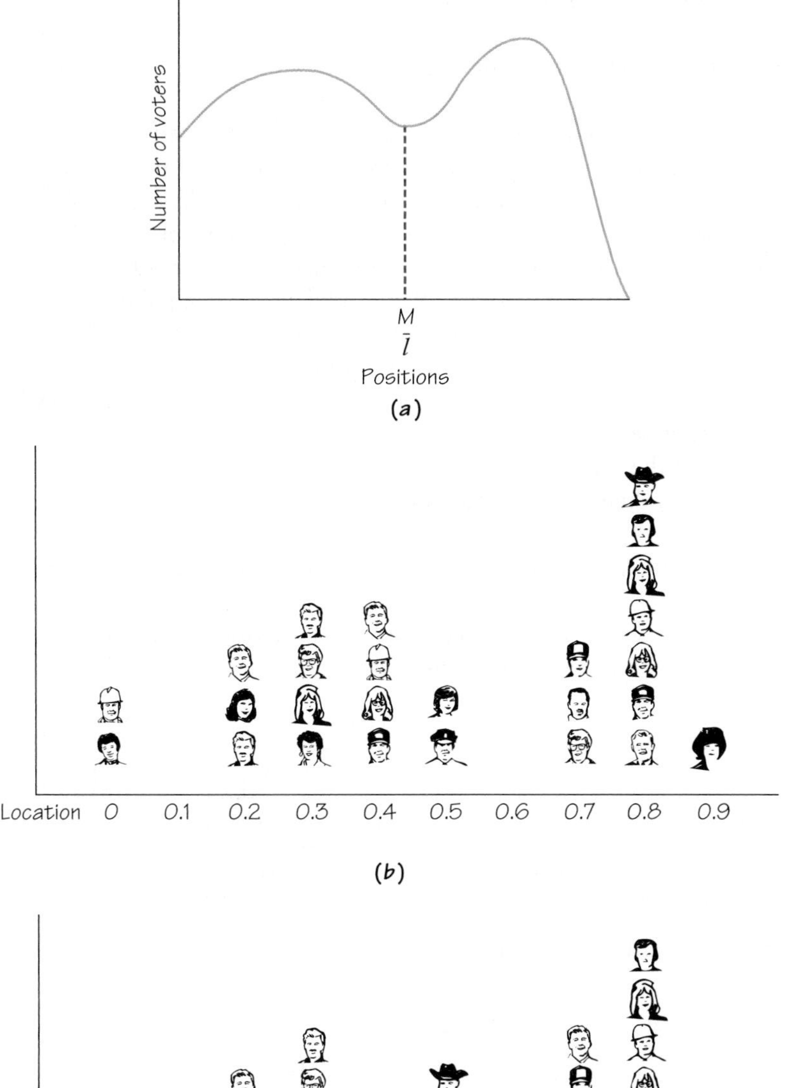

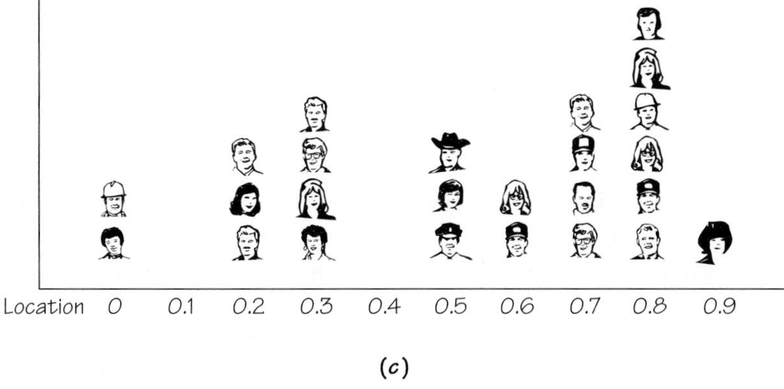

FIGURE 12.3 Unimodal distribution in which median and mean are the same. It is not necessary for a distribution to be symmetric for the median M and mean $\bar{l}$ to coincide.

We begin by calculating the mean $\bar{l}$:

$$\bar{l} = \left(\frac{1}{26}\right)[2(0) + 3(0.2) + 4(0.3) + 4(0.4) + 2(0.5) + 3(0.7) + 7(0.8) + 1(0.9)] = 0.5$$

The median M is not 0.5. For an even number of voters, as in this example, M is the average of the two middle positions; at this average, 13 voters lie to the left and 13 to the right. The two middle voters are the 13th and 14th voters when they are lined up in the order of their positions from left to right. The 13th voter is at position 0.4, and the 14th voter is at position 0.5, so M is 0.45. Thus, this discrete distribution does not mimic the continuous distribution in Figure 12.3a, in which the median and mean coincide.

Note that if both candidates position themselves at the median M, their positions will be in equilibrium, as we showed earlier. But it is not the unique equilibrium pair. In this example, there are many other pairs of equilibrium positions. For instance, the same reasoning shows that any pair of positions between 0.4 and 0.5 will be in equilibrium. Moreover, it is easy to show that the position 0.4 for one candidate, and the position 0.5 for the other candidate, are in equilibrium. The candidate at 0.4 will get the support of the 13 voters at or to his left, and the candidate at 0.5 will get the support of the 13 voters at or to her right. If either candidate takes a different position, either to the left of 0.4 or to the right of 0.5, he or she will not be assured of 13 votes.

In general, if the number of voters is even, and the two middle voters adopt different positions, then if the candidates adopt those positions or any pair of positions in between, then they will be in equilibrium.

It is possible that, for either an odd or even number of voters, there may not be a median position such that half the voters lie to the left and half to the right of this position. As a case in point, consider the following discrete distribution of $n = 25$ voters at $k = 8$ different positions over the interval [0, 1], which is illustrated in Figure 12.3c.

Position, i	1	2	3	4	5	6	7	8
Location (l_i) of position i	0	0.2	0.3	0.5	0.6	0.7	0.8	0.9
Number of voters (n_i) at position i	2	3	4	3	2	4	6	1

We begin by calculating the mean:

$$\bar{l} = \left(\frac{1}{25}\right)[2(0) + 3(0.2) + 4(0.3) + 3(0.5) + 2(0.6) + 4(0.7) + 6(0.8) + 1(0.9)] = 0.52$$

At 0.6, 12 voters lie to the left of 0.6 and 11 voters lie to the right. At 0.5 and 0.7, the imbalances on the left and the right are even more lopsided. Moreover, there is no position, including the mean $\bar{l} = 0.52$, such that exactly half the voters lie to the left and half lie to the right.

In the absence of such a median position, is there an equilibrium? It is easy to show that 0.6 is indeed the equilibrium for two candidates. It is somewhat more difficult to show that if a distribution is discrete and there is no median position, there is still a unique position for both candidates that is in equilibrium. We call this the **extended median** because it extends the median-voter theorem to the discrete case, in which there may be no median position.

Extended Median

DEFINITION

The **extended median** is the equilibrium position of two candidates in the discrete case when there is no median position.

Given the stability of the median or the extended median in a two-candidate, single-issue election, is it any wonder that candidates who want to win try to avoid extreme positions? As shown in Figures 12.2a and 12.3a, even when the greatest concentration of voters does not lie at M but instead at the mode, a candidate would be foolish to adopt this modal position. For although the right-leaning voters would be very pleased, the candidate's opponent would win the votes of a majority by sidling up to this position but staying just to the left.

Voters on the far left may not be particularly pleased to see both candidates situate themselves at M, which is nearer the mode in Figure 12.2a. But in a two-candidate race, they would have nobody else to turn to. Of course, if left-leaning voters felt sufficiently alienated by both candidates, they might decide not to vote at all, which has implications we explore further later.

EXAMPLE 1 ■ Location of Department Stores

There is a rather different application of the foregoing analysis to business, which in fact was the first substantive area to which spatial modeling was applied. Consider two competitive retail businesses, such as department stores, that consider locating their stores somewhere along the main street that runs through a city. Assume that, because transportation is costly, people will buy at the department store closer to them. Then the analysis says that no matter how the population is distributed along or near the main street, the best location is the median.

(*Andria Patino/Corbis.*)

Thus, if the city's population is symmetrically distributed–that is, not skewed toward one end or the other of the main street–then this location will, of course, be at the center of the main street. Indeed, clusters of similar stores are frequently bunched together near the center of many main streets, although these stores may not be particularly convenient to people who live far from the city's center. Consequently, their location seems not to be in the public interest. Wouldn't it be better to have some of the same kinds of stores near one end of the main street and some near the other, so no people are discriminated against? In an election, by contrast, not every voter can so easily be satisfied if only one candidate is to be elected, so the median seems the most attractive location in this context. ■

12.2 Spatial Models for Multicandidate Elections

Primary elections, in which candidates seek the nomination of one of the major parties, tend to attract more than two candidates. In presidential primaries, in particular, many candidates are likely to jump into the fray, especially in the states that go early in the season, if the incumbent president or vice president is not running (as was the case in 2008).

Under what conditions is entry into a multicandidate race attractive? If no positions offer a potential candidate any possibility of success, then it will not be rational for him or her to enter the primary in the first place. Therefore, the rationality of entering a race, and the rationality of the positions he or she might take once there, are really two aspects of the same decision.

Suppose that two candidates have already entered a primary, and they both take positions at M. Is there any room for a third candidate?

EXAMPLE 2

Entry of a Third Candidate in a Two-Candidate Race

Look at Figure 12.4, where A and B are both at M and therefore split the vote. Now if a third candidate, C, enters and takes a position on either side of M (say, to the right), the area under the distribution to C's right may encompass less than $\frac{1}{3}$ of the total area and still enable C to win a plurality of votes.

To show why this is so, consider the portion of the electorate's vote that A/B will receive and the portion that C will receive. If C's area (yellow) is greater than half of A/B's area (blue), C will win more votes than A or B, because C's area includes not only the votes to the right of his or her position but also some votes to the left. More precisely, C will attract voters up to the point midway between his or her position on the horizontal axis and that of A/B; A and B will split the votes to the left of this midway point. Because C picks up some votes to the left of his or her position, less than $\frac{1}{3}$ of the electorate may lie to the right and still enable C to win a plurality of more than $\frac{1}{3}$ of the total vote.

By similar reasoning, it is possible to show that a fourth candidate, D, could take a position to the left of A/B and further chip away at the total of the two centrists. Indeed, D could beat candidate C, as well as A and B, by moving closer to A/B from the left than C moves from the right.

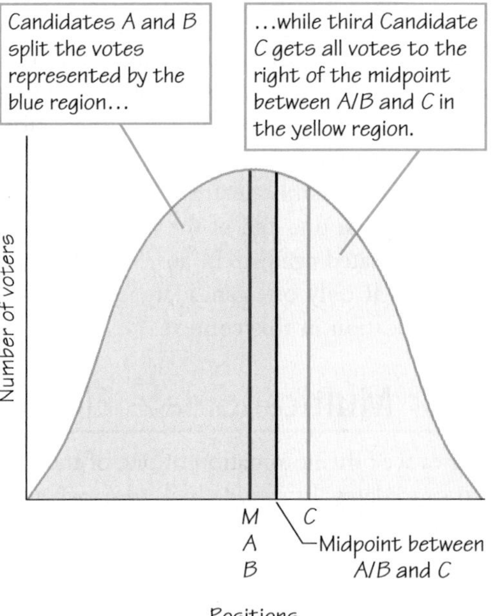
Candidates A and B split the votes represented by the blue region...
...while third Candidate C gets all votes to the right of the midpoint between A/B and C in the yellow region.
Number of voters
M
A
B
C
Midpoint between A/B and C
Positions

FIGURE 12.4 Unimodal distribution with three candidates. Candidate C can take a position with less than $\frac{1}{3}$ of the voters to his or her right and still win if candidates A and B at the median M and mean $\bar{l}$ split the remainder of the vote.

Clearly, M has little appeal, and in fact is quite vulnerable, to a third or fourth candidate contemplating a run against two centrists. Indeed, it is not difficult to show that *whatever* positions two candidates adopt–the same or different–at least one of these candidates will be vulnerable to a third candidate.

This is not to say, however, that a third Candidate C will necessarily win against *both* A and B. There are both obstacles and opportunities for C, which are summarized in Figure 12.5 (the reasoning behind these is explored in Exercises 18 and 19).

1/3-Separation Obstacle

DEFINITION

The **1/3-separation obstacle** occurs when there is little room in the middle, enabling *C* to beat *A* or *B* but, in so doing, causing him or her to lose to the other.

The 1/3-Separation Obstacle and the 2/3-Separation Opportunity

THEOREM

The 1/3-separation obstacle. If *A* and *B* are distinct positions that are equidistant from the median of a symmetric distribution and separated from each other by no more than $\frac{1}{3}$ of the area under the curve (so that no more than $\frac{1}{3}$ of the voters lie between *A* and *B*), *C* can take no position that will displace both *A* and *B* and enable *C* to win (see below).

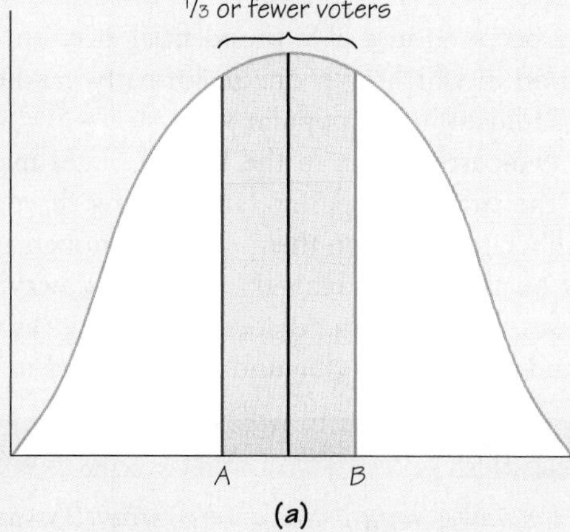

The 2/3-separation opportunity. If *A* and *B* are distinct positions that are equidistant from the median of a symmetric unimodal distribution and separated from each other by at least $\frac{2}{3}$ of the area under the curve (so that at least $\frac{2}{3}$ of the voters lies between *A* and *B*), *C* can defeat both *A* and *B* by taking a position at *M* (exactly between them, as shown below).

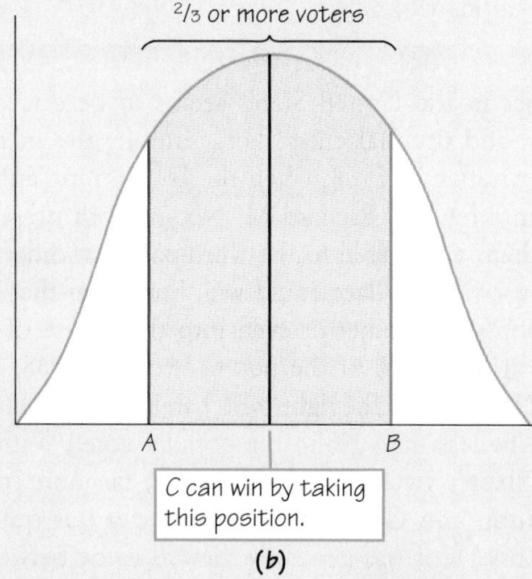

FIGURE 12.5 The obstacles and opportunities for a third candidate, *C*, to enter a race.

Theodore Roosevelt
(Hulton Getty/Liaison.)

This occurred in the 1912 presidential election, when Theodore Roosevelt ran as the Progressive ("Bull Moose") Party candidate after losing the Republican nomination to William Howard Taft. (Roosevelt had previously been president but had lost favor with his party after sitting out one term.) In the general election, Roosevelt received 27% of the popular vote and Taft, 24%. Both candidates were handily defeated by the Democratic candidate, Woodrow Wilson, who got 41% of the popular vote. There was a fourth candidate in this race, socialist Eugene V. Debs, but he received only 6% and was never a serious threat to Wilson on the left. Wilson was also the overwhelming winner in the Electoral College.

2/3-Separation Opportunity — DEFINITION

The **2/3-separation opportunity** occurs when there is a wide separation between A and B, giving enough room in the middle for C to win.

This event has never occurred in a U.S. presidential election. In fact, the 1912 election is the only election in which even one major-party candidate has been defeated by a third-party candidate in the popular vote.

The stability of the two-party system in the United States may be partially explained by the fact that the two major parties, anticipating the possible entry of a third-party candidate, deliberately position themselves far enough away from the median to discourage entry on the left or right–but not so far away as to make entry in the middle advantageous. The following theoretical result gives some insight into how this can be done (and the reasoning behind it is explored in Exercise 20).

Optimal Entry — THEOREM

The optimal entry of two candidates, anticipating a third entrant. Assume A and B are the first candidates to enter an election and anticipate the later entry of C. Assume that the distribution of voters is uniform (rectangular) over $[0, 1]$. Then the optimal positions of A and B are to enter at $\frac{1}{4}$ and $\frac{3}{4}$, whereby $\frac{1}{4}$ of the voters lie to the left of one candidate (say, A) and $\frac{1}{4}$ to the right of the other (B). Then C can do no better than win 25% of the vote: He or she will be indifferent among entering just to the left of A, just to the right of B, or at any position in between. At none of these positions will C win.

Presidential politics in the United States seems to be a reflection of both the median-voter theorem and optimal entry. For example, the median-voter theorem seems to have been operative in 1968, when the Democratic and Republican nominees, Hubert H. Humphrey and Richard M. Nixon, both presented themselves as centrists. This made them vulnerable to the third-party candidate that year, George Wallace–not in the sense that Wallace could win, but rather that he could throw the election in one direction or the other or even into the House of Representatives. In fact, while Wallace won only 14% of the popular vote in 1968, he attracted mostly supporters of Richard N. Nixon on the right, who barely defeated Hubert H. Humphrey that year. (Nixon won by less than 1% in the popular vote.) Without Wallace in the race, polls show that Nixon's victory would have been far more substantial.

In 1992, Bill Clinton and George Bush were viewed as quite far apart on the left–right spectrum. Ross Perot was generally viewed to be between Clinton on the left and Bush on the right, leaving considerable room in the middle that Perot could better exploit than by trying to displace one of the major-party nominees on the left or right. In winning 19% of the popular vote, Perot drew almost equally from each candidate. However, he did not come close to winning, which the optimal entry of the candidates makes difficult. (Clinton won decisively, with 43% of the popular vote to Bush's 38%.)

12.3 Narrowing the Field

Up to now, we have looked at the spatial game that candidates play as they vie to position themselves optimally in two-candidate and multicandidate races so as to (1) maximize their vote totals or (2) deter new candidates from entering. We will continue to assume that candidates take positions along a left–right spectrum, but now we consider the game from the point of view of the voters. More specifically, we ask the following question in a multicandidate race: When one candidate drops out, perhaps because of performing below expectations in an early primary, to whom will the dropout's supporters shift their votes?

Beating expectations is the name of the game in the early primaries. Thus, when Senator Edmund S. Muskie from Maine ran for the Democratic Party nomination in 1972, he was expected to do well in the neighboring state of New Hampshire. To "win" in this first primary, he had to exceed these expectations, whereas other candidates, who were not expected to do so well, could afford more mediocre performances. As in a horse race, to which primary elections are often compared, the contenders are handicapped; that is, they must beat expectations about their performances if they are to gain momentum.

Momentum, or what George H. W. Bush (the father) called the "Big Mo," can start a **bandwagon**, or a presumption that a candidate will win.

Bandwagon Effect DEFINITION

The **bandwagon effect** induces voters to vote for the presumed winner, independent of his or her merit.

Assume three candidates take positions, from left to right, as follows: *A–B–C*. Clearly, if *A* or *C* drops out, their supporters mostly likely will switch to *B*, giving the centrist a boost. But what if *B* is the first to drop out? Then it is unclear whether *A*, *C*, or neither will benefit–it depends on the number of *B*'s supporters who prefer *A* next, *C* next, or neither (and hence may not vote at all). In any event, with *B* out of the race, the winner must be one of the candidates on the extremes.

The possibilities become more interesting when there are four candidates arrayed from left to right as follows: *A–B–C–D*. If one of the extremists, *A* or *D*, drops out, then one of the two centrists, *B* or *C*, will benefit. But what if a centrist, say *C*, drops out? Does this benefit one of the extremists, or does the other centrist (*B*) benefit? At first glance, one might think that, with only one centrist remaining, he or she will surely benefit.

This will not be the case, however, if most *C* supporters prefer *D* to *B*, which is certainly possible. Then the extremist *D* will benefit, which will be most upsetting to *A*'s supporters. Conceivably, *A*'s supporters might encourage *A* to withdraw so they can throw their support to *B*, whom they definitely prefer to *D*.

Does this sound implausible? Think back to the 2000 election, in which our four hypothetical candidates are replaced by the following ordering from left to right: Nader–Gore–Bush–Buchanan. (Ralph Nader was the Green Party candidate on the left and Pat Buchanan the Reform Party candidate on the right.) Just before the election, the polls were showing that Buchanan was not much of a threat to Bush, but Nader–who ended up with 2.7% of the popular vote nationwide (Buchanan got only 0.4%)–was definitely a threat to Gore. Despite pleas from some of his supporters, Nader refused to withdraw and, consequently, gave Bush a victory in Florida and a few other close states that won him the presidency.

This 2000 scenario is not the same as the previous four-candidate hypothetical scenario, in which we argued that the extremist on the right, *D*, might win if one of the centrists, *C*, dropped out. In the 2000 scenario, the extremist on the left, Nader, could have dropped out to "save" the centrist closer to him, Gore.

Unfortunately for Gore, Nader not only refused to make this sacrifice but contended afterward that Gore's loss was due to Gore's own poor performance, not Nader's presence in the race. We will return to this issue when we discuss the effects of approval voting and the abolition of the Electoral College as possible remedies to the so-called **spoiler** problem.

Spoiler DEFINITION

A **spoiler** is a candidate who cannot win but "spoils" the election for a candidate who otherwise would win.

12.4 What Drives Candidates Out?

So far we have considered the possibility that candidates drop out, but not why they do so. Presumably, they do so because of their poor performance in the polls or early primaries, where performance depends in part on expectations of how well they will do. But expectations change over time, and this change in turn affects how voters perceive a race–in particular, who is ahead and who is behind. On this basis, voters choose voting strategies that are likely to benefit their favorites.

Polls make public the standing of candidates in a race, as do the returns from presidential primaries. To be sure, the electorate is different in each primary state, whereas a poll is a sample from the entire electorate. While polls and primaries both provide a glimpse of the "state of the electorate," here we will focus on the change of polls over time. But our results are just as applicable to primaries, whose winnowing-out effects are evident in almost every election in which an incumbent is not running for reelection.

Suppose the election procedure is plurality voting.

Plurality Voting PROCEDURE

Plurality voting is a voting procedure in which each voter votes for one candidate, and the candidate with the most votes wins.

Assume that voters rank candidates. For example, $A\ B\ C$ indicates that a voter prefers A to B to C. Before a poll, we assume that each voter votes **sincerely**–for his or her favorite candidate–because, in the absence of poll information, there is no reason to do otherwise.

EXAMPLE 3 ▪ Poll for Three Candidates

Nine voters, who can be divided into three classes, have the following preferences for candidates *A*, *B*, and *C*:

I. 4: *A C B*
II. 3: *B C A*
III. 2: *C A B*

Because the four voters in class I prefer *A* to *C* to *B*, the poll would indicate *A* to have 4 votes (44%), whereas *B* and *C* would have 3 and 2 votes (33% and 22%), respectively, if voters vote sincerely.

After the poll, we make the following assumption:

Poll Assumption RULE

Voters adjust, if necessary, their sincere voting strategies to differentiate between the top two candidates revealed in the poll, voting for the one they prefer.

Since class I and class II voters chose one of the top two candidates, only the two class III voters, who voted for *C*, would change their votes. Because they prefer *A* to *B*, it would be in their interest to vote insincerely for *A* (instead of *C*), thereby distinguishing *A* from the other top candidate, *B*, by giving *A* their votes. This would result in *A*'s winning with 6 votes, *B*'s getting 3 votes, and *C*'s getting no votes.

Paradoxically, it is *C* who is the Condorcet winner.

Condorcet Winner DEFINITION

A **Condorcet winner** is a candidate who can defeat each of the other candidates in pairwise contests.

C is preferred to *A* by class II and III voters (5 to 4) and to *B* by class I and III voters (6 to 3). Hence, if there were a series of pairwise contests (in any order), *C* would win. Yet the poll not only does not make *C* victorious but instead magnifies *A*'s plurality victory (4 votes) by inducing *C*'s supporters, thinking that their candidate is out of the running, to throw their support to *A*, giving *A* a $\frac{2}{3}$-majority victory (6 out of the 9 votes). ■

This example can be generalized to yield the following:

Condorcet Winner Unsuccessful THEOREM

Given the poll assumption, a Condorcet winner will always lose if he or she is not one of the top two candidates identified by the poll.

Why this is true is considered in Exercise 28. Here we note that even if *C* were given serious consideration in a tight race, *A* would still win with a plurality of votes.

One might think that C's problem is due solely to the fact that the poll assumption puts him or her out of the running by presuming that his or her supporters "jump ship"–desert C for the apparently more viable candidates, A and B. But, surprisingly, even when a Condorcet winner is on top before a poll that distinguishes the top three candidates (instead of the top two), the Condorcet winner may be hurt by the poll after strategic adjustments are made by the voters, as we next illustrate.

EXAMPLE 4 ▪ Poll for Four Candidates

Add a fourth candidate, D, to the three in the preceding example, and assume that there are 12 voters with the following preferences for the four candidates:

I. 3: $A\ C\ B\ D$
II. 3: $B\ C\ A\ D$
III. 4: $C\ A\ B\ D$
IV. 2: $D\ A\ B\ C$

After the poll establishes that A, B, and C are the top three candidates, the two class IV voters would be motivated to switch to their second choice, A. A would thereby increase his or her total from 3 to 5 votes–and win after the poll. Yet C is again the Condorcet winner. In staying the same at 4 votes, C is hurt, relative to A, by the poll. In fact, C would lose to A after the poll (because D's supporters would continue to vote for A), even though C was the winner before the poll.

Thus, the poll assumption, which induces strategic adjustments that favor the top two candidates, may hurt the Condorcet winner when a larger number of candidates (possibly including the Condorcet winner) are considered to be contenders. However, when one of the top two contenders distinguished by the poll assumption is the Condorcet winner, we have the following result:

Condorcet Winner Successful THEOREM

Given the poll assumption, a Condorcet winner will always win if he or she is one of the top two candidates identified by the poll.

Why this is true is considered in Exercise 29. Here we note that a Condorcet winner need not be the winner in the poll but instead can place second and be successful. If a Condorcet winner places second, and the poll has the effect of turning him or her into a majority winner, then it is proper to say that the poll is instrumental in electing this candidate.

We conclude that a poll may either hurt or help a Condorcet winner. If the poll assumption is modified to distinguish more than two candidates, a Condorcet winner may be hurt *even if he or she is among those distinguished by the poll*, as the previous example showed.

12.5 Election Reform: Approval Voting

The furor caused by the divided outcome in the 2000 presidential election, in which George W. Bush won the electoral vote and Al Gore won the popular vote, has spurred efforts for reform of the election system. But except for calls for abolition of the Electoral College, whose effects we will analyze in the next section, most of the discussion has centered on making balloting more accurate and reliable and eliminating election irregularities, especially those that discriminate among different classes of voters.

Unfortunately, such reforms ignore a fundamental problem that plagues *multi-candidate elections*–elections with three or more candidates–namely, that the candidate who wins under plurality voting may not be a Condorcet winner. Indeed, there will be no Condorcet winner if each candidate can be beaten by at least one other candidate. In such a situation, who is the rightful winner?

Chapter 9 presents alternatives to plurality voting. Most of these alternatives allow a voter to rank candidates from best to worst; analysts have investigated their ability to elect a Condorcet winner if one exists. Here, however, we will examine in depth a simple election reform, approval voting, that does not require voters to rank candidates.

Approval Voting — DEFINITION

Under **approval voting**, voters can vote for as many candidates as they like or find acceptable. Each candidate approved of receives one vote, and the candidate with the most approval votes wins the election.

What if approval voting had been used in the 2000 presidential election? Nader supporters, knowing that voting for Nader would be only a protest vote because Nader had no chance of winning, might also have voted for Gore. In fact, because polls show that Gore was the second choice of most Nader voters, Gore almost certainly would have won in Florida if there had been approval voting.

To be sure, Bush would have benefited from the approval votes of Buchanan supporters, but the number of these votes would not have come close to matching the number of votes Gore would have received from Nader supporters. There is therefore little doubt that Gore was the Condorcet winner in this election, because polls show that he could have defeated Bush (with help from Nader voters) as well as each of his less popular opponents in pairwise contests.

Arguments for an election reform like approval voting, however, should not be based on the outcome of only one election. Moreover, even in this one election, we cannot be entirely sure that Gore would have won under approval voting, because the nature of the campaign almost surely would have changed if this reform had been in use.

For example, John McCain, the Republican senator from Arizona who defeated George Bush in the New Hampshire primary but ultimately lost the Republican nomination to him, might have run as an independent candidate if there had been approval voting. As a centrist, he would have been attractive to both Democrats and Republicans and, conceivably, could have won under approval voting. But even if McCain had not run, it is likely that the candidates would have pitched their campaign appeals somewhat differently to try to attract as much approval as possible, especially from Nader and Buchanan supporters.

Although we cannot make precise predictions of the effects of approval voting in a presidential election like that of 2000, we can say what voters will do in certain *types* of situations:

Voting Only for a Second Choice — THEOREM

In a three-candidate election under approval voting, it is never rational for a voter to vote only for a second choice. If a voter finds a second choice acceptable, he or she should also vote for a first choice.

This is certainly not true under plurality voting. If you were a Nader supporter *and* found Gore also acceptable as a second choice, you should have voted for Gore rather than Nader if (1) you thought Gore could win but Nader could not and (2) electing an acceptable candidate was important to you. (Indeed, some Nader supporters switched to Gore for these reasons.) By comparison, because you lose nothing by voting for *both* Nader and Gore under approval voting–and may gain by doing so (if Nader cannot win, you at least help Gore)–you should never vote for just your second choice (Gore in this example).

Why voting only for a second choice is never rational is considered in Exercise 35. In effect, a voter whose favorite candidate in a three-candidate race seems to be out of the running can have his or her cake and eat it too, by casting a *sincere* vote for a favorite candidate and a *strategic* vote for a second choice (to try to prevent a worst choice from winning). Roughly speaking, **strategic voting** (for instance, by Nader supporters for Gore in the 2000 presidential election) is voting that is not sincere but nevertheless has a strategic purpose–to elect an acceptable candidate if one's first choice is not viable.

Another general result about approval voting that is helpful to know uses the concept of dichotomous preferences:

Dichotomous Preferences — DEFINITION

A voter has **dichotomous preferences** if he or she divides the set of candidates into two subsets–a preferred subset and a nonpreferred subset–and is indifferent among all candidates in each subset.

In other words, a dichotomous voter sees the world in two colors, white and black, and there is nothing in between. True, most of us see grays, but it is useful to analyze the dichotomous case. In this case, voters have a **dominant strategy**, which is a strategy that is at least as good as, and sometimes better than, any other strategy they might choose. With this definition, we can show a general condition under which Condorcet winners will be elected under approval voting:

Effect of Dichotomous Preferences — THEOREM

A Condorcet winner will always be elected under approval voting if all voters have dichotomous preferences and choose their dominant strategies.

A dichotomous voter's dominant strategy is to vote for all candidates in his or her preferred subset and no others. This strategy is dominant because the preferred candidates are all assumed to be equally good, so a voter has no reason to distinguish among them. Furthermore, the voter has no reason to vote for a nonpreferred candidate or candidates because they are all equally bad and voting for any one of them could help that candidate win. As an illustration of the effect of dichotomous preferences, consider the following example.

EXAMPLE 5 ■ Dichotomous Preferences

For each class of voters, the preferred subset of candidates is enclosed in the first set of parentheses, the nonpreferred subset in the second set of parentheses. Thus, the four class I voters prefer A and B, between whom they are indifferent, to C and D, between whom they are also indifferent:

I. 4: $(A\ B)\ (C\ D)$
II. 3: $(C)\ (A\ B\ D)$
III. 2: $(B\ C\ D)\ (A)$

Assuming that each class of voters chooses its dominant strategy, B wins with 6 votes to 5 votes for C and 4 votes for A.

In pairwise contests, notice that B is preferred to A by the two class III voters (class I and II voters are indifferent between these two candidates), so B would defeat A by 2 to 0 votes. (We assume that indifferent voters express no preference.) Because B is preferred to C by the four class I voters, and C is preferred to B by the three class II voters (class III voters are indifferent between these two candidates), B would defeat C by 4 to 3 votes. Thus, B, the approval-vote winner, is also the Condorcet winner, which must always be the case when voter preferences are dichotomous and voters vote for all their approved candidates.

Insofar as voters in the 2000 presidential election thought equally well (or badly) of Bush and Buchanan on the one hand, and Gore and Nader on the other, they would have preferences like those of class I voters in the previous example. Of course, most voters probably made finer distinctions, which is allowed by "range voting" (see Suggested Web Sites). In this case, there is no guarantee that approval voting will elect a Condorcet winner.

12.6 The Electoral College

As we have noted, the Electoral College had a decisive effect in the 2000 presidential election. In winning the popular vote in Florida by the slimmest of margins, George Bush captured all 25 of Florida's electoral votes, which gave him a majority in the Electoral College. This won him the presidency even though he lost the popular vote.

What is the justification for the Electoral College? Its original purpose was to place the selection of a president in the hands of a body that, while its members would be chosen by the people, would be sufficiently removed from them that it could make more deliberative choices. As for its composition, each state gets 2 electoral votes for its two senators (total for all states: 100). In addition, a state receives 1 electoral vote for each of its representatives in the House of Representatives, whose numbers are based on population (see Chapter 11) and range from 1 representative for the seven smallest states to 53 representatives for the largest state, California. The House has a total of 435 representatives. The District of Columbia, like the smallest states, is given 3 electoral votes. Altogether, there are 538 electoral votes, and a candidate needs 270 to win. In 2000, George W. Bush got 271 electoral votes.

Although there is nothing in the U.S. Constitution mandating that the popular-vote winner in a state receive all its electoral votes, this has been the tradition almost

CHAPTER 12 EXERCISES

■ Challenge ◆ Discussion

12.1 Spatial Models for Two-Candidate Elections

1. Why do M and $\bar{l}$ not coincide if a distribution is skewed?

2. Show that 0.6 is the equilibrium in Figure 12.3b.

■ 3. Prove that if a distribution is discrete and there is no median position, there is always an *extended median*. (*Hint:* Show that there is always one position at which a majority of voters lies neither to the left nor to the right, and neither candidate would have an incentive to depart from this position.)

4. Assume that the one voter at 0.1 in Figure 12.2b decides not to vote because he is "too far away" from the two candidates who take the median position at 0.6. Would either candidate depart from $M = 0.6$ to try to do better if he or she knew that this voter had decided not to vote–but he or she would vote for the closer of two candidates less than a distance of 0.5 away? What if the candidates knew that the three voters at 0.2 had also decided not to vote–but they would vote for the closer of two candidates less than a distance of 0.4 away?

5. If you are a far-left or a far-right voter, are you helping your cause when you announce, like the voters in Exercise 4, that you will not support candidates who are too far away?

6. Consider the two most extreme voters at 0 in Figure 12.3b (those who are farthest from the extended median of 0.6). Would their nonvoting change the extended median? How about, as well, the nonvoting of the somewhat less extreme voter at 0.9? Show when, if at all, M or the extended median will change as fewer and fewer extreme voters decide not to vote in this example?

◆ 7. In Figure 12.2a, $\bar{l}$ is not in equilibrium–one candidate would do better if he or she moved from 0.56 to 0.6. But is 0.6 really a better reflection of the views of the electorate than 0.56?

8. Define an outcome to be in equilibrium if, given that one candidate chooses it, the other candidate cannot do better than take the same position. Show that this definition is equivalent to the text's definition of being in equilibrium.

■ 9. Consider a trimodal distribution (three peaks). When will taking a position at the middle peak be in equilibrium? Is it possible that one of the other peaks can ever be in equilibrium? If so, give a discrete-distribution example.

10. Define A's position in a two-candidate race to be *opposition-optimal* if, given that the position of B is fixed, it maximizes A's vote total. Show that A's opposition-optimal position must be adjacent to B's position and closer to M, except when B is at the median. (Roughly speaking, being "adjacent" means being a very small distance away.)

◆ 11. Assume the population along a main street is uniformly distributed over [0, 1], so there are equal numbers of people located at all equally spaced intervals from $M/\bar{l}$. (This makes the distribution rectangular, or "flat.") It has been argued that the "social optimum" for the location of two stores are at the points $\frac{1}{4}$ and $\frac{3}{4}$, because then no person would have to travel more than $\frac{1}{4}$ of the length of the street to buy at one store. Is this desirable if the population is not uniformly distributed?

◆ 12. What is a social optimum in an election if only one candidate is to be elected? How about five candidates to a city council? Is it better that the city council members' positions all be centered around 0.5, or should they be more spread out?

◆ 13. Which is better for consumers: (a) to minimize the maximum distance they must travel to a store; or (b) to foster price competition, which would presumably be encouraged if two stores are located at $M = \frac{1}{2}$?

■ 14. Assume a city comprises three equal-sized districts, each of which elects a candidate to the city council. The mayor is elected by the entire city. Show with an example that the median or extended median for the mayor need not be the median or extended median for any of the three city council districts. Does this explain why mayors and city council members often disagree?

■ 15. In Exercise 14, must the median or extended median for the mayor be between the leftmost and rightmost medians, or extended medians, of the three districts? How about the mean $\bar{l}$?

12.2 Spatial Models for Multicandidate Elections

16. Assume that A and B take the *same* nonmedian position. What position should C take to maximize his or her vote total? Is C's position always a winning one?

■ 17. Assume that A and B take *different* positions, with one possibly being at M. What position should C take to maximize his or her vote total? Is C's position always a winning one?

18. Is there a 1/3-separation obstacle if the distribution is not symmetric but no more than $\frac{1}{6}$ of the area under the curve separates *A* (on the left) from *M*, and no more than $\frac{1}{6}$ of the area separates *B* (on the right) from *M*? What if these $\frac{1}{6}$-or-less areas on the left and the right are not the same?

19. Is there a 2/3-separation opportunity if the distribution is not unimodal but at least $\frac{1}{3}$ of the area under the curve separates *A* (on the left) from *M*, and at least $\frac{1}{3}$ separates *B* (on the right) from *M*? (*Hint:* Start by assuming that the distribution is uniform between *A* and *B*–and hence not unimodal–and that exactly $\frac{2}{3}$ of the voters lie between *A* and *B*. Can *C* always win by taking a position at *M*? If not, is there a distribution that affords *C* this opportunity?)

20. Show that *C* cannot win under the conditions for the optimal entry of two candidates, anticipating a third entrant (page 382). (*Hint*: Indicate which candidate will win when *C* enters to the left of *A*, to the right of *B*, or in between.)

21. It is known that *A*, *B*, and *C* will enter an election in that order, with *A* announcing his position first, then *B*, and finally *C*. If the distribution is uniform over [0, 1], what position should each candidate take to maximize his or her vote total, anticipating–in the case of *A* and *B*–the entry of future candidates? [*Hint:* Start by assuming that *A* takes a position at $\frac{1}{4}$. Is *B*'s position at $\frac{3}{4}$ optimal, anticipating the entry of *C*? Or can *B* do better at some other position (perhaps by influencing *C*'s choice of a maximizing position)?]

■ **22.** If *A* and *B* are equidistant from the median of a symmetric distribution and separated from each other by exactly $\frac{1}{2}$ of the area under the curve, under what conditions is this separation an obstacle and under what conditions is it an opportunity? (*Hint:* Start by constructing examples of symmetric distributions in which *C* would either win or lose by taking a position at *M*.)

■ **23.** What are the vote-maximizing positions for four candidates to take if it is known that they will enter in the order *A*, *B*, *C*, *D*?

12.3 Narrowing the Field

24. Assume that the four candidates in the 2000 presidential election can be arrayed from left to right as follows: Nader–Gore–Bush–Buchanan. Suppose a poll reveals Gore at 48%, Bush at 47%, Nader at 3%, and Buchanan at 2%. Would Bush be well advised to offer Buchanan a cabinet position to drop out of the race (as Adams offered Clay the secretary-of-state post after the 1824 election)? What if Bush knew that, after Buchanan dropped out, only half of Buchanan's supporters would switch to him, with most of the remainder not voting, except for a few who would switch to Gore?

25. Assuming the same poll results as in Exercise 24, now suppose that Gore offered the same deal to Nader, knowing that only one-third of Nader supporters would switch to him and the rest would not vote. However, suppose Gore also thought that if Nader dropped out, so would Buchanan, and all Buchanan supporters would vote for Bush. Should Gore set off this train of events?

◆ **26.** Is there any evidence that the four presidential candidates in 2000 might have contemplated "deals" of the kind indicated in Exercises 24 and 25? If you cannot find any evidence, do you think this is because the candidates found such ploys unethical or because they thought they might be found out if they tried to engage in them?

◆ **27.** One tactic that was considered by Nader supporters who thought that their votes for Nader might kill Gore's chances in some states was to swap votes: In close states that Gore might lose if Nader supporters stuck with their candidate, these supporters would switch to Gore if Gore supporters in less contested states, where Gore would almost surely win, would switch to Nader. Thereby the popular-vote totals for the two candidates would not change overall, but Gore would be able to win in the close states he might otherwise lose. Is this a sensible way of dealing with problems created by the Electoral College, which puts a premium on winning in large states?

12.4 What Drives Candidates Out?

28. Show why the Condorcet-winner-unsuccessful result (page 385) is true. Is it true that if the poll assumption was modified to differentiate the top three (rather than the top two) candidates from the rest, and the Condorcet winner was not among the top three, that he or she would still lose?

◆ **29.** Show why the Condorcet-winner-successful result (page 386) is true. Is it proper that the candidate who comes in second in the poll should win after the results of the poll are announced? Why?

◆ **30.** In Example 4 (page 386), after class IV voters switch from *D* to *A*, the vote totals for the top three candidates are *A*–5, *B*–3, and *C*–4. Now assume a second poll is taken, differentiating *A* and *C*, the top two contenders, from *B*. If *B* supporters switch at this point to their second choice, which candidate will win? Do you consider this a desirable outcome?

◆ **31.** Assume there are four classes of voters that rank four candidates as follows:

I. 4: *A D B C*

II. 3: *B D A C*

III. 2: *C D B A*

IV. 1: *D C B A*

Which candidate is the Condorcet winner? Do you find this result strange in the light of what a poll would tell the voters?

32. Assume there is a poll that differentiates the top two candidates in Exercise 31. Which candidate will win the election after the poll?

◆ **33.** Assume there is a poll that differentiates the top three candidates in Exercise 31. Which candidate will win the election after the poll? Comment on the different outcomes in this exercise and the previous one.

◆ **34.** Assume there are three classes of voters who rank three candidates as follows:

I. 4: *A B C*

II. 3: *B C A*

III. 2: *C A B*

Show that there is no Condorcet winner. Applying the poll assumption to this example, which candidate will win? Is this fair, given the preferences of class II and III voters, who, together, are a majority?

12.5 Election Reform: Approval Voting

■ **35.** Prove the voting-only-for-a-second-choice result (page 388).

36. In a three-candidate election, show that your strategy of voting for your top two choices under approval voting is not always better than voting only for your top choice.

37. Consider a four-candidate election under approval voting. Is there ever a situation in which a voter would vote for a first and a third choice without also voting for a second choice? [*Hint:* Assume a voter ranks the four candidates *A B C D* and believes that one of two things can happen: The electorate will favor either liberals (say, *A* and *B*) or conservatives (say, *C* and *D*) but never favor each side equally.]

38. Is there ever a situation under approval voting in which a voter would vote for a worst choice?

39. Is there ever a situation under approval voting in which a voter would *not* vote for a first choice if he or she finds acceptable one or more lower-ranked candidates? (*Note:* This question asks whether the voting-only-for-a-second-choice result can be generalized to more than three candidates.)

■ **40.** Prove the effect-of-dichotomous-preferences result (page 388). (*Hint:* If all voters have dichotomous preferences and vote for all candidates in their preferred subsets, which candidate will get the most approval votes? What does this say about the preferences of voters for the approval-vote winner, compared to their preferences for each of the other candidates?)

41. In the following example, class I and II voters have dichotomous preferences, but the class III voter has *trichotomous preferences* (he or she divides the four candidates into three indifference subsets):

I. 2: (*A B*) (*C D*)

II. 2: (*C*) (*A B D*)

III. 1: (*D*) (*C*) (*A B*)

Is it rational for the class III voter to vote only for his or her top choice, *D*? If not, who else should he or she approve of? Which class of voters will be most unhappy if the class III voter does not vote just for *D*? Can voters in this class, by voting strategically, do anything about their situation?

◆ **42.** Assume the class III voter's preferences change to a different trichotomous ordering:

III. 1: (*D*) (*A C*) (*B*)

Suppose, as in Exercise 41, that the class III voter indicates in an initial poll that he or she intends to vote only for *D* but then, in response to the poll, switches to voting for the candidates in his or her second-choice subset as well. If there is a new poll, based on these results, what will be the outcome? What if there is a third poll, fourth poll, and so on? Do you regard this result as desirable? Why?

43. In Exercise 31, we saw that under plurality voting the Condorcet winner, *D*, comes in fourth in a poll and, therefore, cannot be helped by subsequent polling, even when the poll distinguishes the top three candidates and voters differentiate among them:

I. 4: *A D B C*

II. 3: *B D A C*

III. 2: *C D B A*

IV. 1: *D C B A*

What are the outcomes under approval voting–both with and without polling–if voters approve of their (i) top-ranked, (ii) two top-ranked, and (iii) three top-ranked candidates initially? [*Note:* In making adjustments to the poll results, assume that voters approve not only of the preferred of their two top-ranked candidates identified by the poll but also of *all* candidates ranked above their preferred candidate. For example, when the poll based on (i) above identifies *A* and *B* as the two top-ranked candidates, with 4 and 3 votes, respectively, the class III and class IV voters after the poll will approve not only of their preferred candidate, *B*, but also of *C* and *D*, because they rank the latter two candidates above *B*.]

◆ **44.** On the basis of your answers to the foregoing problems, do you think approval voting would be beneficial in finding Condorcet winners–either with or without polling–in multicandidate elections?

12.6 The Electoral College

45. Assume there are three states with 3, 7, and 9 voters, and that they are all toss-up states. If both the Democratic and Republican candidates choose strategies that maximize their expected popular vote (the proportional rule), and they have the same total resources ($D = R$), what is the expected number of votes that each will receive?

46. Assume the Republican knows in advance what allocations, d_i, to each state i the Democrat will make in Exercise 45. Then the Republican's optimal response can be shown to be

$$r_i = \frac{\sqrt{n_i d_i}}{\sum_{i=1}^{t} \sqrt{n_i d_i}} (R + D) - d_i$$

Suppose that the Democrat ignores the smallest state and makes proportional allocations to the two largest states. (For concreteness, assume both candidates have 100 units of resources.) What is the Republican's optimal response? What if the Democrat makes proportional allocations to all three states?

■ **47.** In Exercise 46, show that if the Democrat makes proportional allocations, and the Republican responds optimally according to the formula given there, this formula simplifies to $r_i = (n_i/N)R$, which does not depend on d_i. What does this say about the proportional rule? [*Hint:* If the Republican finds out (say, through a spy) that the Democrat is making proportional allocations, does the Democrat have anything to worry about?]

48. In Exercise 47, if the Democrat has only half the resources of the Republican, would you recommend that he or she behave differently from proportional allocations to maximize his or her expected popular vote? Why? If the Republican allocates his or her resources proportionally to the three states, is there any way the Democrat can allocate his or her resources to win a majority of votes in states with more than half the votes?

49. Instead of maximizing their expected popular vote, assume the candidates in Exercise 45 want to win in states with more than half the votes. Suppose the candidate who allocates more resources to a state wins that state. Is there any state to which a candidate should not consider allocating resources? Should the states that receive allocations receive equal allocations?

50. In Exercise 45, assume you can choose specific voters in each state to whom you can allocate resources. Suppose the candidate who allocates more resources to a voter wins that voter's vote. If your goal is to win the votes of a majority of voters in states that have more than half the votes, which voters would you target, and how much would you spend on each? (*Hint:* First show which states you would target; then show that these states should receive equal allocations, which in turn should be divided equally among a certain set of voters.)

51. Assume there are three toss-up states, A, B, and C, with, respectively, 2, 3, and 4 voters, which are also the number of electoral votes of each state. In the text, we gave the formulas for the probabilities, P_A, P_B, and P_C, that the Democrat wins a majority of popular votes in each state and, therefore, wins all the electoral votes of that state. Show that the formula for the probability that the Democrat *wins the election* under the Electoral College, PWE_D, is

$$PWE_D = P_A P_B(1 - P_C) + P_A P_C(1 - P_B) + P_B P_C(1 - P_A) + P_A P_B P_C$$

(*Hint:* Winning in any two states is sufficient to win the election.)

■ **52.** Compare the formula for PWE_D with the formula for EEV_D (in the text). Which quantity is it better to maximize? What would be a good resource-allocation strategy for maximizing PWE_D?

53. Is the square root in the formulas for the EEV maximizing strategies of the Democratic and Republican candidates related to the square-root rule for the Electoral College discussed in Chapter 11?

WRITING PROJECTS

1. Do you think polling is useful in helping voters choose the "best" candidate? Or would it be better, as in some countries, to ban the publication of polls before an election? In one to two pages, discuss these questions in light of the theoretical effects polling has when voters react to polls and possibly change their voting strategies. Is there empirical evidence that voters behave in this way?

2. How serious a problem do you think the large-state bias of the Electoral College is? How would you explain the fact that some of the strongest advocates of the Electoral College come from small states? Has the theoretical bias been a reality in the campaign behavior of candidates in recent presidential elections? Discuss in one to two pages.

SUGGESTED READINGS

BRAMS, STEVEN J. *The Presidential Election Game,* 2nd ed., A K Peters, Wellesley, Mass., 2008. Focuses on the strategic aspects of presidential elections–from primaries to conventions to general elections–and also includes an analysis of the "game" played between President Richard Nixon and the Supreme Court over the release of the Watergate tapes that led to Nixon's resignation in 1974 (Nixon has been the only president to resign the presidency). Approval voting and direct popular-vote election of a president are recommended as election reforms.

BRAMS, STEVEN J., and PETER C. FISHBURN. *Approval Voting,* 2nd ed., Springer, New York, 2007. An in-depth analysis of approval voting, which includes several case studies.

BRAMS, STEVEN J. *Mathematics and Democracy: Designing Better Voting and Fair-Division Procedures*, Princeton University Press, Princeton, N.J., 2008. Shows how mathematics can be used to analyze the properties of different democratic procedures and can help to identify those with the most desirable properties.

HINICH, MELVIN J., and MICHAEL C. MUNGER. *Analytical Politics,* Cambridge University Press, Cambridge, U.K., 1997. Extends spatial modeling to more than one dimension, analyzes probabilistic voting, and introduces game-theoretic solution concepts relevant to the study of elections.

SAARI, DONALD G. *Chaotic Elections! A Mathematician Looks at Voting,* AMS [American Mathematical Society], Providence, R.I., 2001. Argues that elections–in particular, the 2000 presidential election, but others as well–have chaotic features that can be understood through mathematics. The mathematics used is an unusual kind of geometry that will be accessible to those with some mathematical background.

SHEPSLE, KENNETH A., and MARK S. BONACHEK. *Analyzing Politics: Rationality, Behavior, and Institutions,* Norton, New York, 1997. Rational strategies in voting and elections are a major component of this text, but it also includes sections on collective action and political institutions, such as courts and legislatures. Several case studies illustrate the theory.

SUGGESTED WEB SITES

www.fec.gov Federal Election Commission–About Elections and Voting.

www.ifes.org International Foundation for Election Systems.

www.rangevoting.org The Center for Range Voting.

wiki.electorama.com/wiki/Election-methods_mailing_list Election-methods mailing list.

Fairness and Game Theory PART IV

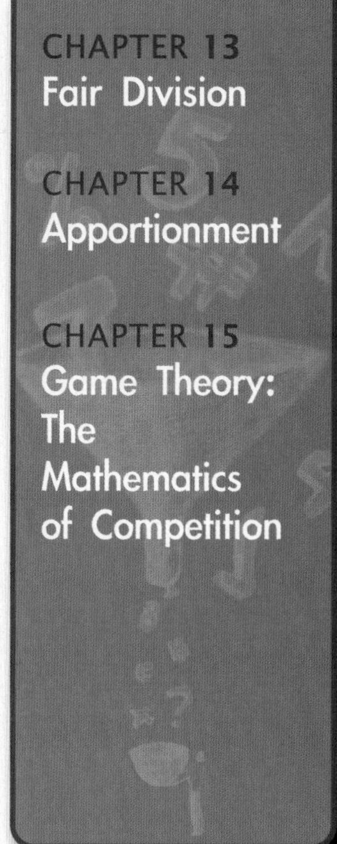

The central thrust of the first two chapters in Part IV is the fair division of divisible and indivisible objects. Whereas a cake or a parcel of land is divisible, the representatives who are apportioned to the different states are indivisible. Sometimes, however, seemingly indivisible objects, like a car, can be shared, rendering them divisible. By contrast, the game theory chapter focuses on what rational players will choose in different strategic situations, which may be highly unfair to some.

Chapter 13 describes fair-decision schemes in which a group of individuals with different values can be assured of each receiving what he or she views as a fair share when dividing objects like cakes or the goods in an estate.

Chapter 14 discusses the apportionment problem, which is to round a set of fractions to whole numbers while preserving their sum; of course, the sum of the original fractions must be a whole number to start. Apportionment problems occur when resources must be allocated in integer quantities–for instance, when legislators allocated seats in the U.S. House of Representatives to the 50 states.

Chapter 15 introduces the mathematical field called game theory, which describes situations involving two or more decision makers having different goals. Game theory provides a collection of models to assist in the analysis of conflict and cooperation as well as strategies for resolution. Interestingly, you will find that the games covered in this chapter provide us with insights into certain social paradoxes that we routinely encounter in our daily lives. ■

CHAPTER 13

Fair Division

When the demands or desires of one party are in conflict with those of another–be it a divorce, a labor–management negotiation, or an international dispute–no one wants to be treated unfairly. And with 1.2 million divorces every year in the United States alone, and crises such as we've seen in the Middle East for decades, it is certainly worth considering how mathematics might help in the search for procedures that can ensure fair and equitable resolutions of such conflicts.

We begin this chapter with one such procedure that was developed in the mid-1990s. The **adjusted winner procedure** allows two parties to settle any dispute involving either issues (as in an international dispute) or objects (as in a divorce or a two-person inheritance) with certain mathematical guarantees of "fairness." Disagreement, it turns out, is both a bad thing and a good thing. On the one hand, disagreement as to how each issue should be resolved typically lies at the heart of a conflict. On the other hand, procedures such as adjusted winner are designed to capitalize on the parties' disagreement as to the importance of each issue, thus allowing each party to end the negotiations thinking it has been met more than halfway.

But adjusted winner is just one of several so-called fair-division procedures that have been developed over the past 65 years. So following our discussion of adjusted winner, we describe a procedure for handling inheritances that was discovered by the Polish mathematician Bronislaw Knaster during World War II. Staying with real-world applications, we next consider the tricky question of finding a priority ranking for potential recipients of an organ that becomes available for transplantation. This is followed by a discussion of an extremely basic fair-division procedure–taking turns–and the question of what the optimal strategy is when taking turns choosing objects.

Bridging the gap between fair-division procedures with obvious real-world potential, such as divorce and inheritance procedures, and procedures that address fundamental mathematical questions of fairness (as do the procedures treated later in this chapter) is the ancient two-person procedure known as divide-and-choose. An application of this procedure to the Law of the Sea Treaty is described.

Divide-and-choose sets the stage for the mathematical investigations of fair division that have gone on for more than half a century. These investigations have often been phrased within the metaphor of "cake cutting." We present three cake-cutting procedures. The first two of these–found by Steinhaus and Banach–Knaster in the 1940s–yield allocations in which each player receives what he or she perceives to be at least his or her fair share of the cake. The last one–found by Selfridge–Conway in 1960–yields allocations in which each player receives what he or she perceives to be a piece at least tied for largest.

13.1 The Adjusted Winner Procedure

To illustrate the *adjusted winner procedure*, we will consider an application to the multibillion-dollar world of business mergers. It turns out that one of the most elusive ingredients in the success of a merger is what deal-makers call *social issues*–how power, position, sacrifice, and status are allocated between the merging companies and their executives.

(*Pascal Plessis/AP Photo.*)

As a case in point, let's revisit the 1998 proposed merger between two giant pharmaceutical companies, Glaxo Wellcome and SmithKline Beecham. While most of the details underlying this aborted deal are still unknown to outsiders, the role of social issues is clearly underscored by reports that the companies "saw nearly 19 billion dollars of stock market value vanish in the clash of two corporate egos."

Exactly what kinds of issues might bring on a "clash of two corporate egos"? While not privy to the details of the Glaxo Wellcome–SmithKline Beecham merger attempt, we can speculate as to their nature. For purposes of illustration, let's assume that the following five social issues were paramount:

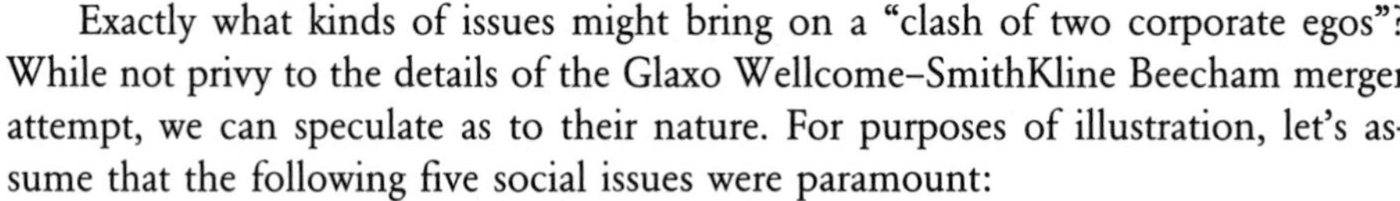

1. The name that the combined company would use
2. The location of the headquarters of the combined company
3. The question of who would serve as chairman of the combined company
4. The question of who would serve as CEO of the combined company
5. The question of where the necessary layoffs would come from

Each of these five social issues is known to have been a major factor in other recent proposed mergers. For example, when Chrysler merged with Daimler-Benz in 1998, the issue of the choice of a name for the combined company was described as a "standoff" before both sides finally agreed to DaimlerChrysler.

So, let's assume that these were the five social issues confronting Glaxo Wellcome and SmithKline Beecham, and let's see how the adjusted winner procedure would have suggested a resolution. The starting point–and something that is quite difficult when dealing with issues (as in a negotiation) as opposed to objects (as in a divorce)–is to have each side quantify the importance it attaches to getting its own way on each of the issues.

With the adjusted winner procedure, quantification is done by having each side–independently and simultaneously–spread 100 points over the issues in a way that reflects the relative worth of each issue to that party. In our present example, let's assume that the companies allocated their 100 points as shown in Table 13.1. Adjusted winner is now used to decide which side gets its way on which issues, but the procedure requires that a compromise of sorts may have to be reached on one of the issues.

TABLE 13.1 Applying the Adjusted Winner Procedure to a Merger of Two Companies

	Point Allocations	
Issue	**Glaxo Wellcome**	**SmithKline Beecham**
Name	5	10
Headquarters	25	10
Chairman	35	20
CEO	15	35
Layoffs	20	25
Total	100	100

Here's how the procedure works. Suppose we have two parties and a list of either issues to be resolved in one party's favor or the other's (as in our merger example) or objects to be awarded either to one party or to the other (as in a divorce or a two-person inheritance). To have a single word covering both issues and objects, we will speak of "items." The adjusted winner procedure follows these basic steps:

Basic Steps in the Adjusted Winner Procedure PROCEDURE

Step 1. As described earlier, each party distributes 100 points over the items in a way that reflects their relative worth to that party.

Step 2. Each item is initially given to the party that assigned it more points. Each party then assesses how many of his or her own points he or she has received. The party with the fewest points is now given each item on which both parties placed the same number of points.

Step 3. Since the point totals are most likely not equal, let A denote the party with the higher point total and B be the other party. Start transferring items from A to B, in a certain order, until the point totals are equal. The point at which equality is achieved may involve a fractional transfer of one item.

Step 4. The order in which this is done is extremely important and is determined by going through the items in order of increasing **point ratio.** An item's point ratio is the fraction

$$\frac{A\text{'s point value of the item}}{B\text{'s point value of the item}}$$

where A is the party with the higher point total.

Let's demonstrate the adjusted winner procedure by continuing with our analysis of the proposed merger between Glaxo Wellcome and SmithKline Beecham. Why step 4 is so important will be explained later.

1. Assume that Glaxo Wellcome and SmithKline Beecham have given us the point assignments shown in Table 13.1.

2. Because Glaxo Wellcome has placed more points on headquarters (25) and chairman (35), it is initially "given" these issues, while SmithKline Beecham is initially given name (10), CEO (35), and layoffs (25). Notice that SmithKline Beecham now has $10 + 35 + 25 = 70$ of its points, whereas Glaxo Wellcome only has $25 + 35 = 60$ of its points.
3. We now start transferring issues from SmithKline Beecham to Glaxo Wellcome until the point totals of the two sides are equal. SmithKline Beecham has initially been given three issues (name, CEO, and layoffs), and step 4 will help us decide in what order to start transferring them.
4. Layoffs has point ratio $25/20 = 1.25$, name has point ratio $10/5 = 2.00$, and CEO has point ratio $35/15 = 2.33$. Because layoffs has the lowest point ratio, we start to transfer that item first.

We now see that transferring the entire layoff item (worth 25 to SmithKline Beecham and 20 to Glaxo Wellcome) gives Glaxo Wellcome more points ($60 + 20 = 80$) than SmithKline Beecham has ($70 - 25 = 45$).

Thus, the entire layoff item cannot be transferred. Glaxo Wellcome and SmithKline Beecham will need to compromise on the issue of layoffs. But compromise may not mean meeting each other half way. Our goal is to equalize points between the two companies, and a little algebra will tell us exactly the extent to which Glaxo Wellcome and SmithKline Beecham should get their way on the issue of layoffs. Conceptually, it's easier to think of SmithKline Beecham retaining some fraction x of the issue in question and Glaxo Wellcome receiving the complementary fraction $1 - x$ of that same issue.

Because x is the fraction of the issue that SmithKline Beecham retains, the number of points SmithKline Beecham gets from this issue is x times 25. The fraction that Glaxo Wellcome gets is $1 - x$, so the number of points it gets from this issue is $1 - x$ times 20. Thus, if we want a fraction that will make SmithKline Beecham's total points and Glaxo Wellcome's total points equal, then x must satisfy the following equation:

$$10 + 35 + 25x = 25 + 35 + 20(1 - x)$$

We use algebra to solve this equation:

$$\begin{aligned} 45 + 25x &= 60 + 20 - 20x \\ 45 + 25x &= 80 - 20x \\ 45x &= 35 \\ x &= \frac{35}{45} = \frac{7}{9} \end{aligned}$$

Inserting $\frac{7}{9}$ back into the equation, we see that

$$45 + 25\left(\frac{7}{9}\right) = 60 + 20\left(\frac{2}{9}\right)$$

or approximately 64 points for each side. In rough terms, equality of points is achieved when SmithKline Beecham gets about three-fourths ($7/9 \cong 3/4$) of its way on the issue of layoffs and Glaxo Wellcome gets about one-fourth of its way.

Having seen how the adjusted winner procedure works, we must now ask the following question: Exactly what is it about the allocation produced by this scheme that would make someone want to use it? To answer this question, we need three definitions.

Equitable DEFINITION

A fair-division procedure, like adjusted winner, is said to be **equitable** if each player believes he or she received the same fractional part of the total value.

Envy-Free DEFINITION

A fair-division procedure is said to be **envy-free** if each player has a strategy that can guarantee him or her a share of whatever is being divided that is, in the eyes of that player, at least as large (or at least as desirable) as that received by any other player, no matter what the other players do.

Pareto-Optimal DEFINITION

A fair-division procedure is said to be **Pareto-optimal** if it produces an allocation with the property that no other allocation, achieved by any means whatsoever, can make any one player better off without making some other player worse off.

The answer to our earlier question is given by the following theorem (whose proof can be found in *Fair Division* by Brams and Taylor, listed in the Suggested Readings):

Properties of the Adjusted Winner Allocation THEOREM

For two parties, the adjusted winner procedure produces an allocation, based on each player's assignment of 100 points over the items to be divided, that has the following properties:

- The allocation is equitable.
- The allocation is envy-free.
- The allocation is Pareto-optimal.

Economists consider Pareto optimality (also named after Vilfredo Pareto) to be an extremely important property, and the order of transfer in step 4 on page 409 of the adjusted winner procedure is so important because it guarantees that the outcome is Pareto-optimal. The fact that the adjusted winner procedure produces an allocation that is efficient in this sense leads us to hope that it can and will play a future role in real-world dispute resolution.

13.2 The Knaster Inheritance Procedure

The adjusted winner procedure can be applied in the case of an inheritance if there are only two heirs. For *more than two heirs*, there is quite a different scheme, the **Knaster inheritance procedure**, first proposed by Bronislaw Knaster in 1945. It has a drawback, though, in that is requires the heirs to have a large amount of cash at their disposal.

EXAMPLE 1 ■ A Four-Person Inheritance

Suppose (for the moment) that there is just one object–a house–and four heirs–Bob, Carol, Ted, and Alice. Knaster's scheme begins with each heir bidding (simultaneously and independently) on the house. Assume, for example, that the bids are

Bob	Carol	Ted	Alice
\$120,000	\$200,000	\$140,000	\$180,000

Carol, being the high bidder, is awarded the house. Her fair share, however, is only one-fourth of the \$200,000 she thinks the house is worth, and so she places \$150,000 (which is three-fourths of the \$200,000 she bid) into a temporary "kitty."

Each of the other heirs now withdraws from the kitty his or her fair share, that is, one-fourth of his or her bid:

Bob withdraws $120,000/4 = $30,000
Ted withdraws $140,000/4 = $35,000
Alice withdraws $180,000/4 = $45,000

Thus, from the \$150,000 kitty, a total of \$30,000 + \$35,000 + \$45,000 = \$110,000 is withdrawn, and each of the four heirs now feels that he or she has the equivalent of one-fourth of the estate. Moreover, there is a \$40,000 surplus (\$150,000 kitty − \$110,000 withdrawn), which is now divided equally among the four heirs (so each receives an additional \$10,000). The final settlement is

Bob	Carol	Ted	Alice
\$40,000	house − \$140,000	\$45,000	\$55,000

(*Transtock/Corbis.*)

This illustrates Knaster's procedure for the simple case in which there is only one object. But what if our same four heirs have to divide an estate consisting of, say, a house (as before), a cabin, and a boat? The easiest answer is to handle the estate one object at a time (proceeding for each object as we just did for the house). To illustrate, assume that our four heirs submit the following bids:

	Bob	Carol	Ted	Alice
House	\$120,000	\$200,000	\$140,000	\$180,000
Cabin	60,000	40,000	90,000	50,000
Boat	30,000	24,000	20,000	20,000

We have already settled the house. Let's handle the cabin the same way. Thus, Ted is awarded the cabin based on his high bid of \$90,000. His fair share is one-fourth of this, so he places three-fourths of \$90,000 (which is \$67,500) into the kitty.

Bob withdraws from the kitty \$60,000/4 = \$15,000. Carol withdraws \$40,000/4 = \$10,000, and Alice withdraws \$50,000/4 = \$12,500. Thus, from the \$67,500 kitty, a total of \$15,000 + \$10,000 + \$12,500 = \$37,500 is withdrawn.

The surplus left in the kitty is thus \$30,000, and this is again split equally (\$7500 each) among the four heirs. The final settlement on the cabin is

Bob	Carol	Ted	Alice
\$22,500	\$17,500	cabin − \$60,000	\$20,000

If we were now to do the same for the boat (we leave the details to you), the corresponding final settlement would be

Bob	Carol	Ted	Alice
boat − $20,875	$7625	$6625	$6625

Putting the three separate analyses (house, cabin, and boat) together, we get a final settlement of

Bob: boat + ($40,000 + $22,500 − $20,875 = $41,625)
Carol: house + (−$140,000 + $17,500 + $7625 = −$114,875)
Ted: cabin + ($45,000 − $60,000 + $6625 = −$8375)
Alice: $55,000 + $20,000 + $6625 = $81,625.

Notice that here, Carol gets the house but must pay $114,875 in cash (and Ted gets the cabin but must put up $8375 in cash). This cash is then disbursed to Bob and Alice. In practice, Carol's having this amount of cash available may be a real problem–the key drawback to Knaster's procedure. Nevertheless, Knaster's procedure shows again that whenever some participants have different evaluations of some objects, there is an allocation in which everyone obtains more than what they would normally consider a fair share.

We summarize Knaster's inheritance procedure as follows.

Basic Steps in Knaster's Inheritance Procedure with *n* Heirs PROCEDURE

For each object, the following steps are performed:
Step 1. The heirs–independently and simultaneously–submit monetary bids for the object.
Step 2. The high bidder is awarded the object, and he or she places all but $1/n$ of his or her bid in a kitty. So, if there are four heirs ($n = 4$), then he or she places all but one-fourth–that is, three-fourths–of his or her bid in a kitty.
Step 3. Each of the other heirs withdraws from the kitty $1/n$ of his or her bid.
Step 4. The money remaining in the kitty is divided equally among the n heirs.

13.3 Fair Division and Organ Transplant Policies

In 1984, the United States Congress passed the National Organ Transplant Act and established a unified transplant network known as the Organ Procurement and Transplantation Network (OPTN). One of the primary goals of the OPTN was to increase the equity in the national system of organ allocation.

Achieving an equitable system of organ allocation is complicated by factors other than demand exceeding supply. For example, should an available organ go to the patient who needs it the most or the one for whom the likelihood of a successful transplant is greatest? Should both of these be taken into consideration, and, if so, how? Questions such as these reveal the extent to which an equitable system of organ allocation is a challenging problem in fair division.

In order to illustrate some of the issues (and paradoxes!) arising in the search for an equitable system for organ allocation, we'll (roughly) follow Peyton Young's synopsis–from his book, listed in Suggested Readings–of the fair division procedure for kidney allocation adopted by the OPTN in the late 1980s.

There were three (main) criteria used in arriving at a final ranking of those needing a kidney, and each potential recipient was awarded points according to a fixed method that we now describe.

- **Criterion 1: Waiting time.** A list of potential recipients was made according to how long they had been waiting for an organ. For each potential recipient, one calculates the fraction of people at or below the spot on the list he or she occupies, and then awards that person a number of points equal to 10 times that fraction. So if there are 5 people on the list, the first (waiting the longest) gets 10 × 1 = 10 points, the second gets 10 × (4/5) = 8 points, the third gets 10 × (3/5) = 6 points, and so on.
- **Criterion 2: Suitability.** The donor and potential recipient each have 6 relevant antigens that are either matched or not matched, with the likelihood of a successful transplant increasing with more matches. Two points are awarded for each match.
- **Criterion 3: Disadvantage.** Each person has antibodies that rule out a certain percentage of the population as being potential donors for that person. For some, only 10% are ruled out, while for others it may be as high as 90%. Those in the latter category are at a serious disadvantage compared to those in the former. Thus, potential recipients are awarded 1 point for each 10% of the population they are "sensitized against."

To illustrate this allocation procedure, let's assume we have 5 potential recipients–A, B, C, D, and E–with the following characteristics:

Potential recipient	Months waiting	Antigens matched	Percent sensitized
A	5	2	10
B	4.5	2	20
C	4	0	0
D	2	3	60
E	1	6	90

According to the procedure we described, points would be allocated as follows:

Potential recipient	Months waiting	Antigens matched	Percent sensitized	Total points
A	10	4	1	15
B	8	4	2	14
C	6	0	0	6
D	4	6	6	16
E	2	12	9	23

Thus, if one kidney became available, it would go to E (with 23 points). Presumably, if two became available at the same time, E would get one and D (with 16 points) would get the other.

But now things get interesting. Peyton Young, being well versed in the paradoxes of voting theory, fair division, and apportionment (among other things), observed the following. In the above scenario, what if two kidneys become available, but one is delayed slightly? Presumably, E gets the first one, and then we redo the chart with only A, B, C, and D. This yields the following:

Potential recipient	Months waiting	Antigens matched	Percent sensitized
A	5	2	10
B	4.5	2	20
C	4	0	0
D	2	3	60

According to the procedure we described, points would be allocated as follows:

Potential recipient	Months waiting	Antigens matched	Percent sensitized	Total points
A	10	4	1	15
B	7.5	4	2	13.5
C	5	0	0	5
D	2.5	6	6	14.5

Thus, A (not D!) now gets the second kidney, having 15 points to 14.5 for D. This is an example of what is called the "priority paradox." For more on this, we invite the reader to consult Peyton Young's book in the Suggested Readings.

13.4 Taking Turns

For many of us, an early lesson in fair division happens in elementary school with the choosing of sides for a spelling bee or when picking teams on the playground. In terms of importance, these pale in comparison with the issue of property settlement in a divorce. Remarkably, however, the same fair-division procedure–*taking turns*–is often used in both.

Taking turns is fairly self-explanatory. With two parties (and that's all we'll consider here), one party selects an object, then the other party selects one, then the first party again, and so on. But in this context, there are several interesting questions that suggest themselves:

1. How do we decide who chooses first?
2. Because choosing first is often quite an advantage, shouldn't we compensate the other party in some way, perhaps by giving him or her extra choices at the next turn?
3. Should a player always choose the object he or she most favors from those that remain, or are there strategic considerations that players should take into account?

The answer to question 1 is often "toss a coin," but there are other possibilities–for example, the two parties could "bid" for the right to go first, as in an auction. The answer to question 2 is less clear, but we outline a discussion of the issue it raises in Writing Project 2.

Question 3, on the other hand, is remarkably interesting, and it is this one that we want to pursue. Let's look at an easy example. Suppose that Bob and Carol are getting a divorce, and their four main possessions, ranked from best to worst by each, are as follows:

	Bob's Ranking	**Carol's Ranking**
Best	Pension	House
Second best	House	Investments
Third best	Investments	Pension
Worst	Vehicles	Vehicles

If Carol knows nothing of Bob's preferences, then we can assume that she will choose sincerely–selecting at her turn whichever item she most prefers from those not yet chosen. Now, if Bob is also sincere, and if he chooses first, the items will be allocated as follows:

First turn:	Bob takes the pension.
Second turn:	Carol takes the house.
Third turn:	Bob takes the investments.
Fourth turn:	Carol is left with the vehicles.

Hence, Bob gets his first and third favorites (the pension and the investments). However, if Bob opens by choosing the house–and bypassing the pension for the moment–then the allocation will be as follows:

First turn:	Bob takes the house.
Second turn:	Carol takes the investments.
Third turn:	Bob takes the pension.
Fourth turn:	Carol is left with the vehicles.

Thus, by being insincere, Bob does better–getting his first and second favorites (the pension and the house).

In general, then, what is the optimal strategy for rational players to use, assuming that both know the preferences of the other? The answer is something called the **bottom-up strategy**, discovered by the mathematicians D. A. Kohler and R. Chandrasekaran in 1969. We will illustrate it with an example.

Suppose we have five objects–*A, B, C, D, E*–and Bob is choosing first. Suppose that Bob and Carol have the following rankings of the objects (called **preference lists** in what follows):

Bob	Carol
A	*C*
B	*E*
C	*D*
D	*A*
E	*B*

It will turn out that Bob should open with *C* (his third choice) followed by Carol's choice of *D* (skipping over *E*, for the moment). Bob will then take *A*, Carol will follow with *E*, and finally Bob will get *B*. Bob gets his first, second, and third choices without selecting his first choice first! Where does this strategy come from?

The intuition here is quite easy. Let's make two assumptions about rational players: A rational player will never willingly choose his or her least preferred alternative, and a rational player will avoid wasting a choice on an object that he or she knows will remain available and thus can be chosen later.

With these assumptions as motivation, let's return to the preceding example and think about the mental calculation Bob will go through in deciding what his first choice will be. Bob knows the eventual sequence of choices will fill in all of the following blanks:

Bob: ______ ______ ______
Carol: ______ ______

Now, working mentally from right to left, Bob knows that Carol will not choose *B*, because it is the bottom thing on her list. Thus, he will get stuck with *B*, and so he will avoid wasting anything but his last choice on alternative *B*. Thus, Bob can pencil in alternative *B* as his last choice:

Bob: ______ ______ *B*
Carol: ______ ______

Bob, placing himself momentarily in Carol's shoes, knows she will reason the same way, and thus he pencils Carol in for the bottom alternative, *E*, on his list:

Bob: ______ ______ *B*
Carol: ______ *E*

Mentally now, Bob reasons as if alternatives *B* and *E* never existed (and the choice sequence had been Bob–Carol–Bob) and continues to pencil in alternatives from right to left, with Bob working from bottom to top on Carol's preference list and Carol working from bottom to top on Bob's preference list. This yields the following sequence of choices mentally penciled in by Bob:

Bob: *C* *A* *B*
Carol: *D* *E*

Remember, this is just a mental calculation that Bob went through to decide upon the actual choice–in this case, *C*–with which he will open. Bob has no guarantee that Carol will, in fact, respond with *D*, so the use of this strategy involves some risk on Bob's part.

This bottom-up strategy can also be viewed as a procedure that a mediator could use to specify a division of several objects between two parties. Given the preference lists of both parties, the mediator could construct a list–exactly as we did for Bob and Carol above–and then offer this to the parties as the suggested allocation. In effect, the mediator is simultaneously playing the role of two rational parties who choose to employ optimal strategies.

13.5 Divide-and-Choose

There are vast mineral resources under the seabed, all of which, one might argue, should be available to both developed and developing countries. In the absence of some kind of agreement, however, what is to prevent the developed countries from mining all of the most promising tracts before the developing countries have reached a technological level where they can begin their own mining operations? Such an agreement went into effect on November 16, 1994, with 159 signatories (including the United States). It was called the **Convention of the Law of the Sea**, and it protects the interests of developing countries by means of the following fair-division procedure.

Whenever a developed country wants to mine a portion of the seabed, that country must propose a division of the portion into two tracts. An international mining company called the Enterprise, funded by the developed countries but representing the interests of the developing countries through the International Seabed Authority, then chooses one of the two tracts to be reserved for later use by the developing countries.

Divide-and-Choose — PROCEDURE

With **divide-and-choose**, one party divides the object into two parts in any way that he desires, and the other party chooses whichever part she wants.

As a fair-division procedure, the origins of divide-and-choose go back thousands of years. The Hebrew Bible tells the story of Abram (later to be called Abraham) and Lot, who settled a dispute over land via a proposed division by Abram–"If you go north, I will go south; and if you go south, I will go north" (Gen. 13:8–9)–and a choice (of the plain of Jordan) by Lot. Divide-and-choose resurfaced later in Hesiod's book *Theogony*. The Greek gods Prometheus and Zeus had to divide a portion of meat. Prometheus began by placing the meat into two piles, and Zeus selected one.

Actually, a fair-division procedure consists of both rules and strategies, and all we have described so far are the rules of divide-and-choose. But the natural strategies here are quite obvious: The divider makes the two parts equal in his estimation, and the chooser selects whichever piece she feels is more valuable.

Rules and strategies differ from each other in the following sense: A referee could determine if a rule is being followed, even without knowing the preferences of the players. Strategies represent choices of how players follow the rules, given their individual preferences (and any other knowledge or goals they may have).

The strategies on which we focus in our discussion of fair-division procedures are those that require no knowledge of the preferences of the other players and yet provide some kind of minimal degree of satisfaction even in the face of collusion by the other players. For example, the strategies just given for divide-and-choose guarantee each player a piece that he or she would not wish to trade for that received by the other.

There are, to be sure, other strategic considerations that might be relevant. For example, in divide-and-choose, would you rather be the divider or the chooser? The answer, given our assumptions that nothing is known of the preferences of the others, is to be the chooser. However, if you knew the preferences of your opponent (and how much she may value spite), then you might want to be the divider.

As a final comment on strategic considerations, we need only look to the origins of the well-known expression "the lion's share." It comes from one of Aesop's fables, as reported by Todd Lowry in *Archaeology of Economic Ideas* (1987, p. 130):

> It seems that a lion, a fox, and an ass participated in a joint hunt. On request, the ass divides the kill into three equal shares and invites the others to choose. Enraged, the lion eats the ass, then asks the fox to make the division. The fox piles all the kill into one great heap except for one tiny morsel. Delighted at this division, the lion asks, "Who has taught you, my very excellent fellow, the art of division?" to which the fox replies, "I learnt it from the ass, by witnessing his fate."

13.6 Cake-Division Procedures: Proportionality

The modern era of fair division in mathematics began in Poland during World War II (see Spotlight 13.1). At this time, Hugo Steinhaus asked what is, in retrospect, the obvious question: What is the "natural" generalization of divide-and-choose to three or more people? The metaphor that has been used in this context, going back at least to the English political theorist James Harrington (1611–1677), is a cake. We picture different players valuing different parts of the cake differently because of concentrations of certain flavors or depth of frosting.

Cake-Division Procedure DEFINITION

A **cake-division procedure** for n players is a procedure that the players can use to allocate a cake among themselves (no outside arbitrators) so that each player has a strategy that will guarantee that player a piece with which he or she is "satisfied," even in the face of collusion by the others.

As we have seen, divide-and-choose is a cake-division procedure for two players, if by "satisfied" we mean either "thinks his piece is of size or value at least one-half" or "does not want to trade what she received for what anyone else received." We define the first notion here; envy-free allocations were defined in Section 13.1.

Proportional Procedure DEFINITION

A cake-division procedure (for n players) will be called **proportional** if each player's strategy guarantees that player a piece of size or value at least $1/n$ of the whole in his or her own estimation.

It turns out that for $n = 2$, a procedure is envy-free if and only if it is proportional; that is, for $n = 2$, the two notions of fair division are exactly the same. For $n > 2$, however, all we can say is that an envy-free procedure is automatically proportional. For example, if a three-person allocation is not proportional, then one player (call him Bob) thinks that he received less than one-third. Bob then feels that the other two are sharing more than two-thirds between them, and thus that at least one of the two (call her Carol) must have more than one-third. But then Bob will envy Carol, and so the allocation is not envy-free. Because all nonproportional allocations fail to be envy-free, it follows that if an allocation is envy-free, then it must be proportional.

Many procedures that are proportional, however, fail to be envy-free, as we shall soon show. Thus, proportional procedures are fairly easy to come by, but envy-free procedures are fairly hard to come by.

(Brand X Pictures/Punchstock.)

EXAMPLE 2 ■ The Steinhaus Proportional Procedure for Three Players (Lone Divider)

Given three players–Bob, Carol, and Ted–we have Bob divide the cake into three pieces, call them X, Y, and Z, each of which he thinks is of size or value exactly one-third. Let's speak of Carol as "approving of a piece" if she thinks it is of size or value at least one-third. Similarly, we will speak of Ted as "approving of a piece" if the same criterion applies. Notice that both Carol and Ted must approve of at least one piece.

If there are distinct pieces—say, X and Y—with Carol approving of X and Ted approving of Y, then we give the third piece, Z, to Bob (and, of course, X to Carol and Y to Ted), and we are done. The problem case is where both Carol and Ted approve of only one piece and it is the *same* piece.

Let's assume that Carol and Ted approve of only one piece, X, and hence (of more importance to us) both *disapprove* of piece Z. Let XY denote the result of putting piece X and piece Y back together to form a single piece. Notice that both Carol and Ted think that XY is at least two-thirds of the cake because both disapprove of Z. Thus, we can give Z to Bob and let Carol and Ted use divide-and-choose on XY. Because half of two-thirds is one-third, both Carol and Ted are guaranteed a proportional share (as is Bob, who approved of all three pieces).

The procedure just described, which guarantees proportional shares but is not necessarily envy-free and is sometimes called the **lone-divider method**, was discovered by Hugo Steinhaus around 1944. Unfortunately, it does not extend easily to more than three players. It was left to Steinhaus's students, Stefan Banach and Bronislaw Knaster, to devise a method for more than three players. Picking up where Steinhaus left off (and traveling in quite a different direction), they devised the proportional procedure that today is referred to as the **last-diminisher method**. Like the lone-divider method, it is proportional but not envy-free. We illustrate it for the case of four players (Bob, Carol, Ted, and Alice), and we include both the rules and the strategies that guarantee each player his or her fair share.

EXAMPLE 3 ▪ The Banach–Knaster Proportional Procedure for Four or More Players (Last Diminisher)

Bob cuts from the cake a piece that he thinks is of size one-fourth and hands it to Carol. If Carol thinks the piece handed her is larger than one-fourth, she trims it to size one-fourth in her estimation, places the trimmings back on the cake, and passes the diminished piece to Ted. If Carol thinks the piece handed her is of size at most one-fourth, she passes it unaltered to Ted.

Ted now proceeds exactly as did Carol, trimming the piece to size one-fourth if he thinks it is larger than this and passing it (diminished or unaltered) on to Alice. Alice does the same, but, being the last player, simply holds onto the piece momentarily instead of passing it to anyone.

Notice that everyone now thinks the piece is of size at most one-fourth, and the last person to trim it (or Bob, if no one trimmed it) thinks the piece is of size exactly one-fourth. Thus, the procedure now allocates this piece to the last person who trimmed it (and to Bob if no one trimmed it).

Assume for the moment that it was Ted who trimmed the piece last, so he takes this piece and exits the game. Bob, Carol, and Alice all think that at least three-fourths of the cake is left, so they can start the process over with (say) Bob beginning by cutting a piece from what remains that he thinks is one-fourth of the original cake. Carol and Alice are both given a chance to trim it to size one-fourth in their estimation, and again, the last one to trim it takes that piece and exits the game. The two remaining players both think that at least half the cake is left, so they can use divide-and-choose to divide it between themselves and thus be assured of a piece that is of size at least one-fourth in their estimation.

SPOTLIGHT 13.1 Sixty Years of Cake Cutting

The modern era of cake cutting began with the investigations of the Polish mathematician Hugo Steinhaus during World War II. His research, and that of dozens of others since, involved dealing with two fundamental difficulties. First, allocation schemes that work in the context of two or three players often do not generalize easily to the context of four or more players. Second, procedures that yield envy-free allocations are considerably harder to obtain than procedures that yield proportional allocations.

The mathematics inspired by these two difficulties constitutes a rather elegant corner of the large and important area of fair division. Steinhaus's investigations in the 1940s led to his observation that there is a rather natural extension of divide-and-choose to the case of three players. This is the "lone-divider procedure" described on page 420. Steinhaus's method was generalized to an arbitrary number of players by Harold W. Kuhn of Princeton University in 1967.

Unable to extend his procedure from three to four players, Steinhaus proposed the problem to some Polish colleagues. Two of them, Stefan Banach and Bronislaw Knaster, solved this problem in the mid-1940s by producing the "last-diminisher procedure" described on page 420.

In addition to the procedures devised by Banach, Knaster, and Kuhn, there are other well-known constructive procedures for obtaining a proportional allocation among four or more players. One of these is by A. M. Fink of Iowa State University and appears in Exercise 29.

Another constructive procedure of note, although different in flavor from the others, is the 1961 recasting by Lester E. Dubins and Edwin H. Spanier of the University of California at Berkeley of the last-diminisher method as a "moving-knife procedure" (illustrated in Exercise 31). The trade-off here involves giving up the "discrete" nature of the last-diminisher method in exchange for the conceptual simplicity of the moving knife.

Although the existence of an envy-free allocation (even for four or more players) was known to Steinhaus in the 1940s, the first constructive procedure for producing an envy-free allocation among three players was not found until around 1960. At that time, John L. Selfridge of Northern Illinois University and, later but independently, John H. Conway of Princeton University found the elegant procedure presented on page 422. Although never published by either, the procedure was quickly and widely disseminated by Richard K. Guy of the University of Calgary and others. Eventually it appeared in several treatments of the problem by different authors.

In 1980, a moving-knife procedure for producing an envy-free allocation among three players was found by Walter R. Stromquist of Daniel Wagner Associates. Then, another procedure, capable of being recast as a moving-knife solution of the three-player case, was found by a law professor at the University of Virginia, Saul X. Levmore, and a former student of his, Elizabeth Early Cook.

In 1992, Steven J. Brams, a political scientist at New York University, and Alan D. Taylor, a mathematician at Union College, succeeded in finding a constructive procedure for producing an envy-free allocation among four or more players. In 1994, Brams, Taylor, and William S. Zwicker (also from Union College) found a moving-knife solution to the four-person envy-free problem. No moving-knife procedure is known that will produce an envy-free allocation among five or more players.

13.7 Cake-Division Procedures: The Problem of Envy

Divide-and-choose has a property that neither of the last two procedures possesses: It can ensure that each player receives a piece of cake he or she considers the largest or tied for the largest. In the case of only two players, this means that each player can get what he or she perceives to be at least half the cake, no matter what the other player does. Thus, divide-and-choose is an envy-free procedure.

Steinhaus's $n = 3$ proportional procedure (the lone-divider method) is not envy-free. For example, consider the case where Carol and Ted both find one piece unacceptable (and this piece is given to Bob). Carol and Ted will not envy each other when one divides and the other chooses, but Bob may think that this is not a 50-50 split. Indeed, if Bob divided the cake initially into what he thought was three equal pieces, an unequal split of the remaining two-thirds of the cake by Carol and Ted means that Bob will prefer the larger of these two pieces to the one-third he got. Consequently, Bob will envy the person who got this larger piece.

Nor is the last-diminisher method envy-free. For example, if Bob initially cuts a piece of cake of size one-fourth, and no one else trims it, then Bob receives this piece and exits the game. If Carol is the one to make the next initial cut, she may well cut a piece from the cake that she thinks is of size one-fourth but that Bob thinks is of size considerably more than one-fourth. But Bob is out of the game. Thus, if Ted and Alice think this piece is of size less than one-fourth, then Carol receives it, and so Bob will envy Carol.

Nevertheless, there do exist cake-division procedures that are envy-free. We present one of these in what follows.

EXAMPLE 4 ■ The Selfridge–Conway Envy-Free Procedure for Three Players

We start with a cake and three people. The point we wish to arrive at is an envy-free allocation of the entire cake among the three people in a finite number of steps. This task may seem formidable, but quite often in mathematics, an important part of solving a problem involves breaking the problem into identifiable parts. In this case, let's call our starting point A and the final point we wish to reach C. Now let's identify an appropriate in-between point B that makes going from A to C–via B– more manageable. Our in-between point B is the following:

> *Point B:* Getting a constructive procedure that gives an envy-free allocation of *part* of the cake.

Can we constructively obtain three pieces of cake, whose union may not be the whole cake, which can be given to the three people so that each thinks he or she received a piece at least tied for largest? This turns out to be quite easy with the solution given by John Selfridge and John Conway. The following process and strategies do the trick:

1. Player 1 cuts the cake into three pieces he considers to be the same size. He hands the three pieces to player 2.
2. Player 2 trims at most one of the three pieces to create at least a two-way tie for largest. Setting the trimmings aside, player 2 hands the three pieces (one of which may have been trimmed) to player 3.

3. Player 3 now chooses, from among the three pieces, one that he considers to be at least tied for largest.
4. Player 2 next chooses, from the two remaining pieces, one that she considers to be at least tied for largest, with the proviso that if she trimmed a piece in step 2, and player 3 did not choose this piece, then she must now choose it.
5. Player 1 receives the remaining piece.

Let's reconsider the five steps of this trimming procedure to assure ourselves that each player experiences no envy. Recall that player 1 cuts the cake into three pieces, and player 2 trims one of these three pieces. Now player 3 chooses, and, as the first to choose, he certainly envies no one. Player 2 created a two-way tie for largest, and at least one of these two pieces is still available after player 3 selects his piece. Hence, player 2 can choose one of the tied pieces she created and will envy no one. Finally, player 1 created a three-way tie for largest and, because of the proviso in step 4, the trimmed piece is not the one left over. Thus, player 1 can choose an untrimmed piece and therefore will envy no one.

So far we have gone from point A to point B: Starting with a cake and three players, we have constructively obtained (in finitely many steps) an envy-free allocation of all of the cake, except the part T that player 2 trimmed from one of the pieces. We will now describe how T can be allocated among the three players in such a way that the resulting allocation of the whole cake is envy-free. (This is the rest of the **Selfridge–Conway envy-free procedure**.)

The key observation for the $n = 3$ case is that player 1 will not envy the player who received the trimmed piece, even if that player were to be given all of T. Recall that player 1 created a three-way tie and received an untrimmed piece. The union of the trimmed piece and the trimmings yields a piece that player 1 considers to be exactly the same size as the one he received. Thus, assume that it is player 3 who received the trimmed piece (it could as well be player 2). Then player 1 will not envy player 3, no matter how T is allocated.

The next step ensures that neither player 2 nor player 3 will envy another player when it comes time to allocate T. Let player 2 cut T into three pieces she considers to be the same size. Let the players choose which of the three pieces they want in the following order: player 3, player 1, player 2.

To see that this yields an envy-free allocation, notice that player 3 envies no one, because he is choosing first. Player 1 does not envy player 2, because he is choosing ahead of her; and player 1 does not envy player 3 because, as pointed out earlier, player 1 will not envy the player who received the trimmed piece. Finally, player 2 envies no one, because she made all three pieces of T the same size.

Hence, for $n = 3$, the Selfridge–Conway procedure will give an envy-free allocation of all the cake except T, followed by an allocation of T that gives an envy-free allocation of all the cake.

■

A naive attempt to generalize to $n = 4$ what we have done for $n = 3$ would proceed as follows: We would begin by having player 1 cut the cake into four pieces he considers to be the same size. Then we would have players 2 and 3 trim some pieces (but how many?) to create ties for the largest. Finally, we would have the players choose from among the pieces–some of which would have been trimmed–in the following order: player 4, player 3, player 2, player 1.

This approach fails because player 1 could be left in a position of envy. To understand how the approach could fail, consider how many pieces player 3 might

(c) Identify a single piece that player 2 and player 3 agree is *not* acceptable. (There are actually two such pieces; for definiteness, find the one on the right.)
(d) Assume that players 2 and 3 give the piece from part (c) to player 1. Suppose they reassemble the rest and players 2 and 3 divide it between themselves using divide-and-choose (with a single vertical cut). Determine what size piece each of the three players will think he or she received (1) if player 2 divides and player 3 chooses, and (2) if player 3 divides and player 2 chooses.

28. Suppose players 1, 2, and 3 view a cake as in Exercise 27. Illustrate the last-diminisher method (still restricting attention to vertical cuts and, in addition, assuming that the piece potentially being diminished is a piece off the left side of the cake) by following steps (a) through (f) below:
(a) Draw a picture showing the third of the cake (6 squares) that player 1 will slice off the cake.
(b) Determine whether player 2 will pass or further diminish this piece. If he or she would further diminish it, make a new drawing.
(c) Determine whether player 3 will pass or further diminish this piece. If he or she would further diminish it, make a new drawing.
(d) Determine who receives the piece cut off the cake and what size or value he or she thinks it is. (Actually, we knew what size the person receiving this first piece would think it was, assuming he or she followed the prescribed strategy. How did we know this?)
(e) Finish the last-diminisher method using divide-and-choose on what remains, with the lowest-numbered player who remains doing the dividing.
(f) Redo step (e) with the other player doing the dividing.

29. The Banach–Knaster last-diminisher method is not the only well-known cake-division procedure that yields a proportional allocation for any number of players. There is also one due to A. M. Fink (sometimes called the *lone-chooser method*). For three players (Bob, Carol, and Ted) it works as follows:

(i) Bob and Carol divide the cake into two pieces using divide-and-choose.
(ii) Bob now divides the piece he has into three parts that he considers to be the same size. Carol does the same with the piece she has.
(iii) Ted now chooses whichever of Bob's three pieces that he (Ted) thinks is largest, and Ted chooses whichever of Carol's three pieces that he thinks is largest.
(iv) Bob keeps his remaining two pieces, as does Carol.

(a) Explain why Ted thinks he is getting at least one-third of the cake.
(b) Explain why Bob and Carol each think they are receiving at least one-third of the cake.
(c) Explain why, in general this scheme is not envy-free.

■ **30.** In A. M. Fink's procedure (described in Exercise 29), suppose that a fourth person (Alice) comes along after Bob, Carol, and Ted have already divided the cake among themselves so that each of the three thinks he or she has a piece of size at least one-third. Mimic what was done in the three-person case to obtain an allocation among the four that is proportional. (*Hint:* Begin by having Bob, Carol, and Ted divide the pieces they have into a certain number–how many?–of equal parts.)

31. There is a moving-knife version of the Banach–Knaster procedure that is due to Dubins and Spanier. To describe it, we picture the cake as being rectangular, and the procedure beginning with a referee holding a knife along the left edge, as illustrated below.

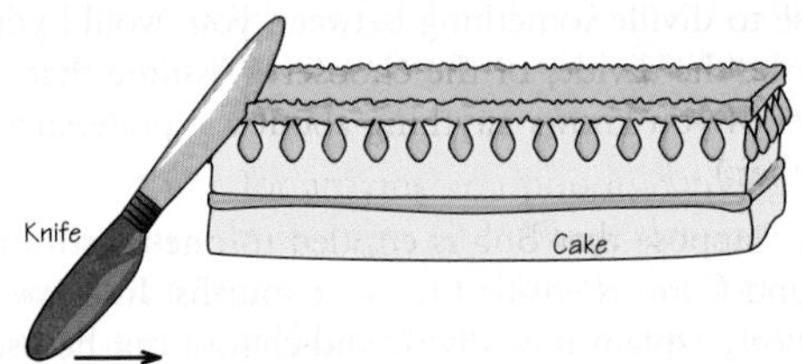

Assume, for the sake of illustration, that there are four players (Bob, Carol, Ted, and Alice). The referee starts moving the knife from left to right over the cake (keeping it parallel to the position in which it started) until one of the players (assume it is Bob) calls "cut." At this time, a cut is made, the piece to the left of the knife is given to Bob, and he exits the game. The knife starts moving again, and the process continues. The strategies are for each player to call "cut" whenever it would yield him or her a piece of size at least one-fourth.
(a) Explain why this procedure produces an allocation that is proportional.
(b) Explain why the resulting allocation is not, in general, envy-free.
(c) Explain why, if you are not the first player to call "cut," there is a strategy different from the one suggested that is never worse for you, and sometimes better.

13.7 Cake-Division Procedures: The Problem with Envy

32. Suppose players 1, 2, and 3 view the cake as in Exercise 27. Illustrate the envy-free procedure for $n = 3$ (yielding an allocation of part of the cake) by following steps (a) through (c) below. Again, restrict attention to vertical cuts.
(a) Provide a total of three drawings to show how each player views a division of the cake by player 1 into three pieces he or she considers to be the same size or value. Label the pieces *A*, *B*, and *C*. (This is the same as Exercise 27a.)
(b) Redraw the picture from player 2's view, and illustrate the trimming of piece *A* that he or she would do. Label the trimmed piece *A′* and the actual trimmings *T*.

Apportionm

14.1 The Apportic

Coach is proud of her field
its record for the season. T
tied one. Table 14.1 is a draft
cated. "Just express the percen
percentages: The winning perc
ing percentage, 17.39%, is rou
is rounded down to 4%. Beca
total is only 99%. The coach n
"Now you have 100%!"

Apportionment Problem

An **apportionment problem**
maintained at its original val
bitrary one, but one that can
cedure is called an **apportion**

When rounding percentag
apportionment problem. The c
it look good for the team," l
unbiased, especially when it co
ing seats in the U.S. House of
The framers of the U.S. C
resentatives "shall be apportion
cording to their respective Nu
apportionment problem that t
it caused trouble right from the

(c) Indicate which piece each player would choose (and what he or she thinks its size is) if the players choose in the following order: player 3, player 2, player 1, according to the envy-free procedure.

■ **33.** There is a two-person moving-knife cake-division procedure due to A. K. Austin that leads to each player receiving a piece of cake that he or she considers to be of size exactly one-half. It begins by having one of the two players (Bob) place two knives over the cake, one of which is at the left edge, and the other of which is parallel to the first and placed so that the piece between the knives (A in the picture below) is of size exactly one-half in Bob's estimation.

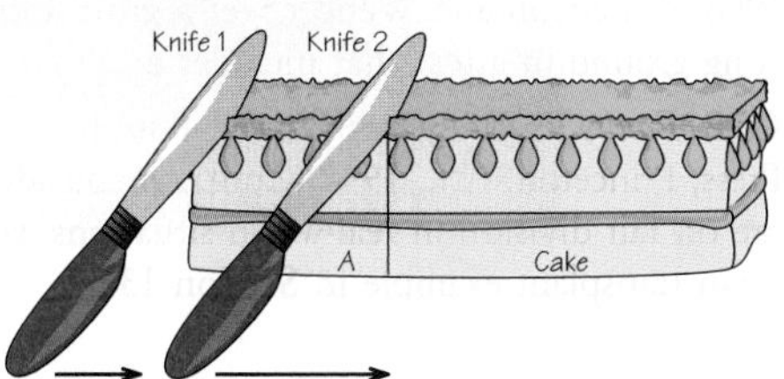

If Carol agrees that this is a 50-50 division, we are done. Otherwise, Bob starts moving both knives to the right–perhaps at different rates–so that the piece between the knives remains of size one-half in his eyes. Carol calls "stop" at the point when she also thinks the piece between the two knives is of size exactly one-half.

(a) If the knife on the right were to reach the right-hand edge, where would the knife on the left be?

(b) Explain why there definitely is a point where Carol thinks the piece between the two knives is of size exactly one-half. (*Hint:* If Carol thinks the piece is too small at the beginning, what will she think of it at the end?)

WRITING PROJECTS

1. It turns out that there is no way to extend the adjusted winner procedure to three or more players. That is, there are point assignments by three players to three objects so that no allocation satisfies the three desired properties of equability (equal points), envy-freeness, and Pareto optimality. On the other hand, there are separate procedures that will realize any two of the three properties. Thus, trade-offs must be made, and these may depend on the circumstances. In a few paragraphs, discuss the relative importance of the three properties and circumstances that may affect the choice of which two of the three properties one might wish to have satisfied.

2. If we use taking turns to divvy up a collection of objects between two people (Bob and Carol), then there is an obvious advantage to going first. Assume that we have decided that Bob will, in fact, choose first (say, by the toss of a coin). Let's think about how Carol might be compensated. First of all, if there are only three objects, then the "choice sequence" Bob–Carol–Carol seems to be the only reasonable one. Do you agree? For four objects, however, there are two choice sequences that suggest themselves: Bob–Carol–Carol–Carol and Bob–Carol–Carol–Bob. Do you think that one of these is obviously more fair than the other? What if there are four identical objects? What if both Bob and Carol value object A twice as much as B, and B twice as much as C, and C twice as much as D? What sequences suggest themselves for five objects? For eight objects? In one page or less, discuss these questions. (For more on this, see *The Win–Win Solution* in the Suggested Readings.)

3. One of the most important differences between the three-person and the n-person envy-free procedures is that the latter procedure may take more than two stages. And, of course, the more stages there are, the more cuts and trimmings may be necessary. Do you consider this a serious practical problem, or is it mainly a theoretical problem? In one paragraph, explain your reasons.

4. One often hears of the importance of "process" versus "product," the latter referring to *what* is achieved and the former referring to *how* it was achieved. In a couple of sentences, comment on the relevance of this to fair division as illustrated by the following rough paraphrasing of an exchange between two old friends, Ralph Kramden (played by Jackie Gleason) and Ed Norton (played by Art Carney) in the 1950s sitcom *The Honeymooners.*

> *Ralph to Ed* (as the two are sitting alone at the dinner table): I can't believe you did that.
>
> *Ed:* Did what, Ralph?
>
> *Ralph:* There were two potatoes there, and you reached right out and took the big one.
>
> *Ed:* What would you have done, Ralph?
>
> *Ralph:* Why, I'd have taken the little one.
>
> *Ed:* You *got* the little one, Ralph.

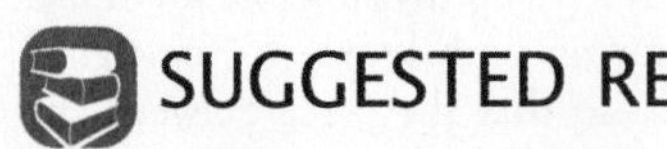

SUGGESTED RE

BRAMS, S. J., and A. D. TAYLO[illegible] *Solution: Guaranteeing Fair Shares t*[illegible] New York, 1999. Brams and Tayl[illegible] adjusted winner, as well as divide-[illegible] turns.

BRAMS, S. J., and A. D. TAYLO[illegible] *Cake-Cutting to Dispute Resolution,* [illegible] Press, Cambridge, 1996. Brams an[illegible] book-length treatment of the kind[illegible] in this chapter, as well as divide-a[illegible] political arena, moving-knife proc[illegible] and fairness as it applies to differe[illegible] election procedures.

TABLE 14.1 The Field Hockey Team's Season

		Percentage
Games won	18	$\frac{18}{23} \times 100\% = 78.26\%$
Games lost	4	$\frac{4}{23} \times 100\% = 17.39\%$
Games tied	1	$\frac{1}{23} \times 100\% = 4.35\%$
Games played	23	100.00%

the least among the 15 states. The total population of the nation at the time was 3,615,920, and the House of Representatives was to have 105 members. Thus, the average congressional district should have a population of 3,615,920 ÷ 105 = 34,437. To obtain its fair share of the House seats, we can divide Delaware's population by the average congressional district population. Ideally, Delaware should have had

$$\frac{55{,}540}{34{,}437} = 1.613 \text{ congressional districts}$$

Table 14.2 displays the apportionment proposed in a bill, written by Alexander Hamilton, that Congress passed and sent to President Washington. Each state's fair share of the 105 congressional districts, or **quota**, was calculated by dividing its population by the average congressional district population of 34,437, as in the case of Delaware, shown above. None of the quotas were whole numbers. The bill gave Delaware two seats in Congress. As you will see in the table, Virginia's quota was 18.310 seats, and Virginia was awarded 18 seats in the House.

TABLE 14.2 The Congressional Apportionment that George Washington Vetoed

State	Population	Quota	Apportionment
Virginia	630,560	18.310	18
Massachusetts	475,327	13.803	14
Pennsylvania	432,879	12.570	13
North Carolina	353,523	10.266	10
New York	331,589	9.629	10
Maryland	278,514	8.088	8
Connecticut	236,841	6.878	7
South Carolina	206,236	5.989	6
New Jersey	179,570	5.214	5
New Hampshire	141,822	4.118	4
Vermont	85,533	2.484	2
Georgia	70,835	2.057	2
Kentucky	68,705	1.995	2
Rhode Island	68,446	1.988	2
Delaware	55,540	1.613	2
Totals	3,615,920	105	105

Although this apportionment may seem fair enough, President Washington vetoed the bill.[1] Washington came from Virginia, a state that would get less than its quota in the apportionment Congress proposed. It is impossible to determine if he was just biased for his home state–as the field hockey coach was in favor of her team–because, as we will discover, there were substantial reasons for rejecting the bill.

First, let's see how to set up an apportionment problem. Many apportionment problems do not involve the House of Representatives; our terminology refers to *states*, *populations*, and a *house size*. In the problem of rounding percentages so that their sum is 100, the house size is 100. The categories (such as wins, losses, and ties for a field hockey team) correspond to the states, and the numbers in each category (in our field hockey story, 18 wins, 4 losses, and 1 tie) correspond to the populations of the states.

Standard Divisor DEFINITION

The quotient of the total population, p, divided by the house size is called the **standard divisor**. If h denotes the house size and s is the standard divisor, then

$$s = \frac{p}{h}$$

Quota DEFINITION

In an apportionment problem, the **quota** is the exact share that would be allocated *if a whole number were not required.* To obtain a state's quota, divide its population by the standard divisor.

In the following course-scheduling problem, the courses to be taught correspond to the states, the numbers of students enrolled in each course correspond to the populations, and the house size is the total number of sections to be scheduled. Thus, in Example 1, there are three states: geometry, pre-calculus, and calculus, with populations 52, 33, and 15, respectively, and the house size is 5.

EXAMPLE 1 ▪ The High School Mathematics Teacher

A high school has one mathematics teacher who teaches all geometry, precalculus, and calculus classes. She has time to teach a total of five sections. One hundred students are enrolled as follows: 52 for geometry, 33 for pre-calculus, and 15 for calculus. How many sections of each course should be scheduled?

(*LWA-Dann Tardif/Corbis.*)

The number of students enrolled in each course is called the *population.* Thus, the populations of geometry, pre-calculus, and calculus are 52, 33, and 15, respectively. There are five sections for the 100 students, so the average section will have $100 \div 5 = 20$ students. We will call this average section size the *standard divisor,* because each quota can be determined by dividing the corresponding population by this number. Table 14.3 displays these calculations.

As shown in the table, the quotas add up to 5. It is tempting to round each quota to the nearest whole number, as in the right column of the table, but this makes 6 sections in all–too many! The purpose of an apportionment method is to

[1]This apportionment bill was the first bill in U.S. history to be vetoed.

find an equitable way to round a set of numbers such as these quotas without increasing or decreasing the original sum.

TABLE 14.3 Calculation of the Quotas for High School Mathematics Courses

Course	Population	Quota	Rounded
Geometry	52	$52 \div 20 = 2.60$	3
Pre-calculus	33	$33 \div 20 = 1.65$	2
Calculus	15	$15 \div 20 = 0.75$	1
Totals	100	5	6

EXAMPLE 2 ■ California's Quota

The Census Bureau recorded the apportionment population[2] of the United States, as of April 1, 2000, to be 281,424,177. There are 435 seats in the House of Representatives; therefore, the standard divisor is

$$s = \frac{281{,}424{,}177}{435} = 646{,}952$$

California's apportionment population was 33,930,798. Its quota is determined by dividing this population by the standard divisor. Thus,

$$\text{California's quota} = \frac{33{,}930{,}798}{646{,}952} = 52.447 \text{ seats}$$

California's apportionment, which is required to be a whole number, was set at 53 seats.

Ideally, each state's apportionment should be close to its quota. It is unrealistic to expect that any state will be apportioned its exact quota because each apportionment is required to be a whole number and the quota is unlikely to be a whole number. In choosing an apportionment method, we must decide what we mean by the phrase "each state's apportionment should be close to its quota."

Apportionment always involves rounding, and there are many ways to round. "Rounding down" means discarding the fractional part of a number q to obtain a whole number that we will denote $\lfloor q \rfloor$. Thus, $\lfloor 7.00001 \rfloor = 7$, $\lfloor 7 \rfloor = 7$, and $\lfloor 6.99999 \rfloor = 6$. "Rounding up" gives the next whole number, $\lceil q \rceil$. Thus, $\lceil 7.00001 \rceil = 8$, but $\lceil 7 \rceil = 7$.

There are numerous different apportionment methods. Each has flaws, and our goal is to understand how to choose a method that is appropriate for a particular apportionment problem.

14.2 The Hamilton Method

The congressional apportionment bill that President Washington vetoed was written by Alexander Hamilton. While it may appear that he simply rounded each quota to the nearest whole number, Hamilton was aware that there would be occasions–analogous to the examples of the field hockey team's percentages of wins, losses, and ties not summing to 100%, or the high school teacher receiving an extra class

[2] The apportionment population includes the resident population and the overseas population.

to teach–when the total number of seats apportioned in this way would be either more or less than the statutory house size. Hamilton called his method *largest fractions.*

Alexander Hamilton's Method and Upper, Lower Quotas DEFINITIONS

With the **Hamilton method**, each state receives either its **lower quota** $\lfloor q \rfloor$, which is its quota rounded down, or its **upper quota,** $\lceil q \rceil$, obtained by rounding the quota up. The states that receive their upper quotas are those whose quotas have the largest fractional parts.

The apportionment method of largest fractions, also known as the Hamilton method, was named for Alexander Hamilton. (*National Portrait Gallery/Art Resource, NY.*)

Implementing the Hamilton method is a three-step procedure:

1. Calculate each state's quota.
2. Tentatively assign to each state its lower quota of representatives. Each state whose quota is not a whole number loses a fraction of a seat at this stage, so the total number of seats assigned at this point will be less than the house size. This leaves additional seats to be apportioned.
3. Allot the remaining seats, one each, to the states whose quotas have the largest fractional parts, until the house is filled.

It is possible that a tie will occur, with the quotas of two states having identical fractional parts, but in practice, this rarely happens when large populations are involved.

EXAMPLE 3 ■ The High School Teacher's Dilemma

Let us use the Hamilton method to determine how many sections of geometry, pre-calculus, and calculus the high school teacher should teach. We have found that the quotas for the three subjects were 2.60, 1.65, and 0.75, respectively (see Table 14.3). The lower quotas are $\lfloor 2.60 \rfloor = 2$, $\lfloor 1.65 \rfloor = 1$, and $\lfloor 0.75 \rfloor = 0$, so we tentatively schedule two sections of geometry and one section of pre-calculus. With three sections apportioned, we have two more to assign to fill the house. These two sections go to the subjects with the largest fractions: Calculus has the largest fraction, 0.75, and gets a section; the second section goes to pre-calculus, whose fraction is 0.65. The final apportionment is as follows: geometry, two sections; pre-calculus, two sections; calculus, one section. ■

EXAMPLE 4 ▪ The Field Hockey Team

In Table 14.1, we saw that the percentages of wins, losses, and ties for the field hockey team were 78.26%, 17.39%, and 4.35%, respectively. These are the quotas that we must round so that they sum to 100%. To start, we apportion $\lfloor 78.26 \rfloor = 78\%$, $\lfloor 17.39 \rfloor = 17\%$, and $\lfloor 4.35 \rfloor = 4\%$ to the three categories. These lower quotas add up to 99%. The remaining 1% to be apportioned goes to the losses because their fraction, 0.39, is the largest. The final apportionment is 78% wins, 18% losses, and 4% ties. Coach will veto this apportionment.

EXAMPLE 5 ▪ Hamilton's Apportionment

The first congressional apportionment involved 15 states, and the House had 105 seats. According to the 1790 census, the U.S. population was 3,615,920. The standard divisor, 3,615,920 ÷ 105 = 34,437, represents the population of the average congressional district.

Table 14.2 displays Alexander Hamilton's proposed apportionment. Each quota shown in the table was calculated by dividing the state's population by this standard divisor. Adding the lower quotas, we find that their sum 97 leaves 8 seats to be apportioned. These go to the 8 states whose quotas had the largest fractional parts.

If you are using a calculator to follow the entries in the table, store the standard divisor in the calculator's memory. Then each quota can be figured by entering the individual state's population and dividing by the divisor recalled from memory.

President Washington's veto message stated that the fractions of seats gained by some states in the third step of the Hamilton apportionment were not related to the states' total populations. You may notice that Hamilton could have achieved the same apportionment by simply rounding each quota to the nearest whole number. He was aware that this doesn't always work, and our field hockey and high school examples confirm this. The famous orator and politician, Congressman Daniel Webster, devised a way to adjust the quotas before rounding, and developed an apportionment method that would have given the same result as Hamilton's apportionment–and would have answered Washington's objection. Webster's method is a **divisor method**, and it and two other divisor methods will be described later in this chapter.

Neither President Washington nor Alexander Hamilton were aware of the **Webster method** of apportionment, but even if they had been, the 1791 apportionment bill could have been dismissed on a technicality, because the Constitution requires each congressional district to have a population of at least 30,000. Delaware, with 55,540 inhabitants, was too small to have two districts. Washington was probably aware of this technicality, making it likely that his objection to the Hamilton method was sincere.

The veto prevented the Hamilton method from being used in 1792, but it was adopted by Congress in 1850 and remained in use until 1900. The half-century of experience with the Hamilton method revealed a paradox.

Paradoxes of the Hamilton Method

A *paradox* is a fact that seems obviously false. The first Hamilton apportionment paradox, called the **Alabama paradox**, was discovered in 1881. As part of the reapportionment procedure mandated by the Constitution, the Census Bureau had supplied Congress with a table of congressional apportionments for a range of different house sizes from 275 to 350, based on the 1880 census. The table revealed a strange phenomenon. Here is a portion of that table, just showing apportionments that changed when the house size increased from 299 to 300:

	House Size	
State	299	300
Alabama	8	7
Illinois	18	19
Texas	9	10

The paradox was that Alabama's apportionment decreased as a result of an increase in the number of seats in the House of Representatives. To see how this could happen, let's look at the quotas, shown in the following table.

	House Size		**Increase in quota**
State	299	300	
Alabama	7.646	7.671	0.025
Illinois	18.640	18.702	0.062
Texas	9.640	9.672	0.032

When the house size increased from 299 to 300, each state's quota increased by a factor of $\frac{300}{299}$. States with larger populations get larger increases, as the table shows. With a 299-seat House of Representatives, Alabama was the last state to get its upper quota. When the house size went to 300, the fractional parts of the quotas of Texas and Illinois advanced past the fractional part of Alabama's quota, so those states were awarded their upper quotas–and that meant Alabama was left with its lower quota. Alabama was left with its lower quota.

The Alabama Paradox DEFINITION

The **Alabama paradox** occurs when a state loses a seat as the result of an increase in the house size.

The Alabama paradox validates President Washington's veto message, issued 90 years before its discovery. If the fractional parts of the quotas, which determine the way the last few seats are apportioned, were related to the populations in any sensible way, the paradox could not have occurred.

EXAMPLE 6

A Mathematics Department Meets the Alabama Paradox

A mathematics department has 30 teaching assistants to cover recitation sections for College Algebra, Calculus I, Calculus II, Calculus III, and Contemporary Mathematics. The enrollments of these courses are given in Table 14.4. The department will use the Hamilton method to apportion the teaching assistants (TA's) to the five subjects. In this problem, the house size is 30 (the number of TA's) and the population is 750. Therefore, the standard divisor is $750 \div 30 = 25$; this represents the average number of students in each recitation section. Each quota shown in the table was determined by dividing the enrollment of the course by this divisor.

TABLE 14.4 Apportioning 30 Teaching Assistants

Course	Enrollment	Quota	Lower Quota	Apportion-ment
College Algebra	188	7.52	7	7
Calculus I	142	5.68	5 ↑	6
Calculus II	138	5.52	5	5
Calculus III	64	2.56	2 ↑	3
Contemporary Mathematics	218	8.72	8 ↑	9
Totals	750	30.00	27	30

The lower quotas add up to 27, so the three courses whose quotas have the largest fractional parts, Calculus I and III and Contemporary Mathematics, were given their upper quotas.

After the TA's were given their teaching assignments, the graduate school authorized the department to hire an additional TA. To determine which course should get the new TA, the department had to recalculate the apportionment. With 31 TA's, the standard divisor was $750 \div 31 = 24.19355$. The new quotas, determined by dividing each population by this new divisor, are shown in Table 14.5. Now the lower quotas add up to 28, so again three additional TA's go to the subjects whose quotas have the largest fractions. The Calculus III fraction, which had been larger than the College Algebra fraction when there were just 30 teaching assistants, has been surpassed. The new TA was placed in College Algebra, and one of the Calculus III TA's had to be reassigned to Calculus II.

TABLE 14.5 Apportioning 31 Teaching Assistants

Course	Enrollment	Quota	Lower Quota	Apportion-ment
College Algebra	188	7.771	7 ↑	8
Calculus I	142	5.869	5 ↑	6
Calculus II	138	5.704	5 ↑	6
Calculus III	64	2.645	2	2
Contemporary Mathematics	218	9.011	9	9
Totals	750	31.000	28	31

■

The size of the House of Representatives has been fixed by statute at 435 members since Arizona and New Mexico became states on February 14, 1912. Therefore, the Alabama paradox cannot occur. A second paradox, called the **population paradox**, is associated with a fixed house size.

The Population Paradox DEFINITION

The **population paradox** occurs when one state's population increases, and its apportionment decreases, while simultaneously another state's population increases proportionally less, or decreases, and its apportionment increases.

EXAMPLE 7 ■ Apportioning Seats in Parliament

A country has four political parties. Its parliament has 100 members, and seats are apportioned by the Hamilton method after each election so that the number of seats each party is awarded is as close as possible to being proportional to the number of votes the party receives.

An election is held but the parties are unable to form a government, so a new election is held. Here are the results of the two elections:

Party	First Election	Repeat Election
Whigs	5,525,381	5,657,564
Tories	3,470,152	3,507,464
Liberals	3,864,226	3,885,693
Centrists	201,203	201,049
Totals	13,060,962	13,251,770

The three major parties, Whigs, Tories, and Liberals, all received more votes in the second election, but the Centrists received fewer. The quotas for each party, shown in the following table, were determined by dividing each party's votes by the the standard divisors

$$13{,}060{,}962 \div 100 = 130{,}609.62$$

for the first election, and 132,517.70 for the second election.
The quotas are given in the following table.

Party	First Election	Repeat Election
Whigs	42.3045	42.6929
Tories	26.5689	26.4679
Liberals	29.5861	29.3221
Centrists	1.5405	1.5171
Totals	100.0000	100.0000

The lower quotas for the results of the first election were 42, 26, 29, and 1, with a sum of 98–thus the Tories and the Liberals, with the largest fractions, get extra seats. The apportionment after the first election was Whigs, 42; Tories, 27; Liberals, 30; and Centrists, 1.

For the repeat election, the lower quotas were the same, but now the largest fractions belong to the Whigs and the Centrists! Therefore the new apportionment is Whigs, 43; Tories, 26; Liberals, 29; and Centrists, 2.

The Centrists have *gained* a seat, although they received fewer votes in the repeat election, while the Liberals lost a seat even though their vote total increased in the repeat election. This is an instance of the population paradox. ■

14.3 Divisor Methods

The Jefferson Method

The Constitution requires congressional districts to be drawn so that the population of each is at least 30,000. President Washington could have vetoed the Hamilton apportionment bill because Delaware's population–only 55,540–was too small for the two congressional districts assigned to it. Thomas Jefferson proposed an apportionment method, now called the **Jefferson method**, to replace the Hamilton method.

Thomas Jefferson favored a method of apportionment biased in favor of states with large populations. (*National Portrait Gallery/Art Resource, NY.*)

In any apportionment, the standard divisor, which we will call s, obtained by dividing the total population by the house size, represents the average district population. In developing his method, Thomas Jefferson specified the population of the *smallest* district in the nation, which we will now call d. Using d, rather than s, as the divisor, each state receives an **adjusted quota** that is *always* rounded *down* to obtain its apportionment. If d is chosen correctly, the apportionments will add up exactly to the statutory house size.

In effect, the Jefferson method apportions to each state the maximum number of congressional districts of population d that will be accommodated by the state's population. Any leftover population is divided among these districts.

Once the divisor d is known, the Jefferson apportionment is easy to compute. The apportionment for state X, with population V, is

$$\left\lfloor \frac{V}{d} \right\rfloor$$

The actual divisor that Jefferson used was $d = 33{,}000$. Thus, Virginia's apportionment was

$$\left\lfloor \frac{\text{population of Virginia}}{33{,}000} \right\rfloor = \left\lfloor \frac{630{,}560}{33{,}000} \right\rfloor = \lfloor 19.108 \rfloor = 19$$

Therefore, Virginia received 19 seats, rather than 18, which Hamilton's bill would have allocated. Delaware's apportionment was 1 seat instead of 2, as the following calculation shows.

$$\left\lfloor \frac{\text{population of Delaware}}{33{,}000} \right\rfloor = \left\lfloor \frac{55{,}540}{33{,}000} \right\rfloor = \lfloor 1.683 \rfloor = 1$$

The Jefferson method is one of a class of apportionment methods called *divisor methods.*

Divisor Methods DEFINITION

A **divisor method** of apportionment determines each state's apportionment by dividing its population by a common divisor d and rounding the resulting quotient. Divisor methods differ in the rule used to round the quotient.

The divisor must be carefully chosen to achieve the correct house size. In the first congressional apportionment, $d = 30{,}000$ would have resulted in larger apportionments for several states and a house size of 112, while $d = 36{,}000$ would have decreased several apportionments, and the house size would have been 91.

Critical Divisors

To implement the Jefferson method, we must determine the divisor, d. Start by determining the standard divisor, s, and the quota for each state, as we did with the Hamilton method. Each state is assigned, as a **tentative apportionment**, its lower quota. Thus, the tentative apportionment of state X is

$$\left\lfloor \frac{\text{Population of } X}{s} \right\rfloor$$

As we noted with the Hamilton method, these tentative apportionments are not enough to fill the house, so the apportionments of some states will have to be increased. To determine which states should receive additional seats, calculate a **critical divisor** for each state.

Critical Divisor DEFINITION

A state's **critical divisor** is the number that can be divided into the state's population to produce a number just on the borderline for changing the state's apportionment.

The formula for the critical divisor for a given apportionment method is determined by the rounding method used. The following theorem gives the critical divisor for the Jefferson method.

Critical Divisor—The Jefferson Method THEOREM

Let N be the tentative apportionment of a state X. The **critical divisor** for X is

$$\frac{\text{Population of } X}{N + 1}$$

EXAMPLE 8

Critical Divisors for Virginia and Delaware

Referring to Table 14.2, the lower quota for Virginia was 18 and the lower quota for Delaware was 1. The populations of the two states were 630,560 and 55,540, respectively, so their critical divisors were

$$\frac{630{,}560}{18 + 1} = 33{,}187 \text{ and } \frac{55{,}540}{1 + 1} = 27{,}770, \text{ repectively.}$$

The significance of the critical divisor is that if the Jefferson divisor d is equal to the critical divisor of state X (whose tentative apportionment was N), then X will contain *exactly* $N + 1$ districts of population d. Thus, with $d = 33{,}187$, Virginia's apportionment would be equal to 19. If we took $d = 27{,}770$, then Delaware would get 2 seats. However, the same divisor must be used for all states, so with this lower value of d, Virginia would get

$$\left\lfloor \frac{630{,}560}{27{,}770} \right\rfloor = 22 \text{ seats}$$

Once the critical divisors are determined, the state with the largest critical divisor is entitled to another seat, because when that divisor is used, no other state will receive a changed apportionment–a state receives additional seats only when a divisor smaller than or equal to its critical divisor is used. The total apportionment is thus increased by 1 in this step.

If there remain additional seats to be apportioned, we recompute the critical divisor for the state whose tentative apportionment has increased, and repeat the process. When the house is filled, the critical divisor most recently used is the divisor d, representing the minimum district population. With the Jefferson method (and any other divisor method), the critical divisors determine the priority of a state for receiving additional seats. *The state with the largest critical divisor gets the next seat, but after receiving that seat, its critical divisor is recomputed, and usually another state will then be first in line.*

EXAMPLE 9 ▪ The Field Hockey Team

The field hockey team had 18 wins, 4 losses, and 1 tie last season. The Jefferson method can be used to express this record as percentages. The house size is 100%, and the total population is the 23 games played. We've previously determined the quotas (see Table 14.1 and Example 4) to be 78.26% for wins, 17.39% for losses, and 4.35% for ties. The lower quotas are the tentative apportionments: 78, 17, and 4, respectively. The critical divisors are $\frac{18}{78+1} = 0.22785$ for wins, $\frac{4}{17+1} = 0.22222$ for losses, and $\frac{1}{4+1} = 0.20000$ for ties.

The category with the largest critical divisor is "wins"; its apportionment is raised to 79%. The house is now full; we have the apportionment that Coach wanted: 79% wins, 17% losses, and 4% ties.

EXAMPLE 10 ▪ The High School Mathematics Teacher

The teacher can be assigned five classes. There are 52 students enrolled in geometry, 33 in pre-calculus, and 15 in calculus. Let's use the Jefferson method to determine her teaching assignment. We have previously determined that the lower quotas for the subjects are 2, 1, and 0, respectively (see Table 14.3). Thus, the critical divisors are $\frac{52}{2+1} = 17\frac{1}{3}$ for geometry, $\frac{33}{1+1} = 16\frac{1}{2}$ for pre-calculus, and $\frac{15}{0+1} = 15$ for calculus.

Geometry, with the largest critical divisor, has first priority for a new section, and its tentative apportionment is increased to 3. Its new critical divisor will be $52 \div (3 + 1) = 13$. Now pre-calculus has top priority, and its tentative apportionment is increased to 2. The house is now full, so the final apportionments are 3 sections of geometry and 2 sections of pre-calculus. The minimum section size will be the $16\frac{1}{2}$, because that is the critical divisor for the subject that was the last to receive an increased apportionment: pre-calculus. In practice, the two sections will have 16 and 17 students. Because the enrollment for calculus is less than the minimum section size, there will be no calculus class.

Examples 7, 8, and 9 demonstrate that different methods may yield different apportionments, because the Hamilton method gave different apportionments in each case.

Here is another distinction between the methods. The Hamilton method gives each state either its upper quota or its lower quota. With the Jefferson method, no state can receive less than its lower quota as its apportionment–because the lower quota is the initial tentative apportionment–but a state can be apportioned more than its upper quota.

EXAMPLE 11 ■ The 1820 Congressional Apportionment

According to the 1820 census, New York had a population of 1,368,775. The total population of the United States was 8,969,878, and the house size was 213. Therefore, the standard divisor was $8{,}969{,}878 \div 213 = 42{,}112$, and New York's quota was $1{,}368{,}775 \div 42{,}112 = 32.503$. The Hamilton method would have apportioned to New York its upper quota, 33 seats. The divisor for the Jefferson method was $d = 39{,}900$. Thus New York's apportionment was $\lfloor 1{,}368{,}775 \div 39{,}900 \rfloor = 34$ seats.

An apportionment method is said to satisfy the **quota condition** if in every situation each state's apportionment is equal to either its lower quota or its upper quota. It takes only one example like the 1820 apportionment to show that the Jefferson method does not satisfy the quota condition. In fact, if the house had continued to use the Jefferson method, it would have violated the quota condition in every apportionment since 1850.

The Hamilton method satisfies the quota condition. This was obvious to Congress in 1850, so it based its apportionment on the Hamilton method.[3]

The Jefferson method, however, is not troubled by the Alabama and population paradoxes. Consider the Alabama paradox, in which a state loses a seat as a result of an increase in the house size. With any apportionment method, the apportionments of some states must increase when the size of the house increases. The Jefferson method awards seats in order of critical divisors. When the house size increases, the next seat will go to the state with the next largest critical divisor. There is no opportunity for a state to lose a seat.

Congress has never used an apportionment method that satisfies the quota condition and avoids the paradoxes. It would be desirable to use such a method, and in the 1970s, the mathematicians Michel L. Balinski and H. Peyton Young set out to find one. They succeeded in finding a method, which they called the *quota method,* that satisfies the quota condition (as the Hamilton method does) and avoids the Alabama paradox (as the Jefferson method does). However, the population paradox remained. They subsequently proved that the only apportionment methods that are free of the population paradox are the divisor methods. It is known that every divisor method is capable of violating the quota condition, so Balinski and Young have proved an impossibility theorem: *No apportionment method that satisfies the quota*

[3]The origins of the Hamilton method had been forgotten in 1850, and the method was named for Congressman Samuel Vinton, who had rediscovered the method.

condition is free of paradoxes. This theorem is like Kenneth Arrow's theorem that there is no completely satisfactory way to decide multicandidate elections based on voter preference schedules (see section 9.4).

The Jefferson method favors the larger states. It is not an accident that in every example that we have considered, the "state" with the largest population fared better with the Jefferson method than it did with the Hamilton method. Virginia got a greater apportionment, and Delaware a smaller apportionment in 1790; the winning percentage for the field hockey team was higher, and the losing percentage lower, when the Jefferson method was used, as compared with the Hamilton method, and there were more sections of geometry, and no sections of calculus when the Jefferson method was substituted for the Hamilton method.

Let's see why the Jefferson method is biased in favor of larger states. In an apportionment problem, let s be the standard divisor and let d be the divisor used in the Jefferson method. The apportionment given to state X is then $\lfloor U \rfloor$, where

$$U = (\text{Population of } X) \div d.$$

We'll call U the state's *adjusted quota.* Comparing U with the quota q for X, we will see that the formulas are similar:

$$q = (\text{Population of } X) \div s.$$

In fact, the state's population neatly cancels out of the ratio $U \div q$:

$$U \div q = s \div d.$$

Thus, while U and q have different values for each state, the ratio of U to q is always the same number, $M = s \div d$. Multiplying both sides of the identity $U \div q = M$ by q, we obtain $U = M \times q$. Therefore the Jefferson apportionment for state X is

$$\lfloor M \times (\text{quota for } X) \rfloor$$

Consider the congressional apportionment of 1820. The standard divisor was $s = 42{,}112$, and the Jefferson divisor was $d = 39{,}900$. Thus, the quotient, M, is $42{,}112 \div 39{,}900 = 1.0554$. Now suppose that a state has a quota of q. The state's adjusted quota is $U = (1 + 0.0554) \times q = q + q \times 5.54\%$. In words, this algebraic formula says the adjusted quota is obtained by giving the state a 5.54% raise on its quota. A state with a large quota will get a greater raise than a small state will.

To see how this works with numbers, consider a state X with $q = 18.96$. The adjusted quota is

$$U = 1.0554 \times 18.96 = 20.01$$

The upper quota for X of 19, but X is awarded $\lfloor 20.01 \rfloor = 20$ seats! This violates the quota condition, and in fact, every state whose quota is 18.96 or more will be guaranteed to get at least its upper quota with this value of M. If a state has a quota of $2 \times 18.96 = 37.92$, an identical calculation shows that it will receive at least its upper quota plus one seat. On the other hand, consider a small state whose lower quota is 1. To increase its apportionment to 2, its quota must be at least $2 \div M = 1.89502$. Thus, a state with quota 18.96 gets more than its upper quota, and a state with quota 1.89 has to settle for its lower quota.

The Webster Method

The Method of Daniel Webster DEFINITION

The **Webster method** is the divisor method that rounds the quota (adjusted if necessary) to the nearest whole number, rounding up when the fractional part is greater than or equal to $\frac{1}{2}$, and rounding down when the fractional part is less than $\frac{1}{2}$.

Statesman and orator Daniel Webster (1782–1852), who developed a divisor method for apportioning the U.S. House of Representatives. (*National Portrait Gallery/Art Resource, NY.*)

The Webster and Jefferson methods are immune to the Alabama and population paradoxes, but neither satisfies the quota condition. However, the Jefferson method favors the large states, the Webster method is neutral, favoring neither the large nor the small states. Furthermore, the Webster method rarely violates the quota condition by giving a state more than its upper quota, or fewer seats than its lower quota, and would not have done so in any of the 22 congressional apportionments that have occurred so far.

Here is the procedure to calculate an apportionment by the Webster method.

1. Determine the standard divisor, and use it to find the quota for each state.
2. Obtain the tentative apportionments by rounding each quota q to $\lfloor q \rfloor$ if the fractional part of q is less than 0.5; otherwise, round to $\lceil q \rceil$. (This is the standard way of rounding numbers.)
3. Add the rounded quotas. If their sum is equal to the house size, the job is finished. The tentative apportionments calculated in step 2 are the final apportionments.

The procedure for implementing the Webster method of apportionment when the rounded quotas of the states don't add up to the house size is similar to the procedure for the Jefferson method. The quotas must be adjusted before rounding by using a divisor that is either larger than the standard divisor (if the sum of the rounded quotas is greater than the house size), or smaller than the standard divisor (if the sum of the quotas is less than the house size). To determine the correct divisor, one has to calculate a critical divisor for each state. We will denote the critical divisor by d^+ if we are looking for a divisor that is smaller than the standard divisor s in order to increase the number of seats apportioned. When a divisor greater than s is needed, to decrease the total apportionment, the critical divisor is denoted d^-.

It is never necessary to compute both critical divisors d^+ and d^-, and when the sum of the rounded quotas is equal to the house size h, no critical divisors are needed at all.

District Population DEFINITION

The **district population** of state X is $V \div A$, where V is the state's population, and A is its apportionment. The district population is the average population of a congressional district in the state.

Apportionments can be evaluated by computing differences in district population. The best apportionment by this standard is the one for which the worst difference in district populations between states is minimal.

EXAMPLE 15 ■ Comparing District Populations

If we consider differences in district population rather than representative share, it was correct to give Michigan 17 seats and Arkansas 7 in the 78th Congress. The district population for Michigan was

$$\frac{\text{population of Michigan}}{\text{Michigan's apportionment}} = \frac{5{,}256{,}106 \text{ people}}{17 \text{ districts}}$$
$$= 309{,}183 \text{ people per district}$$

The district population for Arkansas was $1{,}949{,}387 \div 7 = 278{,}484$. The Arkansas average district population was 30,699 less than Michigan's. If Michigan had 18 seats and Arkansas had 6, Michigan would have the lesser district population, 292,006, while Arkansas's would have increased to 324,898. This adjustment in apportionment would have increased the inequity between the two states, because now Arkansas would be worse off than Michigan by 32,892 in district population.

For state X, the representative share is $\frac{A}{V}$ and the district population is $\frac{V}{A}$, where A is the apportionment of X and V is its population. Thus:

$$\text{representative share for state } X = \frac{1}{\text{district population for state } X}$$

It may be surprising that these two ways of evaluating the fairness of an apportionment could disagree, but we have just seen that they can.

A mathematician, Edward V. Huntington, pointed out that if **relative differences** are compared instead of absolute differences, then either district population and representative share would give identical comparisons of apportionments–and he suggested a compromise.

Absolute and Relative Differences DEFINITION

Given two positive numbers A and B, with $A > B$, the **absolute difference** is $A - B$ and the **relative difference** is the quotient $\frac{(A-B)}{B} \times 100\%$.

For any two states, it turns out that the *relative difference* in district populations is equal to the relative difference in representative share (see Exercise 40). Therefore, an apportionment method that minimizes *relative* difference in representative shares will also minimize the relative difference in district populations.

EXAMPLE 16 ■ Relative Inequity in the 78th Congress

Recall that Michigan was given 17 seats in the 78th Congress and had a representative share of 3.234 seats per million. Arkansas had 7 seats and a representative share of 3.591 seats per million. The relative difference was

$$\frac{3.591 - 3.234}{3.234} \times 100\% = 11.0\%$$

In terms of district populations, recall that Michigan had a district population of 309,183, and Arkansas had a district population of 278,484. The relative difference in district populations was

$$\frac{309{,}183 - 278{,}484}{278{,}484} \times 100\% = 11.0\%$$

Thus, by either measure, Arkansas was 11.0% better represented in the 78th Congress than Michigan was.

If Michigan had 18 seats and Arkansas had 6, the relative difference in representative shares would be found by subtracting the smaller representative share (Arkansas's) from the larger (Michigan's), and expressing the result as a percentage of the smaller representative share. Thus, the relative difference would have been

$$\frac{3.425 - 3.078}{3.078} \times 100\% = 11.3\%$$

in Michigan's favor. The same relative difference would be found if the district populations were compared. Because the relative inequity was less when Michigan had 17 seats and Arkansas had 7, the 1941 apportionment was preferred from the point of view of Professor Huntington's compromise. ■

To optimize apportionment by the relative difference criterion for equity, Professor Huntington and a statistician from the Bureau of the Census, Joseph Hill, designed a new divisor method. It has been used to apportion seats in the U.S. House of Representatives after each decennial census since 1940.

The Hill–Huntington Method

Like the Jefferson and Webster methods, the apportionment is calculated by rounding the quotas, after adjusting them if necessary. The only difference between the three divisor methods is in the rounding procedure.

The Hill–Huntington rounding procedure is related to the **geometric mean.**

Geometric Mean DEFINITION

The **geometric mean** of two positive numbers A and B is equal to the square root of their product, $\sqrt{A \times B}$.

Consider the rectangle $\mathcal{R}$ displayed in Figure 14.1. The area of $\mathcal{R}$ is the product of the lengths A and B, or $A \times B$. The geometric mean of A and B is equal to the length E of the edge of a square $\mathcal{S}$ with the same area as $\mathcal{R}$, because the area of $\mathcal{R}$ is E^2, and thus $E^2 = A \times B$. Taking square roots, $E = \sqrt{A \times B}$.

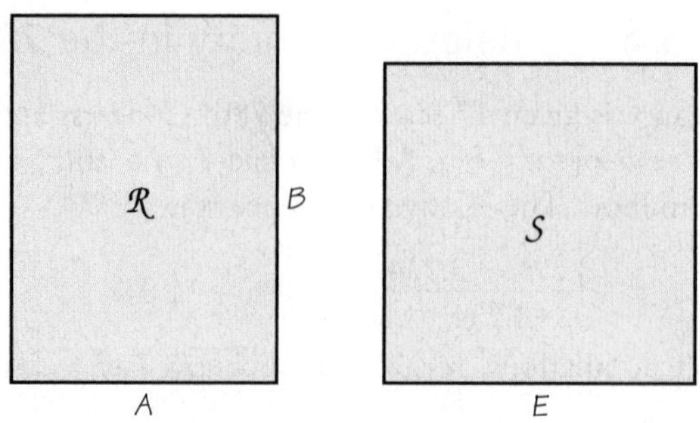

FIGURE 14.1 The edge of the square is the geometric mean of the edges of the rectangle, because the two figures have the same area.

Given a positive number q, let q^* be the geometric mean of $\lfloor q \rfloor$ and $\lceil q \rceil$. This q^* is called the *rounding point* for q. The Hill–Huntington rounding of a number q is equal to $\lfloor q \rfloor$ if $q < q^*$, where q is the rounding point, and $\lceil q \rceil$ if $q \geq q^*$.

EXAMPLE 17 ■ Hill–Huntington Rounding

Suppose that $q = 7.485$. Jefferson and Webster would round 7.485 down to 7. Because $\lfloor 7.485 \rfloor = 7$ and $\lceil 7.485 \rceil = 8$, the rounding point is $q^* = \sqrt{7 \times 8} = 7.48331\ldots$, which is less than 7.485. Therefore, the Hill–Huntington rounding of 7.485 is $\lceil 7.485 \rceil = 8$.

Hill–Huntington apportionment calculations follow the general plan of the Jefferson and Webster methods. Round each state's quota the Hill–Huntington way to obtain a first tentative apportionment. If the sum of the tentative apportionments is equal to the house size, the job is finished. If not, a list of critical divisors must be constructed, each chosen to be just sufficient to change the corresponding state's apportionment by one seat in the desired direction.

Critical Divisors for the Hill–Huntington Method — THEOREM

If N is the number of seats apportioned tentatively to state X, and the total apportionment is too small, the critical divisor for that state is

$$d^+ = \frac{V}{\sqrt{N(N+1)}}$$

where V is the state's population. The state with the largest critical divisor gains a seat. If the house is still not full, recompute the critical divisor for the state whose apportionment changed, and repeat the process.

If the total apportionment is too large, the critical divisor for state X is

$$d^- = \frac{V}{\sqrt{N(N-1)}}$$

In this case, the state with the smallest critical divisor loses a seat.

A zero apportionment is impossible with the Hill–Huntington method because the rounding point for quotas between 0 and 1 is $\sqrt{0 \times 1} = 0$. Any quota less than 1 will be rounded to 1.

EXAMPLE 18 ■ Percent Effort

Faculty members at a certain university must state the percentage of their time spent in several activities. To comply as accurately as possible, Professor Worktorule has requisitioned five stopwatches to keep track of her activities. Here is what she recorded over the course of one week:

Instruction	300 minutes
Instructional support	705 minutes
Independent study	31 minutes
Research	2475 minutes
Committee work	89 minutes
Total	3600 minutes

The professor is too busy to covert the data into percentages–which the university requires in whole numbers with sum 100%–so we'll do it, using the Hill–Huntington method. As with any percentage apportionment problem, the house size is 100, so the standard divisor–one percentage unit–is 36 minutes. Table 14.6 shows the quotas, the rounding points, and the tentative apportionments, obtained by rounding the quotas up or down, depending on whether the quota exceeds the rounding point or not. The tentative apportionments add up to 101%, so we must determine the critical divisors in order to decide which tentative apportionment to reduce. These are calculated by dividing the minutes of effort in each category by $\sqrt{N(N-1)}$, where N is the tentative apportionment. Research has the smallest critical divisor, $2475 \div \sqrt{69 \times 68} = 36.13$. Therefore the percentage effort in research is reported as 68%.

TABLE 14.6 Apportioning Professor Worktorule's Effort by the Hill–Huntington Method

Effort Category	Effort (min.)	Quota	Rounding Point	Tentative Apportionment	Critical Divisor	Final Apportionment
Instruction	300	8.33%	8.49	8%	40.09	8%
Instructional Support	705	19.58%	19.49	20%	36.17	20%
Independent Study	31	0.86%	0	1%	∞	1%
Research	2475	68.75%	68.5	69%	36.13	68%
Committees	89	2.47%	2.45	3%	36.33	3%
Totals	3600	100%	-	101%	-	100%

EXAMPLE 19 ■ The 435th Seat

When Congress was apportioned as a result of the 1940 census, the last seat in the house was in play. Michigan had 17 seats for its population of 5,256,106. Arkansas, with a population of 1,949,387, had 6 seats. With the Webster method, Michigan's critical divisor was $5{,}256{,}106 \div 17.5 = 300{,}349$, while Arkansas's was $1{,}949{,}387 \div 6.5 = 299{,}906$. Thus the Webster method apportions the 435th seat to Michigan.

By the Hill–Huntington method, the critical divisors for Michigan and Arkansas were $5{,}256{,}106 \div \sqrt{17 \times 18} = 300{,}472$ and $1{,}949{,}387 \div \sqrt{6 \times 7} = 300{,}797$, respectively. Because the Hill–Huntington divisor for Arkansas exceeded Michigan's, the seat went to Arkansas. The politics behind the decision to use the Hill–Huntington method rather than the Webster method is the subject of Spotlight 14.2.

SPOTLIGHT 14.2 Mathematics and Politics: A Strange Mixture

Walter F. Willcox

(Department of Manuscripts and University Archives, Cornell University Libraries.)

Edward V. Huntington

(Courtesy of Harvard University Archives.)

The first American to consider apportionment from a theoretical point of view was Walter Willcox (1861–1964), who strongly advocated the Webster method and had computed the apportionment of 1902. His arguments convinced Congress to use the Webster method again in 1912. In 1911, Joseph Hill, a statistician at the Census Bureau, proposed the Hill–Huntington method—with the strong endorsement of Edward V. Huntington, a mathematics professor at Harvard.

In 1920, the two methods were in competition. There were significant differences in the apportionments determined by the two methods, and the result was Washington gridlock: No apportionment bill passed during the decade, and the 1912 apportionments were retained throughout the 1920s. In preparation for the 1930 census results, the National Academy of Sciences formed a committee to study apportionment. In 1929, the committee endorsed the Hill–Huntington method.

The 1930 census was remarkable in that the apportionments calculated by the Webster method were the same as the Hill–Huntington apportionments. The House was therefore reapportioned, but the method used could be claimed to be either one of the competing methods. The coincidence was almost repeated in the 1940 census, but there was one difference. The Hill–Huntington method gave the last seat to Arkansas, while Webster's method gave it to Michigan (see page 453). At the time, Michigan was a predominantly Republican state, and Arkansas was in the Democratic column. The vote on the apportionment bill split strictly along party lines, with Democrats supporting the Hill–Huntington method and Republicans voting for the Webster method. Because the Democrats had the majority, the Hill–Huntington method became the law.

We have seen that the Jefferson method is biased in favor of populous states and that the Webster method is not biased in regard to state population size. It's natural to ask if the Hill–Huntington method exhibits any bias with respect to state population.

A divisor method will show bias in favor of large states when the quotas are adjusted by using a divisor that is smaller than the standard divisor. If the quotas must be adjusted downward–that is, a divisor larger than the standard divisor is used–small states are favored. Because the rounding point for the Webster method is halfway between whole numbers, it is just as likely for the divisor to be smaller than the standard divisor as it is for it to be larger.

For any positive number q, the rounding point used by the Hill–Huntington method is closer to $\lfloor q \rfloor$ than to $\lceil q \rceil$ (see Exercise 36). This means that a random number q is more likely to be above the rounding point and thus rounded up to $\lceil q \rceil$ than it is to be less and thus rounded down to $\lfloor q \rfloor$. The difference between the Webster and Hill–Huntington ways of rounding is not significant for relatively large numbers. For example, the Hill–Huntington rounding point between 50 and 51 is 50.498. Therefore, a number q between 50 and 51 will be rounded up to 51 by Hill–Huntington if it is larger than 50.498. The Webster method would round q to 51 if $q >$ 50.500. The differences are more significant when rounding smaller numbers. Hill–Huntington rounds all numbers between 0 and 1 up to 1; Webster rounds only the numbers in the range 0.500–1 up to 1. When the Hill–Huntington method is used for apportionment, the sum of the tentative apportionments is more likely to exceed the house size than it is to be less, especially if there are many states with small populations. Therefore, the Hill–Huntington method is likely to use a divisor larger than the standard divisor. This favors the less populous states.

In conclusion, the Webster method is the best divisor method for general use. It is the only divisor method that is unbiased regarding population size, and it minimizes differences between representative share. Although the Webster method is capable of violating the quota condition, it is the divisor method least likely to do so.

For apportionment of seats in the U.S. House of Representatives, a slight modification is needed, because no state can receive a zero apportionment. The rounding point for quotas less than 1 is set to zero, rather than 0.5.

There are situations where other apportionment methods could be considered. See Exercise 44 to explore the problem of teaching assignments, and Exercise 34 for apportionment of seats in a parliament.

REVIEW VOCABULARY

$\lfloor q \rfloor$ The result of rounding a number q down; for example, $\lfloor \pi \rfloor = 3$. (p. 437)

$\lceil q \rceil$ The result of rounding a number q up to the next integer; for example, $\lceil \pi \rceil = 4$. (p. 437)

Absolute difference The result of subtracting a smaller number from a larger number. (p. 452)

Adjusted quota The result of dividing a state's quota by a divisor other than the standard divisor. The purpose of adjusting the quotas is to correct a failure of the rounded quotas to sum to the house size. (p. 442)

Alabama paradox A state loses a representative solely because the size of the House is increased. This paradox is possible with the Hamilton method but not with divisor methods. (p. 438)

Apportionment method A systematic way of computing solutions of apportionment problems. (p. 433)

Apportionment problem To round a list of fractions to whole numbers in a way that preserves the sum of the original fractions. (p. 433)

Critical divisor The number closest to the standard divisor that can be used as a divisor of a state's population to obtain a new tentative apportionment for the state. The following table lists formulas for critical divisors for some divisor methods. In the table, V stands for the state's population, and N is its tentative apportionment. (p. 443)

Method	Critical Divisor Causing Tentative Apportionment: To Increase	To Decrease
Jefferson	$\frac{V}{N+1}$	Not necessary
Webster	$\frac{V}{N+\frac{1}{2}}$	$\frac{V}{N-\frac{1}{2}}$
Hill–Huntington	$\frac{V}{\sqrt{N(N+1)}}$	$\frac{V}{\sqrt{N(N-1)}}$

District population A state's population divided by its apportionment. (p. 452)
Divisor method One of many apportionment methods in which the apportionments are determined by dividing the population of each state by a common divisor to obtain adjusted quotas. The apportionments are calculated by rounding the adjusted quotas. Divisor methods differ in the way that the rounding of the quotas is carried out. The methods of Jefferson, Webster, and Hill–Huntington are divisor methods. (p. 443)
Geometric mean For positive numbers A and B, the geometric mean is defined to be $\sqrt{A \times B}$. (p. 453)
Hamilton method An apportionment method that assigns to each state either its lower quota or its upper quota. The states that receive their upper quotas are those whose quotas have the largest fractional parts. (p. 437)
Hill–Huntington method A divisor method that minimizes relative differences in both representative shares and district populations. (p. 453)
Jefferson method A divisor method based on rounding all fractions down. Thus, if U is the adjusted quota of state X, the state's apportionment is $\lfloor U \rfloor$. (p. 442)
Lower quota The integer part $\lfloor q \rfloor$ of a state's quota q. (p. 437)
Population paradox A situation in which the apportionment of one state, A, decreases, although its population has increased; while another state, B, loses population (or increases population proportionally less than state A) and gains a seat. This paradox is possible with all apportionment methods *except* divisor methods. (p. 440)
Quota The quotient $V \div s$ of a state's population divided by the standard divisor s. The quota is the number of seats a state would receive if fractional seats could be awarded. (p. 435)
Quota condition A requirement that an apportionment method should always assign to each state either its lower quota or its upper quota in every situation. The Hamilton method satisfies this condition, but none of the divisor methods do. (p. 445)
Relative difference The relative difference between two positive numbers is obtained by subtracting the smaller number from the larger, and expressing the result as a percentage of the smaller number. Thus, the relative difference of 120 and 100 is 20%. (p. 452)
Representative share A state's representative share is the state's apportionment divided by its population. It is intended to represent the amount of influence a citizen of that state would have on his or her representative. (p. 451)
Standard divisor The ratio $p \div h$ of the total population p to the house size h. In a congressional apportionment problem, the standard divisor represents the average district population. (p. 435)
Tentative apportionment The result of rounding a state's quota or adjusted quota to obtain a whole number. (p. 443)
Upper quota The result of rounding a state's quota *up* to a whole number. A state whose quota is q has an upper quota equal to $\lceil q \rceil$. (p. 437)
Webster method A divisor method of apportionment that is based on rounding fractions the usual way. The Webster method minimizes the *absolute* differences of representative share between states. (p. 438)

✓ SKILLS CHECK

1. A county is divided into 3 districts with the following populations: Southern, 3600; Western, 3100; Northeastern, 1600. There are 6 seats on the county council to be apportioned. What is the quota for the Southern district?

(a) 2.6
(b) 2.8
(c) 3

2. Two calculus teachers can teach a total of 8 classes. Enrollments are as follows: Calculus I, 200; Calculus II, 100; Calculus III, 52. In this apportionment problem, the population is ______, the standard divisor is ______, and the quotas are ______ for Calculus I, ______ for Calculus II, and ______ for Calculus III.

3. A, B, and C are arguing about fractions of a cent. On a project, they worked exactly 33, 34, and 35 minutes, respectively, and were paid $100. Use the Hamilton method to see who gets his upper quota (in cents!).

(a) A
(b) B
(c) C

4. Round each number in the sum 13.62 + 12.58 + 17.51 + 16.77 + 19.52 = 80 to a whole number.

____ + ____ + ____ + ____ + ____ = 80

5. The population paradox occurs when

(a) A state's apportionment decreases because the house size increased.
(b) A state's apportionment decreases, and its apportionment increases, while another state's apportionment decreases, even though its population has increased.
(c) The Jefferson method is used.

6. The Alabama paradox occurred when it was noticed that Alabama would lose a seat, in apportionment by the Hamilton method, if the house size was changed from 299 to ______________ .

7. When rounding the numbers in the sum $20.45 + 30.30 + 49.25 = 100$ by the Jefferson method, the largest critical divisor belongs to

(a) 20.45
(b) 30.30
(c) 49.25

8. Use the Jefferson method to apportion the sum $0.8 + 0.9 + 98.3 = 100$ as a sum of whole numbers.
____ + ____ + ____ = 100

9. The Jefferson method frequently

(a) gives the smallest state less than its lower quota.
(b) gives the largest state more than its upper quota.
(c) gives a state a lesser apportionment if the house size increases.

10. Use the Webster method to apportion the sum $0.8 + 0.9 + 98.3 = 100$ as a sum of whole numbers.
____ + ____ + ____ = 100

11. When rounding the numbers in the sum $20.45 + 30.30 + 49.25 = 100$ by the Webster method, the largest critical divisor belongs to

(a) 20.45
(b) 30.30
(c) 49.25

12. States *A* and *B* have populations of 1 million and 2 million, respectively. If their respective apportionments are 2 and 3, then the absolute difference in representative share is ____ per million.

13. If the apportionments of the two states in Skills Check 12 were changed to 1 for *A* and 4 for *B* then

(a) the absolute difference in representative share would increase.
(b) the absolute difference in representative share would decrease.
(c) the absolute difference in representative share would be unchanged.

14. If the criterion is absolute difference in district population, the most equitable apportionment of 5 seats to states *A* and *B* in Skills Check 12 is ____ for *A*, and ____ for *B*.

15. If the initial calculations leading to the Hill–Huntington apportionment result in a sum that is too large, what happens next?

(a) A seat is taken from the state with the smallest critical divisor.
(b) The largest apportionment is reduced.
(c) A different method must be used.

16. The ________ method has been used since 1941 to apportion seats in the U.S. House of Representatives.

17. Which divisor method never apportions to a state fewer seats than its lower quota?

(a) Hill–Huntington
(b) Webster
(c) Jefferson

18. A school principal is apportioning sections of the school's mathematics classes. She wants to set a minimum section size, and to adjust it so that a total of 32 sections are open. She should use the ____ method.

19. The Hill–Huntington minimizes relative differences in

(a) district population.
(b) representative share.
(c) both district population and representative share.

20. The divisor method that shows the least bias in favor of either large states or small states is the _______ method.

CHAPTER 14 EXERCISES

■ Challenge ◆ Discussion

14.1 The Apportionment Problem

1. Jane has decided to track her daily expenses, and finds them to be as listed in the table at right.

Express these as percentages. If rounded to whole numbers, do the percentages add up to 100%?

Rent	$31
Food	16
Transportation	7
Gym	12
Miscellaneous	5

2. A mathematics department uses 20 teaching assistants to aid in its four-semester calculus course. The number of teaching assistants assigned to each level of the course depends on enrollment. Here are the fall enrollments:

Calculus I	500
Calculus II	100
Calculus III	350
Calculus IV	175
Total	1125

How many teaching assistants should be assigned to each level of the course?

3. Should the mathematics department in Exercise 2 revise the assignments for its TA's? Grades have been posted for the previous semester, and some students need to repeat the previous level of the course. Forty-five students move from Calculus II to Calculus I, 41 students move from Calculus III to Calculus II, and 12 students move from Calculus IV to Calculus III.

◆ **4.** Here is a typical apportionment problem. Round the numbers in the sum to integers:

$$8.37 + 10.33 + 12.38 + 5.47 + 3.45 = 40$$

The rounded numbers must add up to 40. How would you approach this?

14.2 The Hamilton Method

5. Use the Hamilton method to round each of the following numbers in the sum to a whole number, preserving the total of 10.

$$0.36 + 1.59 + 0.99 + 2.33 + 2.38 + 2.35 = 10$$

6. *The 37th pearl.* Three friends have bought a bag guaranteed to contain 36 high-quality pearls for $14,900 at an auction. Abe contributed $5900, Beth's contribution was $7600, and Charles supplied the remaining $1400. After taking the bag to your house, they pour the 36 pearls from the bag onto the kitchen table.

(a) How many should each friend get if the Hamilton method is used to apportion the pearls according to the size of the contributions?

(b) Charles has noticed the bag isn't empty! Another pearl comes out, and you are asked to recalculate the apportionment.

◆ **(c)** How do you explain the result to Charles?

◆ **7.** A country has three political parties, and apportions seats in its 102 seat parliament by the Hamilton method proportionately to the number of votes each receives. In a recent election, the Pro-UFO party received 254,000 votes, the Anti-UFO party got 153,000 votes, and the Who Cares party polled 103,000 votes. Show that two of the parties are tied.

8. A small high school has one mathematics teacher who can teach a total of five sections. The subjects that she teaches, and their enrollments, are as follows: geometry, 52; algebra, 33; calculus, 12. Use the Hamilton method to apportion sections to the subjects.

9. Repeat Exercise 8 using the following enrollments: geometry, 77; algebra, 18; calculus, 20.

10. Use the Hamilton method to express the summands of the following expression as whole number percentages of the total:

$$2746 + 1725 + 1921 + 100 = 6492.$$

Repeat the calculation for the sum:

$$2814 + 1745 + 1933 + 99 = 6591.$$

Do you see a paradox?

11. Abe, Beth, Charles, and David have decided to invest in rare coins. A dealer has offered to sell them a parcel containing 100 identical coins for $10,000. Each person invests all that he or she can afford, but there is not quite enough money, so Charles asks his Aunt Esther to join the group. The coins will be apportioned by the Hamilton method. Here are the amounts contributed:

Abe	$3,619
Beth	1,862
Charles	2,258
David	2,010
Esther	251
Total	$10,000

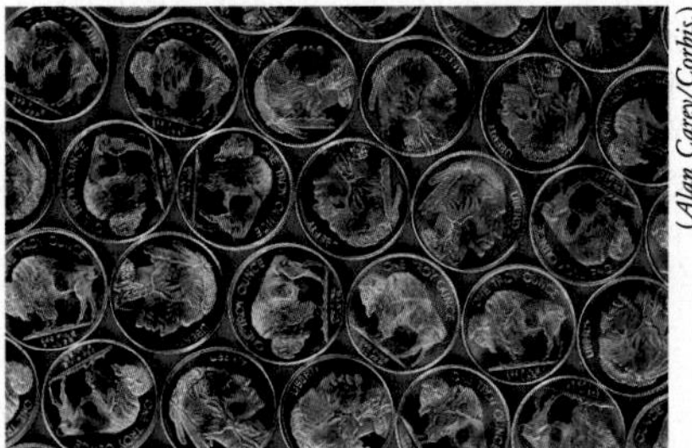
(Alan Carey/Corbis.)

(a) How should the coins be apportioned among the five contributors?

(b) After the coins are distributed, the dealer mentions that there will be $50 in excise tax! Everyone empties their wallet: Abe finds $16 more, Beth has $2, Charles has $1, and David finds $32. This adds up to $51, so a dollar is returned to Aunt Esther. The apportionment is recalculated and one of the coins changes hands. Who has to give a coin to whom?

◆ **(c)** Explain what happened.?

To see how this situation works out with a different apportionment method, refer to Exercise 21.

12. A country has five political parties. Here are the numbers of votes each received in a recent election:

5,576,330; 1,387,342; 3,334,241; 7,512,860; and 310,968. Seats in its parliament are apportioned by the Hamilton method. Calculate the apportionments for house sizes of 82, 83, and 84. Does the Alabama paradox occur?

13. Repeat the apportionments in Exercise 12 for house sizes of 89, 90, and 91.

14.3 Divisor Methods

◆ **14.** Explain why the tentative Webster apportionment of a state with quota q is $\lfloor q + 0.5 \rfloor$.

15. Reapportion the classes in Exercise 9 using the Jefferson method.

16. Reapportion the classes in Exercise 8 using the Webster method.

◆ **17.** The three friends who bought the pearls (see Exercise 6) ask you to suggest a different apportionment method to distribute their purchase. Before answering, determine the apportionments given by the Jefferson and Webster methods for the 36- and 37-pearl house sizes. Then make your suggestion.

◆ **18.** The three friends have bought a lot of 36 identical diamonds, at a total cost of $36,000; Abe's investment was $15,500, Beth's was $10,500, and Charles's was $10,000. They decided to apportion the diamonds using the Webster method, and they can't make it work out. Can you help?

19. Round the following to whole percentages using the methods of Hamilton, Jefferson, and Webster:

$$87.85\% + 1.26\% + 1.25\% + 1.24\% + 1.23\% + 1.22\% + 1.21\% + 1.20\% + 1.19\% + 1.18\% + 1.17\% = 100\%$$

Do any of these methods violate the quota condition?

20. Round the following percentages to whole numbers, using the methods of Hamilton, Jefferson, and Webster.

$$92.15\% + 1.59\% + 1.58\% + 1.57\% + 1.56\% + 1.55\% = 100\%$$

Do any of these methods violate the quota condition?

21. Recalculate the apportionment of the coins in Exercise 11 by the Webster method. Again, after the excise tax is paid, a coin changes hands. Who gives it to whom?

22. Reapportion the parliament of the country in Exercise 12 using the Jefferson method for each of the proposed house sizes. Is there a paradox?

23. Reapportion the parliament of the country in Exercise 13 using the Webster method, for each of the proposed house sizes. Is there a paradox?

24. A country has two political parties, the Liberals and the Tories. The seats in its 99-seat parliament are apportioned to the parties according to the number of votes it receives in the election. If the Liberals receive 49% of the vote, how many seats do the Liberals get with the Hamilton method? With the Webster method? With the Jefferson method?

■ **25.** A country with a parliamentary government has two parties that capture 100% of the vote between them. Each party is awarded seats in proportion to the number of votes received.

(a) Explain why the Webster and Hamilton methods will always give the same apportionment in this two-party situation.

(b) Explain how to use the result of (a) to show that the Alabama and population paradoxes cannot occur when the Hamilton method is used to apportion seats between two parties or states.

(c) Explain why the result of (a) implies that the Webster method satisfies the quota condition when the seats are apportioned between two parties or states.

◆ **(d)** Will the Jefferson and Hill–Huntington methods also yield the same apportionments as the Hamilton method?

14.4 Which Divisor Method is Best?

26. Determine the relative difference between the numbers 5 and 7.

27. Jim is 72 inches tall and Alice is 65 inches tall. What is the relative difference of their heights?

28. In the 2001 apportionment of Congress, the average congressional district in North Carolina was given 13 seats, and its average district population was 620,590. Montana, with a population of 905,316, was the most populous state to receive only one district.

(a) Which state is the more favored in this apportionment?

(b) What are the relative differences in the district sizes?

(c) If North Carolina was apportioned 12 seats, and Montana was apportioned 2, determine the absolute and relative differences in district sizes.

(d) Does this apportionment minimize absolute differences in district size between these two states?

29. The data from Exercise 28 imply that the representative share for North Carolina was 1.6114 representatives per million, and for Montana the representative share was 1.1046 representatives per million.

(a) What are the absolute and relative differences in representative share?

(b) If North Carolina was apportioned 12 seats, and Montana was apportioned 2, determine the absolute and relative differences in representative share.

(c) Does the apportionment minimize absolute differences in representative share between these two states?

30. According to the 2000 census, the population of California was 33,930,798 and the population of Utah was 2,236,714. California was apportioned 53 House seats, and Utah received 3 House seats.

(a) Determine the average congressional district sizes for these states.

(b) Determine the absolute and relative differences in these district sizes.

(c) Suppose a seat were transferred from California to Utah, giving Utah 4 seats and California 52. What would now be the absolute and relative differences in district sizes?

(d) Does this evidence indicate that the apportionment following the 2000 census fulfilled the criterion of keeping absolute differences in district populations as small as possible? What about relative differences?

31. Find the Hill–Huntington rounding points for numbers between 0 and 1; between 1 and 2; between 2 and 3; and between 3 and 4.

32. A high school has one math teacher, who can teach five sections. Fifty-six students have enrolled in the algebra class, 28 have signed up for geometry, and 7 students will take calculus. Use the Hill–Huntington method to decide how many sections of each course to schedule.

33. One year later, the high school described in Exercise 32 still has just one math teacher who teaches five sections. The enrollments are algebra, 36; geometry, 61; and calculus, 3. Apportion the classes by the Webster and Hill–Huntington methods. Which apportionment do you think the school principal would prefer?

◆ **34.** Seats in parliament are apportioned according to votes received by parties.

(a) What would be the drawback of using the Hill–Huntington method?

(b) If it is considered undesirable to give small parties much representation in parliament, which divisor method would be preferable–Jefferson or Webster?

■ **35.** Suppose that in 2000, the governor of Utah believed that the population of his state was undercounted. What increase in population would be large enough to entitle Utah to take a seat from California if the apportionment is by the Hill–Huntington method? The data needed for this problem are given in Exercise 30.

◆ **36. (a)** Show that for any positive numbers A and B, the geometric mean is less than the arithmetic mean,[4] except when $A = B$; then the two means are equal. (*Hint:* Show that the triangle in Figure 14.2 is a right triangle.)

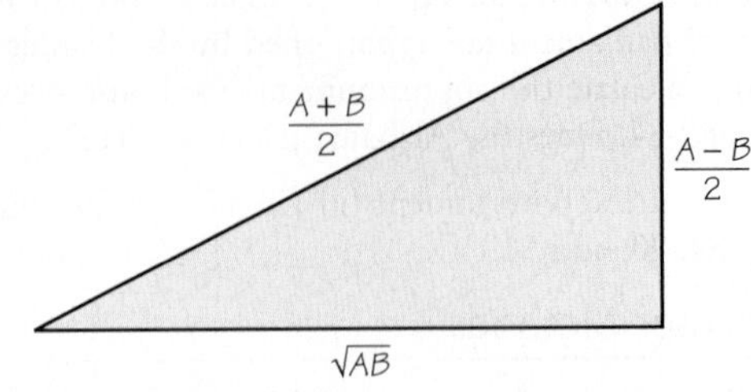

FIGURE 14.2 Is this a right triangle?

(b) Compare the Webster and Hill–Huntington roundings: Show that for any positive number q, if the two roundings differ, then the Hill–Huntington rounding of q is equal to $\lceil q \rceil$ and the Webster rounding is equal to $\lfloor q \rfloor$.

(c) Explain why the fact established in part (a) implies that the Hill–Huntington method is more favorable to small states than the Webster method.

37. A city has three districts with populations of 100,000; 600,000; and 700,000. Its council has 20 members, and seats on the council are apportioned according by the Hill–Huntington method to the districts. Show that there is a tie. Would a tie occur with any of the other apportionment methods that we have considered?

38. In a 1991 federal lawsuit *Massachusetts* v. *Mosbacher,* Massachusetts claimed that the Hill–Huntington method of apportionment is unconstitutional, because it does not reflect the "one person, one vote" principle as well as the Webster method does. Would Massachusetts have gained a seat from Oklahoma if the Webster method had been used to apportion the House of Representatives in 1991? Use the following populations and Hill–Huntington apportionments:

State	Population	Apportionment
Massachusetts	6,029,051	10
Oklahoma	3,145,585	6

◆ **39.** The following apportionment method was invented by Congressman William Lowndes of South Carolina in 1822. Lowndes starts, as Hamilton does, by giving each state its lower quota. But where Hamilton apportions the remaining seats to the states whose quotas have the largest fractional parts–in other words, the states for which the *absolute difference* between q_i and $\lfloor q_i \rfloor$ is greatest–Lowndes gives the extra seats to the states where the *relative difference* between q_i and $\lfloor q_i \rfloor$ is greatest, raising as many as necessary to their upper quotas to fill the House.

(a) Would this method be more beneficial to states with large populations or small populations, as compared with the Hamilton method?

(b) Does the Lowndes method satisfy the quota condition?

[4]The arithmetic mean of A and B is equal to $(A + B)/2$.

(c) Would there be any trouble with paradoxes with the Lowndes method?
(d) Use the method to apportion the 1790 House of Representatives.

◆ **40.** Let the populations of states A and B be p_A and p_B, respectively. The apportionments will be a_A and a_B. Assuming that district populations for state A are larger than district populations for state B, show that the relative difference in district populations is

$$\left(\frac{a_B p_A}{a_A p_B} - 1\right) \times 100\%$$

Also show that this expression is equal to the relative difference in representative share. Hence the relative difference in district populations is equal to the relative difference in representative shares.

◆ **41.** John Quincy Adams, the sixth president of the United States, proposed that the House of Representatives should be apportioned by a divisor method based on the rounding rule that rounds each fraction up to the next whole number.

(a) Is it likely that the initial tentative apportionment will be final?
■ **(b)** Find a formula for a state's critical divisor in terms of the state's tentative apportionment n_i and its population p_i.
(c) Does the method favor small states or large states?
(d) Is it possible for a state to be apportioned zero seats by using this method?

◆ **42.** The U.S. Constitution requires that each state be apportioned at least one seat in the House of Representatives.

(a) Show that the Hill–Huntington method is consistent with this requirement. What about the methods of Hamilton, Jefferson, and Webster?
(b) Does the Adams method (see Exercise 41) meet the requirement?
■ **(c)** The Dean method (see Writing Project 4) is the divisor method that minimizes absolute differences in district population. Using this information, explain why the Dean method will never give any state zero seats, unless the house size is less than the number of states.

43. The Marquis de Condorcet, who proposed a criterion for deciding elections (see Chapter 9), also designed a divisor method for apportionment. His rounding rule was to round down numbers whose fractional parts are less than 0.4 and to round up otherwise.

(a) Explain why the Condorcet rounding of a number q is $\lfloor q + 0.6 \rfloor$.
(b) Does the method favor large states, small states, or is it neutral?
(c) Find a formula for a state's critical divisor in terms of the state's tentative apportionment n_i and its population p_i.

◆ **44.** The choice of a divisor method for apportioning classes to subjects according to enrollments, as in the senior high school example, depends on what the school principal considers most important.

(a) The principal wants to set a minimum class size. For example, if the minimum class size is 20, and 39 students are signed up for English III, there would be one section, because there are not enough students for two sections with enrollment of at least 20. If there were 40 students, there would be two sections. The minimum class size is adjusted so that as many sections as possible are running. What apportionment method should she use, and what will the minimum class size be?
■ **(b)** The principal prefers to set a maximum class size. For example, if the maximum class size is 33, and 67 students are taking History I, there will be three sections, because there are too many students to fit in two 33-student sections. If there were only 66 taking History I, there would be two sections. The maximum class size is adjusted so that as many sections as possible are running. What apportionment method should she use, and what will the maximum class size be? (*Hint:* This divisor method is not described in the text but is mentioned in one of the previous exercises.)
(c) The principal wants to treat students as equitably as possible, so that the differences between students' share of teachers vary as little as possible from course to course. What apportionment method should she use now?
(d) The principal wants to minimize relative difference in class size. What divisor method would work best for her?
(e) The principal wants to cancel any class that has an enrollment of just one student. Which apportionment methods should she avoid using?

■ **45.** Let $q_1, q_2, \ldots, q_n$ be the quotas for n states in an apportionment problem, and let the apportionments assigned by some apportionment method be denoted $a_1, a_2, \ldots, a_n$. The *absolute deviation* for state i is defined to be $|q_i - a_i|$; it is a measure of the amount by which the state's apportionment differs from its quota. The *maximum absolute deviation* is the largest of these numbers. Explain why the Hamilton method always gives the least possible maximum absolute deviation.

APPLET EXERCISES

To do these exercises, go to www.whfreeman.com/fapp8e.

A bus company has three lines–*A*, *B*, and *C*–and a total of 48,000 riders. *A* has 21,700 riders daily, *B* has 17,200, and *C* has 9100. The company has 40 buses to allocate to the three lines. Use the applet *Apportionment* to help you find the standard divisor and determine the allocation of the buses according to the methods of apportionment of Hamilton, Jefferson, Webster, and Hill–Huntington.

WRITING PROJECTS

1. Does the Hill–Huntington method best reflect the intentions of the Founding Fathers, as these intentions were set down in the Constitution and in the debate during the 1787 Constitutional Convention? Good sources of information here include all of the publications listed in the Suggested Readings. This writing project requires that you state your answer to the question and make a case for it.

2. Suppose that in 2000, Congress had reverted to its nineteenth-century habit of increasing the size of the House of Representatives so that no state would have a decrease in the size of its delegation. How many seats would have been added, and which states would have gotten them? (*Warning:* The apportionments of some states might *increase* as a result of this practice.) As the first step of this project, obtain the populations and apportionments for the 50 states from the Census Bureau Web site (www.census.gov/population/www/censusdata/apportionment.html).

3. In 2004, the state of Colorado considered an amendment to its constitution regarding the way electors representing Colorado in the Electoral College would be selected. Until 2000, Colorado's electors were selected by the president/vice president ticket that received a plurality of the votes. The proposed amendment, which would have taken effect in 2004, apportioned electoral votes to each ticket in proportion to the number of popular votes received. The apportionment method specified in the amendment was as follows: Determine each ticket's quota, and round to the nearest whole number. Tickets with quotas less than 0.5 receive no electors. If the number of electors thus apportioned is less than Colorado is entitled to have, give the remaining electors to the ticket that received the most votes. If the number of electors is more than Colorado is entitled to have, take electors from that ticket that, among all who received electors, received the smallest number of popular votes. If more electors need to be removed, take from the ticket that received the next smallest number of popular votes, and so on.

Write an essay exploring the implications of this apportionment method, and compare it to others that Colorado could have chosen. Include a discussion of the consequences that would occur if a third-party candidate received some electoral votes as a result of this procedure.

4. The Dean method. The Webster method was proposed in 1832 after New York received an apportionment in excess of its upper quota. Two other apportionment methods were proposed in the same year: the method of John Quincy Adams, which is biased in favor of small states as badly as the Jefferson method is biased in favor of large states, and the Dean method. The latter method, invented by James Dean, a professor of mathematics and astronomy at Dartmouth College, gives the most equitable apportionment when the measure of inequity is absolute difference in district population. Suppose that state A has population p and its tentative apportionment is n, while state B has population q and tentative apportionment m. If another seat is to be given to one of these states,

(a) Calculate the absolute difference in district populations if A gets the seat, and repeat the calculation for the situation when B gets the seat.

(b) Show that the difference between the two results in (a) is equal to $\frac{p}{n^\#} - \frac{q}{m^\#}$, where $n^\#$ denotes the *harmonic mean* (you may have to look this up) between n and $n + 1$.

(c) Explain the mechanics of the Dean method, including why it is a divisor method that rounds a number r down to $\lfloor r \rfloor$ if r is less that the Dean rounding point, and up to $\lceil r \rceil$ otherwise. It's up to you to figure out the Dean rounding point.

(d) Is the Dean method biased in favor of large states or small states? Is it possible for any state to get a zero apportionment? Compute the apportionment of the House of Representatives according to the latest decennial census by the Dean method. Is the quota condition satisfied?

 SUGGESTED READINGS

BALINSKI, M. L., and H. P. YOUNG. *Fair Representation: Meeting the Ideal of One Man, One Vote,* Yale University Press, New Haven, Conn., 1982. In the 1970s, Balinski and Young analyzed apportionment methods in depth. Their approach was to postulate the desirable properties of an apportionment method as axioms and to deduce from the axioms which method is best. This book combines an account of the history of apportionment of the U.S. House of Representatives with the results of their research.

ERNST, LAWRENCE R. Apportionment methods for the House of Representatives and the court challenges, *Management Science,* 40 (1994): 1207–1227. Ernst, who wrote briefs for the government in both the *Montana* and the *Massachusetts* cases, reviews the apportionment problem and the arguments in favor of and against each of the divisor methods. The article includes a summary of the arguments used by both sides in the two court cases.

YOUNG, H. PEYTON. *Equity,* Princeton University Press, Princeton, N.J., 1994. Chapter 3 covers apportionment and focuses on which apportionment method is the most equitable.

 SUGGESTED WEB SITES

www.census.gov/population/www/censusdata/apportionment.html This site contains a two-page history of apportionment of the Congress.

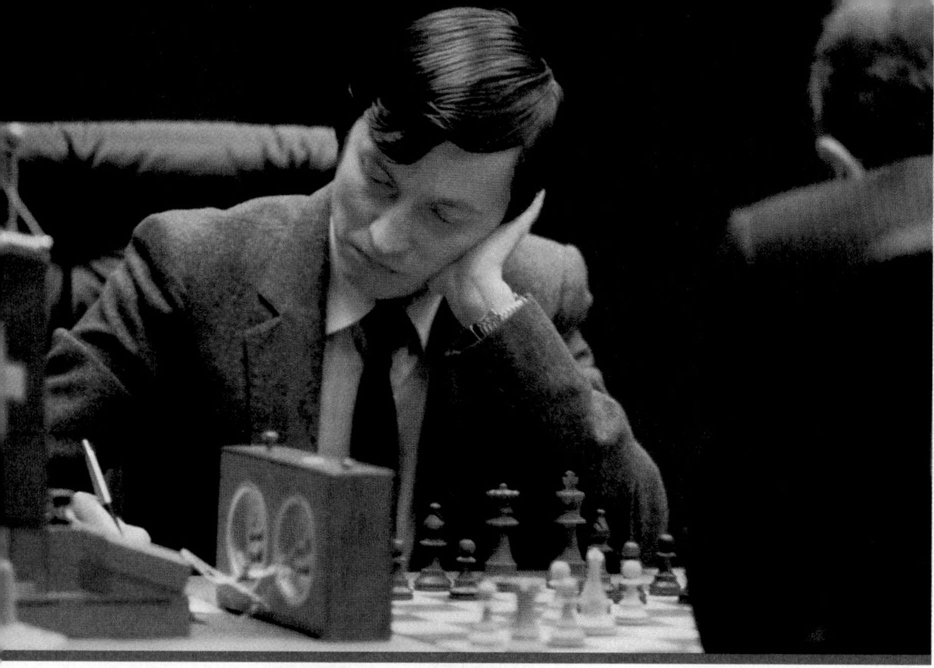

CHAPTER 15

Game Theory: The Mathematics of Competition

Conflict has been prevalent throughout human history. It arises whenever two or more individuals, with different values or goals, compete to try to control the course of events. *Game theory* uses mathematical tools to study situations, called games, involving both conflict and cooperation. Its study was greatly stimulated by the publication in 1944 of the monumental *Theory of Games and Economic Behavior* by John von Neumann and Oskar Morgenstern (see Spotlight 15.1).

The *players* in a game, who may be people, organizations, or even countries, choose from a list of options available to them–that is, courses of action they may take–that are called **strategies**. The strategies chosen by the players lead to *outcomes,* which describe the consequences of their choices. We assume that the players have *preferences* for the outcomes: They like some more than others.

Game theory analyzes the **rational choice** of strategies–that is, how players select strategies to obtain preferred outcomes. Among areas to which game theory has been applied are bargaining tactics in labor–management disputes, resource-allocation decisions in political campaigns, military choices in international crises, and the use of threats by animals in habitat acquisition and protection.

Unlike the subject of *individual* decision making, which researchers in psychology, statistics, and other disciplines study, game theory analyzes situations in which there are at least two players, who may find themselves in conflict because of different goals or objectives. The outcome depends on the choices of *all* the players. In this sense, decision making is *collective,* but this is not to stay that the players necessarily cooperate when they choose strategies. Indeed, many strategy choices are noncooperative, such as those between combatants in warfare or competitors in sports. In these encounters, the adversaries' objectives may be at cross-purposes: A gain for one means a loss for the other. But in many activities, especially in economics and politics, there may be joint gains that can be realized from cooperation.

SPOTLIGHT 15.1 The Early History of Game Theory

John von Neumann
(Bettmann/UPI/Corbis.)

Oskar Morgenstern
(Courtesy of the Institute for Advanced Study, Princeton University Archives.)

As early as the seventeenth century, such outstanding scientists as Christiaan Huygens (1629–1695) and Gottfried W. Leibniz (1646–1716) proposed the creation of a discipline that would apply the scientific method to the study of human conflict and interactions. Throughout the nineteenth century, several leading economists created simple mathematical models to analyze particular examples of competitive encounters. The first general mathematical theorem on this subject was proved for games of perfect information by the distinguished logician Ernst Zermelo (1871–1953) in 1912. A game is said to have *perfect information* if at each stage of the play, every player is aware of all past moves (by itself and others) as well as all future choices that are possible. The theorem stated that any finite game with perfect information, such as checkers or chess, has an optimal solution in *pure* strategies; that is, no randomization or secrecy is necessary. This theorem is an example of an *existence theorem*: It demonstrates that there must exist a best way to play such a game, but it does not provide a detailed plan for playing a complex game, like chess, to achieve victory.

The famous mathematician F. E. Émile Borel (1871–1956) introduced the notion of a *mixed*, or randomized, strategy when he investigated optimal strategies in duels around 1920. The fact that every two-person *zero-sum* game must have a solution in optimal mixed strategies was proved by the Hungarian-American mathematician John von Neumann (1903–1957) in 1928. Von Neumann's result was extended to the existence of equilibrium outcomes in mixed strategies for multiperson games that are either *constant-sum* or *variable-sum* in 1951 by John F. Nash, Jr. (b. 1928), who was portrayed in the movie, *A Beautiful Mind* (2001).

Modern game theory dates from the publication in 1944 of *Theory of Games and Economic Behavior* by John von Neumann and the Austrian-American economist Oskar Morgenstern (1902–1977). They introduced the first general model and solution concept for multiperson *cooperative games*, which are primarily concerned with coalition formation (by economic cartels, voting blocs, or military alliances) and the resulting distribution of gains or losses. Several other suggestions for a solution to such games have since been proposed. These include the value concept of Lloyd S. Shapley (b. 1923), which relates to fair allocation and serves also as index of voting power (see Chapter 11).

The French artist Georges Mathieu designed a medal for the Musée de la Monnaie in Paris in 1971 to honor game theory. It was the seventeenth medal to "commemorate 18 stages in the development of Western consciousness." Game theory also has a mascot, the tiger, arising from the Princeton University tiger and the Russian abbreviation of the term *game theory* (ТЕОРИЯ ИГР), where the underlined letters correspond to the sounds of the English *T*, *G*, and *R*, respectively).

Many interactions involve a delicate mix of cooperative and noncooperative behavior. In business, for example, firms in an industry cooperate to gain tax breaks even as they compete for shares in the marketplace.

Game theory has provided important theoretical foundations in economics, starting with microeconomics but now extending to macroeconomics, industrial organization, and international economics. It also has been increasingly applied in political science, especially in the study of voting, elections, and international relations. In addition, game theory has contributed major insights in biology, particularly in understanding the evolution of species and conditions under which animals–humans included–fight each other for territory or act altruistically. It has also illuminated certain fields in philosophy, including ethics, the philosophy of religion, and political philosophy, and inspired many experiments in social psychology.

In the next two sections, we present several simple examples of two-person **total-conflict games**, in which what one player wins the other player loses, so cooperation never benefits the players. We distinguish two different kinds of solutions to such games. Then we analyze two well-known **partial-conflict games**, in which the players can benefit from cooperation but may have strong incentives not to cooperate. We next turn to the analysis of a larger three-person voting game, in which we show how to eliminate undesirable strategies in stages. Finally, we offer some general comments on solving games and discuss different applications of game theory.

15.1 Two-Person Total-Conflict Games: Pure Strategies

For some games with two players, determining the best strategies for the players is straightforward. We begin with such a case.

EXAMPLE 1 ■ A Location Game

Two young entrepreneurs, Henry and Lisa, plan to locate a new restaurant at a busy intersection in the nearby mountains. They agree on all aspects of the restaurant except one. Lisa likes low elevations, whereas Henry wants greater heights–the higher, the better. In this one regard, their preferences are diametrically opposed. What is better for Henry is worse for Lisa, and likewise what is good for Lisa is bad for Henry.

The layout for their location problem is shown in Figure 15.1. Observe that three routes, Avenue A, Boulevard B, and County Road C (blue lines), run in an east–west direction, and that three highways, numbered 1, 2, and 3 (red lines), run in a north–south direction. Table 15.1 shows the altitudes at the nine corresponding intersections. The same information is shown in three dimensions in Figure 15.2.

To maximize the number of customers, Henry and Lisa agree that the restaurant should be at a location where one of the east–west routes intersects one of the three highways. But they cannot agree on which intersection, so they decide to turn their decision into the following competitive game: Henry will select one of the three routes, *A*, *B*, or *C*, and Lisa will simultaneously choose one of the three highways, 1, 2, or 3. Because their choices will be made simultaneously, neither one can predict beforehand what the other will do.

Henry, worried that Lisa will choose a low elevation, tries to determine the highest altitude he can guarantee by picking one of the three routes. For each choice of a route, this means considering the worst-case (lowest) elevation on each route. These

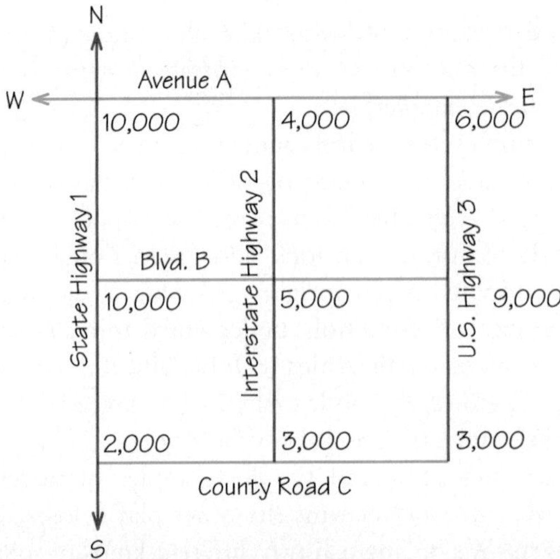

FIGURE 15.1 The road map for the location of Henry and Lisa's restaurant in Example 1. (The elevations in feet are shown at each intersection.)

TABLE 15.1 Heights (in thousands of feet) of the Nine Intersections

	Highways		
Routes	**1**	**2**	**3**
A	10	4	6
B	6	5	9
C	2	3	7

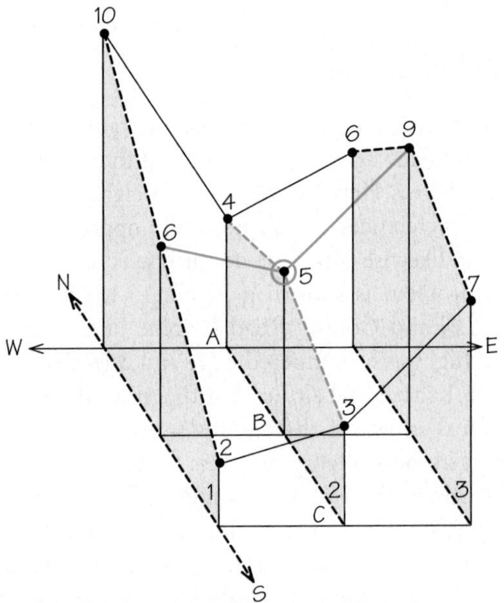

FIGURE 15.2 Three-dimensional road map showing Henry's and Lisa's possible choices (in thousands of feet).

are the numbers 4, 5, and 2, which are the respective *row minima,* indicated in the right-hand column of Table 15.2. He notes that the highest of these values is 5. By choosing the corresponding route, *B,* Henry can guarantee himself an altitude of at least 5000 feet.

TABLE 15.2 Heights (in thousands of feet) in Table 15.1, with the Row Minima (maximum circled) and Column Maxima (minimum circled)

	Routes	Lisa Highways 1	2	3	Row Minima
Henry	*A*	10	4	6	4
	B	6	5	9	⑤
	C	2	3	7	2
	Column Maxima	10	⑤	9	

Maximin DEFINITION

The **maximin** is the maximum value of the minimum numbers in the rows in a table. The strategy that corresponds to the maximin is called the **maximin strategy**. The number 5 in the right-hand column of Table 5.2, which is circled, is the maximin. Route *B* is Henry's maximin strategy.

Lisa likewise does a worst-case analysis and lists the highest–for her, the worst–elevations for each highway. These numbers, 10, 5, and 9, are the column maxima and are listed in the bottom row of Table 15.2. From Lisa's point of view, the best of these outcomes is 5. If she picks Interstate Highway 2, then she is assured of an elevation of no more than 5000 feet.

Minimax DEFINITION

The **minimax** is the minimum value of the maximum numbers in the columns. The strategy that corresponds to the minimax is called the **minimax strategy**. The number 5 at the bottom row of Table 5.2, which is circled, is the **minimax**. Highway 2 is Lisa's minimax strategy.

To summarize, Henry has a strategy that will ensure the height is 5000 or higher, and Lisa has a strategy that will ensure the height is 5000 or lower. The height of 5000 at the intersection of route *B* and Highway 2 is, simultaneously, the lowest value along Boulevard B and the highest along Interstate Highway 2. In other words, the maximin and the minimax are both equal to 5000 for the location game.

Saddlepoint DEFINITION

When a row minimum and a column maximum are the same, the resulting outcome is called a **saddlepoint**.

The reason for the term **saddlepoint** should be clear from the saddle-shaped payoff surface shown in Figure 15.2. The middle point on a horse saddle is simultaneously the lowest point along the spine of the horse and the highest point between the rider's legs. (In Figure 15.2, the rider would be facing leftward or rightward.) In our example, one might also think of the saddlepoint as a mountain pass:

As one drives through the pass, the car is at a high point on a highway (in the north–south direction) and at a low point on a route (in the east–west direction).

The resolution of this contest is for Henry to pick *B* and Lisa to pick 2. This puts them at an elevation of 5000, which is simultaneously the maximin and minimax.

Value DEFINITION

In total-conflict games, the **value** is the best outcome that both players can guarantee. If a game has a saddlepoint, it gives the value (5 in our example): Players can guarantee this outcome by choosing their maximin and minimax strategies.

Total-conflict games without saddlepoints also have a "value," as we shall see later.

There is no need for secrecy in a game with a saddlepoint. Even if Henry were to reveal his choice of *B* in advance, Lisa would be unable to use this knowledge to exploit him. In fact, both players can use the height information in our example to compute the optimal strategy for their opponent as well as for themselves. In games with saddlepoints, players' worst-case analyses lead to the best *guaranteed* outcome–in the sense that each player can ensure that he or she does not do worse than a certain amount (5 in our example) and may do better (if the opponent deviates from a maximin or minimax strategy).

Another well-known game with a saddlepoint is tic-tac-toe. Two players alternately place an X or an O, respectively, in one of the unoccupied spaces in a 3 × 3 grid with 9 cells. The winner is the first player to have three X's, or three O's, in the same row, in the same column, or along a diagonal; if no player does this when all cells are filled in, the game ends in a tie.

An explicit list of all strategies for either the first- or second-moving player in tic-tac-toe is long and complicated, because it specifies a complete plan for all possible contingencies that can arise. For the first-moving player, for example, a strategy might say "put an X in the middle cell, then an X in the corner if your opponent puts an O in a noncorner cell," and so on. While young children initially find this game interesting to play, before long they discover that each player can always prevent the other player from winning by forcing a tie, making the game quite boring. Unlike the game between Henry and Lisa, the list of possible strategies in tic-tac-toe is huge, but only those that force a tie, it turns out, are a saddlepoint.

EXAMPLE 2 ▪ The Restricted-Location Game

Assume in our location game that Henry and Lisa are informed by county officials that it is against the law to locate a restaurant on either Boulevard B or Interstate Highway 2. These two choices, which provided our earlier solution, are now forbidden. The resulting location game without these two strategies is given in Table 15.3 (with payoffs again expressed in thousands of feet).

As before, Henry and Lisa can each do a worst-case analysis. Henry is worried about the minimum number in each row, and Lisa is concerned with the maximum number in each column. These are listed in the right column and bottom row, respectively, in Table 15.4.

Henry sees that his maximin is 6, so he can guarantee a height of 6000 feet or more by choosing route *A*. Likewise, Lisa observes that her minimax is 7, so she can keep the elevation of the restaurant down to 7000 feet or less by selecting Highway 3.

TABLE 15.3 Heights (in thousands of feet) without Boulevard B and Interstate Highway 2

	Highways	
Routes	1	3
A	10	6
C	2	7

TABLE 15.4 Heights (in thousands of feet) in Table 15.3, with Row Minima (maximum circled) and Column Maxima (minimum circled)

		Lisa Highways		
	Routes	1	3	Row Minima
Henry	*A*	10	6	(6)
	C	2	7	2
	Column Maxima	10	(7)	

There is a gap of $7 - 6 = 1$ between the minimax and maximin. When the maximin is less than the minimax, as in this case, then a game does *not* have a saddlepoint, but it does have a value (described in the next section).

If Henry plays his maximin strategy, route *A*, and Lisa plays her minimax strategy, Highway 3, then the resulting payoff is 6. However, Henry may be motivated to gamble in this case by playing his other strategy, route *C*. If Lisa sticks to her conservative strategy, Highway 3, then the payoff is 7. Henry will have gained one unit (1000 feet), going from 6 to 7.

This is, however, a risky move. If Lisa suspected it, she might counter by selecting Highway 1. The payoff would then be 2, the best for Lisa and the worst for Henry. So Henry's gamble to gain 1 unit (6 to 7) by moving has the risk that he might lose 4 units (6 to 2) if Lisa also moves.

But then there is no incentive for Lisa to play her nonminimax strategy (that is, to play Highway 1) if she believes Henry, in turn, will move back to his maximin strategy (route *A*), leading to a payoff of 10. This is worse than 6 from her viewpoint. ■

In two-player games that have saddlepoints, like our original 3 × 3 location game and tic-tac-toe, each player can calculate the maximin and minimax strategies for both players before the game is even played. Once the solution has been determined by either mathematical analysis or practical experience (as was probably true of tic-tac-toe), there may be little interest in actually playing the game.

But this is decidedly not the case for much more complex games, like chess, whose solution has not yet been determined–and is unlikely to be in the foreseeable future. Even though computers are able to beat world champions, the computer's winning moves will not necessarily be optimal against those of *all* other opponents. Nevertheless, we know that chess, like tic-tac-toe, has a saddlepoint. (All games of perfect information, in which the players know each other's moves at every step, have a saddlepoint.) What we do not know is whether it yields a win for white, a win for black, or a draw.

Unlike chess, many games, like the 2 × 2 restricted-location game, do not have an outcome that can always be guaranteed. These games, which include poker, involve uncertainty and risk. In such games, one does not want to have one's strategy detected in advance, because this information can be exploited by an opponent. It is no surprise, then, that poker players are told to keep a "poker face," revealing nothing about their likely choices. But this advice is not very helpful in telling the players what actually to do in the game, such as how many cards to ask for in draw poker.

We will show that there are optimal ways to play two-person total-conflict games without a saddlepoint so as not to reveal one's choices. But their solution is by no means as straightforward as that of games with a saddlepoint.

15.2 Two-Person Total-Conflict Games: Mixed Strategies

Probably most competitive games do not have a saddlepoint like the one we found in our first location-game example. Rather, as is illustrated in our restricted-location game–in which the maximin and minimax are not the same–players must try to keep secret their strategy choices, lest their opponent use this information to his or her advantage.

In particular, players must take care to conceal the strategy they will select until the encounter actually takes place, when it is too late for the opponent to alter his or her choice. If the game is repeated, a player will want to *vary* his or her strategy in order to surprise the opponent.

In parlor games like poker, players often use the tactic of *bluffing*. This tactic involves a player's sometimes raising the stakes when he has a low hand so that opponents cannot guess whether or not his hand is high or low–and may, therefore, miscalculate whether to stay in or drop out of the game (a player would prefer opponents to stay in when he has a high hand and drop out when he has a low hand). In military engagements, too, secrecy and even deception are often crucial to success.

In many sporting events, a team tries to surprise or mislead the opposition. A pitcher in baseball will not signal the type of pitch he or she intends to throw in advance, varying the type throughout the game to try to keep the batter off balance. In fact, we next consider a confrontation between a pitcher and batter in more detail.

EXAMPLE 3 ■ A Duel Game

The pitcher and the batter use mixed strategies. (*Alan Schein/ The Stock Market.*)

Assume that a particular baseball pitcher can throw either a blazing fastball or a slow curve into the strike zone and so has two strategies: *fast* (denoted by F) and *curve* (C). The pitcher faces a batter who attempts to guess, before each pitch is thrown, whether it will be a fastball or a curve, giving the batter two strategies also: guess F and guess C. Assume that the batter has the following batting averages, which are known by both players.

- 0.300 if the batter guesses fast (F) and the pitcher throws fast (F)
- 0.200 if the batter guesses fast (F) and the pitcher throws curve (C)

- 0.100 if the batter guesses curve (C) and the pitcher throws fast (F)
- 0.500 if the batter guesses curve (C) and the pitcher throws curve (C)

A player's batting average is the number of times he hits safely divided by his number of times at bat. If a batter hit safely 3 times out of 10, for example, his average would be 0.300.

This game is summarized in Table 15.5. We see from the right-hand column in the table that the batter's maximin is 0.200, which is realized when he selects his first strategy, F. Thus, the batter can "play it safe" by always guessing a fastball, which will result in his batting 0.200, hardly enough for him to remain on the team.

We see from the bottom row of the table that the pitcher's minimax is 0.300, which is obtained when he throws fast (F). Note that the batter's maximin of 0.200 is less than the pitcher's minimax of 0.300, so this game does not have a saddlepoint. There is a gap of $0.300 - 0.200 = 0.100$ between these two numbers.

Each player would like to play so as to win for himself as much of the 0.100 payoff in the gap as possible. That is, the batter would like to average more than 0.200, whereas the pitcher wants to hold the batter down to less than 0.300.

TABLE 15.5 Batting Averages in a Baseball Duel

		Pitcher		
		F	C	Row Minima (maximum circled)
Batter	F	0.300	0.200	(0.200)
	C	0.100	0.500	0.100
	Column Maxima (minimum circled)	(0.300)	0.500	

A Flawed Approach

If the batter and pitcher in our example consider how they might outguess each other, they may reason along the following lines:

1. *Pitcher* (to himself): If I choose strategy F, I hold the batter down to 0.300 (the minimax) or less. However, the batter is likely to guess F because it guarantees him at least 0.200 (his maximin), and it actually provides him with 0.300 against my F pitch. In this case, the batter wins all the 0.100 payoff in the gap.
2. *Batter* (to himself): Because the pitcher will try to surprise me with C by reasoning as in step 1, I should fool him and guess C. I would thus average 0.500, which will show him up for trying to gamble and outguess me!
3. *Pitcher* (to himself): But if the batter is thinking as in step 2–that is, guessing C–I, on second thought, should really throw F. This will lead to an average of only 0.100 for the batter and teach him to not try to outguess me!

This type of cyclical reasoning can go on forever: "I think that he thinks that I think that he thinks. . . ." It provides no resolution to the players' decision problem.

Clearly, there is no pitch, or guess, that is best under all circumstances. Nevertheless, both the pitcher and the batter *can* do better, but not by trying to anticipate each other's choices. The answer to their problem lies in the notion of a **mixed strategy**.

A Better Idea

The play of many total-conflict games requires an element of surprise, which can be realized in practice by making use of a mixed strategy.

Pure Strategy DEFINITION

Each of the definite courses of action that a player can choose is called a **pure strategy**.

All the choices of players–Henry and Lisa, the batter and the pitcher–that we have considered so far are **pure strategies**.

Mixed Strategy DEFINITION

A **mixed strategy** is a strategy in which the course of action is randomly chosen from one of the pure strategies in the following way: Each pure strategy is assigned some probability, indicating the relative frequency with which that pure strategy will be played. The specific strategy used in any given play of the game can be selected using some appropriate random device.

Note that a pure strategy is a special case of a mixed strategy, with the probability of 1 assigned to just one pure strategy and 0 to all the rest. When a player resorts to a mixed strategy, the resulting outcome of the game is no longer predictable in advance. (For example, if a pitcher throws a curve ball or a fastball with probability 0.5 each, the batter cannot predict which pitch he or she is about to receive.) Rather, the outcome must be described in terms of the probabilistic notion of an **expected value**.

Expected Value *E* DEFINITION

If each of the n payoffs, $s_1, s_2, \ldots, s_n$, will occur with the probabilities $p_1, p_2, \ldots, p_n$, respectively, then the average, or **expected value *E***, is given by

$$E = p_1 s_1 + p_2 s_2 + \cdots + p_n s_n$$

We assume that the probabilities sum to 1 and that each probability p_i is never negative. That is, we assume that $p_1 + p_2 + \cdots + p_n = 1$, and $p_i \geq 0$ $(i = 1, 2, \ldots, n)$.

To see how mixed strategies and expected values are used in the analysis of games, we turn to what is perhaps the simplest of all competitive games without a saddlepoint.

EXAMPLE 4 ■ Matching Pennies

In matching pennies, each of two players simultaneously shows either a head H or a tail T. If the two coins match, with either two heads or two tails, then the first player (player I) receives both coins (a win of 1 for player I). If the coins do not match, that is, if one is an H and the other is a T, then the second player (player II) receives the two coins (a loss of 1 for player I). These wins and losses for player I are shown in Table 15.6.

TABLE 15.6 Wins and Losses for Player I in Matching Pennies

		Player II	
		H	*T*
Player I	*H*	1	−1
	T	−1	1

Payoff Matrix

DEFINITION

A **payoff matrix** (illustrated by Table 15.6) is a table whose rows and columns correspond to the strategies of the two players. The numerical entries give the payoffs to player I when these strategies are chosen.

Although the entries in our earlier tables for the location game also gave payoffs, they were not monetary, as here. A game represented by a payoff matrix is called *a game in strategic form.*

The two rows in Table 15.6 correspond to player I's two pure strategies, H and T, and the two columns to player II's two pure strategies, also H and T. The numbers in the table are the corresponding winnings for player I and losses for player II. If two H's or two T's are played, player I wins 1 from player II. When one H and one T are played, player I pays out 1 to player II.

It is fruitless for one player to attempt to outguess the other in this game. They should instead resort to mixed strategies and use expected values to estimate their likely gains or losses.

The best thing for player I to do is randomly to select H half the time and T half the time. This mixed strategy can be expressed as

$$(p_H, p_T) = (p_1, p_2) = (p, 1 - p) = \left(\frac{1}{2}, \frac{1}{2}\right)$$

Note that the probability p of choosing H, and $(1 - p)$ of choosing T, do indeed sum to 1, as required; in particular, when $p = \frac{1}{2}$, $1 - p = 1 - \frac{1}{2} = \frac{1}{2}$.

This mixture can be realized in practice by the flip of a coin. Player I's resulting expected value is

$$E_H = \frac{1}{2}(1) + \frac{1}{2}(-1) = 0$$

whenever player II plays H (first column of Table 15.6). Whenever player II plays T (second column), player I's resulting expected value is

$$E_T = \frac{1}{2}(-1) + \frac{1}{2}(1) = 0$$

Mixed-Strategy Value

DEFINITION

A player's expected value is the **mixed-strategy value** of the game. Unlike the use of this notion in games with a saddlepoint, the value here can be realized only by the use of mixed strategies.

The value of 0 in matching pennies is really an expected value and so must be understood in a statistical sense. That is, in a given play of the game, player I will either win 1 or lose 1. However, his or her expectation over many plays of this game is 0. The optimal mixed strategy for player II is likewise a 50-50 mix of H and T, which also leads to an expectation of 0, making the game **fair**.

Fair Game DEFINITION

A **fair game** has a value of 0 and, consequently, it favors neither player when at least one player uses an *optimal* (mixed) strategy–one that guarantees that the resulting expected payoff is the best that this player can obtain against all possible strategy choices (pure or mixed) by an opponent.

Player II gains nothing by knowing that player I is using the optimal mixed strategy $(\frac{1}{2}, \frac{1}{2})$. However, player I must not reveal to player II whether H or T will be displayed *in any given play* of the game before player II makes his or her own choice of H or T. Even without this information, if player II knew that player I was using a particular *nonoptimal* mixed strategy $(p_1, p_2) = (p, 1 - p)$, where $p \neq \frac{1}{2}$ (that is, not choosing a 50-50 mixture between H and T), then player II could take advantage of this knowledge and increase his or her average winnings over time to something greater than the value of 0. (See Exercise 14.) ■

EXAMPLE 5 ■ Nonsymmetrical Matching

In this game, players I and II can again show either heads H or tails T. When two H's appear, player II pays \$5 to player I. When two T's appear, player II pays \$1 to player I. When one H and one T are displayed, then player II collects \$3 from player I. Note that although the sum of player I's gains (\$5 + \$1 = \$6) when there are two H's or two T's, and the sum of player II's gains (\$3 + \$3 = \$6) otherwise, are the same, the game is **nonsymmetrical**.

Nonsymmetrical Game DEFINITION

A two-person total-conflict **nonsymmetrical game** is one in which the row player's gains are different from the column player's gains. Note that the row player's gains (\$5 and \$1 in our example) are different from the column player's gains (always \$3). In the original matching pennies, on the other hand, the payoff for winning is the same for each player, so that game is *symmetrical.*

The game just described is given by the payoff matrix in Table 15.7, which shows the payoffs that player I receives from player II. A worst-case analysis, like that which solved our initial location game, is of little help here. Player I may lose \$3 whether he plays H or T, making his maximum -3. Player II can keep her losses down to \$1 by always playing T (and thus avoiding the loss of \$5 when two H's appear), so player II's minimax is 1. However, if player II chooses T and player I knows this, then player I will also play T and collect \$1 from player II. Can player II do better than lose \$1 in each play of the game?

Consider the situation where player I uses a mixed strategy $(p_H, p_T) = (p, 1 - p)$, which involves playing H with probability p and playing T with

TABLE 15.7 Payoffs for Player I in a Nonsymmetrical Matching Game

		Player II	
		H	*T*
Player I	*H*	5	−3
	T	−3	1

probability $1 - p$, where $0 \leq p \leq 1$. Against player II's pure strategy H, player I's expected value is

$$E_H = (5)(p) + (-3)(1 - p) = 8p - 3$$

Against player II's pure strategy T, player I's expected value is

$$E_T = (-3)(p) + (1 - p) = -4p + 1$$

These two linear equations in the variable p are depicted in Figure 15.3. Note that the four points where these two lines intersect the two vertical lines, $p = 0$ and $p = 1$, are the four payoffs appearing in the payoff matrix.

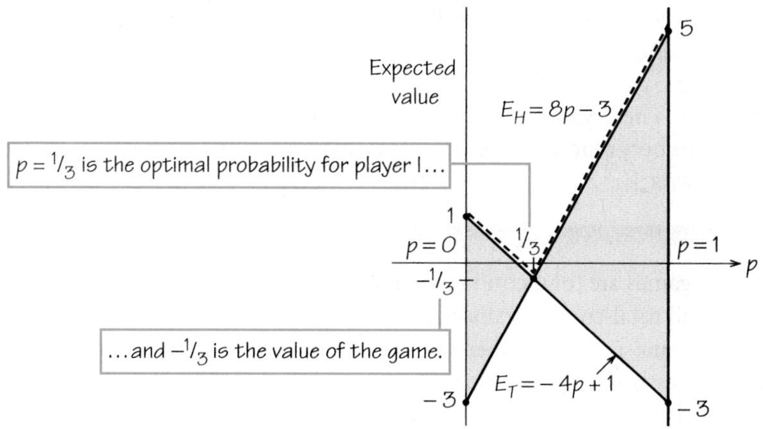

FIGURE 15.3 Solution to the nonsymmetrical matching pennies.

The point at which the lines given by E_H and E_T intersect can be found by setting $E_H = E_T$, yielding

$$8p - 3 = -4p + 1$$
$$12p = 4$$

so $p = \frac{1}{3}$. To the left of $p = \frac{1}{3}$, $E_T > E_H$, and to the right, $E_H > E_T$; at $p = \frac{1}{3}$, $E_H = E_T$. If player I chooses $(p_H, p_T) = (p, 1 - p) = (\frac{1}{3}, \frac{2}{3})$, he can ensure

$$E_H = 8\left(\frac{1}{3}\right) - 3 = E_T = -4\left(\frac{1}{3}\right) + 1 = -\frac{1}{3}$$

regardless of what player II does.

In other words, player I's optimal mixed strategy is to pick H and T with probabilities $\frac{1}{3}$ and $\frac{2}{3}$, respectively, which gives player I an expected value of $-\frac{1}{3}$. As can be seen from Figure 15.3, $-\frac{1}{3}$ is the highest expected value that player I can guarantee against *both* strategies H and T of player II. Although T yields player I a higher expected value for $p < \frac{1}{3}$, and H yields him a higher expected value for $p > \frac{1}{3}$, player I's choice of $p_H, p_T = (\frac{1}{3}, \frac{2}{3})$ protects him against an expected loss greater than $-\frac{1}{3}$, which neither of his pure strategies does (each may produce a maximum loss of -3). Put another way, the intersection of E_H and E_T at $p = \frac{1}{3}$ is the minimum of the function given by E_T to the left and E_H to the right (shown by the dashed line in Figure 15.3). If player II had more than two strategies, this approach to finding a minimum that puts a floor on player I's expected loss can be extended.

A similar calculation for player II results in the same optimal mixed strategy $(\frac{1}{3}, \frac{2}{3})$ and expected value $-\frac{1}{3}$. But because the payoffs for player II are losses, $-\frac{1}{3}$ means that she gains $\frac{1}{3}$ on the average.

Therefore, this game is unfair, even though the sum of the amounts ($6) that player I might have to pay player II when he loses is the same as the sum that player II might have to pay player I when she loses. Interestingly, player II, who will win an average of $33\frac{1}{3}$ cents each time the game is played, is favored, even though she may have to pay more to player I when she loses (a maximum of $5) than player I will ever have to pay her (a maximum of $3).

The symmetrical and nonsymmetrical matching games are examples of what are called **zero-sum games.**

Zero-Sum Game — DEFINITION

A **zero-sum game** is one in which the payoff to one player is the negative of the corresponding payoff to the other, so the sum of the payoffs to the two players is always zero. These games can be completely described by a payoff matrix, in which the numbers represent the payoffs to player I, while their negatives are the payoffs to player II.

Zero-sum games are total-conflict games in which what one player wins the other loses. But not all total-conflict games are zero-sum–in particular, the sum of the payoffs could be some constant other than zero. Nevertheless, the strategic nature of these latter games is the same as that of zero-sum games: What one player wins, the other player still loses. This was true in our location game, in which Henry's payoff was greater the higher the altitude, and Lisa's greater the lower the altitude.

Scoring in professional chess tournaments usually assigns a payoff of 1 for winning, 0 for losing, and $\frac{1}{2}$ to each player for a tie, making the sum of the payoffs to the two players always 1. Such games, called **constant-sum games**, can readily be converted to zero-sum games. Thus, chess could as well be scored -1 for a loss, $+1$ for a win, and 0 for a tie, making the constant 0 in this case. Although constant-sum and zero-sum games have the same strategic nature, constant-sum games are a more general class because the constant need not be zero.

The solution in the symmetrical version of matching pennies illustrated how the mixed strategy of $(\frac{1}{2}, \frac{1}{2})$ guarantees each player the value of 0, but we did not give a *solution technique* for finding optimal mixed strategies. In the nonsymmetrical version of matching pennies, we illustrated a procedure that can be applied to *every* payoff matrix in which each player has only two strategies.

We must use more complex methods, which we will not describe here, to find mixed-strategy solutions when one or both players have more than two strategies. However, one should always check first to see whether a game has a saddlepoint before employing any technique for finding optimal mixed strategies.

In our next example, which is the earlier duel between the pitcher and the batter given by the 2×2 payoff matrix in Table 15.5, we already showed that there is no saddlepoint. Thus, the solution will necessarily be in mixed strategies. We now proceed to find what mix is optimal.

EXAMPLE 6 ■ The Duel Game Revisited

In Table 15.8, we add probabilities, which we explain next, to Table 15.5, where F indicates fastball and C indicates curve ball. The pitcher should use a mixed strategy $(p_1, p_2) = (p_F, p_C) = (p, 1 - p)$. The probabilities p and $1 - p$ (where $0 \leq p \leq 1$) are indicated below the game matrix and under the corresponding strategies, F and C, for the pitcher. If the pitcher plays a mixed strategy $(p, 1 - p)$ against the two pure strategies, F and C, for the batter, he realizes the respective expected values:

$$E_F = (0.3)p + 0.2(1 - p) = 0.1p + 0.2$$
$$E_C = (0.1)p + 0.5(1 - p) = -0.4p + 0.5$$

TABLE 15.8 A Baseball Duel with Probabilities

		Pitcher		
		F	***C***	
Batter	***F***	0.300	0.200	q
	C	0.100	0.500	$1 - q$
		p	$1 - p$	

As in the nonsymmetrical matching-pennies game, the solution to this game occurs at the intersection of the two lines given by E_F and E_C. Setting the equations of these lines equal to each other yields $p = 0.6$, giving $E_F = E_C = E = 0.260$.

Thus, the pitcher should use his optimal mixed strategy, which selects F with probability $p = \frac{3}{5}$ and C with probability $1 - p = \frac{2}{5}$. This choice will hold the batter down to a batting average of 0.260, which is the value of the game. We stress that 0.260 is an average and must be interpreted in a statistical manner. It says that about one time in four the batter will get a hit, but not what will happen on any particular time at bat.

Assume that the batter uses a mixed strategy $(q_1, q_2) = (q_F, q_C) = (q, 1 - q)$, as indicated to the right of the game matrix in Table 15.8. This mixed strategy, when played against the pitcher's pure strategies, F and C, results in the following expected values:

$$E_F = (0.3)q + 0.1(1 - q) = 0.2q + 0.1$$
$$E_C = (0.2)q + 0.5(1 - q) = -0.3q + 0.5$$

The intersection of these two lines occurs at the point $q = 0.8$, giving $E_F = E_C = E = 0.260$. The batter's optimal mixed strategy is therefore $(q_F, q_C) = (\frac{4}{5}, \frac{1}{5})$, which gives him the same batting average of 0.260. ■

We have seen that the outcome of 0.260, which is the value of the game, occurs when either the pitcher selects his optimal mixed pitching strategy $(\frac{3}{5}, \frac{2}{5})$ or the batter selects his optimal mixed guessing strategy $(\frac{4}{5}, \frac{1}{5})$. This particular result holds true for every two-person zero-sum game; it is the fundamental theorem for such games and is known as the **minimax theorem**.

Minimax Theorem DEFINITION

The **minimax theorem** guarantees that there is a unique game value and an optimal strategy for each player, so that either player alone can realize at least this value by playing this strategy, which may be pure or mixed.

The unique value in our example is 0.260.

15.3 Partial-Conflict Games

The 2 × 2 matrix games (two players, each with two strategies) presented so far have been total-conflict games: One player's gain was equal to the other player's loss. Although most parlor games, like chess or poker, are games of total conflict, and therefore constant-sum, most real-life games are surely not. (Elections, in which there are usually a clear-cut winner and one or more losers, probably come as close to being games of total conflict as we find in the real world.) We will consider two games of partial conflict, in which the players' preferences are not diametrically opposed, that have often been used to model real-world conflicts.

Variable-Sum Games DEFINITION

Games of partial conflict are **variable-sum games**, in which the sum of payoffs to the players at the different outcomes varies.

There is some mutual gain to be realized by both players if they can cooperate in partial-conflict games, but this may be difficult to do in the absence of either good communication or trust. When these elements are lacking, players are less likely to comply with any agreement that is made. *Noncooperative games* are games in which a binding agreement cannot be enforced. Even if communication is allowed in such games, there is no assurance that a player can trust an opponent to choose a particular strategy that he or she promises to select.

In fact, the players' self-interests may lead them to make strategy choices that yield both lower payoffs than they could have achieved by cooperating. Two partial-conflict games illustrate this problem.

EXAMPLE 7 ■ Prisoners' Dilemma

Prisoners' Dilemma is a two-person variable-sum game. It provides a simple explanation of the forces at work behind arms races, price wars, and the population problem. In these and other similar situations, the players can do better by cooperating. But there may be no compelling reasons for them to do so unless the players have credible threats of retaliation for not cooperating. The name *Prisoners' Dilemma* was first given to this game by Princeton mathematician Albert W. Tucker (1905–1994) in 1950.

Before defining the formal game, we introduce it through a story.

Prisoners' Dilemma STORY

Prisoners' Dilemma involves two persons, accused of a crime, who are held incommunicado. Each has two choices: to maintain his or her innocence, or to sign a confession accusing the partner of committing the crime. It is in each suspect's interest to confess and implicate the partner, thereby trying to receive a reduced sentence. Yet if both suspects confess, they ensure a bad outcome–namely, they are both found guilty. What is good for the prisoners as a pair–to deny having committed the crime, leaving the state with insufficient evidence to convict them–is frustrated by their pursuit of their own individual interests.

The game of Prisoners' Dilemma, as we already noted, has many applications, but we will use it here to model a recurrent problem in international relations: arms races between antagonistic countries, which earlier included the superpowers but more recently have included such countries as India and Pakistan and Israel and some of its Arab neighbors. Other countries, such as Iran, may be antagonistic to more than one other country (Israel and the United Staes).

For simplicity, assume there are two nations, Red and Blue. Each can independently select one of two policies:

A: Arm in preparation for a possible war (noncooperation).
D: Disarm, or at least try to negotiate an arms-control agreement (cooperation).

There are four possible outcomes:

(D, D): Red and Blue disarm, which is *next best* for both because, while advantageous to each, it also entails certain risks.
(A, A): Red and Blue arm, which is *next worst* for both because they spend needlessly on arms and are comparatively no better off than at (D, D).
(A, D): Red arms and Blue disarms, which is *best for Red* and *worst for Blue*, because Red gains a big edge over Blue.
(D, A): Red disarms and Blue arms, which is *worst for Red* and *best for Blue*, because Blue gains a big edge over Red.

This situation can be modeled by means of the matrix in Table 15.9, which gives the possible outcomes that can occur. Here, Red's choice involves picking one of the two rows, whereas Blue's choice involves picking one of the two columns.

We assume that the players can rank the four outcomes from best to worst, where 4 = best, 3 = next best, 2 = next worst, and 1 = worst. Thus, the higher the number, the greater the payoff, making the resulting game an **ordinal game**: It indicates an ordering of outcomes from best to worst but says nothing about the *degree* to which a player prefers one outcome over another. To illustrate, if a player despises the outcome that he or she ranks 1 but sees little difference among the outcomes ranked 4, 3, and 2, the "payoff distance" between 4 and 2 will be less than that between 2 and 1, even though the numerical difference between 4 and 2 is greater.

The ordinal payoffs to the players for choosing their strategies of A and D are shown in Table 15.10, where the first number in the pair indicates the payoff to the row player (Red), and the second number the payoff to the column player (Blue). Thus, for example, the pair (1, 4) in the second row and first column signifies a payoff of 1 (worst outcome) to Red and a payoff of 4 (best outcome) to Blue. This outcome occurs when Red unilaterally disarms while Blue continues to arm, making Blue, in a sense, the winner and Red the loser.

TABLE 15.9 The Outcomes in an Arms Race, as Modeled by Prisoners' Dilemma

		Blue	
		A	*D*
Red	*A*	Arms race	Favors Red
	D	Favors Blue	Disarmament

TABLE 15.10 Ordinal Payoffs in an Arms Race, as Modeled by Prisoners' Dilemma

		Blue	
		A	*D*
Red	*A*	(2, 2)	(4, 1)
	D	(1, 4)	(3, 3)

Let's examine this strategic situation more closely. Should Red select strategy *A* or *D*? There are two cases to consider, which depend on what Blue does:

- If Blue selects *A*: Red will receive a payoff of 2 for *A* and 1 for *D*, so it will choose *A*.
- If Blue selects *D*: Red will receive a payoff of 4 for *A* and 3 for *D*, so it will choose *A*.

In both cases, Red's first strategy (*A*) gives it a more desirable outcome than its second strategy (*D*). Consequently, we say that *A* is Red's **dominant strategy**, because it is always advantageous for Red to choose *A* over *D*.

In Prisoners' Dilemma, *A dominates D* for Red, so we presume that a rational Red would choose *A*. A similar argument leads Blue to choose *A* as well–that is, to pursue a policy of arming. Thus, when each nation strives to maximize its own payoffs independently, the pair is driven to the outcome (*A, A*), with payoffs of (2, 2). The better outcome for both, (*D, D*), with payoffs of (3, 3), appears unobtainable when this game is played noncooperatively.

The outcome (*A, A*), which is the product of dominant strategy choices by both players in Prisoners' Dilemma, is a **Nash equilibrium**.

Nash Equilibrium DEFINITION

When no player can benefit by departing unilaterally (by itself) from its strategy associated with an outcome, the strategies of the players constitute a **Nash equilibrium**. Technically, while it is the set of strategies that define the equilibrium, the choice of these strategies leads to an outcome that we shall also refer to as the equilibrium.

Note that in Prisoners' Dilemma, if either player departs from (A, A), the payoff for the departing player who switches to D drops from 2 to 1 at (D, A) and (A, D). Not only is there no benefit from departing, but there is actually a loss, with the D player punished with its worst payoff of 1. These losses would presumably deter each nation from moving away from the Nash equilibrium of (A, A), assuming the other nation sticks to A.

Even if both nations agreed in advance jointly to pursue the socially beneficial outcome, (D, D), (3, 3) is unstable. This is because if either nation alone reneges on the agreement and secretly arms (as North Korea did when it developed nuclear weapons), it will benefit, obtaining its best payoff of 4. Consequently, each nation would be tempted to go back on its word and select A. Especially if nations have no great confidence in the trustworthiness of their opponents, they would have good reason to try to protect themselves against the other side's defection from an agreement by arming.

Prisoners' Dilemma

DEFINITION

Prisoners' Dilemma is a two-person variable-sum game in which each player has two strategies, cooperate or defect (not cooperate). Defect dominates cooperate for both players, even though the mutual-defection outcome, which is the unique Nash equilibrium in the game, is worse for both players than the mutual-cooperation outcome.

Note that if 4, 3, 2, and 1 in Prisoners' Dilemma were not just ranks but numerical payoffs, their sum would be $2 + 2 = 4$ at the mutual-defection outcome and $3 + 3 = 6$ at the mutual-cooperation outcome. At the other two outcomes, the sum, $1 + 4 = 5$, is still different, illustrating why Prisoners' Dilemma is a variable-sum game.

In real life, of course, people often manage to escape the noncooperative Nash equilibrium in Prisoners' Dilemma. Either the game is played within a larger context, wherein other incentives are at work, such as cultural norms that prescribe cooperation (though this is just another way of saying that defection from (D, D) is not rational, rendering the game not Prisoners' Dilemma), or the game is played on a repeated basis–it is not a one-short affair–so players can induce cooperation by setting a pattern of rewards for cooperation and penalties for noncooperation.

In a repeated game, factors like reputation and trust may play a role. Realizing the mutual advantages of cooperation in costly arms races, players may inch toward the cooperative outcome by slowly phasing down their acquisition of weapons over time, or even destroying them (the United States and Russia have been doing exactly this). They may also initiate other productive measures, such as improving their communication channels, making inspection procedures more reliable, writing agreements that are truly enforceable, or imposing penalties for violators when their violations are detected (as has occurred through reconnaissance or spy satellites).

Prisoners' Dilemma illustrates the intractable nature of certain competitive situations that blend conflict and cooperation. The standoff that results at the Nash equilibrium of (2, 2) is obviously not as good for the players as that which they could achieve by cooperating–but they risk a good deal if the other player defects.

While saddlepoints are Nash equilibria in total-conflict games, they can never be worse for *both* players than some other outcome (as in partial-conflict games like

CHAPTER 16

Identification Numbers

Modern identification numbers serve at least two functions. An identification number should unambiguously identify the person or thing with which it is associated. Less obvious is a "self-checking" aspect of the number.

16.1 Check Digits

Look at the 13-digit **International Standard Book Number (ISBN)** printed on the back cover of this book. The number 978-1-4292-0900-7 (978-1-4292-1506-0 for the paperback version) is a **code**. It distinguishes this book from all others. The last digit 7 is there solely to detect errors that may occur when the ISBN is entered into a computer. Grocery items, credit cards, overnight mail, magazines, personal checks, traveler's checks, soft-drink cans, automobiles, and many other items you encounter daily have identification numbers that code data and a digit called a **check digit** for error detection. In this chapter, we examine some of the methods used to assign identification numbers and check digits.

Let's begin by considering the U.S. Postal Service money order shown in Figure 16.1. The first 10 digits of the 11-digit number 63024383845 simply identify the money order. The last digit, 5, serves as an **error-detecting** mechanism. Let's see how this mechanism works. The eleventh (last) digit of a Postal Service money order number is the remainder obtained when the sum of the first 10 digits of the number is divided by 9. In our example, the last digit is 5 because $6 + 3 + 0 + 2 + 4 + 3 + 8 + 3 + 8 + 4 = 41$ and the remainder when 41 is divided by 9 is 5.

Now suppose that instead of the correct number, the number 63054383845 (an error in the fourth position) was entered into a computer programmed for error detection in money orders. The machine would divide the sum of the first 10 digits of the entered number, 44, by 9 and obtain a remainder of 8. Since the last digit of the entered number is 5 rather than 8, the entered number cannot be correct. This crude method of error detection will not detect the mistake of

FIGURE 16.1 Money order with identification number 6302438384 and appended check digit 5. The check digit is the remainder after dividing the sum of the digits by 9.

replacing a 0 with a 9, or vice versa. Nor will it detect the transposition of digits, such as 63204383845 instead of 63024383845 (the digits in positions three and four have been transposed).

American Express traveler's checks, VISA traveler's checks, and Euro banknotes also use a check digit determined by division by 9. In these cases, the check digit is chosen so that the sum of the digits, including the check digit, is evenly divisible by 9.

EXAMPLE 1 ■ The American Express Travelers Cheque

The American Express Travelers Cheque with the identification number 387505055 has check digit 7 because $3 + 8 + 7 + 5 + 0 + 5 + 0 + 5 + 5 = 38$ and $38 + 7$ is evenly divisible by 9.

The scheme used on airline tickets, UPS packages, and Avis and National rental cars assigns the remainder after division by 7 of the number itself as the check digit rather than dividing the sum of the digits by 7. For example, the check digit for the number 540047 is 4 because $540047 = 7 \times 77149 + 4$. This method will not detect the substitution of 0 for a 7, 1 for an 8, 2 for a 9, or vice versa. However, unlike the Postal Service method, it will detect transpositions of adjacent digits with the exceptions of the pairs 0, 7; 1, 8; and 2, 9. For example, if 5400474 were entered into a computer as 4500474 (the first two digits are transposed), the machine would determine that the check digit should be 3 since $450047 = 7 \times 64292 + 3$. Because the last digit of the entered number is not 3, the error has been detected.

The scheme used on grocery products, the **Universal Product Code (UPC)**, is more sophisticated. Consider the number 0 38000 00127 7 found on the bottom of a box of corn flakes. The first digit identifies a broad category of goods, the next five digits identify the manufacturer, the next five identify the product, and the last is a check digit. Suppose this number were entered into a computer as 0 58000 00127 7 (a mistake in the second position). How would the computer recognize the mistake?

For any UPC number $a_1a_2a_3a_4a_5a_6a_7a_8a_9a_{10}a_{11}a_{12}$, the computer is programmed to carry out the following computation: $3a_1 + a_2 + 3a_3 + a_4 + 3a_5 + a_6 + 3a_7 + a_8 + 3a_9 + a_{10} + 3a_{11} + a_{12}$. If the result doesn't end with a 0, the computer knows the entered number is incorrect.

For the incorrect corn flakes number, we have $3 \cdot 0 + 5 + 3 \cdot 8 + 0 + 3 \cdot 0 + 0 + 3 \cdot 0 + 0 + 3 \cdot 1 + 2 + 3 \cdot 7 + 7 = 62$. Since 62 doesn't end with 0, the error is detected. Notice that had we used the correct digit 3 in the second position instead of 5, the sum would have ended in a 0 as it should. This simple scheme detects *all* single-position errors and about 89% of all other kinds of errors.

Beginning in January 2005, U.S. retailers were required to have software that could read the 12-digit UPC code used in the United States and the 13-digit European Article Number (EAN) code used in Europe. This change paves the way for the 13-digit EAN to become the worldwide standard. Existing UPC numbers will be converted to EAN numbers by adding an extra 0 at the beginning. The check digit for a 13-digit EAN number $a_1a_2a_3a_4a_5a_6a_7a_8a_9a_{10}a_{11}a_{12}a_{13}$ is selected so that $a_1 + 3a_2 + a_3 + 3a_4 + a_5 + 3a_6 + a_7 + 3a_8 + a_9 + 3a_{10} + a_{11} + 3a_{12} + a_{13}$ ends with 0. Adding an extra 0 in the front of a UPC number does not affect the check digit.

The U.S. banking system uses a variation of the UPC scheme that appends check digits to the numbers assigned to banks. Each bank has an eight-digit routing number $a_1a_2 \cdots a_8$ together with a check digit a_9 so that a_9 is the last digit of $7a_1 + 3a_2 + 9a_3 + 7a_4 + 3a_5 + 9a_6 + 7a_7 + 3a_8$. The numbers 7, 3, and 9 used in this formula are called the **weights.** (The weights for the UPC scheme are 3 and 1.) The weights were carefully chosen so that all single-digit errors and most transposition errors are detected. The use of different weights in adjacent positions permits the detection of most transposition errors.

EXAMPLE 2 ■ Bank Identification Number

The First Chicago Bank has the routing number 071000013 on the bottom of all its checks (see Figure 16.2). The check digit 3 is the last digit of $7 \cdot 0 + 3 \cdot 7 + 9 \cdot 1 + 7 \cdot 0 + 3 \cdot 0 + 9 \cdot 0 + 7 \cdot 0 + 3 \cdot 1 = 33$. The first four digits of a nine-digit bank routing number identify the bank's Federal Reserve District, office, state or special collection arrangement; the next four digits are the bank's identification number; the ninth digit is the check digit. The block of numbers 22 63378 shown in Figure 16.2 is the account number. The last block, 0134, is the check number.

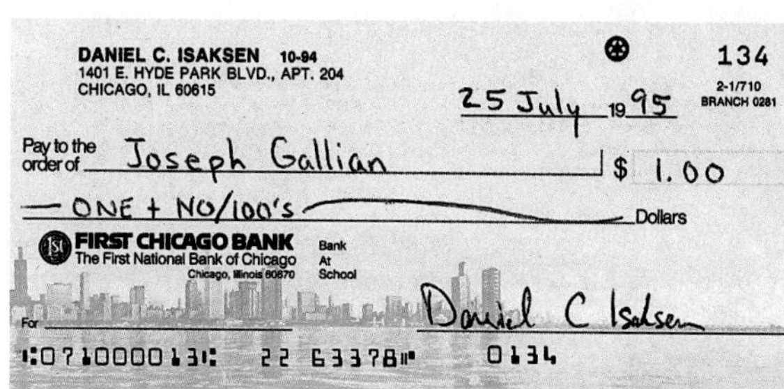

FIGURE 16.2 A bank check with routing number 071000013. The 3 is the check digit.

You may wonder if there is any advantage to using three weights as opposed to using two, as is the case for the UPC error-detection scheme. The answer is yes. While both the UPC scheme and the bank scheme detect 100% of single-position errors and the same transposition errors involving adjacent digits, the bank scheme will detect most transposition errors of the form $\cdots abc \cdots \rightarrow \cdots cba \cdots$, whereas the UPC scheme does not detect such errors.

For example, say we look at a number that begins with 241. In the UPC scheme, these digits contribute $3 \cdot 2 + 4 + 3 \cdot 1 = 13$ toward the total calculation, while the string 142 (the first and third digits are transposed) also contributes $3 \cdot 1 + 4 + 3 \cdot 2 = 13$ toward the total calculation. So the error is not detected. In contrast, using the bank scheme, 241 contributes $7 \cdot 2 + 3 \cdot 4 + 9 \cdot 1 = 35$ toward the total, while 142 contributes $7 \cdot 1 + 3 \cdot 4 + 9 \cdot 2 = 37$ toward the total. Since the total for the correct number ends with 0, the total for the number that had the transposition error would end with the digit 2. Thus, the error is detected.

One of the most efficient error-detection methods is used by all major credit-card companies as well as by many libraries, blood banks, photofinishing companies, and the South Dakota driver's license department. Say a bank intends to issue a credit card with the identification number 312560019643001. It must then add an extra digit for error detection. This is done as follows. Add the digits in positions 1, 3, 5, 7, 9, 11, 13, and 15 and double the result: $(3 + 2 + 6 + 0 + 9 + 4 + 0 + 1) \times 2 = 50$. Next, count the number of digits in positions 1, 3, 5, 7, 9, 11, 13, and 15 that exceed 4 and add this to the total. For our example, only 6 and 9 exceed 4, so the count is 2 and our running total is 52. Finally, take the sum of 52 and the digits in the even-numbered positions: $52 + (1 + 5 + 0 + 1 + 6 + 3 + 0) = 68$.

The check digit is whatever is needed to bring the final tally to a number that ends with 0. Because $68 + 2 = 70$, the check digit for our example is 2. This digit is appended to the end of the number the bank issues for identification purposes. Errors in input data are detected by applying the same algorithm to the input, including the check digit. If the correct number is entered into a computer, the result will end in 0. If the result doesn't end with 0, a mistake has been made. The credit card shown in Figure 16.3 is reproduced from an ad promoting the Citibank VISA card. Notice that the check digit on the card is not valid because the algorithm yields

$$(4 + 2 + 0 + 1 + 3 + 5 + 7 + 9) \times 2 + 3 + (1 + 8 + 0 + 2 + 4 + 6 + 8) + 0 = 94$$

which does not end in 0. This method allows computers to detect 100% of single-position errors and about 98% of other common errors.

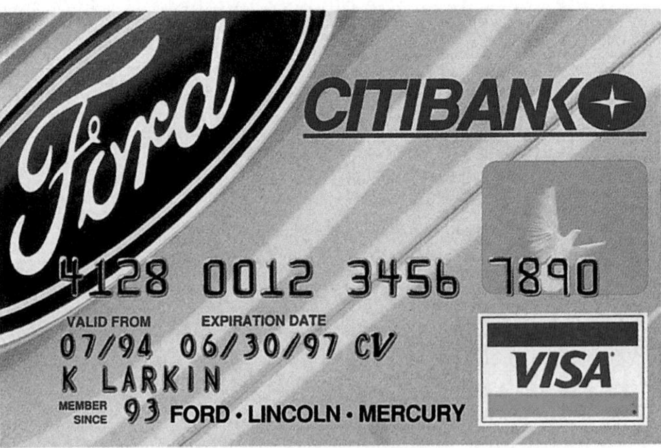

FIGURE 16.3 VISA card with an invalid number.

Besides detecting errors, the check digit offers partial protection against fraudulent numbers. A person who wanted to create a phony credit card, bank account number, or driver's license number would have to know the appropriate check-digit scheme for the number to go unchallenged by the computer.

Thus far we have not discussed any schemes that detect 100% of single errors and 100% of transposition errors. As seen on the back of this book and most others published since 2007, there are two identifications numbers–a 13-digit number called the *13-digit International Standard Book Number (ISBN-13)* and a 10-digit number, called the *10-digit ISBN (ISBN-10)*. The 10-digit ISBN detects 100% of single digit errors and 100% of transposition errors.

A correctly coded 10-digit ISBN $a_1 a_2 \cdots a_{10}$ has the property that $10a_1 + 9a_2 + 8a_3 + 7a_4 + 6a_5 + 5a_6 + 4a_7 + 3a_8 + 2a_9 + a_{10}$ is evenly divisible by 11. Consider the 10-digit ISBN of the book you are now reading: 1-4292-0900-3 (1-4292-1506-2 for paperback version). The 1 at the beginning indicates that the book is published in an English-speaking country, the next block of digits, 4292, identifies the publisher, W. H. Freeman and Company. The third block for the hardback edition, 0900, is assigned by the publisher and identifies this particular book. The last digit 3, for the hardback version, is the check digit. Let's verify that this number is a legitimate possibility. We must compute $10 \cdot 1 + 9 \cdot 4 + 8 \cdot 2 + 7 \cdot 9 + 6 \cdot 2 + 5 \cdot 0 + 4 \cdot 9 + 3 \cdot 0 + 2 \cdot 0 + 3 = 176$. Because $176 = 11 \cdot 16$, it is evenly divisible by 11, so no error has been detected.

How can we be sure that this method detects 100% of the single-position errors? Well, let's say that a correct number is $a_1a_2a_3a_4a_5a_6a_7a_8a_9a_{10}$ and that a mistake is made in the second position. (The same argument applies equally well in every position.) We may write this incorrect number as $a_1a_2'a_3a_4a_5a_6a_7a_8a_9a_{10}$, where $a_2' \neq a_2$. For this error to go undetected, it must be the case that $10 \cdot a_1 + 9 \cdot a_2' + 8 \cdot a_3 + 7 \cdot a_4 + 6 \cdot a_5 + 5 \cdot a_6 + 4 \cdot a_7 + 3 \cdot a_8 + 2 \cdot a_9 + a_{10}$ is evenly divisible by 11. Then, since both $10a_1 + 9a_2 + 8a_3 + 7a_4 + 6a_5 + 5a_6 + 4a_7 + 3a_8 + 2a_9 + a_{10}$ and $10a_1 + 9a_2' + 8a_3 + 7a_4 + 6a_5 + 5a_6 + 4a_7 + 3a_8 + 2a_9 + a_{10}$ are divisible by 11, so is their difference:

$$(10 \cdot a_1 + 9 \cdot a_2 + 8 \cdot a_3 + \cdots + a_{10}) - (10 \cdot a_1 + 9 \cdot a_2' + 8 \cdot a_3 + \cdots + a_{10}) = 9 \cdot (a_2 - a_2')$$

Because a_2 and a_2' are distinct digits between 0 and 9, their difference must be one of $\pm 1, \ldots, \pm 9$. Thus, the only possibilities for the number $9 \cdot (a_2 - a_2')$ are ± 9, ± 18, ± 27, ± 36, ± 45, ± 54, ± 63, ± 72, ± 81, and none of these is divisible by 11. So a single-position error cannot go undetected.

To verify that the method detects all adjacent transposition errors, let's suppose that the first two digits are transposed (the same argument applies to all positions). Say the correct number is $a_1a_2a_3 \cdots a_{10}$. As before, for the incorrect number $a_2a_1a_3 \cdots a_{10}$ to go undetected, it must be the case that the difference of the correct number and the incorrect number is evenly divisible by 11. That is,

$$(10a_1 + 9a_2 + 8a_3 + \cdots + a_{10}) - (10a_2 + 9a_1 + 8a_3 + \cdots + a_{10})$$

is evenly divisible by 11. This reduces to $a_1 - a_2$ is divisible by 11. But the only possible differences of two numbers between 0 and 9 are plus or minus the numbers between 0 and 9. Thus $a_1 - a_2 = 0$. But then $a_1 = a_2$ and there is no error.

With a bit more work, we could prove that every transposition error is detected, not just the transpositions of adjacent digits. This is possible because 11 is prime.

Since this method, in contrast to the other methods we have described, detects all single-position errors and all transposition errors, why is it not used more? Well, it does have a drawback. Say the next title published by W.H. Freeman is to have 0902 for the third block. (The 10-digit ISBN for all W.H. Freeman books begins with 1-4292.) What check digit should be assigned? Call it a. Then the weighted sum is $10 \cdot 1 + 9 \cdot 4 + 8 \cdot 2 + 7 \cdot 9 + 6 \cdot 2 + 5 \cdot 0 + 4 \cdot 9 + 3 \cdot 0 + 2 \cdot 2 + a = 177 + a$. Because the next integer after 177 that is divisible by 11 is 187, we see that $a = 10$. But appending 10 to the existing 9-digit number would result in an 11-digit number instead of a 10-digit one. This is the only flaw in the 10-digit ISBN scheme. To avoid this flaw, publishers use an X to represent the check digit 10. As a result, not all 10-digit ISBNs consist solely of digits–some end with X. Publishers could avoid this inconsistency by simply refraining from using numbers that require an X.

To expand the inventory of ISBNs and make them compatible with the UPC/EAN numbering scheme for other retail items worldwide, publishers began using a 13-digit ISBN in 2007. The 13-digit ISBN is the same as the 10-digit ISBN number except for a prefix of 978 or 979 and the check digit. The check digit for the 13-digit ISBN is calculated so that the weighted sum using the weights 1, 3, 1, 3, … , 1, 3, 1 ends with the 0. Thus the 13-digit ISBN and the 13-digit UPC/EAN numbers used for retail products employ the same check digit method. During a phase-in period publishers will use both the 10-digit and 13-digit numbers.

After single digit errors and adjacent transposition errors, the third most common error is one of the form $\cdots abc \cdots \to \cdots cba \cdots$. In practice, these kinds of errors commonly occur in phone numbers that have matching digits separated by another digit such as 727 5856. A likely mistake when writing or dialing this number is to switch the 8 and the 6, resulting in the number 727 5658. Such an error is called a *jump transposition*. Remarkably, there is a simple way to encode identification numbers so that the three most common errors are detected 100% of the time without having to introduce an alphabetic character as is done for the 10-digit ISBN numbers. To illustrate the method, suppose that a math instructor wants to publicly post student grades without revealing any information about the students' ID numbers. Assuming the last four digits of each student ID number are different, she could assign each student a six-digit number by multiplying the last four digits of their identification numbers by 13 (adding leading 0s when necessary). For example, a student with an ID number that ends with 8912 is assigned $115856 = 8912 \times 13$. (To preserve confidentiality of the original four digits, students are not informed of the encoding method). Of course, the instructor can recapture the original four-digit numbers by dividing the encoded numbers by 13. Since all encoded numbers are divisible by 13, the jump transposition error 115856 →115658 is detected because 115658 is not divisible by 13.

The arguments for verifying that encoding identification numbers as multiples of 13 detect 100% of all single digit errors, all transposition errors involving adjacent digits, and all jump transposition errors are similar to those used to show that the 10-digit ISBN numbers detect errors. In particular, a single-digit error in the number $a_n a_{n-1} \cdots a_i \cdots a_0$ of the form $a_n a_{n-1} \cdots a'_i \cdots a_0$ where $a'_i \neq a_i$ is not detected if and only if $a_n a_{n-1} \cdots a'_i \cdots a_0$ is a multiple of 13. But if both $a_n a_{n-1} \cdots a_i \cdots a_0$ and $a_n a_{n-1} \cdots a'_i \cdots a_0$ are multiples of 13, then so is their difference $(a_n a_{n-1} \cdots a_i \cdots a_0) - (a_n a_{n-1} \cdots a'_i \cdots a_0) = (a_i - a'_i)10^i$. But 13 does not divide the term on the right when $a_i \neq a'_i$. Similarly, the transposition of adjacent digits a_i and a_{i-1} is undetected if and only if $9(a_i - a_{1-1})10^{i-1}$ is divisible by 13, which happens only when $a_i = a_{i-1}$. And the jump transposition $\cdots a_i a_{i-1} a_{i-2} \cdots \to \cdots a_{i-2} a_{i-1} a_i \cdots$ is undetected if and only if $99(a_i - a_{i-2})10^{i-2}$ is divisible by 13, which happens only when $a_i = a_{i-2}$. Incidentally, the arguments just given reveal why we used multiplication by 13 rather than some smaller positive integer. For example, if multiplication by 11 were used to transform the identification numbers instead of 13, then all single digit errors and all adjacent transposition errors are detected, but not all jump transpositions are since 11 divides 99.

Many identification numbers use both alphabetic and numerical characters. When a check digit is included, the alphabetic characters are assigned numerical values. The vehicle identification number (VIN) used to identify cars and trucks is one such example, as explained in Spotlight 16.1

SPOTLIGHT 16.1 The VIN System

Automobiles and trucks are given a vehicle identification number (VIN) by the manufacturer. A typical VIN has 17 alphanumeric characters that code information, such as country where the vehicle was built, manufacturer, make, body style, engine type, plant where the vehicle was built, model year, model, type of restraint, a check digit, and a production sequence number. The check digit is calculated by converting the 26 consecutive letters of the alphabet, respectively, to the numbers 1, 2, 3, 4, 5, 6, 7, 8, 9, 1, 2, 3, 4, 5, 6, 7, 8, 9, 2, 3, 4, 5, 6, 7, 8, 9 (note the skipped digit after the second 9) to obtain a 16-digit number $a_1a_2 \cdots a_{15}a_{16}$ that is weighted with 8, 7, 6, 5, 4, 3, 2, 10, 9, 8, 7, 6, 5, 4, 3, 2. The check digit is the remainder when the weighted sum $8 \cdot a_1 + 7 \cdot a_2 + \cdots + 3 \cdot a_{15} + 2 \cdot a_{16}$ is divided by 11 unless the remainder is 10, in which case an X is used instead. The check digit is inserted in position 9.

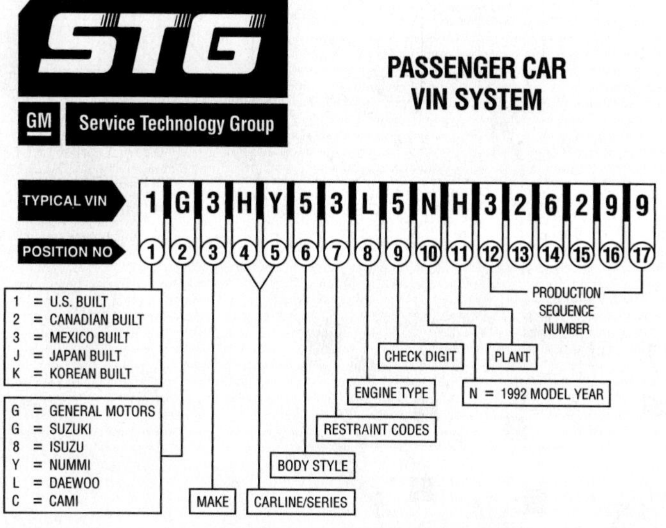

16.2 The ZIP Code

Identification numbers sometimes **encode** geographic data. The ZIP code, Social Security numbers, and telephone numbers are the foremost examples. In 1963, the U.S. Postal Service numbered every American post office with a five-digit **ZIP code.** (ZIP is an acronym for Zone Improvement Plan.) The numbers begin with 0's at the points farthest east–00601 for Adjuntas, Puerto Rico–and work up to 9's at the the points farthest west–99950 for Ketchikan, Alaska (see Figure 16.4).

Let's use one of the ZIP codes for Duluth, Minnesota, as an example:

55812

5 The first digit represents one of 10 geographic areas, usually a group of states. The numbers begin at the points farthest east (0) and end at the points farthest west (9).

58 The second two digits, in combination with the first, identify a central mail-distribution point known as a sectional center. The location of a sectional center is based on geography, transportation facilities, and population density. Although just four centers serve the entire state of Utah, there are six of them to take care of New York City.

12 The last two digits indicate the town or local post office. The order is often alphabetical for towns within a delivery area–for example, towns with names beginning with A usually have low numbers. There are many exceptions to this, such as towns that came into existence after the ZIP code scheme was created. In many cases, the largest city in a region will be given the digits 01 and surrounding towns assigned succeeding digits alphabetically.

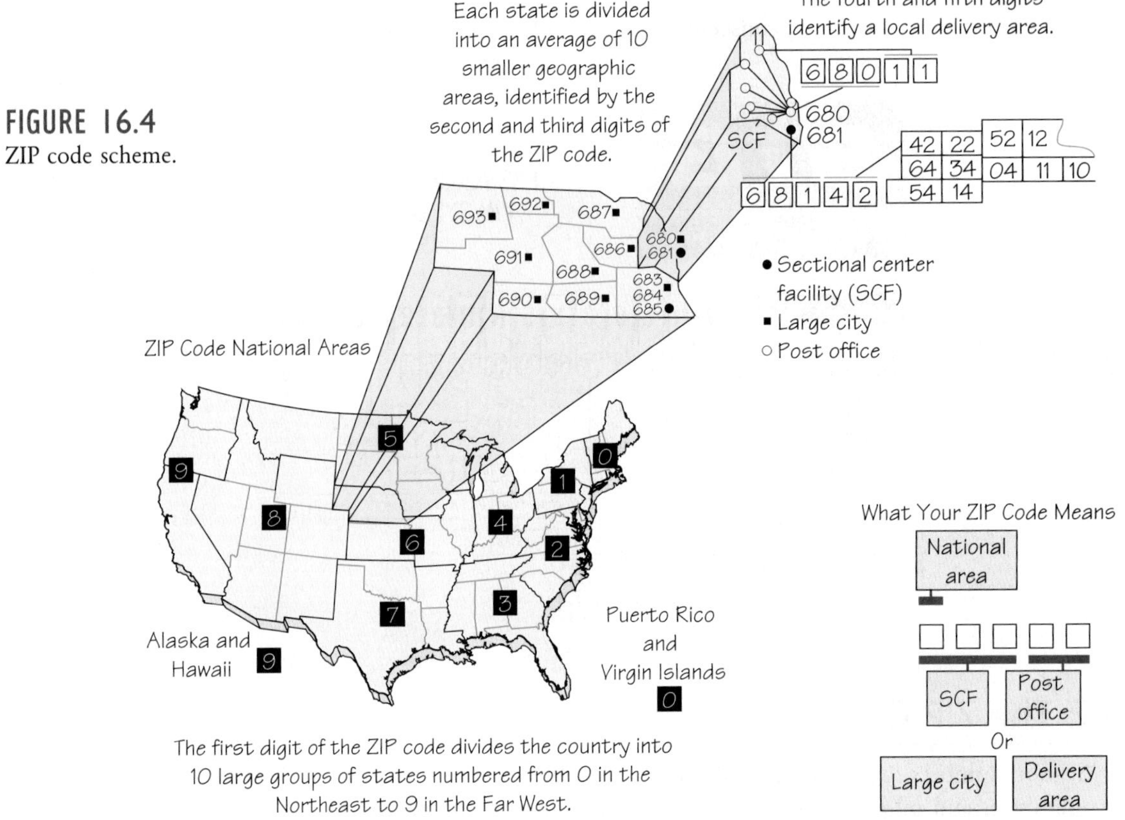

FIGURE 16.4
ZIP code scheme.

In 1983, the U.S. Postal Service added four digits to the ZIP code. When four digits are added after a dash–for example, 68588-1234–the number is called the **ZIP + 4 code**. Mail with ZIP + 4 coding is eligible for cheaper bulk rates, being easier to sort with automated equipment. It's also helpful for businesses that wish to sort the recipients of their mailings by geographic location. The first two numbers of the four-digit suffix represent a delivery sector, which may be several blocks, a group of streets, several office buildings, or a small geographic area. The last two numbers narrow the area further. They might denote one floor of a large office building, a department in a large firm, or a group of post office boxes.

For businesses that receive an enormous volume of mail, the ZIP + 4 code permits automation of in-house mailroom sorting. For example, the first seven digits of all mail sent to the University of Minnesota Duluth, are 55812-24. The school has designated nine pairs of digits for the last two positions to direct the mail to the appropriate dormitory or apartment complex.

16.3 Bar Codes

In modern applications, bar codes and identification numbers go hand in hand. Bar coding is a method for automated data collection. It is a way to transmit information rapidly, accurately, and efficiently to a computer.

Bar Code DEFINITION

A **bar code** is a series of dark bars and light spaces that represent characters.

To **decode** the information in a bar code, a beam of light is passed over the bars and spaces via a scanning device, such as a handheld wand or a fixed-beam device. The dark bars reflect very little back to the scanner, whereas the light spaces reflect much light. The differences in reflection intensities are detected by the scanner and converted to strings of 0's and 1's that represent specific numbers and letters. Such strings are called a **binary coding** of the numbers and letters.

Binary Code DEFINITION

Any system for representing data with only two symbols is a **binary code.**

ZIP Code Bar Code

The simplest bar code is the **Postnet code** used by the U.S. Postal Service and commonly found on business reply forms (see Figure 16.5). For a ZIP + 4 code there are 52 vertical bars of two possible lengths (long and short). The long bars at the beginning and end are called *guard bars* and together provide a frame for the remaining 50 bars. In blocks of five, the 50 bars within the guard bars represent the ZIP + 4 code and a tenth digit for error correction. Each block of five is composed of exactly two long bars and three short bars, according to the pattern shown below:

Decimal Digit	Bar Code
1	ııı||
2	ıı|ı|
3	ıı||ı
4	ı|ıı|
5	ı|ı|ı
6	ı||ıı
7	|ııı|
8	|ıı|ı
9	|ı|ıı
0	||ııı

Handheld scanner reading the shipping bar code on a crate. (*Stewart Cohen/Tony Stone Images.*)

The tenth digit of a Postnet code number is a check digit chosen so that the sum of the nine digits of the ZIP + 4 code and the tenth one is evenly divisible by 10. That is, the check digit C for the ZIP + 4 code $a_1a_2\cdots a_9$ is the digit with the property that the sum $a_1 + a_2 + \cdots + a_9 + C$ ends with 0. For example, the ZIP + 4 code 80321-0421 has the check digit 9, because $8 + 0 + 3 + 2 + 1 + 0 + 4 + 2 + 1 = 21$ and $21 + 9 = 30$ ends with 0.

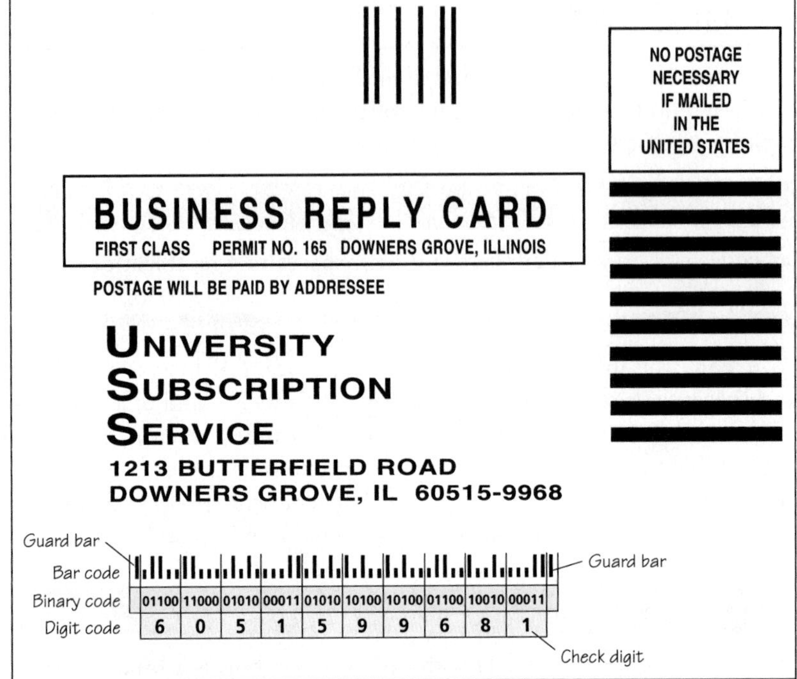

FIGURE 16.5
ZIP + 4 bar code.

Because each digit is represented by exactly two long bars and three short ones, any error in reading or printing a single bar would result in a block of five with only one long bar or three long bars. In either case, the error is detected. This is the reason behind the choice of five bars to code each digit rather than four bars. With five bars per digit, there are exactly 10 arrangements composed of two long bars and three short bars. Any misreading of a single bar in such a block is therefore recognizable, because it does not match any other of the blocks for the 10 digits. And because the block location of the error is known, the check digit permits the correction of the error. Let's look at an example of an incorrectly printed bar code and see how the error is correctable.

EXAMPLE 3 ■ Detecting and Correcting an Error

The scanner ignores the guard bars at the beginning and the end and reads the remaining bars in blocks of five, as shown below. (We have inserted dashed dividing lines for readability.)

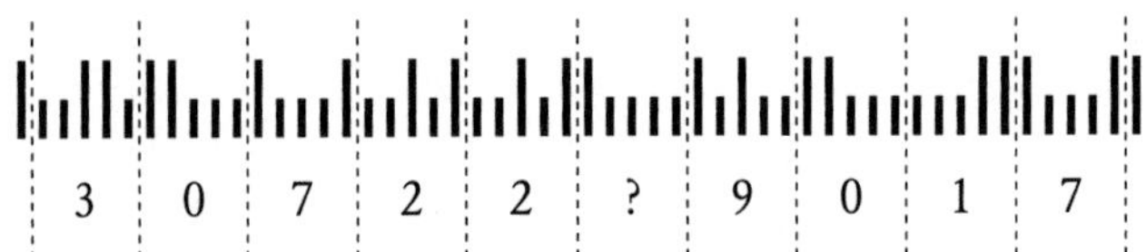

The sixth block is an incorrect one because it has only one long bar. To correct the error, the computer linked with the bar-code scanner sums the remaining 9 digits to obtain 31. Because the sum of all 10 digits ends with 0, the correct value for the sixth digit must be 9.

Beginning in 1993, large organizations and businesses that wanted to receive reduced rates for ZIP + 4 bar-coded mail were required to use a 12-digit bar code called the *delivery-point bar code.* This code permits machines to sort a letter into the order in which it will be delivered by the carrier. Mail for the first location on a mail route occurs first, mail for the second location on a route occurs second, and so on.

The 12-digit bar code uses the Postnet bar scheme to code the 12-digit string composed of the 9-digit ZIP + 4 number followed by the last two digits of the street address or box number and a check digit chosen so that the sum of all 12 digits is evenly divisible by 10. For example, a letter addressed to 1738 Maple Street with ZIP + 4 code 55811-2742 would have the Postnet bar code for the digits 558112742384 (38 is from the street address and 4 is the check digit).

Effective January 1, 2009, businesses that desire to receive discounted postage rates from the U.S. postal service must use a new bar code called the *Intelligent Mail Barcode.* The new bar code converts 31 digits of data into 65 vertical bars that encode the type of service, the mail owner, a unique serial number that enables the user to track the letter at every step from arrival at the post office to delivery, and the delivery-point ZIP code. The code uses bars of three lengths and multiple levels to create four states as shown below. For human use, the four states are denoted by the letters T, A, D, and F. The Intelligent Mail Barcode will appear above the printed address.

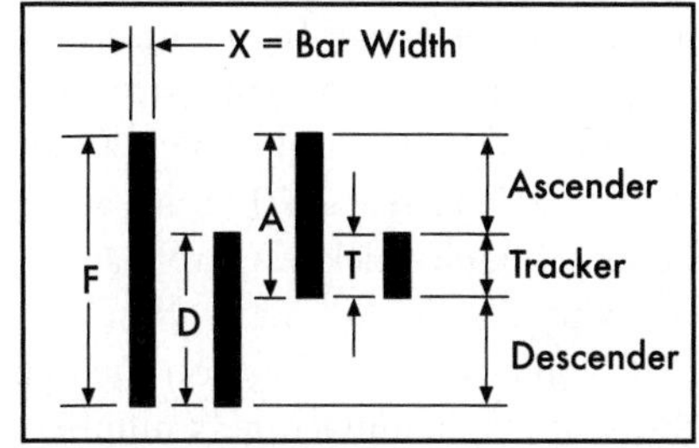

FIGURE 16.6
Entomologist Stephen Buchmann developed a reliable, inexpensive way to track bees using the same technology that supermarkets use to speed up the checkout lines and keep track of inventory. He glued bar-code labels onto the backs of 100 bees and placed a laser scanner above the hive. In the past, researchers marked bees with paint or tags, but monitoring activity required the presence of a human observer. (*Scott Camazine/Sue Trainor.*)

The UPC Bar Code

The bar code that we encounter most often is the UPC, which was first used on grocery items in 1974 and has since spread to most retail products. As Figure 16.6 shows, it has other applications as well. The UPC bar code translates 12-digit UPC identification numbers discussed earlier into bars that can be quickly and accurately read by a laser scanner. The number has four components–two five-digit numbers sandwiched between two single digits–as shown in Figure 16.7.

Here is what the four components represent:

5 The first digit identifies the kind of product. For example, a 2 signals random-weight items, such as cheese and meat; a 3 means drug and certain other health-related products; a 4 means products marked for price reduction by the retailer; a 5 signals cents-off coupons (see Figure 16.7).

13000 The next five digits identify the manufacturer.

22020 The next five digits, assigned by the manufacturer to identify the product, can include size, color, or other important information (but not price).

5 The final digit is the check digit. This digit is often not printed, but it is always included in the bar code.

FIGURE 16.7 UPC identification number 5 13000 22020 5. The initial 5 indicates that the number is a manufacturer's coupon. The block 13000 identifies the manufacturer as Heinz. The block 22020 identifies the product. The last digit, 5, is a check digit.

Each digit of the UPC code is represented by a space divided into seven modules of equal width, as illustrated in Figure 16.8. How these seven modules are filled depends on the digit being represented and whether the digit being represented is part of the manufacturer's number or the product number. In every case, there are two light spaces and two dark bars of various thicknesses that alternate. A UPC code has on each end two long bars of one-module thickness separated by a light space of one-module thickness. These two modules are called the *guard bar patterns* (Figure 16.9). The guard bar patterns define the thickness of a single module of each type. They are not part of the identification number. The manufacturer's number and the product number are separated by a center bar pattern consisting of the following five modules: a light space, a (long) dark bar, a light space, a (long) dark bar, and a light space (see Figure 16.9). The center bar pattern is not part of the identification number but merely serves to separate the manufacturer's number and product number. Figure 16.8 shows how the digits 6 and 0 in a manufacturer's number are coded.

Observe the following pattern in Figure 16.8: a light space of one-module thickness, a dark bar of one-module thickness, a light space of one-module thickness, a dark bar of four-module thickness. Symbolically, such a pattern of light spaces and dark bars is represented as 0101111. Here each 0 means a one-module-thickness light space and each 1 means a one-module-thickness dark bar.

Table 16.1 shows the binary code for all digits. Notice that the code for the digits in the product number (the block of five digits on the right side) can be obtained from the code for the digits in the manufacturer's number (the block of digits on the left side), and vice versa, by replacing each 0 by a 1 and each 1 by a 0. Thus, the code 0111011 for 7 in a manufacturer's number becomes 1000100 in the product number. Also notice that each manufacturer's number has an odd number of 1's, whereas each product number has an even number of 1's. This permits a computer linked with an optical scanner to determine whether the bar code was scanned

TABLE 16.1 Binary UPC Coding

Digit	Manufacturer's Number	Product Number
0	0001101	1110010
1	0011001	1100110
2	0010011	1101100
3	0111101	1000010
4	0100011	1011100
5	0110001	1001110
6	0101111	1010000
7	0111011	1000100
8	0110111	1001000
9	0001011	1110100

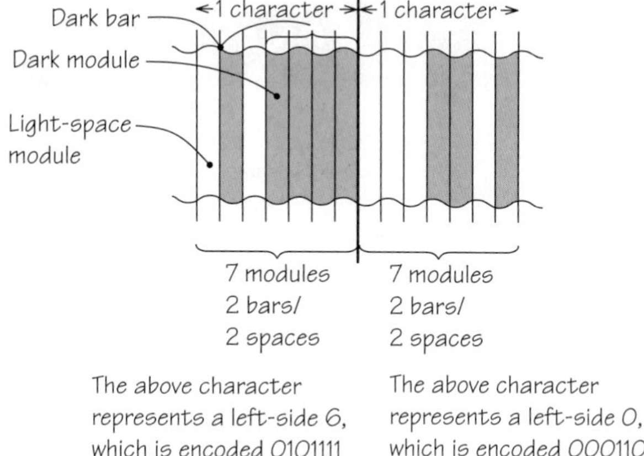

FIGURE 16.8 UPC bar coding for a left-side 6 and left-side 0, part of the manufacturer's number.

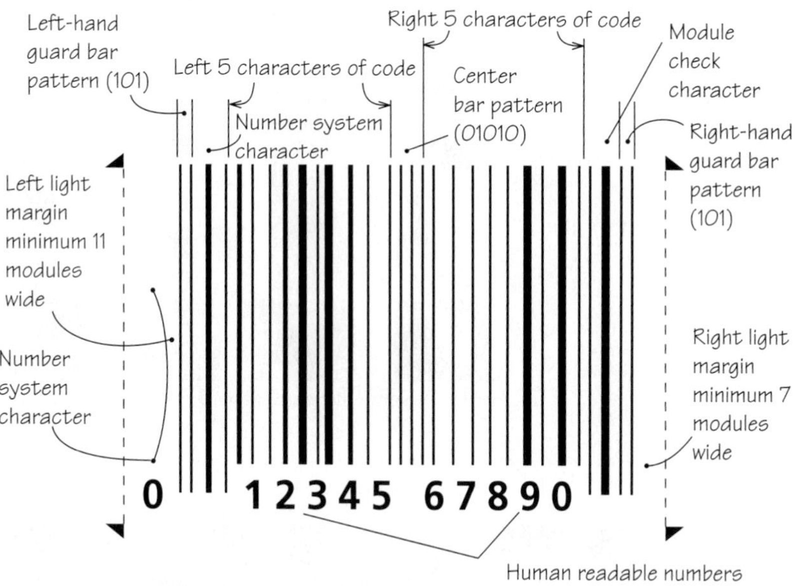

FIGURE 16.9 UPC bar-code format.

left to right or right to left. (If the first block of digits has an even number of 1's for each digit, the scanning is being done right to left.) Thus, scanning can be done in either direction without ambiguity.

New Applications of Bar Coding

New applications of bar coding continue to be found. In 2003 a method of bar coding genetic information about animal species that provides a convenient, inexpensive way to identify species was introduced. (See Spotlight 16.2.) In Japan a new generation of bar codes uses mosaics of black and white rectangles that encode much more information than traditional bar codes (see Figure 16.10). These bar codes can be read by specially equipped cell phones to display video, music or text on the screen or link the cell phone to a Web page. Users can point their cell phone at the bar code in a magazine, on a billboard or on the side of a building to receive information about a product or service. A bar code for a movie will allow the viewer to watch a trailer. Scanning the wrapper of a hamburger will provide nutrition information. This new technology is currently in development for use in the United States.

FIGURE 16.10 Bar code on a building in Japan that can be read by a properly equipped cellphone. (*Ko Sasaki/ The New York Times/Redux.*)

SPOTLIGHT 16.2 New Frontier: Bar Coding DNA

In 2003 Paul Hebert from the University of Guelph in Canada proposed the compilation of a public library of DNA bar codes for animal species. Rather than scanning an animal's entire genome, which is expensive and time consuming, Hebert pinpointed a short piece of a section of a single gene that could be used to distinguish one animal species from another cheaply and quickly. For about $2 per sample the genetic sequence of this tiny gene section can be converted to a four color bar code that corresponds to the four nucleotides that make up the genetic code. The bar code identifies the species of its source in the same way that the UPC bar code identifies a retail item. By 2007 more than 31,000 species were bar coded and a new field of science was born. The technique has already resulted in improved food safety, disease prevention, and better environmental monitoring. The Consortium for the Barcode of Life has set a goal of bar coding 500,000 species by 2012.

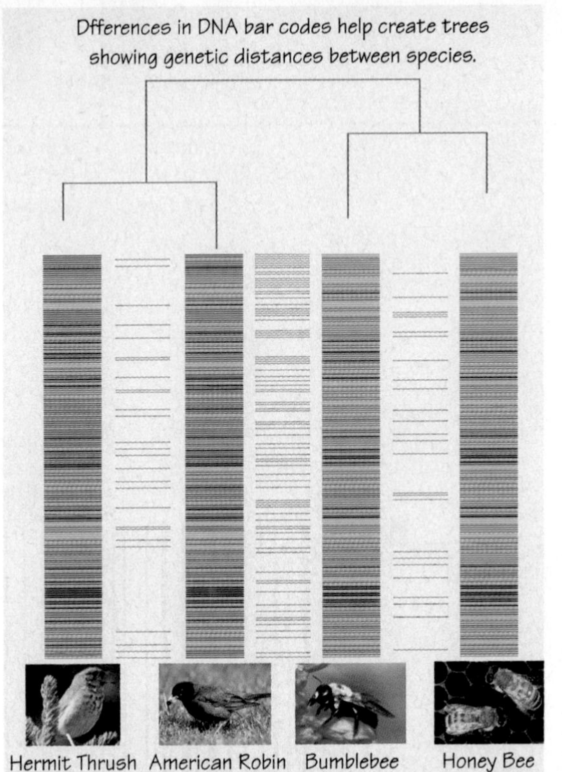
Dfferences in DNA bar codes help create trees showing genetic distances between species.
Hermit Thrush
American Robin
Bumblebee
Honey Bee

Hermit Thrush (*George Jameson*) American Robin (*Jeffrey Lepore/ Photo Researchers, Inc.*) Bumblebee (*Mark Stoeckle/ The Rockefeller University*) Honey Bee (*Scott Camazine/Photo Researchers, Inc.*) DNA bar code (*Mark Stoekle/The Rockefeller University*)

SPOTLIGHT 16.3 History of the Bar Code

1948 Graduate students Norman Joseph Woodland and Bernard Silver at Drexel Institute of Technology begin working on a bar code.

October 7, 1952 Woodland and Silver receive a U.S. patent.

October 10, 1967 The Association of American Railroad adopted an optical bar code.

December 1971 The Uniform Code Council, originally called the Uniform Grocery Product Code Council, is formed to administer the UPC.

1972 U.S. Supermarket Ad Hoc Committee on a Uniform Grocery Product Code recommends the adoption of the 1972 UPC.

April 3, 1973 An ad-hoc committee composed of grocery executives chooses the linear bar code with 11 digits and a 12th check digit.

June 26, 1974 A 10-pack of Wrigley's Juicy Fruit chewing gum was the first product with a bar code scanned at a checkout counter in Troy, Ohio. Today, the pack of gum is on display at the Smithsonian Institution's National Museum of American History.

1974 95% of the railroad fleet was labeled with a bar code.

February 1977 European Article Numbering Association formed in Belgium.

September 1, 1981 United States Department of Defense adopted the use of bar codes for marking all products sold to the United States military.

1992 Norman Joseph Woodland was awarded the 1992 National Medal of Technology by President George Bush.

16.4 Encoding Personal Data

Consider this Social Security number: 189-31-9431. What information about the holder can be deduced from the number? Only that the holder obtained it in Pennsylvania (see Spotlight 16.4). Figure 16.11 shows an Illinois driver's license number: M200-7858-1644. What information about the holder can be deduced from this number? This time we can determine the date of birth, sex, and much about the person's name.

These two examples illustrate the extremes in coding personal data. The Social Security number has no personal data encoded in the number. It is entirely determined by the place and time it is issued, not the individual to whom it is assigned. In contrast, in some states the driver's license

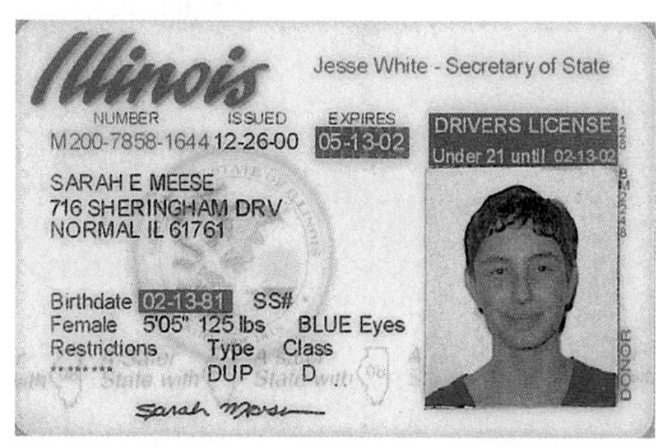

FIGURE 16.11
Illinois driver's license.

SPOTLIGHT 16.4 Social Security Numbers

The first three digits of Social Security numbers show where the number was applied for. Changes in population have forced some numbers to be moved or assigned out of sequence over the years.

001–003	New Hampshire	400–407	Kentucky	526–527	Arizona
004–007	Maine	408–415	Tennessee	& 600–601, 764–765	
008–009	Vermont	& 756–763		528–529	Utah
010–034	Massachusetts	416–424	Alabama	& 646–647	
035–039	Rhode Island	425–428	Mississippi	530, 680	Nevada
040–049	Connecticut	& 587–588, 752–755		531–539	Washington
050–134	New York	429–432	Arkansas	540–544	Oregon
135–158	New Jersey	& 676–679		545–573	California
159–211	Pennsylvania	433–439	Louisiana	& 602–626	
212–220	Maryland	& 659–665		574	Alaska
221–222	Delaware	440–448	Oklahoma	575–576	Hawaii
223–231	Virginia	449–467	Texas	& 750–751	
& 669-699		& 627–645		577–579	District of Columbia
232–236	West Virginia	468–477	Minnesota		
232, 237–246	North Carolina	478–485	Iowa	580	Virgin Islands
& 681–690		486–500	Missouri	580–584	Puerto Rico
247–251	South Carolina	501–502	North Dakota	& 596–599	
& 654–658		503–504	South Dakota	586	Guam
252–260	Georgia	505–508	Nebraska	586	American Samoa
& 667–675		509–515	Kansas	586	Philippine Islands
261–267	Florida	516–517	Montana	700–728	*through July 1, 1963, reserved for railroad employees*
& 589–595, 766–772		518–519	Idaho		
268–302	Ohio	520	Wyoming		
303–317	Indiana	521–524	Colorado		
318–361	Illinois	& 650–653			
362–386	Michigan	525 & 585	New Mexico		
387–399	Wisconsin	& 648–649			

Source: Social Security Administration.

numbers are entirely determined by personal information about the holders. It is no coincidence that the unsophisticated Social Security numbering scheme predates computers. Agencies that have large databases that include personal information such as names, sex, and dates of birth find it convenient to encode these data into identification numbers. Examples of such agencies are the National Archives (where census records are kept), genealogical research centers, the Library of Congress, and state motor vehicle departments.

There are many methods in use to encode personal data such as name, sex, and date of birth. These methods are perhaps most widely used in assigning driver's license numbers in some states. Coding license numbers solely from personal data enables automobile insurers, government entities, and law enforcement agencies to determine the number from the personal data.

Many states encode the surname, first name, middle initial, date of birth, and sex by quite sophisticated schemes.

In one scheme that is based on sound, the first four characters of the license number are obtained by applying the **Soundex Coding System** to the surname as follows.

1. Delete all occurrences of *h* and *w*. (For example, *Schworer* becomes *Scorer* and *Hughgill* becomes *uggill.*)
2. Assign numbers to the remaining letters as follows:

$a, e, i, o, u, y \rightarrow 0$	$l \rightarrow 4$
$b, f, p, v \rightarrow 1$	$m, n \rightarrow 5$
$c, g, j, k, q, s, x, z \rightarrow 2$	$r \rightarrow 6$
$d, t \rightarrow 3$	

3. If two or more letters with the same numeric value are adjacent, omit all but the first. (For example, *Scorer* becomes *Sorer* and *uggill* becomes *ugil*).
4. Delete the first character of the original name if still present. (*Sorer* becomes *orer*).
5. Delete all occurrences of *a, e, i, o, u,* and *y*.
6. Retain only the first three digits corresponding to the remaining letters; append trailing 0's if fewer than three letters remain; precede the three digits obtained in step 6 with the first letter of the surname.

Figure 16.12 shows three examples.

What is the advantage of this method? It is an error-correcting scheme. Indeed, it is designed so that likely misspellings of a name nevertheless result in the correct coding of the name. For example, frequent misspellings of the name *Erickson* are *Ericksen, Eriksen, Ericson,* and *Ericsen.* Observe that all of these yield the same coding as *Erickson.* If a law enforcement official, a genealogical researcher, a librarian, or an airline reservation agent wanted to pull up the file from a data bank for someone whose name was pronounced "Erickson," the correct spelling isn't essential because the computer searches for records that are coded as E-625 for all spelling variations. The search feature of a Web site where many mathematicians post their research papers uses the Soundex Coding System. This system was designed for the U.S. Census Bureau when much census information was obtained orally.

Step 1 Step 2
Schworer → Scorer → Scorer
220606

Step 3 Step 4 Step 5 Step 6
→ Sorer → orer → rr → S-660
20606 0606 66

Step 1 Step 2
Hughgill → uggill → uggill
022044

Step 3 Step 4 Step 5 Step 6
→ ugil → ugil → gl → H-240
0204 0204 24

Step 1 Step 2
Schmidlapper → Scmidlapper → Scmidlapper
22503401106

Step 3 Step 4 Step 5 Step 6
→ Smidlaper → midlaper → mdlpr → S-534
250340106 50340106 53416

FIGURE 16.12 The Soundex Coding System.

There are many schemes for encoding the date of birth and the sex in driver's license numbers. For example, the last five digits of Illinois and Florida driver's license numbers capture the year and date of birth as well as the sex. In Illinois, each day of the year is assigned a three-digit number in sequence beginning with 001 for January 1. However, each month is assumed to have 31 days. Thus, March 1 is given the number 063 because both January and February are assumed to have 31 days. These numbers are then used to identify the month and day of birth of male drivers. For females, the scheme is identical except that 600 is added to the number. The last two digits of the year of birth, separated by a dash (probably to obscure the fact that they represent the year of birth), are listed in the fifth and fourth positions from the end of the driver's license number. Thus, a male born on October 13, 1940, would have the last five digits 4-0292 ($292 = 9 \cdot 31 + 13$), whereas a female born on the same day would have 4-0892. The scheme to identify birth date and sex in Florida is the same as in Illinois except that each month is assumed to have 40 days and 500 is added for women. Moreover, a dash occurs between the two digits for the year and the three digits for the day. For example, the five digits 49-585 belong to a woman born on March 5, 1949.

In this chapter, we have investigated how mathematics is used to append a check digit to an identification number for error detection. In the next chapter, we will show how codes consisting of 0's and 1's can be devised so that errors can be corrected.

REVIEW VOCABULARY

Bar code A code that employs bars and spaces to represent information. (p. 517)

Binary code A coding scheme that uses two symbols, usually 0 and 1. (p. 517)

Check digit A digit included in an identification number for the purpose of error detection. (p. 509)

Code A group of symbols that represent information together with a set of rules for interpreting the symbols. (p. 509)

Decoding Translating code into data. (p. 517)

Encoding Translating data into code. (p. 515)

Error-detecting code A code in which certain types of errors can be detected. (p. 509)

International Standard Book Number (ISBN) An identification number used on books throughout the world that contains a check digit for error detection. (p. 509)
Postnet code The bar code used by the U.S. Postal Service for ZIP codes. (p. 517)
Soundex Coding System An encoding scheme for surnames based on sound. (p. 524)
Universal Product Code (UPC) A bar code and identification number that is used on most retail items. The UPC code detects 100% of all single-digit errors and most other types of errors. (p. 510)
Weights Numbers used in the calculation of check digits. (p. 511)
ZIP code A five-digit code used by the U.S. Postal Service to divide the country into geographic units to speed sorting of the mail. ZIP stands for Zone Improvement Plan. (p. 515)
ZIP + 4 code The nine-digit code used by the U.S. Postal Service to refine ZIP codes into smaller units. (p. 516)

✓ SKILLS CHECK

1. When a single incorrect digit is entered, an error-detecting code

(a) will sometimes detect the error.
(b) will always detect the error but may not be able to correct the error.
(c) will always detect and correct the error.

2. If a U.S. Postal Service money order is numbered 1012065994X, where X indicates that the last digit is obliterated, X is ______________________.

3. If the first five digits of a valid U.S. Postal Service money order are rearranged, the resulting number will have the same check digit as the original number

(a) Always
(b) Sometimes
(c) Never

4. The sum of the digits of a correctly coded American Express Travelers Cheque identification number is evenly divisible by ______________________.

5. Is the number 105408970012 a legitimate airline ticket number?

(a) Yes.
(b) No, but if the final digit is changed to a 5, the resulting number 105408970015 is legitimate.
(c) No, but if the final digit is changed to a 3, the resulting number 105408970013 is legitimate.

6. If an American Express Travelers Cheque is numbered X425036790, where X indicates that the first digit is obliterated, X is ______________________.

7. If the first two digits of a valid airline ticket identification number are transposed, the resulting number will be valid

(a) Always
(b) Sometimes
(c) Never

8. A correctly coded UPC number has a weighted sum that is evenly divisible by ______________________.

9. The bank routing number error detection scheme detects

(a) all transportation and most single-digit errors.
(b) all single-digit errors and most transpositions.
(c) all single-digit errors and all transpositions.

10. The check digit that should be appended to the UPC code 0-14300-25433 is ______________________.

11. The bank routing number error detection scheme

(a) detects the same errors as the UPC scheme.
(b) detects fewer errors than the UPC scheme.
(c) detects more errors than the UPC scheme.

12. The check digit that should be appended to the bank routing number 01500085 is ______________________.

13. Suppose the 10-digit ISBN 0-1750-3549-0 is incorrectly reported as 0-1750-3540-1. Which of the following statements is true?

(a) This error will not be detected by the check digit.
(b) While this particular error will be detected, the check digit does not detect all two-digit errors in ISBNs.
(c) All two-digit errors in a 10-digit ISBN are detectable by the check digit.

14. A correctly coded 10-digit ISBN has a weighted sum that is evenly divisible by ______________________.

15. If an error in an identification number is made by transposing the first and third digits, the error is

(a) usually detected by the UPC scheme.
(b) always detected by the bank scheme.
(c) always detected by the ISBN-10 scheme.

16. The ISBN-10 error detection scheme detects __________ percent of single digits errors and __________ percent of transposition errors.

17. As far as the ability to detect single-position errors and adjacent-digit transposition errors

(a) the credit card scheme is superior to the UPC scheme.

(b) the credit card scheme is superior to the ISBN-10 scheme.
(c) the UPC scheme is superior to the bank scheme.

18. If the sixth digit of the Postnet code 20001-5800-7 is incorrect, the correct Postnet code is ______________ .

19. If a scanner misreads exactly one bar of a Postnet code the computer will

(a) not always detect the error.
(b) always detect the error but will not always be able to correct it.
(c) always detect the error and correct it.

20. If the identification 48945 has been encoded by multiplying the original number by 13, the original number is ______________________________ .

CHAPTER 16 EXERCISES

■ Challenge ◆ Discussion

16.1 Check Digits

1. Determine the check digit for a money order with identification number 3953981640.

2. Determine the check digit for a money order with identification number 7234541780.

3. Determine the check digit for the United Parcel Service (UPS) identification number 873345672.

4. Suppose a money order with the identification number and check digit 21720421168 is erroneously copied as 27750421168. Will the check digit detect the error? Explain your reasoning.

5. Determine the check digit for the airline ticket number 30860422052.

6. Determine the check digit for the Avis rental car with identification number 540047.

7. Determine the check digit for the UPC number 38137009213.

8. If the packaging of a retail item were damaged in such a way that the first digit of a UPC code was scratched off, but the remaining digits were 88072303584, determine the first digit.

9. Determine the check digit for the ISBN 0-669-19493.

10. When this edition was in preparation, the publisher sent the author of this section the following ISBNs for the book: ISBN-10: 1-4292-0900-3; ISBN-13: 978-1-4292-0890-0. How did the author know the ISBN-13 was wrong and how did he know how to correct it? (This really happened!)

11. When calculating the check digit for a 13-digit ISBN, why can you disregard the first two digits? (Try it for Exercise 10.)

12. Determine the check digit for the bank routing number 09100001.

13. Determine the check digit for the American Express Travelers Cheque with identification number 461212023.

14. Suppose a check digit is assigned to a four-digit number by appending the remainder after division by 7. If the number 96806 has a single-digit error but the check digit is correct, determine the possibilities for the correct number.

15. Determine whether the Master Card number 3541 0232 0033 2270 is valid.

16. Suppose that the digit indicated by a question mark in the Master Card number 426452002177?337 is unreadable. What is the unreadable number?

17. Create a check digit for the UPC number 38137009213 using the weights 7, 1, 7, 1, 7, 1, . . . , 7, 1, instead of 3, 1, 3, 1, 3, 1, . . . , 3, 1. Test to see whether this check digit will detect single-digit errors by trying several examples.

18. Create a check digit for the UPC number 38137009213 using the weights 2, 1, 2, 1, 2, 1, . . . , 2, 1, instead of 3, 1, 3, 1, 3, 1, . . . , 3, 1. Is the error caused by replacing the 3 in the first position with an 8 detected? What about the error caused by replacing the 1 in the third position with a 6? Explain why or why not.

19. If the weights 5, 1, 5, 1, 5, 1, . . . , 5, 1 were used for the UPC code, which single-digit errors would go undetected?

20. Exercises 17, 18, and 19 reveal that using the weights 1, 3, or 7 for a particular position detects all errors in that position, whereas using weights 2 or 5 in a position does not detect all errors. Using this observation, make a guess about error-detection capability using weights 9, 4, 6, or 8.

21. Use the credit card scheme to determine the check digit for the number 300125600196431.

22. Determine the check digit for the VIN JM1GD222J1581570 (see Spotlight 16.1 for a description of the method to be used).

23. For some products, such as soft-drink cans and magazines, an 8-digit UPC number called Version E is used instead of the 12-digit number. The method of calculating the eighth digit, which is the check digit, depends on the value of the seventh digit. The check digit a_8 for a UPC Version E identification number $a_1a_2a_3a_4a_5a_6a_7$, where a_7 is 0, 1, or 2, is chosen so that $a_1 + a_2 + 3a_3 + 3a_4 + a_5 + 3a_6 + a_7 + a_8$ is divisible by 10. Use this fact to determine the check digit for the following Version E numbers:

(a) 0121690
(b) 0274551
(c) 0760022
(d) 0496580

(georgphotos/Alamy.)

24. The check digit a_8 for a UPC Version E identification number $a_1a_2a_3a_4a_5a_6a_7$, where a_7 is 4, is chosen so that $a_1 + a_2 + 3a_3 + a_4 + 3a_5 + 3a_6 + a_8$ is divisible by 10. Use this fact to determine the check digit for the following Version E numbers:

(a) 0754704
(b) 0774714
(c) 0724444

■ **25.** The 10-digit ISBN 0-669-03925-4 is the result of a transposition of two adjacent digits not involving the first or last digit. Determine the correct ISBN.

26. Explain why the bank scheme will detect the error 751 ··· → 157 ··· but the UPC scheme will not.

27. Suppose the check digit a_9 for the bank routing number was chosen to be the last digit of $3a_1 + 7a_2 + a_3 + 3a_4 + 7a_5 + a_6 + 3a_7 + 7a_8$ instead of the way described in this chapter. How would this compare with the actual check digit?

28. Explain why an error caused by transposing the first two digits of a Postal Service money order is not detected by the check-digit scheme. Explain why the same is true for the second and third digits. What about the last two digits?

29. Suppose a company assigns an extra digit to every employee Social Security number by appending a 0 if the sum of the digits is even and a 1 if the sum is odd. If a 2 were mistakenly read as a 7 would the error be detected? What if a 2 were mistakenly read as an 8? Try a few other experiments with single-digit errors (for experiments you can use three-digit numbers instead of nine-digit numbers). Determine which errors are detected by this method. Explain your reasoning.

30. Explain why an error caused by transposing any two digits of an American Express Travelers Cheque is not detected by the check-digit scheme.

31. When using the traveler's check, credit card or UPC number algorithms for detecting errors does the computer have to know which digit is the check digit?

32. Explain why the Postal Service money order check-digit scheme does not detect the mistake of substituting a 0 for a 9, or vice versa.

33. Which digit never appears as a check digit on a Postal Service money order?

34. Which digit never appears as a check digit on an American Express Travelers Cheque?

35. Which digits never appear as a check digit for an airline identification number?

36. Suppose four-digit numbers $a_1a_2a_3a_4$ are assigned a check digit a_5 so that $a_1 + 2a_2 + a_3 + 2a_4 + a_5$ is evenly divisible by 10. Test the number 43216 created in this way to see whether the method detects adjacent-digit transposition errors.

37. Starting with the 10-digit ISBN 0-7167-4782-0, create three new numbers by transposing any two different digits. (They need not be adjacent.) Are these errors detected by the scheme?

38. Suppose in a UPS number an 8 is mistaken for a 5. Is the error detected? What if a 9 is mistaken for a 2?

39. Give an argument to show that the 10-digit ISBN error-detection method will detect a transposition error involving the first and third digits. Does the same argument work for the fourth and sixth digits?

■ **40.** Suppose the check digit a_{10} of 10-digit ISBNs were chosen so that $a_1 + 2a_2 + 3a_3 + 4a_4 + 5a_5 + 6a_6 + 7a_7 + 8a_8 + 9a_9 + 10a_{10}$ is divisible by 11 instead of the way described in the chapter. How would this compare with the actual check digit?

■ **41.** Consider a UPC number in which the digits 7 and 2 appear consecutively (that is, the number has the form ··· 72 ···). Will the error caused by transposing these digits (that is, the number is taken as ··· 27 ···) be detected? What if the digits 6 and 2 were transposed instead? State the general criterion for the detection of an error of the form ··· *ab* ··· → ··· *ba* ··· by the UPC scheme.

42. If the first three digits of a routing number for a checking account are 537 and the 5 and 3 are transposed, will the error be detected? If the first three numbers are 237 and the 2 and 7 are transposed, will the error be detected?

■ **43.** State a general criterion for the detection of an error of the form $\cdots abc \cdots \rightarrow \cdots cba \cdots$ for the routing number of a checking account.

■ **44.** The state of Utah appends a ninth digit a_9 to an eight-digit driver's license number $a_1a_2 \cdots a_8$ so that $9a_1 + 8a_2 + 7a_3 + 6a_4 + 5a_5 + 4a_6 + 3a_7 + 2a_8 + a_9$ is divisible by 10.

(a) If the first eight digits of a Utah driver's license number are 14910573, what is the ninth digit?
(b) Suppose a legitimate Utah driver's license number 149105767 is miscopied as 149105267. How would you know a mistake was made? Is there any way you could determine the correct number? Suppose you know the error was in the seventh position. Could you correct the mistake?
(c) If a legitimate Utah driver's license number 149105767 were miscopied as 199105767, would you be able to tell a mistake was made? Explain.
(d) Explain why any transposition error involving adjacent digits of a Utah driver's license number would be detected.

■ **45.** The Canadian province of Quebec assigns a check digit a_{12} to an 11-digit driver's license number $a_1a_2 \cdots a_{11}$ so that $12a_1 + 11a_2 + 10a_3 + 9a_4 + 8a_5 + 7a_6 + 6a_7 + 5a_8 + 4a_9 + 3a_{10} + 2a_{11} + a_{12}$ is divisible by 10. Criticize this method. Describe all single-digit errors that are undetected by this scheme.

46. Speculate on the reason why telephone numbers, Social Security numbers, and serial numbers on most currency do not have check digits.

47. Suppose a company uses a check-digit scheme similar to the UPC scheme, except that instead of using the UPC weights 3, 1, 3, 1, . . . it uses w, 1, w, 1, ... If two of the ID numbers used by the company are 73215674 and 73215661, determine w.

48. If a publishing company has headquarters in both the United States and Germany and publishes the same book in both countries, it is likely that the 10-digit ISBN for the book will be identical except for the first and last digits (because the first digit for U.S. publications is 0 and the first digit for German publications is 3). If the last digit of the U.S. edition is 1, what is the last digit for the German publication?

16.2 The Zip Code

49. Determine the ZIP + 4 code and check digit for each of the following Postnet bar codes:

(a)
(b)
(c)

50. Determine the ZIP + 4 code and check digit for each of the following Postnet bar codes:

(a)
(b)
(c)

51. In each of the following Postnet bar codes, exactly one mistake occurs (that is, a long bar appears instead of a short one, or vice versa). Determine the correct ZIP code.

(a)
(b)
(c)

52. Below is a 12-digit delivery-point bar code. Determine the ZIP + 4 number, the last two digits of the street address, and the check digit.

53. Explain why any two errors in a particular block of five bars in a Postnet code are always detectable. Explain why not all such errors can be corrected.

54. Change 173 into a Postnet code.

55. Form all possible strings consisting of exactly three a's and two b's and arrange the strings in alphabetical order (for example, the first two possibilities are *aaabb* and *aabab*). Do you see any relationship between your list and the Postnet code?

16.3 Bar Codes

56. Many recently published books include a bar code on the back cover that has the 10-digit ISBN above the bars and a 13-digit identification number below the bars. Examine several books with a bar code on the back cover. How does the number below the bar code differ from the UPC code? How is the number below the bar code related to the ISBN? Given the fact that the last digit in the number below the bar code is a check digit, determine how it is calculated.

57. Suppose the first block of a UPC bar code following the guard bar pattern that a scanner reads is 1000100. Is the scanner reading left to right or right to left?

58. The following is an actual identification number and bar code from a roll of wallpaper. What appears to be wrong with them? Speculate on the reason for the apparent violation of the UPC format.

Building Regulations: 1985 Class 0
FINE ART WALLCOVERINGS LTD.
HOLMES CHAPEL, CHESHIRE
MADE IN ENGLAND
FABRIQUE EN ANGLETERRE

16.4 Encoding Personal Data

59. Judging from the information in Spotlight 16.4, which three states were most likely to have had the smallest populations when Social Security numbers were allocated to the states?

60. What geographical information was used in allotting the first three digits of Social Security numbers to the states?

61. What demographic information was used in allotting the first three digits of Social Security numbers to the states?

62. As of 2007 there were seven states with a population under 1 million. Use the data in Spotlight 16.4 to identify those seven states.

◆ **63.** The Canadian postal system has assigned each geographic region a six-character code composed of alternating letters and digits, such as P7B5E1 and K7L3N6. Discuss the advantages this scheme has over the five-digit ZIP code used in the United States.

64. Determine the Soundex code for Smith, Schmid, Smyth, and Schmidt.

65. Determine the Soundex code for Skow, Sachs, Lennon, Lloyd, Ehrheart, and Ollenburger.

66. In Florida, the last three digits of the driver's license number of a female with birth month m and birth date b are $40(m - 1) + b + 500$. For both males and females, the fourth and fifth digits from the end give the year of birth. Determine the last five digits of a Florida driver's license number for a female born on July 18, 1942.

67. Explain why an Illinois driver's license number that ends with the last five digits 99817 cannot be valid.

68. Determine the last five digits of an Illinois driver's license number for a male born on June 18, 1942.

69. In Illinois, one obtains the last three digits of the driver's license number for a female by adding 600 to the number for a male with the same birthday. In Florida 500 is added to the number for a male. Why can't Florida use 600?

70. Explain why an Illinois driver's license number that ends with 77061 cannot be valid.

71. Determine the birth date of a person whose Illinois driver's license number ends with 58818.

72. In Florida, the last three digits of the driver's license number of a male with birth month m and birth date b are $40(m - 1) + b$. For both males and females, the fourth and fifth digits from the end give the year of birth. Determine the birth dates of people with numbers whose last five digits are 42218 and 53953.

73. Provide three names that share the same Soundex code as Gallihan.

74. Another math book describes the Soundex algorithm for the surname code as follows:

(i) Leave the first letter alone, then cross off all occurrences of the letters a, e, i, o, u, y, h and w.
(ii) Cross off the second of any double letters.
(iii) Leave the first letter alone, and replace each of the other letters with the appropriate number (using the same assignment as given on page 524).
(iv) The code is the first letter of the surname followed by the first three numbers.

Compare the codes for Jackson, Mnack and Shaw using this method and the method given on page 524. (The method on page 524 is the correct one.)

75. For driver's license numbers issued in New York before September 1992, the last two digits were the year of birth. The three digits preceding the year encoded the sex and the month and day of birth. For a woman with birth month m and birth date b, the three digits were $63m + 2b + 1$ (insert a 0 in front for numbers less than 100). For a man with birth month m and birth date b, the three digits were $63m + 2b$. Determine the birth months, birth dates, and sexes of drivers with the three digits 248 and 601 preceding the year.

76. The state of Washington encodes the last two digits of the year of birth into driver's license numbers (in positions 8 and 9) by subtracting the two-digit number from 100. For example, a person born in 1942 has 58 in positions 8 and 9, whereas a person born in 1971 has 29 in positions 8 and 9. Speculate on the reason for subtracting the birth year from 100.

77. Driver's license number-assignment schemes that use personal data sometimes produce the same number for different people. Speculate about circumstances under which this is more likely to occur.

78. Apply the Soundex code to common ways to misspell your name. Do they give the same code as your name does?

79. Why would the Soundex system of coding last names be a poor method for encoding names in China?

WRITING PROJECTS

1. Prepare a two-page report on coded information in your location. Possibilities for investigation include driver's license numbers in your state, student ID numbers and bar codes at your school, and bar codes used by your school library and city library. Identify the coding schemes and, when possible, determine whether a check digit is employed. Include samples. The Suggested Readings for this chapter contain information that will assist you.

2. Prepare a two-page report on the driver's license coding schemes used by Michigan, Maryland, and Washington (Michigan and Maryland use the same method). J. Gallian's "Assigning Driver's License Numbers" has the information you will need (see the Suggested Readings).

3. Use the Web to find material for a two-page report on the history of the bar code.

4. Use the Web to find material for a two-page report on the Barcode of Life project.

5. Use the Web to find material for a two-page report on Smart Card technology.

SUGGESTED READINGS

GALLIAN, J. The mathematics of identification numbers, *College Mathematics Journal,* 22 (1991): 194–202. A survey of check-digit schemes associated with identification numbers.

GALLIAN, J. Assigning driver's license numbers, *Mathematics Magazine,* 64 (1992): 13–22. Discusses various methods used by the states to assign driver's license numbers. Several of these methods include check digits for error detection.

GALLIAN, J. Error detection methods, *ACM Computing Surveys,* 28 (1996): 504–517. A detailed description of many error-detection methods.

GALLIAN, J., and S. WINTERS. Modular arithmetic in the marketplace, *American Mathematical Monthly,* 95 (1988): 548–551. A detailed analysis of the check-digit schemes presented in this chapter. In particular, the error-detection rates for the various schemes are given.

KIRTLAND, J. *Identification Numbers and Check Digit Schemes,* Mathematical Association of America, Washington, D.C., 2001. Provides more examples and exercises for the check-digit schemes discussed in this chapter.

SUGGESTED WEB SITES

www.d.umn.edu/~jgallian/fapp7 This Web site enables users to calculate check digits using the various methods discussed in this chapter.

CHAPTER 17

Information Science

With the enormous volume of email, faxes, Internet traffic, and cellular phone calls, the Digital Revolution has brought about many mathematical challenges. One is how to correct errors in data transmission. Another is how to electronically send and store information economically. A third is how to ensure security of transmitted data. Yet another is how to improve Web search efficiency. In this chapter we illustrate some of the ways in which mathematicians and engineers have responded to these challenges.

17.1 Binary Codes

A system for coding data made up of two states (or symbols) is called a **binary code**. Binary codes are the hidden language of computers. The Postnet code (short and long bars) and the Universal Product Code (UPC) bar code (white spaces and dark bars) are two examples of binary codes. Morse code (dots and dashes) and Braille (bumps and flat) are two more. The Ebert and Roeper "thumbs up/thumbs down" rating of films is a binary code with four messages. CDs (compact disks), fax machines, DVDs (digital video disks), high definition television signals, cell phones, and space probes represent data as strings of 0's and 1's rather than the usual digits 0 through 9 and letters *A* through *Z*. In this section we will illustrate one way binary codes can be devised so that errors in the transmission of the code can be corrected.

The idea behind error-correction schemes is simple and one you often use. To illustrate, suppose you are reading the employment section of a newspaper and you see the phrase "must have a minimum of bive years experience." Instantly you detect an error because *bive* is not a word in the English language. Moreover, you are fairly confident that the intended word is *five*. Why so? Because *five* is a word derived from *bive* by changing a single letter and it makes the phrase understandable. In other phrases, words such as *bike* or *give* might be sensible alternatives to *bive*. Using the extra information provided by the context, we are often able to infer the intended meaning when errors occur.

To demonstrate the way error-correcting schemes work, suppose that NASA sends a spacecraft to land at one of 16 possible landing sites on Mars. The spacecraft orbits Mars while surveying the sites for the most favorable landing conditions. NASA officials have coded the 16 landing sites with four-digit strings of 0's and 1's such as 0000, 0001, 0010, 0100 (see Table 17.1 for the complete list). Once the best site has been selected, NASA will inform the spacecraft where to land by sending the code for the site. However, signals sent through space are subject to interference called *noise*. The noise might cause the spacecraft to interpret the signal as 0001 when the signal actually sent was 1001. Fortunately, over the past 60 years mathematicians and engineers have devised highly sophisticated schemes to build extra information into messages composed of 0's and 1's that often permits one to infer the correct message even though the message may have been received incorrectly (see Spotlights 17.1 and 17.2).

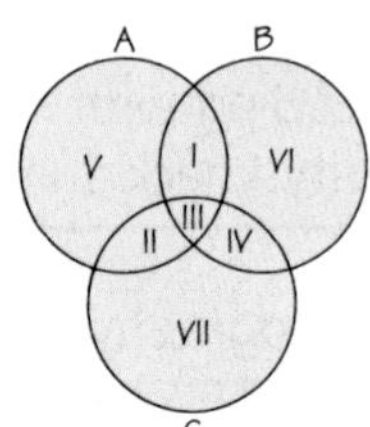

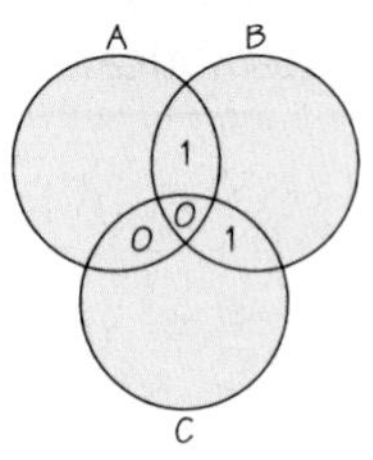

FIGURE 17.1 Diagram for message 1001.

As a simple example, let's assume our message is 1001. We will build extra information into this message with the aid of the diagram in Figure 17.1. Begin by placing the four message digits in the four overlapping regions I, II, III, IV, with the digit in the first position (starting at the left of the sequence) in region I, the digit in the second position in region II, and so on. For regions V, VI, and VII, assign 0 or 1 so that the total number of 1's in each circle is even. See Figure 17.2.

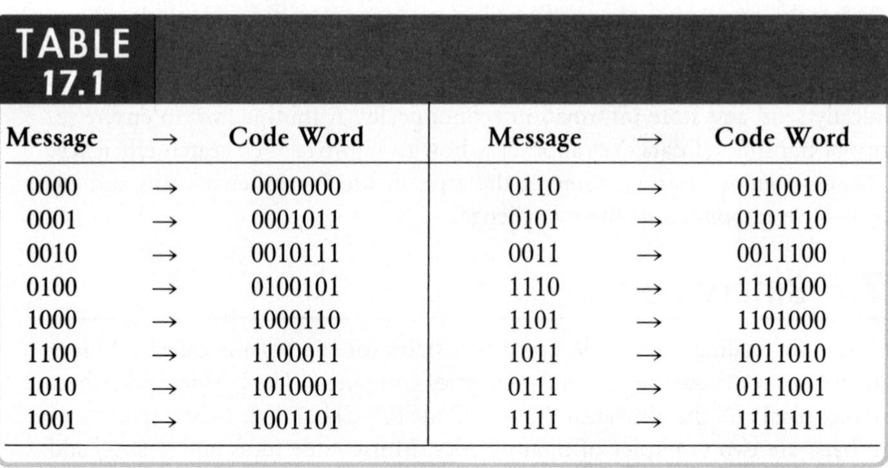

TABLE 17.1

Message	→	Code Word	Message	→	Code Word
0000	→	0000000	0110	→	0110010
0001	→	0001011	0101	→	0101110
0010	→	0010111	0011	→	0011100
0100	→	0100101	1110	→	1110100
1000	→	1000110	1101	→	1101000
1100	→	1100011	1011	→	1011010
1010	→	1010001	0111	→	0111001
1001	→	1001101	1111	→	1111111

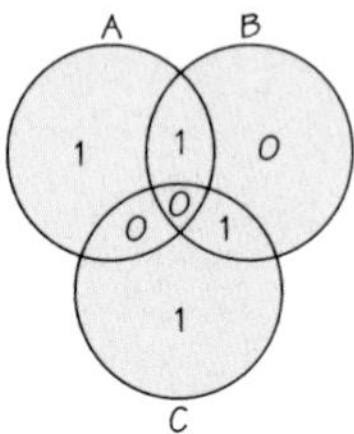

FIGURE 17.2 Diagram for encoded message 1001101.

We have now encoded our message 1001 using the diagram as 1001101. Now suppose that this encoded message is received as 0001101 (an error in the first position). How would we know an error was made? We place each digit from the received message in its appropriate region, as in Figure 17.3.

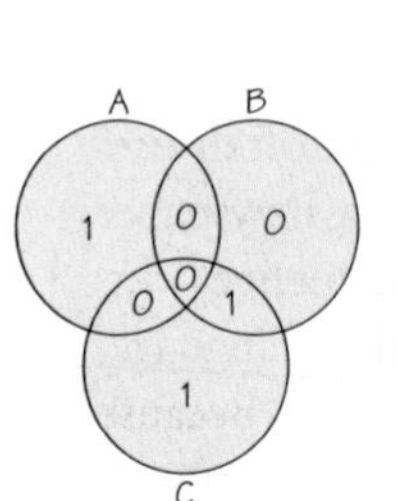

FIGURE 17.3 Diagram for received message 0001101.

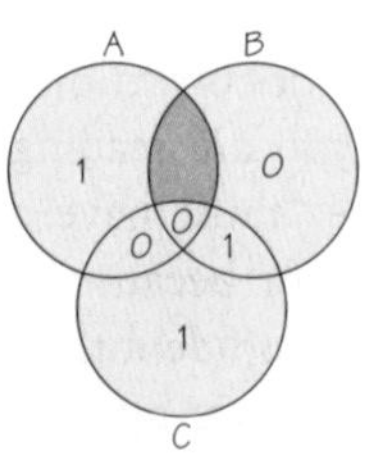

FIGURE 17.4 Circles A and B but not C have wrong parity.

SPOTLIGHT 17.1 The Ubiquitous Reed-Solomon Codes

One of the mathematical ideas underlying current error-correcting techniques for everything from computer hard-disk drives to CD players was first introduced in 1960 by Irving Reed and Gustave Solomon. Reed–Solomon codes made possible the stunning pictures of the outer planets sent back by the space probes *Voyager 1* and *2*. They make it possible to scratch a compact disc and still enjoy the music.

"When you talk about CD players and digital audio tape and now digital television, and various other digital imaging systems that are coming—all of those Reed–Solomon [codes] are an integral part of the system," says Robert McEliece, a coding theorist at Caltech.

Irving Reed and Gustave Solomon
At the Jet Propulsion Laboratory in 1989 to monitor the encounter of *Voyager 2* with Neptune. (*Rex Ridenhouse.*)

Why? Because digital information consists of 0's and 1's, and a physical device may occasionally confuse the two. *Voyager 2*, for example, was transmitting data at incredibly low power over billions of miles. Error-correcting codes are a kind of safety net, mathematical insurance against the vagaries of an imperfect material world.

In 1960, the theory of error-correcting codes was only about a decade old. Through the 1950s, a number of researchers began experimenting with a variety of error-correcting codes. But the Reed–Solomon paper, McEliece says, "hit the jackpot." "In hindsight it seems obvious," Reed later said. However, he added, "Coding theory was not a subject when we published the paper." The two authors knew they had a nice result; they didn't know what impact the paper would have.

Four decades later, the impact is clear. The vast array of applications, both current and pending, has settled the questions of the practicality and significance of Reed–Solomon codes. Billions of dollars in modern technology depend on ideas that stem from Reed and Solomon's original work.

Source: Adapted from an article by Barry Cipra, with permission from *SIAM News*, January 1993, p. 1.

Noting that in both circles A and B there is an odd number of 1's, we instantly realize that something is wrong, because the intended message had an even number of 1's in each circle. How do we correct the error? Because circles A and B have the wrong parity (parity refers to the oddness or evenness of a number–even integers have **even parity**; odd integers have **odd parity**) and C does not, the error is located in the portion of the diagram in circles A and B, but not in circle C; that is, region I (see Figure 17.4). Here we also see the advantage of using only 0's and 1's to encode data. If you have only two possibilities and one of them is incorrect, then the other one must be correct. Because the 0 in region I is incorrect, we know 1 is correct.

This technique can be used to encode all 16 possible binary messages of length 4, as shown in Table 17.1. The encoded messages are called *code words*. The three digits appended to each string of length 4 provide the "extra information" that is sufficient to infer the intended four-digit message as long as the received seven-digit message has at most one error. If a received message has two or more errors, this method will never yield the correct message.

SPOTLIGHT 17.2 Vera Pless

Vera Pless was born on March 5, 1931, to Russian immigrants on the West Side of Chicago. The neighborhood was intellectually stimulating, and there was a tradition of teaching each other things. At age 12, Pless was taught some calculus by a mathematics graduate student. She accepted a scholarship to attend the University of Chicago at age 15. The program at Chicago emphasized great literature but paid little attention to physics and mathematics. At age 18, with no more than one precalculus course in mathematics, she entered the prestigious graduate program in mathematics at Chicago, where, at that time, there were no women on the mathematics faculty nor even women colloquium speakers. After receiving her master's degree, Pless took a job as a research associate at Northwestern University while pursuing a Ph.D. there. In the midst of writing her thesis, she moved to Boston with her husband and continued to work on her thesis at home. She defended her thesis two weeks before her daughter was born.

Vera Pless
A leader in the field of coding theory.
(*Courtesy of Vera Pless.*)

Over the next several years, Pless stayed at home to raise her children and taught part time at Boston University. When she decided to work full time, she found that women were not welcome at most colleges and universities. Some people told her outright, "I would never hire a woman." Fortunately, there was an Air Force lab in the area that had a group working on error-correcting codes. Although she had never even heard of coding theory, she was hired because of her background in algebra. When the lab discontinued basic research, she took a position as a research associate at MIT. In 1975, she went to the University of Illinois–Chicago, where she remained until her retirement. Having written more than 100 research papers and a widely used book on coding theory, Pless is a leader in the field.

17.2 Encoding with Parity-Check Sums

Strings of 0's and 1's with extra digits for error correction can be used to send full-text messages. A simple way to do this is to assign a space the string 00000, the letter *a* the string 00001, *b* the string 00010, *c* the string 00100, and so on. Because there are 32 possible binary strings of length 5, the five unassigned strings can be used for special purposes, such as indicating uppercase letters or numerals. For example, we might use the string 11111 to indicate a "shift" from lowercase to uppercase when it precedes the code for a letter (1111100010 represents *B*). This is like the shift key on a keyboard. Similarly, we could use 11110 to indicate we are "shifting" from letters to numerals. Here 11110 followed by the code for *a* represents the numeral 0, 11110 followed by the code for *b* represents the numeral 1, and so on up to 9. Punctuation marks could be handled in the same fashion. However, our diagram method for assigning extra digits does not work for strings with five or more digits. Rather, the messages are encoded by appending extra digits determined by the parity of various sums of certain portions of the messages. We illustrate this method for the 16 messages shown in the left-hand column of Table 17.1. (See also Spotlight 17.3.)

Our goal is to take any binary string $a_1a_2a_3a_4$ and append three check digits $c_1c_2c_3$ so that any single error in any of the seven positions can be corrected. This is done as follows: Choose

$$c_1 = 0 \text{ if } a_1 + a_2 + a_3 \text{ is even}$$
$$c_1 = 1 \text{ if } a_1 + a_2 + a_3 \text{ is odd}$$
$$c_2 = 0 \text{ if } a_1 + a_3 + a_4 \text{ is even}$$
$$c_2 = 1 \text{ if } a_1 + a_3 + a_4 \text{ is odd}$$
$$c_3 = 0 \text{ if } a_2 + a_3 + a_4 \text{ is even}$$
$$c_3 = 1 \text{ if } a_2 + a_3 + a_4 \text{ is odd}$$

The sums $a_1 + a_2 + a_3$, $a_1 + a_3 + a_4$, and $a_2 + a_3 + a_4$ are called **parity-check sums**. They are so named because their function is to guarantee that the sum of various components of the encoded message is even. Indeed, c_1 is defined so that $a_1 + a_2 + a_3 + c_1$ is even. (Recall that this is precisely how the value in region V in Figure 17.2 was defined.) Similarly, c_2 is defined so that $a_1 + a_3 + a_4 + c_2$ is even, and c_3 is defined so that $a_2 + a_3 + a_4 + c_3$ is even.

Let's revisit the message 1001 we considered in Figure 17.1. Then $a_1a_2a_3a_4 = 1001$ and

$$c_1 = 1 \text{ because } 1 + 0 + 0 \text{ is odd}$$
$$c_2 = 0 \text{ because } 1 + 0 + 1 \text{ is even}$$

and

$$c_3 = 1 \text{ because } 0 + 0 + 1 \text{ is odd}$$

So, because $c_1c_2c_3 = 101$, we have $1001 \rightarrow 1001101$.

Now how is the intended message determined from a received encoded message? This process is called **decoding.** Say, for instance, that the message 1000, which has been encoded using parity-check sums as $u = 1000110$, is received as $v = 1010110$ (an error in the third position). We simply compare v with each of the 16 code words (that is, the possible correct messages) in Table 17.1 and decode it as the one that differs from v in the fewest positions. (Put another way, we decode v as the code word that agrees with v in the most positions.) This method works even if the error in the message is one of the check digits rather than one of the digits of the original message string. In a situation when there is more than one code word that differs from v in the fewest positions, we do not decode. To carry out this comparison, it is convenient to define the distance between two strings of equal length.

Distance Between Two Strings — DEFINITION

The **distance between two strings** of equal length is the number of positions in which the strings differ.

For example, the distance between $v = 1010110$ and $u = 1000110$ is 1, because they differ in only one position (the third). In contrast, the distance between 1000110 and 0111001 is 7, because they differ in all seven positions. Thus, our decoding procedure is simply to decode any received message v as the code word v' that is "nearest" to v in the sense that among all distances between v and code words, the distance between v and v' is a minimum. If there is more than one possibility for v',

SPOTLIGHT 17.3 Neil Sloane

In the middle of Neil Sloane's office, which is in the center of AT&T Bell Laboratories, which in turn is at the heart of the Information Age, there sits a tidy little pyramid of shiny steel balls stacked up like oranges at a neighborhood grocery. Sloane has been pondering different ways to pile up balls of one kind or another for most of his professional life. Along the way he has become one of the world's leading researchers in the field of sphere packing, a field that has become indispensable to modern communications. Without it, we might not have modems or compact discs or satellite photos of Neptune. "Computers would still exist," says Sloane. "But they wouldn't be able to talk to one another."

To exchange information rapidly and correctly, machines must code it. As it turns out, designing a code is a lot like packing spheres: Both involve cramming things together into the tightest possible arrangement. Sloane, fittingly, is also one of the world's leading coding theorists, not least because he has studied the shiny steel balls on his desk so intently.

Here's how a code might work. Imagine, for example, that you want to transmit a child's drawing that uses every one of the 64 colors found in a jumbo box of Crayola crayons. For transmission, you could code each of those colors as a number—say, the integers from 1 to 64. Then you could divide the image into many small units, or pixels, and assign a code to each one based on the color it contains. The transmission would then be a steady stream of those numbers, one for each pixel.

In digital systems, however, all those numbers would have to be represented as strings of 0's and 1's. Because there are 64 possible combinations of 0's and 1's in a six-digit string, you could handle the entire Crayola palette with 64 different six-digit

Neil Sloane
At work, wearing his famous "Codemart" T-shirt (952 points in a sphere). (*Courtesy of Neil Sloane/AT&T Labs.*)

"code words." For example, 000000 could represent the first color, 000001 the next color, 000010 the next, and so on.

But in a noisy signal, two different code words might look practically the same. A bit of noise, for example, might shift a spike of current to the wrong place, so that 001000 looks like 000100. The receiver might then wrongly color someone's eyes. An efficient way to keep the colors straight in spite of noise is to add four extra digits to the six-digit code words. The receiver, programmed to know the 64 permissible combinations, could now spot any other combination as an error introduced by noise and it would automatically correct the error to the "nearest" permissible color.

In fact, says Sloane, "If any of those ten digits were wrong, you could still figure out what the right crayon was."

Source: Adapted from an article by David Berreby, *Discover*, October 1990.

we do not decode. Table 17.2 shows the distance between $v = 1010110$ and all 16 code words. From this table, we see that v will be decoded as u, because it differs from u in only one position, whereas it differs from all others in the table in at least two positions. This method is called **nearest-neighbor decoding**.

Assuming that errors occur independently, the nearest-neighbor method decodes each received message as the one it most likely represents.

TABLE 17.2

v	1010110	1010110	1010110	1010110	1010110	1010110	1010110	1010110
Code word	0000000	0001011	0010111	0100101	1000110	1100011	1010001	1001101
Distance	4	5	2	5	1	4	3	4
v	1010110	1010110	1010110	1010110	1010110	1010110	1010110	1010110
Code word	0110010	0101110	0011100	1110100	1101000	1011010	0111001	1111111
Distance	3	4	3	2	5	2	6	3

Nearest-Neighbor Decoding — DEFINITION

The **nearest-neighbor decoding** method decodes a received message as the code word that agrees with the message in the most positions provided that there is only one such message.

The scheme we have just described was first proposed in 1948 by Richard Hamming, a mathematician at Bell Laboratories. It is one of a family of codes that are called the Hamming codes.

Strings obtained from all possible messages of a given length of 0's and 1's by appending extra 0's and 1's using parity-check sums, as illustrated earlier, are called **binary linear codes**. The strings with the appended digits are called **code words**.

Binary Linear Code — DEFINITION

A set of words composed of 0's and 1's obtained from all possible messages of a given length by using parity-check sums to append check digits to the messages is called a **binary linear code**. The resulting strings are called **code words**.

Think of a binary linear code as a set of n-digit strings in which each string is composed of two parts: the message part, consisting of the original messages, and the remaining check-digit part.

The longer the messages are, the more check digits are required to correct errors. For example, binary messages consisting of six digits require four check digits to ensure that all messages with one error can be decoded correctly.

Given a binary linear code, how can we tell whether it will correct errors and how many errors it will detect? It is remarkably easy. We examine all the code words to find one that has the fewest number of 1's, excluding the *zero code word* consisting entirely of 0's. Call this minimum number of 1's in any nonzero code word the *weight* of the code and denote it by t.

Weight of a Binary Code — DEFINITION

The **weight of a binary code** is the minimum number of 1's that occur among all nonzero code words of that code.

If t is odd, the code will correct any $(t - 1)/2$ or fewer errors. If t is even, the code will correct any $(t - 2)/2$ or fewer errors. If we prefer simply to detect errors rather than to correct them (as is often the case in applications), the code will detect any $t - 1$ or fewer errors.

Applying this test to the code in Table 17.1, we see that the weight is 3, so it will correct any $(3 - 1)/2 = 1$ error or it will detect any $3 - 1 = 2$ errors. Be careful here. We must decide *in advance* whether we want our code to correct single errors or detect any two errors. It can do whichever we choose, but not both. If we decide to detect errors, then we will not decode any message that was not among our original list of encoded messages (just as *bive* is not a word in the English language). Instead, we simply note that an error was made and, in most applications, request a retransmission. An example of this occurs when a bar-code reader at the supermarket detects an error and therefore does not emit a sound (in effect, requesting a rescanning). On the other hand, if we decide to correct errors, we will decode any received message as its nearest neighbor.

Here is an example of another binary linear code. Let the set of messages be {000, 001, 010, 100, 110, 101, 011, 111} and append three check digits c_1, c_2, and c_3 using

$$c_1 = 0 \text{ if } a_1 + a_2 + a_3 \text{ is even}$$
$$c_1 = 1 \text{ if } a_1 + a_2 + a_3 \text{ is odd}$$
$$c_2 = 0 \text{ if } a_1 + a_3 \text{ is even}$$
$$c_2 = 1 \text{ if } a_1 + a_3 \text{ is odd}$$
$$c_3 = 0 \text{ if } a_2 + a_3 \text{ is even}$$
$$c_3 = 1 \text{ if } a_2 + a_3 \text{ is odd}$$

For example, if we take $a_1a_2a_3$ as 101, we have

$$c_1 = 0 \text{ because } 1 + 0 + 1 \text{ is even}$$
$$c_2 = 0 \text{ because } 1 + 1 \text{ is even}$$
$$c_3 = 1 \text{ because } 0 + 1 \text{ is odd}$$

So we encode 101 by appending 001, that is, $101 \rightarrow 101001$. The entire code is shown in Table 17.3.

TABLE 17.3

Message	→	Code Word	Message	→	Code Word
000	→	000000	110	→	110011
001	→	001111	101	→	101001
010	→	010101	011	→	011010
100	→	100110	111	→	111100

Because the minimum number of 1's of any nonzero code word is three, this code will either correct any single error or detect any two errors, whichever we choose.

It is natural for you to ask how the method of appending extra digits with parity-check sums enables us to detect or even correct errors. Error detection is obvious. Think of how a computer spell-checker works. If you type *bive* instead of *five*, the spell-checker detects the error because the string *bive* is not on its list of valid words. On the other hand, if you type *give* instead of *five*, the spell-checker will not detect the error because *give* is on its list of valid words.

Our error-detection scheme works the same way, except that if we add extra digits to ensure that our code words differ in many positions–say, t positions–then even as many as $t - 1$ mistakes will not convert one code word into another code word. And if every pair of code words differ from each other in at least three positions, we can correct any single error because the incorrect received word will differ from the correct code word in one position, but it will differ from all others in two or more positions. Thus, in this case, the correct word is the unique "nearest neighbor." So the role of the parity-check sums is to ensure that code words differ in many positions. For example, consider the code in Table 17.1. The messages 1000 and 1100 differ in only the second position. But the two parity-check sums $a_1 + a_2 + a_3$ and $a_2 + a_3 + a_4$ will guarantee that encoded words for these messages will have different values in positions 5 and 7 as well as in position 2. It is the job of mathematicians to discover the appropriate parity-check sums to correct several errors in long, complicated codes.

Data Compression

Binary linear codes are fixed-length codes. In a fixed-length code, each code word is represented by the same number of digits (or symbols). In contrast, the Morse code (see Spotlight 17.4), designed for the telegraph, is a **variable-length code**, that is, a code in which the number of symbols for each code word may vary.

Notice that in the Morse code the letters that occur most frequently have the shortest coding, whereas the letters that occur the least frequently have the longest coding. By assigning the code in this manner, telegrams could convey more information per line than would be the case for fixed-length codes or a randomly assigned variable-length coding of the letters. The Morse code is an example of data compression.

Data Compression — DEFINITION

Data compression is the process of encoding data so that the most frequently occurring data are represented by the fewest symbols.

SPOTLIGHT 17.4 Morse Code

The Morse code is a ternary code consisting of short marks, long marks and spaces (see figure on the right). It was invented in the early 1840s by Samuel Morse as an efficient way to transmit messages using electronic pulses through telegraph wires. The code enabled operators to send strings of short pulses, long pulses and pauses representing characters into indentations on paper tape that could be easily converted back to characters. Although it was widely used up until the mid-twentieth century, it has gradually been supplanted by more machine-friendly codes. Because the Morse code uses data compression, sending messages using Morse code is faster than text messaging. Many Nokia cellphones can convert text messages to Morse code.

A	·—	N	—·
B	—···	O	———
C	—·—·	P	·——·
D	—··	Q	——·—
E	·	R	·—·
F	··—·	S	···
G	——·	T	—
H	····	U	··—
I	··	V	···—
J	·———	W	·——
K	—·—	X	—··—
L	·—··	Y	—·——
M	——	Z	——··

Morse Code

Figure 17.5 shows a typical frequency distribution for letters in English-language text material.

FIGURE 17.5 A widely used frequency table for letters in normal English usage.

	A	B	C	D	E	F	G	H	I	J	K	L	M
Percentage:	8	1.5	3	4	13	2	1.5	6	6.5	0.5	0.5	3.5	3
	N	O	P	Q	R	S	T	U	V	W	X	Y	Z
Percentage:	7	8	2	0.25	6.5	6	9	3	1	1.5	0.5	2	0.25

Data compression provides a means to reduce the costs of data storage and transmission. A **compression algorithm** converts data from an easy-to-use format to one optimized for compactness. Conversely, an uncompression algorithm converts the compressed information back to its original form or approximately its original form. Downloaded files in the ZIP format are an example of a particular kind of data compression. When you "unzip" the file, you return the compressed data to its original state. In some applications, such as data sets that represent images, the original data need only be recaptured in approximate form. In these cases, there are algorithms that result in a great saving of space. Graphics Interchange Format (GIF) encoding returns compressed data to its exact original form, while JPEG encoding and MPEG encoding return data only approximately to its original state.

EXAMPLE 1 Data Compression

Let's illustrate the principles of data compression with a simple example. Biologists are able to describe genes by specifying sequences composed of the four letters A, T, G, and C, which represent the four nucleotides adenine, thymine, guanine, and cytosine, respectively. One way to encode a sequence such as AAACAGTAAC in fixed-length binary form would be to encode the letters as

$$A \rightarrow 00 \qquad C \rightarrow 01 \qquad T \rightarrow 10 \qquad G \rightarrow 11$$

The corresponding binary code for the sequence AAACAGTAAC is then

00000001001110000001

On the other hand, if we knew from experience that A occurs most frequently, C second most frequently, and so on, and that A occurs much more frequently than T and G together, the most efficient binary encoding would be

$$A \rightarrow 0 \qquad C \rightarrow 10 \qquad T \rightarrow 110 \qquad G \rightarrow 111$$

For this encoding scheme, the sequence AAACAGTAAC is encoded as

0001001111100010

Notice that this binary sequence has 20% fewer digits than our previous sequence, in which each letter was assigned a fixed length of 2 (16 digits versus 20 digits). However, to realize this savings, we have made decoding more difficult. For the binary sequence using the fixed length of two symbols per character, we decode the sequence by taking the digits two at a time in succession and converting them to the corresponding letters. For the compressed coding, we can decode by examining the digits in groups of three.

EXAMPLE 2 ■ Decode 0001001111100010

Consider the compressed binary sequence 0001001111100010. Look at the first three digits: 000. Since our code words have one, two, or three digits and neither 00 nor 000 is a code word, the sequence 000 can represent only the *three* code words 0, 0, and 0. Now look at the next three digits: 100. Again, because neither 1 nor 100 is a code word, the sequence 100 represents the *two* code words 10 and 0. The next three digits, 111, can represent only the code word 111 because the other three code words all contain at least one 0. Next consider the sequence 110. Because neither 1 nor 11 is a code word, the sequence 110 can represent only 110 itself. Continuing in this fashion, we can decode the entire sequence to obtain AAACAGTAAC.

The following observation can simplify the decoding process for compressed sequences. Note that 0 occurs only at the end of a code word. Thus, each time you see a 0, it is the end of the code word. Also, because the code words 0, 10, and 110 end in a 0, the only circumstances under which there are three consecutive 1's is when the code word is 111. So, to quickly decode a compressed binary sequence using our coding scheme, insert a comma after every 0 and after every three consecutive 1's. The digits between the commas are code words.

EXAMPLE 3 ■ Code AGAACTAATTGACA and Decode the Result

Recall: A → 0, C → 10, T → 110, and G → 111. So

$$\text{AGAACTAATTGACA} \rightarrow 01110010110001101101110100$$

To decode the encoded sequence, we insert commas after every 0 and after every occurrence of 111 and convert to letters:

0,	111,	0,	0,	10,	110,	0,	0,	110,	110,	111,	0,	10,	0
A,	G,	A,	A,	C,	T,	A,	A,	T,	T,	G,	A,	C,	A

Delta Encoding

For data sets of numbers that fluctuate little from one number to the next, the method of compression called the *delta function* works well. Consider the following closing prices (rounded to the nearest integer) of the Standard & Poor's index of the stock prices of 500 companies in September 2007.

1489 1472 1479 1454 1452 1471 1472 1484 1484 1477 1520 1529
1519 1526 1518 1517 1525 1531 1527 1547 1547 1540 1543 1558

These numbers use 96 characters in all. To compress this data set using the delta method, we start with the first number and continue by listing only the change from each entry to the next. So our list becomes

1489	−17	7	−25	−2	19	1	12	0	−7	43	9
−10	7	−8	−1	8	6	−4	20	0	−7	3	15

This time we have used only 44 characters, counting the minus signs, to represent the same data–a savings of 54%.

SPOTLIGHT 17.5 David Huffman

Large networks of IBM computers use it. So do high-definition televisions, modems, and a popular electronic device that takes the brainwork out of programming a videocassette recorder. All these digital wonders rely on the results of a 58-year-old term paper by an MIT graduate student—a data-compression scheme known as Huffman encoding.

In 1951, David Huffman and his classmates in an electrical engineering graduate course on information theory were given the choice of a term paper or a final exam. For the term paper, Huffman's professor had assigned what at first appeared to be a simple problem. Students were asked to find the most efficient method of representing numbers, letters, or other symbols using binary code. Huffman worked on the problem for months, developing a number of approaches, but none that he could prove to be the most efficient. Finally, he despaired of ever reaching a solution and decided to start studying for the final. Just as he was throwing his notes in the garbage, the solution came to him. "It was the most singular moment of my life," Huffman says. "There was the absolute lightning of sudden realization. It was my luck to be there at the right time and also not have my professor discourage me by telling me that other good people had struggled with the problem," he says. When presented with his student's discovery, Huffman recalls, his professor exclaimed: "Is that all there is to it!"

"The Huffman code is one of the fundamental ideas that people in computer science and data communications are using all the time," says Donald Knuth of Stanford University. Although others have used Huffman's code to help make millions of dollars, Huffman's main compensation was dispensation from the final exam. He never tried to patent an invention from his work and experienced only a twinge of regret at not having used his creation to make himself rich. "If I had the best of both worlds, I would have had recognition as a scientist, and I would have gotten monetary rewards," he says. "I guess I got one and not the other."

David Huffman

(Matthew Mulbry.)

But Huffman received other compensation. A few years ago an acquaintance told him that he had noticed that a reference to the code was spelled with a lowercase h. Remarked his friend to Huffman, "David, I guess your name has finally entered the language."

David Huffman died October 7, 1999.

Source: Adapted from an article by Gary Stix, *Scientific American*, September 1991, pp. 54, 58.

Huffman Coding

The methods we have shown previously are too simple for general use, but in 1951 a graduate student named David Huffman (see Spotlight 17.5) devised a scheme for data compression that became widely used. As was the case for the first scheme we discussed, Huffman coding assigns short code words to those characters with high probabilities of occurring and long code words to those with low probabilities of occurring. A Huffman code is made using a so-called *code tree* by arranging the characters from top to bottom according to increasing probability; it proceeds by combining, at each stage, the two least probable combinations and repeating this process until there is only one combination remaining. To illustrate the method, say we have a data set of six letters that occur with the following probabilities.

A	0.125
B	0.051
C	0.215
D	0.173
E	0.210
F	0.226

Rearranging them in increasing order, we have:

B	0.051
A	0.125
D	0.173
E	0.210
C	0.215
F	0.226

Because *B* and *A* are the two least likely to occur, we begin our tree by merging them with the one with the smallest probability on the left (that is, *BA* rather than *AB*), adding their probabilities, and rearranging the resulting items in increasing order:

D	0.173
BA	0.176
E	0.210
C	0.215
F	0.226

This time *D* and *BA* are the two least likely remaining entries so we merge them with *D* on the left since it has smallest probability, add their probabilities, and re-sort from smallest to largest. This gives:

E	0.210
C	0.215
F	0.226
DBA	0.349

Next we combine *E* and *C* with *E* on the left and re-sort to get:

F	0.226
DBA	0.349
EC	0.425

Then we combine *F* and *DBA* with *F* on the left and re-sort:

EC	0.425
FDBA	0.575

And finally,

ECFDBA	1.000

To assign a binary code word to each letter, we work our way back from the end of the tree to each letter by assigning, at each merging juncture, 0 to the branch with the lower probability, as shown in Figure 17.6. (There is more than one way to draw the tree. Moreover, we can assign the 0's and 1's to the branches in any fashion, but we do it in this specific way for convenience.)

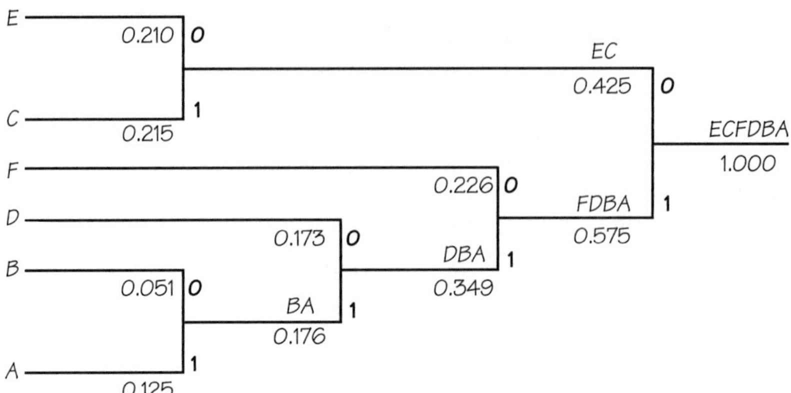

FIGURE 17.6
A Huffman tree.

The path to each letter determines the code word for that letter. So we have:

A	1111
B	1110
C	01
D	110
E	00
F	10

Notice that the letters that occur least often have the longest codes and the letters that occur most often have the shortest codes. Decoding codes created from a Huffman tree is possible because at each stage there is only one way a particular string could have occurred. Here is an example. Consider the Huffman code created using the code words given in the previous display:

1110100001101001111010010

How can we determine the corresponding string of letters that has this Huffman code? The method is quite simple. Look at the first two digits. If they correspond to a code word, then decode them as that letter. If not, then look at the next digit. If the three digits correspond to a code word, then decode it as the corresponding letter. If not, then these three digits and the next one are a four-digit code word. Looking at our example 1110100001101001111010010 we see that neither 11 nor 111 is a code word but 1110 is the code word for B. So, replacing 1110 with B, we have B100001101001111010010. The next possibility is 10, which is the code word for F, so we have BF0001101001111010010. Next we have 00, which is the code word for E, giving us BFE01101001111010010. Continuing in this way, we obtain BFECFFCACEF. Of course, in practice, coding and decoding are done by computers.

17.3 Cryptography

Thus far, we have discussed ways in which data can be encoded to detect errors or correct errors in transmission. In many situations, there is also a desire for security against unauthorized interpretation of coded data (that is, a desire for secrecy). The process of disguising data is called **encryption**. **Cryptology** is the study of methods to make and break secret codes.

(*Corbis.*)

Historically, encryption was used primarily for military and diplomatic transmissions. Today, encryption is essential for securing electronic transactions of all kinds. Cryptography is what allows you to have a Web site safely receive your credit-card number. Cryptographic schemes prevent hackers from charging calls to your cellular phone. Cryptography is also used for authenticating electronic transactions. In September 1998, history was made when former President Bill Clinton and Ireland's Prime Minister Bertie Ahern used digital signatures to sign an intergovernmental document. Each leader had a unique signing code and a digital certificate that served as a "digital ID," thereby ensuring that the document was approved by them. Although modern encryption schemes are extremely complex, we will illustrate the fundamental concepts involved with a few simple examples.

Among the first known cryptosystems is the so-called **Caesar cipher** used by Julius Caesar to send messages to his troops. To encrypt a message with the method employed by Caesar, we use the following table to replace each letter in the top row with the letter below it.

A B C D E F G H I J K L M N O P Q R S T U V W X Y Z
D E F G H I J K L M N O P Q R S T U V W X Y Z A B C

For example, the message ATTACK AT DAWN is encrypted as DWWDFN DW GDZQ.

To decrypt the message, replace each letter with the letter above it in the table. Obviously, it would not require much effort for someone to "crack" this code.

To describe more sophisticated schemes for transmitting messages secretly, it is convenient to introduce a special kind of arithmetic used in cryptography. For any positive integers a and n, we define a mod n (read: "a modulo n" or just "a mod n") to be the remainder when a is divided by n. Thus,

$$3 \bmod 2 = 1 \text{ because } 3 = 1 \cdot 2 + 1$$
$$6 \bmod 2 = 0 \text{ because } 6 = 3 \cdot 2 + 0$$
$$5 \bmod 3 = 2 \text{ because } 5 = 1 \cdot 3 + 2$$
$$37 \bmod 10 = 7 \text{ because } 37 = 3 \cdot 10 + 7$$
$$38 \bmod 26 = 12 \text{ because } 38 = 1 \cdot 26 + 12$$
$$342 \bmod 85 = 2 \text{ because } 342 = 4 \cdot 85 + 2$$
$$62 \bmod 85 = 62 \text{ because } 62 = 0 \cdot 85 + 62$$

Arithmetic involving mod n is called **modular arithmetic**. Although this arithmetic may appear unfamiliar, you often unconsciously use it. For example, if it is now September, what month will it be 25 months from now? Of course, you answer "October," but the interesting fact is that you didn't arrive at the answer by starting with September and counting off 25 months. Instead, without even thinking about it, you simply observed that $25 = 2 \cdot 12 + 1$ so that 25 mod 12 = 1, and you added one month to September. Similarly, if it is now Wednesday, you know that in 23 days it will be Friday. This time, you arrived at your answer by noting that $23 = 3 \cdot 7 + 2$ (that is, 23 mod 7 = 2), so you added 2 days to Wednesday instead of counting off 23 days. An application of modular arithmetic to genetics is described in Spotlight 17.6.

With modular arithmetic, we can easily describe the Caesar cipher as follows. Begin by saying that the letter A is in position 0, B is in position 1, C is in position 2, and so on. Then the Caesar cipher replaces the letter in position i with the

TABLE 17.5

ATTACK AT DAWN	0	19	19	0	2	10	0	19	3	0	22	13
MATHMA TH MATH	12	0	19	7	12	0	19	7	12	0	19	7
MTMHOK TA PAPU	12	19	12	7	14	10	19	0	15	0	15	20

Encrypting Credit-Card Data on the Web

Suppose that you want to purchase a compact disc from Amazon.com. Should you be concerned that a hacker will intercept your credit-card number during the transaction? As you might expect, your credit-card number is sent to Amazon in encrypted form to protect the data.

To describe one way that this encryption can be done, we need to perform addition of binary strings. We add two binary strings $a_1a_2 \cdots a_n$ and $b_1b_2 \cdots b_n$ as follows:

$$\begin{array}{r} a_1a_2 \cdots a_n \\ +\ b_1b_2 \cdots b_n \\ \hline c_1c_2 \cdots c_n \end{array}$$

where $c_i = 0$ if $a_i = b_i$ and $c_i = 1$ if $a_i \neq b_i$. Equivalently, $c_i = (a_i + b_i) \bmod 2$. (Add a_i and b_i in the ordinary way, but replace 2 by 0.)

EXAMPLE 6 ■ Sum of Binary Strings

$$\begin{array}{r} 11000111 \\ +\ 01110110 \\ \hline 10110001 \end{array} \qquad \begin{array}{r} 00111011 \\ +\ 01100101 \\ \hline 01011110 \end{array} \qquad \begin{array}{r} 10011100 \\ +\ 10011100 \\ \hline 00000000 \end{array}$$

We can now explain one way to send credit-card numbers over the Web securely. When you place an order with Amazon, the company sends your computer a randomly generated string of 0's and 1's called a **key**. This key has the same length as the binary string corresponding to your credit-card number, and the two strings are added (think of this process as "locking" the data). The resulting sum is then transmitted to Amazon. Amazon in turn adds the same key to the received string, which then produces the original string corresponding to your credit-card number (adding the key a second time "unlocks" the data).

To illustrate the idea, say you want to send an eight-digit binary string such as $s = 10101100$ to Amazon (actual credit-card numbers have very long strings), and Amazon sends your computer the key $k = 00111101$. Your computer returns the string $s + k = 10101100 + 00111101 = 10010001$ to Amazon, and Amazon adds k to this string to get $10010001 + 00111101 = 10101100$, which is the string representing your credit-card number. If someone intercepts the number $s + k = 10010001$ during transmission, it is of no value without knowing k. This method works because of the property of binary addition that $a_1a_2 \cdots a_n + b_1b_2 \cdots b_n = 00 \cdots 0$ if and only if the two strings are identical. Thus, $(s + k) + k = s + (k + k) = s + 00 \cdots 0 = s$. The method is secure because the key sent by Amazon is randomly generated and used only one time.

You can tell when you are using an encryption scheme on a Web transaction by looking to see if the Web address begins with "https" rather than the customary "http." You will also see a small padlock in the status bar at the bottom of the browser window.

Public Key Cryptography

In the mid-1970s Ronald Rivest, Adi Shamir, and Leonard Adleman devised an ingenious method that permits each person who is to receive a secret message to publicly tell how to scramble messages sent to him or her. Even though the method used to scramble the message is known publicly, only the person for whom it is intended will be able to unscramble the message.

To illustrate their method for transmitting messages secretly, we need the following property of modular arithmetic:

$$(ab) \bmod n = ((a \bmod n)(b \bmod n)) \bmod n$$

This property allows you to replace integers that are greater than or equal to n with integers that are less than n to simplify calculations. You should think of it as saying, "mod before you multiply."

EXAMPLE 7
Multiplication Property for Modular Arithmetic

$$\begin{aligned}(17 \cdot 23) \bmod 10 &= ((17 \bmod 10)(23 \bmod 10)) \bmod 10\\ &= (7 \cdot 3) \bmod 10 = 21 \bmod 10 = 1\\ (22 \cdot 19) \bmod 8 &= ((22 \bmod 8)(19 \bmod 8)) \bmod 8\\ &= (6 \cdot 3) \bmod 8 = 18 \bmod 8 = 2\\ (100 \cdot 8) \bmod 85 &= ((100 \bmod 85)(8 \bmod 85)) \bmod 85\\ &= (15 \cdot 8) \bmod 85 = 120 \bmod 85 = 35\end{aligned}$$

We now describe the Rivest, Shamir, and Adleman method by way of a simple example. Say we wish to send the message "IBM." We convert the message to digits by replacing A by 1, B by 2, . . . , and Z by 26. So the message IBM becomes 9213. The person to whom the message is to be sent has picked two primes p and q, say, $p = 5$ and $q = 17$. (Recall that a *prime* is an integer greater than 1 whose only divisors are 1 and itself.) The receiver has also picked a number r, such as 3, that has no divisors in common with the least common multiple m of $(p - 1) = 4$ and $(q - 1) = 16$ other than 1, and published $n = pq = 85$ and $r = 3$ in a public directory. To decode our message, the receiver must find a number s so that $r \cdot s = 1 \bmod m$ (this is where knowledge of p and q is necessary). That is, $3 \cdot s = 1 \bmod 16$. This number is 11. (The number s can be found by calculating successive powers of $r \bmod m$. When 1 is reached, the previous power of r is s. In our example, we have $3 \bmod 16 = 3$, $3^2 \bmod 16 = 9$, $3^3 \bmod 16 = 11$, $3^4 \bmod 16 = 1$, so $s = 3^3 \bmod 16 = 11$.)

To send our message to this person, we consult the public directory to find $n = 85$ and $r = 3$, then send the "scrambled" numbers $9^3 \bmod 85$, $2^3 \bmod 85$, and $13^3 \bmod 85$ rather than 9, 2, and 13, and the receiver will unscramble them. Thus, we send:

$$\begin{aligned}9^3 \bmod 85 &= 49\\ 2^3 \bmod 85 &= 8\\ 13^3 \bmod 85 &= 72\end{aligned}$$

Now the receiver must take the numbers he or she receives–49, 8, and 72–and convert them back to 9, 2, and 13 by calculating 49^{11} mod 85, 8^{11} mod 85, and 72^{11} mod 85.

The calculation of 49^{11} mod 85 can be simplified as follows:[1]

$$49 \bmod 85 = 49$$
$$49^2 \bmod 85 = 2401 \bmod 85 = 21$$
$$49^4 \bmod 85 = 49^2 \cdot 49^2 \bmod 85 = 21 \cdot 21 \bmod 85 = 441 \bmod 85 = 16$$
$$49^8 \bmod 85 = 49^4 \cdot 49^4 \bmod 85 = 16 \cdot 16 \bmod 85 = 1$$

So,

$$\begin{aligned} 49^{11} \bmod 85 &= (49^8 \bmod 85)(49^2 \bmod 85)(49 \bmod 85) \\ &= (1 \cdot 21 \cdot 49) \bmod 85 \\ &= 1029 \bmod 85 \\ &= 9 \end{aligned}$$

Thus, the receiver has correctly determined the code for I. The calculations for 8^{11} mod 85 and 72^{11} mod 85 are left as exercises. Notice that without knowing how $n = pq$ factors, we cannot find the least common multiple of $p - 1$ and $q - 1$ (in our case, 16), and therefore the s that is needed to determine the intended message.

The procedure just described is called the **RSA public key encryption scheme** in honor of Rivest, Shamir, and Adleman, who discovered it. The method is practical and secure because efficient methods exist for finding very large prime numbers (say, about 100 digits long) and for multiplying large numbers, but no one knows an efficient algorithm for factoring large integers (say, about 200 digits long).

The algorithm is summarized below. In practice, the messages are not sent one letter at a time. Rather, the entire message is converted to decimal form, with A represented by 01, B by 02, . . . , and a space by 00. The message is then broken up into blocks of uniform size and the blocks are sent. See step 2 under Sender below.

Receiver

1. Pick very large primes p and q and compute $n = pq$.
2. Compute the least common multiple of $p - 1$ and $q - 1$; let's call it m.
3. Pick r so that it has no divisors in common with m other than 1 (any such r will do).
4. Find s so that $rs = 1$ mod m. (To find s, simply compute r^2 mod m, r^3 mod m, r^4 mod m, . . . until you reach r^t mod $m = 1$. Then $s = r^{t-1}$ mod m.)
5. Publicly announce n and r, but keep p, q, and s secret.

Sender

1. Convert the message to a string of digits.
2. Break up the message into uniformly sized blocks of digits, appending 0's in the last block if necessary. Call them $M_1, M_2, \ldots, M_k$. For example, for a

[1]To determine 49^2 mod 85 with a calculator, enter 49 × 49 to obtain 2401, then divide 2401 by 85 to obtain 28.247058. Finally, enter 2401 − (28 × 85) to obtain 21. Provided that the numbers are not too large, the search engine Google at www.google.com will do modular arithmetic by simply entering the number and the mod value in the format 49^4 mod 85. Be careful, however, because entering 49^11 mod 85 yields 0, which is incorrect (m^r mod n is never 0 when the greatest common divisor of m and n is 1). Instead, we can use Google to compute smaller powers such as 49^4 mod 85 = 16 and 49^7 mod 85 = 59, then compute (16 × 59) mod 85 = 9.

string such as 2105092315, we would use $M_1 = 2105$, $M_2 = 0923$, and $M_3 = 1500$.

3. Check to see that the greatest common divisor of each M_i and n is 1. If not, n can be factored and the code is broken. (In practice, the primes p and q are so large that they exceed all M_i, so this step may be omitted.)
4. Calculate and send $R_i = M_i^r \bmod n$.

Receiver

1. For each received message R_i, calculate $R_i^s \bmod n$.
2. Convert the string of digits back to a string of characters.

Let's do another example step by step with $p = 7$, $q = 11$, and the message "HI."

Receiver

1. $n = 77$.
2. The least common multiple m of $7 - 1 = 6$ and $11 - 1 = 10$ is 30.
3. We pick $r = 7$.
4. Since $7^4 = 1 \bmod 30$, we have $s = 7^3 \bmod 30 = 13$.
5. Make public $n = 77$ and $r = 7$.

Sender

1. HI converts to 89.
2. We will send 8 and 9 individually (that is, our blocks have size 1).
3. The greatest common divisor of 8 and 77 is 1, and the greatest common divisor of 9 and 77 is 1, so we can proceed.
4. Send $8^7 \bmod 77 = 57$ and $9^7 \bmod 77 = 37$.

Receiver

1. $57^{13} \bmod 77 = 8$; $37^{13} \bmod 77 = 9$.
2. 89 converts to HI.

This method works because of a basic property of modular arithmetic and the choice of r. As a result of choosing the number m as we described, it has the property that for each positive integer x having no common divisors with n except 1, we have $x^m = 1 \bmod n$. So, in the case of our first example with $n = 85$, $m = 16$, and $r = 3$, for the original message 9 and the received message 49, we have mod 85

$$49^{11} = (9^3)^{11} = 9^{33} = 9 \cdot 9^{32} = 9 \cdot (9^{16})^2 = 9 \cdot 1^2 = 9$$

In 2002, Rivest, Shamir, and Adleman received the Association for Computing Machinery A. M. Turing Award, which is considered to be the "Nobel Prize of Computing," for their seminal contribution to public key cryptography.

17.4 Web Searches and Mathematical Logic

With the number of Web pages indexed by large Internet search engines such as Google numbering in billions, computer scientists and mathematicians attempt to manage massive data sets by taking advantage of the associated network structure, which represents the interrelations of the data. The algorithm used by the Google search engine, for instance, ranks all pages on the Web using these interrelations to determine their relevance to the user's search. Factors such as the frequency, location near the top of the page of key words, font size, and number of links are taken into account. Spotlight 17.7 discusses the "Kevin Bacon" game that illustrates the so-called small-world phenomenon of the interconnectedness of data. In this section we will show how a branch of mathematics called **Boolean logic**, after the nineteenth-century mathematician George Boole (1815–1864), can be used to make search engine queries more efficient.

SPOTLIGHT 17.7 Six Degrees of Kevin Bacon

The actor Kevin Bacon is probably better known for a game named after him than for any of the movies he has been in. The game works like this. Every actor who has been in a film with Kevin Bacon has Bacon number 1. Every actor who does not have Bacon number 1, but has been in a film with someone with Bacon number 1, has Bacon number 2. Any actor who does not have Bacon number 1 or 2, but has been in a film with someone who has Bacon number 2, has Bacon number 3, and so on. For example, Nicole Kidman has not been in a film with Bacon but was in *The Interpreter* with Sean Penn, and Sean Penn was in *Mystic River* with Kevin Bacon. So Kidman's Bacon number is 2.

This game was conceived by three college students who first explained it to the public on an MTV show hosted by Jon Stewart. It sometimes goes by the name "Six Degrees of Kevin Bacon" because nearly every actor has a Bacon number that is at most 6. In fact, it is a challenge to think of an actor with a Bacon number exceeding 3. The game is an example of what scientists call the "small-world phenomenon," by which they mean that every person is connected to every other person by a surprisingly short number of links. You can play the Bacon game at www.cs.virginia.edu/oracle/.

The first experiment involving the small-world phenomenon occurred in 1967, when social psychologist Stanley Milgram mailed a series of traceable letters from points in Kansas and Nebraska to "targets" in Boston. The letters could be sent only to someone whom the holders knew on a first-name basis, and who they thought was more likely to know the target than they were themselves. The data revealed a median chain length of about 6.

Mathematicians have their own version of the Bacon game called the "Erdös number," where coauthors of research papers with Paul Erdös have Erdös number 1. The author of this spotlight has Erdös number 3. Ironically, Erdös's Bacon number is 3. In April 2004, a person with Erdös number 4 auctioned on eBay the opportunity for someone to get an Erdös number of 5 by offering to write a joint paper with the highest bidder. The auction was halted when someone bid $1 million as a protest to the idea of selling coauthorships. Before the auction was halted the highest bid was $1031. That bidder also refused to pay as a protest. Interestingly, the actress Danica McKellar, who played Winnie Cooper on the TV series *The Wonder Years*, has a Bacon number 2 and an Erdös number 4.

An *expression* in Boolean logic is simply a statement that is either true or false. For example, the expression, "There are infinitely many prime numbers," is either true or false. For purposes of mathematical reasoning, it is not necessary that we know whether this statement is true or false, but simply that it is one or the other. (It was proved to be true by Euclid more than 2000 years ago.) The expression, "The integer 51 is prime," is an example of an expression that is false, since $51 = 3 \cdot 17$. When combining expressions in logic, we avoid statements that are subject to opinion or various interpretations, such as "math is cool." When we enter a phrase such as "college football" as a query to a search engine, the search engine automatically interprets it as the expression, "This Web page contains the phrase 'college football'." The search engine then returns the list of Web pages for which this expression is true.

In this section we discuss how complex expressions can be constructed by connecting individual expressions with the *connectives* AND, OR, and NOT. For example, to obtain Web pages containing the phrase "football" but not pages that also contain "NFL" or "college," we could formulate the query using the expression "football AND (NOT NFL) AND (NOT college)." The search engine interprets this as, "This Web page contains the phrase 'football' AND it is NOT the case that this Web page contains the word 'NFL' AND it is NOT the case that this Web page contains the word 'college'." The parentheses are not necessary but are sometimes useful, as we will see shortly. Most search engines are not *case sensitive*. That is, no distinction is made between uppercase and lowercase letters.

Each search engine has slightly different conventions for formulating queries. Virtually every search engine has a hyperlink on its Web page that explains how to formulate queries. Although we use traditional terminology and notation from logic for our connectives, most popular search engines employ a more user-friendly format for advanced searches. For Google, "Find results with all of the words" is a substitute for our AND connective; "with at least one of the words" plays the role of our OR connective; and "without the words" is the same as our NOT connective. Some search engines use + for our AND connective and − for our NOT connective. Despite the differences in format, the logic is the same.

Our interest is finding out how we can use Boolean logic to decide whether two different expressions have the same meaning. For example, is the expression, "football AND (NOT NFL) AND (NOT college)," equivalent to "football AND NOT (NFL AND college)"? Or is it equivalent to the expression "football AND NOT (NFL OR college)"? To answer these questions, we will now take a closer look at the connectives AND, OR, and NOT.

The NOT connective allows us to take an expression P and create a new expression NOT P, called the *negation* of P. If P is true, then NOT P is false. If P is false, then NOT P is true. Rather than writing NOT P, we will use the more standard mathematical notation $\neg P$. The negation relationship can be summarized in the following format, known as a **truth table**:

P	$\neg P$
T	F
F	T

Notice that T and F are used here as shorthand for *true* and *false*, respectively. The left column of the truth table shows the two possible values of P: T and F. The right column shows the values of $\neg P$ for each of the corresponding values of P.

The AND connective allows us to combine two expressions, P and Q, into a new expression P AND Q called the *conjunction* of P and Q. The new expression is true when both statements P and Q are true and is otherwise false. The mathematical notation for P AND Q is $P \wedge Q$. (You can remember this by noting that $\wedge$ has a shape like the first letter of AND.) This relationship can also be summarized in a truth table:

P	Q	$P \wedge Q$
T	T	T
T	F	F
F	T	F
F	F	F

Here, the first two columns are used to show all possible values of P and Q, and the right column shows the value of $P \wedge Q$.

Finally, the OR connective allows us to combine two expressions P and Q into a new expression P OR Q, called the *disjunction* of P and Q, which is true if either P or Q, or both, are true and is otherwise false. The mathematical notation for P OR Q is $P \vee Q$. This relationship is summarized by the truth table

P	Q	$P \vee Q$
T	T	T
T	F	T
F	T	T
F	F	F

In everyday circumstances the word "or" is used in two distinct ways. In some situations "P or Q" means either P is valid, or Q is valid, or both are valid, while in other situations "P or Q" means exactly one of P or Q is valid. A typical example of the former is the criterion for admission to an entertainment event that states "Must be at least 18 years old or accompanied by an adult." One the other hand, a menu entry that says "Price includes soup or salad" is an example of the latter. To distinguish between these two usages mathematicians call the first the *inclusive or* and the second the *exclusive or*. When you encounter a mathematical statement of the form P OR Q, the inclusive or is meant. In nonmathematical, ambiguous situations some people use the term "and/or" to mean the inclusive or.

A statement involving three expressions P, Q, and R such as $P \wedge Q \wedge R$ appears to be ambiguous. Does this expression mean that P and Q are first combined into a new expression $P \wedge Q$ and this new expression is then combined with R? In other words, should we interpret this expression as $(P \wedge Q) \wedge R$? Perhaps the intention was to connect P with the single expression $Q \wedge R$. In this case, the expression is interpreted as $P \wedge (Q \wedge R)$. Just as with the case for arithmetic statements such as $5 + 3 + 6$, which we can interpret to mean $(5 + 3) + 6$ or $5 + (3 + 6)$, it turns out that both interpretations are the same. For example, consider the three requirements for the office of president of the United States. A candidate for president must be at least 35 years old, must be a natural-born U.S. citizen, and must have lived in the United States for at least 14 years. Let P be the statement "a candidate must be at least 35 years old," let Q be the statement "a candidate must be a natural-born U.S. citizen," and let R be the statement "a candidate must have lived in the United States for at least 14 years."

The expression $(P \wedge Q) \wedge R$ can then be interpreted as "a candidate must be at least 35 years old and a natural-born U.S. citizen and also must have lived in the United States for at least 14 years." The expression $P \wedge (Q \wedge R)$ can be interpreted as "a candidate must be at least 35 years old and also a natural-born U.S. citizen who has lived in the United States for at least 14 years." Both of these descriptions are effectively the same. Both expressions are true only in the event that each of P, Q, and R is true. One way to verify this is to construct the truth table for expression $(P \wedge Q) \wedge R$ and the truth table for $P \wedge (Q \wedge R)$ and show that they give the same values for every possible value of P, Q, and R. Thus, since the order of operations does not matter in this case, we can simply write $P \wedge Q \wedge R$ without worrying about any possible ambiguity. The same thing is true for $P \vee Q \vee R$. The connectives $\wedge$ and $\vee$ have the associative property. Notice this terminology is consistent with the "associative property" of real-number addition and multiplication: $(a + b) + c = a + (b + c)$ and $(ab)c = a(bc)$.

In some cases, however, the ambiguity is not easy to resolve. For example, consider the expression $P \wedge Q \vee R$, which can be interpreted as either $(P \wedge Q) \vee R$ or as $P \wedge (Q \vee R)$. Using the statements P, Q, and R as before, $(P \wedge Q) \vee R$ can be interpreted as "a candidate must be at least 35 years old and a natural-born U.S. citizen, or must have lived in the United States at least 14 years." On the other hand, $P \wedge (Q \vee R)$ can be interpreted as "a candidate must be at least 35 years old, and be a natural-born citizen or have lived in the United States at least 14 years." These two descriptions are certainly not the same! For example, since Arnold Schwarzenegger is not a natural-born U.S. citizen, the first expression excludes him as a candidate for president, whereas the second one includes him. One way to see exactly how the two statements differ is to compare the truth table for $(P \wedge Q) \vee R$ to the truth table for $P \wedge (Q \vee R)$. The truth table for $(P \wedge Q) \vee R$ is:

P	Q	R	$(P \wedge Q)$	$(P \wedge Q) \vee R$
T	T	T	T	T
T	T	F	T	T
T	F	T	F	T
T	F	F	F	F
F	T	T	F	T
F	T	F	F	F
F	F	T	F	T
F	F	F	F	F

The truth table for $P \wedge (Q \vee R)$ is:

P	Q	R	$(Q \vee R)$	$P \wedge (Q \vee R)$
T	T	T	T	T
T	T	F	T	T
T	F	T	T	T
T	F	F	F	F
F	T	T	T	F
F	T	F	T	F
F	F	T	T	F
F	F	F	F	F

Because the last columns of these two truth tables differ for some values of P, Q, and R, the two expressions are not equivalent. For example, notice that if P is false and Q and R are both true, then $(P \wedge Q)$ is false. Because R is true, however,

$(P \wedge Q) \vee R$ is true. On the other hand, $P \wedge (Q \vee R)$ is false regardless of whether $(Q \vee R)$ is true or false, because P is false. To avoid ambiguity, it is often best to use parentheses.

A way to avoid ambiguity without using parentheses is to adopt a convention on the order of operations. For example, in arithmetic, the convention is that multiplication takes precedence over addition. Therefore, $3 + 4 \times 5$ is determined by first evaluating 4×5 and then adding 3. Of course, we could have written $3 + (4 \times 5)$ to avoid the ambiguity altogether. Similarly, in Boolean logic we adopt the convention that $\wedge$ (AND) takes precedence over $\vee$ (OR). Therefore, the expression $P \wedge Q \vee R$, by convention, is to be interpreted as $(P \wedge Q) \vee R$. Furthermore, the convention states that $\neg$ (NOT) takes the highest precedence of all. Thus, $\neg P \wedge \neg Q \wedge R$ is interpreted as $((\neg P) \wedge (\neg Q)) \vee R$.

EXAMPLE 8 ■ Applying Boolean Logic to a Web Search

Let's revisit the Web queries mentioned before. Let P represent the query "football," which corresponds to the expression, "This Web page contains the phrase 'football'." Let Q represent the expression, "This Web page contains the word 'NFL'." Let R represent the expression, "This Web page contains the word 'college'." We now translate the query "football AND (NOT NFL) AND (NOT college)" as $P \wedge (\neg Q) \wedge (\neg R)$ and write its truth table:

P	Q	R	$\neg Q$	$\neg R$	$P \wedge (\neg Q) \wedge (\neg R)$
T	T	T	F	F	F
T	T	F	F	T	F
T	F	T	T	F	F
T	F	F	T	T	T
F	T	T	F	F	F
F	T	F	F	T	F
F	F	T	T	F	F
F	F	F	T	T	F

For every possible value of P, Q, and R, the truth table gives us the value of our expression. The fourth and fifth columns of the table are not strictly necessary, but they are helpful in determining the values in the last column. As expected, this table tells us that the expression $P \wedge (\neg Q) \wedge (\neg R)$ is true precisely when P is true, Q is false, and R is false.

■

Two expressions are said to be **logically equivalent** if they have the same value, true or false, for each possible assignment of the Boolean variables. To decide whether two expressions are logically equivalent, we construct the truth tables for each one and then check if they have the same values for each of the possible assignments of the Boolean variables. If they do, the expressions are logically equivalent. If they differ for even one case, however, then the expressions are not equivalent.

So, to determine whether the expression, "football AND (NOT NFL) AND (NOT college)," is equivalent to the expression, "football AND NOT (NFL AND college)," we need only compare their corresponding truth tables.

EXAMPLE 9 ■ Logically Equivalent Expressions

We first determine the truth table for the expression, "football AND NOT (NFL AND college)," which is represented as $P \wedge \neg(Q \wedge R)$. Its truth table is:

P	Q	R	$Q \wedge R$	$\neg(Q \wedge R)$	$P \wedge \neg(Q \wedge R)$
T	T	T	T	F	F
T	T	F	F	T	T
T	F	T	F	T	T
T	F	F	F	T	T
F	T	T	T	F	F
F	T	F	F	T	F
F	F	T	F	T	F
F	F	F	F	T	F

This truth table differs in the last column from the truth table for $P \wedge (\neg Q) \wedge (\neg R)$ in the previous example. For instance, when P is true, Q is true, and R is false, we see that $P \wedge (\neg Q) \wedge (\neg R)$ is false but $P \wedge \neg(Q \wedge R)$ is true. Therefore, we must conclude that $P \wedge (\neg Q) \wedge (\neg R)$ is not logically equivalent to $P \wedge \neg(Q \wedge R)$.

On the other hand, the expression, "football AND NOT (NFL OR college)," is represented by $P \wedge \neg(Q \vee R)$. Its truth table is:

P	Q	R	$Q \vee R$	$\neg(Q \vee R)$	$P \wedge \neg(Q \vee R)$
T	T	T	T	F	F
T	T	F	T	F	F
T	F	T	T	F	F
T	F	F	F	T	T
F	T	T	T	F	F
F	T	F	T	F	F
F	F	T	T	F	F
F	F	F	F	F	F

Because the last column of this table agrees with the last column for the expression $P \wedge (\neg Q) \wedge (\neg R)$, we know the expression, "football AND (NOT NFL) AND (NOT college)," is equivalent to the expression, "football AND NOT (NFL OR college)." ■

Applying Logic to Message Routing

The AND operator in the truth table on page 556 is also used by computers to deliver messages over the Internet with a device called a *router*. Recall that if P and Q are expressions then the statement $P \wedge Q$ is true when both P and Q are true and false otherwise. Computer scientists use an analogous operation on 0 and 1 by allowing P and Q to represent 0 or 1 and defining that $P \wedge Q = 1$ when P and Q are 1 and $P \wedge Q = 0$ otherwise. When $\wedge$ is used in this way it is called the *bitwise AND*. Notice that we can obtain an operation table for the bitwise AND from the table for the logical AND on page 556 by substituting 1 for T and 0 for F. In particular, we have

P	Q	$P \wedge Q$
1	1	1
1	0	0
0	1	0
0	0	0

The bitwise AND operation can be extended to binary strings of equal length by applying it individually to corresponding entries. Thus $11001001 \wedge 01101101 = 01001001$ since both strings have a 1 only in positions 2, 5 and 8. In general, for any binary string s we can use the bitwise AND to copy whichever entries of s we desire while converting all the other entries of s to 0. For example, if we have a list of binary strings of length 8 and we wish to modify these strings by copying the entries in positions 2, 7, and 8 and changing all other entries to 0, we simply take each string in the list and combine it with 01000011 using the bitwise AND operator. Thus we have, $11101001 \wedge 01000011 = 01000001$; and $00011110 \wedge 01000011 = 00000010$. When doing the bitwise AND operation on binary strings it is convenient to put one string directly above the other. In those positions where both entries are 1 the result is 1; otherwise the result is 0.

s	11101001	s	00011110
t	<u>01000011</u>	t	<u>01000011</u>
$s \wedge t$	01000001	$s \wedge t$	00000010

The bitwise AND operation is used by computers to determine when certain entries of two binary strings match. Say, for example, that a computer would take a particular action if two binary strings s and t of equal length match in the first three positions. This happens precisely when $s \wedge 11100000 = t \wedge 11100000$. The reason why this works is that $x \wedge 1 = 1$ only when $x = 1$. Thus, if s and t both begin with 1 then both $s \wedge 11100000$ and $t \wedge 11100000$ will begin with 1; if s and t both begin with 0 then both $s \wedge 11100000$ and $t \wedge 11100000$ will begin with 0; if s and t begin with different digits then $s \wedge 11100000$ and $t \wedge 11100000$ will begin with different digits. The same reasoning applies to positions 2 and 3. The string of five 0's at the end of 11100000 ensures that $s \wedge 11100000$ and $t \wedge 11100000$ will both end with five 0's. So, checking that $s \wedge 11100000 = t \wedge 11100000$ checks whether s and t agree in the first three positions while disregarding the other positions.

In order for devices on a network to communicate with each other each one must be given a unique identifier. This is done with an **Internet Protocol address.**

Internet Protocol Address — DEFINITION

An **Internet Protocol (IP) address** is a sequence of four numbers between 0 and 255 separated by dots assigned to routers, computers, printers and fax machines that allows those linked electronically to uniquely identify and communicate with each other.

Each computer on the Internet is assigned an IP address. From the IP address, we can determine the **network address** of the computer, which specifies the network or subnet that the computer is a part of. While each computer on the network or subnet will have a different IP address, they will all have the same network address. To demonstrate how computers determine the network address from an IP address, we must first explain how to convert the decimal form of each component of an IP address such as 131.212.66.17, which is convenient for humans, to their binary forms of length 8, which are convenient for computers. To convert an IP address in decimal to binary, we express each decimal number as a sum of distinct powers of 2 ranging from $128 = 2^7$ to $1 = 2^0$. For example, $213 = 128 + 64 + 16 + 4 + 1 = 2^7 + 2^6 + 2^4 + 2^2 + 2^0$. Now, make a row of the powers of 2 from 128 to 1 and beneath each one place a 1 if that power of 2 appears in the sum and a 0 if it does not. For the number 213, we have

128	64	32	16	8	4	2	1
1	1	0	1	0	1	0	1

Reading off the sequence of 0's and 1's, we have that the binary form of 213 is 11010101. Since 00000000 is the binary form of length 8 of 0, and 11111111 is the binary form of $255 = 128 + 64 + 32 + 16 + 8 + 4 + 2 + 1$, all the integers between 0 and 255 can be written as binary strings of length 8 using leading 0's as needed.

Each IP address is assigned a companion number called a **subnet mask** that also consists of four numbers, each of which range from 0 to 255. To determine the network address from the IP address, one performs the bitwise AND operation to the binary forms of the IP address and its companion subnet mask. The resulting number is the network address.

EXAMPLE 10 ▪ Network Address for IP Address 131.212.66.17 With Subnet Mask 255.255.255.0

We determine the network address corresponding to the IP address 131.212.66.17 with subnet mask 255.255.255.0. Since the binary form of 131.212.66.17 is 10000011.11010100.01000010.00010001 and the binary form of the subnet mask is 11111111.11111111.11111111.00000000, combining them using the bitwise AND gives

IP address	10000011.11010100.01000010.00010001
Subnet mask	11111111.11111111.11111111.00000000
Bitwise AND	10000011.11010100.01000010.00000000.

Thus, the network address is 10000011.11010100.01000010.00000000 = 131.212.66.0 (some authors omit one or more 0's when they appear at the end of a network address). ▪

From Example 10 it appears that the network address corresponding to a IP address with subnet mask 255.255.255.0 is simply the same as the IP address with the last number changed to 0. This is correct in this case but for other subnet masks, the network address is not readily apparent from the IP address.

EXAMPLE 11 ▪ Network Address for IP Address 131.212.66.56 With Subnet Mask 255.255.255.240

We determine the network address corresponding to the IP address 131.212.66.56 with subnet mask 255.255.255.240. Since the binary form of 131.212.66.56 is 10000011.11010100.01000010.00111000 and the binary form of the subnet mask is 11111111.11111111.11111111.11110000, combining them using the bitwise AND gives

IP address	10000011.11010100.01000010.00111000
Subnet mask	11111111.11111111.11111111.11110000
Bitwise AND	10000011.11010100.01000010.00110000.

In this example, the network address is 10000011.11010100.01000010.00110000 = 131.212.66.48. Since the subnet mask for this IP address ends with four 0's, even if we changed the last four bits in the IP address, we would still get the same network address. Any decimal number whose binary form starts with 0011 is a number between 48 and 63, so any device with a IP address that begins with 131.212.66 and ends with any number between 48 and 63 will be in the network whose address is 131.212.66.48. We note that a device with an IP address that begins with 131.212.66 and ends with a number that is not between 48 and 63 is not on the same network as any device that starts with 131.212.66 and ends with a number between 48 and 63 since the devices will have different network addresses.

The network addresses allow routers to deliver messages to IP addresses in much the same way the postal service delivers mail to home addresses. That is, the postal service first checks to see if the mail is to be sent to a local address and if not, it is relayed to a larger mail center. Likewise, if a message is to be sent from one computer to another, a local router checks to see if both computers are on the same network. If so, the message is sent over that network. Otherwise it is routed to a larger network. In this case the local router searches its memory for the network address that most closely matches that of the destination network address and relays the message to a router in a larger network that has that address in its memory. A postal analogy might be a letter from Duluth, Minnesota addressed to Rochester, Minnesota being routed through Minneapolis.

REVIEW VOCABULARY

Binary linear code A code consisting of words composed of 0's and 1's obtained by using parity-check sums to append check digits to messages. (p. 539)
Boolean logic Logic attributed to George Boole that uses operations such as ∧, ∨, and ¬ to connect statements. (p. 554)
Caesar cipher A cryptosystem used by Julius Caesar whereby each letter is shifted the same amount. (p. 547)
Code word A string of digits composed of a message and check digits. (p. 539)
Compression algorithm A procedure for converting data from one format to another one optimized for compactness. (p. 542)
Cryptology The study of how to make and break secret codes. (p. 546)
Data compression The process of encoding data so that the most frequently occurring data are represented by the fewest symbols. (p. 541)
Decimation cipher A cryptosystem that uses multiplication by a fixed value to shift each letter. (p. 548)
Decoding The process of translating received data into code words. (p. 537)
Distance between two strings The distance between two strings of equal length is the number of positions in which they differ. (p. 537)
Encryption The process of encoding data to protect against unauthorized interpretation. (p. 546)
Even parity Even integers are said to have even parity. (p. 535)
IP address A sequence of four numbers that uniquely identifies a device on a network. (p. 560)
Key A string used to encode and decode data. (p. 550)
Key word A word used to determine the amount of shifting for each letter while encoding a message. (p. 549)
Logically equivalent Two expressions are said to be logically equivalent if they have the same values for all possible values of their Boolean variables. (p. 558)
Modular arithmetic Addition and multiplication involving modulo n. (p. 547)

Nearest-neighbor decoding A method that decodes a received message as the code word that agrees with the message in the most positions. (p. 538)
Network address The portion of an IP address that identifies a local network. (p. 560)
Odd parity Odd integers are said to have odd parity. (p. 535)
Parity-check sums Sums of digits whose parities determine the check digits. (p. 537)
RSA public key encryption scheme A method of encoding that permits each person to announce publicly the means by which secret messages are to be sent to him or her. (p. 552)
Subnet mask A companion number to an IP address that allows a router to determine the network portion of an IP address. (p. 561)
Truth table A tabular representation of an expression in which the variables and the intermediate expressions appear in columns and the last column contains the expression being evaluated. (p. 555)
Variable-length code A code in which the number of symbols for each code word may vary. (p. 541)
Vigenère cipher A cryptosystem that utilizes a key word to determine how much each letter is shifted. (p. 549)
Weight of a binary code The minimum number of 1's that occur among all nonzero code words of a code. (p. 539)

SKILLS CHECK

1. Using the circular diagram method to encode the message 1011, the encoded message is

(a) 1011001.
(b) 1011010.
(c) 1010001.

2. A four-digit binary message was encoded using Table 17.1 and the message 1010010 was received. Using the nearest neighbor method, the decoded four-digit message is ______________.

3. Using the nearest-neighbor method and the code in Table 17.2, the word 1110011 decodes as

(a) 0110010.
(b) 1100011.
(c) 1010001.

4. The distance between received words 1011001 and 1000101 is ______________.

5. The weight of the binary linear code {0000000, 0011111, 0101011, 0110100} is

(a) 3.
(b) 4.
(c) 5.

6. If the two messages 0 and 1 are encoded as 000 and 111, respectively, the number of errors the code can correct is ______________.

7. If every pair of code words differs in at least five positions, then nearest-neighbor decoding can accurately decode words that have

(a) two mistakes.
(b) three mistakes.
(c) four mistakes.

8. If a binary linear code has weight 4, the maximum number of errors that it will detect is ______________.

9. Using the encoding scheme $A \to 0$, $B \to 10$, $C \to 11$, the string 010110 decodes as

(a) ABCB.
(b) ABCA.
(c) ABACA.

10. The sum of the binary string 1011001 and 1001101 is ______________.

11. The Caesar cipher encrypts GO HOME NOW as

(a) JR KRPH QRZ.
(b) DL ELJB KLT.
(c) Neither of these.

12. Using modular arithmetic, 3^5 mod 20 is equal to ______________.

13. Using the Vigenère cipher and the key word ADAM to decrypt EIEIO, we obtain

(a) ELELR.
(b) EFEFL.
(c) EFEWO.

14. If the message EAPL was encrypted using the decimation cipher with the key 9, the message is _____.

15. Using the RSA scheme with $n = 91$ and $s = 5$, the message 4 decodes as

(a) 11.
(b) 20
(c) 23.

16. If we use $p = 7$, $q = 17$, and $r = 5$ in the RSA scheme, the value of s is ______________.

17. Which messages are the hardest to break?

(a) Messages encrypted with the Vigenère cipher.
(b) Messages encrypted with the decimation cipher.
(c) Messages encrypted with the RSA scheme.

18. The statement "P OR Q" is true if and only if

____ .

19. When the statement "P AND NOT Q" is true, it must be the case that

(a) P is true.
(b) P is false.
(c) either P or Q is false.

20. $00111001 \wedge 11111001 =$ ____________________ .

CHAPTER 17 EXERCISES

■ Challenge ◆ Discussion

17.1 Binary Codes

1. Use the diagram method shown in Figures 17.1 and 17.2 to verify the code words in Table 17.1 for the messages 0101, 1011, and 1111.

2. Use the diagram method to decode the received messages 0111011 and 1000101.

3. Find the distance between each of the following pairs of words:

(a) 11011011 and 10100110
(b) 01110100 and 11101100

4. Referring to Table 17.1, use the nearest-neighbor method to decode the received words 0000110 and 1110100.

5. If the code word 0110010 is received as 1001101, how is it decoded using the diagram method?

6. Suppose a received word has the Venn diagram arrangement shown here:

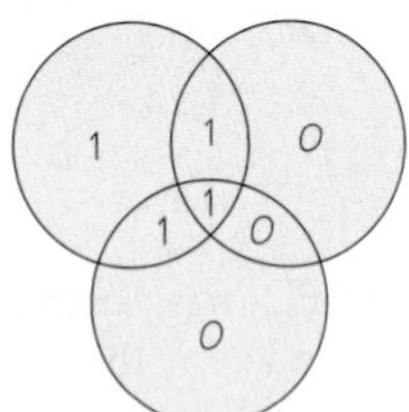

What can we conclude about the received word?

17.2 Encoding with Parity-Check Sums

7. Determine the binary linear code that consists of all possible three-digit messages with three check digits appended using the parity-check sums $a_2 + a_3$, $a_1 + a_3$, and $a_1 + a_2$. (That is, $c_1 = 0$ if $a_2 + a_3$ is even, $c_1 = 1$ if $a_2 + a_3$ is odd, and similarly for c_2 and c_3.)

8. Let C be the code

$$\{0000000, 1110100, 0111010, 0011101, \\ 1001110, 0100111, 1010011, 1101001\}$$

What is the error-correcting capability of C? What is the error-detecting capability of C?

9. Find all code words for binary messages of length 4 by adding three check digits using the parity-check sums $a_2 + a_3 + a_4$, $a_2 + a_4$, and $a_1 + a_2 + a_3$. Will this code correct any single error?

10. Consider the binary linear code

$$C = \{00000, 10011, 01010, 11001, \\ 00101, 10110, 01111, 11100\}$$

Use nearest-neighbor decoding to decode 11101 and 01100. If the received word 11101 has exactly one error, can you determine the intended code word? Explain your reasoning.

11. Construct a binary linear code using all eight possible binary messages of length 3 and appending three check digits using the parity-check sums $a_1 + a_2$, $a_2 + a_3$, and $a_1 + a_3$. Decode each of the received words below by the nearest-neighbor method:

001001, 011000, 000110, 100001

12. Extend the code words listed in Table 17.1 to eight digits by appending a 0 to words of even weight and a 1 to words of odd weight. What are the error-detecting and error-correcting capabilities of the new code?

13. Extend the code words listed in Table 17.2 to eight digits by appending a 0 to words of even weight and a 1 to words of odd weight. What are the error-detecting and error-correcting capabilities of the new code?

14. Suppose the weight of a binary linear code is 6. How many errors can the code correct? How many errors can the code detect?

15. How many code words are there in a binary linear code that has all possible messages of length 5 with three check digits appended? How many possible received words are there with this code?

■ **16.** Explain why no binary linear code with all possible three-digit messages together with three check digits can correct all possible errors involving two digits.

■ **17.** A *ternary* code is formed by starting with all possible strings of a fixed length composed of 0's, 1's, and 2's and appending extra digits that are also 0's, 1's, or 2's. Form a ternary code by appending to each message a_1a_2 the check digits c_1c_2 using:

$$c_1 = (a_1 + a_2) \bmod 3$$
$$c_2 = (2a_1 + a_2) \bmod 3$$

■ **18.** Use the ternary code in the preceding exercise and the nearest-neighbor method to decode the received word 1211.

■ **19.** Suppose a ternary code is formed by starting with all possible strings of 0's, 1's, and 2's of length 4 and appending two extra digits that are also 0's, 1's, and 2's. How many code words are there in this code? How many possible received words are there in this code?

17.3 Data Compression

20. Suppose we code a four-symbol genetic set {A, C, T, G} into binary form as follows:

A → 0 C → 10 T → 110 G → 111

Convert the sequence ACAAGTAAC into binary code.

21. Use the code in the previous exercise to determine the sequence of symbols represented by the binary code 001100001111000.

22. Suppose we code a five-symbol set $\{A, B, C, D, E\}$ into binary form as follows:

$A \to 0$ $B \to 10$ $C \to 110$
$D \to 1110$ $E \to 1111$

Convert the sequence of *AEAADBAABCB* into binary code. Determine the sequence of symbols represented by the binary code 01000110100011111110.

23. Use the code in the previous exercise to convert the sequence *EABAADABB* into binary code. Determine the sequence of letters represented by the binary code 0010001100111110111010.

■ **24.** Devise a variable-length binary coding scheme for a six-symbol set $\{A, B, C, D, E, F\}$. Assume the A is the most frequently occurring symbol, B is the second most frequently occurring symbol, and so on.

25. Judging from the Morse code, what are the three most frequently occurring consonants in English text material? What is the most frequently occurring vowel?

26. In English, the letter H occurs more often than D, G, K, and W, but in Morse code, H has a longer code than D, G, K, and W. Speculate on the reason for this apparent violation of data-compression principles.

27. Explain why the Morse code must include a space after each letter but fixed-length codes do not.

28. Guess the percentage of the occurrences of spaces in typical English text material.

29. Following are the closing values (rounded to the nearest integer) of the Dow Jones Industrial Average Index stock market values for the period September 17, 2007–September 28, 2007. Use the delta function method to compress these values. What percentage reduction in characters is there?

13403 13739 13816 13767 13820
13759 13779 13878 13913 13896

30. The following numbers were encoded using delta function encoding. Determine the original numbers.

1207 373 −57 −97 −234 −105 178 −73 275
79 −183 −146 −94 129

31. Decode the binary string 11100100001001110110011010, which has been encoded using the Huffman code given on page 544.

32. Use a Huffman tree code to assign a binary code to the letters that occur with the probabilities

A	0.025
B	0.150
C	0.015
D	0.170
E	0.200
F	0.225
G	0.215

33. Suppose a Huffman tree has been used to create a binary code for the letters A through J, and the results include $B = 111110$, $J = 111111$, and $G = 11110$. If the code has only two code words of length 6 and one of length 5, what can you say about the probability of the occurrence of the letters B, J, and G?

17.3 Cryptography

34. For each part below, explain how modular arithmetic can be used to answer the question.

(a) If today is Wednesday, what day of the week will it be in 16 days?
(b) If a clock (with hands) indicates that it is now four o'clock, what will it indicate in 37 hours?
(c) If a military person says it is now 0400, what time would it be in 37 hours? (Instead of A.M. and P.M., military people use 1300 for 1:00 P.M., 1400 for 2:00 P.M., and so on.)
(d) If it is now July 20, what day will it be in 65 days?
(e) If the odometer of an automobile reads 97,000 now, what will it read in 12,000 miles?

35. Use the Caesar cipher to encrypt the message RETREAT. Determine the intended message corresponding to the message DGYDQFH that was encrypted using the Caesar cipher.

36. Suppose you take a message and repeatedly apply the Caesar cipher to it until you return to the original message. How many iterations must be done before this occurs?

37. Using 0, 1, 2, . . . , 25 to label the positions of the letters $A, B, C, \ldots, Z$, suppose we create a cipher by replacing the letter in position i with the letter in position $(i + 8) \bmod 26$. How many iterations of this cipher must be done before a message will return to its original state?

38. The message ADDAOS was encrypted using decimation cipher with the key 7. Decrypt it.

39. Use the decimation cipher with the key 5 to encrypt RETREAT.

40. If you attempted to use the decimation cipher with the key 13, how would the word MESSAGE be encrypted?

41. Explain why 2 cannot be used for the key in a decimation cipher.

42. Use the Vigenère cipher with the key word HELP to encrypt the message PHONE HOME.

43. Given that the BEATLES was used as the key word for the Vigenère cipher to encrypt SSLETRY TXOGPW, decrypt the message.

44. Use the Vigenère cipher with the key word CLUE to encrypt the message THE WALRUS WAS PAUL.

45. Add the following pairs of binary strings:
(a) 10111011 and 01111011
(b) 11101000 and 01110001

46. All binary linear codes have the property that the sum of two code words is another code word. Use this fact to determine which of the following sets cannot be a binary linear code
(a) {0000, 0011, 0111, 0110, 1001, 1010, 1100, 1111}
(b) {0000, 0010, 0111, 0001, 1000, 1010, 1101, 1111}
(c) {0000, 0110, 1011, 1101}

47. Use the RSA scheme with $p = 5$, $q = 17$, and $r = 3$ to determine the numbers sent for the message VIP.

48. Use the RSA scheme with $p = 5$, $q = 17$, and $r = 3$ to decode the received numbers 52 and 72.

49. In the RSA scheme with $p = 5$, $q = 17$, and $r = 5$, determine the value of s.

50. Why can't we use the RSA scheme with $p = 7$, $q = 11$, and $r = 3$?

51. Explain why we can't employ the RSA scheme to send the message "NO" with $p = 7$ and $q = 11$ using blocks of length 2, but we can send it if we use blocks of length 4.

52. Use the search box at www.google.com to compute $13^9 \bmod 77$, $13^6 \bmod 77$, and $13^{15} \bmod 77$. (To compute $13^9 \bmod 77$ enter 13^9 mod 77.)

17.4 Web Searches and Mathematical Logic

53. Show that $P \vee (P \wedge Q)$ is logically equivalent to P.

54. Show that $\neg(P \vee Q)$ is logically equivalent to $\neg P \wedge \neg Q$.

55. Show that $\neg(P \wedge Q)$ is logically equivalent to $\neg P \vee \neg Q$. This relationship and the one in the previous exercise are known collectively as *De Morgan's Laws.*

56. Show that $P \vee (Q \wedge R)$ is logically equivalent to $(P \vee Q) \wedge (P \vee R)$.

57. Show that $P \wedge (Q \vee R)$ is logically equivalent to $(P \wedge Q) \vee (P \wedge R)$.

58. A patron at a restaurant tells the waiter to bring her the chef's recommendation as long as it has "lots of anchovies or is not spicy and in addition the portion must be large." The waiter goes to the kitchen and tells the chef to prepare a dish that has "lots of anchovies and is also large or is spicy and is also large." Did the waiter communicate the patron's wishes correctly to the chef? Use truth tables to support your answer.

59. The *implication connective* is defined by the following truth table:

P	Q	$P \to Q$
T	T	T
T	F	F
F	T	T
F	F	T

Use truth tables to show that $P \to Q$ is logically equivalent to $\neg P \vee Q$.

60. The Minnesota Vikings football coach tells his team before the last game of the regular season that if the team wins, they will be in the playoffs. Use the truth table given in the previous exercise to verify that if the Vikings lose and are still in the playoffs, the coach made a truthful statement to the team.

61. Using the implication connective and other connectives, variables, and truth tables, determine whether the statement "If it snows, there will be no school" is logically equivalent to the statement "It is not the case that it snows and there is school."

62. Suppose s and t are binary strings of length 8. How would you use the bitwise operator $\wedge$ to determine if the last three digits of s and t match? How would you determine if s and t match in positions 2, 4, 6, and 8?

63. Using $\wedge$ to denote the bitwise AND operator compute:

(a) 11110001 $\wedge$ 00101110
(b) 01110001 $\wedge$ 10111110

64. Explain why 01100000 $\wedge$ 1011111 is undefined.

65. In practice, a computer checks to see if $s \wedge 11100000 = t \wedge 11100000$ by checking if $s \wedge 11100000 + t \wedge 11100000 = 00000000$ where addition is done mod 2 in each component. Explain why this works.

66. If s is an 8-digit binary string, determine $s \wedge 11111111$ and $s \wedge 00000000$.

67. If s is a binary string of length 8 and $s \wedge 01010101 = 00000001$ what is the most that you can say about s?

68. Given a binary string s of length 8, how could you use the bitwise AND operator to determine if the digits in positions 1, 3, and 5 are 0's?

69. Find four binary strings s that satisfy $s \wedge 11100111 = 01100010$.

70. How many binary strings s of length 8 are there that satisfy $s \wedge 11100011 = 10000001$?

71. Determine the network address for the IP address 8.20.15.1 with subnet mask 255.000.000.000.

72. Determine the network address for the IP address 8.20.15.1 with subnet mask 255.255.000.000.

73. Determine if the IP address 172.16.17.30 with subnet mask 255.255.255.240 has the same network address as the IP address 172.16.17.15 with the subnet mask 255.255.255.240.

WRITING PROJECTS

1. Prepare a two-page report on applications of modular arithmetic. Explain the calculation of the check digits described in Exercises 7, 9, and 11 with modular arithmetic. Use modular arithmetic to describe the error-detection schemes used in Chapter 16.

2. Use the Web to find information for a two-page report on the Braille system of coding.

3. Use the Web to find information for a two-page report on the Morse code.

4. Use the Web to find information about Smart Card technology and write a two-page report on your findings.

SUGGESTED READINGS

DENEEN, L. Secret encryption with public keys. *UMAP Journal,* 8 (1987): 9–29. Describes several ways in which modular arithmetic can be used to code secret messages.

PETZOLD, C. *Code,* Microsoft Press, Redmond, Wash., 1999. The first three chapters of this book provide an excellent explanation of the Morse code and the Braille system of coding.

SUGGESTED WEB SITE

www.d.umn.edu/~jgallian/fapp7e This site implements the nearest-neighbor decoding method for seven-digit binary strings using the code given in Table 17.1.

On Size and Growth PART VI

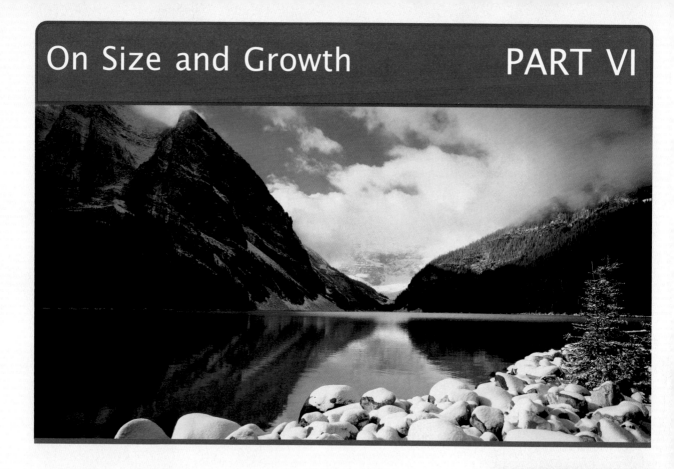

Mathematics is the study of patterns and relationships. It can explain why there are no King Kongs, analyze designs on ancient pottery, and suggest new and beautiful artistic designs. Mathematicians search for and classify numerical, geometric, and even abstract patterns. In these chapters, we follow some of those searches. We concentrate on geometric patterns but find that those lead to numerical considerations, too.

In Chapter 18, Growth and Form, we look at how the sizes of objects influence their forms. We investigate some big things, such as King Kong, tall trees, mile-high buildings, and mountains. Seeing the underlying principles of scaling will help you to appreciate why objects in the world have the shapes and sizes that they do.

We start with a simple numerical pattern in Chapter 19, Symmetry and Patterns, which leads to questions about esthetically pleasing proportions and the importance of bilateral symmetry. We expand our notion of symmetry and discover surprising limitations that even broader notions of symmetry face. We examine the beauty of fractal patterns, ones that resemble themselves at finer and finer scales, in nature and in traditional art from Africa and elsewhere.

Chapter 20, Tilings, answers the question of how to arrange objects symmetrically on a surface. What shapes can we use? What patterns can arise if the objects themselves are symmetrical, or if we allow irregular shapes but demand that they all face the same way? Most curious of all, you can arrange shapes in a pattern that does not repeat but is nevertheless systematic. ■

CHAPTER 18

Growth and Form

Fantasy films have made us familiar with giant creatures, including King Kong, Godzilla, and the auliphants in *The Lord of the Rings.* We also find supergiants in literature, such as the giant of "Jack and the Beanstalk," the Big Friendly Giant in Roald Dahl's *The BFG,* and the Brobdingnagians of *Gulliver's Travels.*

Even from an early age, though, we don't really believe in monsters and giants. But could such beings ever exist? What problems would their enormous size cause them? How would they have to adapt to cope? (See Figure 18.1.)

Every species adapts to its environment. In particular, it faces the **problem of scale**: how to adapt and survive at the different sizes from the beginning of life to the final size of a mature adult. For example, the giant panda ranges from barely 1 pound (lb) at birth to 275 lb in adulthood. A baby panda could be crushed by its mother; an adult panda needs to eat a lot.

For contrast, consider the horse. A newborn foal that weighed as little as a newborn panda would be too small to keep up with the herd and could not survive. An adult horse weighs much more than a panda and has to consume far more food, but the horse can move much more quickly and cover greater distances to find sustenance.

There have been large land mammals (mammoths) and huge sea mammals (the blue whale)–not to mention the dinosaurs. But the tallest humans have been only 9 to 10 feet (ft) tall. The largest mammoth was 16 ft at the shoulder (about twice as tall as an elephant). Even the tallest dinosaur, *Supersaurus,* stood only 40 ft high.

What about supergiants and utterly huge monsters? That they have never existed suggests physical limits to size. In fact, with a few simple principles of geometry, we can show that no objects or living beings could exist, unchanged in shape, on a vastly different scale, larger or smaller.

FIGURE 18.1 Could King Kong actually exist? (*Bettmann/Corbis.*)

TABLE 18.1 Units of the U.S. Customary Systems

Distance:

1 mile (mi) = 1760 yards (yd) = 5280 feet (ft) = 63,360 inches (in.)
1 yard (yd) = 3 feet (ft) = 36 inches (in.)
1 foot (ft) = 12 inches (in.)

Area:

1 square mile = 1 mi × 1 mi = 5280 ft × 5280 ft
= 27,878,400 $ft^2 \approx 2.8 \times 10^7$ ft^2
= 63,360 in. × 63,360 in.
= 4,014,489,600 in.$^2 \approx 4 \times 10^9$ in.^{2}
= 640 acres
1 acre = 43,560 ft^2

Volume:

1 cubic mile = 1 mi × 1 mi × 1 mi
= 5280 ft × 5280 ft × 5280 ft
= 147,197,952,000 ft^3
$\approx 147 \times 10^9$ ft^3
= 63,360 in. × 63,360 in. × 63,360 in.
$\approx 2.5 \times 10^{14}$ in.^{3}
1 U.S. gallon (gal) = 4 U.S. quarts (qt) = 231 in.^{3}, exactly

Mass:

1 ton (t) = 2000 pounds (lb)

TABLE 18.2 Units of the Metric System

Distance:

1 meter (m) = 100 centimeters (cm)
1 kilometer (km) = 1000 meters (m)
= 100,000 centimeters (cm) = 1×10^5 cm

Area:

1 square meter (m^2) = 1 m × 1 m
= 100 cm × 100 cm = 10,000 (cm^2) = 1×10^4 cm^2
1 hectare (ha) = 10,000 m^2

Volume:

1 liter (L) = 1000 cm^3 = 0.001 m^3
1 cubic meter (m^3) = 1 m × 1 m × 1 m
= 100 cm × 100 cm × 100 cm
= 1,000,000 cm^3 = 1×10^6 cm^3 (or cc)

Mass:

1 kilogram (kg) = 1000 grams (g) = 1×10^3 g

All other units of length, area, and volume are *defined* in terms of the meter. For example, a centimeter (cm) is a hundredth of a meter.

Mass is quantity of matter. The metric unit of mass, the *kilogram* (kg), is defined as the mass of a platinum–iridium standard kept in Paris. Since you can't determine the mass of a sack of potatoes by comparing it to that, we measure the mass indirectly by seeing how much force gravity exerts on it–that is, we weigh it, on a scale calibrated in pounds or kilograms. However, a mass of 1 kg would "weigh" (register on the scale) only one-sixth as much on the Moon.

Table 18.2 lists the units of the metric system.

Converting Between Systems

What are the conversion factors between the U.S. Customary System and the metric system? Since 1959, the fundamental units of the U.S. Customary System, the yard (for length) and the pound (for mass), have been *defined* in terms of metric units, so that we have

$$1 \text{ yd} = 0.9144 \text{ m, exactly}$$
$$1 \text{ lb} = 0.45359237 \text{ kg, exactly}$$

Table 18.3 illustrates the conversion factors. In the following examples we explain how to convert measurements between systems.

TABLE 18.3 Conversions Between the U.S. Customary System and the Metric System

Distance:

1 in. = 2.54 cm, exactly
1 ft = 12 in. = 12×2.54 cm = 30.48 cm = 0.3048 m, exactly
1 yd = 0.9144 m, exactly
1 mi = 5280 ft = 5280×30.48 cm
= 160,934.4 cm, exactly $\approx$ 1.61 km
1 cm $\approx$ 0.393701 in. $\approx$ 0.4 in.
1 m $\approx$ 39.37 in. $\approx$ 3.281 ft
1 km $\approx$ 0.621 mi

Area:

1 ft^2 $\approx$ 0.09290 m^2 = 929.0 cm^2
1 m^2 $\approx$ 10.76 sq ft
1 hectare (ha) $\approx$ 2.47 acres

Volume:

1 ft^3 $\approx$ 28.32 liters (L)
1 gallon $\approx$ 3.785 liters (L)
1 cubic meter (m^3) = 1000 liters $\approx$ 264.2 U.S. gallons $\approx$ 35.31 ft^3
1 liter (L) = 1000 cm^3 $\approx$ 1.057 U.S. quarts (qt) $\approx$ 0.2642 U.S. gallons

Mass:

1 lb = 0.45359237 kg, exactly
1 kg $\approx$ 2.205 lb

EXAMPLE 2 ■ What's That in Feet?

An international student tells her American student friends that she is 160 cm tall. They ask how much that is in feet and inches.

We approach this conversion by using the scaling factor 1 cm = $\frac{1}{2.54}$ in.:

$$160 \text{ cm} = 160 \text{ cm} \times 1 \approx 160 \text{ cm} \times \frac{1 \text{ in.}}{2.54 \text{ cm}}$$

$$\approx 63.0 \text{ in.} = 63.0 \text{ in.} \times \frac{1 \text{ ft}}{12 \text{ in.}} \approx \frac{63.0}{12} \text{ ft} \approx 5.25 \text{ ft}$$

However, because we normally give height in feet and a whole number of inches, the height is

$$63.0 \text{ in.} = 5 \times (12 \text{ in.}) + 3.0 \text{ in.} = 5 \text{ ft} + 3.0 \text{ in.} \approx 5 \text{ ft } 3 \text{ in.}$$

Another way to approach the problem is by means of a proportion:

$$\frac{\text{height in in.}}{\text{height in cm}} = \frac{\text{length of 1 in. in inches}}{\text{length of 1 in. in cm}} = \frac{1 \text{ in.}}{2.54 \text{ cm}}$$

so that

$$\text{height in in.} = \text{height in cm} \times \frac{1 \text{ in.}}{2.54 \text{ cm}}$$

$$= 160 \text{ cm} \times \frac{1 \text{ in.}}{2.54 \text{ cm}} \approx 63.0 \text{ in.}$$

■

EXAMPLE 3 ■ Got Gas?

In the United States, we measure the efficiency of cars in miles per gallon (mpg); most of the rest of the world measures it in liters per 100 kilometers. The conversion between these two measures is more complicated than other conversions, because the U.S. measure has distance (mi) in the numerator and quantity of fuel (gal) in the denominator, while the other measure has quantity of fuel (L) in the numerator and distance (km) in the denominator. We need to take this difference into account when doing the conversion.

For example, according to the Environmental Protection Administration, the 2008 Toyota Camry hybrid gets 34 mpg on the highway. What is the equivalent in liters per 100 km?

SOLUTION

$$34 \text{ mpg} = 34 \times \frac{1 \text{ mi}}{1 \text{ gal}}$$

$$\approx 34 \times \frac{1.609 \text{ km}}{3.785 \text{ L}} = 34 \times \frac{1.609}{3.785} \times \frac{\text{km}}{\text{L}} \approx 14.45 \frac{\text{km}}{\text{L}} = \frac{14.45}{1} \times \frac{100 \text{ km}}{100 \text{ L}}$$

$$= \frac{1}{\frac{1}{14.45}} \times \frac{100 \text{ km}}{100 \text{ L}}$$

$$\approx \frac{100 \text{ km}}{6.9 \text{ L}},$$

or 6.9 L per 100 km. The key steps in the solution are to multiply both units by 100, then divide both numerator and denominator of the fraction by 14.45, so as to get exactly 100 km in the numerator of the result.

18.3 Scaling a Mountain

Gravity exerts an enormous effect on the size and shape that objects and beings can assume. **Weight** (force under Earth's gravity) is the reading at sea level on a scale (such as your bathroom scale) *calibrated in pounds or kilograms of mass.*

Suppose that the two cubes in Figure 18.3 on p. 573 are made of steel and that the first is 1 ft on a side and the second is 3 ft on a side. A cubic foot of steel weighs about 500 lb; we say that the **density** of steel is 500 lb per cubic foot, or 500 lb/ft^3. The cube 1 ft on a side weighs 1 ft^3 × 500 lb/ft^3 = 500 lb. The weight W of an object of volume V and uniform density D is

$$W = DV$$

Each cube's bottom face supports the weight of the entire cube. **Pressure** is the force per unit area, so the pressure exerted on the bottom face by the weight of the cube is equal to the weight of the cube divided by the area of the bottom face, or

$$P = \frac{W}{A}$$

The first cube weighs 500 lb and has a bottom face with area 1 ft^2, so the pressure exerted on this face is 500 lb/ft^2.

The second cube is 3 ft on a side. The area of the bottom face increases with the square of the linear scaling factor, so it is $3^2 \times 1$ ft^2 = 9 ft^2. As we saw earlier, volume goes up with the cube of the linear scaling factor. So this larger cube has a volume of $3^3 \times 1$ ft^3 = 27 ft^3. Because both cubes are made of the same steel, the larger cube has 27 times as much steel as the smaller one. Hence, it weighs 27 times as much as the smaller cube, or 27 × 500 lb = 13,500 lb.

When we divide this weight by the area of the bottom face (9 ft^2), we find that the pressure exerted on the bottom face is 1500 lb/ft^2, or three times the pressure on the bottom face of the original cube. This makes sense because over each 1 ft^2 area stands 3 ft^3 of steel. In general, if the linear scaling factor for the cube is L, the pressure on the bottom face is L times as much. Using the notation of proportionality, we have $A \propto L^2$ and $W \propto V \propto L^3$, so

$$P = \frac{W}{A} \propto \frac{L^3}{L^2} \propto L$$

EXAMPLE 4 ▪ What About a 10-Foot Cube?

SOLUTION If we scale the original cube of steel up to a cube 10 ft on a side, then the dimensions are

$$10 \text{ ft} \times 10 \text{ ft} \times 10 \text{ ft}$$

The total volume is

$$\begin{aligned} V &= \text{length} \times \text{width} \times \text{height} \\ &= 10 \text{ ft} \times 10 \text{ ft} \times 10 \text{ ft} = 1000 \text{ ft}^3 \end{aligned}$$

The weight of the cube is

$$W = D \times V$$
$$= \frac{500 \text{ lb}}{\text{ft}^3} \times 1000 \text{ ft}^3 = 500{,}000 \text{ lb}$$

The area of the bottom face is

$$A = \text{length} \times \text{width}$$
$$= 10 \text{ ft} \times 10 \text{ ft} = 100 \text{ ft}^2$$

The pressure on the bottom face is

$$P = \frac{W}{A} = \frac{500{,}000 \text{ lb}}{100 \text{ ft}^2} = 5000 \text{ lb/ft}^2$$

This is 10 *times*–not "10 times *more* than"–the pressure on the bottom face of the original 1-ft cube.

At some scale factor, the pressure on the bottom face will exceed the steel's ability to withstand that pressure–and the steel will deform under its own weight. That point for steel is reached for a cube about 3 miles (mi) on a side–the pressure exerted by the cube's weight exceeds the **crushing strength** of steel, which under Earth's gravity is about 7.5 million lb/ft^2. Because 3 mi = 3 × 5280 ft = 15,840 ft, a 3-mi-long cube of steel would be more than 15,000 times as long as the original 1-ft cube; that is, the linear scaling factor is more than 15,000. The pressure on the bottom face of the cube would therefore be more than 15,000 times as much as for the 1-ft cube, or more than 15,000 × 500 lb/ft^2 = 7.5 million lb/ft^2.

EXAMPLE 5 ■ What About Taipei 101?

Taipei 101 in Taiwan, completed in 2004 and named for its 101 floors (see Spotlight 18.1), is the world's tallest skyscraper, at 1671 ft (how much is that in meters?), not counting radio and television antennas. What is the pressure at the bottom of its walls?

SOLUTION The building is made of reinforced concrete, which weighs 160 lb/ft^3. Although the building tapers a bit toward the top, we are not far off if we model it as straight up and down, that is, as a rectangular solid. Consider one of its supporting walls. The volume of the wall is its height H times the area A of its base, or $V = HA$. The weight of the wall is $W = DV = DHA$. The pressure at the bottom is

$$P = \frac{W}{A} = \frac{DV}{A} = \frac{DHA}{A} = DH$$
$$= \frac{160 \text{ lb}}{\text{ft}^3} \times 1671 \text{ ft} = \frac{267{,}360 \text{ lb}}{\text{ft}^2}$$

or approximately 270,000 lb/ft^2.

So the pressure at the bottom of the wall from the wall's weight alone is about 270,000 lb/ft^2. That's not counting the contents of the tower, which also must be supported!

Could we have a Super Taipei Tower 10 times as high? The bottom of its walls would have to support about 10 × 270,000 lb/ft^2 = 2.7 million lb/ft^2. The crushing strength of reinforced concrete under Earth's gravity is 8.5 million lb/ft^2, which would leave some safety margin.

If built as planned, the Freedom Tower in New York City, whose cornerstone was laid July 4, 2004, and which was scheduled to be completed in 2009, would be taller still–1776 ft–but contain only 60 floors, and be topped by a 276-ft spire.

Burj Dubai ("Dubai Tower"), a skyscraper in the United Arab Emirates scheduled for occupancy in late 2009, is projected to rise to 818 m (2,684 ft) and have 160 floors, including a spire at the top.

SPOTLIGHT 18.1 A Mile-High Building?

In 1956, the famous American architect Frank Lloyd Wright (1867–1959) proposed a mile-high tower for the Chicago lakefront. In the text, we focus on the problem of holding up the weight of such a structure.

But there are other limits to the height of a building. For example, the bending of the building in the wind, which can go up dramatically with height, can be controlled by making the building stiffer.

The terrorist destruction of the World Trade Center towers in 2001—resulting not from the aircraft impacts but from the subsequent fires—revealed a vulnerability in the towers' structure.

Even if designed to better resist fires and impacts, however, a mile-high building might not be practical. For example, the enormous number of people (perhaps 100,000) living, working, or visiting in such a building would create enormous traffic problems (pedestrian, parking, deliveries) for blocks around.

Cost per square foot of usable area is an important consideration. Even if the building did not taper, the space in the upper floors might not justify their additional expense. With increasing height, an increasingly larger proportion of the cross-sectional area of all floors must be devoted to services, such as elevators, plumbing, and conduits for heat and air-conditioning. But everyone entering the building and going to any floor needs to start in an elevator on the ground floor, so there must be more elevators and more elevator shafts. In an emergency evacuation, the people must walk down!

Some architects, however, maintain that the main limit on height of a building is human physiology. Differences in air pressure between the top and bottom of a building limit how fast elevators can rise or drop without discomfort to passengers, thereby enforcing long travel times for "vertical commuters." Human psychology might also present some limits.

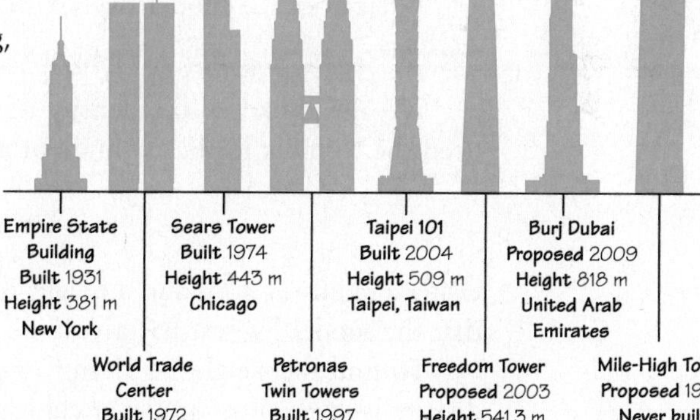

EXAMPLE 6 ■ How High Can a Mountain Be?

The height of mountains too is limited, by gravity, their composition, and their shape. How tall can a mountain be?

SOLUTION We build a simple mathematical model of a mountain. Suppose that it is made of granite, a common material, with uniform density. Granite weighs 165 lb/ft^3 and has a crushing strength of about 4 million lb/ft^2.

In the interests of both realism and simplicity, we assume that the mountain is a solid cone whose width at the base is the same as its height. Let's model Mount Everest, the tallest mountain on Earth, at about 6 mi high. The base, then, is a circle with diameter (distance across) 6 mi. The radius (half the diameter) is 3 mi (Figure 18.4). Because we took round numbers (6 mi) for the height and width, we record as significant only the first digit or two of the results of the calculations.

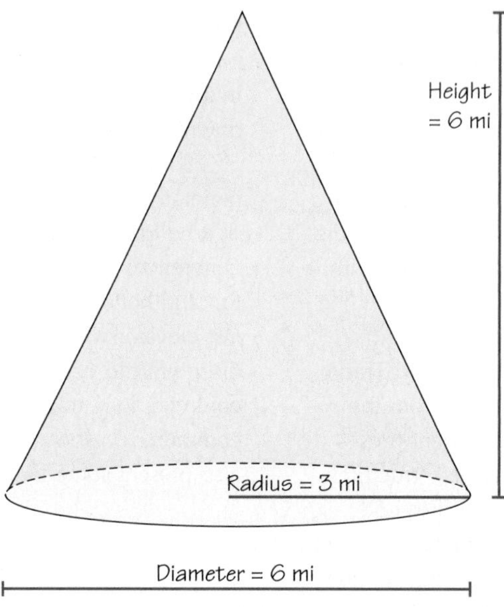

FIGURE 18.4 Model of Mount Everest as a cone of granite.

What does the model Everest weigh? The relevant formula is $W = DV$, or

$$\text{weight} = \text{density} \times \text{volume}$$

We already know the density of granite (165 lb/ft^3), so to find the weight we need the formula for the volume of a cone of radius r and height h:

$$V = \text{volume} = \frac{1}{3}\pi r^2 h$$

Using a radius of 3 mi and a height of 6 mi and π (pi) approximately 3.14, we find that the model Everest has a volume of about 57 mi^3.

To find the weight of 57 mi^3 of granite, we need to convert units, because the density is given in pounds per cubic foot (lb/ft^3). Let's convert to units of feet:

$$\begin{aligned} 1 \text{ mi}^3 &= 1 \text{ mi} \times 1 \text{ mi} \times 1 \text{ mi} \\ &= 5280 \text{ ft} \times 5280 \text{ ft} \times 5280 \text{ ft} \\ &\approx 1.5 \times 10^{11} \text{ ft}^3 \end{aligned}$$

Thus

$$57 \text{ mi}^3 \approx 57 \times 1.5 \times 10^{11} \text{ ft}^3 \approx 8.6 \times 10^{12} \text{ ft}^3$$

So we have

$$\begin{aligned} W = \text{weight of mountain} &= \frac{165 \text{ lb}}{\text{ft}^3} \times 8.6 \times 10^{12} \text{ ft}^3 \\ &\approx 1.4 \times 10^{15} \text{ lb} \\ &\approx 1.4 \text{ quadrillion lb} \end{aligned}$$

Now that we know the weight of the mountain, we want to find the pressure on the base of the cone and compare it with the crushing strength of granite. (Everest is standing, so if our model is any good, that pressure will be below the crushing strength.) Physics tells us that the weight of the mountain is spread evenly over the base of the cone (though we are oversimplifying the geology underlying mountains). Because

$$P = \frac{W}{A} \quad \left(\text{pressure} = \frac{\text{weight}}{\text{area}}\right)$$

we need to calculate the area of the base of the cone. The shape is a circle, and the familiar formula

$$A = \text{area} = \pi r^2$$

gives an area of 28 mi^2 for a radius of 3 mi.

Once again, we need to convert units to express the pressure in pounds per square foot, the units in which the crushing strength is expressed. We get

$$A = \text{area} = 28\ \text{mi}^2 = 28 \times 5280\ \text{ft} \times 5280\ \text{ft} \approx 7.8 \times 10^8\ \text{ft}^2$$

Then

$$\begin{aligned} P &= \frac{W}{A} \\ &= \frac{1.4 \times 10^{15}\ \text{lb}}{7.8 \times 10^8\ \text{ft}^2} \\ &= 1.8 \times 10^6\ \text{lb/ft}^2 \end{aligned}$$

This number is about half the crushing strength of granite under Earth's gravity, 4×10^6 lb/ft^2.

For a mountain to come close to the limitation of the crushing strength of granite, it would have to be only about 10 mi high, not quite twice as high as Everest. Other physical considerations suggest a maximum height of at most 15 mi. That no current mountains are that high may be a consequence of Earth's high amount of volcanic activity and the structural deformation of Earth's crust. ■

What about mountains made of other materials–glass, ice, wood, old cars? They couldn't be nearly as high. The pressure would cause glass to flow, ice to melt, and old cars to compact. What about mountains on another planet? Or on an asteroid? Their potential height depends on the gravity there.

18.4 Sorry, No King Kongs

Unfortunately, the resistance of bone to crushing is not nearly as great as that of steel or granite. This fact helps to explain why there couldn't be any King Kongs (unless they were made of steel or granite!). A King Kong scaled up by a factor of, say, 20 would weigh $20^3 = 8000$ times as much. Though the weight increases with the cube of the linear scaling factor, the ability to support the weight–as measured by the cross-sectional area of the bones, like the area of the bottom face of the cube in Figure 18.3–increases only with the square of the linear scaling factor.

These simple consequences of the geometry of scaling apply not only to supermonsters but also to other objects, such as trees.

FIGURE 18.5 Even giant sequoias can grow no taller than their form and materials allow. (*Michael Rothman.*)

EXAMPLE 7 ■ How Tall Can a Tree Be?

Galileo suggested that no tree could grow taller than 300 ft (see Spotlight 18.2). The world's tallest trees are giant sequoias (Figure 18.5), which grow only on the West Coast of the United States and hence were unknown to Galileo. The tallest known today is 379 ft.

What limits the height of a tree? If the roots do not adequately anchor it, a tall tree can blow over. (This happened in 1990 to the world's then-tallest tree, the Dyerville Giant, a giant sequoia in Humboldt Redwoods State Park in California.) The tree could buckle or snap under its own weight and the force of a strong wind. The wood at the bottom will crush if there is too much weight above. Finally, there is a limit to how far the tree can lift water and minerals from the roots to the leaves.

SOLUTION Could a tree be a mile high? To make a rough estimate of the pressure at the base of the tree due to gravity, let's model the tree as a perfectly vertical cylinder. Over each square foot at the bottom, there is 5280 ft^3 of cells of wood, which weighs about half as much as water. A convenient fact of the metric system is that water weighs just about 1 gram (g) per cubic centimeter. So, to calculate the weight, we first translate 1 ft^3 into metric measurement:

$$1 \text{ ft}^3 = (12 \text{ in.})^3 = (12 \times 2.54 \text{ cm})^3 \approx 28{,}317 \text{ cm}^3$$

So, 1 ft^3 of water weighs about

$$28{,}317 \text{ g} = 28.317 \text{ kg} = 28.317 \times 2.205 \text{ lb} \approx 62.44 \text{ lb}$$

Consequently, 5280 ft^3 of water weighs about 5280×62.4 lb $\approx$ 330,000 lb. The weight of the same volume of wood is about half as much, or about 165,000 lb. Therefore, the pressure at the bottom of the tree would be about 165,000 lb/ft^2.

This is an overestimate, because we assumed that the tree does not taper. A tree that tapers steadily looks like an elongated cone; using a more realistic cone model (as we did in the last section for a mountain), you would find that the pressure at the bottom of the tree would be one-third of 164,000 lb/ft^2, or about 55,000 lb/ft^2.

A biological organism needs a safety factor of at least two to four times the absolute minimum physical limits, so a mile-high tree would need from 110,000 to 220,000 lb/ft^2 of upward pressure for water and minerals. Tension in the string of water molecules from root to leaf ranges from 80,000 to 3.2 million lb/ft^2, for different kinds and heights of trees, so this consideration does not rule out mile-high trees.

At more than about 500 lb/in.^{2} = 70,000 lb/ft^2, though, the bottom of the tree would begin to crush under the weight above. On this basis, a mile-high tree is barely feasible, with little margin of safety. However, researchers who hauled themselves up to the top of the tallest trees in 2004 found a much lower limit, at least for giant sequoias. With increasing height, leaves are smaller, dryer, and less efficient at photosynthesis. The researchers estimated that trees can't top out higher than 400 to 427 ft. The tallest reliably measured tree was a North American Douglas fir, measured at 413 ft in 1902.

There are other considerations. The taller the tree, the greater the area from which it must draw water and minerals, for which nearby trees also compete. Moreover, for a tree to grow very tall, it would have to live for a very long time. Evolution and time may select against extremely tall trees, or maybe, for no reason at all, they have just never evolved.

SPOTLIGHT 18.2 Galileo and the Problem of Scale

The Italian physicist Galileo Galilei (1564–1642) was the first to describe the problem of scale, in 1638, in his *Dialogues Concerning Two New Sciences* (in which he also famously discussed the idea of Earth revolving around the Sun):

> You can plainly see the impossibility of increasing the size of structures to vast dimensions either in art or in nature; likewise, the impossibility of building ships, palaces, or temples of enormous size in such a way that their oars, yards, beams, iron-bolts, and, in short, all their other parts will hold together; nor can nature produce trees of extraordinary size because the branches would break down under their own weight, so also would it be impossible to build up the bony structures of men, horses, or other animals so as to hold together and perform their normal functions if these animals were to be increased enormously in height; for this increase in height can be accomplished only by employing material which is harder and stronger than usual, or by enlarging the size of the bones, thus changing their shape until the form and appearance of the animals suggests a monstrosity.
>
> To illustrate briefly, I have sketched a bone whose natural length has been increased three times and whose thickness has been multiplied until, for a correspondingly large animal, it would perform the same function which the small bone performs for its small animal. From the figures shown here you can see how out of proportion the enlarged bone appears. Clearly then if one wishes to maintain in a great giant the same proportion of limb as that found in an ordinary man he must either find a harder and stronger material for making the bones, or he must admit a diminution of strength in comparison with men of medium stature; for if his height be increased inordinately he will fall and be crushed under his own weight. Whereas, if the size of a body be diminished, the strength of that body is not diminished in proportion; indeed the smaller the body the greater its relative strength. Thus a small dog could probably carry on his back two or three dogs of his own size; but I believe that a horse could not carry even one of his own size.

Translated by Henry Crew and Alfonso De Salvo, and published by Macmillan, 1914, and Northwestern University, 1946.

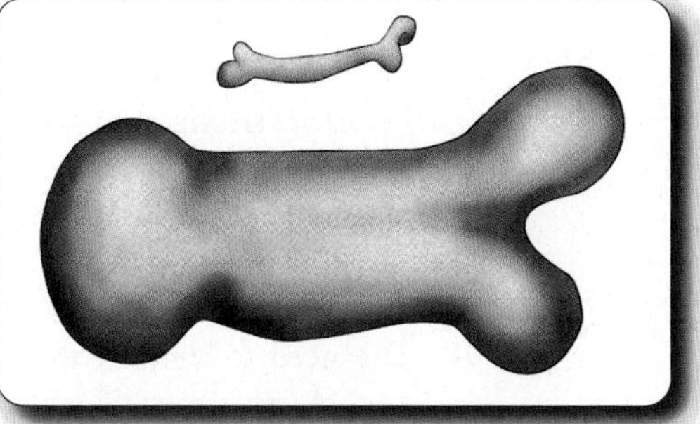

One bone, with another three times as long and thick enough to perform the same function in a scaled-up animal.

18.5 Dimension Tension

A large change in scale forces a change in either materials or form. A major manifestation of the scaling problem is the tension between weight and the need to support it. For example, a real building or machine must differ from a scale model: The balsa wood or plastic of the model would never be strong enough for the real thing, which would need aluminum, steel, or reinforced concrete.

Another way to compensate is to redesign the object to distribute its weight better. Let's go back to the original cube. It supports all its weight on its bottom face. In the version scaled up by a factor of 3, each small cube of the bottom layer has a bottom face supporting that cube's weight plus the weight of the two cubes piled on top of it.

Let's redesign the scaled-up cube, concentrating for simplicity only on the front face, with its nine small cubes. We take the three cubes on top and move them to the bottom, alongside the three already there. We take the three cubes on the second level, cut each in half, and put a half-cube over each of the six ground-level cubes (see Figure 18.6). We have the same volume and weight that we started with, but now there is less pressure on the bottom face of each small cube. Of course, the new design is not geometrically similar to the object that we started with–it's no longer a cube. We have solved the scaling problem by changing the proportions.

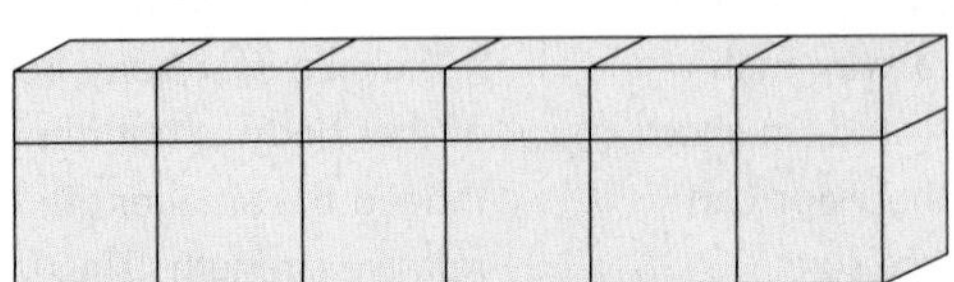

FIGURE 18.6 Nine small cubes rearranged to support greater weight.

We observe in nature both strategies for scaling: change of materials and change of form. Small animals (such as insects) do not have bony internal skeletons. Larger animals generally do. Animals made of similar materials but differing greatly in size, such as a mouse and an elephant, must differ in shape. If a mouse were scaled up to the size of an elephant, it would need the disproportionately thicker legs of the elephant to support its weight and the elephant's thick hide to contain its tissue.

Some dinosaurs, like *Supersaurus* (which weighed 30 tons), had special adaptations to lighten their weight, such as hollow bones, just as some birds have. Hollow bones are stronger, a paradox that Galileo analyzed. Of two bones of the same weight and length, the hollow one is wider across at its midpoint because of the air it contains; and the greater the width, the greater the resistance to fracture.

Falls, Jumps, and Flight

The need to support weight can be thought of as a tension between volume and area. As an object is scaled up, its volume and weight go up together, as long as the density remains constant (for example, no air bubbles introduced into the steel to make it into a Swiss cheese!). At the same time, the ability to support the weight goes up with the cross-sectional area, like the bottom face of the steel cube.

Area-Volume Tension DEFINITION

Area–volume tension is a result of the fact that as an object is scaled up, the volume increases faster than the surface area and faster than areas of cross sections.

Because volume V is proportional to the cube of the linear scaling factor L, we have $V \propto L^3$; taking each side to the one-third power, $L \propto V^{1/3}$. The fact that surface area A is proportional to the square of the linear scaling factor becomes

$$A \propto L^2 \propto (V^{1/3})^2 = V^{2/3}$$

so that surface area scales as the two-thirds power of volume.

In any crowded city, you can observe tension between length, area, and volume. Consider an apartment building that spans a city block. The area of parking spaces on the adjacent streets is proportional to the perimeter of (length around) the building. But the number of cars belonging to people in the building is proportional to the number of apartments, which is proportional to the volume of the building. So the higher the building, the greater the parking tension!

In some cities, zoning tries to help the situation by putting shops on the ground floor, which cuts out one floor of apartments. If the residents' cars are away during the day, customers and employees of the shops can park where the apartment dwellers do at night. A more common solution is an underground garage, usually with several levels (with an area for cars proportional to the volume of the building). However, garages that were designed for one car per apartment have proven inadequate now that families tend to have more than one car.

Other examples of dimensional tension solutions include the old-fashioned diner, with its serving counter in the form of S-shapes to expand its effective length, and your small intestine, which coils its 20-ft length to fit into your abdomen.

Area–volume tension has many other practical consequences, some of them related to our childhood fantasies. We can forget about humans "leaping tall buildings in a single bound," "soaring like an eagle," or diving miles below the sea. Consider the following examples.

EXAMPLE 8 ■ Falls

Area–volume tension affects how animals respond to falling, another of gravity's effects. A mouse may be unharmed by a 10-story fall, and a cat by a two-story fall, but many humans are injured by falling while running, walking, or even just standing.

What is the explanation? The energy acquired in falling is proportional to the weight of the falling object, hence to its volume. This energy must be absorbed either by the object or by what it hits, or must be otherwise dissipated at impact–for example, as sound. The fall is absorbed over part of the surface area of the object, just as the weight of the cube was distributed over its base. With scaling up, volume–hence weight, hence falling energy–goes up much faster than area. As size increases, the hazards of falling from the same height increase. ■

EXAMPLE 9 ■ Jumps

A flea can jump about 2 ft vertically, many times its own height. Many people believe that if a flea were as large as a person, it could jump 1000 ft into the air. Imagining–against our earlier arguments–that there could be so large a flea, we know its limits: A scaled-up flea could jump about the same height as a small flea. The strength of a muscle is proportional to its cross-sectional *area* (see Spotlight 18.3). A jump involves suddenly contracting the muscle through its length, so it turns out that the ability to jump is proportional to the *volume* of muscle. But the volume of the flea and the volume of its leg muscles go up in proportion.

Let's say that a real flea's leg muscles account for 1% of its body. If we scale the flea up to the size of a person (without any change in its form), the enlarged flea's leg muscles will still make up 1% of its body. For either flea, each bit of muscle has the same power: In a jump, it propels 100 times its own weight, and it can do so to the same height. Both the weight of the flea and the power of its legs go up proportionately. In fact, the maximum heights that people, fleas, grasshoppers, and kangaroos can jump are all within a factor of 3 of each other. ■

SPOTLIGHT 18.3 Scaled to Fit

Big isn't always beautiful when it comes to the U.S. military's physical fitness tests.

Paul Vanderburgh, chair of the University of Dayton Dept. of Health and Sport Science, has spent more than a dozen years researching how a person's body mass affects performance on such tests, which consist of distance runs, push-ups, sit-ups and abdominal crunches. The Arnold Schwarzeneggers of the world actually tend to score *lower.*

Vanderburgh emphasizes that some larger people (like Schwarzenegger) have more muscle, not more fat. Nevertheless, he and fellow researcher Todd Crowder found that scores for larger and heavier (though muscular) men and women are 15–20% lower than for their smaller and lighter counterparts. "A person's strength doesn't increase as fast as their size," explains his student Liz Trouten. "The extra muscle that big people have doesn't make up for their size."

Vanderburgh noticed at the U.S. Military Academy that, even at similar fitness levels, smaller cadets tend to score higher on physical fitness tests than larger cadets. "Fitness testing is a big part of cadets' grade point averages, and the stakes are pretty high. The test results affect class rank and even a cadet's first assignment, so it matters a lot how well a cadet does."

For example, a larger cadet with a fitness test score of 256, which Vanderburgh compares to a grade of C+, may not be eligible for certain awards and assignments. However, a smaller cadet with a perfect score of 300 "would get lots of attention."

(Cynthia Johnson/Time Life Pictures/ Getty Images.)

But that doesn't mean the C+ cadet isn't worthy. "In fact, if these two cadets were scale models of each other, these two performances would be biologically the same, and they should receive the same score."

Vanderburgh gives another example using the scale-model approach. Take a woman who is 5 feet 5 inches tall, weighs 130 pounds, and scores a perfect 300 on the fitness test. If she were 5 feet 8 inches tall and 30% heavier, she would score only 250.

To compensate for this body mass "penalty," Vanderburgh and Crowder developed a correction factor, which multiplies the score by a number based on weight, "to place everybody on an even playing field."

This formula is similar to the Flyer Handicap, developed by Vanderburgh and colleague Lloyd Laubach. The handicap adjusts a runner's race time based on age and body weight. "A higher body weight is definitely a handicap for performance, whether it be running a marathon or military physical fitness tests," Laubach said. (A Web calculator for the Flyer Handicap is at http://academic.udayton.edu/PaulVanderburgh/weight_age_grading_calculator.htm.)

Source: adapted from an article by Kristen Wicker in the *University of Dayton Quarterly* (Winter 2006–07) 21–22.

EXAMPLE 10 ▪ Flight

Wouldn't it be nice to be able to fly? Well, you have to be able to stay up. The power necessary for sustained flight is proportional to the **wing loading**, which is the weight supported divided by the area of the wings. We know that in scaling up, weight grows with the cube of the length of the bird or plane, and wing area with the square of the length. So the wing loading is proportional to the length of the flying object.

For example, if a bird or plane is scaled up proportionally by a linear scaling factor of 4, it will weigh $4^3 = 64$ times as much but will have only $4^2 = 16$ times as much wing area. So each square foot of wing must support 4 times as much weight.

Once you're up, you have to keep moving. To stay level, an airborne object must fly fast enough to maintain the lift on the wings. The minimum necessary speed is proportional to the square root of the wing loading. Combining this fact with the first consideration, we conclude that the minimum speed goes up with the square root of the length. A bird scaled up by a factor of 4 must fly $\sqrt{4} = 2$ times as fast. (Hovering helicopters, hummingbirds, and insects maintain lift by moving their wings directly rather than through forward motion.)

Take, for instance, a sparrow, whose minimum speed is about 20 miles per hour (mph). An ostrich is 25 times as long as a sparrow, so the minimum speed for an ostrich would be $\sqrt{25} \times 20 = 100$ mph. Have you seen any flying ostriches lately? Heavy birds have to fly fast or not at all!

Of course, ostriches are not just scaled-up sparrows, nor are eagles. Larger flying birds have disproportionately larger wings than a sparrow to keep the wing loading down. The largest animal ever to take to the air was *Quetzalcoatlus northropi,* a flying reptile of 65 million years ago, with a wingspan of 36 ft and a weight of about 100 lb.

You have to stay up, you have to keep moving–and you have to get up there. Here basic aerodynamics imposes further limits. Paleontologists originally thought that *Q. northropi* weighed 200 lb and had a 50-ft wingspan. Even though that works out to about the same wing loading as for 100 lb and a wingspan of 36 ft, other considerations from aerodynamics show that at the larger size, the reptile couldn't have gotten off the ground.

Keeping Cool (and Warm)

Area–volume tension is also crucial to an animal's thermal equilibrium. Both warm-blooded and cold-blooded animals gain or lose heat from the environment in proportion to body surface area.

Warm-Blooded Animals

A warm-blooded animal's basal metabolism, or rate of food intake needed to maintain body heat, depends primarily on its surface area, the temperature of its environment, and the insulation provided by its coat or skin. Other factors being equal, a scaled-up mammal scales up its food consumption with *surface area* (proportional to the square of the linear scaling factor), *not with volume* (proportional to its cube). For example, a mouse eats about half of its weight in food every day, while a human consumes only about one-fiftieth of its own weight, because the mouse has more surface area per unit volume.

Thus, the metabolic rate should be proportional to the surface area. Using proportionality notation, we can find how the metabolic rate changes with the mass of the animal. We know that mass is proportional to volume, which in turn is proportional to the cube of length, or

$$M \propto V \propto L^3$$

Taking each side to the one-third power, we have

$$M^{1/3} \propto V^{1/3} \propto L \qquad \text{or} \qquad L \propto M^{1/3}$$

Meanwhile, the metabolic rate (call it R) is proportional to surface area, so

$$R \propto A \propto L^2 \propto (M^{1/3})^2 = M^{2/3}$$

So, based on area–volume tension, we would expect metabolic rate to scale as the two-thirds power of body mass. But it doesn't–instead, it scales as the *three-quarters* power of body mass, that is, $R \propto M^{3/4}$. The least-squares line (see section 6.4) through the points in the "mouse-to-elephant" curve of Figure 18.7 has a slope of 0.74, very close to three-quarters. (The logarithmic coordinates used in this graph are explained in section 18.6.)

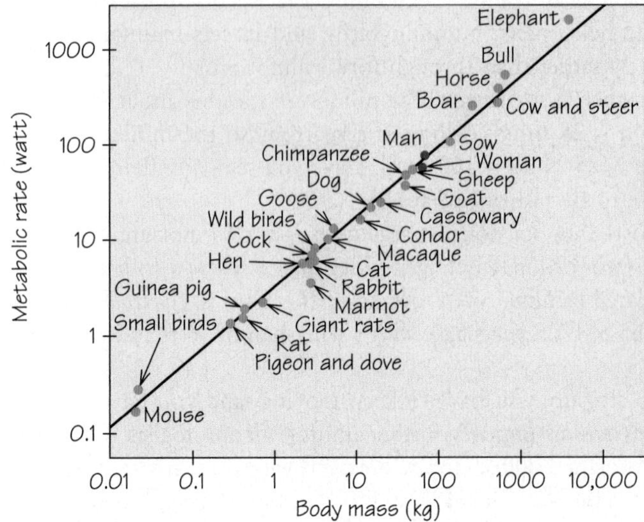

FIGURE 18.7 Metabolic rates for mammals and birds, when plotted against body mass on logarithmic coordinates, tend to fall along a single straight line. (*Adapted from Benedict, 1938.*)

Why the difference from the two-thirds that area–volume tension would predict? And does the small difference between two-thirds and three-quarters matter? The answers lie in further considerations from geometry, physiology, and physics. A plant or animal needs a network of vessels (like the blood system) to transport resources to, and wastes away from, every part of the animal's tissues. The terminal branches (in the blood system, capillaries) tend to be just about the same size in all species, for reasons of the physics involved. To minimize the energy involved in transport, the network of vessels needs to be organized as a fractal-like tree, with smaller and smaller vessels branching off (see section 19.5 to learn about fractal patterns). With same-size smallest branches at the ends, minimization of energy demands that the metabolic rate scale as the three-quarters power of body mass. Fractal branching makes it possible for the circulatory system of a whale, with 10^7 times the mass of a mouse, to have only 70% more branches than the mouse has.

EXAMPLE 11 ■ Dives

Sperm whales (and some other species) regularly hold their breath and stay under water for an hour. Why can't we? In part, because we aren't as large as whales. A mammal's breath-holding ability depends on how much air it can hold in its lungs, which is proportional to its mass. It also depends on how fast it uses up air–in other words, on its metabolic rate, which is proportional to the three-quarters power of its mass. Hence, the limit of duration of a dive should be proportional to

$$\frac{M}{M^{3/4}} = M^{1/4}$$

For a 90,000-lb sperm whale, this limit is proportional to $90{,}000^{1/4} = 17.3$, while the corresponding figure for a 150-lb human is $150^{1/4} = 3.5$. So the sperm whale should be able to hold its breath for about $17.3/3.5 \approx 5$ times as long. However, humans cannot hold their breath for one-fifth of an hour (12 minutes)! This fact tells us that the whale has special adaptations to make long dives possible. The stars of the 2005 film *The March of the Penguins*, emperor penguins, weigh 80–90 lb, but can dive for as long as 20 minutes. Their special adaptations are more blood per pound of body weight, an abundance of myoglobin (which can store oxygen) in their tissues, and slowing their heart rate during dives.

Cold-Blooded Animals

Mammals and birds regulate their metabolism and maintain a constant internal body temperature. Cold-blooded animals, such as alligators and lizards, have a somewhat different problem. They absorb heat from the environment for energy, but they must also dissipate any excess heat to keep their temperatures below unsafe levels. The amount of heat that must be gained or lost is proportional to total volume, because the entire animal must be warmed or cooled. But the heat is exchanged through the skin, so the rate is proportional to surface area.

Dimetrodon was a large mammal-like reptile that roamed present-day Texas and Oklahoma 280 million years ago (see Figure 18.8). *Dimetrodon* had a great "sail," or fan, on its back. As an individual grew, and as the species evolved, the sail grew. But it did not grow according to *geometric similarity,* the kind of growth we refer to as **proportional growth**.

Proportional Growth — DEFINITION

Proportional growth is growth according to geometric similarity, where the length of every part of the organism enlarges by the same linear scaling factor.

FIGURE 18.8 *Dimetrodon* may have evolved a sail to absorb and dissipate heat efficiently. (*Robert F. Walters.*)

Instead, the area of *Dimetrodon*'s sail grew in proportion to the volume of the animal, a fact that strongly suggests to paleontologists that the sail was a temperature-regulating organ. Larger specimens of *Dimetrodon* didn't look like scaled-up smaller ones. We would say that the sail grew disproportionately compared to the rest of the animal. An individual twice as long would have eight (2^3) times as much weight and volume and a sail with eight times as much area. If it had grown according to geometric similarity, the sail would have been twice as high and twice as wide, and hence would have had only four times as much area.

Dimetrodon was a large animal, but heat regulation is even more important for small animals–like human babies, they can lose heat quickly because of their high ratio of surface area to volume. Paleontologists believe that birds evolved from dinosaurs and that feathers are modified reptilian scales. The wings of birds and insects may have evolved not for flight but as temperature-control devices.

Some scientists have speculated that African Pygmies are small in part because a small body can better lose heat in the hot, humid climate of the Ituri Forest in the Congo, where Pygmies live. The discovery announced in late 2004 of "hobbit-sized" people (1 m tall) who lived on the island of Flores in Indonesia 13,000 years ago, suggests another explanation. Being marooned on the island with a limited food supply (they hunted pygmy elephants) made large size–and a corresponding need for more calories–a disadvantage.

Other scientists have suggested that ancestors of human beings began walking on two legs in part to keep cool in a hot climate. Walking upright exposes less body area to the rays of the sun than walking on all fours and also reduces the amount of water needed by about one-half.

18.6 How to Grow

A large change of scale forces adaptive changes in materials or form. However, within narrow limits–in most cases, up to a factor of 2–creatures can grow according to geometric similarity. That is, they can grow proportionally, so that their shape is preserved. A striking example of such growth by a far greater factor is the chambered nautilus (*Nautilus pompilius*). Each new chamber that it adds to its shell is larger than, but geometrically similar to, the previous chamber and also similar to the shape of the shell as a whole–an *equiangular,* or *logarithmic,* spiral (see Figure 18.9).

FIGURE 18.9 A chambered nautilus shell. (*Photodisc/Punchstock.*)

Most living things grow over the course of their lives by a factor greater than 2. We've seen with *Dimetrodon* that a big specimen was not just a scaled-up small one. Nor is a human adult simply a scaled-up baby: A baby's head is relatively much larger than an adult's, and its arms are disproportionately shorter. In growth from baby to adult, the body does not scale up as a whole. Different parts of the body scale geometrically, each with a different linear scale factor. That is, a baby's eyes grow to perhaps twice their original size, while the arms grow by a factor of about four.

Although the laws for growth can be much more complicated than for proportional growth (or even for the allometric growth that we discuss later), more sophisticated mathematics–for example, differential geometry, the geometry of curves and surfaces–permits analysis of complex and interlocking scalings. For a model of the process in which a baby's head changes shape to grow into an adult head, we can use graph paper: First, we put a picture of the baby's skull on graph paper. Then we determine how to deform the grid until the pattern matches an adult skull (see Figure 18.10 and Spotlight 18.4). The same idea lies at the heart of computerized "morphing," the process in which the face of one person can be changed smoothly into the face of another, with different scalings for different parts of the face.

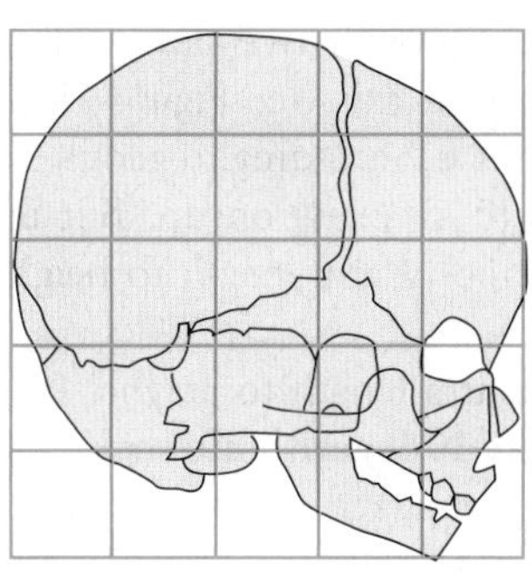

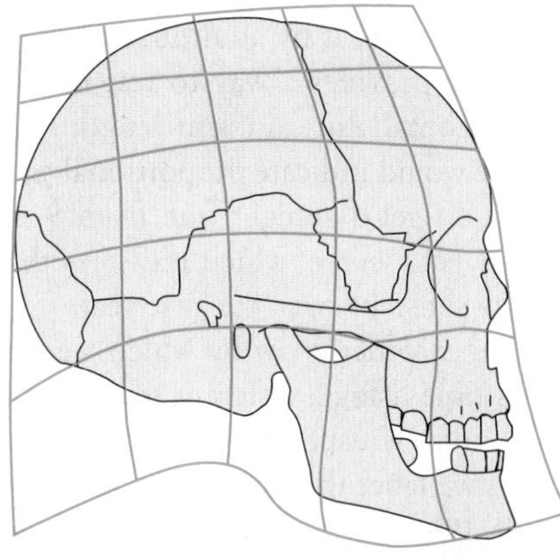

FIGURE 18.10 Modeling the changes in the shape of a human head from infancy to adulthood.

SPOTLIGHT 18.4 Helping to Find Missing Children

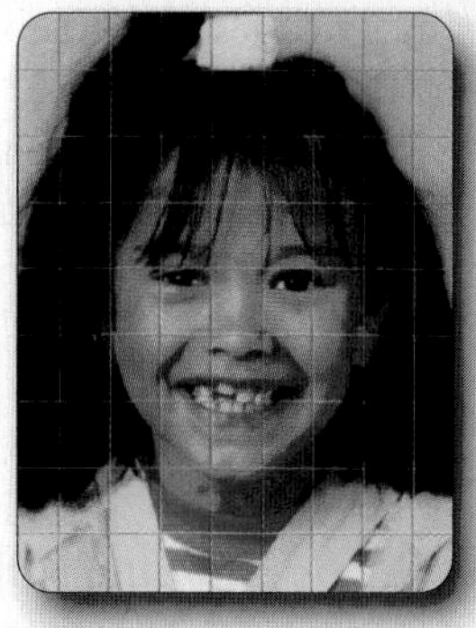

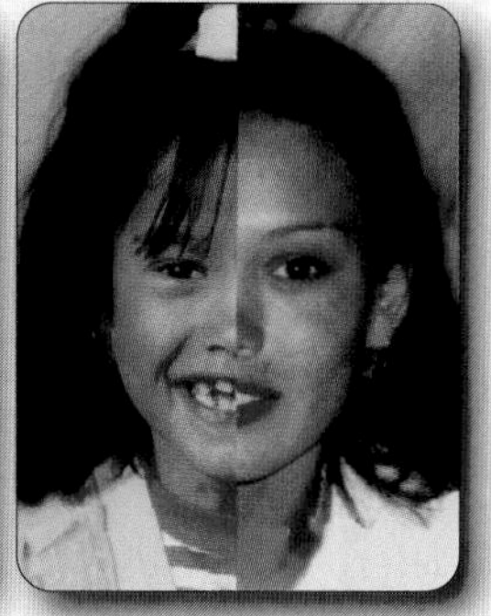

Photograph at age 7; photo showing one side of the child's face and one side of her mother's, to which the child's photo will be stretched.

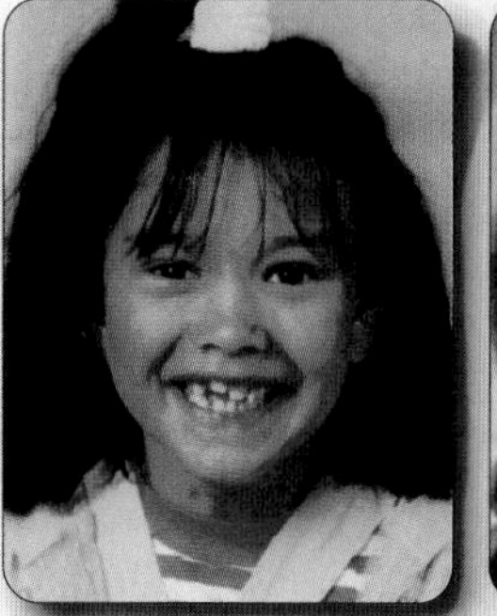

Stretching, merging, and age progression to age 17. *(National Center for Missing and Exploited Children, www.missingkids.com.)*

What does a child who was kidnapped at age 3 look like now, six years later?

At the National Center for Missing and Exploited Children (NCMEC) in Arlington, Virginia, a computer and a more sophisticated version of the graph-paper technique are used to answer such questions. Computer age-progression specialists scan photographs of both the missing child at age 3 and an older sibling or a biological parent at age 9 into a computer. Then the face of the 3-year-old is stretched, depending on age, to reflect craniofacial growth and merged with the image of the sibling or parent at 9 years old. The result is a rough idea of what the missing child may look like. As mathematicians and biologists refine their models of how faces change over time, this technique will improve. It may even become possible to gain an idea of how a child may look at age 40 or 65.

Allometric Growth

If we measure the arm length or head size for humans of different ages and compare these measurements with body height, we observe that humans do not grow proportionally, that is, in a way that maintains geometric similarity. The head of a newborn baby may be one-third of the baby's length, but an adult's head is usually close to one-seventh of the individual's height. The arm, which at birth is one-third as long as the body, is by adulthood closer to two-fifths as long (see Figure 18.11a).

Graphing provides a way to test for differential growth. We plot body height on the horizontal axis and arm length on the vertical axis (see Figure 18.11b). A straight line would indicate proportional growth, that is, according to geometric similarity. We do get a straight line from 9 months (0.75 years) on up. But up to 9 months, we get a curve, which indicates that the ratio of arm length to height does not remain constant over the first year.

Is there an orderly law by which we can relate arm length to height? Let's plot again, this time using a different scale. For this **base-10 logarithmic scale,** we mark off equal units, as usual. But instead of labeling the marked points with 0, 1, 2, 3, and so on, we label them with the corresponding powers of 10: $10^0 = 1$, $10^1 = 10$, $10^2 = 100$, $10^3 = 1000$, and so on, which are also called **orders of magnitude.** Plotting a point on such a scale is not easy, because the point midway between 1 and 10 is not 5.5, but instead is closer to 3. Special graph paper (available in most college bookstores) marks smaller divisions and makes it easier to plot; paper marked with log scales on both axes is called **log-log paper,** while **semilog paper** has a logarithmic scale on just one axis. Also, many computer plotting packages can produce logarithmic scales.

We could use a logarithmic scale for either height or arm length, or for both. Using logarithmic scales for both, as in Figure 18.11c, the data plot closely to a straight line. Looking carefully, we can discern two different straight lines: a steeper one that fits early development (we will see shortly that it has slope 1.2), and a less steep one (with slope 1.0) that fits development after 9 months of age.

The change from one line to another after 9 months indicates a change in pattern of growth. The pattern after 9 months, characterized by the straight line with slope 1, is indeed proportional growth (sometimes called **isometric growth**). For the pattern before 9 months, the slope is 1.2. The fact that it is greater than 1 means that arm length is increasing relatively faster than height. This early growth also follows a definite pattern, called **allometric growth.**

Allometric Growth — DEFINITION

Allometric growth is growth of the length of one feature at a rate proportional to a power of the length of another.

In geometric scaling, area grows according to the square (second power), and volume according to the cube (third power) of length, so they grow allometrically with length.

If we denote arm length by y and height by x, a straight-line fit on log-log paper corresponds to the algebraic relation

$$\log_{10} y = B + a \log_{10} x$$

where a is the slope of the line and B is the point where the graph crosses the vertical axis. If we raise 10 to the power of each side, we get

$$y = bx^a$$

where $b = 10^B$. This equation describes a **power curve**: y is a constant multiple of x raised to a certain power.

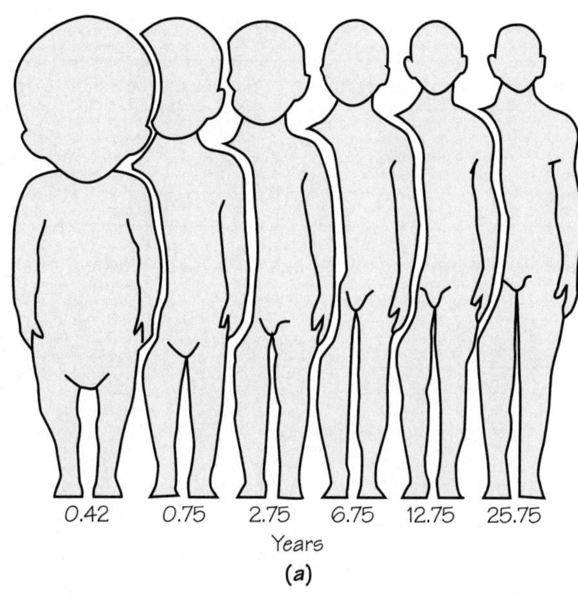

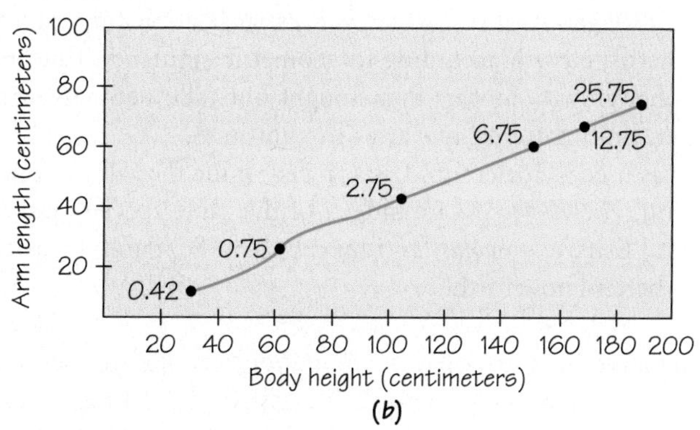

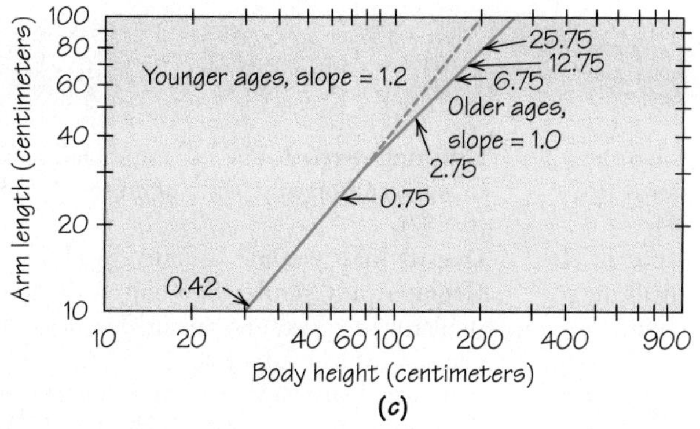

FIGURE 18.11
(a) The proportions of the human body change with age. (b) A graph of human body growth on ordinary graph paper. The numbers shown beside the points indicate the age in years; they correspond to the stage of human development shown in part (a). (c) A graph of human body growth on log-log paper.

EXAMPLE 12 ■ Finding the Power of Growth

How do we arrive at those slopes of 1.0 and 1.2 in Figure 18.11?

SOLUTION You could use statistical software or your calculator to find the equation of the least-squares regression line (section 6.4) through the points on the log-log plots. Here we find approximate values for the slope a for each line from the coordinates of the points at the ends of the lines, for ages 0.42, 0.75, and 25.75. The observations and the corresponding logarithms are:

Age	Height	Log (Height)	Arm Length	Log (Arm Length)
0.42	30.0	1.48	10.7	1.03
0.75	60.4	1.78	25.1	1.40
25.75	180.8	2.26	76.9	1.89

The slope for the line from age 0.42 to age 0.75 is the vertical change over the horizontal change in terms of log units:

$$\frac{\log 25.1 - \log 10.7}{\log 60.4 - \log 30.0} = \frac{1.40 - 1.03}{1.78 - 1.48} = \frac{0.37}{0.30} \approx 1.2$$

The slope for the line from age 0.75 to age 25.75 is

$$\frac{\log 76.9 - \log 25.1}{\log 180.8 - \log 60.4} = \frac{1.89 - 1.40}{2.26 - 1.78} = \frac{0.49}{0.48} \approx 1.0$$

So $a = 1.2$ up to 9 months, and $a = 1.0$ after 9 months. Up to 9 months, arm length grows according to $(\text{height})^{1.2}$. After 9 months, arm length grows according to $(\text{height})^{1.0}$, and we get $y = bx^{1.0}$, which is a linear relationship describing proportional growth, that is, growth according to geometric similarity. On ordinary graph paper, proportional growth appears as a straight line, allometric growth as a curve. On log-log paper, both patterns appear as straight lines.

Allometry was used by paleontologists to determine that all specimens (just six!) of the earliest bird, *Archaeopteryx*, are indeed of the same species, and that the puzzling minute fossil fish *Palaeospondylus* (found only in Scotland) is probably just the larval stage of a better-known fish.

In this chapter we have explored the limitations on life imposed by dwelling in three dimensions. In Chapters 19 and 20, we will see that dimensionality also imposes surprising limits on artistic creativity in devising patterns.

REVIEW VOCABULARY

Allometric growth A pattern of growth in which the length of one feature grows at a rate proportional to a power of the length of another feature. (p. 594)

Area–volume tension A result of the fact that as an object is scaled up, the volume increases faster than the surface area and faster than areas of cross sections. (p. 586)

Base-10 logarithmic scale A scale on which equal divisions correspond to powers of 10. (p. 594)

Crushing strength The maximum ability of a substance to withstand pressure without crushing or deforming. (p. 580)

Density Mass per unit volume. (p. 579)

Geometrically similar Two objects are geometrically similar if they have the same shape, regardless of the materials of which they are made. They need not be the same size. Corresponding linear dimensions must have the same factor of proportionality. (p. 572)

Isometric growth Proportional growth. (p. 594)
Linear (length) scaling factor The number by which each linear dimension of an object is multiplied when it is scaled up or down; that is, the ratio of the length of any part of one of two geometrically similar objects to the length of the corresponding part of the second. (p. 572)
Log-log paper Graph paper on which both the vertical and the horizontal scales are logarithmic scales; that is, the scales are marked in orders of magnitude 1, 10, 100, 1000, . . . , instead of 1, 2, 3, 4, ... (p. 594)
Orders of magnitude Powers of 10. (p. 594)
Power curve A curve described by an equation $y = bx^a$, so that y is proportional to a power of x. (p. 595)
Pressure Force per unit area. (p. 579)
Problem of scale As an object or being is scaled up, its surface and cross-sectional areas increase at a rate different from its volume, forcing adaptations of materials or shape. (p. 571)
Proportional growth Growth according to geometric similarity, where the length of every part of the organism enlarges by the same linear scaling factor. (p. 591)
Semilog paper Graph paper on which only one of the scales is a logarithmic scale. (p. 594)
Weight Force under gravity. (p. 579)
Wing loading Weight supported divided by wing area. (p. 588)

✓ SKILLS CHECK

1. A penny and a nickel are

(a) not geometrically similar, because they are made of different materials.
(b) not geometrically similar, because they are of different sizes.
(c) geometrically similar, because they have the same shape and proportional dimensions.

2. A scale model of a carillon stands 10 in. tall, and the actual carillon stands 100 ft tall. The linear scaling factor of the carillon compared to its model is ________ .

3. If a model car is built to a scale of 1 to 40, and the actual car has a turning circle of 37 ft, what should be the turning circle of the model?

(a) 11.1 in.
(b) 0.925 in.
(c) 1480 ft

4. You want to enlarge a 3-in. by 5-in. photograph to a 12-in. by 20-in. copy. Assuming that the cost of photographic paper is proportional to its area, and that 3-in. by 5-in. reprints cost 40 cents each, you would expect to pay ____________ for the large copy.

5. If a medium 10-in. pizza costs \$8 and a similar 14-in. pizza costs \$14, which is the better buy?

(a) The 10-in. pizza
(b) The 14-in. pizza
(c) They are about the same price per square inch.

6. An artist plans to melt 100 pennies and re-form a larger penny proportional in all dimensions to an ordinary penny. The linear scaling factor of the large penny compared with the ordinary penny is ________ .

7. The actor Elijah Wood, who plays Frodo Baggins in the movie of Tolkien's *Lord of the Rings,* is 5 ft 6 in. tall, but his character is barely 4 ft tall. Put correctly, how much shorter is Frodo than Wood?

(a) 138% shorter
(b) 73% shorter
(c) 27% shorter

8. The distance for the marathon race was established in 1921 as 42.195 km. Converted to the U.S. Customary System, the distance is ____________ .

9. A kilometer is approximately equal in length to

(a) 5 mi.
(b) 3 mi.
(c) 3/5 mi.

10. A weight of 130 lb is approximately the same as ________ kg.

11. A 2-liter bottle contains approximately

(a) 2 quarts.
(b) 1 gallon.
(c) 10 pints.

12. A common speed limit in European neighborhoods is 30 km/h, which is about ________ mph.

13. Coffee costs about \$8 per pound in the United States. If a Canadian dollar (Cdn\$) exchanges for U.S.\$1.02, what is the approximate cost in Canadian dollars of 500 g of coffee?

(a) Cdn\$2
(b) Cdn\$4
(c) Cdn\$8

14. A sculpture weighs 140 lb and is supported by three legs, each of which is 0.5 in. by 0.5 in. by 2 in. high. The legs exert a pressure of ________ lb/in.^{2} on the floor.

15. In comparing flight speeds of birds, an analysis of wing loading leads to the conclusion that

(a) light birds fly faster than heavy birds.
(b) heavy birds fly faster than light birds.
(c) heavy and light birds fly at about the same speed.

16. If an object is scaled linearly so that its volume grows to 8 times its original volume, its surface area is scaled to ______ times its original surface area.

17. In comparing the heights that large and small animals jump, analysis of the impact on scaling leads to the conclusion that

(a) smaller jumping animals can jump much higher than larger jumping animals.
(b) larger jumping animals can jump much higher than smaller jumping animals.
(c) all jumping animals jump to about the same height.

18. Assuming that a catfish maintains the same shape and proportions as its grows, and a catfish 8 in. long weighs about 1 lb, a 2-lb catfish is about _____ in. long.

19. The population of Mexico grew by 2.4% in 2002, by another 2.4% (of the 2002 population) in 2003, and by another 2.4% (of the 2003 population) in 2004. The population was growing

(a) proportionally.
(b) allometrically.
(c) by a constant amount each year.

20. The base-10 logarithm of 72 is approximately ____ .

CHAPTER 18 EXERCISES

■ Challenge ◆ Discussion

Most of these exercises require a calculator; one with square roots will suffice.

18.1 Geometric Similarity

1. Your digital camera likely takes pictures with an aspect ratio of 4 to 3, meaning that the longer side is 4/3 times as long (in pixels) as the shorter side. For example, you probably can take a "small" picture with 640 pixels by 480 pixels, or perhaps a "large" picture with 2592 pixels by 1944 pixels (for a total of 2592 × 1944 pixels, or just a little more than 5 megapixels). Photographic prints from your digital camera are available in various sizes of paper, quoted in inches: 4 × 6, 5 × 7, and 8 × 10.

(a) Which of the paper sizes, if any, is geometrically similar to the original digital image?
(b) If a 4 × 6 print is instead made by scaling the shorter side of the digital image to be exactly 4 in., how long should the longer side of the image be on the print?
(c) If a 4 × 6 print is made by scaling the longer side of the digital image to be exactly 6 in., how long should the shorter side of the image be on the print? (*Hint*: The paper isn't wide enough!)

2. The area of a circle of radius r is πr^2. Expressed in terms of the diameter, $d = 2r$, the area is $\frac{1}{4}\pi d^2$. If we apply a linear scaling factor L to the diameter, then the area of the scaled circle–as in the case of the square that we considered in the text–changes with L^2, the square of the linear scaling factor. A natural application of this idea is with pizza. The prices at Domenico's pizza restaurant near Beloit College are \$7, \$8, \$9, and \$10, respectively, for small (10-in.), medium (12-in.), large (14-in.), and extra-large (16-in.) cheese pizzas.

(a) What is the linear scaling factor for an extra-large pizza compared with a small one?
(b) How many times as large in area is the extra-large pizza compared with the small one?
(c) How much pizza does each size give per dollar? What "hidden" assumptions are you making about how the pizzas are scaled up?
(d) The corresponding prices for a pizza with "the works" are \$9.95, \$11.95, \$13.95, and \$15.95. Is there any size of these for which you get more pizza per dollar than some size of the cheese pizzas?
(e) All of the prices are, to the nearest cent, exactly 5% lower than 3 years ago! What kind of scaling is that?

3. The human figures in Lego sets are 4 cm tall (without hats or helmets).

(a) What is the linear scaling factor of a Lego figure if it represents a human who is 160 cm tall?
(b) How does the volume of a real human compare with the volume of a Lego figure?
(c) The car in one Lego set is 10 cm long. Using the linear scaling factor in part (a), how long would a real car be?

4. Dollhouses and their furnishings are usually built to a scale of exactly 1 in. to 1 ft, meaning that an item 1 ft long in a real house is 1 in. long in a dollhouse.

(a) What is the linear scaling factor for a dollhouse?
(b) If a dollhouse were made of the same materials as a real house, how would their weights compare?

5. According to *Time* (March 7, 2005), men's brains on average are 10% larger than women's, even though men on average are only 8% taller. (The article mainly discusses the many differences in brain structure that likely outweigh any size differences.) If the brain scales linearly with height, and men are 8% taller, what percentage larger would you expect their brains to be?

6. At our house, we have some 10-in. frying pans and a 12-in. one; the 12-in. one weighs a lot more, never gets

as hot, and cooks food more slowly. Suppose that a 10-in. frying pan weighs 1 lb, apart from its handle. How much would a geometrically similar 12-in. frying pan weigh? How much would it weigh if it had the same thickness of metal as the 10-in. pan?

7. A famous geometry problem of Greek antiquity was *duplication of the cube.* Our knowledge of the history of the problem comes from the third century B.C. from Eratosthenes of Cyrene, famous for his estimate of the circumference of the earth. According to him, the citizens of Delos were suffering from a plague. They consulted the oracle, who told them that to rid themselves of the plague, they must construct an altar to a particular god that would be geometrically similar to the existing one but double the volume.

(a) How would the volume of the new altar compare with that of the old if each of its linear dimensions were doubled?
(b) What should the linear scaling factor be for the new altar? (The problem intended by the oracle was to construct with straightedge and compasses a line segment equal in length to this particular linear scaling factor. Not until the nineteenth century was the task shown to be impossible. Eratosthenes relates that the Delians interpreted the problem in this sense, were perplexed, and asked Plato about it. Plato told them that the god didn't really want an altar of double the volume but wished to shame them for their "neglect of mathematics and their contempt for geometry.")

8. Unfortunately, recent dollar coins (Susan B. Anthony, Sacagawea) have been a failure–the public found them too small and light. It is not yet known if the Presidential series of dollar coins begun in 2007 will be any more successful (they are the same size). Suppose that you are to design a new *$5* coin (whom should it depict?). The sole requirement is that it be made of the same material as the quarter but weigh four times as much. A quarter can be described geometrically as a circular cylinder approximately $\frac{15}{16}$ in. in diameter and $\frac{1}{16}$ in. thick. Because your new dollar should weigh four times as much, it needs to have four times the volume of a quarter. [The formula for the volume of a cylinder is $\pi \times (\text{diameter}/2)^2 \times$ height.]

Susan B. Anthony dollar.

Presidential series dollar.

Coins shown actual size (26.5 mm diameter).
(United States Mint.)

(a) A member of your public advisory panel suggests just doubling the diameter and doubling the thickness. What do you tell this individual, in the most diplomatic terms?
(b) If you double the diameter, how thick does the coin need to be?
(c) Another member feels that the result of part (b) would be inconveniently large and proposes instead to scale up the quarter proportionality (she has studied an earlier edition of this book). What would the dimensions be for this coin? (Incidentally, nothing in U.S. law requires a dollar coin to be round. The Canadian $1 coin introduced in 1989 has been very popular; it has the shape of an 11-sided polygon.)

9. Criticize the following statement and write a correct version.

"[The Apple Web browser] Safari loads pages up to 2 times faster than Internet Explorer 7 and up to 1.7 times faster than Firefox 2." [http://www.apple.com/safari/, November 2007.]

10. Criticize the following statement and write a correct version.

"Wii uses 10 times less power consumption than PS3 & Xbox 360."
[http://www.maxconsole.net/?mode=news&newsid=14580, February 22, 2007.]

11. Criticize the following statement and write a correct version.

"The stock of Countrywide Financial Corporation [then the country's largest mortgage lender] has fallen 150% since the start of the year." [Heard on National Public Radio News, September 12, 2007.]

12. Criticize the following statement and write a correct version.

"Waiting time for surgery down 500%." [*Evening Standard* (Edinburgh), 5 July 2005.]

13. Abuses of the language of comparison aren't hard to find. For example, the phrase "35 times less than" occurs in more than 1900 Internet documents. Search on the Internet and find either an abuse of *times* and *less than* together, or else an abuse of percentages. Figure out what the author meant to say, and write it in correct language.

14. Criticize the following statement and write a correct version.

"The average full-time [real estate] agent working in Steubenville [Ohio] sells more than 22 houses per year, whereas the same agent in San Francisco sells fewer than 4–5.7 times less." [Austan Goolsebee, "Bubble-lusions: Why most real-estate agents aren't getting rich," http://www.slate.com/id/2124506, August 26, 2005.]

18.2 How Much Is That in...?

15. The cost of mailing a lightweight airmail letter from the United States to most of western Europe in 2008 was \$0.90. How much was that in euros (€), the currency of the European Union, when the exchange rate was €1 = \$1.47? (For comparison, the cost then of an airmail letter to the United States varied from country to country in the European Union, ranging from €0.65 to €1.70.)

16. The cost of mailing a lightweight letter from the United States to Canada in 2008 was US\$0.69. How much was that in Canadian dollars when the exchange rate was US\$1 = Cdn\$1.02? (The postage cost from Canada to the United States was Cdn\$0.93.)

17. In Germany the fuel efficiency of cars is measured in liters of fuel per 100 km (L/100 km). A typical average in a compact station wagon is 7.3 L/100 km. What is that in miles per gallon (mpg)?

18. According to Environmental Protection Agency ratings, the highest-mileage 2008 car was the gasoline/electric hybrid Toyota Prius at 48 mpg in the city. How many liters of gasoline does such a Prius use to travel 100 km in the city?

19. Consider a real locomotive that weighs 88 tons and an HO-gauge scale model of it, for which the linear scaling factor is 1/87.

(a) How much would an exact scale model weigh in tons?
(b) What assumptions are involved in your answer to part (a)?
(c) How much would an exact scale model weigh in pounds?
(d) In kilograms?
(e) In metric tonnes (1 metric tonne = 1000 kg)?

20. What's wrong in the following quotations?

(a) "President Bush visited California, where 12 forest fires have charred more than 700,000 square miles." [Steve Stadelman, WTVO television news, Channel 17, Rockford, Illinois, October 2007.] (Curiously, the same number 700,000 also appeared in news reports for California fires in 2003, 2000, and 1987.)
(b) "The population of the USA has topped 300 million.... If current trends continue, it is expected to reach 400 billion by 2043. This makes it an acceleration of growth...." [*Significance* 3 (4) (December 2006) p. 146.]

21. Gasoline is sold in the United States by the U.S. gallon and in Europe by the liter (1 U.S. gal = 231 in.^{3}; 1 L = 1000 cm^3). What was the equivalent cost, in U.S. dollars per U.S. gallon, for gasoline in Germany priced in euros at €1.42 per liter, when €1 = \$1.25, in September 2005?

22. In 1991, Edward N. Lorenz, a meteorologist who was an early researcher into chaos and dynamical systems (discussed in Chapter 23), received the Kyoto Prize in Basic Sciences, consisting of a gold medal and 45 million Japanese yen (¥). If US\$1 = ¥125 at the time, what was the value of the cash award in 1991 U.S. dollars? (In Chapter 21, we show how to convert such an amount to its value in today's dollars.)

23. The year 2008 marked the 50th anniversary of the installation of length markers on the sidewalk of the Harvard Bridge across the Charles River between Boston and Cambridge, Massachusetts (where in 1908 Harry Houdini performed one of his "escapes"). The bridge is marked at 10-smoot intervals, where 1 smoot was the height of MIT fraternity pledge, Oliver Smoot. The length of the bridge is 620 m or 364.4 smoots and one ear. How long is a smoot in feet and inches? (Oliver Smoot later became the head of the International Standards Organization.)

18.3 Scaling a Mountain

24. The two fastest elevators in Taipei 101 go up (or down) at 55 ft/s.

(a) With no stops, how long would it take to get to the top of Taipei 101?
(b) With no stops, how long would it take to get to the top of the Petronas Towers, which are 1483 ft high and whose elevators run at 41 ft/s?
(c) The answers to parts (a) and (b) suggest that building designers find half a minute a reasonable standard for getting to the top of a building (or else elevators don't come any faster). How fast would the elevators in a mile-high building have to be to achieve that standard?

25. Calculate the speed, in miles per hour, of the elevators in the three buildings in parts (a), (b), and (c) of Exercise 24.

For Exercises 26–31, refer to the following.

The Canadian dollar for many years fell steadily in value in terms of the U.S. dollar, but that trend has been reversed in the past few years. In January 2002, a U.S. dollar was worth Cdn\$1.60. In January 2008, US\$1 = Cdn\$1.02. How should you measure how much one currency has depreciated (lost value) against another? Let the *home currency* be the one whose change in value you are interested in, and let the *target currency* be the one in terms of which you will measure the change. In our example, we want to track the change in value of the U.S. dollar (home currency) in terms of the Canadian dollar (target currency).

There are two competing practices. Both begin by calculating the change as measured in the target currency, the new value minus the old value–here, Cdn\$1.02 − Cdn\$1.60 = Cdn−\$0.58.

- Option A, used by the International Monetary Fund and the British periodical *The Economist*, divides this difference by the new trading value (Cdn$1.02) and multiplies by 100 to get a result in percent–here, −57%.
- Option B, sometimes called the "popular method," divides instead *by the old trading value*, Cdn$1.60, here getting −36% as the percentage change in value of the U.S. dollar in terms of the Canadian dollar.

So, depending on the method used, we can say that the U.S. dollar lost either 57% or 36% in value against the Canadian dollar.

26. To get a feeling for how the results of these two options come out, we consider an artificial example where the numbers are simple. Suppose that in January 2008 the imaginary currency of the imaginary country Middle Earth, the Middie ($\mathcal{M}$), traded at US$2 but in January 2009 its value was only $1.

(a) Using Option A, calculate how much percentage value the Middie lost against the dollar.
(b) Calculate how much percentage value the Middie lost against the dollar using Option B.

27. We use the same data as in Exercise 26 but look at matters from the perspective of an American entrepreneur importing rings of power from Middle Earth. In January 2008, $1 = $\mathcal{M}$0.50; in January 2009, $\mathcal{M}$1 = $1.

(a) Using Option A, calculate how much percentage value the dollar gained against the Middie.
(b) Using Option B, calculate how much percentage value the dollar gained against the Middie.

28. In January 2002, when European Union euro (€) currency coins and bills were introduced, the conversion to U.S. dollars was US$1 = €1.160. Six years later, in January 2008, the dollar had declined severely, so that $1 = €0.693.

(a) Using Option A, calculate how much percentage value the dollar lost against the euro.
(b) Using Option B, calculate how much percentage value the dollar lost against the euro.

29. We use the same data as in Exercise 28 but look at matters from the perspective of a European considering a vacation in the United States. In January 2002, €1 = $0.862; in January 2008, €1 = $1.442.

(a) Using Option A, calculate how much percentage value the euro gained against the dollar.
(b) Using Option B, calculate how much percentage value the euro gained against the dollar.

◆ **30.** Using your results from either Exercises 26–27 or 28–29:

(a) Why don't the numbers agree? If the dollar loses a certain percentage against another currency, shouldn't that currency gain the same percentage against the dollar? Why or why not?
(b) Which option, A or B, seems to give a better sense of the effect of the change in relative values of currencies?

◆ **31.** For both Options A and B:

(a) Can either give a percentage loss that is more than 100%? Would it make sense to speak of a currency declining more than 100%?
(b) Is the percentage from Option A–loss or gain–always higher than the Option B percentage? Always lower? Neither, but do you see any pattern?
(c) Which option would you expect a person to use who wants to make a decline seem large? To make a gain seem large?

18.4 Sorry, No King Kongs

32. The weight of a 1-ft cube of steel is 500 lb. What is the pressure on the bottom face in

(a) pounds per square inch?
(b) atmospheres (1 atm = 14.7 lb/sq in.)?

33. In an article on adding organic matter to soil, the magazine *Organic Gardening* (March 1983) said, "Since a 6-inch layer of mineral soil in a 100-square-foot plot weighs about 45,000 pounds, adding 230 pounds of compost will give you an instant 5% organic matter."

(a) What is the density of the mineral soil, according to the quotation?
(b) How does this density compare with that of steel?
(c) How do you think the quotation should be revised to be accurate?

For Exercises 34 and 35, refer to the following.

A mature gorilla weighs 400 lb and stands 5 ft tall; its two feet combined have an area of about 1 ft^2.

34. (a) Give an estimate of the gorilla's weight when it was half as tall.
(b) What assumptions are involved in your estimate?
(c) When the gorilla is standing, what is the pressure on its feet in pounds per square inch?

35. Suppose King Kong is a gorilla scaled up with a linear scaling factor of 10.

(a) How much does King weigh?
(b) What is the pressure on King's feet in pounds per square inch?

36. You may want a waterbed, but waterbeds are not allowed in your building. Apart from the danger of flood if the bed should puncture or leak, the weight is an issue.

(a) Suppose that a queen-size waterbed mattress is 80 in. long by 60 in. wide by 12 in. high, and water weighs 1 kg/L. How much does the water in the mattress weigh in pounds?

(b) If the weight of the mattress and frame is carried by four legs, each 2 in. by 2 in., what is the pressure, in pounds per square inch, on each leg?
(c) How does the pressure on the legs of the waterbed compare with the pressure that a person exerts on his or her feet–for example, a 130-lb person with a total foot area of about one-quarter of a square foot in contact with the ground?

37. If you aren't allowed to have a waterbed, how about a spa (hot tub)? Find the weight of the water in a spa that is in the shape of a cylinder 6 ft in diameter and 3.5 ft deep. (*Hint:* The volume of a cylinder is $\pi r^2 h$, where r is the radius and h is the height.)

38. What does the largest giant sequoia tree (named "Hyperion") weigh? Model the tree as a (very elongated) cone. Assume that the tree is 379 ft high, has a circumference of 40 ft at the base, and that the density of the wood is 31 lb/ft^3. (The volume of a cone of height h and radius r is $\frac{1}{3}\pi r^2 h$.)

39. A 6-ft-tall indoor holiday tree needs four strings of lights to decorate it. How many strings of lights are needed for an outdoor tree that is 30 ft high? (Contributed by Charlotte Chell of Carthage College, Kenosha, Wisconsin.)

For Exercises 40 and 41, refer to the following.

An ancient measure of length, the *cubit,* was the distance from the elbow to the tip of the middle finger of a person's outstretched arm. So the length of a cubit depended on the person, though there was some attempt at standardization. Most estimates place the cubit between 17 and 22 in.

40. According to classical Greek sources, Pythagoras (sixth century B.C.) used geometric scaling to model the height of Hercules, the heroic figure of classical mythology. Pythagoras compared the lengths of two racecourses, one (according to tradition) paced off by Hercules and the other by a man of average height. Both were 600 "paces" long, but the one by Hercules was longer because of his longer stride. A normal man in the time of Pythagoras would have been about 5 ft tall.

(a) If the distance paced off by Hercules was 30% longer than the other racecourse, how tall was Hercules? What does your calculation assume?
(b) The sources give two conflicting answers, that Hercules was 4 cubits tall and 4 cubits 1 "foot" tall. What range does this give for his height in feet and inches? In centimeters? (Assume that a Greek "foot" was the length of a modern foot.)

41. Goliath [of David and Goliath, as related in the Bible (I Samuel 17:4)] was "six cubits and a span." A "span" was originally the distance from the tip of the thumb to the tip of the little finger when the hand is fully extended, about 9 in. What range of heights would this indicate for Goliath in feet and inches? In centimeters?

For Exercises 42–45, refer to the following.

The *body mass index* (BMI) is the basis for the National Heart, Lung and Blood Institute's weight guidelines. BMI is body weight (in kilograms) divided by the square of height (in meters). A BMI of 25 through 29 is considered "overweight"; a BMI of 30 or over is considered "obese." Some 55% of American adults have a BMI of 25 or above. (*Note:* BMI is not a useful measure for young children, pregnant or breast-feeding women, the frail elderly, or very muscular people.) For practice with this concept, calculate your own BMI.

(moodboard/Corbis.)

42. Calculate the BMI for a woman 160 cm tall who weighs 65 kg. Is she overweight according to the institute's guidelines?

43. How much in kilograms must a man weigh who is 190 cm tall if he is not to be considered overweight according to the institute's guidelines?

44. Suppose that weight and height are measured instead in U.S. Customary units of pounds and inches. We can still calculate body weight divided by the square of height using these units. What conversion factor is necessary to convert this number to the BMI?

◆ **45.** Because body weight is average density times body volume, BMI is average density times a quantity that has units of length. Discuss whether BMI makes sense as a measure of being overweight. Would dividing by a different power of height make for a better measure?

18.5 Dimension Tension

◆ **46.** Jonathan Swift's Gulliver traveled to Lilliput, where the Lilliputians were human-shaped but only about 6 in. tall. In other words, they were geometrically similar in shape to ordinary human beings but only one-twelfth as tall. What would a Lilliputian weigh?

Are Lilliputians ruled out by the size–shape and area–volume considerations in this chapter? If you think they are, what considerations do you find convincing? If not, why not?

47. **(a)** What would you expect an individual *Quetzalcoatlus northropi* to weigh if it had half the wingspan of an adult?
(b) If an individual weighed half as much as an adult, what would you expect its wingspan to be?

48. In the children's story *Peter Pan*, Peter and Wendy can fly. We may suppose that they are 4 ft tall, so they are about 12 times as tall as a sparrow is long. What should their minimum flying speed be?

49. Icarus of Greek legend escaped from Crete with his father, Daedalus, on wings made by Daedalus and attached with wax. Against his father's advice, Icarus flew too close to the Sun; the wax melted, the wings fell off, and he fell into the sea and drowned. What must have been his minimum cruising speed? What assumptions does your answer involve?

◆ **50.** Recent years have seen the beginnings of human-powered controlled flight, in the *Gossamer Condor* and other superlightweight planes, which have disproportionately large wings compared with geometric scaling up of birds. The *Gossamer Condor* is far longer than an ostrich but it flies at only 12 mph. How can it?

51. Justify the claim on p. 589 that a *Q. northropi* weighing 200 lb with a wingspan of 50 ft would have had the same wing loading as one weighing 100 lb with a wingspan of 36 ft.

52. The largest and heaviest aircraft in service today is the An-225–and we mean "The" because there is only one! (A second was scheduled to enter service in 2008.) It has been used to bring humanitarian equipment to Iraq, as well as–in a single flight–216,000 meals for American military personnel. The plane has a wing area of 905 m^2 and a maximum takeoff weight of 1.3 million lb. What is its wing loading, in kg/m^2?

53. The cult movie *Them* (1954) features enormous ants (8 m long by 3 m wide). We can investigate the feasibility of such a scaled-up insect by considering its oxygen consumption. A common ant, 1 cm long, needs 24 milliliters (mL) of oxygen per second for each cubic centimeter of its volume. Because an ant has no lungs, it absorbs oxygen through its "skin" at a rate of 6.2 mL per second per square centimeter. Suppose that the tissues of a scaled-up ant would have the same need for oxygen for each cubic centimeter, and that its skin could absorb oxygen at the same rate, as a common ant.

(a) Compared with a common ant, how many times as large is an enormous ant's

- **(i)** length?
- **(ii)** surface area?
- **(iii)** volume?

(b) What proportion of such an ant's oxygen need could its skin supply?

(c) What can you conclude about the existence of such insects? (Adapted from George Knill and George Fawcett, Animal form or keeping your cool, *Mathematics Teacher,* May 1982, 395–397.)

For Exercises 54–57, refer to the following.

Maybe some trees could grow to a mile high, but they just don't live long enough to have the chance. In this problem, we try to determine how fast the height of a tree increases. We can measure indirectly how much mass the tree adds in a year by the area of the annual tree ring added. Here are two relevant facts:

- As you may have noticed from stumps, as a tree grows older, its annual rings get less wide. Although the width of the ring varies somewhat from year to year with the amount of rainfall and other factors, the total *area* of each annual ring is roughly the same over the years, meaning that *the tree adds roughly the same amount of mass each year.* Call that amount a; then the mass M of the tree is $M = at$, where t is its age in years.
- Over a large range of tree sizes and tree species, the diameter d of a tree of a species is approximately proportional to the three-halves power of the height h of the tree (different species have different constants of proportionality). Thus, $d \propto h^{3/2}$ (this is shown in Exercise 54).

Now, if we assume that the bulk of the mass of the tree is in the trunk, and if we model the trunk either as a long cylinder or as a thin cone, the mass is proportional to the volume, so $M \propto d^2h$. Then

$$at = M \propto d^2h \propto (h^{3/2})^2h = h^4$$

so $h \propto t^{1/4}$. In other words, *the tree grows in height as the fourth root of its age.*

54. Suppose that a tree grows to 20 m in 30 years. How tall will it be (if it lives long enough) when it is 60 years old?

55. How long would it take the tree in Exercise 54 to grow to be 40 m tall?

56. Giant sequoias can reach 100 m after about 1000 years. If it could keep on growing at the same rate of addition of mass, how long would it take a giant sequoia 100 m tall to grow to 200 m?

■ **57.** The branching of trees is similar to the branching of systems in the bodies of animals. For similar reasons, the area of the cross section of the tree at its base scales as the three-fourths power of the tree's mass, that is, $A \propto M^{3/4}$. Assume that most of the mass is in the trunk and model the tree either as a tall cylinder ($V = \pi r^2 h$) or as a cone ($V = \pi r^2 h/3$). Show that the diameter d of a tree is approximately proportional to the three-halves power of the height, that is, $d \propto h^{3/2}$.

◆ **58.** Some humans, such as the Bushmen of the Kalahari Desert in Africa, live in desert environments, where it is important to be able to do without water for periods of time. Would you expect such an environment to favor short people or tall ones? (Adapted from A. Zherdev, Horseflies and flying horses, *Quantum,* May–June 1994, 32–37, 59–60.)

◆ **59.** Smaller birds and mammals generally maintain higher body temperatures than do larger ones. Explain why you would expect this to be so. (Adapted from A. Zherdev, Horseflies and flying horses, *Quantum*, May–June 1994, 32–37, 59–60.)

18.6 How to Grow

60. Listed below are the numbers of species of reptiles and amphibians on some Caribbean islands, together with the approximate areas of the islands. (Suggested by Florence Gordon of the New York Institute of Technology, with contributions from Kevin Mitchell and James Ryan of Hobart and William Smith Colleges, Geneva, N.Y. This table is adapted from Tables 15 and 16 in P. J. Darlington, *Zoogeography: The Geographic Distribution of Animals,* Wiley, New York, 1957, pp. 483–484).

Island	**Area (mi²)**	**Species**
Redonda	1	3
Saba	4.9	5
Montserrat	40	9
Trinidad	2,000	80
Puerto Rico	3,400	40
Jamaica	4,500	39
Hispaniola	30,000	84
Cuba	40,000	76

(a) Plot number of species versus area on ordinary graph paper and then on log-log graph paper. If you don't have log-log paper available, use a calculator or spreadsheet to take the logarithms ($\log_{10}$) of all the numbers and graph logarithm of number of species versus logarithm of area on ordinary graph paper. (*Note:* Trinidad is an outlier from the general pattern–see Chapter 6).
(b) Is the relationship that you graphed in part (a) proportional? Allometric?
(c) What would be the expected number of species on an island of 400 mi²?
(d) For each 10-fold increase in the island's size, what happens to the number of species, approximately?

61. Listed below are the weights and wingspans of some birds and of some fully loaded airplanes. (Idea and most data contributed by Florence Gordon of the New York Institute of Technology.)

Bird	**Weight (lb)**	**Wingspan (ft)**
Crow	1	2.9
Harris hawk	2.6	3.2
Blue-footed booby	4	3
Red-tailed hawk	4	4
Horned owl	5	5
Turkey vulture	6.5	6
Eagle	12	7.5
Golden eagle	13	7.3
Whooping crane	16.1	7.5
Vulture	18.7	9.3
Condor	22	9.9
Quetzalcoatlus northropi	100	36
Plane		
Boeing 737	117,000	93
DC9	121,000	93.5
Boeing 727	209,500	108
Boeing 757	300,000	156.1
Boeing 707	330,000	145.7
DC8	350,000	148.5
DC10	572,000	165.4
Boeing 747	805,000	195.7
Boeing 747-400	895,000	212.6
Anton An-225	1,323,000	290.2

(a) Use a calculator or spreadsheet to take the logarithms ($\log_{10}$) of all the numbers and then graph logarithm of weight versus logarithm of wingspan on ordinary graph paper.
(b) For the birds, is the relationship that you graphed in part (a) proportional? Allometric? How about for the planes?
(c) Does the same relationship of wingspan to weight seem to hold for birds and planes?

WRITING PROJECTS

1. A human infant at birth usually weighs between 5 and 10 lb and has a height (length) between 1 and 2 ft, with the shorter babies having the lesser weight. Considering the weight and height of an adult human, write a paragraph arguing that human growth must not be just proportional growth.

2. The principle that area scales with the square of length, and volume with the cube, has important consequences for the depiction and interpretation of data in graphic form. Suppose we wish to indicate in an artistic way that the weekly income of a U.S. carpenter is twice that of a carpenter in (mythical) Rotundia. We draw one moneybag for the Rotundian and another one "twice as large" for the American. (Illustration from Darrell Huff, *How to Lie with Statistics,* Norton, New York, 1954, p. 69.)

What's the problem? Well, first, people tend to respond to graphics by comparing areas. Because the larger moneybag is twice as high and twice as wide as the smaller one, its image has four times the area. Second, we are used to interpreting depth and perspective in drawings in terms of three-dimensional objects. Because the larger bag is also twice as thick as the smaller, it has eight times the volume. The graphic leaves the subconscious impression that the U.S. carpenter earns eight times as much, instead of twice as much. With these ideas in mind, evaluate–in a paragraph each–the following data depictions.

(a) Percentages of Ph.D.s earned by women in three fields. (From *Science*, 260, April 16, 1993, 409, as reproduced in Jessica Utts, *Seeing Through Statistics*, Duxbury, Belmont, Calif., 1996, p. 142.)

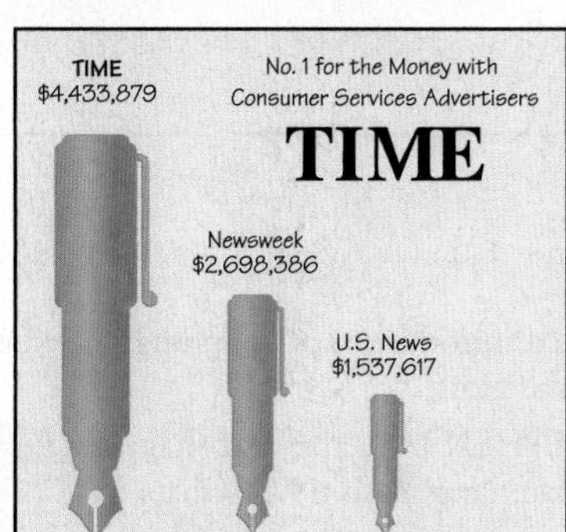

(b) Advertising spending in three prominent newsmagazines. (From *Time* magazine, as reproduced in David S. Moore, *Statistics: Concepts and Controversies*, 4th ed., W. H. Freeman, New York, 1997, p. 207.)

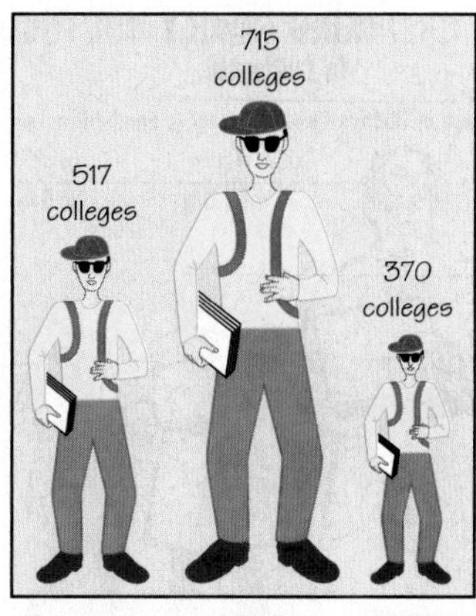

(c) U.S. colleges as classified by enrollment. (From David S. Moore, *Statistics: Concepts and Controversies*, 4th ed., W. H. Freeman, New York, 1997, p. 217.)

3. Evaluate in a paragraph each of the following depictions (a–c). (Illustrations from Edward R. Tufte, *The Visual Display of Quantitative Information*, Graphics Press, 1983, pp. 70, 69, and 57.)

1958—Eisenhower

1973—Nixon

1978—Carter

1963—Kennedy

1984—Reagan

1990—Bush

1968—Johnson

1994—Clinton

2006—Bush

(a) Value of the dollar.

(b) The shrinking family doctor.

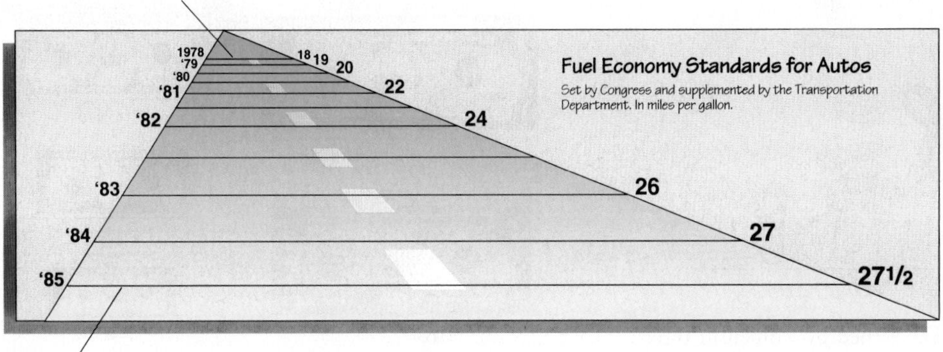

(c) Fuel economy standards for autos. (From *The New York Times*, August 9, 1978, p. D2.)

4. With the ideas of Writing Projects 2 and 3 in mind, collect and evaluate similar depictions of data from magazines and newspapers.

5. Dolls and human figures are usually scaled to be geometrically similar to actual humans. But are dolls designed to represent babies or adult humans? Go to a toy store and measure the height, the vertical height of the head, and the arm length of some dolls and other figures. Scale your measurements to compare them with Figure 18.11; from that comparison, try to estimate the ages of the humans that the figures resemble. Write up your procedure, data, calculations, and conclusions in a page or two.

SUGGESTED READINGS

ADAM, JOHN A. *Mathematics in Nature: Modeling Patterns in the Natural World*, Princeton University Press, Princeton, N.J., 2003.

BONNER, JOHN TYLER. *Why Size Matters: From Bacteria to Blue Whales*, Princeton University Press, Princeton, N.J., 2006.

DUDLEY, BRIAN A. C. *Mathematical and Biological Interrelations*, Wiley, New York, 1977. Excellent and gentle extended introduction to graphing, scale factors, and logarithmic plots.

GOULD, STEPHEN JAY. Size and shape. In *Ever Since Darwin,* Norton, New York, 1977, Chap. 21.

HALDANE, J. B. S. On being the right size. In *Possible Worlds and Other Papers,* Harper, New York, 1928. Reprinted in James R. Newman (ed.), *The World of Mathematics,* vol. 2, Simon & Schuster, New York, 1956, pp. 952–957. Also reprinted in John Maynard Smith (ed.), *On Being the Right Size and Other Essays by J. B. S. Haldane,* Oxford University Press, Oxford, 1985, pp. 1–8. Succinctly surveys area–volume tension, flying, the size of eyes, and even the best size for human institutions.

McMAHON, T. A., and J. T. BONNER. *On Size and Life,* Scientific American Library, New York, 1983. Astonishingly beautiful and informative book on the effects of size and shape on living things.

SCHMIDT-NIELSEN, KNUT. *Scaling: Why Is Animal Size So Important?* Cambridge University Press, New York, 1984.

WEIBEL, EWALD R. *Symmorphosis: On Form and Function in Shaping Life*, Harvard University Press, Cambridge, Mass., 2000.

SUGGESTED WEB SITES

physics.nist.gov/cuu/Units/index.html In-depth information on SI, the modern metric system.

www.missingkids.com National Center for Missing and Exploited Children.

www.usmint.gov U.S. Mint.

www.thusness.com/bmi.t.html Body mass index calculator and further links.

CHAPTER 19

Symmetry and Patterns

"The senses delight in things duly proportional." So said the famous philosopher-theologian Thomas Aquinas more than 700 years ago. In this chapter, we examine elements of esthetic appreciation, particularly *symmetry*.

What is symmetry and what does mathematics have to do with it? Symmetry, like beauty, is hard to define. Dictionaries talk about "correspondence of form on opposite sides of a dividing line or plane or about a center or an axis," "correspondence, equivalence, or identity among constituents of an entity," and "beauty as a result of balance or harmonious arrangement" (*American Heritage Dictionary,* 3rd ed.).

In the narrowest sense, symmetry refers to mirror-image correspondence between parts of an object. Crystals, in both their appearance and their atomic structure, provide examples of symmetry in this sense. Taken in a wider sense, though, symmetry includes notions of *balance, similarity,* and *repetition.*

Our wider sense of symmetry leads us to appreciate patterns. *Mathematics is the study of patterns,* and it gives important insights into symmetry.

You are already familiar with mirror-image symmetry, which mathematicians call **reflection symmetry**.

Another kind of symmetry that you know well is **rotation symmetry**, in which rotation of an object about its center leaves it looking the same. A snowflake is a familiar example of both reflection symmetry and rotational symmetry.

Beautiful examples of rotational symmetry arise in nature in the shoots, leaves, and seeds of plants that grow from a central stem. For instance, the scales of a pineapple or a pinecone (Figure 19.1) and the stickers on a cactus follow such a pattern, as do the seeds of a sunflower (Figure 19.2a) and the petals of a daisy. In this pattern, known as **phyllotaxis**, the spirals and their elements are geometrically similar to one another.

The chambered nautilus of Figure 19.2b may stretch your notion of symmetry, as will other examples in this chapter. It has neither reflection symmetry nor rotational symmetry. However, although the successive sections of the nautilus are of different sizes, they are geometrically similar to one another, and the resulting spiral has the same shape at any size: A photographic enlargement superimposed on it would fit exactly. There is balance, similarity, and repetition–the characteristics of symmetry that we identified above.

FIGURE 19.1 Spirals of scales on a pinecone: 8 right, 13 left. (*From Verner E. Hoggatt, Jr.,* Fibonacci and Lucas Numbers, *Houghton Mifflin, New York, 1969, p. 81.*)

(*Don Hammond/Design Pics/Corbis.*)

FIGURE 19.2 (a) This sunflower has 55 spirals in one direction and 89 spirals in the other direction. (*Harvey Lloyd/The Stock Market.*) (b) A chambered nautilus shell. (*James Randkler/Tony Stone Images.*)

This chapter explores and classifies the fundamentally different ways in which a two-dimensional design can be symmetrical. What will be surprising is that there are so few such patterns.

19.1 Fibonacci Numbers and the Golden Ratio

Fibonacci Numbers

Associated with the geometric symmetry of phyllotaxis is a kind of *numeric symmetry*, with a "proportion" in the sense of a ratio of numbers. Strangely, the number of spirals in plants with phyllotaxis is not just any whole number but always comes from a particular sequence of numbers called the **Fibonacci numbers** (see Spotlight 19.1).

Fibonacci Numbers (Fibonacci Sequence) DEFINITION

Fibonacci numbers occur in the sequence

1, 1, 2, 3, 5, 8, 13, 21, 34, 55, 89, 144, 233, 377, . . .

This sequence begins with the numbers 1 and 1 again, and each next number is obtained by adding the two preceding numbers together.

SPOTLIGHT 19.1 Leonardo of Pisa ("Fibonacci")

Born in Pisa in 1170, Leonardo of Pisa has been known as "Fibonacci" for the past century and a half. This nickname, which refers to his descent from an ancestor named Bonaccio, is modern, and there is no evidence that he was known by it in his own time.

Leonardo was the greatest mathematician of the Middle Ages. His stated purpose in his book *Liber abbaci* (1202) was to introduce calculation with Hindu-Arabic numerals into Italy, to replace the Roman numerals then in use. Other books of his treated topics in geometry, algebra, and number theory.

We know little of Leonardo's life apart from a short autobiographical sketch in the *Liber abbaci*:

> I joined my father after his assignment by his homeland Pisa as an officer in the customhouse located at Bugia [Algeria] for the Pisan merchants who were often there. He had me marvelously instructed in the Arabic-Hindu numerals and calculation. I enjoyed so much the instruction that I later continued to study mathematics while on business trips to Egypt, Syria, Greece, Sicily, and Provence and there enjoyed discussions and disputations with the scholars of those places.

Leonardo of Pisa ("Fibonacci")
A portrait of unlikely authenticity.
(From Columbia University, D. E. Smith Collection.)

(*Source*: L. E. Sigler, *Leonardo Pisano Fibonacci, The Book of Squares: An Annotated Translation into Modern English*, Academic Press, New York, 1987, p. xvi.)

The *Liber abbaci* contains a famous problem about rabbits, whose solution is now called the Fibonacci sequence. Leonardo did not write further about it.

Sometimes a sequence of numbers is specified by stating the value of the first term or first several terms and then giving an equation to calculate succeeding terms from preceding ones. This is called a *recursive rule,* and the sequence is said to be defined by **recursion**. Let's denote the nth Fibonacci number by F_n; then the Fibonacci sequence can be defined by

Recursion for the Fibonacci Sequence PROCEDURE

$$F_1 = 1, F_2 = 1, \quad \text{and} \quad F_{n+1} = F_n + F_{n-1} \quad \text{for } n \geq 2$$

The recursive rule just expresses in algebraic form that the next Fibonacci number is the sum of the previous two.

Look at the sunflower in Figure 19.2a. You see a set of spirals running in the counterclockwise direction and another set in the clockwise direction. It is (just barely) possible to count the number of spirals in both directions. In the sunflower there are 55 in one direction and 89 in the other–two consecutive Fibonacci numbers. In the case of the pineapple, there are three sets of spirals, one each along the three directions through each hexagonally shaped scale. For the common grocery pineapple (*Ananas comosus*), there are always 8 spirals to the right, 13 to the left, and 21 vertically–again, consecutive Fibonacci numbers.

Why are the numbers of spirals in plants the same numbers that appear next to each other in a purely mathematical sequence? There is no easy answer. There are several intricate theories about the dynamics of the plant's growth.

The Golden Ratio

During the last several centuries, an attractive myth has arisen that the ancient Greeks considered a specific numerical proportion essential to beauty and symmetry. Known in modern times as the **golden ratio**, **golden mean**, or even **divine proportion**, this proportion was investigated by Euclid in Book II of his *Elements*. Recent research reveals little evidence connecting this proportion to Greek esthetics, but let's pursue the golden ratio briefly because of its intimate connection to the Fibonacci sequence and because it does have appeal as a standard for beautiful proportion.

Golden Ratio — DEFINITION

The value of the **golden ratio**, which is usually denoted by the Greek letter phi (ϕ), is

$$\phi = \frac{1 + \sqrt{5}}{2} \approx 1.618034 \ldots$$

The basic esthetic claim is that a **golden rectangle**–one whose height and width are in the ratio of 1 to ϕ–is the most pleasing of all rectangles. The Greeks treated lengths geometrically, so for them it was important to construct lengths using straightedge and compass. In Spotlight 19.2 we show how to construct a golden rectangle that is 1 unit by ϕ units.

SPOTLIGHT 19.2 How the Greeks Constructed a Golden Rectangle

In constructing a golden rectangle, the Greeks started from a one-by-one square (shown in black in the figure), which they made by constructing perpendiculars at the two ends of a horizontal segment of unit length. To extend the square to a golden rectangle, they bisected the original segment, getting a new point that divides it into two pieces of length one-half each. Using this new point and a compass opening equal to the distance from it to a far corner of the square (shown by the blue line in the figure), they could add the blue length to the length one-half to get an interval (in red at bottom) with total length ϕ.

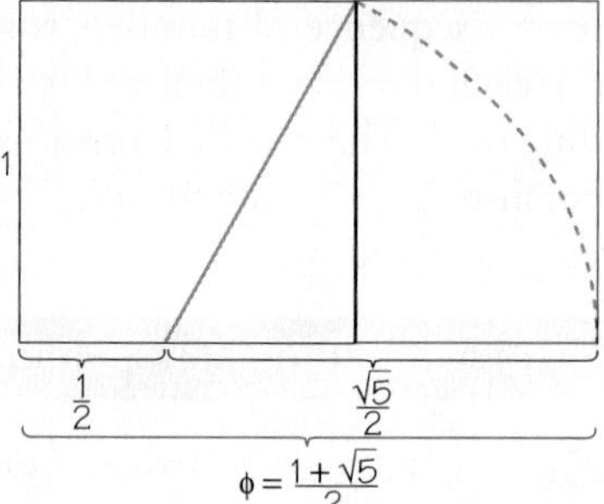

A golden rectangle has the pleasing property that if you cut a square-shaped piece off one end of it, the rectangle that remains is again a golden rectangle.

Why would anyone think that this is an attractive ratio? And where did it come from? The answer lies not in Fibonacci numbers but in the Greeks' pursuit of balance in their study of geometry.

Given two line lengths, one way to find a length that "strikes a balance" between the two is to average them. For lengths l (the larger) and w (the smaller), their average, or *arithmetic mean,* is $m = (l + w)/2$, and it satisfies

$$l - m = m - w$$

The length m strikes a balance between l and w, in terms of a common *difference* from the two original lengths. More generally, the arithmetic mean of n numbers or lengths is their sum divided by n. (See Chapter 5 for its use in statistics.)

The Greeks, however, preferred a balance in terms of *ratios* rather than differences. They sought a length s, the **geometric mean**, that gives a common ratio

$$l \div s = s \div w \qquad \text{or} \qquad \frac{l}{s} = \frac{s}{w}$$

Hence $lw = s^2$, which expresses the geometric fact that s is the side of a square whose area equals the area of an l by w rectangle (the Greeks thought in terms of geometric objects). In geometry, the geometric mean s is called the *mean proportional* between l and w (see Figure 19.3).

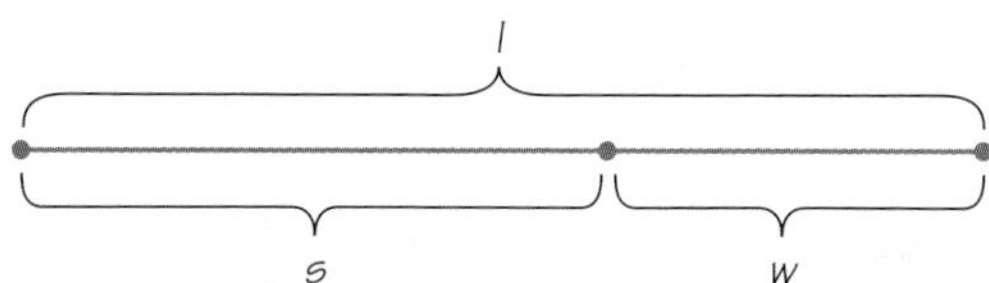

FIGURE 19.3 The line segment of length l is divided so that the length of s is the geometric mean between l and $w = l - s$. The dividing point divides the length l in the golden ratio.

Geometric Mean

DEFINITION

The quantity $s = \sqrt{lw}$ is the **geometric mean** of l and w. More generally, the geometric mean of n numbers is the nth root of the product of all n factors: The geometric mean of $x_1, \ldots, x_n$ is $\sqrt[n]{x_1 \times \cdots \times x_n}$. For example, the geometric mean of 1, 2, 3, and 4 is $\sqrt[4]{1 \times 2 \times 3 \times 4} = \sqrt[4]{24} = 24^{1/4} \approx 2.213$.

The Greeks found symmetry and proportion in the geometric mean, but the geometric mean also has important practical applications (see Spotlight 19.3).

The Greeks were interested in cutting a single line segment of length l into lengths s and w, where $l = w + s$, so that s would be the mean proportional between w and l. Surprisingly, the ratio ϕ arises, as we show. Denote by x the common ratio

$$\frac{l}{s} = \frac{s}{w} = x$$

Substituting $l = s + w$, we get

$$x = \frac{l}{s} = \frac{s + w}{s} = \frac{s}{s} + \frac{w}{s} = 1 + \frac{w}{s}$$

But w/s is just $1/x$, so we have

$$x = 1 + \frac{1}{x}$$

Multiplying through by x gives

$$x^2 = x + 1 \qquad \text{or} \qquad x^2 - x - 1 = 0$$

This is a quadratic equation of the form

$$ax^2 + bx + c = 0$$

SPOTLIGHT 19.3 The Consumer Price Index: An Application of the Geometric Mean

The Bureau of Labor Statistics (BLS) uses the geometric mean—not the arithmetic mean—to calculate the Consumer Price Index (CPI), which tracks changes in the cost of the goods and services that people buy.

The geometric mean takes into account substitutions that consumers make when prices change. For example, if the price of beef goes up but the price of chicken doesn't, then consumers may buy less beef and substitute the cheaper chicken for some beef.

Suppose that, overall, U.S. families consume equal dollar values of beef and chicken. A typical family might consume weekly 5 lb of beef at \$4/lb and 10 lb of chicken at \$2/lb, for \$20 each and a total cost of \$40. We say that beef and chicken each have a relative *market share* of 0.5 (50% beef, 50% chicken, by dollar value).

What if beef goes up to \$6/lb but chicken stays at \$2/lb? The *relative price change* in beef is \$6/\$4 = 1.5 and the relative price change in chicken is \$2/\$2 = 1.00 (no change). If the average family continues to eat just as much beef and chicken as before, the cost is now \$50, an increase of 25%. Because \$30 goes for beef and \$20 for chicken, the relative market shares (0.6 and 0.4) have changed. The *relative price change* for the family's meat is \$50/\$40 = 1.25, which is just the arithmetic mean of the two relative price changes (1.50 and 1.00). A more general formulation is:

relative price change

$$= (\text{old market share of beef})\,\frac{\text{new cost of beef}}{\text{old cost of beef}} + (\text{old market share of chicken}) \times \frac{\text{new cost of chicken}}{\text{old cost of chicken}}$$

$$= 0.5 \times \frac{6.00}{4.00} + 0.5 \times \frac{2.00}{2.00}$$

$$= \frac{1.50 + 1.00}{2} = 1.25$$

A family that eats no beef sees no increase. A family that eats only beef sees an increase of 50%. The CPI is an average over *all* families, weighted by the dollar value that each consumes.

If instead we use the geometric mean, we get a relative price change of $\sqrt{1.50 \times 1.00} = 1.225$. The more general formulation is

relative price change

$$= \left(\frac{\text{new cost of beef}}{\text{old cost of beef}}\right)^{(\text{old market share of beef})} \times \left(\frac{\text{new cost of chicken}}{\text{old cost of chicken}}\right)^{(\text{old market share of chicken})}$$

$$= \left(\frac{6.00}{4.00}\right)^{0.5} \times \left(\frac{2.00}{2.00}\right)^{0.5}$$

$$= \sqrt{1.50 \times 1.00} = 1.225$$

This relative price change, a 22.5% increase, is less than the 25% using the arithmetic mean.

The intention of the CPI is to measure the change in the cost of goods and services that still yield the same level of satisfaction to consumers. Use of the arithmetic mean presumes that a family buys the same amount of beef and chicken (5 lb beef, 10 lb chicken) as before. Use of the geometric mean presumes that a family buys the same *relative dollar value* of each meat as before, hence \$24.50 (12.25 lb) of chicken and \$24.50 (4.08 lb) of beef, for a total of \$49 = 1.225 × \$40. Buying 2.25 lb more chicken and 0.92 lb less beef is supposed to yield the "same satisfaction" as before.

Because the geometric mean is always less than or equal to the arithmetic mean (see Exercise 17), the geometric mean gives a lower figure for inflation than using the arithmetic mean would produce.

Social Security payments, some wage increases, and income tax rates are all automatically geared to the CPI, which we treat in detail in Chapter 21.

with $a = 1$, $b = -1$, and $c = -1$. We apply the famous quadratic formula,

$$x = \frac{-b \pm \sqrt{b^2 - 4ac}}{2a}$$

to get the two solutions

$$x = \frac{1 + \sqrt{5}}{2} \approx 1.618034 \ldots \quad \text{and} \quad \frac{1 - \sqrt{5}}{2} \approx -0.618034 \ldots$$

The negative solution does not correspond to a length. The first solution is the golden ratio ϕ. It occurs often in other contexts in geometry; for example, ϕ is the ratio of a diagonal to a side of a regular pentagon (see Figure 19.4).

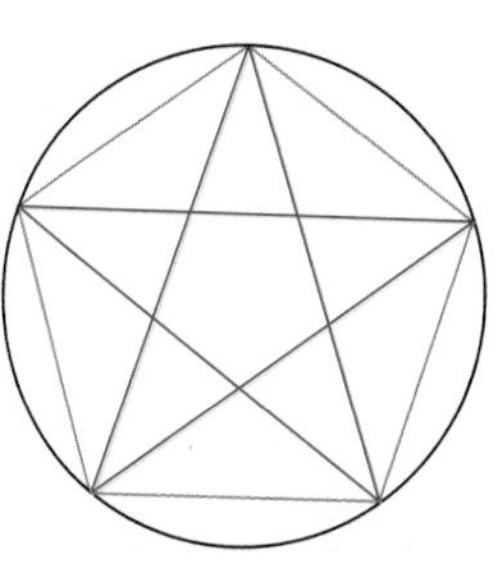

FIGURE 19.4 In a pentagon with equal sides, ϕ is the ratio of a diagonal to a side. The five-pointed star formed by the diagonals was the symbol of the followers of the ancient Greek mathematician Pythagoras.

Thanks to Roger Herz-Fischler (Wilfrid Laurier University) and George Markowsky (University of Maine), we know that the term *golden ratio* was not used in antiquity and that there is no evidence that the Great Pyramid was designed to conform to ϕ, nor that the Greeks used ϕ in the proportions of the Parthenon, nor that Leonardo da Vinci used ϕ in proportions for the human figure (Figure 19.5a). The area from the top of the head of the "Mona Lisa" to the top of her bodice may form a golden rectangle, as claimed by Bulent Atalay in his *Math and the Mona Lisa: The Art and Science of Leonardo da Vinci* (2004), but Leonardo left no documents saying that was his intention or design principle. Others have claimed that the impressionists Gustave Caillebotte (1848–1894) and Georges Seurat (1859–1891) used the golden ratio to design some of their paintings, but the painters themselves left no word about it. Wolfgang Amadeus Mozart (1756–1791), who was fascinated by mathematics as a student, may have constructed the lengths of parts of some of his piano sonatas with an eye to the golden ratio; but we do not have evidence that this was his intention.

Moreover, experiments show that people's preferences for dimensions of rectangles cover a wide range, with golden rectangles not holding any special place.

It is true that human bodies exhibit ratios close to the golden ratio, as you can see by comparing your overall height to the height of your navel. The Swiss-born architect Le Corbusier (Charles-Edouard Jeanneret [1887–1965]) used the golden ratio (including a navel-height feature) as the basis for his "Modulor" scale of proportions (Figure 19.5b).

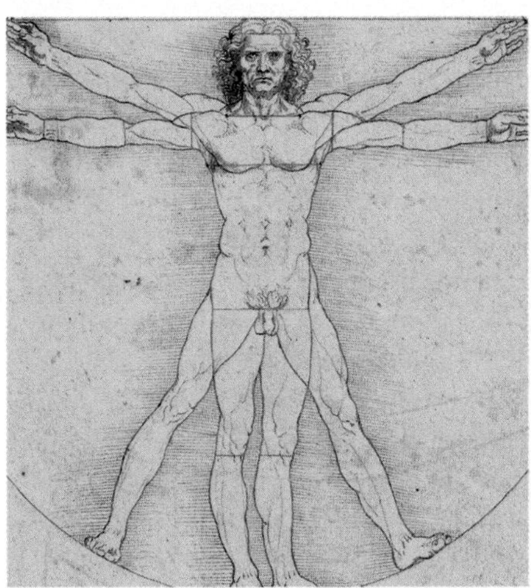

(a)

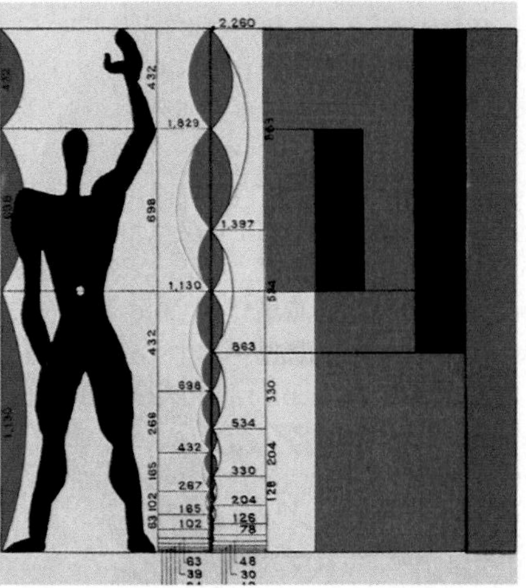

(b)

FIGURE 19.5 (a) Leonardo da Vinci's "Vitruvian Man" (ca. 1490), based on body proportions by Vitruvius (architect and engineer, first century B.C.). Despite claims on the Web and in the thriller *The Da Vinci Code,* neither Vitruvius nor Leonardo suggested using ϕ for human proportions or anything else. (*Accademia, Venice, Italy/Scala/Art Resource, New York.*) (b) Le Corbusier, however, did use ϕ in his "Modulor" scale of proportions. (*Le Corbusier, "Le Modulor," 1945. © 2000 Artists Rights Society [ARS], New York/ADAGP, Paris/FLC.*)

The spirals of the sunflower are approximations to an equiangular, or logarithmic spiral (Figure 19.6). The mathematical reason for this connection is that the ratios of consecutive Fibonacci numbers

$$\begin{array}{ccccccc} \frac{1}{1} & \frac{2}{1} & \frac{3}{2} & \frac{5}{3} & \frac{8}{5} & \frac{13}{8} & \frac{21}{13} \ldots \\ 1.0 & 2.0 & 1.5 & 1.666\ldots & 1.6 & 1.625 & 1.615\ldots \end{array}$$

provide alternately under- and overapproximations to $\phi \approx 1.618034\ldots.$

FIGURE 19.6 A logarithmic spiral determines a sequence of golden rectangles and corresponding squares.

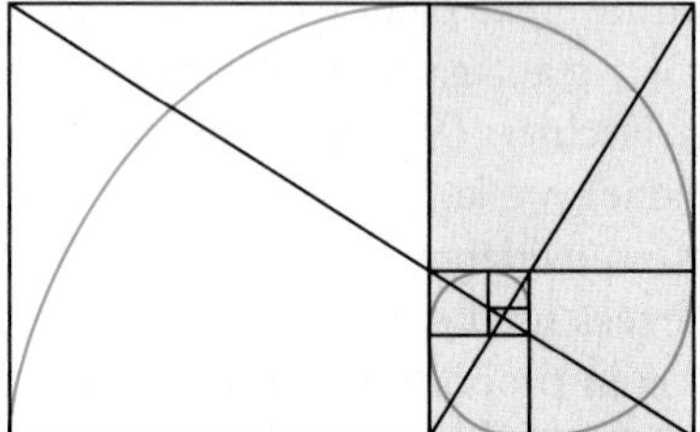

(The spiral of the chambered nautilus shell of Figure 19.2b (p. 610) is also an equiangular spiral, but the rectangles formed have a ratio near 1.3 rather than the golden ratio.)

For reasons that we do not understand, some ratios in the DNA molecule are close to the golden ratio; for example, the length of one full cycle of a strand in the double helix is about 1.62 times its width. Perhaps the most surprising appearance of the golden ratio is in connection with black holes, regions of space in which the gravitational field is so strong that nothing can escape (even light). A rotating black hole loses energy and up to a point heats up as it does so; after that point–when the mass of the hole equals its angular momentum times the square root of ϕ–the hole starts to cool down instead.

19.2 Rosette, Strip, and Wallpaper Patterns

The spiral distribution of the seeds in a sunflower head and the spiraling of leaves around a plant stem are instances of *similarity* and *repetition,* two key aspects of symmetry. They also illustrate *balance,* which refers to regularity in *how* the repetitions are arranged. In considering patterns with repetition, we distinguish the individual element or figure of the design (sometimes called the *motif*) from the *pattern* of the design–*how the copies of the motif are arranged.*

The problem that we focus on in this chapter is how to explore and classify the fundamentally different ways in which a flat design can be symmetrical. The ideas that we discuss were used by scientists to discover what crystalline forms are possible. Although there is a limitless number of chemical structures, and of motifs that people can make, what is quite surprising is that there is only a limited number of ways to arrange atoms in a structure or motifs in a design in a symmetrical way.

How can we enumerate the ways that designs can be put together without counting all the actual designs themselves? The key mathematical idea is to look at what you can *do* to the pattern without changing its appearance.

Rigid Motions

Mathematicians describe various kinds of symmetry by using the geometric notion of a **rigid motion**, also known as an **isometry** (which means "same size"). A rigid motion is a specific kind of variation on the original pattern: We pick it up and move it, perhaps rotate it, possibly flip it over–but we *don't change its size or shape.* (The original figure and its image are not just geometrically similar, in the language of section 18.1, but also the same size.)

Rigid Motion DEFINITION

A **rigid motion** is one that preserves the size and shape of figures. In particular, any pair of points is the same distance apart after the motion as before.

Figure 19.7 shows the results of various motions applied to the rectangle and its interior of Figure 19.7a. In Figure 19.7b, each side is shrunk by 50%–not a rigid motion, because the size of the rectangle changes. For Figure 19.7c, we shear ("squash") the rectangle–again, this is not a rigid motion because the shape of the rectangle changes. In Figure 19.7d we rotate the rectangle 90° (a quarter-turn) clockwise around the center of the rectangle: This is a rigid motion. Similarly, in Figure 19.7e, rotating by 180° (a half-turn) is a rigid motion.

In Figure 19.7f we reflect the rectangle along a vertical mirror down the middle: Could you tell? The right and left halves exchange places.

Figure 19.7g shows the result of reflecting across a diagonal of the rectangle. All reflections and all rotations are rigid motions. So are all **translations**, which move every point in the plane a certain distance in the same direction.

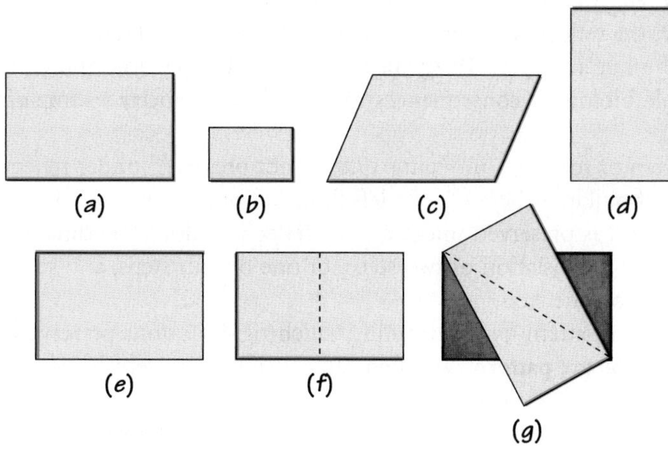

FIGURE 19.7 Results of various motions applied to a blue-edged rectangle and its interior: (a) the original rectangle and interior; (b) 50% reduction (not a rigid motion); (c) shearing (not a rigid motion); (d) quarter-turn; (e) half-turn; (f) reflection along the vertical line down the middle; (g) reflection along a diagonal line.

The only remaining kind of rigid motion in the plane is a combination of reflection and translation. **Glide reflection** is the kind of pattern that your footprints make as you walk: Each successive element of the design (footprint) is a reflection of the previous one (Figure 19.8). The motion combines, in an integral way, translation ("glide") with a reflection across a line parallel to the direction of the translation.

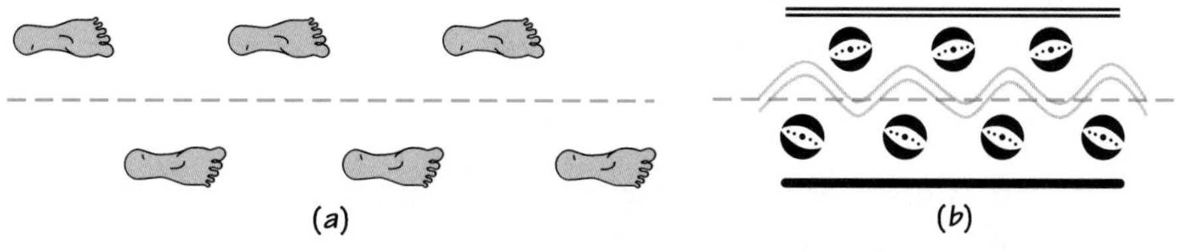

FIGURE 19.8 Glide reflection of (a) footprints; (b) design elements on a pot from San Ildefonso Pueblo, in New Mexico.

Any rigid motion of the plane must be one of:

- Reflection (across a line)
- Rotation (around a point)
- Translation (in a particular direction)
- Glide reflection (across a line)

Performing one rigid motion after another results in a rigid motion that (surprisingly) must be one of the four types that we have just explored.

Preserving the Pattern

In terms of symmetry, we are especially interested in rigid motions like those of Figures 19.7e and 19.7f that **preserve the pattern**–that is, ones for which the pattern looks exactly the same, *with all the parts appearing in the same relative places,* after the motion is applied.

You might enjoy thinking of applying these motions as "The Pattern Game": You turn your back, I apply a transformation, then you turn back and see if you can tell whether anything is changed.

The 90° rotation of Figure 19.7a into Figure 19.7d does not preserve the pattern. The moved rectangle doesn't fit exactly over the original rectangle. On the other hand, the 180° rotation in Figure 19.7e does preserve the pattern. It's true that the top of the original rectangle is now on the bottom of the transformed version, but you can't tell. A rotation by any multiple of 180° would also preserve the pattern.

Similarly, the reflection across the vertical line in Figure 19.7f preserves the pattern, while the one in Figure 19.7g, along a diagonal, does not. Spotlight 19.4 discusses possible biological consequences of reflection symmetry or imperfections in it.

The pattern of footsteps in Figure 19.8a is not preserved under reflection along the direction of walking–there is not a left footprint directly across from a right footprint. The pattern is preserved under a glide reflection along the direction of walking, as well as by a translation of two steps, or one of four steps, and so on–but not by a translation of one step.

We analyze a pattern by determining which rigid motions preserve it; they are the **symmetries of the pattern**. We then can classify the pattern by which rigid motions preserve it.

We may think of a pattern as a recipe for repeating a figure (motif) indefinitely. Of course, any pattern in nature or art has only finitely many copies of the figure. If the recipe for repetition is clear, we can imagine that we are looking at just a part of a pattern that extends indefinitely.

Patterns in the plane can be divided into those that have indefinitely many repetitions in

- no direction–the **rosette patterns**
- exactly one direction (and its reverse)–the **strip patterns**
- more than one direction (and their reverses)–the **wallpaper patterns**

Rosette Patterns

A rosette pattern describes the possible symmetries for a single flower. The repetition aspect of asymmetry consists of the repetition of the petals around the stem. Translations and glide reflections do not come into play. The pattern is preserved under a rotation by certain angles corresponding to the number of petals. There may

SPOTLIGHT 19.4 "Strive Then to Be Perfect"

Is there no such thing as objective and universal beauty, as claimed by American feminist Naomi Wolf in her book *The Beauty Myth*?

Stand in front of a mirror and look at yourself. Do your left and right sides look exactly symmetrical? What about the part in your hair, freckles on your face, evenness of your shoulders, bending of your ears?

Symmetry may be a proxy for fitness. Symmetrical racehorses tend to run faster; male lions with lopsided facial whisker-spot patterns die younger. The more symmetrical a flower is, the more nectar it produces, making it a better food source for pollinating insects; and correspondingly, insects prefer symmetrical flowers, giving such flowers a better chance of being pollinated.

Perhaps because of association with fitness, symmetry may affect mate selection among animals. Female zebra finches prefer males with symmetrical leg bands. Fruit flies and female barn swallows prefer males with symmetrical tails; a particular parasite can lead to an uneven tail.

What about people? Both male and female Britons, as well as Tanzanian hunter-gatherers, find facial symmetry more attractive than asymmetry. Perfectly symmetrical female faces that are computer-generated from composites of individual photos appear more attractive to men than photos of actual women's faces. Studies also indicate that "symmetrical" men tend to have an earlier first sexual experience, more sexual partners, and more extramarital affairs, while asymmetry of the hands is associated with low sperm count and poor sperm motility. Finally, women with symmetrical breasts tend to be more fertile and less susceptible to breast cancer.

Michalangelo's David

(Roger Antrobus/Getty Images.)

or may not be reflections that preserve the pattern, depending on whether the petal itself has reflection symmetry. Most flowers do (Figure 19.9a), but some do not. An everyday example of the rosette pattern–a human-made one–that does not have reflection symmetry is a pinwheel (Figure 19.9b). If there is no reflection symmetry, the motif of the pattern (the element that is repeated) is an entire petal. If there is reflection symmetry, the motif is just half a petal, because the entire pattern can be generated by rotation and reflection of a half petal. The fact that these are the only possibilities is sometimes called *Leonardo's theorem,* after Leonardo da Vinci, who, in the course of planning the design of churches, needed to decide if chapels and niches could be added without destroying the symmetry of the central design.

Leonardo realized that there are two different classes of rosettes, the ones without reflection symmetry (*cyclic rosettes*) and those with it (*dihedral rosettes*) (see Figure 19.9). The respective notations for the patterns are *cn* and *dn*, where *n* is the number of times that the rosette coincides with its original position in one complete turn around the center. A cyclic pattern has no lines of reflection symmetry, while the dihedral pattern *dn* has *n* different lines of reflection symmetry. The flower in Figure 19.9a has dihedral pattern *d34*, because each petal has reflection symmetry, while the pinwheel in Figure 19.9b has pattern *c8*.

(a) (b)

FIGURE 19.9 (a) Flower; each petal has reflection symmetry. (*Gregory G. Dimijian/Photo Researchers, Inc.*) (b) Pinwheel with eight "leaves," each asymmetric symmetrical, hence pattern *c8* (*Bloomimage/Corbis.*)

Strip Patterns

We illustrate the different kinds of strip patterns, and their "ingredient" symmetries, with patterns in the art of the Bakuba people of the Democratic Republic of the Congo, who are noted for their fascination with pattern and symmetry (see Spotlight 19.5).

All the strip patterns offer repetition and **translation symmetry** along the direction of the strip. For simplicity, we always position the pattern so that its repetition runs horizontally.

It may be that the pattern has no other rigid motions that preserve it apart from translation, as in Figure 19.10a.

The simplest other rigid motion to check is reflection across a line. For a strip pattern, the center line of the strip may be a reflection line, as in Figure 19.10b; we say that the pattern has symmetry across a horizontal line. There may instead be reflection across a *vertical* axis, such as the vertical lines through or between the V's in Figure 19.10c.

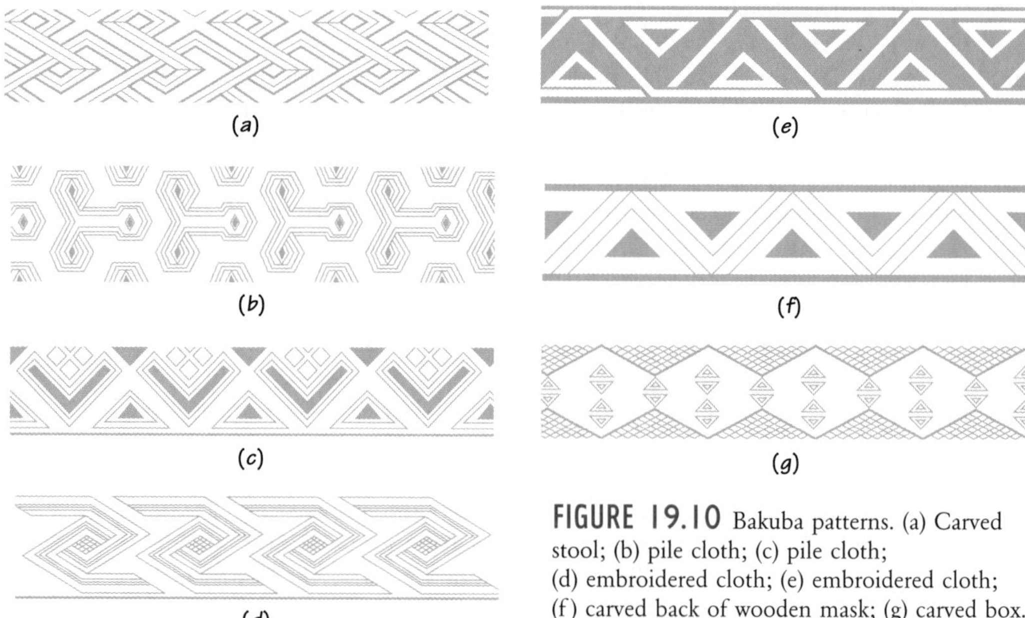

FIGURE 19.10 Bakuba patterns. (a) Carved stool; (b) pile cloth; (c) pile cloth; (d) embroidered cloth; (e) embroidered cloth; (f) carved back of wooden mask; (g) carved box.

SPOTLIGHT 19.5 Patterns Created by the Bakuba People

Among the Bakuba people of the Democratic Republic of the Congo (shaded area of map), it is considered an achievement to invent a new pattern, and every Bakuba king had to create a new pattern at the outset of his reign. The pattern was displayed on the king's drum throughout his reign and, for some kings, on his dynastic statue.

When missionaries first showed a motorcycle to a Bakuba king in the 1920s, he showed little interest in it. But the king was so enthralled by the novel pattern the tire tracks made in the sand that he had it copied and gave it his name.

Source: Adapted from Jan Vansina, *The Children of Woot*, University of Wisconsin Press, Madison, 1978, p. 221.

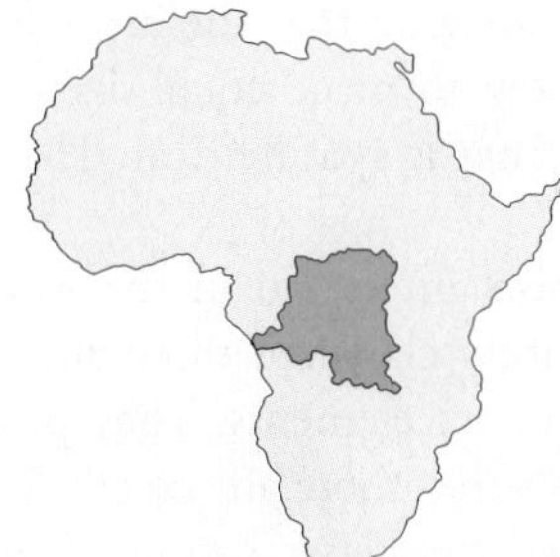

Two women with raffia cloths from the Bakuba village of Mbelo, July 1985. *Left*: Mpidi Muya with embroidered raffia (a kind of fiber) cloth. *Right*: Muema Kenye with plush and embroidered raffia cloth. *(Dorothy K. Washburn.)*

The pattern made by tire tracks fascinated the Bakuba people. *(Travis Amos.)*

What kind of rotational symmetry can a strip pattern have? The only possibility for a strip pattern is a rotation by 180° (a half-turn), because any other angle won't even bring the strip back into itself. (We don't count rotations of 360° or integer multiples [full turns], because any pattern is preserved under these.) Figure 19.10d shows a strip pattern that is unchanged by a 180° rotation about any point at the center of the small crosshatched regions.

What about glide reflections? A row of alternating p's and b's has glide reflection:

Glide: p p p p p p p p p

Reflection: p p p p p p p p p
----
b b b b b b b b b

Glide reflection: p b p b p b p b p (on a dashed line)

For glide reflection, a p is translated as far as the next b and is then reflected upside down. Figure 19.10e shows a Bakuba pattern whose only symmetry (except for translation) is glide reflection.

Having examined symmetries on strip patterns, we can ask: What *combinations* of the four are possible? It turns out that apart from the five kinds of patterns we have already seen, there are only two other possibilities: We can have vertical line reflection, half-turns, and glide reflection, either with horizontal line reflection (Figure 19.10g) or without (Figure 19.10f).

Mathematical analysis reveals the following.

There Are Only Seven Ways to Strip RULE

There are only seven ways to repeat a pattern along a strip.

That this number is so small is quite surprising, because there are myriad different design elements (motifs). Two designs may look entirely different yet share the same pattern of reproducing their design elements.

Wallpaper Patterns and Crystal Structures

So far we have classified the patterns that have no translation repetition (the rosette patterns) and those with repetition in one direction (the strip patterns). What about repetitions in more than one direction–say, in two different directions across a plane? It turns out that there are exactly 17 ways to do so, called *wallpaper patterns.* We give illustrations, notation, and a flowchart in Spotlight 19.6.

We emphasize again that "pattern" does not refer to the basic design but to how its repetition is structured across the plane. There is an infinite variety of possible designs that artists can devise. You should imagine that the artist has created one copy of the design and is contemplating how to place equal-sized copies of it in other parts of the (infinite) plane, in a way that is symmetrical. There are very few (17) strategies possible for doing so.

Crystallographers (physicists and chemists interested in the ways that crystals can occur or be built) in the nineteenth century classified three-dimensional crystal structures in terms of combinations of symmetry elements. They proved–after several years of coming up with different totals!–that there are exactly 230 patterns for crystals. Mathematicians have further refined the classification of patterns to take into account colors that are repeated in a symmetrical way.

SPOTLIGHT 19.6 The 17 Wallpaper Patterns

There are exactly 17 wallpaper patterns. Here we give an example of each, together with a flowchart for identifying them. Crystallographers have standard notations and abbreviations for the patterns. The full notation consists of four symbols:

1. The first symbol is *c* (for "centered") if all rotation centers lie on the reflection lines, or *p* (for "primitive") otherwise.
2. The second symbol indicates rotational symmetry. It is either *1, 2, 3, 4,* or *6,* corresponding to rotational symmetry of, respectively, 360°, 180°, 120°, 90°, or 60°. The symbol is the largest applicable number. For example, if symmetries of 360°, 120°, and 60° are present, the symbol is 6.
3. The third symbol is either *m, g,* or *1,* corresponding to the presence of "mirror," "glide," or no reflection symmetry.
4. The fourth symbol (*m, g,* or *1*) is for describing symmetry relative to an axis at an angle to the symmetry axis of the third symbol.

(*Note*: The patterns *p31m* and *p3m1* are exceptions to this scheme.)

Below each pattern illustration, we give both the standard abbreviation (on top) and the full notation (below).

(continued on page 623)

The 17 Wallpaper Patterns *(continued)*

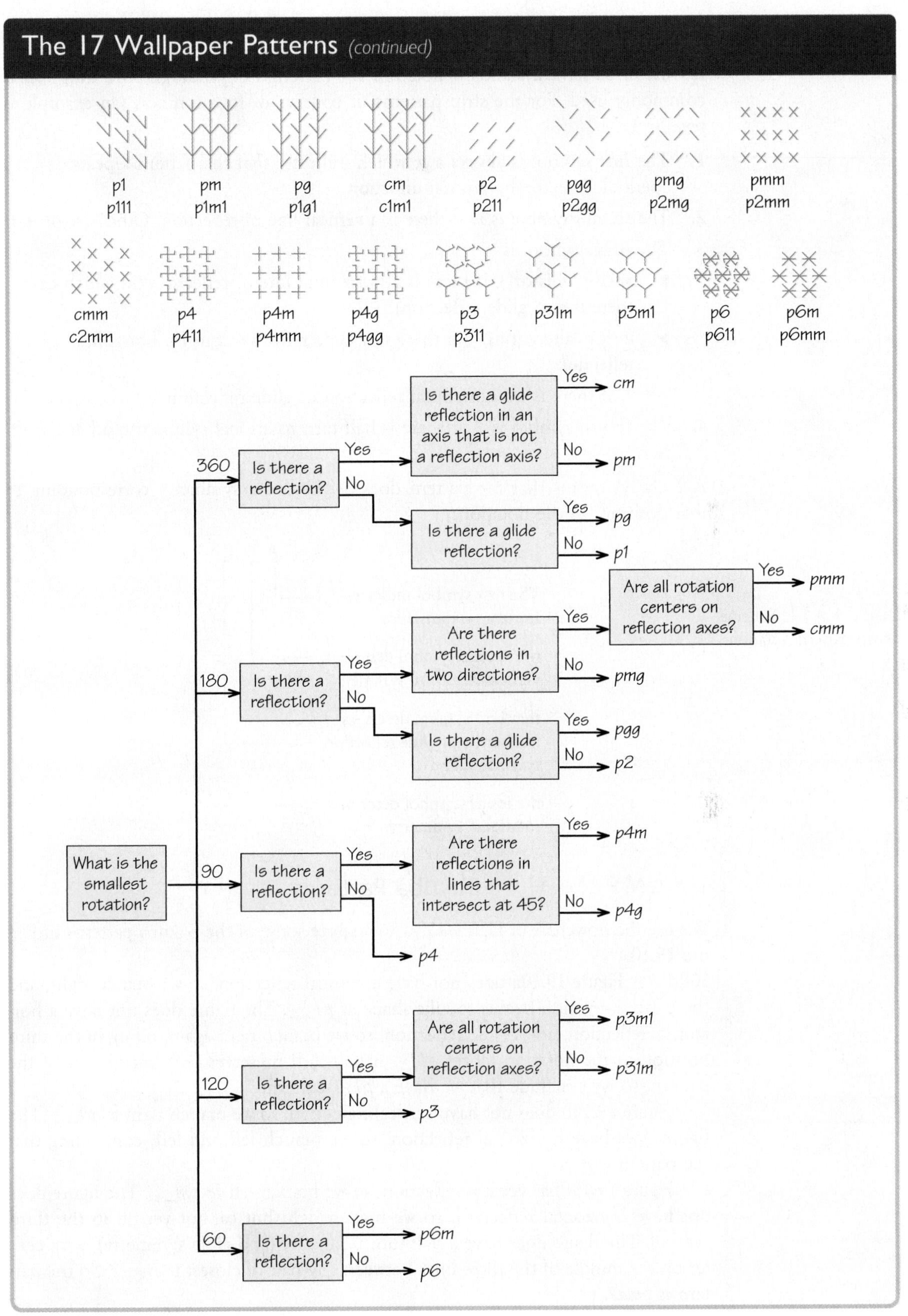

19.3 Notation for Patterns

It's useful to have a standard notation for patterns. **Crystallographic notation** is commonly used. For the strip patterns, it consists of four symbols (an example is *pma2*):

1. The first symbol is always a *p*, which indicates that the pattern repeats (is "periodic") in the horizontal direction.
2. The second symbol is *m* if there is a vertical line of reflection. Otherwise, it is *1*.
3. The third symbol is
   - *m* (for "mirror"), if there is a horizontal line of reflection (in which case there is also glide reflection)
   - *a* (for "alternating"), if there is a glide reflection but no horizontal reflection
   - *1*, if there is no horizontal reflection or glide reflection
4. The fourth symbol is *2*, if there is half-turn rotational symmetry; otherwise, it is *1*.

A *1* always means that the pattern does *not* have the symmetry corresponding to that position. In the notation:

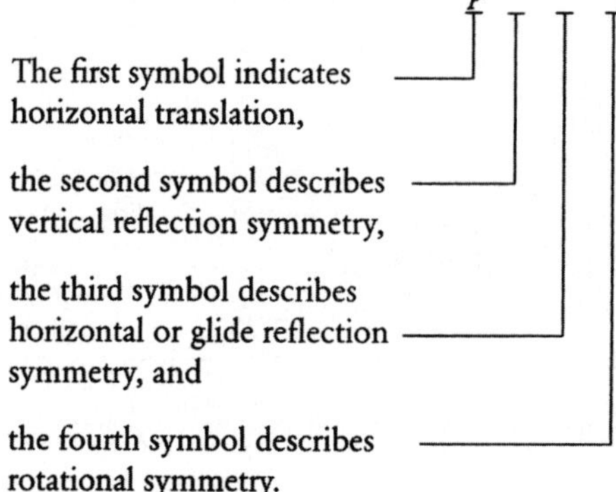
p
The first symbol indicates horizontal translation,
the second symbol describes vertical reflection symmetry,
the third symbol describes horizontal or glide reflection symmetry, and
the fourth symbol describes rotational symmetry.

FIGURE 19.11 Scheme for strip pattern notation.

EXAMPLE 1 ■ Bakuba Patterns

We use the flowchart of Figure 19.11 to analyze some of the Bakuba patterns of Figure 19.10.

SOLUTIONS Figure 19.10a does not have a vertical reflection, so we branch right, and the pattern notation begins to take shape as *p1_ _*. The figure does not have a horizontal reflection, nor a glide reflection, so we branch right again, filling in the third position in the notation to get *p11_*. A half-turn preserves part but not all of the pattern, so we conclude that we have a *p111* pattern.

Figure 19.10b does not have vertical reflection, so we branch right to *p1_ _*. The figure does have horizontal reflection, so we branch left and left, concluding that the pattern is *p1m1*.

Figure 19.10f has vertical reflection, so we branch left to *pm_ _*. The figure does not have horizontal reflection, so we branch right but cannot yet fill in the third symbol. The figure does have a half-turn symmetry (and glide symmetry), with center on the middle of the three lines between any pair of closest triangles. So the pattern is *pma2*.

Remember the Bakuba people's fascination with tire treadmarks? Apart from esthetic value, certain symmetries are important for practical purposes. The Museum of Transport in Glasgow, Scotland, includes all kinds of vehicle tires. However, only five of the seven strip patterns appear among treads of all the tires there. Examining Figure 19.12, can you guess which two patterns do not appear, what they have in common, and the practical reason why they are not used in tire treads?

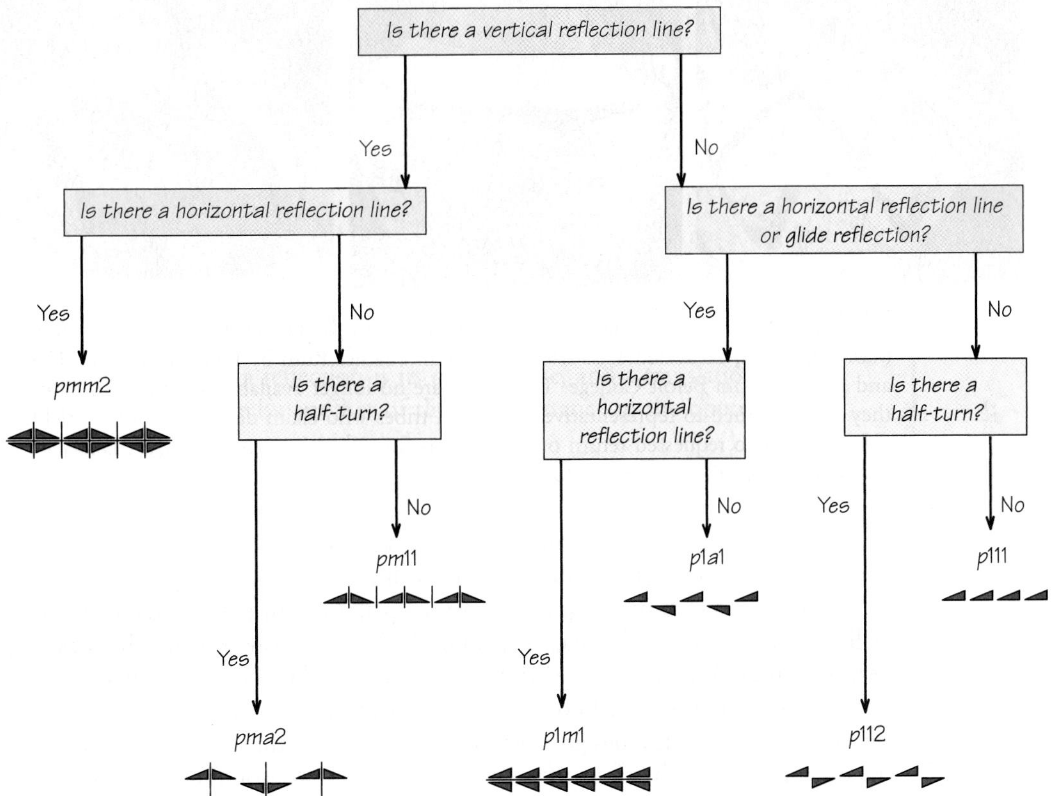

FIGURE 19.12 A flowchart for identifying the seven strip patterns and classifying them according to crystallographic notation.

Imperfect Patterns

In applying these classification schemes to patterns on real objects, we need to take into account that the pattern itself may not be perfectly rendered. Also, patterns that are not on flat surfaces–for example, the pattern around the rim of a bowl or around the body of a jar–require some latitude in interpretation.

EXAMPLE 2 ▪ Patterns on Pueblo Pottery

The pitchers in Figure 19.13 are from a thousand-year-old Pueblo site at Starkweather Ruin near Reserve, New Mexico. Consider the patterns on the bodies of the pitchers, which continue on the back sides. Let's suppose that they could be unwrapped and continued as strip patterns, but we'll disregard the patterns on the spouts and handles.

We immediately come up against the question of the perfectness of the patterns. In Figure 19.13a the "teeth" (represented by the zigzagging of lightning bolts)

observe which other motif it is carried into under the symmetry. For instance, T^2 takes each motif into the motif two to the right. The symmetry T has an inverse T^{-1} among the symmetries of the pattern: the smallest translation to the *left* that preserves the pattern. Moreover, $T^{-1} \circ T^{-1}$ (which we write as T^{-2}), $T^{-1} \circ T^{-1} \circ T^{-1} = T^{-3}$, and so forth are also symmetries. The entire collection of symmetries of the pattern is

$$\{\ldots, T^{-3}, T^{-2}, T^{-1}, I, T, T^2, T^3, \ldots\}$$

From this listing, you see that it is natural to think of the identity I as being T^0. All the strip patterns are preserved by translations, so the symmetry group of each includes the *subgroup* consisting of all translations in this list. We say that the group is **generated by** T, and we write the group as $\langle T\rangle$, where between the angle brackets we list symmetries (**generators**) that, in combination, produce all of the group elements.

The symmetry group of Figure 19.10e includes, in addition, a glide reflection G and all combinations of the glide reflection with the translations. Doing two glide reflections is equivalent to doing a translation, which we express as $G^2 = T$. The glide is only "half as far" as the shortest translation that preserves the pattern. Check that $G \circ T = T \circ G$. The symmetry group of the pattern is

$$\{\ldots, G^{-3}, G^{-2} = T^{-1}, G^{-1}, I, G, G^2 = T, G^3, \ldots\} = \langle G\rangle$$

The pattern of Figure 19.10c is preserved by vertical reflections at regular intervals. If we let V denote reflections at a fixed particular location, the other reflections can be obtained as combinations of V and T. To get a handle on what each of the symmetries does, it helps to make a "simplified" copy of the strip (we use V's), number fixed positions on the page, and identify individual copies of the V's with letters, as in the following:

1	2	3	4	5
V_a	V_b	V_c	V_d	V_e

The symmetries move the V's among the numbered positions. Let V be the reflection across the vertical line through the middle of position 3, and let T be the translation that moves each V one square to the right. To familiarize yourself with the symmetries, write out the result for each of V, T, and VT (V followed by T). (For convenience, we can omit the operation sign between the two symmetries.)

The symmetry group of the pattern, the list of all of the symmetries, is

$$\{\ldots, T^{-3}, T^{-2}, T^{-1}, I, T, T^2, T^3, \ldots; \\ \ldots, T^{-3}V, T^{-2}V, T^{-1}V, V, TV, T^2V, T^3V, \ldots\}$$

This group is notable because not all of its elements satisfy the *commutative property* that $A \circ B = B \circ A$, which you are used to for numerical operations ($a + b = b + a$; $a \times b = b \times a$). In fact, we do not have $VT = TV$, but instead $VT = T^{-1}V$. Verify this fact by working out the effect of $T^{-1}V$, using your simplified strip from above, and compare with what you got for VT earlier.

We can express this group compactly as $\langle T, V | VT = T^{-1}V\rangle$, where we list the generators and indicate what relations hold among them. ■

We have made a transition from thinking about patterns in geometrical terms to reasoning about them in algebraic notation–in effect, applying one branch of mathematics to another. This kind of cross-fertilization is characteristic of contemporary mathematics.

The concept of a group is a fundamental one in the mathematical field of abstract algebra. The generality ("abstractness") is exactly why groups and other algebraic structures arise in so many applications, in areas ranging from crystallography, quantum physics, and cryptography, to error-correcting codes (see Chapters 16 and 17) and anthropology (describing kinship systems).

19.5 Fractal Patterns and Chaos

We noted earlier that similarity and repetition are key aspects of symmetry, as are balance and proportionality. In most of our examples, the repetitions of a motif have been at the same size. Exceptions were the chambers of the nautilus in Figure 19.2b and the varying sizes of leaves and seeds in plants that feature the spiral pattern of phyllotaxis (Figures 19.1 and 19.2a). These exhibit a kind of "proportion," or numeric symmetry–symmetry with changes of scale. Another example of similarity with changes of scale are the nested dolls ("matrioshka") shown in Figure 19.15. They feature a linear scaling factor (see section 18.1) between one doll and another. Each part of one doll (face, arm, and so forth) has the same proportion (scaling factor) to the corresponding part of a second doll.

Fractals

Fractals are another example of symmetry in which linear scaling is used. The word **fractal** was invented in 1975 by Benoit Mandelbrot from the Latin word *fractus* meaning "broken into fragments" (of varied sizes), from which we get *fragment* and also *fracture* and *fraction*. Mandelbrot defined a fractal in strict mathematical terms that we formulate more informally as follows:

Fractal DEFINITION

A **fractal** is a pattern that exhibits similarity at ever finer scales.

The scaling is usually by a *linear scaling factor.*

We show various fractals in Figure 19.16. Figure 19.16a looks to us like an orchid with pronounced "bee guides" to the pollen. With its vertical mirror line, the overall pattern has *d1* rosette symmetry. However, the basic motif of the lacy wings is repeated at an infinite number of scales. In Figure 19.16b, the "suckers" on the "tentacles" appear in smaller and smaller sizes as the "tentacles" wind their way toward the point at the center. The pattern in Figure 19.16c has overall rosette symmetry of type *c2*, but the "seahorse" motif, with two large "seahorses" foot-to-foot in the center, is repeated in diminishing sizes throughout. Figure 19.16d features (to our imagination) "spikey snowmen," with smaller ones growing out of the sides of larger ones. What do they look like to you? And does the overall pattern as a rosette have symmetry *c1*, *c2*, *d1*, or *d2*?

A famous example of a fractal pattern is Maurits Escher's print "Circle Limit IV" (Figure 19.17). As you examine the angels and devils closer and closer to the boundary of the circle, you notice that they are not necessarily geometrically similar to the ones at the center. However, if you imagine that the print is the image of a hemisphere, then figures farther away from your viewpoint should indeed appear smaller.

FIGURE 19.15 Nested "matrioshka" dolls from Russia exhibit symmetry at different scales. (*Photodisc Green/Getty Images.*)

(a) "Paradise." (b) "Purgatory." (c) "r-crest." (d) "Scarab 2."

FIGURE 19.16 Various fractals. (*Courtesy of Noel Giffin and the Spanky Fractal Database.*)

FIGURE 19.17 M. C. Escher's fractal pattern "Circle Limit IV." (*© 2005 M. C. Escher Company–Holland. All rights reserved. www.mcescher.com.*)

Apart from their beauty and the opportunity that they offer as an art form (there is even "fractal music"!), fractals have two major applications:

- Fractals with very simple rules for replicating the motif mimic very well certain natural phenomena, such as the structure of a leaf, a tree, or a mountain (see Figure 19.18). This fact not only allows us to model leaves, trees, and so on, using fractals but also suggests that such natural phenomena are produced by corresponding simple "rules of nature." Moreover, computer special effects in films can use fractals to mimic nature very closely, as in *Star Trek II: The Wrath of Khan,* for landscapes on the Genesis planet, and in *Return of the Jedi,* for the moons of Endor and the Death Star. In Chapter 23, we investigate in detail one particularly simple replication rule, called an **iterated function system (IFS)**; this complicated-sounding name hides the fact that such a system is just a recursive rule, like the one for forming the Fibonacci numbers. Figure 19.19 shows a simple geometric IFS.

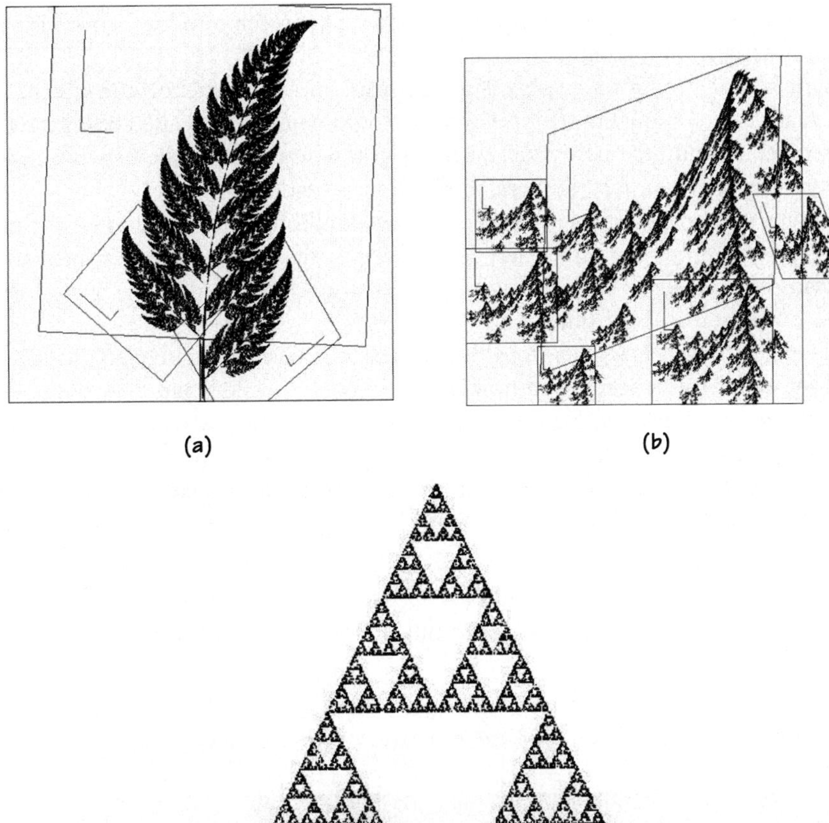

FIGURE 19.18 (a) Barnsley's fractal fern and (b) a snowy mountain landscape, both with templates showing how they were formed using reflections and linear "distortions" in addition to linear scaling. Each leaf of the fern and each mountain peak is in fact just a smaller copy of the entire image. (*Fred Solomon, Warren Wilson College.*)

FIGURE 19.19 A Sierpiński "triangle." Start with the big triangle, and remove its middle triangle. Then do the same for the three remaining smaller triangles. Recursively do the same for each subsequent smaller triangle. Can you guess the area of the resulting figure? (*Hint:* It may be less than you think.) (*Annalisa Crannell, Franklin and Marshall College.*)

▶ Fractals form the basis for an important method of image compression, simliar in efficiency to the better-known JPEG algorithm. The key idea is to store not the millions of bits that make up an image but instead a much smaller number (maybe thousands) of rules for generating patterns that can be found in the image. A simple example (which doesn't use fractals) is that you can compress a checkerboard of a million pixels that alternate between black and white to just two simple rules: If the current pixel is white, the next one is black, and if the current one is black, the next one is white. A more realistic example is Microsoft's original 1992 Encarta Encyclopedia. It contained thousands of articles and photographs, plus color animations and hundreds of maps–all on one CD-ROM, thanks to fractal data compression.

Symmetry in Chaos

While the patterns in rosettes, strips, wallpaper patterns, and some fractals can be produced from very simple rules for the symmetries involved, you may be surprised to learn that symmetry can also arise from apparently random behavior.

We think of symmetry as referring to order, and chaos to disorder and randomness. Scientists use the word *chaos* in a technical sense to describe systems whose behavior over time is inherently unpredictable. We explore chaotic systems in Chapter 23 (section 23.5). Here we investigate how chaos can produce astonishingly beautiful designs on a computer screen.

One way to produce a graphic is to start with an initial pixel on the screen, apply a mathematical function (formula) to its coordinates to generate coordinates of a new pixel, light up the new pixel, then repeat the process with the new pixel. The process is recursive, in fact, an iterated function system.

Iterate the process a large number of times–millions or even hundreds of millions of times. Since the screen has many fewer pixels than that, by what mathematicians call the *pigeon-hole principle,* some pixels must be visited more than once–maybe even thousands of times.

The clue to producing art from this process is "color by number": Choose the color for each pixel according to how many times it is visited, and choose the colors with an eye to beauty. Figure 19.20 shows an example with *d5* symmetry that was produced by 30,000,000 iterations. The scale on the right in Figure 19.20 shows the colors for the number of times that pixels were hit; unhit pixels stay black. The order in which pixels are visited appears to be completely chaotic and is irrelevant to the final image.

- If you ignore the first thousand or so pixels visited, *it doesn't matter what pixel you start from–you get the same image!* But the pixels are visited in different orders.
- The formulas are variations on the *logistic map,* an iterated function system that we discuss in Chapter 23 in connection with biological populations.

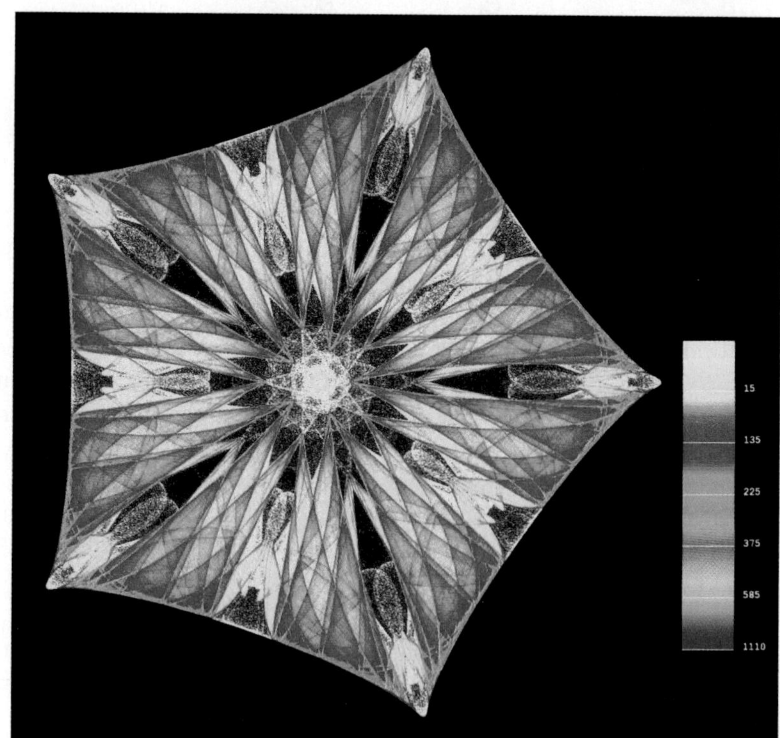

FIGURE 19.20
"Emperor's Cloak," with *d5* symmetry. This work of art was produced by iterating a chaos-producing function, starting at one point and successively generating new points according to a fixed rule. The color bar shows the coloring of pixels according to how often they are visited by the iterations. (*Figure 1.13, p. 20, of Michael Field and Martin Golubitsky,* Symmetry in Chaos: A Search for Pattern in Mathematics, Art and Nature, *New York, Oxford University Press, 1992.*)

SPOTLIGHT 19.7 The Father of Fractals

Benoit Mandelbrot (1924–) was born in Poland, grew up partly in France, and came to the United States to work for IBM. He found that many phenomena feature both repeating patterns and power curves (see section 18.6). The patterns are repeated at a change of scale, as Barnsley's fern in Figure 19.18 and the Sierpiński triangle in Figure 19.19 show, and can be described by very simple rules. In 1975 he coined the term *fractals,* and the systems of rules became known as iterated function systems. Now retired from IBM, he teaches a course in Fractal Geometry at Yale University. The Web site for the course (see Suggested Web Sites) lists 100 or so examples of fractal phenomena in nature and society. Mandelbrot's book, *The (mis)Behavior of Markets: A Fractal View of Risk,* applies fractals to describe how stock market prices vary.

Benoit Mandelbrot

(Roger Ressmeyer/Corbis.)

REVIEW VOCABULARY

Crystallographic notation A four-symbol notation used by crystallographers (and mathematicians) to classify strip patterns and wallpaper patterns. (p. 624)

Divine proportion Another term for the golden ratio. (p. 612)

Fibonacci numbers The numbers in the sequence 1, 1, 2, 3, 5, 8, 13, 21, 34, ... Each number after the second is obtained by adding the two preceding numbers. (p. 610)

Fractal A pattern that exhibits similarity at ever-finer scales. (p. 631)

Generated, generators A group is generated by a particular set of elements (they are the generators) if composing them and their inverse in combinations can produce all elements of the group. (p. 630)

Geometric mean The geometric mean of two numbers a and b is $\sqrt{ab}$. (p. 613)

Glide reflection A combination of translation (= glide) and reflection in a line parallel to the translation direction. Example: pbpbpb. (p. 617)

Golden ratio, golden mean The number

$$\phi = \frac{1 + \sqrt{5}}{2} = 1.618034\ldots$$ (p. 612)

Golden rectangle A rectangle the lengths of whose sides are in the golden ratio. (p. 612)

Group A group is a collection of elements with an operation on pairs of them such that the collection is closed under the operation, there is an identity for the operation, each element has an inverse, and the operation is associative. (p. 628)

Isometry Another word for *rigid motion.* Angles and distances, and consequently shape and size, remain unchanged by a rigid motion. For plane figures there are only four possible isometries: reflection, rotation, translation, and glide reflection. (p. 616)

Iterated function system (IFS) A sequence of elements (numbers or geometric objects) in which each successive element is determined recursively by applying the same function (rule) to the previous element. (p. 632)

Phyllotaxis The spiral pattern of shoots, leaves, or seeds around the stem of a plant. (p. 609)

Preserves the pattern A transformation preserves a pattern if all parts of the pattern look exactly the same after the transformation has been performed. (p. 618)

Recursion A method of defining a sequence of numbers, in which the next number is given in terms of previous ones. (p. 611)

Reflection symmetry Mirror-image symmetry. (p. 609)

Rigid motion A motion that preserves the size and shape of figures. In particular, any pair of points is the same distance apart after the motion as before. (Also called *isometry.*) (p. 616)

Rosette pattern A pattern whose only symmetries are rotations about a single point and reflections through that point. (p. 618)

Rotational symmetry A figure has rotational symmetry if a rotation about its "center" leaves it looking the same, like the letter S. (p. 609)

Strip pattern A pattern that has indefinitely many repetitions in one direction. (p. 618)

Symmetry of the pattern A transformation of a pattern is a symmetry of the pattern if it preserves the pattern. (p. 618)
Symmetry group of the pattern The group of symmetries that preserve the pattern. (p. 629)
Translation A rigid motion that moves everything a certain distance in one direction. (p. 617)
Translation symmetry An infinite figure has translation symmetry if it can be translated (slid, without turning) along itself without appearing to have changed. Example: AAA (p. 620)
Wallpaper pattern A pattern in the plane that has indefinitely many repetitions in more than one direction. (p. 618)

✓ SKILLS CHECK

1. Symmetry includes notions of
(a) balance
(b) similarity
(c) repetition
(d) all of the above.

2. Many people think that mathematics is just about numbers, but in fact mathematics is the study of _____ .

3. Which of the following rectangles is an approximate golden rectangle?
(a) 10 by 16
(b) 6 by 13
(c) 8 by 11

4. The geometric mean of 4 and 36 is ____________ .

5. Which artist claimed to use the golden ratio in his work?
(a) Leonardo da Vinci
(b) Wolfgang Amadeus Mozart
(c) Neither

6. In the Fibonacci sequence ______ follows 13 and 21.

7. A rigid motion always moves any pair of points
(a) in the same direction.
(b) to another pair of points the same distance apart.
(c) to their mirror images.

8. The capital letters ___, ___, ___, ___, ___, ___, and ___ each have a rotation isometry.

9. Assume that the following two patterns continue in both directions. Which of these patterns has a reflection isometry?

ZZZZZZZZZ
UUUUUUUUU

(a) ZZZZZZZZZ only
(b) UUUUUUUUU only
(c) Neither

10. This strip pattern

╜╒ ╜╒ ╜╒ ╜╒

has ___________ and ___________ isometries.

11. What isometries does this wallpaper pattern have?

(a) Translation and reflection only
(b) Translation and rotation only
(c) Translation, rotation, and reflection

12. This wallpaper pattern

has ___________ and ___________ isometries.

13. If a horizontal strip pattern has a glide reflection isometry, then
(a) it always has a horizontal reflection isometry.
(b) it may also have a horizontal reflection isometry.
(c) it cannot have a horizontal reflection isometry.

14. If a strip pattern has both vertical and horizontal reflection isometries, then it always has a ___________ isometry.

15. Consider the strip pattern in the raffia cloth held by the woman in the photo on the right. What isometries does it have?

(Dorothy K. Washburn.)

(a) Vertical reflection
(b) Horizontal reflection
(c) Glide reflection

16. The symbol *p* indicates that a strip pattern has ____________ symmetry.

17. The symbol *2* indicates that a pattern has
(a) rotational symmetry.
(b) reflection symmetry.
(c) too much symmetry.

18. The symbol m indicates that a wallpaper pattern has ____________ symmetry.

19. The symmetry group of a rectangle has how many elements?

(a) 4
(b) 6
(c) 8

20. The symmetry group of the strip pattern *pmm2* has ________ elements.

CHAPTER 19 EXERCISES

■ Challenge ◆ Discussion

19.1 Fibonacci Numbers and the Golden Ratio

1. Examine the "scales" on the surface of a pineapple, which are arranged in spirals around the fruit. Note that there are spirals in three distinct directions. For each direction, how many spirals are there?

2. Repeat Exercise 1, but for a pinecone from your area.

3. Repeat Exercise 1, but for a sunflower.

4. Here are two primitive models of natural increase of biological populations, similar to those Fibonacci hypothesized around the year 1200. A pair of newborn male and female rabbits is placed in an enclosure to breed.

(a) Suppose that the rabbits start to bear young one month after their own birth. This may be unrealistic for rabbits, but we could substitute another species for which it is realistic. Fibonacci used rabbits. At the end of each month, they have another male–female pair, which in turn mature and start to bear young one month later. Assuming that none of the rabbits die, how many pairs of rabbits will there be at the end of six months from the start (just before any births for that month)? (*Hint:* Draw a month-by-month chart of the situation at the end of the month, just before any births.)
(b) Repeat part (a), but assume instead that the rabbits start to bear young exactly two months after their own birth.

5. New houses are to be built along one side of a street ("Leonardo's Lane"), divided into equal-sized lots. Each house is either a single-family detached house, taking up one lot, or a duplex, taking up two lots. Suppose that there are n lots on the street. How many different arrangements (orderings) of houses are there, for $n = 1$, 2, 3, 4, 5, and in general? (This exercise was inspired by a puzzle by Paul Dixon at the Suggested Web Site by Ron Knott.)

6. My wife works in a school district with about 900 faculty and staff. When it snows in the winter, the school district superintendent must decide by 5 A.M. whether to declare a "snow day" and cancel school. The faculty and staff are notified through a binary "telephone tree," in which the superintendent calls two people and each person who receives a call calls two others. Suppose that each call takes exactly one minute.

(a) Draw the telephone tree of calls for, and determine how many calls take place in, the first 1 minute, 2 minutes, 3 minutes, 4 minutes, and 5 minutes.
(b) How many calls does it take to notify all the faculty and staff? How long does that take?
(This exercise was inspired by a puzzle at the Suggested Web Site by Ron Knott.)

7. Here is a trick to "prove" that you can calculate faster than a person with a calculator. Turn your back and ask a friend to write down any two positive integers, then add them to get a third, then add the second and third to get a fourth, and so on, adding each time the last two integers until there are 10 numbers. Have your friend show you the list, whereupon you write down right away the total of all 10, while your friend begins to add them up on the calculator (to prove that you're right). The secret: The total is always 11 times the seventh number, and multiplying by 11 is pretty easy to do in your head–just add each pair of neighboring digits, carrying if necessary. Suppose that your friend writes down m and n as the first two numbers. Show that indeed the total of all 10 numbers is 11 times the seventh number. (Adapted from Martin Gardner, *Mathematical Circus*, Knopf, New York, 1979.)

8. The game of Fibonacci Nim begins with n counters. Two players take turns removing at least one counter, but no more than twice as many as the opponent just did. The winner is the player who takes the last counter. One other rule: The first player may not win immediately by taking all the counters on the first turn! (Adapted from Martin Gardner, *Mathematical Circus*, Knopf, New York, 1979.)

(a) Play this game taking turns with an opponent and starting with different numbers n of counters and try to come up with a strategy for one player or the other to win. (*Hint:* The key is that any positive integer can be represented uniquely as a sum of Fibonacci numbers.)
(b) Proceed as in part (a), but with the rule changes that the player who takes the last counter loses and the first player may not take all but one counter.

9. Put the golden ratio $\phi = (1 + \sqrt{5})/2$ into the memory of your calculator.

(a) Look at the value of ϕ. Now square it (either use the $\boxed{x^2}$ button or multiply it by itself). What do you observe?
(b) Back to ϕ. Now take its reciprocal (either use the $\boxed{1/x}$ button or divide it into 1). What do you observe?
(c) What formula explains what you saw in part (a)?
(d) What formula explains what you saw in part (b)?

10. The golden ratio satisfies the equation $x^2 = x + 1$. Show that $(1 - \phi)$ also satisfies the equation, so that $(1 - \phi) = (1 - \sqrt{5})/2$ is the other solution to $x^2 - x - 1 = 0$.

11. The geometric mean has interpretations in both arithmetic and geometry.
(a) Find the geometric mean of 3 and 27.
(b) Find the length of a side of a square that has the same area as a rectangle that is 4 by 64.

12. Here's further practice on arithmetic and geometric interpretations of the geometric mean.
(a) Find the geometric mean of 4 and 9.
(b) You are to make a golden rectangle with 6 inches of string. How wide should it be, and how high?

13. What is the geometric mean of 3, 6, and 12?

14. What is the geometric mean of 2, 4, 8, 16, and 32? (Such a sequence, in which each successive number is the same constant times the previous one, is called a *geometric sequence.*)

15. Another sequence closely related to the Fibonacci sequence is the *Lucas sequence,* which is formed using the same recursive rule but different starting numbers. The nth Lucas number L_n is given by

$$L_1 = 1,\ L_2 = 3, \text{ and } L_{n+1} = L_n + L_{n-1} \text{ for } n \geq 2$$

(a) Calculate L_3 through L_{10}.
(b) Calculate the ratio of successive terms of the Lucas sequence:

$$\frac{L_2}{L_1}, \frac{L_3}{L_2}, \ldots, \frac{L_{10}}{L_9}$$

What do you notice?

16. For a sequence specified by a recursive rule, finding an explicit expression for the nth term is not easy, nor is the form necessarily simple. An exact expression for the nth term of the Fibonacci sequence is given by the Binet formula:

$$F_n = \frac{1}{\sqrt{5}}\left(\frac{1+\sqrt{5}}{2}\right)^n - \frac{1}{\sqrt{5}}\left(\frac{1-\sqrt{5}}{2}\right)^n$$

(a) Verify the formula for $n = 1$ and $n = 2$ by multiplying out, not by using a calculator.
(b) Use the Binet formula and your calculator to find F_5.
(c) In fact, the second term on the right of the equation gets closer and closer to 0 as n gets large. Because we know that the Fibonacci numbers are integers, we can just round off the result of calculating the first term. Find F_{13} by calculating the first term with your calculator and rounding.

17. For two positive numbers x and y, show that the arithmetic mean $(x + y)/2$ is always greater than or equal to the geometric mean $x^{1/2}y^{1/2} = \sqrt{xy}$. Try some values for x and y and convince yourself, then demonstrate algebraically that it is true in general. When does equality hold? [*Hint:* Suppose that the claim is false, so that $(x + y)/2 < \sqrt{xy}$.) Square both sides of the inequality, bring all terms to one side, factor, and observe a contradiction.]

■ **18.** You may remember having to work problems like, "If Joe can dig a ditch in 3 days, and Sam can dig it in 4, how long will it take the two of them working together?" The answer is related to the *harmonic mean* of 3 and 4. The formula for the harmonic mean of two numbers x and y is

$$\frac{2}{1/x + 1/y}$$

(a) Calculate the answer for Joe and Sam, which is *one-half* of the harmonic mean of 3 and 4. Explain why this is the correct answer.
(b) Show that the harmonic mean of two positive numbers is always less than or equal to the geometric mean. (Thus, in light of Exercise 17, we have the general conclusion that $H \leq G \leq A$, where H stands for the harmonic mean, G for the geometric mean, and A for the arithmetic mean.) (*Hint:* Suppose that the claim is false. Simplify the fraction that is the harmonic mean, square both sides of the inequality, and proceed as in Exercise 17.)
(c) Show once more that the harmonic mean of two positive numbers is always less than the geometric mean, but this time do it with less work: let $A = 1/x$ and $B = 1/y$, and discover one connection (equation) between the harmonic mean of x and y and the arithmetic mean of A and B, and a second connection between the geometric mean of x and y and the geometric mean of A and B. Then use Exercise 17 on A and B.
(d) What should be the formula for the geometric mean of three numbers? Of n numbers?
(e) Proceed as in part (d), but for the harmonic mean.

19. Shari Lynn Levine, a high school student, published an article in *The Fibonacci Quarterly* that investigated the "Beta-nacci" sequence that results if instead of bearing one pair of baby rabbits per month, mature rabbits bear two pairs every month, starting when they reach two months of age. Here we ask you to rediscover some of Shari's results.

(a) How many rabbits will there be each month for the first 12 months?

(b) What is the recursive rule for the nth Beta-nacci number B_n?
(c) For the terms of the sequence in part (a), calculate the ratios B_{n+1}/B_n of successive terms. (*Motivating hint:* It's not the golden ratio this time.)
(d) Suppose that the ratio of successive terms approaches a number x. We show how to find x exactly. For very large n, we have $B_{n+1} \approx xB_n \approx x^2B_{n-1}$. Substituting these values into the recursive rule for the sequence and dividing by B_{n-1} gives the equation $x^2 = x + 2$. Solve this equation for x (you can use the quadratic formula). Make a table of values of $3B_n$ versus 2^n. From the evidence, can you suggest a formula for B_n?

20. Generalize Exercise 19, parts (a) through (d):

(a) to the case of each pair of rabbits having three pairs of rabbits (the "Gamma-nacci" sequence).
(b) to the case of each pair of rabbits having q pairs of rabbits.

For Exercises 21 and 22, refer to the following.

We have seen that the golden ratio is a positive root of the quadratic polynomial $x^2 - x - 1$. We can generalize this polynomial to $x^2 - mx - 1$ for $m = 1, 2, 3, \ldots$ and consider the positive roots of those polynomials as generalized means–the "metallic means family," as they are sometimes known. In particular, for $m = 2, 3, 4$, and 5, we have respectively the silver, bronze, copper, and nickel means. It is surely surprising that these numbers arise both in connection with quasicrystals (investigated in Chapter 20) and in analyzing the behavior of some dynamical systems (a topic investigated in Chapter 23) as the systems evolve into chaotic behavior.

21. Use the quadratic formula to find expressions in terms of square roots for the silver, bronze, copper, and nickel means, and approximate these to three decimal places. Find a general expression in terms of a square root for the mth metallic mean.

22. Just as the golden mean arises as the limiting ratio of consecutive terms of the Fibonacci sequence, each of the metallic means arises as the limiting ratio of consecutive terms of generalized Fibonacci sequences. A generalized Fibonacci sequence G can be defined by

$$G_1 = 1,\ G_2 = 1, \qquad \text{and} \qquad G_{n+1} = pG_n + qG_{n-1}$$

where p and q are positive integers. The Fibonacci sequence itself is the case $p = q = 1$.

(a) Try various small values of p and q and determine which mean they lead to.
(b) Divide the equation for G_{n+1} by G_n. Assume that G_{n+1}/G_n and G_n/G_{n-1} both tend toward the same number x as n gets large, replace those quantities by x, and simplify the resulting equation. What must be the value of x?
(c) What happens to the sequence and to the mean if we allow one or both of p and q to be negative integers?

19.2 Rosette, Strip, and Wallpaper Patterns

23. Determine whether each of the following statements is always true or sometimes false. Drawing some sketches may be helpful.

(a) A line reflection preserves collinearity of points. That is, if the points A, B, and C are in a straight line (collinear), then their images reflected in some other line also lie in a straight line.
(b) A line reflection preserves betweenness. That is, if the collinear points A, B, and C (with B between A and C) are reflected about a line, then the image of B is between the images of A and C.
(c) The image of a line segment under a line reflection is a line segment of the same length.
(d) The image of an angle under a line reflection is an angle of the same measure.
(e) The image of a pair of parallel lines under a line reflection is a pair of parallel lines.

24. Determine whether each of the following statements is always true or sometimes false. Drawing some sketches may be helpful.

(a) The image of a pair of perpendicular lines under a line reflection is a pair of perpendicular lines.
(b) The image of a square under a line reflection is a square.
(c) Label the vertices of a square A, B, C, and D in a clockwise direction. Then their images A', B', C', and D' under a line reflection also follow a clockwise direction.
(d) The length of the perimeter of a geometric figure is equal to the length of the perimeter of its image under a line reflection.
(e) The image of a vertical line under a line reflection is always a vertical line.

25. Which of the capital letters of the alphabet, when drawn in the most symmetrical way, has the following symmetries? For example, assume that the upper and lower loops of B are the same size.

(a) A horizontal line of reflection symmetry
(b) A vertical line of reflection symmetry
(c) A rotational symmetry

26. Repeat Exercise 25 for the lowercase letters.

27. In *The Complete Walker III* (3rd ed., Knopf, New York, 1984, p. 505), Colin Fletcher's answer to "What games should I take on a backpacking trip?" is the game he calls "Colinvert": "You strive to find words with meaningful mirror (or half-turn) images." Some of the words he found are

MOM WOW pod MUd bUM

(a) Which of his words reflect into themselves?
(b) Which of his words rotate into themselves?
(c) Find some more words or phrases of these various types–the longer, the better.

28. Repeat Exercise 27, but for words written vertically instead of horizontally.

29. For each of the following patterns, identify the rigid motions that preserve the pattern:

(a) CCCCCCCCCC
(b) GGGGGGGGGG
(c) HHHHHHHHHH
(d) MMMMMMMMMM

30. Repeat Exercise 29, but for

(a) SSSSSSSSSS
(b) bdbdbdbdbd
(c) dbpqdbpqdbpq

19.3 Notation for Patterns

31. What is the notation (such as *d4* or *c5*) for the symmetry pattern of a regular pentagon (which has all five sides equal)?

32. What is the notation for the symmetry pattern of a snowflake?

33. Give the notation (such as *d4* or *c5*) for the symmetry patterns of the rosettes in hubcaps (a) through (c) below, disregarding the logos in the centers. (Can you identify the make of car for each hubcap?)

34. Repeat Exercise 33 for hubcaps (d) through (f).

35. Repeat Exercise 33 for corporate logos (a) through (c) below. (Can you identify the corporations?)

(a)

(b)

(c)

36. Repeat Exercise 33 for automobile logos (d) through (f) below.

(d)

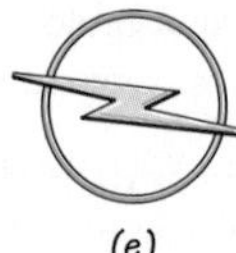
(e)

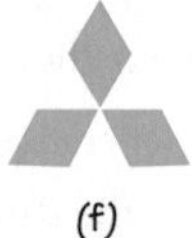
(f)

For Exercises 37–38, refer to the following.

Step patterns are found in Celtic illuminated manuscripts, metal work, and stone crosses. Square ones were constructed by first designing on a square lattice one quarter of the pattern (say, the top right), using horizontal and diagonal lines to produce a prototype such as the following:

(a) (b) (c)

(d) (e) (f)

(*All hubcap photos courtesy of Joe Gallian, University of Minnesota, Duluth.*)

Then three copies were added, either by (1) rotating the original successively by 90° [as in accompanying illustration (a)], or else by (2) reflecting it across its right and bottom edges [as in illustration (b)]. (Based on research by Mark A. M. Lynch of Glasgow Caledonian University, Scotland.)

(a)

(b)

37. Identify the rosette pattern for:

(a) step pattern (a).
(b) step pattern (b).

38. Which rosette pattern would result if the prototype, unlike the one above, has reflection symmetry across its diagonal from top left to lower right and

(a) strategy (1) is used.
(b) strategy (2) is used.

39. Use the flowchart in Figure 19.11 to identify the notation for the types of strip patterns from the pottery and basketry, shown in the illustrations below.

40. In each of the four accompanying examples, two adjacent triangles of an infinite strip are shown. (Contributed by Margaret A. Owens, California State University, Chico.)

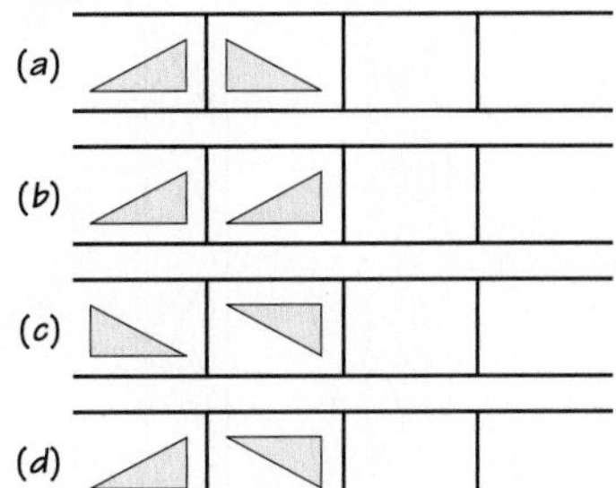

For each example:

(a) Determine a motion (translation, reflection, rotation, or glide reflection) that takes the first (= left) triangle to the second (= right) one.
(b) Draw the next four triangles of the infinite strip that would result if the second triangle is moved to the next space by another motion of the same kind, and so on.
(c) Identify (by notation) the resulting strip as one of the seven possible strip patterns.

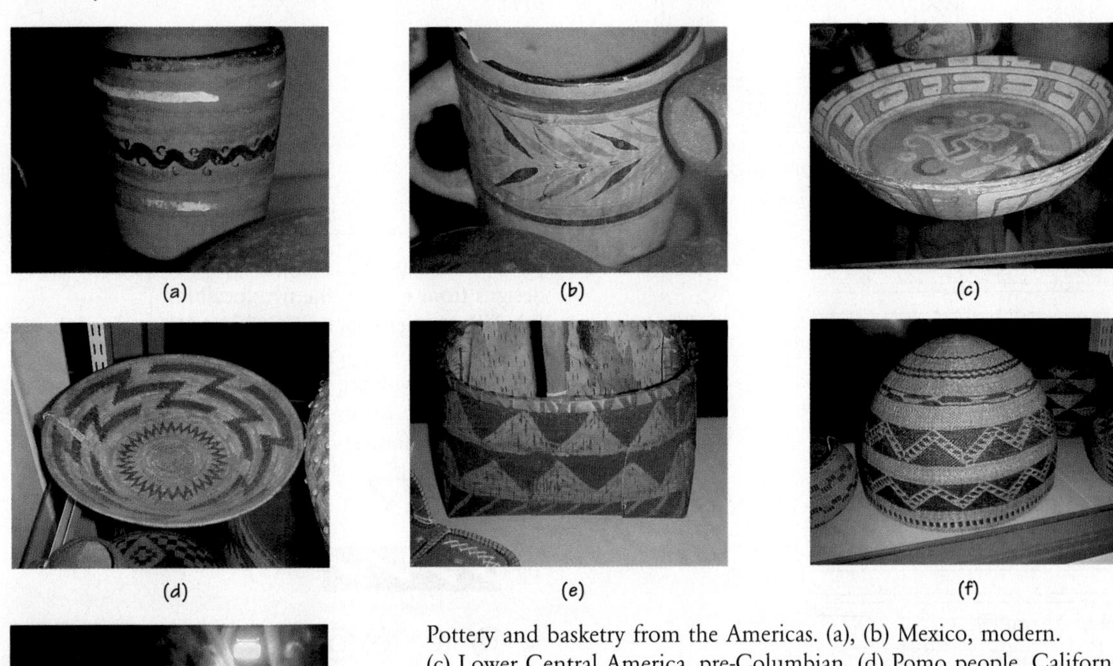

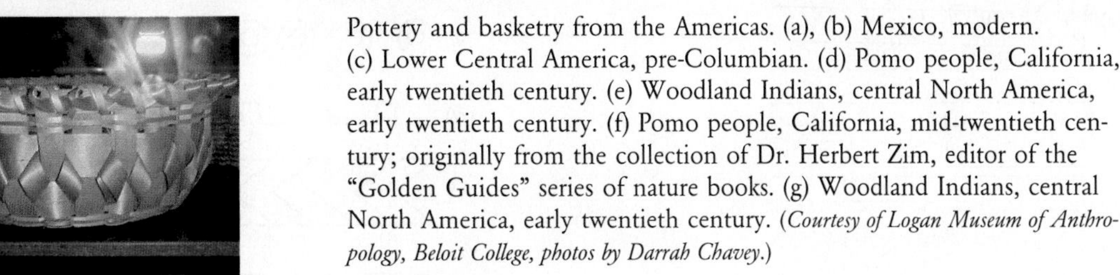

Pottery and basketry from the Americas. (a), (b) Mexico, modern. (c) Lower Central America, pre-Columbian. (d) Pomo people, California, early twentieth century. (e) Woodland Indians, central North America, early twentieth century. (f) Pomo people, California, mid-twentieth century; originally from the collection of Dr. Herbert Zim, editor of the "Golden Guides" series of nature books. (g) Woodland Indians, central North America, early twentieth century. *(Courtesy of Logan Museum of Anthropology, Beloit College, photos by Darrah Chavey.)*

41. Repeat Exercise 39 for the accompanying eight strip patterns, all of which appear on the brass straps for a single lamp from nineteenth-century Benin in West Africa. (From H. Ling Roth, *In Great Benin.*)

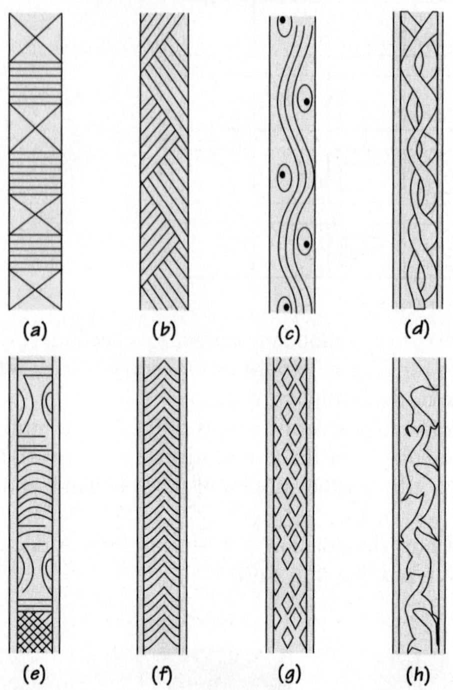

Note that the patterns are roughly carved, so you will need to discern the intent of the artist.

42. Repeat Exercise 39 for the accompanying patterns from San Ildefonso Pueblo, New Mexico.

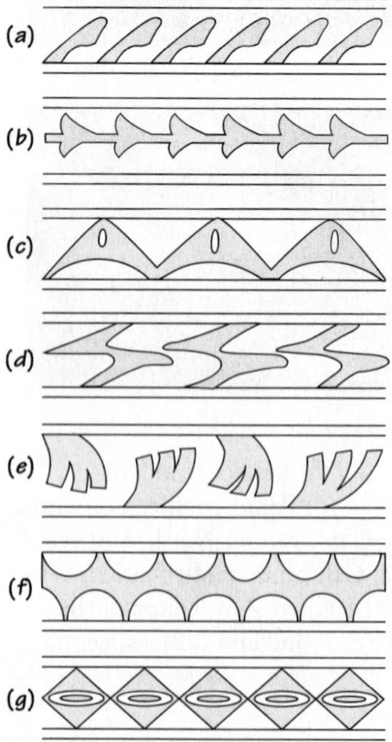

43. The following table shows comparative data about the frequency of occurrence of strip designs of various types on pottery (Mesa Verde, Colorado, United States) and smoking pipes (Begho, Ghana, Africa) from two continents.

Frequency of Strip Designs on Mesa Verde Pottery and Begho Smoking Pipes

	Mesa Verde	
Strip Type	Number of Examples	Percentage of Total
p111	7	4
p1m1	5	3
pm11	12	7
p112	93	53
p1a1	11	6
pma2	27	16
pmm2	19	11
Total	174	

	Begho	
Strip Type	Number of Examples	Percentage of Total
p111	4	2
p1m1	9	4
pm11	22	10
p112	19	8
p1a1	2	1
pma2	9	4
pmm2	165	72
Total	230	

(a) Which types of motions appear to be preferred for designs from each of the two localities?
(b) What other conclusions do you draw from the data of this table?
(c) On the evidence of the table alone, in which locality is each of the following strip patterns most likely to have been found?

(i) 

(ii) 

(iii)

(iv) 

(v) 

(vi) 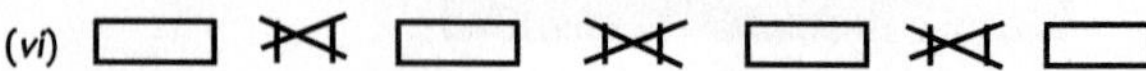

(vii) 

(viii) 

(ix)

44. For the Nigerian Yoruba cloths (a) and (b) in the following illustration, use the flowchart in Spotlight 19.6 to identify (by notation) the type of wallpaper pattern.

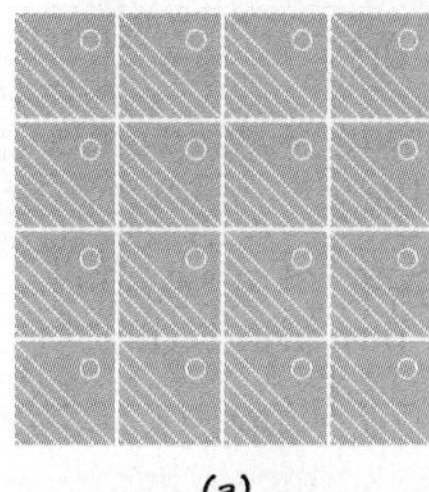

(a)

(b)

Patterns on Yoruba (West Africa) *adire* cloth, made by starching a pattern onto white cloth, then dyeing the cloth before rinsing out the starch, so that the starched portion remains as a white design against a colored background.

45. For the Nigerian Yoruba cloths (c) and (d) in the accompanying illustration, use the flowchart in Spotlight 19.6 to identify (by notation) the type of wallpaper pattern.

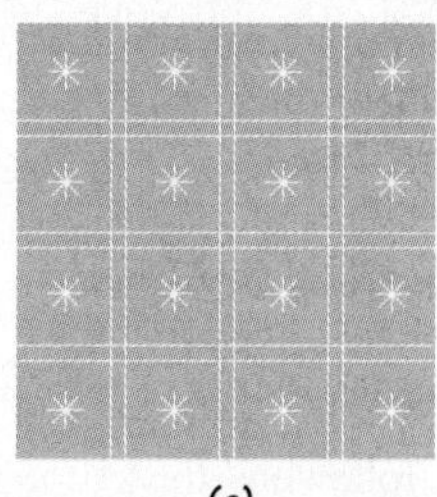

(c)

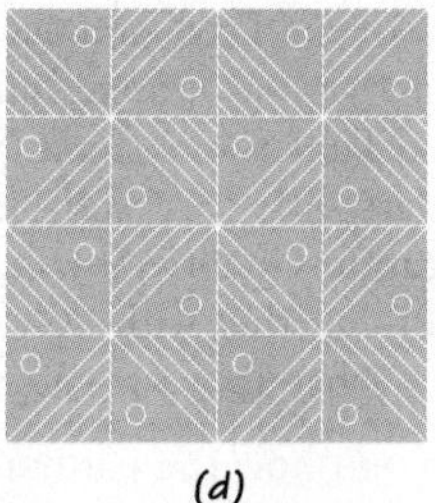

(d)

46. The triangles in the grid at the top of the following figure show beginning steps in forming instances of several of the wallpaper patterns by putting together a vertical motion and a horizontal motion.

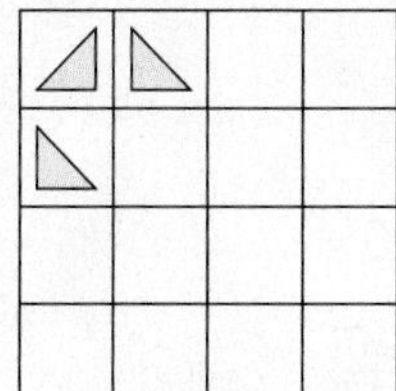

(a) Identify the horizontal motion.
(b) Identify the vertical motion.
(c) Fill in the remaining empty squares.
(d) Identify the wallpaper pattern.

■ **47.** Which of the 17 wallpaper patterns can be formed by the technique of Exercise 46?

For Exercises 48–53, refer to the following.

In Chapter 20, we study both repeating and nonrepeating plane patterns, from the point of view of their basic building blocks (tiles). Here we ask you to analyze the repeating patterns from figures in that chapter according to wallpaper type, using the flowchart of Spotlight 19.6. Identify all the symmetries and give the notational type for the wallpaper pattern of:

48. Figure 20.10.

49. Figure 20.11.

50. Figure 20.15.

51. The figure in Spotlight 20.2.

52. The hexagonal regular tiling at upper right in Figure 20.5.

53. The convex hexagon tiling of type 3 in Figure 20.9.

54. Visit the Web site escher.epfl.ch/escher/, which features an interactive Java program called Escher Web Sketch. Experiment with choosing wallpaper patterns using crystallographic notation. For each, draw on the screen a colored design for the motif; the program will reproduce the motif using the pattern.

19.4 Symmetry Groups

55. For positive integers a and n, the expression a mod n means remainder when a is divided by n. Thus, 23 mod 4 = 3, because $23 = 5 \cdot 4 + 3$, and we say that "23 is equivalent to 3 modulo 4" (see Chapter 17 for further details about this *modular arithmetic*). Every positive integer is equivalent to 0, 1, 2, or 3 modulo 4. Consider the collection of elements {0, 1, 2, 3} and the operation $\oplus$ on them defined by $a \oplus b = (a + b) \bmod 4$. Show that under this operation, the collection forms a group.

56. Explain, by referring to the properties of a group, whether the collection of all real numbers is a group under the operation of **(a)** addition; **(b)** multiplication.

57. Explain why the table for the operation * below shows that the elements indicated do not form a group under *.

*	A	B	C
A	B	B	B
B	B	C	A
C	C	A	B

58. Consider the table for the operation # below.

#	A	B	C	D	E	F
A	A	B	C	D	E	F
B	B	A	D	C	F	E
C	C	E	A	F	B	D
D	D	F	B	E	A	C
E	E	C	F	A	D	B
F	F	D	E	B	C	A

(a) Explain why the elements form a group under #. (Don't bother to check associativity.)
(b) What do you notice about F # E vs. E # F? (This is the smallest example of a group that is noncommutative.)

59. For the traditional North American beadwork shown below:

(Courtesy of Dr. Ron Eglash, RPI. See www.rpi.edu/~Eglash/csdt/na/loom/loom_symm4.html.)

(a) Which rosette pattern does it have?
(b) Specify two rigid motions that are generators of the group of the pattern.
(c) List the elements of the group.

60. Repeat Exercise 59 for the Plains Indian embroidery shown below.

(Courtesy of Dr. Ron Eglash, RPI. See www.rpi.edu/~Eglash/csdt.html.)

61. What is the group of symmetries of a square?

62. What is the group of symmetries of:
(a) an equilateral triangle (all three sides equal)?
(b) an isosceles triangle (two equal sides) that is not equilateral?
(c) a scalene triangle (no pair of sides equal)?

63. (a) Give a numerical example to show that the operation of subtraction on the integers is not associative.
(b) Repeat part (a), but for division on the positive real numbers.

64. What are the elements of the group of symmetries of **(a)** Figure 19.10b? **(b)** Figure 19.10f?

65. What are the elements of the group of symmetries of **(a)** Figure 19.10d? **(b)** Figure 19.10g?

66. What are the elements of the group of symmetries of the dihedral pattern *d8?* (See the flower in Figure 19.9a.)

67. What is the group of symmetries of the cyclic pattern *c8*?

■ **68.** What is the group of symmetries of a cube?

■ **69.** What is the group of symmetries of a general rectangular solid (its length, width, and height are all unequal)?

19.5 Fractal Patterns and Chaos

◆ **70.** Explore Sprott's Fractal Gallery at sprott.physics.wisc.edu/fractals.htm, which features a "Fractal of the Day" and accompanying fractal music. There are various "rooms" in the gallery–including "Iterated Function Systems," "Natural Fractals" (I particularly like "Broccoli" and "Trees"), and "Publication Quality Attractors" ("SMKBNZQA" is our favorite)–together with PC programs for generating such fractals. What are your favorites, and why?

71. Explain how the pattern of the following illustration is fractal.

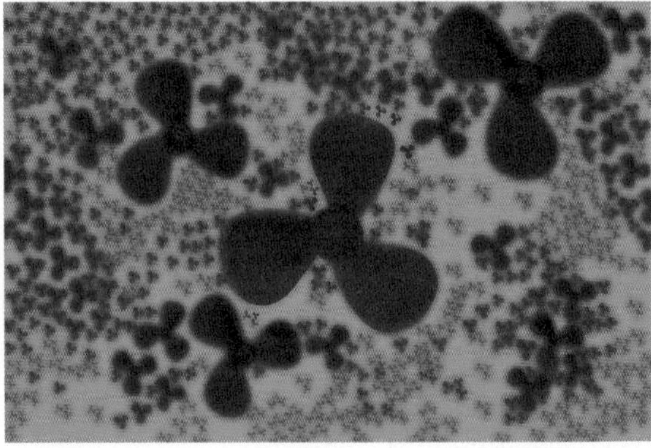

Exercises 72–76 use applets that require a computer with a Web browser equipped with Java and Flash plug-ins. These are available at links from the Web site www.rpi.edu/~eglash/csdt.html.

72. The Web site www.ccd.rpi.edu/Eglash/csdt/african/MANG_DESIGN/culture/mang_homepage.html has information about a fractal-patterned ivory hatpin from the Mangbetu culture in Africa. The site includes a tutorial on producing similar designs using reflection, rotation, translation, and scaling. Work your way through the tutorial and then create a Mangbetu-style artifact.

73. Cornrow hairstyles are fractal in nature. At the Web site www.ccd.rpi.edu/Eglash/csdt/african/CORNROW_CURVES/, you can see how and why, including a tutorial on designing cornrow hairstyles using reflection, rotation, translation, and scaling. Work your way through the tutorial and then create a hairstyle. The Web site also includes instructions for actual braiding, with a short video.

74. The architecture of some African villages follows a fractal pattern–see www.ccd.rpi.edu/Eglash/csdt/african/archi/afractal/afarch.htm. At the Web site www.ccd.rpi.edu/Eglash/csdt/african/archi/mangbetuUpdate2_3_modified_final.swf is an applet, similar to the ones in Exercises 72–73 but without a tutorial. Decide on a plan and use the applet to create a simulated African village.

75. The Yupik (Western Eskimo, along the coast of Alaska) make their parka coats with black-and-white patterns. At the Web site www.ccd.rpi.edu/Eglash/csdt/na/yupik/yupik.html is an applet, similar to the ones in Exercises 72–73 but without a tutorial. Yupik designs emphasize reflection symmetry, but the applet allows you to use also rotation, translation, and scaling of a basic motif. Experiment with one of the motifs available and use the applet to create a pattern for a parka.

76. Download fractal-creation software and accompanying documentation and use the software to create your own fractal. Recommended software:

For Windows: Fractint, from spanky.triumf.ca/www/fractint/fractint.html

For Macintosh: FractaSketch, from www.info.ucl.ac.be/~pvr/fracta.html with draft of manual (shows how to make fractal trees and leaves).

WRITING PROJECTS

1. Generations of children have enjoyed the popular toy Spirograph®, which allows the user to trace out symmetric patterns. A pencil or pen is placed in a hole in one of several plastic circular disks with teeth on the outside rim. The disk is then meshed in the teeth of another plastic circle and rotated around its inside or outside. Each plastic piece is labeled with the number of teeth that it has on its circumference.

Either obtain a copy of Spirograph® or a closely related toy, or else visit the Web site www.wordsmith.org/~anu/java/spirograph.html, which offers an interactive Java application (which you can download) that mimics what the Spirograph® toy does.

(a) Experiment to determine, from the numbers of teeth on the rotating circular disk and the fixed circle, what symmetry pattern the result will have.
(b) Choose a rotating circular disk and a fixed circle for which the ratio of the number of teeth reduces to a whole number. For each of several "offsets" (holes to choose for the pencil or pen), trace overlapping designs. What symmetry pattern do you get for the design taken as a whole? Repeat this experiment for other pairs of pieces and try to reach a general conclusion.
(c) Write up, in a page or so, a description of your experiments and what conclusions you reached.

2. (Project for a team of 2 or 3) Explore your campus looking for symmetrical patterns in decorative elements of walls, floor, carpets, and ceilings. Find one example each of a rosette pattern, a strip pattern, and a wallpaper pattern. Take a digital photo of each and incorporate your photos into a document of three pages or so that explains to the reader where the pattern can be found, what symmetries (translation, rotation, reflection, glide reflection) it has, and how you identified the notation for it.

3. (Project for a team of 2 or 3) Visit a store that sells wallpaper and ask for a few old samples. Identify three that have different patterns according to the flowchart in Spotlight 19.6. Write in a page or two your explanations of how you identified the patterns, and attach the wallpaper samples to your report.

 SUGGESTED READINGS

BELCASTRO, SARAH-MARIE, and THOMAS C. HULL. Classifying frieze patterns without using groups, *College Mathematics Journal,* 33 (March 2002): 93–98. Elementary analysis of why there are only seven ways to repeat a pattern along a strip.

CROWE, DONALD W. *Symmetry, Rigid Motions and Patterns,* High School Mathematics and Its Applications (HiMAP) Module 4, COMAP, Lexington, Mass., 1987. Reprinted in smaller format in *The UMAP Journal,* 8(3) (1987): 207–236. Instructional module on rigid motions of the plane, strip patterns, and wallpaper patterns, with worksheets.

LEE, KEVIN D. KaleidoMania!: Interactive Symmetry, Windows/Macintosh program, Key Curriculum Press, 1999. Lets the user construct rosette, strip, and wallpaper patterns.

LIVIO, MARIO. *The Golden Ratio: The Story of Phi, the World's Most Astonishing Number.* Broadway Books, New York, 2002.

POSAMENTIER, ALFRED S., and INGMAR LEHMANN. *The (Fabulous) Fibonacci Numbers,* Prometheus Books, Amherst, N.Y., 2007.

WASHBURN, DOROTHY K., and DONALD W. CROWE. *Symmetries of Culture: Theory and Practice of Plane Pattern Analysis,* University of Washington Press, Seattle, 1988. An introduction to the mathematics of symmetry, splendidly illustrated with photographs of patterns from cultures all over the world. Includes a complete analysis of patterns with two colors, and proofs that there are only four rigid motions in the plane and exactly seven strip patterns.

 SUGGESTED WEB SITES

www.geom.umn.edu/software/tilings Tessellation resources. Lists programs for various platforms that allow the user to create designs featuring the rosette, strip, and wallpaper patterns.

escher.epfl.ch/escher/ Interactive Escher Web Sketch program that allows a user to design repeating patterns. Choose a wallpaper pattern using crystallographic notation and draw on the screen a colored design for the motif; the program then reproduces the motif using the pattern. The software (for Windows, Macintosh, and Unix) can also be downloaded.

www.geom.uiuc.edu/java/Kali/ Interactive Java Kali Web program that lets the user draw pictures under the action of rosette, strip, or wallpaper groups. Versions for various platforms can be downloaded.

www.wordsmith.org/~anu/java/spirograph.html Interactive Spirograph Java application (which you can download) that lets you do electronically what the Spirograph toy does.

www.rpi.edu/~eglash/eglash.dir/afractal/afractal.htm African fractals site.

classes.yale.edu/fractals/index.html Web site for Mandelbrot's course in fractals at Yale. Features many applets for different kinds of IFS (for example, incorporating randomness), including a fractal music composer.

maven.smith.edu/~phyllo/index.html Phyllotaxis: An interactive site for the study of plant pattern formation, by Pau Atela and Christophe Golë.

www.mcs.surrey.ac.uk/Personal/R.Knott/Fibonacci/fib.html Fibonacci numbers and the golden section. Splendidly illustrated extensive Web pages by Ron Knott about Fibonacci numbers and the golden ratio: their occurrences in nature, their applications, puzzles, and much more.

CHAPTER 20

Tilings

When our ancestors covered the floors and walls of their houses with stones, they selected shapes and colors to form pleasing designs. We can see the artistic impulse at work in mosaics, from Roman dwellings to Muslim religious buildings (see Figure 20.1). The same intricacy and complexity arise in other decorative arts–on carpets, fabrics, baskets, and even linoleum.

Such patterns have one feature in common: They use repeated shapes to cover a flat surface, without gaps or overlaps. If we think of the shapes as tiles, we can call the pattern a **tiling**, or *tessellation.* Even when efficiency is more important than esthetics, designers value clever tiling patterns. In manufacturing, for example, stamping components from a sheet of metal is most economical if the shapes of the components fit together without gaps–in other words, if the shapes form a tiling. We are interested in patterns that could be extended indefinitely far in any direction.

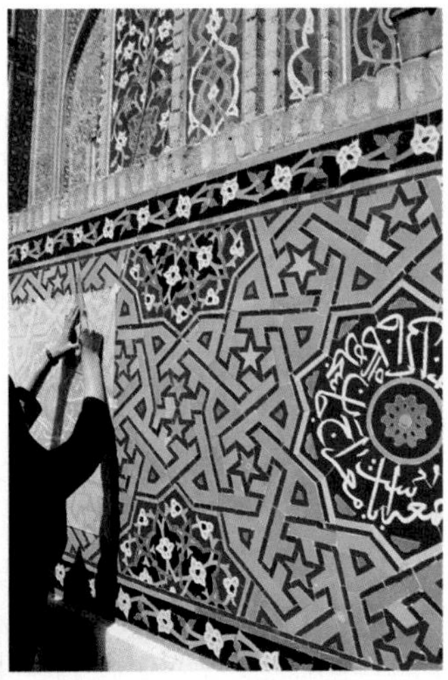

FIGURE 20.1 Arab mosaic. (*Jose Fuste Raga/Corbis.*)

Tiling (Tessellation) DEFINITION

A **tiling (tessellation)** is a covering of the entire infinite plane by nonoverlapping figures.

Some of the tilings in this chapter can be analyzed as wallpaper patterns, as in Spotlight 19.6. But in this chapter, we look at patterns "from the ground up." That is, we start with the basic ingredients of a type or types of tile and ask if or how they can fit together in a pattern. If the result repeats in all directions, it must be one of the wallpaper patterns. We will see, though, that there are other surprising possibilities.

The major mathematical question about tilings is: Given one or more shapes (in specific sizes) of tiles, can they tile the plane? And, if so, how?

The surprising answer to the first question is that it is undecidable. For some particular sets of tiles, we can exhibit tilings; for others, we can prove that there can't be any tiling. In this chapter we will see examples of both situations. But mathematicians have proved that there is no algorithm (mechanical step-by-step process) that can tell for every conceivable set of tile shapes which of the two situations holds. (See Chapter 9, pp. 356–359, for other examples of "unattainable ideals" in regard to voting.)

Given this sobering (and puzzling) limitation, we begin our investigation by considering the simplest kinds of tiles and tilings.

20.1 Tilings with Regular Polygons

The simplest tilings use only one size and shape of tile. They are known as **monohedral tilings**.

Monohedral Tiling DEFINITION

A **monohedral tiling** is a tiling that uses only one size and shape of tile.

We are particularly interested in tiles that are **regular polygons**, figures all of whose sides are the same length and all of whose angles are equal. A square is a regular polygon with four equal sides and four equal interior angles; a triangle with all sides equal (an **equilateral triangle**) is also a regular polygon. A polygon with five sides is a pentagon, one with six sides is a hexagon, and one with n sides is an ***n*-gon**. Regular polygons are especially interesting because of their high degree of symmetry. Each has the reflection and rotation symmetries of a dihedral rosette pattern (see section 19.2). In three dimensions, the corresponding highly symmetrical figures are called *regular polyhedra* (see Spotlight 20.1).

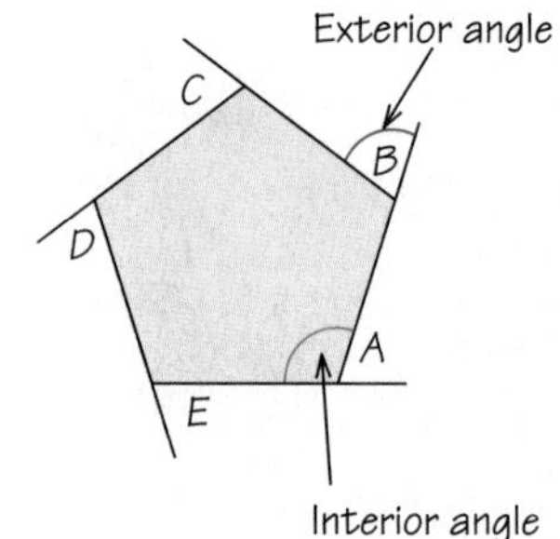

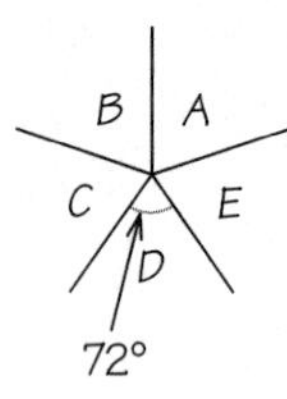

FIGURE 20.2 The exterior angles of a regular pentagon, like those of any regular polygon, add up to 360°. Each exterior angle measures 72°.

An **exterior angle** of a polygon is an angle formed by one side and the extension of an adjacent side (Figure 20.2). At each vertex of the polygon, there are two exterior angles, depending on which side we extend; but we will consider only one of them. Let us agree to extend the sides consistently in turn as we proceed counterclockwise around the polygon, as in Figure 20.2, producing the set of exterior angles A through E, one at each vertex.

By a convention dating back to the ancient Babylonians, angles are measured in degrees, with the total of angles around a point being 360°. If we bring a set of exterior angles together at a point, we can see that they add up to 360° (see Figure 20.2). Hence for a regular polygon with n sides, each exterior angle must measure $360°/n$. For example, a square, with $n = 4$ sides, has 4 exterior angles in a set, each measuring 90°; a regular pentagon, with $n = 5$ sides, has 5, each measuring 72°; a regular hexagon, with 6 sides, has 6, each measuring 60°.

Each exterior angle is paired with a corresponding **interior angle** (the angle inside the polygon formed by the two adjacent sides), and the pair adds up to a straight line, or 180°. For a regular polygon with more than six sides, each interior angle is between 120° and 180°. This last consideration will shortly prove to be crucial to determining how regular polygons can fit together to form tilings.

SPOTLIGHT 20.1 Regular Polyhedra and Buckyballs

The three-dimensional analogue of a regular polygon is a regular polyhedron, a convex solid whose faces are regular polygons all alike (same number of sides, same size), with each vertex surrounded by the same number of polygons. Although there are infinitely many regular polygons, there are only five regular polyhedra, a fact proved by Theaetetus (414–368 B.C.). They were called the *Platonic solids* by the ancient Greeks.

If the restriction that the same number of polygons meet at each vertex is relaxed, five additional convex polyhedra are obtained, all of whose faces are equilateral triangles. If we allow more than one kind of regular polygon, 13 further convex polyhedra are obtained, known as the *semiregular polyhedra* or *Archimedean solids* (although there is no documented evidence that Archimedes studied them—but Kepler in the early 1600s catalogued them all). The truncated icosahedron, whose faces are pentagons and hexagons, is known throughout the world (once inflated) as a regulation soccer ball. Drawings of it appear in the work of Leonardo da Vinci.

The truncated icosahedron is also the structure of C_{60}, a form of carbon known as buckminsterfullerene and, more familiarly, the "buckyball." Sixty carbon atoms lie at the 60 vertices of this molecule, which was discovered in 1985. It is named after R. Buckminster Fuller (1895–1983), inventor and promoter of the geodesic dome. The molecule resembles a dome.

The buckyball is part of a family of carbon molecules, the *fullerenes*, in which each carbon atom is joined to three others. Thirty years before the discovery of fullerenes, mathematicians had shown that a convex polyhedron in which every vertex has three edges must have 12 pentagon faces and may have any number of hexagon faces, from 0 on up, except for 1.

That there must be 12 pentagons follows from a famous equation due to Leonhard Euler (1707–1783). For any convex polyhedron, it must be true that $v - e + f = 2$, where v is the number of vertices, e is the number of edges, and f is the number of faces of the polyhedron.

In 2003 astronomers and mathematicians advanced a remarkable new theory about the shape of the universe, in an effort to explain why it does not show as much historic fluctuation in temperature as other models predict. This lack of fluctuation could be explained by the universe being in the shape of a dodecahedron (the figure shown here with 12 pentagonal sides), with opposite faces coinciding. This theory harks back to Kepler, who had conceived of the universe in terms of the five regular polyhedra nested within one another.

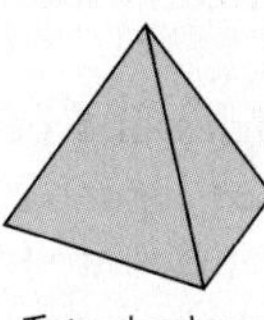
Tetrahedron

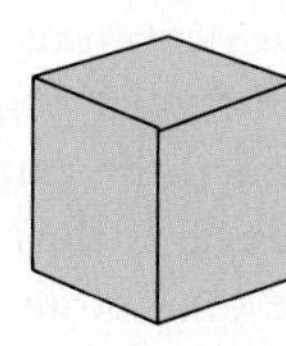
Cube

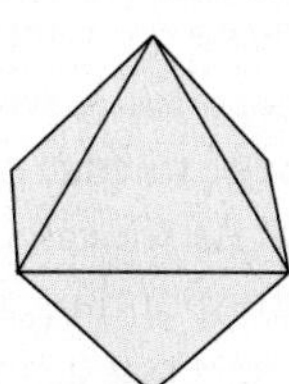
Octahedron

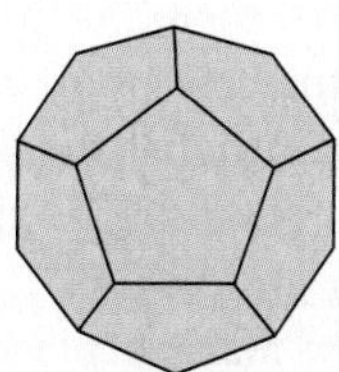
Dodecahedron

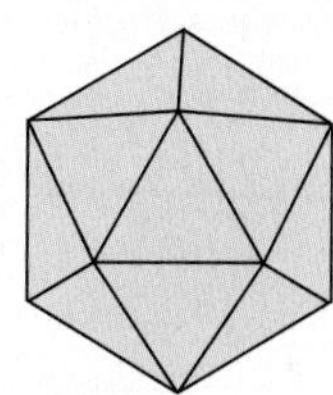
Icosahedron

Regular Tilings

Regular Tiling DEFINITION

A monohedral tiling whose tile is a regular polygon is called a **regular tiling**.

A square tile is the simplest case. Apart from varying the size of the square, which would change the scale but not the pattern of the tiling, we can get different tilings by offsetting one row of squares some distance from the next (Figure 20.3a).

However, there is only one tiling using a square that is **edge-to-edge**.

Edge-to-Edge DEFINITION

In an **edge-to-edge tiling**, the edge of a tile coincides entirely with the edge of a bordering tile.

Figure 20.3 shows one tiling that is not edge-to-edge and another that is.

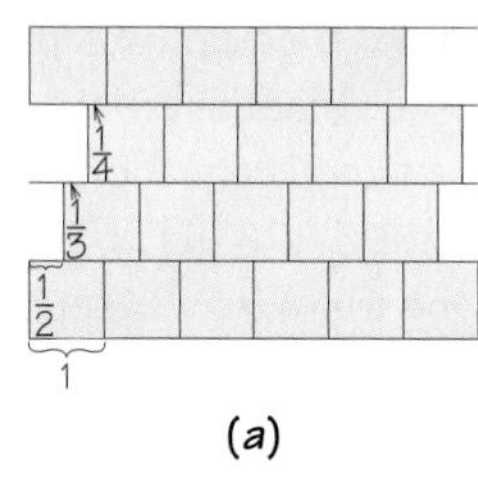

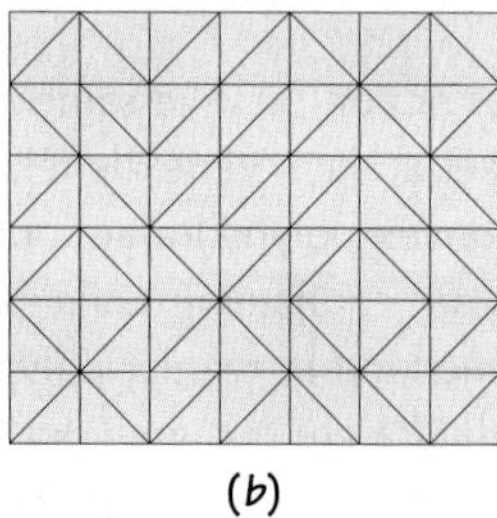

FIGURE 20.3 (a) A tiling that is not edge-to-edge. The horizontal edges of two adjoining squares do not exactly coincide. (b) A tiling by right triangles that is edge-to-edge.

For simplicity, from now on by *tiling* we mean "edge-to-edge tiling." The edges themselves may be curvy.

Any tiling by squares can be refined to one by triangles by drawing a diagonal of each square, but these triangles are not regular (equilateral). Equilateral triangles can be arranged in rows by alternately inverting triangles. As with squares, there is only one pattern of equilateral triangles that forms an edge-to-edge tiling.

What about tiles with more than four sides? An edge-to-edge tiling with regular hexagons is easy to construct (see the upper right pattern in Figure 20.5).

However, if we look for a tiling with regular pentagons, we won't find one. How do we know whether we're just not being clever enough or there really isn't one to be found? This is the kind of question that mathematics is uniquely equipped to answer. In the other sciences, phenomena may exist even though we have not observed them. Such was the case for bacteria before the invention of the microscope. In the case of an edge-to-edge tiling with regular pentagons, we can conclude with certainty that there is no edge-to-edge tiling with regular pentagons.

The proof is very easy. As we calculated earlier, the exterior angles of a pentagon are each 72°; each corresponding interior angle is thus 108° (see Figure 20.2). How many pentagons can meet at a point? The total of all of the angles around a point must be 360°. As you can see in Figure 20.4, four pentagons at a point would be too many (their angles would add to $4 \times 108° = 432°$, so they'd have to overlap), and three would be too few (their angles would add to $3 \times 108° = 324°$, so some of the area wouldn't be covered). Because 108 does not evenly divide 360, *regular pentagons can't tile the plane.*

With this argument, we can do something that is a favorite with mathematicians—we can *generalize* it to a criterion for when a regular polygon can tile the plane: when the size of its interior angles divides 360 evenly. We can apply this criterion to determine exactly which other regular polygons can tile the plane.

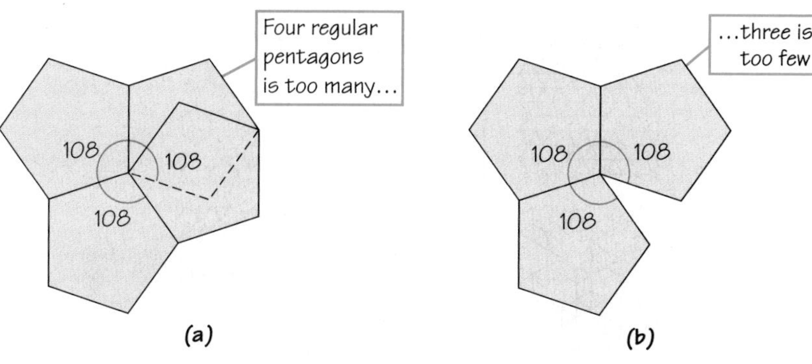

FIGURE 20.4 Polygons that come together at a vertex in a tiling must have interior angles that add up to 360°–no more, no less.

EXAMPLE 1 ■ Identifying the Regular Tilings

SOLUTION A regular hexagon has interior angles of 120°; 120 divides 360 evenly, and three regular hexagons fit together exactly around a point. A regular 7-gon (heptagon)–or any regular polygon with more than six sides–has interior angles that are larger than 120° but smaller than 180°. Now 360 divided by 120 gives 3, and 360 divided by 180 gives 2–and there aren't any other possibilities in between. Angles between 180° and 120° divided into 360° will give a result between 2 and 3, and consequently not an integer. So there are no regular tilings of the plane with polygons of more than six sides. ■

Only Three Regular Tilings — THEOREM

The only regular tilings are the ones with equilateral triangles, with squares, and with regular hexagons.

The follow-up question, of course, is which *combinations* of regular polygons of different numbers of sides can tile the plane edge-to-edge.

Vertex Type — DEFINITION

In an edge-to-edge tiling by regular polygons, the **vertex type** of a vertex is the arrangement of the polygons around the vertex.

To describe a vertex type, we list the sizes of polygons, separated by periods, in either clockwise or counterclockwise order starting from the smallest number of sides. For example, 4.4.4.4 (or 4^4 for short) denotes four squares meeting at a vertex. Similarly, 4.6.12 denotes a square followed by a hexagon then by a dodecagon (12-gon); see the tiling in the middle of the bottom row of Figure 20.5. Two vertices have the same type even if one has the polygons in clockwise order and the other has them in counterclockwise order; both versions of the 4.6.12 type occur in that tiling in Figure 20.5.

Semiregular Tiling — DEFINITION

A systematic tiling that uses a mix of regular polygons with different numbers of sides but in which all vertex types are alike–the same polygons in the same order–is called a **semiregular tiling** (see Figure 20.5).

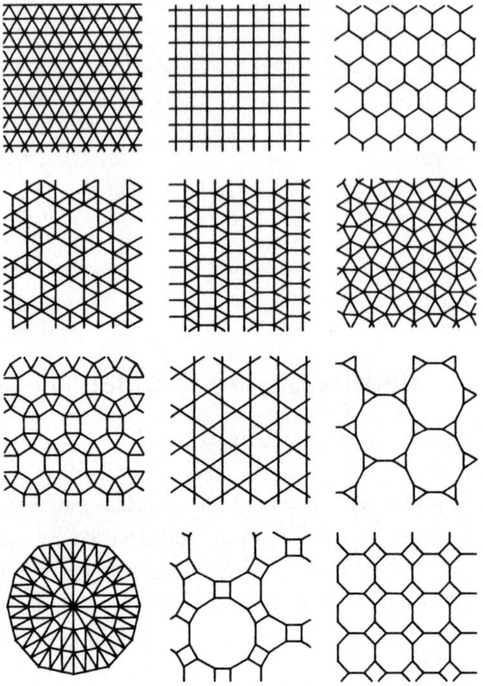

FIGURE 20.5 The three regular tilings . . . and the eight semiregular tilings, plus a "mystery" tiling that does not belong to either group. Can you identify it? (*Hint:* It uses just one tile, which isn't regular.)

As before, the technique of adding up angles at a vertex (to be 360°) can eliminate some impossible combinations, such as "pentagon, pentagon, pentagon" (Figure 20.4). Once we have found an arrangement that is numerically possible, we must confirm the actual existence of each tiling by constructing it (showing that it is geometrically possible). For example, even though a possible arrangement of regular polygons around a point is "triangle, square, square, hexagon," it is not possible to construct a tiling with that vertex figure at every vertex.

The result of such an investigation is that in a semiregular tiling, no polygon can have more than 12 sides. In fact, polygons with 5, 7, 9, 10, or 11 sides do not occur either. Figure 20.5 exhibits all of the semiregular tilings.

If we abandon the restriction about the vertex types being the same at every vertex, then there are *infinitely many* systematic edge-to-edge tilings with regular polygons, even if we continue to insist that all polygons with the same number of sides have the same size.

20.2 Tilings with Irregular Polygons

What about edge-to-edge tilings with irregular polygons, which may have some sides longer than others or some interior angles larger than others? We will look just at monohedral tilings (in which all tiles have the same size and shape) and investigate in turn which triangles, **quadrilaterals** (four-sided polygons), hexagons, and so forth, can tile the plane.

The most general shape of a triangle has all sides of different lengths and all interior angles of different sizes. Such a triangle is called a **scalene triangle**, from the Greek word for "uneven." We can always take two copies of a scalene triangle and fit them together to form a **parallelogram**, a quadrilateral whose opposite sides are parallel (Figure 20.6a). It's easy to see that we can then use such parallelograms to tile the plane by making strips and then fitting layers of strips together edge-to-edge (Figure 20.6b).

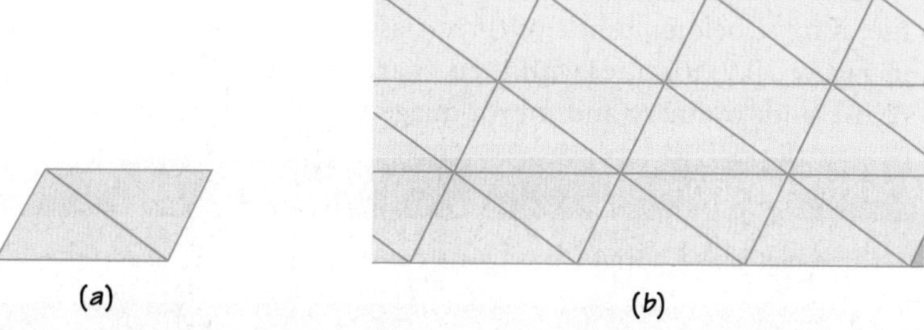

(a) (b)

FIGURE 20.6 (a) Two scalene triangles form a parallelogram. (b) Every scalene triangle tiles the plane.

Tiling with Triangles THEOREM

Any triangle can tile the plane.

What about quadrilaterals? We have seen that squares tile the plane, and rectangles certainly will, too. We have just noted that any parallelogram will tile. What about a quadrilateral (four-sided polygon) with its opposite sides not parallel, as in Figure 20.7a? The same technique as for triangles will work. We fit together two copies of the quadrilateral, forming a hexagon whose opposite sides are parallel. Such hexagons fit next to each other to form a tiling, as in Figure 20.7b.

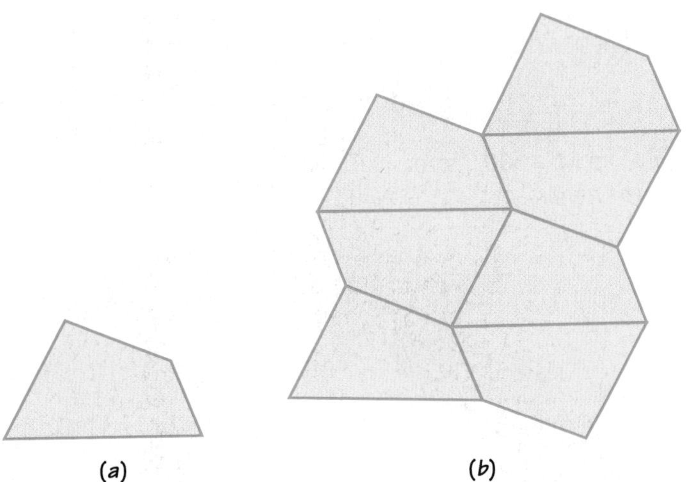

(a) (b)

FIGURE 20.7 (a) A general quadrilateral. (b) Any quadrilateral tiles the plane.

FIGURE 20.8 (a) A general nonconvex quadrilateral. (b) Any quadrilateral, convex or not, tiles the plane.

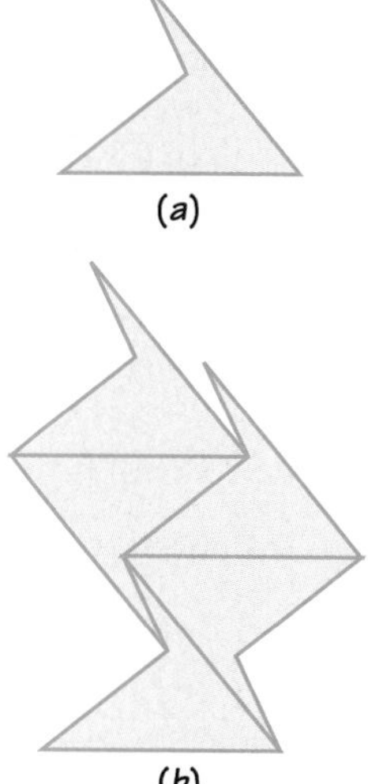

(a) (b)

The quadrilaterals shown in Figure 20.7 are all **convex**. If you take any two points on the tile (including the boundary), the line segment joining them lies entirely within the tile (again, including the boundary). The quadrilateral of Figure 20.8a is not convex, but the same approach works for using it to form a tiling (Figure 20.8b).

Tiling with Quadrilaterals THEOREM

Any quadrilateral, even one that is not convex, can tile the plane.

We could hope that such success would extend to irregular polygons with any numbers of sides, but it doesn't. The situation for convex hexagons was determined by Karl Reinhardt in his 1918 doctoral thesis. He showed that for a convex hexagon to tile, it must belong to one of three classes. Examples of the three classes are shown in Figure 20.9, together with their characterizations. Tilings with a hexagon of type 2 use both ordinary and mirror-image versions of the hexagon.

Tiling with Hexagons THEOREM

Exactly three classes of convex hexagons can tile the plane.

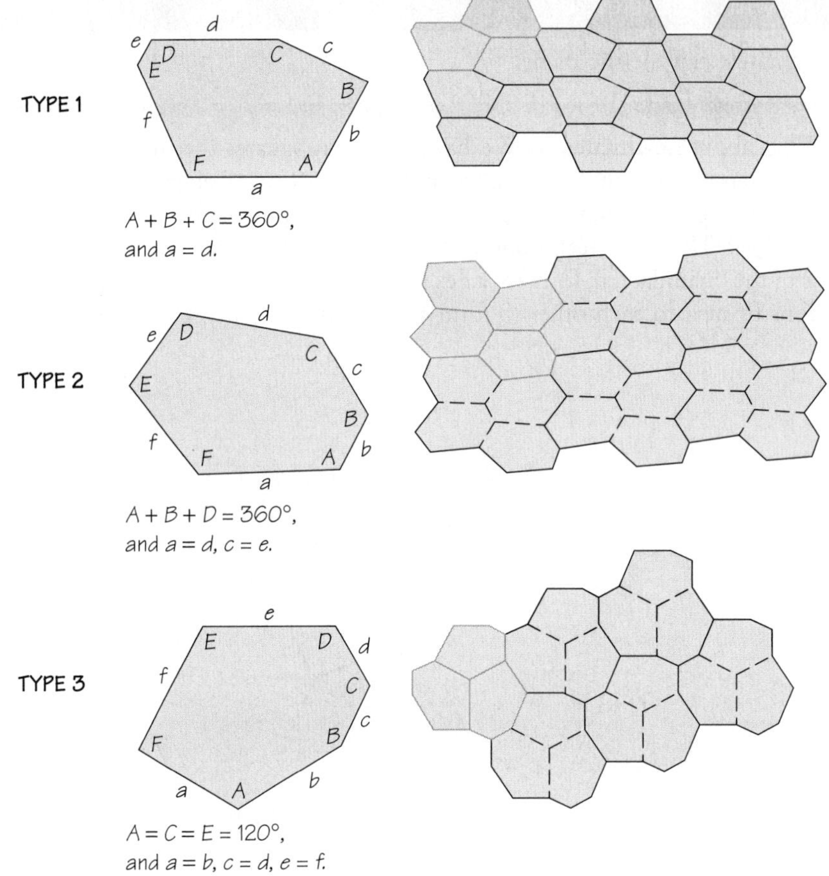

FIGURE 20.9 The three types of convex hexagon tiles.

Reinhardt also explored convex pentagons and found five classes that tile. For example, any pentagon with two parallel sides will tile. Reinhardt did not complete the solution, as he did for hexagons, by proving conclusively that no other pentagons could tile. He claimed that it would be very tedious to finish the analysis. Still, he felt that he had found them all. In 1968, after 35 years of working on the problem on and off, R. B. Kershner, a physicist at Johns Hopkins University, discovered three more classes of pentagons that will tile. Kershner was sure that he had found all pentagons that tile, but, like Reinhardt, he did not offer a complete proof, which "would require a rather large book."

When an account of the "complete" classification into eight types appeared in *Scientific American* (July 1975), the article provoked an amateur mathematician to discover a ninth type! A second amateur, Marjorie Rice, a housewife with no formal education in mathematics beyond high school "general mathematics" 36 years earlier, devised her own mathematical notation and found four more types (see Spotlight 20.2). A fourteenth type was found by a mathematics graduate student in 1985. Since then, no new types have been discovered, yet no one knows if the classification is complete.

With the situation so intricate for convex pentagons, you might think that it must be still worse for polygons with seven or even more sides. In fact, however, the situation is remarkably simple, as Reinhardt proved in 1927.

Tiling with Polygons with More Sides THEOREM

A convex polygon with seven or more sides cannot tile.

M. C. Escher and Tilings

The Dutch artist M. C. Escher (1898–1972) was inspired by the great variety of decoration in tilings in the Alhambra, a fourteenth-century palace built during the last years of Islamic dominance in Spain. He devoted much of his career of making prints to creating tilings with tiles in the shapes of living beings (a practice forbidden to Muslims). Those prints of interlocking animals and people have inspired awe and wonder among people all over the world. Figures 20.10–20.13 illustrate a few of his drawings and finished works. Like Marjorie Rice, he, too, developed his own mathematical notation for the different kinds of patterns for the tilings.

20.3 Using Translations

You may wonder just how much liberty can be taken in shaping a tile and how you might be able to design an Escher-like tiling yourself.

The simplest case is when the tile is just *translated* in two directions; that is, copies are laid edge-to-edge in rows, as in Figure 20.10. Each tile must fit exactly into the ones next to it, including its neighbors above and below. We say that each tile is a **translation** of each other one, because we can move one to coincide with another without doing any rotation or reflection.

When is it possible for a tile to cover the plane in this manner? The boundary of the tile must be divisible into matching pairs of opposing parts that will fit together. Figures 20.10 and 20.11 illustrate two basic ways that this can happen. In the first, two opposite pairs of sides match; in the second, three opposite pairs of sides match.

Translation Criterion RULE

A tile can tile the plane by translations if either

1. There are four consecutive points A, B, C, and D on the boundary such that
   (a) the boundary part from A to B is congruent by translation to the boundary part from D to C, and
   (b) the boundary part from B to C is congruent by translation to the boundary part from A to D (see Figure 20.12a) or

2. There are six consecutive points A, B, C, D, E, and F on the boundary such that the boundary parts AB, BC, and CD are congruent by translation, respectively, to the boundary parts ED, FE, and AF (see Figure 20.12b).

SPOTLIGHT 20.2 In Praise of Amateurs

Marjorie Rice

(Courtesy Sharon Whittaker.)

R. B. Kershner's claim to have found all convex pentagons that tile was read by many puzzle enthusiasts, including Richard James III and Marjorie Rice. James found a tiling that Kershner had missed.

Rice, a San Diego housewife and mother of five, read about James's new tile. "I thought I would see if I could find still another type. It was a delightful new puzzle to me."

With no formal education in mathematics beyond a high school general mathematics course, she not only worked out her own method of attack but invented her own notation.

"I began drawing little diagrams on my kitchen counter when no one was there, covering them up quickly if someone came by, for I didn't wish to have to explain what I was doing. I was searching for a new type and a few weeks later, I found it." Over the next two years, she found three additional new tilings.

What makes a person pursue a problem so patiently and persistently? She was not trained for it nor paid, but she gained great personal satisfaction.

She was born in 1923 in St. Petersburg, Florida, and went to a one-room country school.

"When I was in the 6th or 7th grade, our teacher pointed out to us the Golden Section in the proportions of a picture frame. This immediately caught my imagination and I never forgot it. I've . . . been especially interested in architecture and the ideas of architects and planners such as Buckminster Fuller. I've come across the Golden Section again in my reading and considered its use in painting and design."

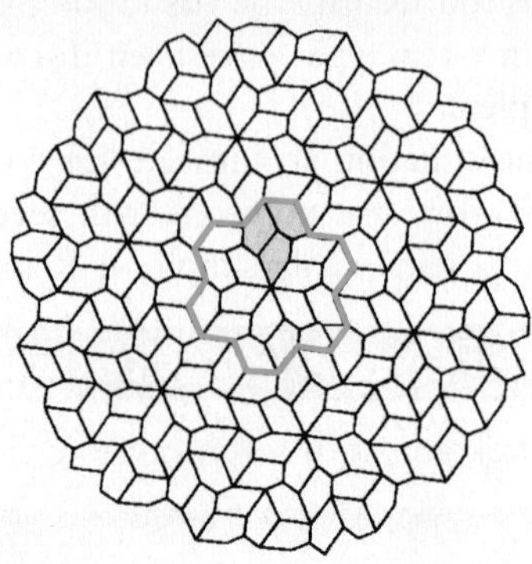

This tiling in the headquarters of the Mathematical Association of America in Washington, D.C., was discovered in 1995 by Marjorie Rice. The angles of each pentagon tile are 60°, 90°, 120°, and 150°—all multiples of 30°. The tiling is periodic, although not every pentagon is surrounded in the same way. Three pentagons form a fundamental block, and the outlined group of 18 pentagons tiles by translation.

After high school, Rice worked until her marriage in 1945. She was drawn back into mathematics by her children, finding solutions to their homework problems "by unorthodox means, since I did not know the correct procedures." She became especially interested in textile design and the works of M. C. Escher. As she pursued the pentagonal tilings, she produced some imaginative Escher-like patterns (see Figure 20.19 and the figure here).

Intense spirit of inquiry and keen perception are the forte of all such amateurs. No formal education provides these gifts. Lack of a mathematical degree separates these "amateurs" from the "professionals," yet their curiosity and ingenious methods make them true mathematicians.

Source: Adapted from Doris Schattschneider, "In Praise of Amateurs," in David A. Klarner (ed.), *The Mathematical Gardner*, pp. 140–166, plus Plates I–III, Wadsworth, Belmont, Calif., 1981.

FIGURE 20.10 Escher No. 128 (*Bird*), from Escher's 1941–1942 notebook. (*© 1967 M. C. Escher Foundation, Baarn, Holland, all rights reserved.*)

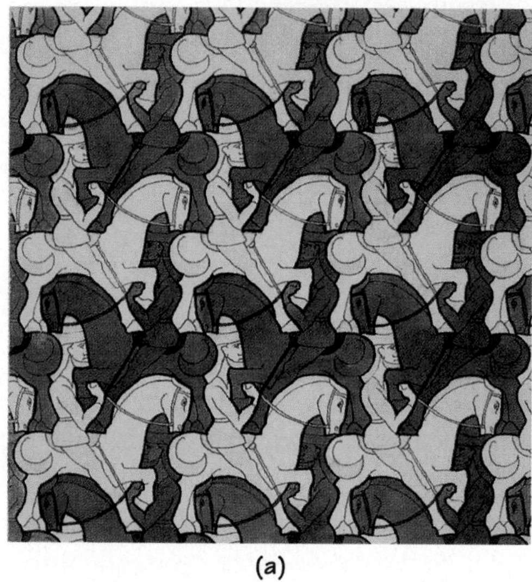

(a)

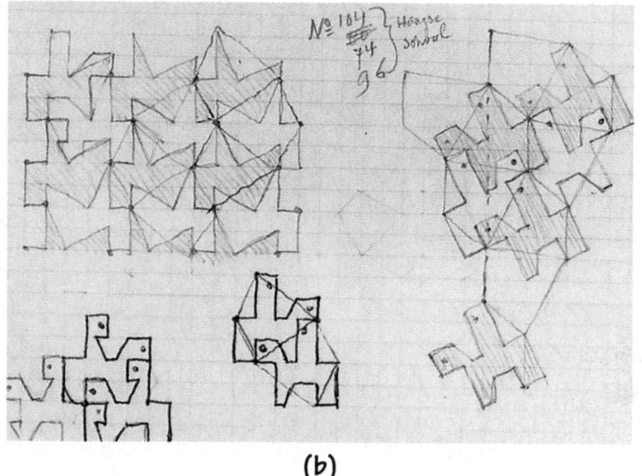

(b)

FIGURE 20.11 (a) Escher No. 67 (*Horseman*), from Escher's 1941–1942 notebook. (*© 1947 M. C. Escher Foundation, Baarn, Holland, all rights reserved.*) (b) Sketch showing the tile design for the *Horseman* print. (*© 1947 M. C. Escher Foundation, Baarn, Holland, all rights reserved. From the collection of Michael S. Sachs.*)

The tiles for Figures 20.10 and 20.11 are shown in outline form in Figure 20.12, together with points marked to show how the tiles fulfill the criterion.

In fact, alternative 1 of the criterion is a special case of alternative 2 (see Exercises 19–22 and 24). Moreover, alternative 2 *completely characterizes* tiles that can tile by translations. That is, not only is it true that *if alternative 2 is true, then the tile can tile by translations,* but also that the criterion works "in reverse": *If a tile can tile by translations, then alternative 2 must be true* (for some choice of six consecutive points).

A nice feature of the translation criterion is that if you can find points as required for alternative 2, then you can join them in order, as in Figures 20.12a and b, to see how to do the tiling.

To create tilings, though, you can proceed exactly as Escher did. His notebooks show that he designed his patterns in just the way that we now describe.

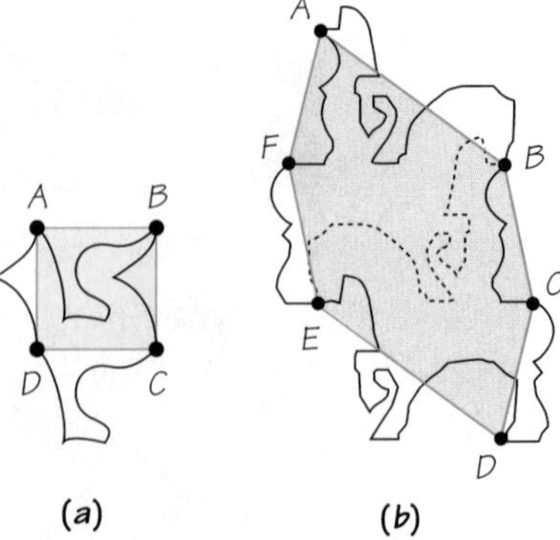

FIGURE 20.12 Individual tiles traced from the Escher prints of Figures 20.10 and 20.11, with points marked to show they fulfill the criteria for tiling by translations. The two knights form a block that tiles by translation, although a single knight can tile by itself if we allow mirror-image reflections too.

EXAMPLE 2 ■ Tiling Starting from a Parallelogram

SOLUTION For the first alternative of the criterion, start from a parallelogram, make a change to the boundary on one side, then copy that change to the opposite side. Similarly, change one of the other two sides and copy that change on the side opposite it (Figure 20.13). Revise as necessary, always making the same change to opposite sides. You might find it useful (as Escher did) to make your designs on graph paper, or you can work by cutting and taping together pieces of heavy paper. ■

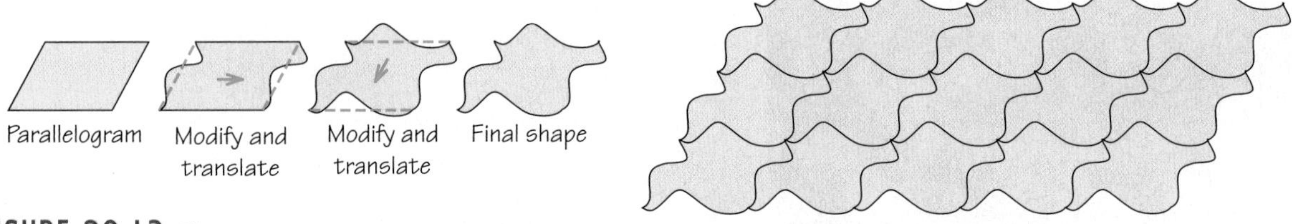

FIGURE 20.13 How to make an Escher-like tiling by translations, from a parallelogram base.

EXAMPLE 3 ■ Tiling Starting from a Hexagon

SOLUTION For the second alternative, start from a **par-hexagon**, a hexagon whose opposite sides are equal and parallel. This is one of the kinds of hexagons that tile the plane. Again, make a change on one boundary and copy the change to the opposite side, and do this for all three pairs of opposite sides (Figure 20.14). ■

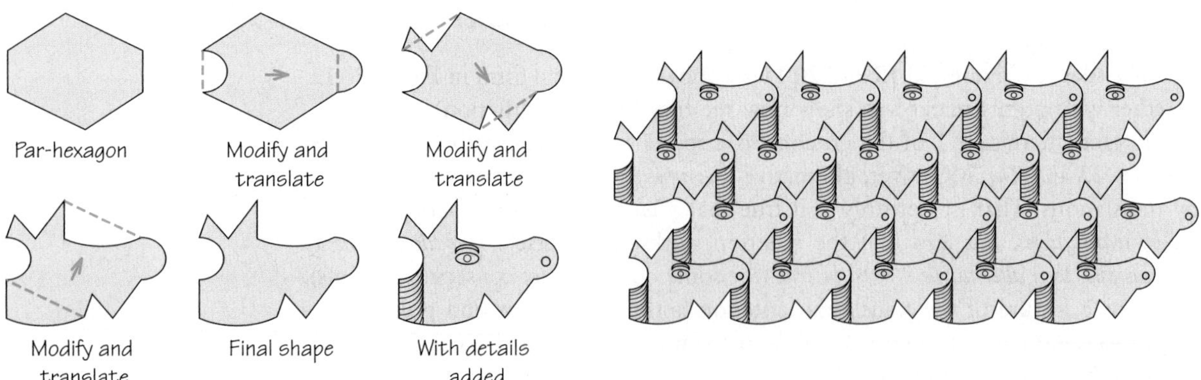

FIGURE 20.14 How to make an Escher-like tiling by translations, from a par-hexagon base.

20.4 Using Translations Plus Half-Turns

If the tiling is to allow half-turns, so that some of the figures are "upside down," the part of the boundary of a right-side-up figure has to match the corresponding part of itself in an upside-down position. For that to happen, that part of the boundary must be **centrosymmetric**, that is, symmetric about (unaltered by) a 180° rotation around its midpoint. The key to some of Escher's more sophisticated monohedral designs, and the fundamental principle behind some further easy recipes for making Escher-like tilings, is the **Conway criterion**, formulated by John H. Conway of Princeton University.

Conway Criterion

RULE

A tile can tile the plane by translations and half-turns if there are six consecutive points on the boundary (some of which may coincide, but at least three of which are distinct)–call them A, B, C, D, E, and F–such that

- the boundary part from A to B is congruent by translation to the boundary part from E to D, and
- each of the boundary parts BC, CD, EF, and FA is centrosymmetric.

The first condition means that we can match up the two boundary parts exactly, curve for curve, angle for angle. The second condition means that each of the remaining boundary parts is brought back into itself by a half-turn around its center. Either condition is automatically fulfilled if the boundary part in question is a straight-line segment.

FIGURE 20.15 Escher No. 6 (*Camel*), from Escher's 1941–1942 notebook. (*© 1937–1938 M. C. Escher Foundation, Baarn, Holland, all rights reserved.*)

The tiles for Figures 20.15 and 20.16 are shown in outline form in Figure 20.17, together with points marked to show how the tiles fulfill the Conway criterion. Figure 20.15 shows that Escher sketched little circles exactly where we have red dots in Figure 20.17a.

FIGURE 20.16 Escher No. 88 (*Sea Horse*). (*© 1947 M. C. Escher Foundation, Baarn, Holland, all rights reserved.*)

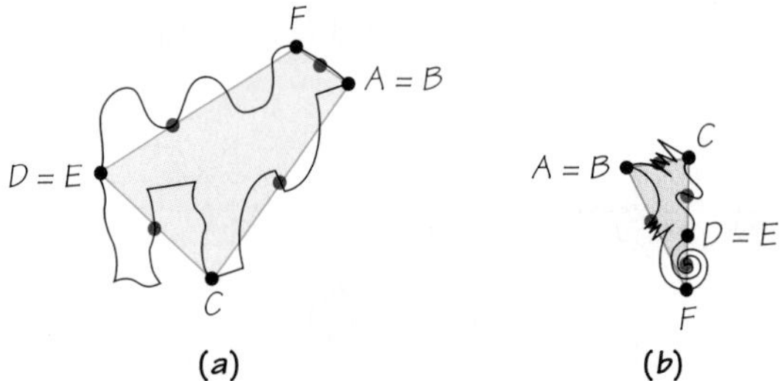

FIGURE 20.17 Individual tiles traced from the Escher prints of Figures 20.15 and 20.16, with points marked to show they fulfill the Conway criterion for tiling by translations and half-turns (around the red dots).

Mathematicians do not know if the Conway criterion completely characterizes tiles that can tile by translations and half-turns. Tiles that fulfill the Conway criterion can tile by translations and half-turns, but not necessarily vice versa: There could be tiles that can tile that way but do not satisfy the criterion–however, nobody knows of any. (The Conway criterion does, however, completely characterize tiles that produce the wallpaper pattern *p2* of Spotlight 19.6: Any tile that satisfies the criterion can be used to make a *p2* pattern, and any tile that can produce that pattern must satisfy the criterion.)

Once again, you can make Escher-like tilings by starting from simple geometric shapes that tile. This time, the starting geometric tile can be any triangle or any quadrilateral.

EXAMPLE 4 ■ Tiling Using a Triangle

SOLUTION For a triangle, modify half of one side, then rotate that side around its center point to extend the modification to the rest of the side, thereby making the new side centrosymmetric. Then you can do the same to the second and third sides (Figure 20.18).

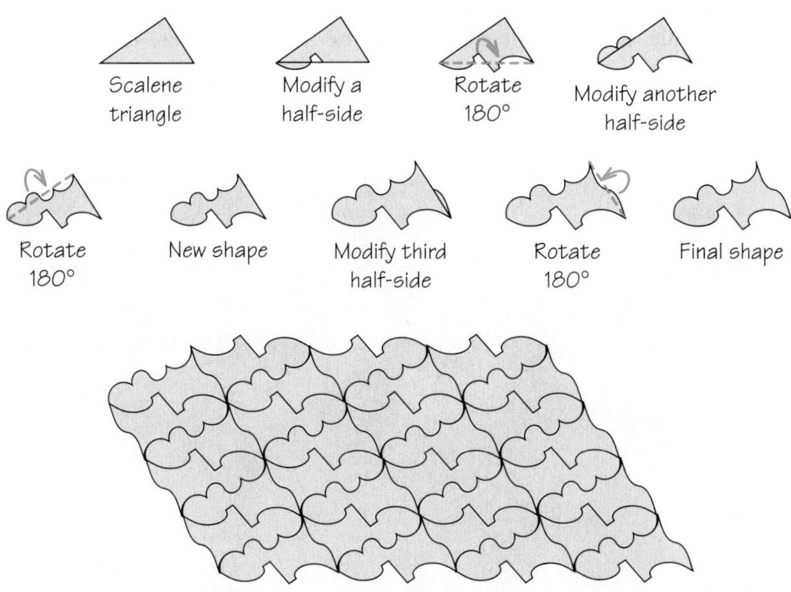

FIGURE 20.18 How to make an Escher-like tiling by translations and half-turns, from a scalene triangle base.

EXAMPLE 5 ■ Tiling Using a Quadrilateral

SOLUTION For the quadrilateral, do the same, modifying each of the four sides, or as many as you wish (Figure 20.19).

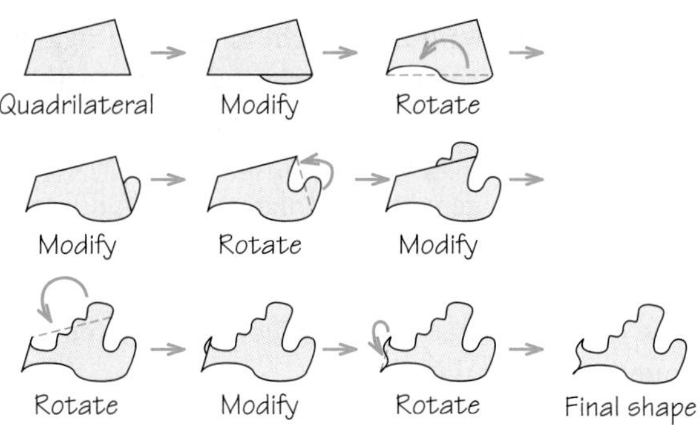

FIGURE 20.19 How to make an Escher-like tiling by translations and half-turns, from a quadrilateral base.

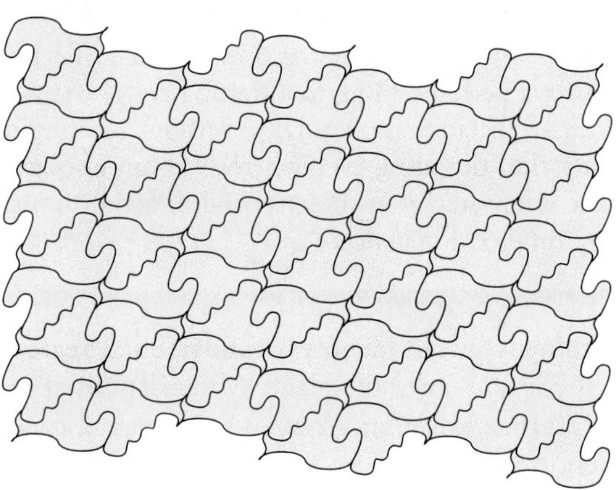

The same approach will work with some of the sides of some pentagons and hexagons that tile. Because not all sides can be modified, there is less freedom for designing tiles, so it is more difficult to make the resulting tiles resemble intended figures. Figure 20.20 shows the beautiful results achieved by Marjorie Rice, using one of the unusual tilings by pentagons that she discovered.

FIGURE 20.20 *Fish,* by Marjorie Rice, based on one of her unusual tilings by pentagons.

Sketches in Escher's notebook indicate how he designed many of his prints. For the bird tiling of Figure 20.10, the single bird below the tiling shows that he modified the sides of a square. For the knights tiling of Figure 20.11a, the sketches in Figure 20.11b show that he modified the pairs of sides of a par-hexagon. We redraw the two fundamental figures more clearly in Figure 20.12. (The knight tiling also has a reflection symmetry, taking a leftward-facing light knight to a rightward-facing dark knight; but we have not discussed criteria for producing a tiling with such a symmetry.)

As can be seen in faint lines in Figure 20.15, Escher used a parallelogram as a base for the camel tiling. In Figure 20.17a, the blue overlay shows how to make the tiling starting from a more general quadrilateral by modifying half of each side. For the seahorse tiling of Figure 20.16, Escher used a triangle base. However, once more he avoided modifying half of every side; instead, he treated the triangle ACF (Figure 20.17b) as a quadrilateral $ACDF$ in which two adjacent sides (CD and DF) happen to continue on in a straight line.

Periodic Tilings — PROCEDURE

All the patterns that we have exhibited and discussed so far have been **periodic tilings**. If we transfer a periodic tiling to a transparency, it is possible to slide the transparency a certain distance horizontally, without rotating it, until the transparency exactly matches the tiling everywhere. We can also achieve the same result by moving the transparency in some second direction (possibly vertically) by a certain (possibly different) distance.

In a periodic tiling, you can identify a **fundamental region**–a tile, or a block of tiles–with which you can cover the plane by translations at regular intervals. For example, in Figure 20.10, a single bird forms a fundamental region. In Figure 20.15, two adjacent camels, one right side up and one upside down, form a fundamental region. In the terminology of Chapter 19, the periodic tilings are ones that are preserved under translations in more than two directions.

20.5 Nonperiodic Tilings

In Figure 20.3a, the second row from the bottom is offset one-half of a unit to the right from the bottom row, the third row from the bottom is offset one-third of a unit further, and so forth. Because the sum $\frac{1}{2} + \frac{1}{3} + \frac{1}{4} + \cdots + \frac{1}{n}$ never adds up to exactly a whole number, there is no direction (horizontal, vertical, or diagonal) in which we can move the entire tiling and have it coincide exactly with itself.

Nonperiodic Tiling DEFINITION

A **nonperiodic tiling** is a tiling in which there is no regular repetition of the pattern by translation.

EXAMPLE 6 ■ A Nonperiodic Tiling through Randomness

SOLUTION Consider the usual edge-to-edge square tiling. For each square, flip a coin. Depending on the result, divide the square into two right triangles by adding either a rising or a falling diagonal (see Figure 20.3b). Because what happens in each individual square is unconnected to what happens in the rest of the tiling, this random tiling by right triangles has no chance of being periodic. ■

Penrose Tiles and Quasicrystals

For all known cases, if a single tile can be used to make a nonperiodic tiling of the plane, then it can also be used to make a periodic tiling. It is still an open question whether this property is true for every possible shape. In 1993, Conway discovered an example in three dimensions of a single convex polyhedron that tiles space nonperiodically but cannot be used to make a periodic tiling.

For a long time, mathematicians also tended to believe the more general assertion that if you can construct a nonperiodic tiling with a set of one *or more* tiles, you can construct a periodic tiling from the same tiles. But in 1964 a set of tiles was found that permits only nonperiodic tiling. It contains 20,000 different shapes! Over the next several years, smaller sets were discovered with the same property, with as few as 100 shapes. But it was still amazing when in 1975 Sir Roger Penrose, a mathematical physicist at Oxford, announced a set that tiles only nonperiodically–consisting of just two tiles! (See Figure 20.21 and Spotlight 20.3.)

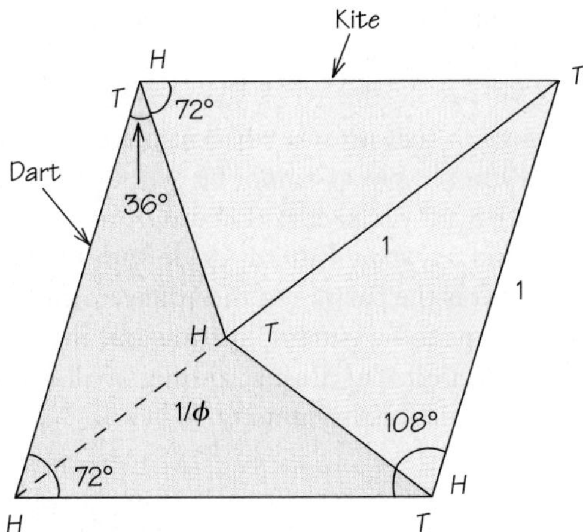

FIGURE 20.21 Construction of Penrose's "dart" (beige area) and "kite" (blue area). The length $1/\phi \approx 0.618$ is the reciprocal of the golden ratio ϕ.

SPOTLIGHT 20.3 Sir Roger Penrose

Sir Roger Penrose
(Anthony Howarth/Photo Researchers.)

Sir Roger Penrose, a professor at the University of Oxford, received a doctorate in mathematics but has been seriously interested in physics for many years; he was one of the first to conjecture the existence of black holes. He discovered what are now called the Penrose pieces in 1973. His latest endeavor has been to try to establish that the mind is not a machine, that is, that the ideas and concepts of artificial intelligence cannot explain human consciousness.

A chance meeting with Maurits Escher resulted in Penrose sending him some of his grandfather's art, which helped inspire some of Escher's prints.

Penrose called his tiles "darts" and "kites," and both of these *Penrose tiles* can be obtained from a single rhombus. A **rhombus** is a quadrilateral with four equal sides and equal opposite interior angles. The particular rhombus from which the Penrose tiles are constructed has interior angles of 72° and 108°. If we cut the longer diagonal in two pieces so that the longer piece is the golden ratio, or $(1 + \sqrt{5})/2 \approx 1.618$ times as long as the shorter (see Chapter 19, p. 712), and connect the dividing point to the remaining corners, we split the rhombus into a dart and a kite (Figure 20.21).

Label the front and back vertices of the dart with H (for head) and its two wing tips with T (for tail), and do the reverse for the kite. Then the rule for fitting the pieces together is that only vertices with the same letter may meet: Heads must go to heads, tails must go to tails. Thus the rules don't allow the pieces to fit together as a rhombus (which would allow them to tile periodically).

A prettier method of enforcing the rules, proposed by Conway, is to draw circular arcs of different colors on the pieces and require that adjacent edges must join arcs of the same color. The result is the pretty patterns of Figure 20.22. In fact, Conway thinks of the darts as children, each with two hands. The rule for fitting the pieces together is that children are required to hold hands. Penrose patterns become dancing circles of children.

Figure 20.23 shows a tiling by a different pair of pieces, both rhombuses, that tile the plane only nonperiodically. Figure 20.24 shows a modification of the Penrose pieces into two bird shapes. Figure 20.25 shows a coloring of one particular tiling with the Penrose pieces so that no two adjacent pieces have the same color.

Although tilings with Penrose's pieces cannot be periodic, the tilings possess unexpected symmetry. As you recall, we have explored our intuitions of symmetry in terms of *balance, similarity,* and *repetition.* Patterns made with the Penrose pieces certainly involve repetition, but it is the balance in the arrangement that we seek. What balance can there be in a nonperiodic pattern? It turns out that some Penrose patterns have a single line of reflection. But most surprising of all is that every Penrose pattern has a kind of fivefold rotational symmetry.

FIGURE 20.22 A Penrose tiling with specially marked tiles, forming what is known as the cartwheel tiling. *(From Sir Roger Penrose.)*

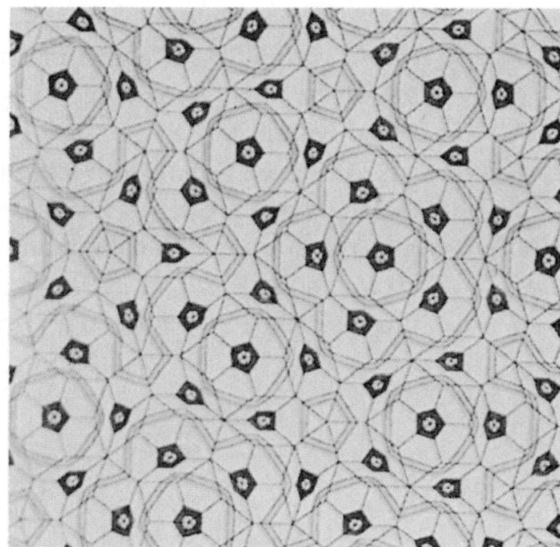

FIGURE 20.23 A Penrose nonperiodic tiling made with two rhombus shapes, one thin and one fatter. The fatter one has a yellow stripe across one end. *(Tiling by Sir Roger Penrose.)*

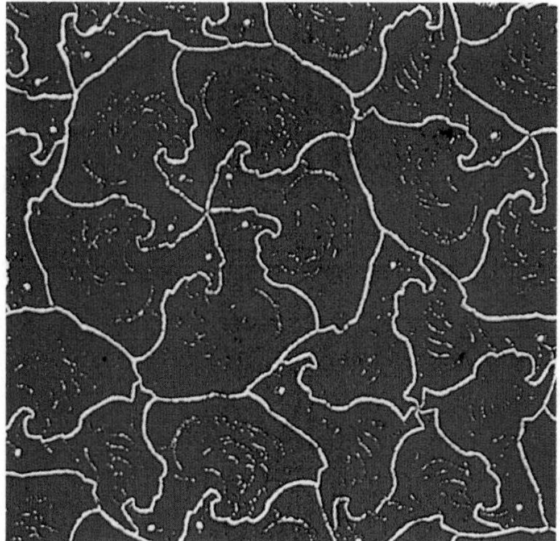

FIGURE 20.24 A modification of a Penrose tiling by refashioning the kites and darts into bird shapes. *(Tiling by Sir Roger Penrose.)*

FIGURE 20.25 A Penrose tiling by kites and darts, colored with five colors. A Penrose tiling can always be colored using four colors, in such a way that two tiles that share an edge have different colors. Whether a Penrose tiling can be colored in such a way using only three colors is an unsolved problem. However, we know that if even one Penrose tiling can be colored using three colors, all Penrose tilings can. *(Tiling by Sir Roger Penrose.)*

EXAMPLE 7

How Does a Penrose Pattern Have Fivefold Symmetry?

SOLUTION Look again at Figure 20.21, which shows how to split a rhombus into the Penrose dart and kite pieces. Except in the recess of the dart and the matching part of the kite, all of the internal angles of the kite and of the dart are either 72° or 36°.

Now, 72° goes into 360° five times, and 36° goes into 360° ten times. If we recall that it is the interior angles that matter in arranging polygons around a point, we see why it might be possible for a Penrose pattern to have fivefold or tenfold rotational symmetry.

A Penrose pattern with tenfold rotational symmetry is impossible, but there are exactly two Penrose patterns that tile the entire plane with fivefold rotational symmetry about one particular point. We show finite parts of these patterns in Figure 20.26. For each pattern, the center of rotational symmetry is at the center of the figure, surrounded by either five darts or five kites.

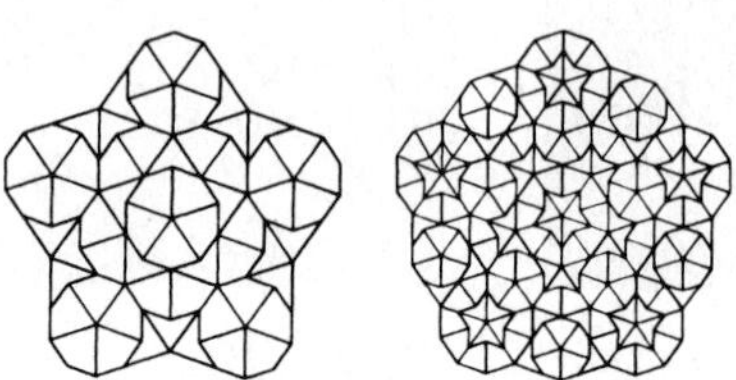

FIGURE 20.26 Successful deflation (that is, the systematic cutting up of large tiles into smaller ones) of patches of tiles of a Penrose nonperiodic tiling.

For any other Penrose pattern, the pattern as a whole does not have fivefold rotational symmetry. However, what is surprising is that the pattern must have arbitrarily large finite regions with fivefold rotational symmetry. You can see this feature in the regions of Figure 20.23 that are enclosed by yellow lines. In Conway's metaphor, whenever a chain of children (darts) closes, the region inside has fivefold symmetry.

Conway invented a process called *inflation* that takes any Penrose pattern into a different Penrose pattern with larger darts and kites. The inflation operation (we don't give the details here) systematically cuts up the darts and kites into triangles and regroups the triangles into larger darts and kites.

We can use inflation to show that a Penrose pattern must be nonperiodic. Suppose (contrary to what we want to establish) that some Penrose pattern is periodic; that is, it has translation symmetry. Let d be the distance along the translation direction to the first repetition. Performing inflation does the same thing to each repetition, so the inflated pattern must still have translation symmetry and a distance d along the translation direction to the first repetition. Keep on performing inflation, time after time, until the darts and kites are so large that they are more than d across. The pattern, as we have just argued, must still have translation symmetry at a distance d, but it can't, because there's no repetition inside a single tile! We reach a contradiction. So what's wrong? Our initial supposition, that the pattern was periodic in the first place, must have been erroneous. We conclude that all Penrose tilings are nonperiodic.

Despite their being nonperiodic, all Penrose patterns are somewhat alike, in the following remarkable sense.

Penrose Inside of Penrose — THEOREM

The subpattern of any finite region in one Penrose pattern is contained somewhere inside every other Penrose pattern. In fact, any subpattern occurs infinitely many times in every Penrose pattern.

The nonperiodicity of Penrose filings found a surprising application in 1997–to bathroom tissue. Quilted bathroom tissue is embossed with a pattern to keep the layers together (Figure 20.27). If the pattern is regular, then the multiple layers on the roll can produce lumpy ridges and grooves. Using a nonrepeating Penrose pattern averts the lumpiness. However, the company used Penrose's pattern without his permission, and Penrose sued successfully.

Penrose tilings have another feature that allows us to characterize them as *quasiperiodic,* or somewhere between periodic and random. (Noting the precise definition of random would take us too far afield.) Robert Ammann introduced onto the two rhombic Penrose pieces used in Figure 20.23 lines now known as *Ammann bars.* In any Penrose tiling, these bars line up into five sets of parallel lines, each set rotated 72° from the next, forming a pentagonal grid (Figure 20.28). The distance between two adjacent parallel bars is one of only two values, either A or B. Do you want to guess what the ratio of the longer A is to the shorter B? You don't think it could possibly be anything but the golden ratio of Chapter 19, do you? And so it is.

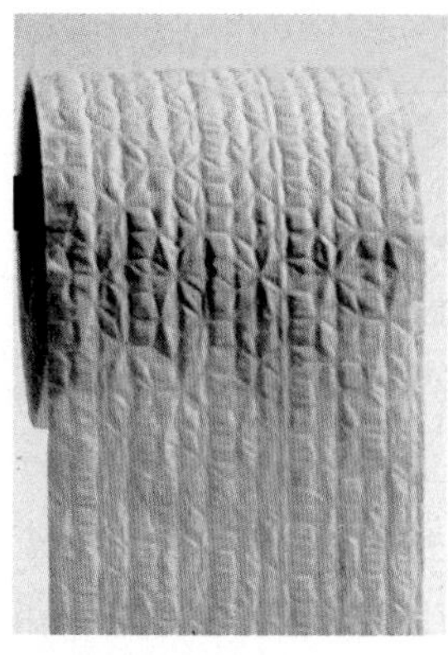

FIGURE 20.27
Penrose toilet paper.
(*Mario Ruiz/Time Magazine.*)

EXAMPLE 8 ■ Musical Sequences

What about the order in which the A's and B's occur, as we move from left to right in Figure 20.28? Is there any pattern to that?

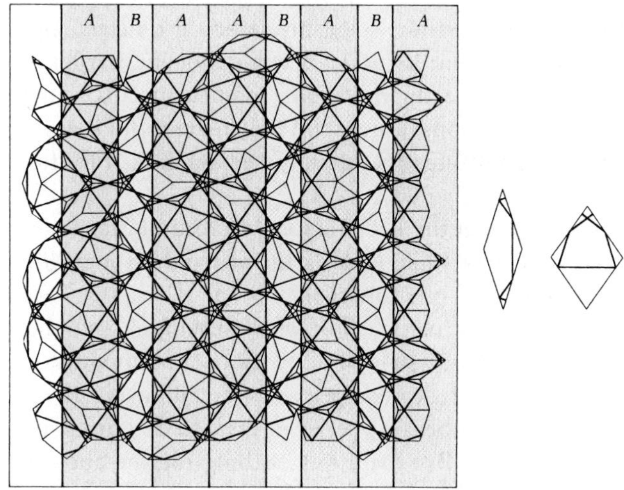

FIGURE 20.28
Penrose tilings with Ammann bars. Specially placed lines on the tiles produce five sets of parallel bars in different directions.

SOLUTION From the limited part of the pattern that we can observe, we see the sequence as

$$ABAABABAABABA$$

You might think from the figure that the pattern continues repeating the group

$$ABAAB$$

indefinitely; after all, there are five symbols in this group. But such is not the case. The sequence of intervals between Ammann bars is nonperiodic–it cannot be produced by repeating any finite group of symbols. We can think of it as a one-dimensional analogue of a Penrose tiling. The notation is reminiscent of the melody pattern of songs: Many popular songs follow the pattern ABA, with the first and the last sections having the same melody but the middle section being different. Consequently, a sequence of intervals between Ammann bars is known as a *musical sequence.*

There is some regularity in musical sequences. Two *B*'s can never be next to each other, nor can we have three *A*'s in a row. Just as any finite part of any Penrose tiling occurs infinitely often in any other Penrose tiling, any finite part of any musical sequence appears infinitely often in any other one. The order of the symbols is neither periodic nor random, but between the two–quasiperiodic.

The ratio of darts to kites in an infinite Penrose tiling, or of *A*'s to *B*'s in a musical sequence, is exactly the golden ratio, approximately 1.618. So if you are going to play with sets of Penrose pieces to see what kinds of patterns you can create, you will need about 1.6 times as many darts as kites.

As pointed out by geometers Marjorie Senechal (Smith College) and Jean Taylor (Rutgers University), Penrose tilings have three important properties:

- They are constructed according to rules that force nonperiodicity.
- They can be obtained from a substitution process (inflation and deflation) that features self-similarity at different scales (like the fractals in Chapter 19).
- They are quasiperiodic.

These properties are somewhat independent, meaning that one or two may be true of a tiling without all three being true.

Quasicrystals and Barlow's Law

Although Penrose's discovery was a big hit among geometers and in recreational mathematics circles in the mid-1970s, few people thought that his work might have practical significance. In the early 1980s some mathematicians even generalized Penrose tilings to three dimensions, using solid polyhedra to fill space nonperiodically. Like the two-dimensional Penrose patterns, these have orderly fivefold symmetry but are nonperiodic.

Yet in 1982 scientists at the U.S. National Bureau of Standards discovered unexpected fivefold symmetry while looking for new ultrastrong alloys of aluminum (mixtures of aluminum with other metals).

Manganese doesn't ordinarily alloy with aluminum, but the experimenters were able to produce small crystals of alloy by cooling mixtures of the two metals at a rate of millions of degrees per second. Following routine procedures, chemist Daniel Shechtman began a series of tests to determine the atomic structure of the special crystals. But there was nothing routine about what he found: The atomic structures of the manganese–aluminum crystals were so startling that it took Shechtman three years to convince his colleagues they were real.

Why did he encounter such resistance? His patterns–and the crystals that produced them–defied one of the fundamental laws of crystallography. Like our discovery that the plane cannot be tiled by regular pentagons, **Barlow's law**, also called the **crystallographic restriction**, says that a crystal must be periodic and hence can have only rotational symmetries that are twofold, threefold, fourfold, or sixfold. If there were a center of fivefold symmetry, there would have to be many such centers. Barlow proved this impossible.

Peter Barlow (1776–1862) argued by contradiction, similar to Conway's proof in which we saw earlier that Penrose patterns are not periodic. Suppose (contrary to what we intend to show) that there is more than one fivefold rotation center. Let *A* and *B* be two of these that are closest together (see Figure 20.29). Rotate the pattern of Figure 20.28 by one-fifth of a turn clockwise around *B*, which carries *A* to some point *A*′. Because the pattern has fivefold symmetry around *B*, the point *A*′, which is the image of the fivefold center A, must itself be a fivefold center. Now use A as a center and rotate the pattern by one-fifth of a turn clockwise, which carries B to some point B'. As we just argued in the case of A', B' must also be a fivefold center. But A' and B' are closer together than A and B, which is a contradiction. Hence our original supposition must be false, and a pattern can have at most one fivefold rotation center (as the patterns in Figure 20.26 in fact do) and so cannot be periodic.

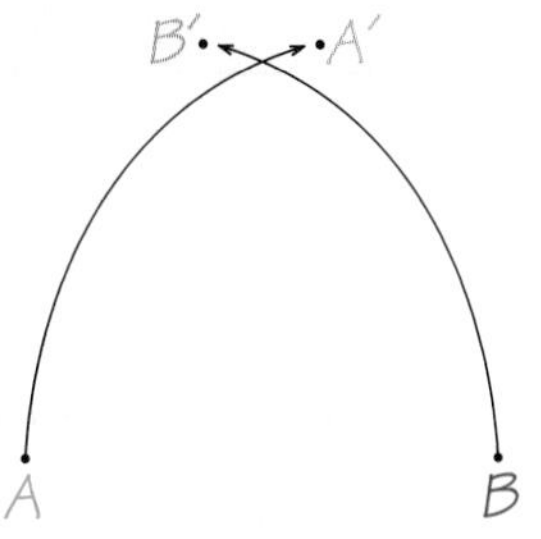

FIGURE 20.29 Barlow's proof that no pattern can have two centers of fivefold symmetry.

For chemists, crystals are modeled well by periodic three-dimensional tilings; an array of atoms with no symmetry whatever would not be considered a crystal. Since Barlow's law shows that fivefold symmetry is impossible in a *periodic* tiling, no one suspected until Penrose's discovery that there could be symmetric *non*periodic tilings, nor until Schechtman's alloys that real atoms could arrange themselves in such a way.

Schechtman's alloys, since they are not periodic, are not crystals, though in other respects they do resemble crystals. It is scientifically more fruitful to extend the concept of crystals to include them than to rule them out. They are now known as *quasicrystals* (Spotlight 20.4).

Once again, as so often happens in history, pure mathematical research anticipated scientific applications. Penrose's discovery, once just a delightful piece of recreational mathematics, has prompted a major reexamination of the theory of crystals. Barlow's law is not refuted, since it applies only to periodic crystals, not to quasicrystals.

SPOTLIGHT 20.4 Quasicrystals

In 1984, working at the University of Pennsylvania, Paul Steinhardt and Don Levine did a computer simulation of what a three-dimensional Penrose pattern would be like. They decided to call such structures *quasicrystals*. Later that fall, their chemist colleague Daniel Shechtman showed that quasicrystals really exist. He produced images of an alloy of aluminum and manganese that were amazingly similar to images from the computer simulations. In short order, sevenfold, ninefold, and other symmetries were also shown to occur in real materials.

In 1991, Sergei Burkov showed that quasiperiodic tilings can be made using only a single kind of 10-sided tile, *provided the tiles are allowed to overlap.* With overlaps, the resulting patterns are no longer tilings. They are called *coverages.* In late 1998, scientists presented electron microscope photos that demonstrated that atoms really do form such coverages.

The current theory is that quasicrystals are packings of copies of a single type of atom cluster, with each cluster sharing atoms with its neighbors, that is, overlapping nearby clusters. The clusters form a quasiperiodic pattern that maximizes their density, thereby minimizing the energy of the atoms involved.

In 2007 Steinhardt and Peter J. Lu announced the discovery of decagonal and Penrose tilings in medieval Islamic architecture in Iran.

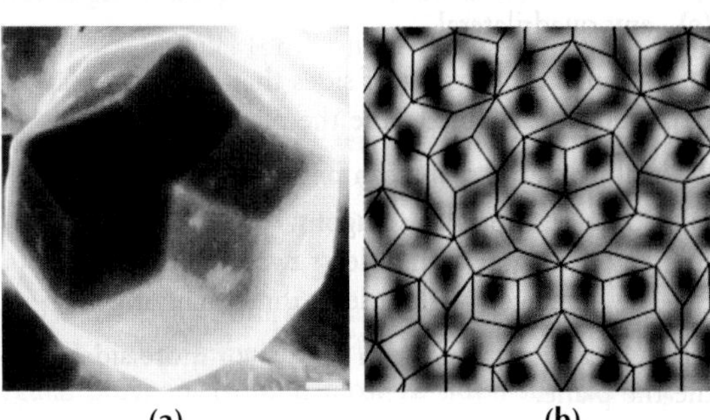

(a) (b)

(a) A scanning electron microscope image of the quasicrystal alloy $Al_{5???}Li_3Cu$ (the question marks indicate uncertainty about how many aluminum atoms are involved). The fivefold symmetry can be seen in the five rhombic faces that meet at a single point in the center of the photograph, forming a starlike shape. (b) This image of the quasicrystal material $Al_{65}Co_{20}Cu_{15}$ was obtained with a scanning tunneling microscope. The resulting image has been overlaid with a nonperiodic tiling to display the local fivefold symmetry. *[Both adapted from Hans C. von Baeyer, "Impossible Crystals,"* Discover *11(2) (February 1990): 69–78, 84.]*

tile the plane? (See Figure 19.4 for a regular pentagon that you can trace.)

12. For each of the tiles below, show how it can be used to tile the plane. (Adapted from *Tilings and Patterns*, by Branko Grünbaum and G. C. Shephard, Freeman, New York, 1987, p. 25.)

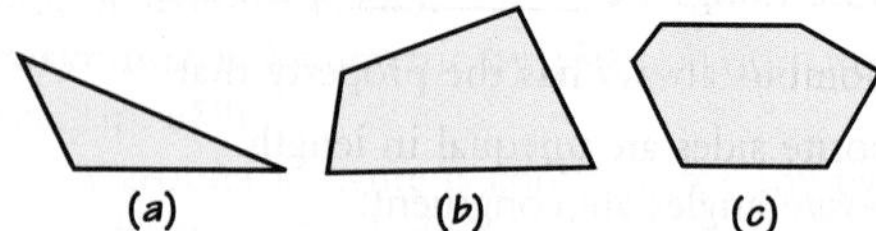

20.3 Using Translations

Refer to tiles (a) through (g) below in doing Exercises 13 and 14.

13. For each of the tiles (a) through (c), determine whether it can be used to tile the plane by translations. (From *Tiling the Plane*, by Frederick Barber et al., COMAP, Lexington, Mass., 1989, pp. 1, 8, 9.)

14. Repeat Exercise 13, but for tiles (d) through (g).

15. Start from a par-hexagon of your choice and modify it to tile the plane by translations. (You will probably find it useful to do your work on graph paper. If you choose a regular hexagon, there is special graph paper, ruled into regular hexagons, that would be particularly useful.) Can you draw a design on the tile so as to make an Escher-like pattern?

16. Start from a parallelogram of your choice and modify it to tile the plane by translations. (You will probably find it useful to do your work on graph paper.) Can you draw a design on the tile so as to make an Escher-like pattern?

Refer to the following information in doing Exercises 17–20.

A particularly simple kind of polygon, called a *polyomino*, is one made of squares joined edge-to-edge. The name is a generalization of "domino"; indeed, there is only one kind of domino (two squares joined at an edge to form a rectangle). There are just two "trominos" (short for "triominos"), the straight tromino and the L-tromino.

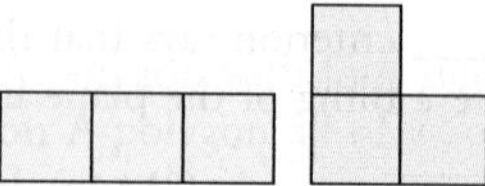

The straight tromino has the shape of a rectangle, so it can tile the plane by translations; and the L-tromino has the shape of a hexagon.

17. Is the L-tromino convex? Does the result about what hexagons can tile the plane (p. 654) give any information about whether the L-tromino can tile the plane or not?

18. Find a tiling of the plane using just the L-tromino and translations of it. Is there more than one way to do the tiling?

19. Show how alternative 2 of the translation criterion can be applied to the L-tromino.

■ **20.** Give an argument why alternative 1 of the translation criterion cannot be applied to the L-tromino. (*Hint:* Label each of the eight corners of the component squares of the tromino with the letters $S, T, \ldots, Z$. Let these be our candidates for the points A, B, C, and D of the criterion.) Each of the sides of the tromino that is two units long has nowhere to go under a translation. Any application of the criterion must divide each side

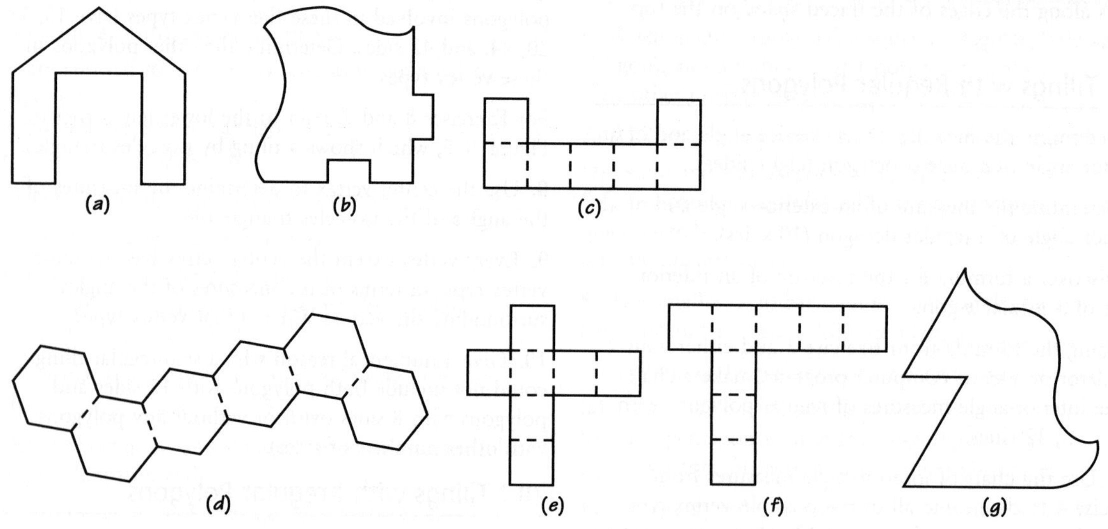

into two pieces, so their midpoints must be two of the points A, B, C, and D. Make a similar argument about two corners of the tromino. Thus, we have four points, which can be labeled consecutively A, B, C, and D, starting at any one of them. Show that none of the four possibilities "works." (This argument can be generalized to show that trying A, B, C, and D at points other than the corners of the squares won't work either.)

Refer to the following information in doing Exercises 21–26.

Demonstrate to your own satisfaction that there are exactly five shapes of tetrominos (each made of four squares joined at edges)–plus differing mirror images of two of them–as shown below. In the order shown, they are called the square, straight, T, L, and skew tetrominos. The straight and the square tetromino certainly can tile the plane by translations.

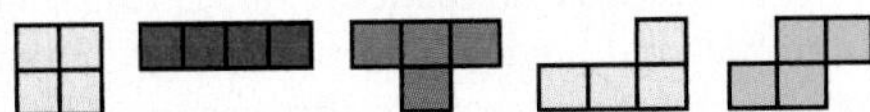

You will definitely find it useful to make yourself several copies of each of the polyominos mentioned below, by cutting them out of graph paper.

21. Apply alternative 1 of the translation criterion to the straight-tetromino and show how it can tile.

22. Show how alternative 2 of the translation criterion applies to the T-tetromino.

23. Apply alternative 1 of the translation criterion to the L-tetromino and show how it can tile.

24. Show how alternative 2 of the translation criterion applies to the skew-tetromino.

25. Show how alternative 2 of the translation criterion can be applied to the skew-tetromino, and show how it can tile.

26. In Exercises 23 and 25, we indulged in what appears to be "overkill," proving the same fact in two different ways. But those exercises should give you the idea that alternative 2 of the translation criterion can reduce to (and hence is more general than) alternative 1 if some points are allowed to coincide. For such a reduction, which pairs of points must coincide? (We are allowed to relabel the remaining four distinct points.)

20.4 Using Translations Plus Half-Turns

For Exercises 27 and 28, refer to tiles (a) through (g) on p. 672.

27. For each of the tiles (a) through (c), determine whether it can be used to tile the plane by translations and half-turns.

28. Repeat Exercise 27, but for tiles (d) through (g).

29. Show how an arbitrary pentagon with two parallel sides can tile the plane.

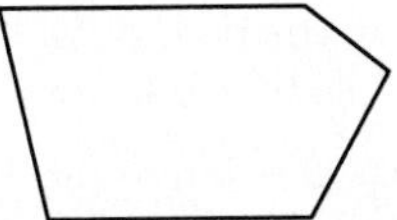

30. The following is a pentagonal tile of type 13, discovered by Marjorie Rice. Show how it can tile the plane. (*Hint:* Carefully trace and cut out a dozen or so copies and try fitting them together.)

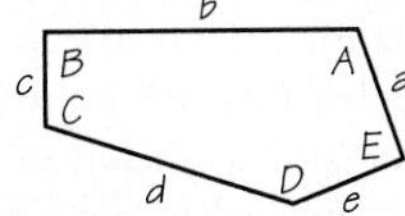

The parts of this pentagon satisfy the following relations: $A = C = D = 120°$, $B = E = 90°$, $a = e$, and $a + e = d$. [Adapted from "In Praise of Amateurs," by Doris Schattschneider, in David A. Klarner (ed.), *The Mathematical Gardner,* Wadsworth, Belmont, Calif., 1981, p. 162.]

31. Start from a triangle of your choice and modify it to tile the plane by translations and half-turns. (You will probably find it useful to do your work on graph paper.) Can you draw a design on the tile so as to make an Escher-like pattern?

32. Start from a quadrilateral of your choice and modify it to tile the plane by translations and half-turns. (You will probably find it useful to do your work on graph paper.) Can you draw a design on the tile so as to make an Escher-like pattern?

Refer to the information about polyominos preceding Exercise 17, and to the following, in doing Exercises 33–36.

We saw earlier that all the dominos, trominos, and tetrominos tile the plane by translations. Here we investigate the 12 pentominos, shown as follows with a letter notation for each (if you allow mirror images to count as different pentominos, there are 18). It will be useful for you to make several copies of each of the pentominos discussed below.

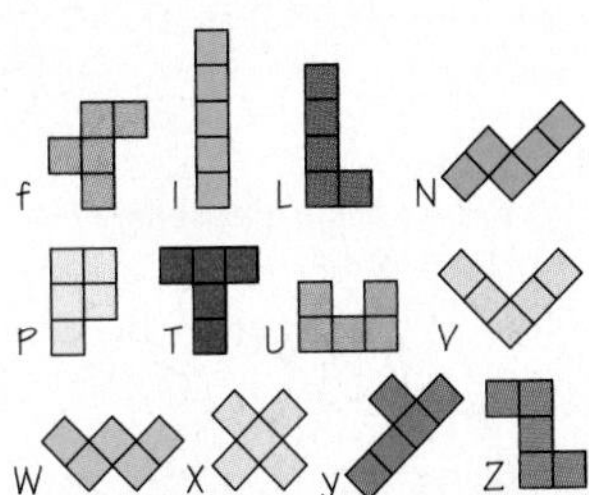

33. Just by experimenting, determine which of the pentominos can tile the plane by translations. (*Hint:* There are nine.)

34. Apply the Conway criterion to the f-pentomino, and show how it can tile by translations and half-turns.

35. Apply the Conway criterion to the U-pentomino, and show how it can tile by translations and half-turns.

36. Apply the Conway criterion to the T-pentomino, and show how it can tile by translations and half-turns.

37. In the text we discuss criteria and methods for generating Escher-like patterns that involve just translations or translations and half-turns. A slight variation on one of those methods allows construction of tilings that feature a tile and its mirror image.

Begin with a parallelogram made from two congruent isosceles triangles, as shown in the figure below. Each of these triangles has two sides equal. Be sure that the two triangles are arranged so that they have one of the equal sides in common, forming a diagonal of the parallelogram.

Make any modification to half of the third side of one of the triangles. Mirror-reflect that modification across the side, then translate the reflection to become the modification of the other half of the side. Take the complete modification of this side, and translate it to become the modification of the opposite side of the parallelogram.

Modify in any way one of the two remaining sides of the parallelogram, and make the same modification to the opposite side (that is, translate the modification, without rotation or reflection). Then reflect this modification across the diagonal of the parallelogram.

The result is a modified parallelogram that tiles by translation and splits into two pieces that are mirror images of each other. Escher used a similar technique, but starting from a par-hexagon made from two quadrilaterals, in his *Horseman* print, as shown in his sketch in Figure 20.11b.

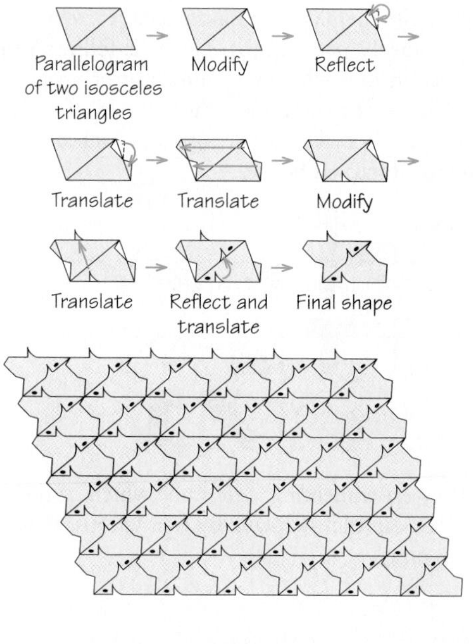

Use this technique to produce a tiling of your own design. Can you draw a design on the tile so as to make an Escher-like pattern?

38. Show that the modified parallelogram in Exercise 37 fulfills the Conway criterion, by identifying the six points of the criterion.

20.5 Nonperiodic Tilings

39. In this chapter we have been concerned mostly with tiling the plane, with some attention to using crystals to fill space. We can also consider a simpler case–tiling the line. For the line, a tile is a line segment of a particular length.

(a) What are the monohedral tilings of the line?
(b) What are the periodic tilings of the line that use two tiles of different lengths?

Exercises 40–42 connect nonperiodic tilings to the Fibonacci numbers and golden ratio of Chapter 19 but do not require any information from that chapter. (Thanks for this idea to David J. Wright, Oklahoma State University.) Just as the plane can be tiled quasiperiodically with Penrose tiles, the line can be tiled quasiperiodically with a pair of tiles, provided that their lengths are in the right proportion. Let the lengths of the tiles be a and b, with $b < a$ and a exactly c times as long as b, so that $cb = a$. Thus, scaling up a tile of length b by a factor of $c > 1$ produces a tile of length a. Similarly, we determine the scale factor c so that scaling up a tile of length a produces a tile of length a followed by a tile of length b, that is, $ca = a + b$.

40. (a) Using substitution, eliminate a and reduce the two equations to a single equation in just c and b alone.
(b) Use the quadratic formula to solve for the two possible values of c. Since we want $c > 1$, we choose the larger value.

41. We can define an inflation process for a line segment consisting of a's and b's: First, replace each original b by an a and each original a by two adjacent segments a and b. For example, we would replace aba by $(ab)(a)(ab)$, where we have inserted parentheses for clarity.

(a) Start with just a single b and repeat the inflation process, showing the stages, until you reach a stage with 21 segments.
(b) How many segments are there at each stage? (If we continue this process forever, we tile a half-line to the right; we tile the entire line by reflecting this right half-line over to cover the left half-line. The result is called a *Fibonacci tiling* of the line.)
(c) If a line segment contains m copies of the a tile and n copies of the b tile, how many tiles will the inflation of the segment contain?

42. We can similarly define a deflation process: Replace each original adjacent pair ab by an a and each remaining original a by a b.

(a) Apply this deflation process repeatedly to the stage in your answer to Exercise 41a that has 21 segments. What do you end up with?
(b) Apply this deflation process repeatedly to the periodic sequence of tiles *abababab*.... What do you end up with?
(c) Devise your own periodic sequence of *a* and *b* tiles. Apply the deflation process to it repeatedly. What do you end up with? What do you conjecture?
(d) In what sense is the Fibonacci tiling quasiperiodic?

For Exercises 43–46, refer to the following.

The rabbit problem in Chapter 19 (Exercise 4) leads us directly into nonperiodic patterns and musical sequences. Let A denote an adult pair of rabbits and B denote a baby pair. We record the population at the end of each month, just before any births, in a particular systematic way–as a string of A's and B's. At the end of their second month of life, a rabbit pair will be considered to be adult and give birth to a baby pair. At the end of the first month, the sequence is just A, and the same is true at the end of the second month. When an adult pair A has a baby pair B, we write the new B immediately to the right of the A. So at the end of the third month, the sequence is AB; at the end of the fourth, it is ABA, because the first baby pair is now adult; at the end of the fifth month we have $ABAAB$.

Mathematicians and computer scientists call this manner of generating a sequence a *replacement system.* At each stage we replace each A by AB and each B by A.

43. What is the sequence at the end of the sixth month?

44. Why can't we ever have two B's next to each other?

45. Why can't we ever have three A's in a row?

46. Show that from the fourth month on, the sequence for the current month consists of the sequence for last month followed by the sequence for two months ago.

For Exercises 47–54, refer to the following.

We can define inflation and deflation of a sequence of A's and B's, and musical sequences themselves, without reference to Penrose patterns, and thereby arrive at an example of a nonperiodic pattern in one dimension. Inflation consists of replacing each A by AB and each B by A, and deflation consists of replacing each AB by A and each A by B; inflation and deflation undo each other on musical sequences. Call a sequence *musical* if it results from applying inflation to the sequence consisting of a single B. Then inflation and deflation preserve musicality: If we inflate or deflate a musical sequence, we get another musical sequence. Another way to think of this relationship is that a musical sequence is self-similar under inflation and deflation.

■ **47.** Let the lone B be considered the first stage of inflation. Show that at the nth stage of inflation, for $n \geq 3$, there are F_n (the nth Fibonacci number, section 19.1) symbols in the sequence, of which F_{n-1} are A's and F_{n-2} are B's. (*Hint:* Check it for $n = 1, 2, 3$, and 4.)

48. Show that no musical sequence contains AAA or BB.

49. Show that no musical sequence ends in AA or in $ABAB$.

50. Show that apart from the lone sequence B, every musical sequence is an initial subsequence of all the succeeding musical sequences.

51. Slightly modified, deflation can be used to check whether a finite block of A's and B's can belong to a musical sequence or not. First, if the block has length greater than one, we may suppose that it begins with an A (why?). So at any stage of the deflation with a block beginning with B, we may add an initial A. Second, we add the additional deflation rule to replace an ending AA with BA. If at any stage of this modified deflation we arrive at two or more B's in a row, or three or more A's in a row, then the original block could not be part of a musical sequence. Otherwise, the original block will eventually deflate to a single symbol, at which point we conclude that the original block is a part of a musical sequence. Check the two blocks $ABAABABAAB$ and $ABAABABABA$.

52. From Exercise 50 we know that each application of inflation to a musical sequence simply extends it. By successive inflation, then, we build an infinite sequence. Show that as we approach this limiting sequence, the ratio of B's to A's tends toward the golden ratio ϕ.

53. Conclude from Exercise 52 that the sequence cannot be periodic, nor settle into a period after a finite "burn-in" period. Thus, the sequence is nonperiodic. (*Hint:* ϕ is not a rational number; that is, it cannot be represented as a ratio m/n of whole numbers m and n.)

54. Show that any finite block of A's and B's that occurs in the infinite sequence must occur over and over again (just as any patch of tiles in a Penrose pattern occurs infinitely often in the pattern). Thus, the infinite sequence is self-similar.

the end of the second year, you again receive only \$100; so at the beginning of the third year, the account contains \$1200. In fact, at the end of each year you receive just \$100 in interest.

The formulas for simple interest are themselves simple.

Simple Interest RULE

For a principal P and an annual rate of interest r, the interest earned in t years is

$$I = Prt$$

and the total amount A accumulated in the account is

$$A = P + I = P + Prt = P(1 + rt)$$

You may find this method for interest rather strange, since you are accustomed from your savings account to a different system of awarding interest–**compound interest**, which we will consider shortly. However, simple interest is often used for:

- private loans between individuals, because it is easy to calculate;
- commercial loans for less than one year–not just because it is easy to calculate–but also because for low interest rates, simple interest differs negligibly from compound interest; and
- financing of corporations and the government through bonds. A bond is a loan with repayment at the end of a fixed term and simple interest in the mean time, paid usually annually or semiannually.

EXAMPLE 1 ■ Simple Interest on a Student Loan

Let's suppose that you have exhausted the amount that you can borrow under federal loan programs and need a private direct student loan for \$10,000.

National City Corporation (headquartered in Cleveland, Ohio) quoted a rate in May 2008 of 5.7% for the 2007–08 school year. It offers an interest-only repayment option, under which you make monthly interest payments while you are in school and pay on the principal only after graduation. Under this plan, National City earns simple interest from you while you are in school.

How much monthly interest would you pay for such a \$10,000 loan?

SOLUTION The principal is $P = \$10{,}000$, the annual interest rate is $r = 5.7\% = 0.057$ per year, and the number of years is $t = \frac{1}{12}$ year. The interest for one month would be $I = Prt = \$10{,}000 \times 0.057 \times \frac{1}{12} = 47.50$. (Actually, National City would charge an "origination fee" of between 3% and 10.5%, added to the principal, so the payment would be greater; and the initial interest rate might not be 5.7% but could be more than 12%, since it would depend on the creditworthiness of you and your cosigner. Finally, the interest rate could increase each year during the term of the loan.) ■

We frequently observe the kind of growth corresponding to simple interest, called **arithmetic growth** or **linear growth**, in other contexts.

Arithmetic Growth DEFINITION

Arithmetic growth (also called **linear growth**) is growth by a constant amount in each time period.

For example, the population of medical doctors in the United States grows arithmetically, because the medical schools graduate the same number of doctors each year (and the number of doctors dying is also fairly constant). The concept of linear growth has appeared already in the discussions of linear programming (Chapter 4) and linear regression (Chapter 6).

21.2 Geometric Growth and Compound Interest

What you probably expected to happen to the savings account discussed in the last section is that during the second year the account would earn interest of 10% not on the *initial* balance of \$1000 (as with simple interest) but on the *new* balance of \$1100. Then, at the end of the second year, 10% of \$1100, or \$110, would be added to the account.

Thus, during the second year you would earn interest on both the principal of \$1000 and on the \$100 interest that was earned during the first year. You receive more interest during the second year than during the first; that is, the account grows by a greater amount during the second year. At the beginning of the third year the account contains \$1210, so at the end of the third year you receive \$121 in interest. Again, this is larger than the amount you received at the end of the preceding year. Moreover, the increase during the third year,

$$\text{third-year interest} - \text{second-year interest} = \$121 - \$110 = \$11$$

is larger than the increase during the second year,

$$\text{second-year interest} - \text{first-year interest} = \$110 - \$100 = \$10$$

Thus, not only is the account balance increasing each year, but the amount added also increases each year.

Compound Interest DEFINITION

Compound interest is interest that is paid on both the original principal and accumulated interest.

Savings institutions usually compound interest and credit it to accounts more often than once a year–for example, quarterly (four times per year). With an interest rate of 10% per year and quarterly compounding, you get one-fourth of the rate, or 2.5%, paid in interest each quarter year. The "quarter" (three months) is the **compounding period**, or the time elapsing before interest is paid.

Consider again a principal of \$1000. At the end of the first quarter, you have the original balance plus \$25 interest, so the balance at the beginning of the second quarter is \$1025. During the second quarter, you receive interest equal to 2.5% of \$1025, or \$25.625, which in posting to your account is rounded up (since the fraction is half a cent or more) to \$25.63. Continuing in this manner, the balance at the end of the first year is \$1103.82. (You should "read" all calculations in this chapter by confirming them on your calculator.)

Even though the account was advertised as paying 10% interest, the interest for the year is \$103.82, which is 10.382% of the principal of \$1000.

Practical note: Without rounding the interest for each quarter, the interest for the year would have been \$1103.81, as shown in Table 21.1. This table shows the results of calculation with rounding done only at the end of the year, while savings institutions must round at each compounding and post the amount to your account. A spreadsheet program could duplicate the results of their computer programs; but in the table and in all later calculations, we take the simpler route of rounding only at the final answer. Any differences will be very small; and if your answers differ just a few cents, that will be OK.

TABLE 21.1 Compound Interest on $1000, at an Interest Rate of 10% Compounded Quarterly

Date	Beginning Balance	Interest on Principal	Interest on Interest	Total Interest Added	Ending Balance
January 1	1000.00				
March 31	1000.00	25.00	0.00	25.00	1025.00
June 30	1025.00	25.00	0.63	25.63	1050.63
September 30	1050.63	25.00	1.27	26.27	1076.90
December 31	1076.90	25.00	1.92	26.92	1103.82

If interest is compounded monthly (12 times per year) or daily (365 or 366 times per year), the resulting balance is even larger, as shown in Table 21.2, (together with the results of continuous compounding, which we discuss later). We will show you shortly the compound interest formula for these calculations.

Terminology for Interest Rates

We have seen that an account at a particular annual rate of interest can produce different amounts of interest, depending on how the compounding is done. To help prevent confusion on the part of consumers, the Truth in Savings Act establishes specific terminology and calculation methods for interest.

A **nominal rate** is any stated rate of interest for a specified length of time, such as a 3% annual interest rate on a savings account or a 1.5% monthly rate on a credit-card balance. By itself, a nominal rate *does not indicate or take into account whether or how often interest is compounded.*

Table 21.2 shows that at an annual interest rate of 10% (a nominal rate) compounded daily for one year, $1000 yields $105.16 in interest, which is 10.516% of the principal. Hence the effective annual rate is 10.516%. In other words, $1000 at *simple* interest of 10.516% for one year would earn exactly the same amount of interest.

Effective Rate and APY DEFINITION

The **effective rate** is the rate of simple interest that would realize exactly as much interest over the same length of time. For a year, the effective rate is called the **annual percentage yield (APY).**

To keep these different rates straight, we use i for a nominal rate for a specified *compounding period*–a day, a month, or a year–*within which no compounding is done.* Because no compounding is done for shorter intervals than this period, the effective rate and the nominal rate are the same for the compounding period.

TABLE 21.2 Comparing Compound Interest: The Value of $1000, at 10% Annual Interest, for Different Compounding Periods*

Years	Compounded Yearly	Compounded Quarterly	Compounded Monthly	Compounded Daily	Compounded Continuously
1	1100.00	1103.81	1104.71	1105.16	1105.17
5	1610.51	1638.62	1645.31	1648.61	1648.72
10	2593.74	2685.06	2707.04	2717.91	2718.28

*Without rounding at posting of interest and neglecting leap years; the difference in most cases is no more than one cent.

Rate Per Compounding Period RULE

For a nominal annual rate r compounded m times per year, the rate per compounding period is

$$\text{periodic rate} = \boxed{i = \frac{r}{m}} = \frac{\text{nominal annual interest rate}}{\text{number of compounding periods per year}}$$

For that \$1000 in savings at 10% compounded quarterly, we have $r = 10\%$ and $m = 4$, so $i = 2.5\%$ per quarter.

We denote the number of compounding periods per year by m. We use r only for an annual rate and t for the number of years.

To avoid confusion, we don't use the terminology *annual percentage rate.* That term has a special meaning just for loans (see Chapter 22, p. 712).

Geometric Growth

Here we look for the underlying mathematical pattern of compounding. For quarterly compounding, you have at the end of the first quarter

$$\text{initial balance} + \text{interest} = \$1000 + \$1000(0.025) = \$1000(1 + 0.025)$$

and at the end of the second quarter

$$\begin{aligned}\text{initial balance} + \text{interest} &= \$1000(1 + 0.025) \\ &\quad + [\$1000(1 + 0.025)](0.025) \\ &= [\$1000(1 + 0.025)] \times (1 + 0.025) \\ &= \$1000(1 + 0.025)^2\end{aligned}$$

The pattern continues in this way, so that you have $\$1000(1 + 0.025)^4$ at the end of the fourth quarter. You use the calculator button marked $\boxed{y^x}$ to evaluate expressions like $(1.025)^4$; on a spreadsheet, use the caret key ^ (Shift-6), as in 1.025^4.

More generally, with initial principal P and interest rate i (= 100 i%) per compounding period, you have at the end of the first compounding period

$$P + Pi = P(1 + i)$$

This amount can be viewed as a new starting balance. Hence, in the next compounding period, the amount $P(1 + i)$ grows to

$$P(1 + i) + P(1 + i)i = P(1 + i)(1 + i) = P(1 + i)^2$$

The pattern continues, and we reach the following conclusion.

Compound Interest Formula — RULE

An initial principal P in an account that pays interest at a periodic interest rate i per compounding period grows after n compounding periods to

$$A = P(1 + i)^n$$

For convenience, we convert the general interest formula into one specific for years and annual rates. An annual rate of interest r with m compounding periods per year gives a rate $i = r/m$ per compounding period, and t years contain $n = mt$ compounding periods.

Compound Interest Formula for an Annual Rate — RULE

An initial principal P in an account that pays interest at a nominal annual rate r, compounded m times per year, grows after t years to

$$A = P\left(1 + \frac{r}{m}\right)^{mt}$$

Notation for Savings — DEFINITION

A	amount accumulated
P	initial principal
r	nominal annual rate of interest
t	number of years
m	number of compounding periods per year
$n = mt$	total number of compounding periods
$i = r/m$	interest rate per compounding period

The amount added each compounding period is proportional to the amount present. This type of growth is called **geometric growth**.

Geometric Growth (Exponential Growth) — DEFINITION

Geometric growth (also called **exponential growth**) is growth proportional to the amount present.

EXAMPLE 2 ■ Compound Interest

Suppose that you have a principal of $P = \$1000$ invested at 10% nominal interest per year. Using the compound interest formula $A = P(1 + i)^n$, you can determine the amount in the account after 10 years–once you know the compounding period.

SOLUTION

- *Annual compounding.* The annual rate of 10% gives $i = 0.10$, and after 10 years the account has

$$\$1000(1 + 0.10)^{1\times 10} = \$1000(1.10)^{10} = \$2593.74$$

- *Quarterly compounding.* Then $i = r/m = 0.10/4 = 0.025$, and after 10 years ($mt = 4 \times 10 = 40$ quarters) the account contains

$$\$1000\left(1 + \frac{0.10}{4}\right)^{4\times 10} = \$1000(1.025)^{40} = \$2685.06$$

- *Monthly compounding.* Then $i = r/m = 0.10/12 = 0.008333$. The amount in the account after 10 years ($mt = 12 \times 10 = 120$ months) is

$$\$1000\left(1 + \frac{0.10}{12}\right)^{12\times 10} = \$2707.04$$

These entries are found in the last row of Table 21.2.

In doing the calculations, use as many decimal places as your calculator or spreadsheet carries and don't round off until the final result. We show intermediate results with enough decimal places to give the final result to the nearest cent.

Effective Rate

For an interest rate i per compounding period, a principal of $1 grows to $(1 + i)^n$ in n periods, so the interest earned on that $1–which is the effective rate of interest for n periods–is given by:

Formula for Effective Rate RULE

$$\text{effective rate} = (1 + i)^n - 1$$

Mostly, we will be interested in the effective rate on an annual basis. For a nominal *annual* interest rate r compounded m times, the interest rate per compounding period is $i = r/m$, and an amount of $1 grows in one year to

$$\left(1 + \frac{r}{m}\right)^m$$

The effective *annual* rate of interest (the annual percentage yield, or APY) is the amount of interest earned

$$\left(1 + \frac{r}{m}\right)^m - 1$$

divided by the original principal. Since that principal is $1, we have:

Formula for Annual Percentage Yield (APY) RULE

$$\text{APY} = \left(1 + \frac{r}{m}\right)^m - 1$$

EXAMPLE 3 ■ Finding the Annual Percentage Yield (APY)

With a nominal annual rate of 6% compounded monthly, what is the APY?

SOLUTION

$$\left(1 + \frac{0.06}{12}\right)^{12} - 1 = 0.0617 = 6.17\%$$

In some cases you know the principal, the current balance, and the interval of time, and you want to learn the interest rate. For example, money market funds typically report earnings to investors each month, based on interest rates that vary from day to day, but often do not report the average interest rate. We find the equivalent average effective *daily* rate, from which we calculate the APY.

The compound interest formula gives the end-of-month balance as $A = P(1 + i)^n$, where P is the balance at the beginning of the month, i is the average daily interest rate, and n is the number of days that the statement covers. So we have

$$\frac{A}{P} = (1 + i)^n$$

Taking the nth root gives

$$1 + i = \left(\frac{A}{P}\right)^{1/n} \qquad i = \left(\frac{A}{P}\right)^{1/n} - 1$$

EXAMPLE 4 ■ Money Market Account

Suppose that the monthly statement from the fund reports a beginning balance (P) of \$7373.93 and a closing balance (A) of \$7382.59 for 28 days (n). What is the effective daily rate?

SOLUTION We thus have

$$i = \left(\frac{7382.59}{7373.93}\right)^{1/28} - 1 = 0.0000419194$$

Thus the average effective daily rate is 0.00419194%. Compounding daily for a year, we would have $(1 + 0.0000419194)^{365} = 1.01542$, for an APY of 1.54%.

Simple Interest Versus Compound Interest

The amounts in accounts paying interest at 10% per year with compound and simple interest are shown in Table 21.3 and in the graph in Figure 21.1, which dramatically illustrate the exponential growth of money at compound interest compared with the linear growth at simple interest.

In some situations, the contrast is not so immediately dramatic at first glance. The amount of carbon dioxide in the atmosphere, which contributes to global warming, has been growing *superexponentially* since 1750 as a result of constantly increasing burning of fuels. The amount is growing faster than exponentially: The "interest rate," or growth rate, increases every year. The current growth rate is about 0.5% per year–a seemingly low rate of "interest." The international Kyoto Protocol that went into effect in early 2005 (without U.S. participation) aims to lower worldwide emissions. Even at a fixed lower level, though, the "population" of carbon dioxide atoms in the atmosphere would still increase–and global warming would intensify–but just arithmetically, instead of superexponentially. We are in effect "saving" carbon dioxide into the atmosphere, at an unknown future cost.

We noted earlier that the population of U.S. medical doctors grows as if it were at simple interest (arithmetic growth) because the same number of doctors graduate from medical school each year. On the other hand, general human populations tend to grow as if they were at compound interest (geometric growth), because the number of children born–the "interest"–increases as the population–the "balance"–increases.

TABLE 21.3 The Growth of $1000: Compound Interest Versus Simple Interest

Years	Amount in Account from Compounded Interest	Amount from Simple Interest
1	1100.00	1100.00
2	1210.00	1200.00
3	1331.00	1300.00
4	1464.10	1400.00
5	1610.51	1500.00
10	2593.74	2000.00
20	6727.50	3000.00
50	117,390.85	6000.00
100	13,780,612.34	11,000.00

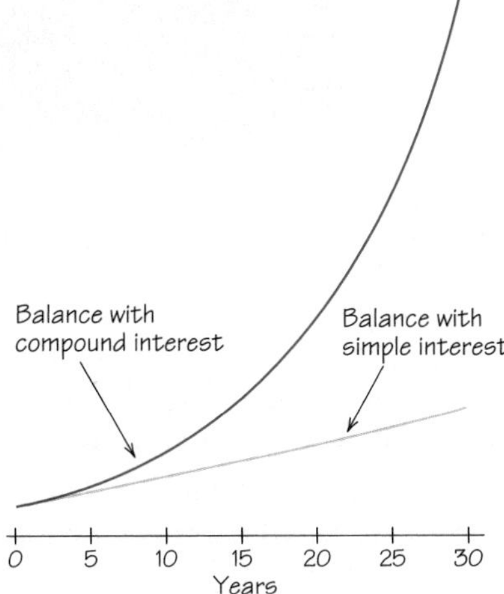

FIGURE 21.1 The growth of $1000: compound interest and simple interest. The straight line explains why growth at simple interest is also known as linear growth.

The distinction between arithmetic growth and geometric growth is fundamental to the major theory of demographer and economist Thomas Robert Malthus (1766–1834). He claimed that human populations grow geometrically but food supplies grow arithmetically, so that populations tend to outstrip their ability to feed themselves (see Spotlight 21.1).

The situation of nuclear waste generated by a nuclear power plant is more complicated. The absolute volume of waste added each year depends on the size and output of the power plant, not on the growing amount of waste in storage. Hence the volume of waste grows arithmetically. What about the total amount of radioactive material in the storage dump? The waste is a mixture of radioactive and nonradioactive substances. Over time, the radioactive ingredients decay very slowly into nonradioactive ones. While the radioactivity of waste already in storage is decreasing, new amounts of radioactive material are being added each year. The situation requires a hybrid model that incorporates positive arithmetic growth (adding to the dump) accompanied by (much smaller) negative geometric growth (radioactive decay). The situation is like turning on the faucet in the bathtub while leaving the drain hole open a little. The faucet determines how fast water runs in, the height of the water determines how fast it runs out, and those two rates determine what happens to the volume of water in the tub.

SPOTLIGHT 21.1 Thomas Robert Malthus

Thomas Robert Malthus (1766–1834), a nineteenth-century English demographer and economist, based a well-known prediction on his perception of the different patterns of growth of the human population and growth of the "population" of food supplies.

He believed that human populations increase geometrically but food supplies increase arithmetically—so that increases in food supplies will eventually be unable to match increases in population. He concluded, however, that over the long run there would be restrictions on the natural growth of human populations, too, including war, disease, and starvation—hardly an optimistic forecast and doubtless responsible for the dreary image associated with his views.

Some observers suggest that the genocide in Rwanda in 1994 was indirectly a result of overpopulation compared to available food resources.

Thomas Robert Malthus
(The Granger Collection, New York.)

21.3 A Limit to Compounding

The rows in Table 21.2 show a trend: More frequent compounding yields more interest. But as the frequency of compounding increases, the interest tends to a limiting amount, shown in the far right column.

Why is this so? Basically, because the extra interest from more frequent compounding is *interest on interest.* For example, in the first row of Table 21.2, the \$3.81 extra interest from compounding quarterly is interest on the \$100 yearly interest. The \$3.81 is less than 10% of the \$100 because the \$100 interest is not on deposit for the whole year. Just part of it is credited to the account (and begins earning interest) at the end of each quarter. As compounding is done more and more often, smaller and smaller amounts of interest on interest are added.

Let's see what happens with the crazy interest rate of 100% per year compounded m times per year. This crazy rate makes the numbers simple (but it is nowhere close to the crazy 66,000 percent rate in Zimbabwe in December 2007, or the incredible 313 million percent rate in Yugoslavia in 1994). Later we examine interest rates closer to those in stable economies. For an initial balance of \$1, the amount at the end of one year–from the compound interest formula, with $P = \$1$ and $i = 100\%$–is

$$A = \$1 \times \left(1 + \frac{100\%}{m}\right)^m = \$\left(1 + \frac{1.00}{m}\right)^m$$

As m increases, this amount, which is just $(1 + 1/m)^m$, gets closer and closer to a special number called $e \approx 2.71828$ (see Spotlight 21.2). This is illustrated in Table 21.4, where the dots (ellipses) indicate that more decimal places follow.

TABLE 21.4 Yield of $1 at 100% Interest, Compounded *m* Times per year

m	$\left(1 + \frac{1}{m}\right)^m$
1	2.0000000 . . .
5	2.4883200 . . .
10	2.5937424 . . .
50	2.6915880 . . .
100	2.7048138 . . .
1,000	2.7169239 . . .
10,000	2.7181459 . . .
100,000	2.7182682 . . .
1,000,000	2.7182804 . . .
10,000,000	2.7182816 . . .

SPOTLIGHT 21.2 The Number e

The number e is similar to the number π in several respects. Both arise naturally, π in finding the area and circumference of circles, and e in compounding interest continuously (e is also the base for the system of "natural" logarithms). In addition, neither is rational (expressible as the ratio of two integers, such as 7/2) or even algebraic (the solution of a polynomial equation with integer coefficients, such as $x^2 = 2$); we say that they are *transcendental* numbers. Finally, no pattern has ever been found in the digits of the decimal expansion of either number.

For a general interest rate r, as m is made larger and larger, the limiting amount is e^r, and the interest method is called **continuous compounding**. The APY, is $(e^r - 1)$. (You can calculate powers of e using the $\boxed{e^x}$ button on your calculator. On some calculators, this button is the 2nd function of the button marked $\boxed{\text{LN}}$ or $\boxed{\ln x}$. For example, to calculate $e^{0.10}$, push $\boxed{\text{2nd}}$, push $\boxed{\ln x}$, and enter 0.10. You get 1.105170918.)

Continuous Compounding DEFINITION

Continuous compounding is the method of calculating interest in which the amount of interest is what compound interest tends toward with more and more frequent compounding.

EXAMPLE 5 ■ Continuous Compounding

For $1000 at an annual rate of 10%, compounded m times in the course of a single year, what is the balance at the end of the year?

SOLUTION It is

$$\$1000\left(1 + \frac{0.10}{m}\right)^m$$

This quantity gets closer and closer to $\$1000e^{0.1} = \$1105.17\ldots$ as the number of compoundings m is increased. No matter how frequently interest is compounded–daily, hourly, every second, infinitely often ("continuously")–the original \$1000 at the end of one year cannot grow beyond \$1105.17. The values after 5 and 10 years are shown in the lower rows of Table 21.2.

Continuous Interest Formula RULE

For a principal P deposited in an account at a nominal annual rate r, compounded continuously, the balance after t years is

$$A = Pe^{rt}$$

We illustrate with \$1000 at 10%. For one year, we have $t = 1$ and

$$A = \$1000e^{0.10} = \$1105.17$$

To find the amount in the account after 5 years, we have $t = 5$:

$$A = \$1000e^{(0.10)(5)} = \$1000e^{0.5} = \$1648.72$$

exactly as shown in the rightmost column of Table 21.2.

It makes virtually no difference whether compounding is done daily or continuously over the course of a year. Most banks apply a daily periodic rate (based on compounding continuously) to the balance in the account each day and post interest daily (rounded to the nearest cent). The daily nominal rate is $r/365$, so each day the balance of the account is multiplied by $e^{r/365}$, the daily effective rate. Except for the rounding in posting interest, the effect is the same as continuous compounding throughout the year, because the compound interest formula gives $A = P(e^{r/365})^{365}$, which is the same as Pe^r from the continuous interest formula.

For example, for a principal of \$1000 and an interest rate of 5%, interest compounded daily over a year yields an amount

$$\$1000\left(1 + \frac{0.05}{365}\right)^{365} = \$1051.2675,$$

while continuous compounding yields $\$1000e^{.05} = \$1051.2711$.

21.4 A Model for Saving

The compound interest formula tells the fate over time of a single deposited amount, but another common question that arises in finance is: What size deposit do you need to make regularly in an account with a fixed rate of interest, to have a specified amount at a particular time in the future?

This question is important in planning for a major purchase in the future or accumulating a retirement nest egg. Later, in Chapter 22, we apply the same concepts and formula to paying off a mortgage and making installment payments on a car.

(*Sean Justice/Corbis.*)

EXAMPLE 6 ■ A Savings Plan

A graduate at her first job saves \$100 per month, deposited directly into her credit union account on payday, the last day of the month. The account earns 1.8% per year, compounded monthly. How much will she have at the end of five years, assuming that the credit union continues to pay the same interest rate?

SOLUTION Note that she makes the first deposit at the end of the first month and the last deposit at the end of the sixtieth month. The monthly interest rate is $i = r/12 = 0.018/12 = 0.0015$.

It's easier to look at the deposits in reverse time order. The last deposit is made on the last day of the five years, so it earns no interest and contributes just \$100 to the total.

The second-last deposit earns interest for one month, contributing $\$100(1 + i)$. Similarly, the third-last contribution is on deposit for two months, contributing $\$100(1 + i)^2$.

Continuing in the same way, we find that the first deposit earns interest for 59 months and contributes $\$100(1 + i)^{59}$. The total of all of the contributions is

$$\$100 + \$100(1 + i)^1 + \$100(1 + i)^2 + \cdots + \$100(1 + i)^{59}$$
$$= \$100[1 + (1 + i)^1 + (1 + i)^2 + \cdots + (1 + i)^{59}]$$

This expression is known as a **geometric series**, because the successive terms have geometric growth: Each succeeding term is a constant–in this case, $(1 + i)$–times the preceding term. For the sum of such a series with general ratio x, we have:

Formula for Sum of a Geometric Series RULE

$$1 + x + x^2 + x^3 + \cdots + x^{n-1} = \frac{x^n - 1}{x - 1}$$

That this formula works for all x (except $x = 1$) can be confirmed by multiplying both sides $(x - 1)$ and watching terms on the left cancel. (You should do this confirmation for $n = 4$.)

In our example, we have $x = 1 + i$, and the formula becomes

$$1 + (1 + i)^1 + (1 + i)^2 + \cdots + (1 + i)^{n-1} = \frac{(1 + i)^n - 1}{i}$$

We have $n - 1 = 59$, or $n = 60$ months, and $i = 0.0015$, the interest rate per month. The total accumulation after five years is

$$A = \$100\left[\frac{(1 + 0.0015)^{60} - 1}{0.0015}\right] = \$6273.37$$

■

For a uniform deposit of d per compounding period (deposited at the end of the period) and an interest rate i per period, the amount A accumulated after n compounding periods is given by the **savings formula**:

Savings Formula RULE

$$A = d\left[\frac{(1 + i)^n - 1}{i}\right] = d\left[\frac{(1 + \frac{r}{m})^{mt} - 1}{\frac{r}{m}}\right].$$

The expression on the right gives the amount accumulated in terms of the nominal annual interest rate r, the number m of compounding periods per year, and the number t of years, using the relations $i = r/m$ and $n = mt$.

The savings formula involves four quantities: A, d, i, and n. If any three are known, the fourth can be found. A common situation is for A, i, and n to be known, with d (the regular payment) to be found, because the practical concern for most people is how much their monthly payment will be.

Since we often want to solve for d, we solve the savings formula algebraically once and for all for d to get the *payment formula*:

Payment Formula RULE

$$d = A\left[\frac{i}{(1+i)^n - 1}\right] = A\left[\frac{\frac{r}{m}}{(1+\frac{r}{m})^{mt} - 1}\right]$$

Sometimes the purpose of saving is to accumulate a fixed sum by a particular date. Such a savings plan is called a **sinking fund**, because you sink money into it. If the same amount is deposited regularly, the sinking fund is called an **annuity**, a term for any series of (usually) equal payments at regular intervals.

Annuity DEFINITION

An **annuity** is a specified number of (usually equal) periodic payments.

Sinking Fund DEFINITION

A **sinking fund** is a savings plan to accumulate a fixed sum by a particular date, usually through equal periodic deposits.

EXAMPLE 7 ■ Sinking Fund

Suppose that your parents had started saving for your college education when you were born. How much would they have had to save each month to accumulate \$15,000 over 18 years with an account earning a steady 5% per year, compounded monthly?

SOLUTION Applying the payment formula with $A = \$15{,}000$, monthly rate $i = r/m = 0.05/12$, and $n = mt = 12 \times 18 = 216$, we get

$$d = \$15{,}000\left[\frac{\frac{.05}{12}}{(1+\frac{.05}{12})^{216} - 1}\right] = \$42.96$$

This sounds like a manageable amount to contribute, but it doesn't take into account inflation, nor costs beyond the first year, nor the higher cost of a private college. In the next section, we investigate how to take inflation into account.

Saving for Retirement (It's Never Too Early to Start)

Financial advisers stress the importance of beginning early to save for retirement. Many firms offer a *401(k) plan* (named after a section of law regulating pensions), which allows an employee to make monthly contributions to a retirement account. The plan has the advantage that income tax on the contributions is deferred until the employee withdraws the money during retirement. That means, for example, that an employee making a \$100 monthly contribution may see a reduction in take-home pay of only \$75 or less, since taxes are not withheld on the contribution.

Sometimes a company's pension plan consists of just contributing company stock to the employee's individual 401(k) account. In 2002, the bankruptcy of Enron Corporation resulted in thousands of its employees losing almost their entire retirement savings. Those savings consisted largely of Enron stock contributed by Enron, which fell from \$90 per share to under \$1 per share in just a couple of months. The Enron bankruptcy illustrated how unwise it is for most of an employee's retirement fund to consist of stock in just one company, particularly if–as was the case for Enron–the employee is not free to sell the stock. Even more people lost retirement savings and jobs when the stock of WorldCom declined more than 99% in 2002, after news of financial fraud by its management.

EXAMPLE 8 ■ Retirement Fund Annuity Savings

Suppose that you start a 401(k) plan when you turn 23 and contribute \$50 at the end of each month until you turn 65 and retire. Suppose (unlike some Enron employees) you put your contributions into a very safe long-term investment that returns a steady 5% annual interest compounded monthly. How much will be in your fund at retirement?

SOLUTION Apply the savings formula with $d = \$50$, $i = 0.05/12$, and $n = mt = 12 \times (65 - 23) = 504$. We get

$$A = \$50 \left[\frac{(1 + \frac{0.05}{12})^{504} - 1}{\frac{0.05}{12}} \right] = \$85{,}567.43$$

At first glance, that may seem like a lot of money, but it is not so much if that's all you have to live off for the rest of your life (of course, there is also Social Security). In the exercises we explore the effects of saving more each month, getting a higher interest rate, saving on taxes, and–especially–having inflation erode the value of your savings. ■

Annuities are a common way for lotteries to pay out grand prizes and for retirees to receive funds saved up for retirement. We examine an example of each in Chapter 22 (pp. 721–722), where we turn the savings formula around to get a formula (the amortization formula) to relate your savings to a regular payout. In Chapter 22 you find out what monthly income for a fixed period–or for life–\$86,000 could buy.

21.5 Present Value and Inflation

Suppose that you want to make a one-time deposit of amount P that will grow to a specific amount A in n compounding periods from now by earning interest at a rate i per period. The quantities A, P, i, and n are related through the compound interest formula, $A = P(1 + i)^n$. The quantity P is called the **present value** of the amount A to be paid n compounding periods in the future.

Present Value DEFINITION

The **present value** P of an amount A to be paid t years in the future, earning a nominal annual rate of interest r compounded m times per year–that is, after $n = mt$ compounding periods at a rate $i = r/m$ per compounding period–is

$$P = \frac{A}{(1 + i)^n} = \frac{A}{(1 + \frac{r}{m})^{mt}}$$

EXAMPLE 9 ■ Certificate of Deposit

A certificate of deposit (CD) pays a fixed rate of interest for a term specified in advance, from a month to 10 years. As I was researching this section, our second son was expecting to enter college in four years. Based on current costs, tuition, room, and board would cost $12,000 for the first year at the public university in our state capital (and more if he goes to a private college). The best rate that I could find for a four-year CD was 4.80% compounded monthly. How much would we need to set aside now in such a CD to have $12,000 in four years?

SOLUTION We find the present value of $12,000 four years from now, with $r = 0.048$, $m = 12$, and $t = 4$. The present value formula gives

$$P = \frac{A}{(1+\frac{r}{m})^{mt}} = \frac{\$12{,}000}{(1+\frac{0.048}{12})^{12\times 4}} = \frac{\$12{,}000}{(1+0.004)^{48}} = \$9907.48$$

In times of economic inflation, prices increase. When the rate of inflation is constant, the compound interest formula can be used to project prices.

EXAMPLE 10 ■ Inflation

Suppose that there is constant 3% annual inflation from mid-2009 through mid-2013. What will be the projected price in mid-2013 of an item that costs $100 in mid-2009?

SOLUTION The compound interest formula applies with $P = \$100$, $r = 3\%$, $m = 1$ and $t = 4$. The projected price is $A = P(1 + r)^t = \$100(1 + 0.03)^4 = \$112.55$.

Inflation and Depreciation as Exponential Decay

During constant-rate inflation, prices grow geometrically (exponentially) and the value of the dollar goes down geometrically: one is growing exponentially; the other is decaying exponentially.

Exponential Decay — DEFINITION

Exponential decay is geometric growth with a negative rate of growth.

Let a (for "additional") represent the annual rate of inflation; what costs $1 now will cost $\$(1 + a)$ this time next year. For example, if the inflation rate were $a = 25\%$, then what costs $1 now would cost $1.25 this time next year. A dollar next year would buy only 0.8 (= 1/1.25) times as much as a dollar buys today. In other words, a dollar next year would be worth only $0.80 in today's dollars–by next year, a dollar would have lost 20% of its purchasing power. We say that the present value of receiving a dollar next year is $0.80. Notice that although the inflation rate is 25%, the loss in purchasing power is 20%. For a general inflation rate a, a dollar a year from now will buy only a fraction of what a dollar today can buy.

Present Value of a Dollar a Year from Now with Inflation Rate a — RULE

$$\frac{\$1}{1+a} = \$1 - \frac{\$a}{1+a}$$

A dollar a year from now is worth $\$[1 - a/(1 + a)]$ today, and the loss in purchasing power is the fraction $a/(1 + a)$. (You should calculate what these expressions become for $a = 25\%$.) The quantity $i = -a/(1 + a)$ behaves like a negative interest rate. We can use the compound interest formula to find the present value of P dollars t years from now as

$$A = P(1 + i)^t = P\left[1 - \frac{a}{1+a}\right]^t.$$

The actual posted price of an item, at any time, is said to be in **current dollars.** That price can be compared with prices at other times by converting all prices to **constant dollars,** dollars of a particular year.

EXAMPLE 11 ■ Deflated Dollars

Suppose that there is 25% annual inflation from mid-2009 through mid-2013. What will be the value of a dollar in mid-2013 in constant mid-2009 dollars? The inflation figure is unrealistic (we hope!), but it makes the calculations easy, so you can focus on the ideas.

We have $a = 0.25$, so $i = -a/(1 + a) = -0.25/1.25 = -0.20$. This, not 25%, is the negative interest rate, the rate at which the dollar is losing purchasing power. We have $t = 4$ years, so the value of \$1 four years from mid-2009 in 2009 dollars will be

$$\$1(1 + i)^4 = \$(1 - 0.20)^4 = (0.80)^4 = \$0.41$$

In Example 11, we may think of the value of the dollar as "depreciating" 20% per year. Depreciation of the value of equipment is similar.

EXAMPLE 12 ■ Depreciation

If you bought a car at the beginning of 2009 for \$12,000 and its value in current dollars depreciates steadily at a rate of 15% per year, what will be its value at the beginning of 2012 in current dollars?

SOLUTION We have $P = \$12{,}000$, $i = -0.15$, and $n = 3$. The compound interest formula gives

$$A = P(1 + i)^n = \$12{,}000(1 - 0.15)^3 = \$7369.50$$

The Consumer Price Index

In our preceding model, we supposed that inflation stayed constant over a period of time. That is not generally the case. However, based on regular measures of inflation, we can determine the equivalent today of a price in an earlier year or how much a dollar in that year would be worth today.

The official measure of inflation is the Consumer Price Index (CPI), determined by the Bureau of Labor Statistics. Here we describe and use the CPI-U, the index for all urban consumers, which covers about 80% of the U.S. population and is the index of inflation that is usually referred to in newspaper and magazine articles.

Each month, the Bureau of Labor Statistics determines the average cost of a "market basket" of goods, including food, housing, transportation, clothing, and other items. It compares this cost with the cost of the same (or comparable) goods in a base period. The base period used to construct the CPI-U is 1982–1984. The index for 1982–1984 is set to 100, and the CPI-U for other years is calculated by using the proportion

$$\frac{\text{CPI for other year}}{100} = \frac{\text{cost of market basket in other year}}{\text{cost of market basket in base period}}$$

For example, the cost of the market basket in 1976 (in 1976 dollars) was 0.569 times the cost in 1982–1984 (in 1982–1984 dollars), so the CPI for 1976 is 100×0.569, or 56.9.

Table 21.5 shows the average CPI for each year from 1913 through 2007, with estimates for 2008 and 2009. This table can be used to convert the cost of an item in dollars for one year to what it would cost in dollars in a different year, using the proportion

$$\frac{\text{cost in year A}}{\text{cost in year B}} = \frac{\text{CPI for year A}}{\text{CPI for year B}}$$

EXAMPLE 13

The Price of Our House and the Value of a Dollar

(*Peter Gridley/Getty Images.*)

Where my family and I live, housing is relatively inexpensive. We bought our house in mid-1992 for \$133,000 (close to the median price of U.S. housing at that time). What would be the equivalent cost in mid-2009 dollars?

SOLUTION We see from Table 21.5 that the CPI for 1992 is 140.3 and the CPI for 2009 is estimated to be 221.7. The table gives the average value for each year, which is very close to the value at midyear. Month-by-month values are available at the Bureau of Labor Statistics Web site.

Using the proportion, we have

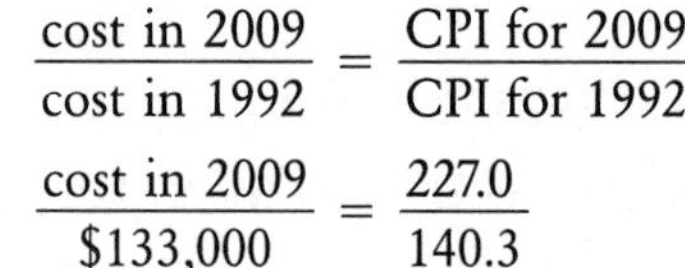

$$\frac{\text{cost in 2009}}{\text{cost in 1992}} = \frac{\text{CPI for 2009}}{\text{CPI for 1992}}$$

or

$$\frac{\text{cost in 2009}}{\$133{,}000} = \frac{227.0}{140.3}$$

so that

$$\text{cost in 2009} = \$133{,}000 \times \frac{227.0}{140.3} \approx \$215{,}000$$

That's what our house would sell for if its price exactly matched inflation.

The ratio $227.0/140.3 = 1.61796$ is the *scaling factor* for converting 1992 dollars to 2009 dollars. What we are observing is a proportion, or *numerical similarity*, between 1992 dollars and 2009 dollars, analogous to the geometric similarity of Chapter 18 (p. 572). To convert from 2009 dollars to 1992 dollars, we would multiply by $1/1.61796 \approx 0.618$.

Spotlight 19.3 (p. 614) describes how the consumer price index is calculated, using the geometric mean (defined and introduced on p. 613).

Real Growth Under Inflation

It's natural to think that if your investment is growing at 6% per year and inflation is at 3% per year, then the real growth in the value (purchasing power) of your investment is $6\% - 3\% = 3\%$. Such, however, is not the case.

Let's suppose that you invest \$500 for a year at 6%. At the beginning of the year, you have \$500, which at \$5 per pound could buy 100 pounds of steak. At the end of

TABLE 21.5 U.S. Consumer Price Index (1982–1984 = 100)

–	–	1931	15.2	1951	26.0	1971	40.5	1991	136.2
–	–	1932	13.7	1952	26.6	1972	41.8	1992	140.3
1913	9.9	1933	13.0	1953	26.7	1973	44.4	1993	144.5
1914	10.0	1934	13.4	1954	26.9	1974	49.3	1994	148.2
1915	10.1	1935	13.7	1955	26.8	1975	53.8	1995	152.4
1916	10.9	1936	13.9	1956	27.2	1976	56.9	1996	156.9
1917	12.8	1937	14.4	1957	28.1	1977	60.6	1997	160.5
1918	15.1	1938	14.1	1958	28.9	1978	65.2	1998	163.0
1919	17.3	1939	13.9	1959	29.1	1979	72.6	1999	166.6
1920	20.0	1940	14.0	1960	29.6	1980	82.4	2000	172.2
1921	17.9	1941	14.7	1961	29.9	1981	90.9	2001	177.1
1922	16.8	1942	16.3	1962	30.9	1982	96.5	2002	179.9
1923	17.1	1943	17.3	1963	30.6	1983	99.6	2003	184.0
1924	17.1	1944	17.6	1964	31.0	1984	103.9	2004	188.9
1925	17.5	1945	18.0	1965	31.5	1985	107.6	2005	195.3
1926	17.7	1946	19.5	1966	32.4	1986	109.6	2006	201.6
1927	17.4	1947	22.3	1967	33.4	1987	113.6	2007	207.3
1928	17.1	1948	24.1	1968	34.8	1988	118.3	2008 (est.)	219.0
1929	17.1	1949	23.8	1969	36.7	1989	124.0	2009 (est.)	227.0
1930	16.7	1950	24.1	1970	38.8	1990	130.7		

Note: This the CPI-U index, which covers all urban consumers, about 80% of the U.S. population. Each index is an average for all cities for the year. The basis for the index is the period 1982–1984, for which the index was set equal to 100. For each year, the figure is the average during the year, hence is usually close to the value at midyear.

Source: http://stats.bls.gov/cpi/

the year, you have $500(1 + 0.06) = $530, but steak now costs $5(1 + 0.03) = $5.15 per pound. How much steak would that buy? $530/$5.15 lb = 102.91 lb. In other words, in terms of purchasing power, or real gain, your investment has grown only 2.91%. This is not a great deal different from 3%, but it *is* different, and the difference is greater for higher rates of interest and inflation.

Consider an investment principal P and a market basket of goods of value m. Let the annual yield (rate of interest) of the investment be r and the rate of inflation be a. We calculate the rate of real growth g of the investment as follows.

At the beginning of the year, the investment would buy quantity $q_{\text{old}} = P/m$ of the market basket. At the end of the year, the investment would buy

$$q_{\text{new}} = \frac{P(1 + r)}{m(1 + a)}$$

market basket. Notice that the gain of r in the investment multiplies the principal by $(1 + r)$, while the erosion due to inflation divides the principal by $(1 + a)$. Here you see directly that the two influences on the investment have directly opposite effects.

The growth of the investment, relative to how many market baskets it could have bought originally, is

$$g = \frac{q_{\text{new}} - q_{\text{old}}}{q_{\text{old}}} = \frac{\frac{P(1+r)}{m(1+a)} - \frac{P}{m}}{\frac{P}{m}} = \frac{1+r}{1+a} - 1 = \frac{r-a}{1+a}$$

In the last expression, the numerator is the difference of the two rates (6% − 3% in our example), which is divided by a quantity greater than 1 if there is inflation. You should confirm that this formula gives 2.91% for $r = 6\%$ and $a = 3\%$.

One way to understand why this is the correct formula is to realize that the gain itself is not in original dollars but in deflated dollars.

The relationship between interest rate, inflation rate, and rate of real growth is called *Fisher's effect,* after the American economist Irving Fisher (1867–1947).

Real Rate of Growth

RULE

The real (effective) annual rate of growth of an investment at annual interest rate r with annual inflation rate a is

$$g = \frac{r-a}{1+a}$$

SPOTLIGHT 21.3 Nobel Prize for a Model in Economics

The 1997 Nobel Memorial Prize in Economics was awarded to Robert C. Merton of Harvard University and Myron S. Scholes of Stanford University for their method of valuing financial derivatives. Together with the late Fischer Black in the 1970s, they formulated a mathematical model with appropriate assumptions and solved the resulting equation. At the time, Black was a mathematician with Arthur D. Little Consultants in Boston, Scholes was a professor of finance at MIT, and Merton was an assistant to the economist Paul Samuelson at MIT. All were under age 30 at the time.

The major achievement of Merton, Black, and Scholes was to incorporate variability of market prices into the formula for the value of an option. They realized that the risk involved in the market is already implicitly taken into account in the stock's current price and its volatility (tendency to vary), and they were able to find the right formulation for incorporating the risk into the value of the option. Black and Scholes actually modeled the rate of change of the option value and then used methods from calculus to work backward to calculate the value and the formula itself.

This fairly complicated formula is based on simplifying assumptions (no stock dividends, no transactions costs, fixed price volatility, fixed interest rate, efficient market) plus a major modeling assumption. That modeling assumption is that the change in the price of the stock, as a percentage of the price, has a fixed component proportional to elapsed time (price trends upward over time) plus a random component deriving from volatility (so the price also can jump around). The random component is modeled using the normal distribution of Chapter 5.

REVIEW VOCABULARY

Annual percentage yield (APY) The effective interest rate per year. (p. 682)
Annuity A specified number of (usually equal) periodic payments. (p. 692)
Arithmetic growth Growth by a constant amount in each time period. (p. 680)
Compound interest Interest that is paid on both the original principal and the accumulated interest. (p. 680)
Compound interest formula Formula for the amount in an account that pays compound interest periodically. For an initial principal P and an effective rate i per compounding period, the amount after n compounding periods is $A = P(1 + i)^n$. (p. 684)

Compounding period The fundamental interval for compounding, within which no compounding is done. Also called simply *period*. (p. 681)
Constant dollars Costs are expressed in constant dollars if inflation or deflation has been taken into account by converting the costs to their equivalent in dollars of a particular year. (p. 695)
Continuous compounding Payment of interest in an amount toward which compound interest tends with more and more frequent compounding. (p. 689)
Current dollars The actual cost of an item at a point in time; inflation or deflation before or since then has not been taken into account. (p. 695)
e The base for continuous compounding, geometric (exponential) growth, and natural logarithms; $e = 2.71828\ldots$ (p. 688)
Effective rate The rate of simple interest that would realize exactly as much interest over the same period of time. (p. 682)
Exponential decay Geometric growth at a negative rate. (p. 694)
Exponential growth Geometric growth. (p. 684)
Geometric growth Growth proportional to the amount present. (p. 684)
Geometric series A sum of terms, each of which is the same constant times the previous term; that is, the terms undergo geometric growth. (p. 691)
Interest Money earned on a savings account or a loan. (p. 679)
Linear growth Arithmetic growth. (p. 680)
Nominal rate A stated rate of interest for a specified length of time; a nominal rate does not take into account any compounding. (p. 682)
Present value The value today of an amount to be paid or received at a specific time in the future, as determined from a given interest rate and compounding period. (p. 693)
Principal Initial balance. (p. 679)
Savings formula Formula for the amount A accumulated after $n = mt$ periods, with a uniform deposit of d at the end of each compounding period and interest rate $i = r/m$ per period:

$$A = d\left[\frac{(1+i)^n - 1}{i}\right] = d\left[\frac{(1+\frac{r}{m})^{mt} - 1}{\frac{r}{m}}\right]. \quad \text{(p. 691)}$$

Simple interest The method of paying interest only on the initial balance in an account, not on any accrued interest. (p. 679)
Sinking fund A savings plan to accumulate a fixed sum by a particular date, usually through equal periodic deposits. (p. 692)

SKILLS CHECK

1. Simple interest is an example of
(a) linear growth.
(b) variable growth.
(c) constant growth.

2. If a savings account pays 3% simple annual interest, a deposit of $250 will earn __________ in 2 years.

3. Which of the following pays more interest?
(a) 6% compounded annually
(b) 6% compounded monthly
(c) 6% compounded continuously

4. If a bond matures in 3 years and will pay $10,000 at that time, the fair value of it today is __________, assuming that the bond has an interest rate of 6% compounded annually and there is no inflation.

5. If a single deposit is made into a compound interest certificate of deposit, the account
(a) earns interest only for the first period.
(b) earns the same amount of interest each period.
(c) earns more interest in each subsequent period.

6. If $800 is invested for one year at 6% compounded quarterly, the amount of interest earned is __________.

7. An 18% annual rate on a credit-card balance is an example of
(a) an effective rate.
(b) a nominal rate.
(c) an adjusted rate.

8. If you deposit $1000 at 6.2% simple interest, the balance after three years is __________.

9. Suppose you invest $250 in an account that pays 4.5% interest compounded quarterly. After 30 months, how much is in your account?
(a) $279.08
(b) $279.59
(c) $279.71

10. Suppose you deposit $15 at the end of each month into a savings account that pays 2.5% interest compounded monthly. After a year, __________ is in the account.

11. Which of the following is the most generous interest rate for a one-year CD?
(a) 6% simple interest
(b) 5.9% compounded annually
(c) 5.9% compounded continuously

12. The APY for 5.90% compounded monthly is ____ .

13. The number e is

(a) irrational.
(b) irrelevant.
(c) irrotational.

14. The APY for 5% compounded daily is ________ .

15. The value of e is approximately

(a) 1.414.
(b) 2.718.
(c) 3.14.

16. When $1000 is invested at 8% compounded continuously for 5 years, the balance is ___________ .

17. An example of exponential decay is

(a) the depreciation of factory equipment.
(b) a retirement annuity.
(c) the Consumer Price Index.

18. Depositing $100 on a child's annual birth date is an example of ________________________________ .

19. If your investment is growing at a rate less than the rate of inflation,

(a) you have a positive real growth in your investment.
(b) you do not have a positive real growth in your investment.
(c) you do not have enough information to determine whether real growth is positive or negative.

20. If a new car costs $18,000 and loses value at a rate of 20% per year, its value after 3 years is ____________ .

CHAPTER 21 EXERCISES

■ Challenge ◆ Discussion

The exercises below require a scientific calculator with buttons for powers y^x, exponential e^x, and natural logarithm ln x.

21.1 Arithmetic Growth and Simple Interest

1. Suppose that you need $30,000 for your last year of college. You could go to a private lending institution and apply for a signature student loan; rates range from 7% to 14%. However, your Aunt Sally is willing to loan you the money from her retirement savings, with no repayment until after graduation. All she asks is that in the meantime you pay her each month the amount of interest that she would otherwise get on her savings (since she needs that to live on), which is 6%. What is your monthly payment to her, and how much interest will you pay her over the year (9 months)? (Aunt Sally will be glad to hear from you every month anyway!)

2. On December 28, 2007, you could buy a 10-year U.S. Treasury note ("T-note," a kind of bond) for $10,000 that pays 4.21% simple interest every year through December 28, 2017. How much total interest would it earn by then?

21.2 Geometric Growth and Compound Interest

3. An often heard claim is that "the amount of information in the world doubles every three days." Presumably the claim refers to the amount of data, which can be quantified in terms of number of bits. (A bit is the smallest unit of storage in a computer.) Show that the claim is absolutely preposterous by doing a little arithmetic and comparing your result with the estimated number of particles in the universe (10^{70}). In particular:

(a) Start with one bit of data and double the number of bits every third day. How long does it take to get past 10^{70}? (*Hint:* Don't just keep multiplying by 2 over and over. Convince yourself that since the amount of data increases by a factor of 2 every 3 days, then it increases by a factor of $2^2 = 4$ every 6 days, a factor of $4^2 = 16$ every 12 days, a factor of $16^2 = 256$ every 24 days, and so forth.)

(b) Part (a) involves a lot of multiplying by 2, even if you do it efficiently. Another approach is to use the fact that $2^{10} = 1024 \approx 1000$. Thus, the amount of data increases by a factor of more than 1000 every $3 \times 10 =$ 30 days, or every month (except February, but the 31-day months make up for it). By when will the total surely be past 10^{70}?

4. In a "Foxtrot" cartoon by Bill Amend (9/10/2006) on the next page, the girl Paige confronts a math problem in which "a math teacher assigns one second of homework the first week of school, two seconds the second week, four seconds the third, and so on." She is asked whether she would agree to this weekly homework doubling for the duration of the 36-week school year. How much homework (in hours) would this plan require in week 36?

5. You deposit \$1000 at 3% per year. What is the balance at the end of one year, and what is the annual yield, if the interest paid is

(a) simple interest?
(b) compounded annually?
(c) compounded quarterly?
(d) compounded daily?

6. Repeat Exercise 5, but for \$1000 at 6% per year.

7. I have a CD with National City through 2010 paying 4.69% interest compounded daily. What is the APY for this rate?

8. I have an account with First Community Credit Union of Beloit, Wisconsin, which pays dividends on independent retirement accounts (IRAs) at 0.75% per year, compounded monthly, for accounts with balances up to \$2000. What is the APY for such a rate?

9. U.S. Savings Bonds are a common form of award; for example, in December 2007 the Rodel Exemplary Teacher Initiative, which addresses the shortage of effective teachers in Arizona's neediest schools, chose 12 teachers each to receive a \$10,000 U.S. Savings Bond. (The interest is exempt from state and local income taxes and may also be exempt from federal income tax if used to pay for college tuition and fees.) Series EE Savings Bonds issued between November 2007 and April 2008 earn 3.00% interest, compounded semiannually, for the 20 years until their maturity. What is the APY? Will such a bond double in value in 20 years? (If not, the U.S. Treasury will make a one-time adjustment at the end of 20 years to ensure doubling in value.)

10. A Paper Series EE Savings Bond is sold at half face value, and the U.S. Treasury Dept. guarantees that it will double in value by 20 years from the issue date. What is the minimum APY for such a bond?

11. Suppose that on the statement for a money market account this month, the initial balance was \$7744.70, the statement was for 34 days, and the final balance was \$7770.84. Calculate the APY.

12. Repeat Exercise 11, but for the previous month, which had an initial balance of \$7722.54, a period of 27 days, and a final balance of \$7744.70.

13. *The rule of 72* is a rule of thumb for finding how long it takes money at interest to double: If r is the annual interest rate, then the doubling time is approximately $72/(100r)$ years.

(a) Calculate the balance at the end of the predicted doubling time for each \$1000, with annual compounding, for the small growth rates of 3%, 4%, and 6%.

(b) Repeat part (a) for the intermediate interest rates of 8% and 9%.

(c) Repeat part (a) for the larger interest rates of 12%, 24%, and 36%.

(d) What do you conclude about the rule of 72?

14. More frequent compounding yields greater interest, but with diminishing returns as the frequency of compounding is increased. For small interest rates, there is little difference in yield for compounding annually, quarterly, monthly, daily, or continuously. Investigating doubling times with continuous compounding leads to understanding why the rule of 72 of Exercise 13 works. Recall that for continuous compounding at annual rate r, the balance A at the end of t years is Pe^{rt} for an initial principal of P. For the initial principal to double, we have $2P = A = Pe^{rt}$, so $e^{rt} = 2$. Taking the natural logarithm of both sides yields $rt = \ln 2$, where ln stands for the natural logarithm, represented on a calculator by a button marked either [ln] or [LN] (not [log] or [log10], which stands for a different kind of logarithm). Using the button gives $\ln 2 = 0.693$. So we have $rt = 0.693$, from which we can determine t if we know r.

Calculate the doubling times for continuous compounding at 3%, 6%, and 9%, and compare them with those predicted by the rule of 72. What do you conclude? Why do you think people prefer a rule of 72 over a rule of 69.3?

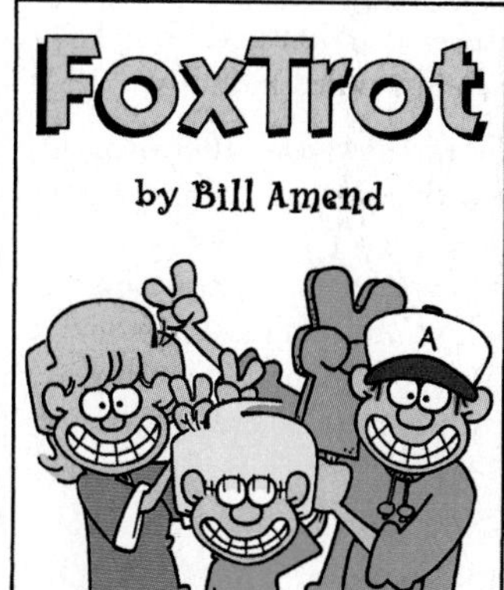

21.3 A Limit to Compounding

15. Use your calculator to evaluate for $n = 1$, 10, 100, 1000, and 1,000,000:

(a) $\left(1 + \frac{1}{m}\right)^m$

(b) $\left(1 + \frac{2}{m}\right)^m$

(c) As m gets larger, what numbers are the expressions in parts (a) and (b) tending toward?

16. (Contributed by John Oprea of Cleveland State University.) Use your calculator to evaluate for $m = 1$, 10, 100, 1000, and 1,000,000:

(a) $\left(1 - \frac{1}{m}\right)^m$

(b) $\left(1 - \frac{2}{m}\right)^m$

(c) As m gets larger, what numbers are the expressions in parts (a) and (b) tending toward?

17. You have $1000 on deposit at your bank at an annual rate of 3%. How much interest do you receive after one year if the bank compounds

(a) continuously?

(b) daily, using 365 days in a year?

18. Suppose that you have a bank account with a balance of $4532.10 at the beginning of the year and $4632.10 at the end of the year. Your bank advertises "continuous compounding," but in fact compounds continuously over each 24-hour day and posts interest to accounts daily.

(a) What effective rate did you receive?

(b) What nominal rate is the calculation based on?

(c) What difference is there between what the bank is doing and true continuous compounding?

19. Suppose that you have an investment that earns 0% in the first year, but 10% in the second year.

(a) What rate of interest, compounded annually, would yield the same return after two years?

(b) What rate of interest, compounded continuously, would yield the same return after two years?

20. Suppose that you have an investment that earns 10% in the first year, 20% in the second year, and 30% in the third year.

(a) What rate of interest, compounded annually, would yield the same return after three years? (The answer here is related to the geometric mean of Chapters 14 and 19, but you do not need to know about that to solve the problem.)

(b) What rate of interest, compounded continuously, would yield the same return after three years?

(Thanks for the idea to Yi Cheng, Indiana University South Bend.)

21. We saw in Example 5 on pp. 689–690 that a nominal rate of 5% compounded continuously yields an effective annual rate of 5.12711%.

(a) What effective annual rate does a nominal rate of 4% yield with continuous compounding?

(b) The difference between the effective rate under continuous compounding, $e^r - 1$, and the nominal rate r is $e^r - 1 - r$. You can't calculate this formula in your head, but you can approximate it closely with one that you can: $e^r - 1 - r \approx \frac{1}{2}r^2$. Thus, for $r = 4\% = 0.04$, the difference is $\frac{1}{2}(0.04)(0.04) = \frac{1}{2}(0.0016) = 0.0008 = 0.08\%$. So the effective rate is about 4.08%. Apply this formula to approximate the difference for a nominal rate of 5%, and compare the result with the 5.12711%.

22. [Suggested by Arthur R. Segal, University of Alabama at Birmingham (retired).] We approximate the smaller difference between regular compounding and continuous compounding. With a nominal annual rate r over t years, continuous compounding yields e^{rt}, while compounding m times per year yields $(1 + \frac{r}{m})^{mt}$, for a difference of $D = e^{rt} - (1 + \frac{r}{m})^{mt}$ that can be approximated by

$$D \approx \frac{r^2 t e^{rt}}{2m + \frac{4r}{3} + \frac{r^2 t}{2}}$$

For a $1000 initial investment, calculate both the true difference and the approximate difference between continuous compounding and

(a) quarterly compounding at a nominal rate of 4% for 10 years.

(b) daily compounding at a nominal rate of 18% for 10 years.

21.4 A Model for Saving

23. Suppose that you want to save up $2000 for a trip abroad two years from now. How much do you have to put away each month in a savings account that earns 5% interest compounded monthly?

24. Repeat Exercise 23, except that you have found a better deal, 7% interest compounded monthly.

25. Parents struggle for the first few years after their child is born but are finally able to start saving toward the child's college education when the child goes to school at age 6 (because the parents stop paying for day care). If they save $400 per month in a credit union account paying 5.5% interest compounded monthly, how much will they have for college expenses 12 years later?

26. Suppose that you save for retirement by contributing the same amount each month from your

23rd birthday until your 65th birthday, in an account that pays a steady 5% annual interest compounded monthly.

(a) How much will be in your fund at age 65 if you save $100 a month?

(b) How much will be in your fund if you get a steady return of 7.5% compounded monthly?

(c) How much will be in your fund if you get a steady return of 10% compounded monthly? (This is comparable to the average annual return of about 11% for all stocks on the New York Stock Exchange from 1950 to 2000.)

27. A colleague feels that he will need $1 million in savings to afford to retire at age 65 and still maintain his current standard of living. A younger colleague, age 30, decides to begin saving for retirement based on that advice. How much does the younger colleague need to save per month to have $1 million at retirement if the fund earns a steady 5% annual interest compounded monthly?

28. The younger colleague of Exercise 27 is not satisfied with 5% return, which he could get with long-term certificates of deposit. Instead, he wants to take the riskier route of investing in the stock market, which has over its history returned an APY of about 10% per year (although for 2001–2007, the APY was only 1.6%). Assuming that over the 35 years until his retirement that the stock market behaves just that way (a big assumption!), how much would he need to invest each month to achieve his goal of $2 million by age 65?

29. Many young people do not start saving right away for retirement, although by the time that they do, they may be earning more and thus be able to afford to save more each month. How much will be in your fund at age 65 if you don't start saving until age 35 and at that age start saving $100 per month in an account paying a steady 6% annual interest compounded monthly?

30. Suppose instead that you have children young, pay for their college expenses, and finally start saving for retirement at age 45. How much do you have to save per month, with a steady return of 7.5% compounded monthly, to accumulate $250,000 by age 65?

■ **31.** Suppose that you are 25, single, and in a 25% bracket for federal income tax and a 7% bracket for state and local income taxes (in 2007 this corresponded to an income, beyond exemptions and deductions, of $32,000–$77,000 for a single person). This means that you pay a total of 32% in income tax on part of your income but a lower rate on the rest (you also pay 7.65% Social Security and Medicare payroll tax). Assume that instead of paying 32% on some income, you put it into a tax-deferred retirement account (TDA) as follows:

(a) Suppose that you are willing to commit to $100 a month less take-home pay. You realize that you don't pay income tax on the money that you put into the retirement plan, so you can actually put in more than $100 per month while reducing your take-home pay by only $100. How much can you put into the retirement fund each month?

(b) How much will be in your fund at age 65 if you can get a steady return of 7.5% compounded monthly?

(c) Suppose that when you turn 65 you withdraw the entire amount in your account and pay the deferred taxes that are owed on it, say a total of 32% (federal, state, and local combined). How much do you net?

■ **32.** We continue the tax-deferral considerations of the previous exercise.

(a) Suppose that instead of contributing to a tax-deferred plan, you take the money as income, pay 32% income tax on it, and deposit what remains into a savings account or safe investment that pays a steady 7.5% compounded monthly. (Note that compared with not putting away any money, your paycheck is reduced by just what you contribute, since you still must pay income tax on the $147.06.) How much will be in your account at age 65?

(b) Under another alternative, you take $100 per month, pay 32% income tax on it, and deposit what remains into a *Roth IRA* (individual retirement account). For this kind of retirement account, the interest earned is not taxed. Assuming the same savings account or safe investment that pays a steady return of 7.5% compounded monthly, how much will be in your account, tax-free, at age 65?

33. Apart from certificates of deposit, returns on investments are rarely the same from year to year, as they vary with prevailing interest rates. How should you calculate an "average" rate of return over several years? Consider a mutual fund that delivers 100% return one year and loses 50% the next year. Calculate just the ordinary average (the arithmetic mean) to get $(100\% + (-50\%))/2 = 25\%$. That sounds good, but check what happens to a $1000 investment: It grows to $2000, then halves back to $1000–for a 0% gain. The customary way used in finance to calculate the "average" return is to use the geometric mean. If the initial value of the portfolio was P, and its value after n years is A, then the average annual rate of return is the value of r that solves $(1 + r)^n = A/P$, or $r = (A/P)^{1/n} - 1$.

(a) Use this formula to determine the average annual rate of return for a portfolio with returns of 15%, 7%, and −20% in three consecutive years.

(b) Is the average rate that the formula finds a nominal rate or an effective rate?

■ **34.** We continue the theme of Exercises 25 and 26 by comparing three kinds of investments for retirement: an ordinary after-tax investment, a tax-deferred investment [such as a tax-deferred annuity or an individual retirement account (IRA)], and a Roth IRA. Let an investment earn interest at a steady annual yield r and let your income (in whatever year you receive it) be taxed at rate τ.

(a) Ordinary after-tax investment: Explain why if you earn $\$E$, pay taxes on it, let what remains earn interest, and pay tax each year on that year's interest, the $\$E$ grows after n years to $\$E(1 - \tau) \times [1 + r(1 - \tau)]^n$.
(b) Ordinary IRA: Explain why if you earn $\$E$, defer taxes on it, let it earn interest, and defer taxes on all the interest, then the $\$E$ grows after n years to $\$E(1 + r)^n(1 - \tau)$.
(c) Roth IRA: Explain why if you earn $\$E$, pay taxes on it, let what remains earn interest, and pay no taxes on all the interest, the $\$E$ grows after n years to $\$E(1 - \tau)(1 + r)^n$.
(d) Which investment gives the best return after n years?
(e) The assumptions of constant interest rate and stable tax rate won't necessarily hold, because interest rates fluctuate (though you can lock in a long-term constant interest rate by buying a long-term bond or certificate of deposit) and the tax rate may change (with your income, your state of residence, and changes in tax laws). If your marginal tax rate (the rate you pay on one more dollar of income) is lower in one year than the tax rate you expect to pay in retirement, what kind of retirement investment is better for you that year? If you have a windfall one year and your marginal tax rate is higher that year than the tax rate that you expect to pay in retirement, what kind of retirement investment is better for you that year?

21.5 Present Value and Inflation

35. Classify the following growth and decay scenarios as linear (arithmetic), exponential (geometric), or neither:

(a) The amount of caffeine in the bloodstream decreases by 10% every hour.
(b) The amount of trash in a landfill increases by 350 tons per week.
(c) The amount of alcohol in the bloodstream decreases by 10 grams (the amount in a standard drink) per hour.
(d) Your age increases every day.

(Adapted from Terence Blows, Northern Arizona University.)

36. Assume the same situation as in Exercise 35, but for

(a) The mean concentration of carbon dioxide in the atmosphere increases by 2 ppm (parts per million) per year.
(b) The mean concentration of carbon dioxide in the atmosphere increases 0.5% per year.
(c) Your knowledge of mathematics and its applications increases with each section of this book that you study.
(d) The number of people in the world increases by 1.3% per year.

(Adapted from Terence Blows, Northern Arizona University.)

37. What is the present value of \$10,000, four years from now, at an APY of 5%?

38. What is the present value of \$150,000, ten years from now, at an APY of 3%?

39. As you will see in Chapter 22, if you have a 30-year \$200,000 mortgage at 8% on a house or apartment, after 22 years of payments you will still owe about \$150,000! What is the present value of \$150,000, 22 years from now, at an interest rate of 8%? (If you put this much more into a down payment, but made the same-size payments as for the 30-year mortgage on \$200,000, you would own the house free and clear after 22 years instead of still owing \$150,000.)

40. If you have a 30-year \$200,000 mortgage at 6.48% on a house or apartment, after 10 years of payments you will still owe about \$170,000. What is the present value of \$170,000, 10 years from now, at an interest rate of 6.48%?

41. Suppose that inflation proceeds at a constant rate of 3% per year from mid-2009 through mid-2012.

(a) Find the cost in mid-2012 of a basket of goods that cost \$1 in mid-2009.
(b) What will be the value of a dollar in mid-2012 in constant mid-2009 dollars?

42. Suppose that you bought a car in mid-2009 for \$10,000. If its value (in current dollars) depreciates steadily at 12% per year, what will its value (in current dollars) be in mid-2012?

43. For the car in Exercise 42, suppose that there is also 3% annual inflation from 2009 through 2015. What will the value of the car be in mid-2015 in inflation-adjusted (mid-2009) dollars?

44. I bought my first vinyl record in 1965, at list price, for \$4.98. How much would that be in 2009 dollars? How does that compare with the list price of a CD today?

(mediacolor's/Alamy.)

45. My first-semester college mathematics book cost \$10.75 in 1962. What would the equivalent price be in 2009 dollars? How does that compare with what you paid for this book? (My book had black-and-white text and figures, with no photographs, color or otherwise.)

46. In 1970, before the OPEC oil embargo, gasoline cost about 25 cents per gallon. In 1974, after the embargo, it cost about 70 cents per gallon. What would the equivalent prices be in 2009 dollars? How do they compare with the price of gasoline today?

Refer to the following in doing Exercises 47 and 48. From Table 21.5, you can determine the average rate of inflation from one year to another. For example, you find the inflation from 1990 to 2000 by subtracting the two index numbers and dividing by the earlier one: $(172.2 - 130.7)/130.7 = 31.752\%$. However, the average rate of inflation is not this number divided by the number of years (10). We must take into account compounding of the rate of inflation. We set $(1 + a)^{10} = 1.31752$ and find $a = (1.31752)^{1/10} - 1 = 2.80\%$.

47. Find the average rate of inflation from 1997 to 2007. Is 3% a good approximation?

48. If inflation had been 3% each year from 1997 to 2007, what would the CPI have been in 2007?

49. (Suggested by Ed Barbeau's column "Fallacies, Flaws, and Flimflam" in *The College Mathematics Journal.*) Suppose that you get a pay raise of 10%, but in the meantime, there has been inflation of 20%–so in effect you have suffered a pay decrease in terms of what your salary will buy. What is the percentage decrease?

50. What is the present value of a \$2000 raise now, which you will enjoy over the course of 40 years more of working, if inflation is a steady 3% per year?

Refer to the following in doing Exercises 51 and 52. A new assistant professor at a typical American liberal arts college starts at age 30 with a salary of \$45,000, while colleagues retiring at age 65 make about twice that. One college gives annual pay raises of inflation plus one percentage point, plus a promotion raise (to associate professor) of \$1500 after (usually) 6 years and another promotion raise (to full professor) of \$1500 after (usually) another 6 years.

51. (Spreadsheet helpful) Can a new assistant professor who starts now expect to be making the equivalent of \$90,000 in today's dollars when she retires 35 years from now if inflation holds steady at

(a) 3%?
(b) 5%?

52. (Spreadsheet helpful) Repeat Exercise 51, but suppose that you are the vice president for academic affairs at the college. Suggest a salary policy that would result in the new assistant professor, when she retires in 35 years, making the equivalent of

(a) \$90,000 in today's dollars.
(b) \$135,000 in today's dollars. (She would prefer that!–and hence she would be more likely to accept an offer to come work at your college.)

53. Surprise! Just for fun, one of your friends wrote your name on an Illinois State Lottery ticket, and you are the sole winner of \$40 million! You discover, however, that you don't get the \$40 million all at once. In fact, it is paid in 20 equal annual installments of \$2 million each. All you get right away is the first installment of \$2 million (minus 20% withheld against federal income tax due and whatever you think your friend deserves for the favor). So, what is the prize really worth to you? That depends on the rate of inflation over the years. Assume a constant rate of 3% inflation over the 19 years until your last payment and calculate the present value of your prize winnings by using the formula for present value combined with the formula for the sum of a geometric series. Do the calculation for a rate of interest of 4%.

Actually, the checks will come not from the state of Illinois but from an insurance company from which Illinois purchases an annuity, whose price depends on current long-term interest rates.

54. Repeat Exercise 53, but for an interest rate of 6%.

55. (Spreadsheet helpful) Your roommate (a business major) has already planned her retirement and will start funding it in 2009. She plans to retire in 2044 at age 57 on \$100,000 per year in 2044 dollars, living on just the interest on her investments. Assume that she realizes a steady 7.2% and assume a steady 3% annual inflation.

(a) What must the size of her nest egg be, and what should her monthly investment be over the 35 years, to achieve this goal?
(b) What will be the value in 2009 dollars of her 2044 income of \$100,000?
(c) What will be the value in 2009 dollars of her income of \$100,000 in 2072 (when she is 85)?

(Suggested by Terence Blows, Northern Arizona University, Flagstaff, Ariz.)

56. (Spreadsheet helpful) You think what your roommate means in Exercise 55 is that she wants to retire in 2044 with a *steady* income of \$100,000 a year in 2009 dollars. You also feel that she should plan to receive that same value of income for 43 years in case she lives to 100 (2% of your classmates will). What is the present value in 2009 of the planned stream of 43 years of retirement income?

In the savings formula, the interest rate i appears twice. The particular ways in which i is involved make it impossible to solve it algebraically to get an explicit

SUGGESTED READINGS

KASTING, MARTHA. *Concepts of Math for Business: The Mathematics of Finance.* UMAP Modules in Undergraduate Mathematics and Its Applications: Module 370–372. COMAP, Inc., Arlington, Mass., 1980.

LINDSTROM, PETER A. *Nominal vs. Effective Rates of Interest.* UMAP Modules in Undergraduate Mathematics and Its Applications: Module 474. COMAP, Inc., Arlington, Mass., 1988. Reprinted in Paul J. Campbell (ed.), *UMAP Modules: Tools for Teaching 1988,* COMAP, Inc., Arlington, Mass., 1989, pp. 21–53. A learning module, requiring no more background than this chapter, that teaches about nominal and effective rates of interest and how to calculate them. Gives real examples of banks using different options for calculating interest.

MILLER, CHARLES D., VERN E. HEEREN, and JOHN HORNSBY. Consumer mathematics. In *Mathematical Ideas,* 11th ed., Pearson Education/Addison Wesley Longman, Boston, 2007.

VEST, FLOYD, and REYNOLDS GRIFFITH. The mathematics of bond pricing and interest rate risk. *Consortium (COMAP),* 59 (Fall 1996): HiMAP Pullout Section 1–6.

SUGGESTED WEB SITES

www.bls.gov/cpi/ Home page for the inflation tables prepared by the Bureau of Labor Statistics.

www.bls.gov/data/inflation_calculator.htm CPI Inflation Calculator. Converts dollar value from any year to its equivalent buying power in any other year.

www.westegg.com/inflation/ Inflation calculators for the United States (1800–2001), Canada, and Italy, by S. Morgan Friedman, with links to sites about the current purchasing power of amounts of currencies of other countries in the past.

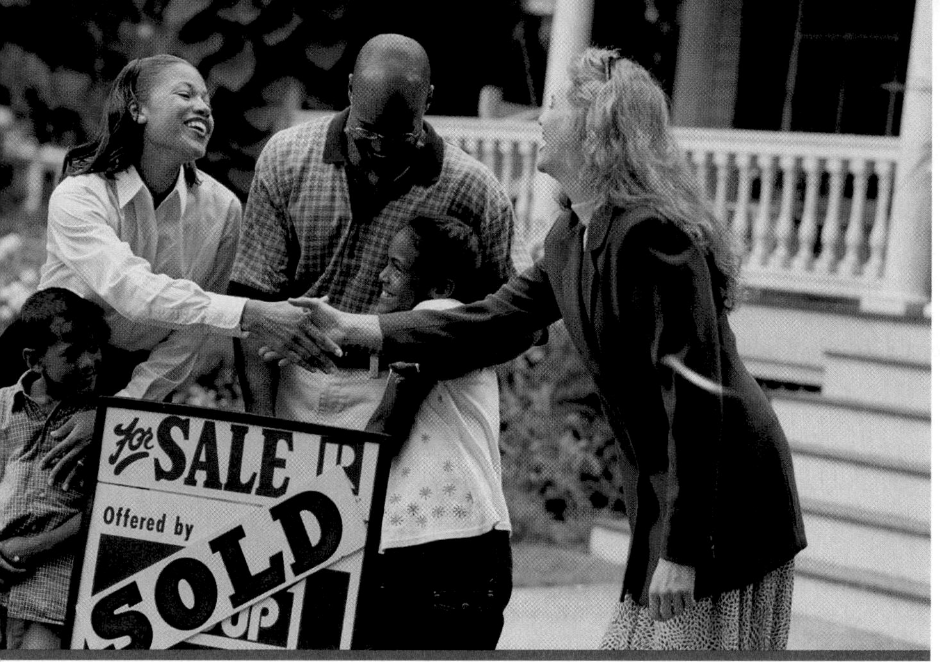

CHAPTER 22

Borrowing Models

In the previous chapter, we looked at consumer financial models for saving and formulas for calculating the amount accumulated. Savings or investments would not earn interest unless they could be loaned to someone to make productive use of the money.

In this chapter, we examine the other side of consumer finance, borrowing. You may have a student loan, you are likely to borrow to buy a car, you will almost certainly borrow if you buy a house or apartment, and you are borrowing if you use a credit card. For any such loan, you pay "finance charges," which include interest and perhaps other "fees" as well. We investigate and compare some common kinds of loans.

We briefly (re)acquaint you with compound interest and a few formulas from Chapter 21. If you have a grasp of the ideas behind compound interest and can use the compound interest formula (p. 684) and the savings formula (p. 691), which we repeat shortly for your convenience, you can proceed with this chapter without first reading Chapter 21.

22.1 Simple Interest

The amount of **interest** charged on a loan is determined by the **principal**, by the amount borrowed, and by the method used to calculate the interest. With **simple interest**, the borrower pays a fixed amount of interest for each period of the loan. The interest rate is usually quoted as an annual rate.

For a principal P and an annual rate of interest r, the interest owed after t years is

$$I = Prt$$

and the total amount A due on the loan is

$$A = P(1 + rt)$$

EXAMPLE 1 ■ Simple Interest on a Federal Student Loan

(*moodboard/Corbis.*)

The federal government offers guaranteed loans through banks to students to pay for tuition, fees, housing, and textbooks (such as this one), with repayment deferred until after graduation. Any eligible student can take out an unsubsidized Stafford loan, in which you are charged interest from the time that you receive the loan until it is repaid in full. The interest rate for new loans after July 1, 2006 is 6.8%. One option is to pay each month the interest due, and defer paying back the principal and interest until six months (the "grace period") after you leave school.

Suppose that you took out a \$3500 Stafford loan on September 1 before your freshman year and begin paying it back on December 1 after graduation (so you will have had the loan for 4 years + 3 months = 51 months).

How much is the monthly interest, how much total interest will you have paid over the 51 months, and how much will you owe when you start to pay back?

SOLUTION We have $P = \$3500$ and $r = 6.8\% = 0.068$, and for one month we have $t = \frac{1}{12}$ years. So the interest for one month is $I = Prt = \$3500 \times 0.068 \times \frac{1}{12} = \$19.8333 \approx \$19.83$. Over the 51 months, you will have paid $51 \times \$19.83 = \$1011.33$. You will still owe the original principal of \$3500.

22.2 Compound Interest

Compounding is the calculation of interest on interest. A common example is the balance on a credit card. As long as there is an outstanding balance owed, the interest owed is calculated on the entire balance, including any part of it that was previously calculated as interest and added to the balance in earlier months.

EXAMPLE 2 ■ Credit-Card Interest

Suppose that you owe \$1000 on your credit card, the company charges 1.5% interest per month, and you just let the balance ride. How much interest do you pay in the first year?

SOLUTION Your interest the first month is 1.5% of \$1000, or $0.015 \times \$1000 = \$15$. The new balance owed is $(1 + 0.015) \times \$1000 = \$1015$. Your interest the second month is not 1.5% of \$1000, or \$15 (as would be the case for simple interest), but 1.5% of \$1015, or $0.015 \times \$1015 = \$15.23$, so the new balance is

$$(1 + 0.015) \times \$1015 = \$(1 + 0.015) \times (1 + 0.015) \times \$1000 = \$1030.23$$

(We neglect the extra charges for your failure to make minimum payments.) After 12 months of letting the balance ride, it has become

$$(1.015)^{12} \times \$1000 = \$1195.62$$

In other words, the actual interest for the year comes to \$195.62, which is 19.562% of \$1000. So, although the quoted rate of interest is 1.5% per month, which seems as if it should amount to $12 \times 1.5 = 18\%$ per year, the interest owed is actually more.

We apply two formulas from Chapter 21: the **compound interest formula** (p. 684) and the **savings formula** (p. 691), phrasing them for loans. Here is the compound interest formula, followed by an example:

Compound Interest Formula RULE

If a principal P is loaned at interest rate i per compounding period, then after n compounding periods (with no repayment) the amount owed is

$$A = P(1 + i)^n$$

This formula just generalizes what we saw happen with the credit-card balance. We give the formula in a slightly more elaborate version below, to make the connection to multiple compoundings per year.

Compound Interest Formula RULE

For a principal P loaned

- at a nominal annual rate of interest rate r
- with m compounding periods per year (so interest rate $i = r/m$ per compounding period), the amount owed
- after t years (hence $n = mt$ compounding periods) with no payment of interest or principal is

$$A = P(1 + i)^n = P\left(1 + \frac{r}{m}\right)^{mt}$$

EXAMPLE 3 ■ Not Repaying Your Student Loan

As noted above, 6.8% annual simple interest accrues on an unsubsidized Stafford loan from the time that you receive the loan until you begin repayment. When you begin repayment, the interest is (in the terminology of the student loan documents) *capitalized*, meaning that it is added to the principal. The interest rate remains 6.8%, for a monthly rate of 6.8%/12 = 0.566667%. However, if you do not make the payments due, the interest continues to accumulate and this additional interest is usually capitalized every quarter. That means that during a quarter of a year, the interest each month is simple interest on the amount still due at the beginning of each month, and at the end of the quarter the three months of interest is added to the amount due–there is no interest on the quarter's interest until after the quarter is ended. In our terminology, the compounding period is one quarter.

Suppose that you owe $5000 on your Stafford loan but you fail to make any payments for 9 months. (This would be very foolish, since after 270 days of nonpayment the loan is in default and all kinds of bad things would happen!) How much would you owe then?

SOLUTION The principal P is $5000. The quarterly interest rate is $i = 6.8\%/4 = 1.7\%$ and there are $n = 3$ compounding periods. The compound interest formula gives the amount owed as $A = \$5000\,(1 + 0.017)^3 = \$5259.36$. ■

Terminology for Loan Rates

The interest on a loan depends on whether or not compounding is done and how the interest is calculated. Just like the Truth in Savings Act mentioned in Chapter 21 (p. 682), the Truth in Lending Act establishes terminology and calculation methods for interest.

A **nominal rate** is any stated rate of interest for a specified length of time. For instance, a nominal rate could be a 1.5% monthly rate on a credit-card balance. By itself, such a rate does not indicate or take into account whether or how often interest is compounded.

The **effective rate** *takes into account compounding*. It is the rate of simple interest that would realize exactly as much interest over the same period of time.

We saw that \$1000 at a yearly interest rate of 18% (a nominal rate), calculated as 1.5% per month compounded monthly, yields \$195.62 in interest owed at the end of the year, which is 19.562% of the original principal. Hence, the effective annual rate is 19.562%. In other words, a \$1000 loan at simple interest of 19.562% for one year would owe exactly the same interest.

Finally, when stated per year ("annualized"), the effective rate is called the **effective annual rate (EAR)**. (The EAR is the same concept as the APY of Chapter 21, p. 682).

To keep the rates straight, we use i for a nominal rate for the specified **compounding period**–such as a day, month, or year–*within which no compounding is done;* this rate is the effective rate for that length of time. For a nominal rate compounded m times per year, we have $i = r/m$. For that \$1000 credit-card balance at 18% compounded monthly, we have $r = 18\%$ and $m = 12$, so $i = 1.5\%$ per month.

The Truth in Lending Act introduced the term **annual percentage rate (APR)**.

Annual Percentage Rate (APR) DEFINITION

The **annual percentage rate (APR)** is the number of compounding periods per year times the rate of interest per compounding period.

$$\text{APR} = m \times i$$

In the example of the credit-card balance, the interest is compounded monthly, or $m = 12$ times per year, and the interest rate for the compounding period is $i = 1.5\%$, so the APR is $12 \times 1.5\% = 18\%$. The APR is the rate that the Truth in Lending Act requires the lender to disclose to the borrower. *The APR is not equal to the EAR* (as we have already seen in the credit-card example), and Spotlight 22.1 explains further.

22.3 Conventional Loans

A common situation that you are likely to encounter is a loan–for a house, a car, or college expenses–to be paid back in equal periodic installments. Your payments are said to **amortize** (pay back) the loan. In these so-called **conventional loans**, each payment pays the current interest and also repays part of the principal. *As the principal is reduced, there is less interest owed, so less of each payment goes to the interest and more toward paying off the principal.*

We remind you of the savings formula from Chapter 21 (p. 691):

Savings Formula RULE

The amount A that is accumulated

- at a nominal annual rate of interest rate r
- with m compounding periods per year (so interest rate $i = r/m$ per compounding period)
- after t years (hence $n = mt$ compounding periods)
- by a uniform deposit d at the end of each compounding period

is

$$A = d\left[\frac{(1+i)^n - 1}{i}\right] = d\left[\frac{(1+\frac{r}{m})^{mt} - 1}{\frac{r}{m}}\right]$$

SPOTLIGHT 22.1 What's the Real Rate?

Financial experts agree that the real, "true" rate of interest for savings or loans is the effective annual rate (EAR).

The 1991 Federal Truth in *Savings* Act requires that savers be told the annual percentage yield (APY), which is the EAR.

The 1968 Federal Truth in *Lending* Act, however, requires that borrowers be told the *annual percentage rate (APR),* which is *not* the same as the EAR. The APR is the rate of interest per compounding period times the number of compounding periods per year. Thus, a credit-card rate of 1.5% per month translates to an APR of 18%. The APR does not take into account compounding. Hence it is not equivalent to—indeed, it understates—the true cost of borrowing, that is, the EAR. For the credit-card loan, with monthly compounding, the EAR is

$$(1 + 0.015)^{12} - 1 \approx 19.6\%$$

The APR also ignores costs that are sometimes involved in borrowing, such as a flat charge for making the loan in the first place ("loan-processing fee"), charges for late payments, and charges for failing to make a minimum payment.

For home mortgage loans, however, the Truth in Lending Act requires that lenders include in the APR some of the upfront costs referred to as *closing costs*: any "loan origination" fee, "loan-processing" fee, and "points" (additional charges to get a reduced interest rate). The APR does not include title insurance, appraisal, credit-report fees, or transaction taxes.

Closing costs are paid at the closing of the sale, while interest is paid over the life of the loan. However, the APR treats the closing costs included in it as if they were amortized over the term of the mortgage, despite the fact that they are paid beforehand. Here, too, the APR understates the true costs.

However, very few people hold a mortgage to its maturity. The median life of a 30-year mortgage is only about 5 years; that is, half of all mortgage holders pay off their mortgage before 5 years are up, usually because they sell their homes and move elsewhere. Thus, for almost all home loans, the APR also includes interest that will never be paid.

Finally, we must take into account inflation. One advantage of buying a home with a fixed-rate mortgage is that your payment stays the same but your earnings and the value of your home are likely to go up with inflation: You are thus paying back the loan with dollars of lesser value. For any loan in a time of inflation, *Fisher's effect* comes into play: If your loan has an EAR of 7% but inflation is running at 3.5% per year, the true cost to you of the loan is not exactly 7% − 3.5% = 3.5%. Instead, for an EAR of r and an inflation rate of a, the cost of the loan at the beginning of the first year is indeed $r - a$ (= 3.5% in our example), but at the end of the first year it is

$$g = \frac{r - a}{1 + a}$$

For $r = 7\%$ and $a = 3.5\%$, we get $g = 3.38\%$. The reason that this is less than the expected 3.5% is that at the end of the first year you are paying back the loan with dollars that have been inflated for a year. As inflation mounts over the term of a mortgage, the cost g goes down steadily each year. For example, at the end of five years of steady inflation at 3.5%, the total inflation has been $a = (1 + 0.035)^5 - 1 = 18.8\%$, and we have $g = 2.95\%$.

A final—and major—consideration is that interest paid on your home mortgage is deductible from taxable income on federal, state, and some local income tax returns. Thus, your home ownership is subsidized by other taxpayers (just as you help subsidize home buyers among them), and the cost to you of the loan is reduced further.

EXAMPLE 4 ■ Buying a House

Let's suppose that you buy a house with a $100,000 loan to be paid off over 30 years in equal monthly installments. Suppose that the interest rate for the loan is 6.00%. How much is your monthly payment?

SOLUTION Imagine changing the setup slightly so that instead of making monthly payments, you are supposed to pay off the entire principal and interest at the end. Meanwhile, you make payments to a savings fund that you're building up to pay off the loan, and the savings fund earns the same rate of interest as the loan costs. The interest rate of 6.00% on the loan is compounded monthly, so the monthly rate is 0.5%. At the end of 30 years, the principal and interest on the loan will (by the compound interest formula) amount to

$$\$100{,}000 \times (1 + 0.005)^{12\times 30} = \$602{,}257.52$$

On the other hand, saving d each month for 30 years at 6.00% interest compounded monthly, we know from the savings formula (p. 691) that you will accumulate

$$d\left[\frac{(1 + 0.005)^{360} - 1}{0.005}\right]$$

To make d just the right amount to pay off the loan exactly, we need to solve the equation

$$d\left[\frac{(1 + 0.005)^{360} - 1}{0.005}\right] = \$100{,}000 \times (1 + 0.005)^{12\times 30} = \$602{,}257.52$$

for the value of d, getting $d = \$599.55$ as your monthly payment. The total of the payments is "only" 360 × $599.55 = $215,838.00–on a loan of just $100,000. (Actually, since the value of d more accurately is $599.5505, your regular monthly payment would be rounded *up* to $599.56 but your last payment would be correspondingly slightly less. We will neglect this fine point here and in later calculations.) ■

We put this idea into a more general setting: *Paying off a conventional loan is like saving*. You can think of paying off the loan as making payments to a savings account that earns interest at the same rate as the loan. At the end of the loan term, the savings balance will exactly equal the principal and interest on the loan. Let the loan amount be P, the effective interest rate per compounding period be i, the number of compounding periods be n, and the loan payment be d. We equate the principal and interest on the loan (from the compound interest formula) with the savings balance (from the savings formula):

$$P(1 + i)^n = d\left[\frac{(1 + i)^n - 1}{i}\right]$$

The quantity P is sometimes called the *present value of an annuity* of n payments of d, each at the end of a compounding period with interest i per period. This terminology is used in the financial mode of some calculators, such as the TI-83.

Solving the above equation for d requires a little algebra. To make things simpler, let $b = (1 + i)^n$, so

$$Pb = d\left[\frac{b - 1}{i}\right]$$

Then

$$d = P\left[\frac{b}{\frac{b-1}{i}}\right] = P\left[\frac{bi}{b-1}\right]$$

Now divide numerator and denominator by b, getting

$$d = P\left[\frac{i}{1-b^{-1}}\right]$$

Substituting $(1 + i)^n$ back for b, we get the usual form of the **amortization payment formula**:

Amortization Payment Formula RULE

A conventional loan amount P

- at a nominal annual rate of interest rate r
- with m compounding periods per year (so interest rate $i = r/m$ per compounding period)
- for t years (hence $n = mt$ compounding periods) can be paid off by uniform payments at the end of each compounding period in the amount

$$d = P\left[\frac{i}{1-(1+i)^{-n}}\right] = P\left[\frac{\frac{r}{m}}{1-(1+\frac{r}{m})^{-mt}}\right]$$

EXAMPLE 5 ■ Repaying Your Student Loan

The standard repayment option for federal student loans is repayment over 10 years with a minimum monthly payment of \$50 at the end of each month, beginning six months after you graduate. For the student loan of Example 1, what will your monthly payments be?

SOLUTION With the amortization payment formula, it's easy to figure out your monthly payment. We have $P = \$4511.33$, monthly interest rate $i = r/m = \frac{0.068}{12} = 0.00566667$, and $n = mt = 12 \times 10 = 120$ months for the payback. We find the payment d as

$$d = P\left[\frac{\frac{r}{m}}{1-(1+\frac{r}{m})^{-mt}}\right] = \$4511.33\left[\frac{\frac{0.068}{12}}{1-(1+\frac{0.068}{12})^{-12\times 10}}\right] = \$51.92$$

So your monthly payment will be \$51.92, just a little more than the minimum payment. (That is for this loan; you may owe more for loans for your other years in college.) Hence over the lifetime of the loan, you will pay 120 × \$51.92 = \$6230.40, of which \$6230.40 − \$4511.33 = \$1719.07 is interest, plus the \$1031.16 interest accrued during the deferment and grace periods, for a total of \$2750.23 in interest. You will pay almost as much in interest as the original principal. If you were to stretch your payments over more years (permissible in some circumstances), an even greater proportion would be interest.

■

EXAMPLE 6 ■ Buying a Car

You decide to buy a new Wheelmobile car. After a down payment, you need to finance (borrow) \$12,000. Comparing interest rates offered by the car dealership, local banks, and your credit union, the best deal you can find is 4.9% compounded monthly over 48 months. What is your monthly payment?

We have $P = \$12{,}000$, monthly interest rate $i = \frac{0.049}{12}$, and $n = 48$. Using the amortization formula, we have

$$d = \$12{,}000\left[\frac{\frac{0.049}{12}}{1-(1+\frac{0.049}{12})^{-48}}\right] \approx \$275.81$$

How much interest do you pay? You make payments totaling $48 \times \$275.81 = \$13{,}238.88$, so the interest is $\$13{,}238.88 - \$12{,}000 = \$1{,}238.88$.

What if you could get a 60-month loan at the same rate? Then your monthly payments would be \$225.91; over 60 months, you would pay \$13,554.60, of which \$1554.60 would be interest.

If you had bought a Plushmobile instead, with \$24,000 to finance, you would have borrowed twice as much and your monthly payment would have been twice as much.

A car loan is usually for 48 or 60 months, but when you buy a home, you usually borrow a great deal more money and pay it off over a much longer period. The usual term for a home mortgage is 30 years.

EXAMPLE 7

Thirty–Year Mortgage on Median-Priced Home

Let's suppose that you are a family with the U.S. median income of about \$59,000 for a family of four, that you want to buy a median-priced home (\$225,000 in August 2007) with a 30-year fixed-rate mortgage at 6.48%, and that you can make a down payment of only \$10,000. Can you afford such a home?

SOLUTION Lenders have "affordability" guidelines that suggest that a family can afford to spend about 28% of its monthly income on housing. Thus, by their guidelines, you can afford $0.28 \times \$59{,}000/12 = \$1376.67$ per month.

What is the monthly payment on the loan? The principal is $P = \$215{,}000$, the monthly interest rate is $i = 0.0648/12 = 0.0054$, and $n = 360$ months. The amortization formula gives a monthly payment of

$$d = \$215{,}000\left[\frac{\frac{0.0648}{12}}{1-(1+\frac{0.0648}{12})^{-360}}\right] \approx \$1356.12$$

Well, that sounds good. But unfortunately there is more to the mortgage than just the amount needed to amortize the loan. Your payment may also have to cover real estate taxes and homeowner's insurance on the property. On a \$225,000 home, these may add \$450 to the monthly payment, which will then total about \$1800.

So, no, the median family can't afford the median-priced home, at least not without a bigger down payment or a lower-interest loan.

A payment on an amortized loan includes both the current interest and a portion toward repaying the principal. You are "building **equity**" in the house or car that you are paying off.

Equity — DEFINITION

Equity is the amount of principal of a loan that has been repaid.

EXAMPLE 8 ■ Home Equity

My wife's parents sold their house in rural Minnesota to move to the town where we live. They had bought their house in 1980 for \$100,000 with a 30-year mortgage at an 8% interest rate. After 22 years, how much *equity* did they have in the house–that is, how much of the principal had been repaid? And how much did they still owe on the house?

(*Norbert Schwerin/ The Image Works.*)

SOLUTION What may shock you is that when they sold their house in May 2002–after 269 months of payments, almost exactly three-quarters of the 30 years of the mortgage–they had only \$50,000 in equity (hence still owed \$50,000 on the house) but had already paid \$147,000 in interest. *Three-quarters of their payments had gone to interest.*

We can use the amortization formula to determine just how much equity they had after 269 months of payments, but first we need to determine their monthly payment. We see $P = \$100{,}000$, $n = 360$ months, and $i = \frac{0.08}{12}$ monthly interest, getting $d = \$733.76$.

Now we use the formula again, this time "in reverse." Knowing $i = \frac{0.08}{12}$ and $d = \$733.76$, we find out how much of the loan would have been paid off by the remaining $n = 360 - 269 = 91$ payments:

$$d\left[\frac{1-(1+i)^{-n}}{i}\right] = \$733.76\left[\frac{1-(1+\frac{0.08}{12})^{-91}}{\frac{0.08}{12}}\right] = \$49{,}940.03$$

This is how much my parents-in-law had yet to pay, so their equity was $\$100{,}000 - \$49{,}940.03 = \$50{,}059.97$.

Figure 22.1 and Table 22.1 show that equity builds up very slowly at first but rapidly later. (The values shown do not take into account possible increase or decrease in the value of the house itself or the effect of inflation.) In fact, the amount of principal in a payment grows by a factor of $1 + i$ from one payment to the next, so the equity at any point is the sum of a geometric series (p. 691) whose common ratio is $1 + i$.

Cumulative equity, 30-year mortgage for $100,000 at 8% interest

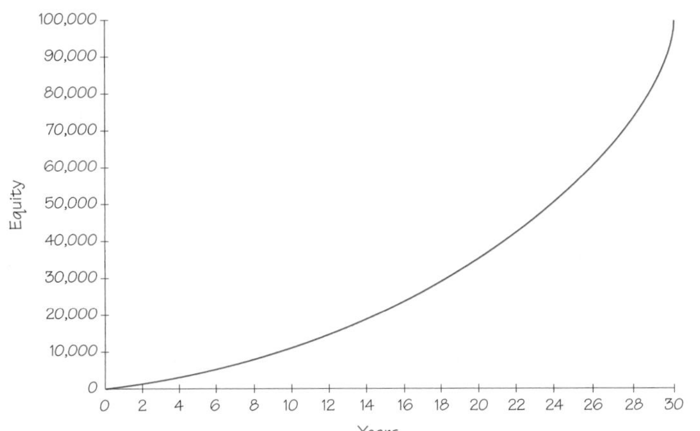

FIGURE 22.1 Equity grows almost exponentially, especially in the later years of a mortgage.

TABLE 22.1 Equity in a 30-year Mortgage for $200,000 at 6.48% Interest

End of Year	1	2	3	4	5	6	7	8	9	10	15	20	25	30
Equity ($\$ \times 10^3$)	2	5	7	10	13	16	19	23	26	30	55	89	135	200

When you buy a home, you have several options for the mortgage: a conventional 30-year mortgage, a conventional 15-year mortgage, or a mortgage for either length of time but with an interest rate that can vary.

You might expect the payment on a 15-year mortgage to be double that of a 30-year mortgage. On the contrary, the payment is only 47% more (for a 5% mortgage) to 26% more (for a 9% mortgage). This range includes the prevailing mortgage rates over the past 20 years. Moreover, over the course of a $200,000 mortgage at 6%, you would pay $328,000 in interest over 30 years but only $144,000 over 15 years. At 9%, the interest totals are $380,000 versus $166,000. (Some financial counselors advise taking a 30-year mortgage and making extra payments when you can afford them, rather than incurring the higher payment obligation of a 15-year loan, which, if you encounter tight personal financial circumstances, you might not be able to afford.) In Spotlight 22.2, we discuss what we did in our own circumstances and mention other options.

Since very few mortgages are held for the full term, it is useful to compare the status of mortgages after five years, the median length of time that Americans remain in a home. Table 22.2 shows the equity after five years for a variety of interest rates. For a 30-year mortgage, the equity after five years may be less than the cost of selling the home through a realtor. Of course, the resale value of the home may also be higher after five years.

A mortgage with an interest rate that can vary is called an **adjustable-rate mortgage (ARM)**. Often such mortgages have a substantially lower interest rate (hence a lower payment) than a fixed-rate mortgage. The ARM's interest rate may go up or down with interest rates in the economy. Usually the rate can be raised or lowered only every year or two, and then by a limited percentage. An ARM may be attractive if you plan to pay off the mortgage after only a few years or because it allows lower payments, it facilitates buying a more expensive home, or you do not plan to keep the home long (hence, you would be selling before the interest rate could rise substantially).

Does it pay to buy a house or apartment? Apart from the joys of ownership, you need to take into account the up-front expenses of closing costs (perhaps $3000), plus the back-end expense of selling the house (usually 6%–7% if through a realtor, so say $12,000). Consulting the table, you might think that you would finally be in the black on your house as an investment after 6 years. However, we have not yet taken into account the ongoing expenses of maintenance, repairs, insurance, and real estate taxes (perhaps $4000–$7000 per year). Of course, if your house is rising in value 5% ($10,000) per year or more, it's a different story. The growth of home ownership, which rose to 73% before the bursting of the housing bubble in 2007, depended on such a steady rise in value. Renting may be attractive if you anticipate moving in just a few years; each year, 20% of Americans move.

SPOTLIGHT 22.2 What We Did with Our House, and What Else You Could Do

We bought our house in 1992. We were offered a choice between an 8.375% fixed-rate 30-year mortgage and an adjustable-rate 30-year mortgage (ARM) at 6.875% whose rate could be raised (or lowered) by up to 2% every year. When we asked, we were also quoted slightly lower rates for corresponding 15-year mortgages.

We were planning to stay in the house much longer than the median of five years, and we were concerned that inflation might force the ARM considerably higher. Also, we did not want the obligation of the higher payments of a 15-year mortgage, in case our circumstances changed (such as through job loss or death). Some loans provide for penalties for paying off the loan early, but in our case (thanks to state law) there was no penalty for making extra payments (if we could afford them).

We chose the 8.375% fixed-rate 30-year mortgage (and made some extra payments). Others in other circumstances, or with a different tolerance for risk, would no doubt have decided otherwise. Had we been sure then that interest rates would not go higher in the 1990s, we would have gone for the ARM. But hindsight is always better than foresight. During the early years of this decade, most homeowners with mortgage interest rates such as ours refinanced at much lower prevailing rates, near 5% for a fixed-rate 30-year mortgage.

Currently, about a third of people take ARMs rather than fixed-rate mortgages as one way to respond to soaring real estate prices in some parts of the country. Newer mortgage "products" include interest-only mortgages and shared appreciation mortgages (SAMs). With an interest-only ARM, payments are (just slightly) lower than for a conventional 30-year mortgage, but you accumulate no equity (still, the market value of the house may rise). After five to seven years, you start also paying off the principal—which means that your payments go up then. In some such loans, the interest rate—and your payments—fluctuates as frequently as every month.

In a SAM, interest payments are lower or absent, but the lender receives a portion of any appreciation (rise in value) when the house is sold. In a nationally reported instance in 2003, a single mother received a no-interest SAM loan to finance the $30,000 down payment on a $223,000 house in Pleasanton, California, through the city's affordable housing program. Four years later, she sold the house for $385,000, and the "affordable housing" lenders got 60% of the $162,000 appreciation, or $97,000. She herself realized $65,000 (minus the cost of the sale) but complained bitterly, saying that she would have been better off to have put the loan on her credit card! Critics have termed SAMs an urban form of sharecropping.

TABLE 22.2 Equity (in thousands of dollars) on a $200,000 Mortgage After Five Years

Term (years)	Interest rate 5%	6%	7%	8%	9%
15	51	48	45	42	40
30	16	14	12	10	8

SPOTLIGHT 22.3 The Mortgage Crisis

Late 2007 saw the development and widening consequences of what has become known as the "mortgage crisis." To understand what that was, why it took place, and how it will have widespread effects for some time, you need to know what happens when you get a mortgage to buy a house—compared to what used to happen. In the "good" old days, you would go to a local bank (or savings and loan, or credit union). If you proved "credit-worthy,"—meaning that after careful consideration, the personnel felt that you could repay the loan—the bank would loan you *its* money, raised from its depositors. The interest rate depended on your credit rating and down payment. You paid back the loan at a fixed rate of interest, usually over 30 years. If interest rates went down, you could refinance the loan at a lower rate by taking out a new loan to pay off the old one; if rates went up, the bank would still receive your regular payments (but wish that it had charged a higher rate of originally). Meanwhile, the value of your house usually went up 5% to 10% per year, your income went up, and your payments stayed the same. (In fact, if you have only 10% equity in your house and it goes up in value 10% in one year, you have made 100% on your investment! This kind of "leveraging" can make real estate investment very profitable—as long as prices keep rising.)

What changed? Efforts to extend home ownership to a wider proportion of the population resulted in banks making more "subprime" loans (loans to people with poorer credit histories who are less likely to repay), with lower down payments but higher rates of interest (because of the greater risk). Some of those loans were "predatory lending," at high rates of interest to people who could not possibly make the payments. Certainly, real estate speculation played a role, as did greed. House prices rose (the "housing bubble"—house prices doubled from 2000 to 2006) but median income (adjusted for inflation) remained stagnant, which made it increasingly hard to afford houses. Banks countered with adjustable-rate and interest-only mortgages. They also realized that they could maximize current income with up-front charges ("origination fee," and "points" paid by the buyer to lower the interest rate) and at the same time minimize risk by immediately "flipping" the mortgages (selling them to bigger banks or other investors, to whom buyers would then make their payments). All of these factors led to banks making more loans that were riskier.

(David H. Wells/ Getty Images.)

What went wrong? Interest rates rose, and people with adjustable-rate mortgages (ARMs) saw their payments rise beyond their ability to pay. At the same time, the "pyramid" of housing prices could not continue with the higher mortgage rates; housing prices fell. When your house becomes worth less than what you would have to pay to keep it, you might be better off just walking away from it (especially since you build up almost no equity in the house in the first few years of the mortgage). Mortgage defaults lowered the value of investments in bundles of mortgages.

Houses are worth perhaps hundreds of billions of dollars less than they were just a year earlier, and big investment banks have mortgages on their hands that are worth hundreds of billions of dollars less than they paid for them. That means that for some time to come, banks have less money to loan and can (should, must) demand more-credit-worthy clients and higher rates of interest.

How does all this affect you? Financial institutions have less money to loan for *any* purpose—buying a home, buying a car, starting a business, etc.—hence the "credit crunch" following the mortgage crisis. We can only hope that the worst will be over before you are in a market for a house, car, or business loan.

Where you won't see direct effect is in the Consumer Price Index (CPI) (see pp. 695–697), which is based on the *rental value* of houses, not their prices. If the doubling of housing prices in 2000–2006 had been taken into account, the CPI would have risen by 5% per year rather than 3%.

22.4 Annuities

We recall from Chapter 21 the concept of an **annuity**:

Annuity DEFINITION

An **annuity** is a specified number of (usually equal) periodic payments.

We restrict our discussion to *ordinary annuities,* for which payments are made at the end of each interval and the interval is also the compounding period.

An annuity can be interpreted as involving borrowing. For example, winners of lotteries are often offered the choice of receiving either the jackpot amount paid as an annuity over a number of years or else a smaller lump sum to be paid immediately. The cost to the lottery administration is the same. If the winner wants an annuity, the administration buys one from an insurance company for the lump sum. You can think of the insurance company as borrowing the lump sum in exchange for making the payments of the annuity. In effect, the insurance company is amortizing the lump sum over the duration of the annuity.

EXAMPLE 9 ▪ Winning the Lottery

On March 6, 2007, two winning tickets shared a record Mega Millions jackpot of \$390 million. Each ticket's share was one-half of the total, or \$195 million. One option was to receive the \$195 million as an annuity in 26 equal annual installments of \$7.5 million each, the first payment being right away. However, each winner chose instead an instant lump sum of \$116,557,083. What was the interest rate of the annuity?

(*Tim Boyle/Getty Images.*)

SOLUTION The insurance company offering the annuity regarded \$116,557,083 as the present value of the annuity. In order to consider it as an ordinary annuity, with payments at the end of each period, we must subtract the first payment, leaving \$195 − \$7.5 million = \$187.5 million to be paid in 25 equal installments at the end of each year, with \$116,557,083 − \$7,500,000 = \$109,057,083 as the present value of the stream of payments.

Solving for P in the amortization payment formula gives

$$P = d\left[\frac{1-(1+i)^{-n}}{i}\right]$$

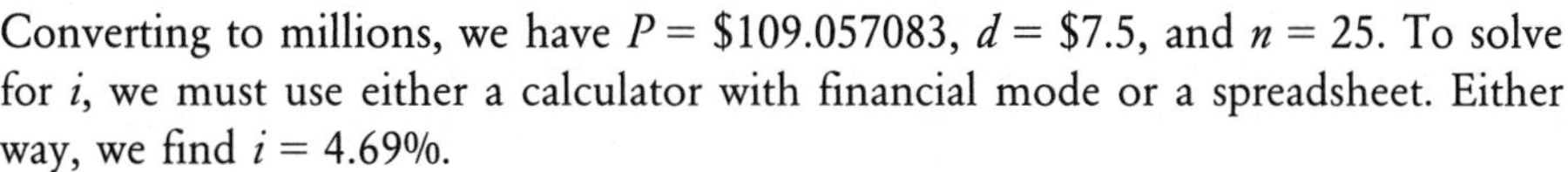

Converting to millions, we have $P = \$109.057083$, $d = \$7.5$, and $n = 25$. To solve for i, we must use either a calculator with financial mode or a spreadsheet. Either way, we find $i = 4.69\%$. ▪

As you save for retirement, it is probably wise to save part of your funds in the form of a tax-deferred annuity. If you do not, at retirement you can still sell all of your holdings in other forms and purchase an annuity.

EXAMPLE 10 ▪ How Much Do You Need to Retire?

Suppose that your father wants to retire at age 65 with an annuity that pays \$1000 per month for 25 years and is willing to assume that at retirement the long-range steady interest rate will be 4% per year compounded monthly. What amount should he expect to have to pay for such a stream of income?

SOLUTION We apply the amortization formula with $d = \$1000$, $r = 0.04$, $m = 12$, and $t = 25$, to find the amount P:

$$P = d\left[\frac{1-(1+\frac{r}{m})^{-mt}}{\frac{r}{m}}\right] = \$1000\left[\frac{1-(1+\frac{0.04}{12})^{-12\times 25}}{\frac{0.04}{12}}\right] = \$189{,}452.48$$

So purchasing such an annuity would cost \$189,452.48.

Such annuities differ in a crucial way from the lottery annuity in Example 9. If you retire at 65 and purchase an ordinary annuity, you would be in trouble if you live longer than the term of the annuity (past 90), because the payments would stop and you would have no further income from the annuity. (About 2% of U.S. children born today can expect to live to age 100.) Similarly, if you die sooner, your estate would still get the payments due after your death, but they wouldn't have helped you meet your living expenses while you were alive.

An approach that avoids these two disadvantages is the *life income annuity*: You receive a fixed amount of income per month for as long as you live. How much you receive per month is based on the life expectancy of people your age, as determined from population data. There are many variations on life annuities, such as payments that increase with anticipated cost-of-living increases, or payments that last until both you and your life partner die (see Spotlight 22.4). But we focus on a simple one-life annuity.

The insurance company that sells you the annuity makes money if you die younger than average and loses money if you die older than average. As in any kind of insurance, over a large number of people, the company can expect gains to balance losses. This is a manifestation of the law of large numbers of Chapter 8. Also, the company's profits vary with the prevailing interest rate during the annuity as compared with the rate built into the annuity.

How much you receive per month depends on your gender. Because women on average live longer than men, the monthly payment to a woman is lower.

EXAMPLE 11 ■ Life Income Annuity

Suppose that you are a 65-year-old male retiring with \$250,000 in a life income annuity. According to the table from one particular insurance company, you would receive \$6.72 per month for every \$1000, so your monthly income would be \$1680. According to the Social Security Administration, your life expectancy at age 65 is about 16.6 years = 199 months. If you lived exactly that long, you would receive a total of 199 × \$1680 = \$334,320. However, simple algebra cannot be used to find the rate of interest that your annuity would need to earn to last that long. We use the RATE function in Excel; entering =RATE(199,1680,−250000) gives a monthly rate of 0.3065%, for an effective annual rate of 3.74%.

If you are female and retire now at the same age with the same \$250,000 savings in a life income annuity, you would receive \$6.26 per month for every \$1000, or \$1565 per month. Your life expectancy would be about 19.6 years = 235 months. If you lived exactly that long, you would receive a total of 235 × \$1565 = \$367,775. The rate of interest that your annuity would need to earn to last that long can be calculated from the amortization formula; using =RATE(235, 1565, −250000) in

Excel gives a monthly rate of 0.3516%, for an effective annual rate of 4.30%. The difference of this figure from that for a man probably reflects the company's use of different values for life expectancy, which vary with region of the country.

Notice that a man and a woman who save the same amount receive different monthly incomes at retirement: The woman receives less but for longer–about 90% as much for 25% longer. Yet their living expenses are likely to be the same. That consideration has resulted in some companies offering "merged gender" rate schedules for annuity payments, so that the individual receives the same monthly payment regardless of gender.

SPOTLIGHT 22.4 What Actuaries Do

The Truth in Savings Act and the Truth in Lending Act specify that the APY for savings and the APR for loans must be calculated "according to the actuarial method."

Actuaries are financial experts who assess the costs of risks and investigate the probability of various contingencies—for example, death, default, or cancellation—that might occur. Actuaries are crucially involved in setting premiums. Their calculations take into account historical rates—such as the percentage of female 85-year-olds who live to be 86, or the percentage of unmarried male drivers under age 25 who have auto accidents—and project those rates and the accompanying costs into the future.

Other actuaries concentrate on setting up and evaluating pension and fringe benefit plans. For example, the city of Beloit, Wisconsin, hired a consulting actuary to estimate the current and future costs of free lifetime medical benefits to families of police and firefighters.

Another major activity of actuaries is managing return on investment. Contrary to popular belief, insurance companies (particularly life insurance companies) do not earn all of their money from premiums paid. In fact, a substantial portion of their income comes from return on investment of financial *reserves*, funds that they are required to have to meet current and future insurance obligations.

(David Young-Wolff/PhotoEdit.)

Becoming an actuary requires training in mathematics, statistics, economics, and finance, and includes a sequence of professional exams taken over several years.

REVIEW VOCABULARY

Adjustable-rate mortgage (ARM) A loan whose interest rate can vary during the course of the loan. (p. 718)

Amortization payment formula Formula for installment loans that relates the principal P, the interest rate i per compounding period, the payment d at the end of each period, and the number of compounding periods n needed to pay off the loan:

$$d = P\left[\frac{i}{1-(1+i)^{-n}}\right], \quad P = d\left[\frac{1-(1+i)^{-n}}{i}\right]. \text{ (p. 715)}$$

Amortize To repay in regular installments. (p. 712)

Annual percentage rate (APR) The number of compounding periods per year times the rate of interest per compounding period. (p. 712)

Annuity A specified number of (usually equal) payments at equal intervals of time. (p. 721)
Compound interest formula Formula for the amount in an account that pays compound interest periodically. For an initial principal A and effective rate i per compounding period, the amount after n compounding periods is $A = P(1 + i)^n$. (p. 710)
Compounding period The fundamental interval for compounding, within which no compounding is done. Also called simply *period.* (p. 712)
Conventional loan A loan in which each payment pays all the current interest and also repays part of the principal. (p. 712)
Effective annual rate (EAR) The effective rate per year. (p. 712)
Effective rate The actual percentage rate, taking into account compounding. (p. 712)
Equity The amount of principal of a loan that has been repaid. (p. 717)
Interest Money charged on a loan. (p. 709)
Nominal rate A stated rate of interest for a specified length of time; a nominal rate does not take into account any compounding. (p. 711)
Principal Initial balance. (p. 709)
Savings formula Formula for the amount in an account to which a regular deposit is made (equal for each period) and interest is credited, both at the end of each period. For a regular deposit of d and an interest rate i per compounding period, the amount A accumulated is

$$A = d\left[\frac{(1 + i)^n - 1}{i}\right]. \quad \text{(p. 710)}$$

Simple interest The method of paying interest on only the initial balance in an account and not on any accrued interest. For a principal P, an interest rate r per year, and t years, the interest I is $I = Prt$. (p. 709)

✓ SKILLS CHECK

1. A nominal rate of interest

(a) takes into account any compounding involved.
(b) is always stated as an annual rate.
(c) neither of the above.

2. (Compound Interest Formula) If you borrow $1000 at 5% interest per year, compounded quarterly, and pay back the principal and interest after four years, the amount that you pay back is ______________ .

3. An effective interest rate

(a) always takes inflation into account.
(b) is the same as the nominal rate.
(c) takes compounding into account.

4. (Savings Formula) If you put $100 at the end of each month for two years in an account that pays 6% annual interest compounded monthly, at the end of the two years you have ______________ .

5. Credit-card interest

(a) is computed using compound interest.
(b) is computed using simple interest.
(c) is included in the late fees.

6. APR stands for ______________ .

7. The nominal rate of interest for a loan is

(a) the same as the effective rate.
(b) less than the effective rate.
(c) never greater than the effective rate.

8. If a store credit account charges 1.5% interest each month, the effective annual rate is ______________ .

9. In a 30-year mortgage, most of the initial payments

(a) go toward reducing the balance.
(b) go toward paying the interest.
(c) pay insurance costs.

10. If a store credit account charges 1.5% interest each month, the APR is ______________ .

11. After 15 years of payments on a 30-year mortgage, the balance remaining is

(a) about one-third of the original balance.
(b) about one-half of the original balance.
(c) about two-thirds of the original balance.

12. Your credit union offers to finance a $6000 conventional loan at 4% to be repaid in four years of monthly payments. Your monthly payment is ______ .

13. An adjustable-rate mortgage

(a) has variable interest rates but maintains a fixed payment amount.
(b) has variable payment amounts.
(c) is always a better alternative to fixed-rate mortgages.

14. If you finance $15,000 for 3 years at 6% compounded monthly, the monthly payments will be ______________ .

15. Equity in a 30-year conventional mortgage grows

(a) linearly.
(b) logarithmically.
(c) exponentially, but slowly.

16. Monthly payments for a 15-year 6% mortgage are about ______ times the payments for a 30-year mortgage of the same amount and the same interest rate.

17. Which of the following arrangements could be an ordinary annuity?

(a) Monthly payments, annual compounding
(b) Annual payments, monthly compounding
(c) Annual payments, annual compounding

18. A convenient rule of thumb is that for a 30-year mortgage at 6%, the monthly payment is about 0.6% of the loan. So, on a $100,000 mortgage, the monthly payment is about $600. About ______ of the first payment goes toward interest.

19. A life income annuity is designed to pay a fixed amount each period until

(a) the annuity runs out of money.
(b) you die.
(c) you reach your life expectancy.

20. If you just won a lottery jackpot paid in 25 equal annual installments of $1 million each at 6% annual effective interest, the present value of the jackpot is ______.

CHAPTER 22 EXERCISES

■ Challenge ◆ Discussion

22.1 Simple Interest

1. Suppose that you take out an unsubsidized Stafford loan on September 1 before your senior year for $5500 and plan to begin paying it back on December 1 after graduation (so you will have had the loan for 15 months, including the six months grace period after leaving school). The interest rate is 6.8%. How much will you owe then, and how much of that will be interest?

2. Assume the same situation as in Exercise 1, but you borrow $5500 on September 1 before your junior year and plan to begin paying it back on December 1 after graduation and grace period 27 months later. How much will you owe then, and how much of that will be interest?

3. Suppose that you borrow $3500 for your first year and $4500 for your second year, as unsubsidized Stafford loans. Suppose that each loan begins on September 1 of its year, that you finish college in four years, and that you begin repayment on December 1 after graduation. What is your total debt then, and how much of that is interest?

4. As in Exercise 3, but you also borrow $5500 for each of your third and fourth years, again on September 1. You finish college in four years, and you begin repayment on December 1 after graduation. What is your total debt then, and how much of that is interest?

22.2 Compound Interest

5. If you borrowed $20,000 to buy a new car at 6% interest per year, compounded annually, and paid back the principal and interest at the end of 5 years, how much would you pay back?

6. Assume the same situation as in Exercise 5, but the interest is compounded monthly.

7. If you borrowed $200,000 to buy a house at 6% interest per year, compounded annually, and paid back the principal and interest at the end of 30 years, how much would you pay back?

8. Assume the same situation as in Exercise 7, but the interest is compounded monthly (this is usually the case).

9. A recent credit-card bill of mine showed a daily interest rate of $i = 0.05819\%$.

(a) What is the APR for this rate?
(b) What is the effective annual rate?

10. I received an offer for a credit card with 0% fixed APR for the first 12 months, followed by one of several rates depending on credit history. The highest was a 22.74% APR (and the company reserves the right to change the APR "at any time for any reason").

(a) What is the corresponding daily rate?
(b) What is the effective annual rate?

22.3 Conventional Loans

11. A credit-card bill of mine showed $500 due, with a minimum payment of $10 and daily interest rate of $r = 0.04932\%$. If I make no more charges on the card and pay $10 a month as soon as I get each bill, how long would it take to pay off the total? (*Hint:* The amortization payment formula can be changed algebraically into the form

$$(1 + i)^n = \frac{d}{d - iP}$$

Note that making the first payment immediately will reduce the principal P to be amortized to \$500 − \$10 = \$490. If I delay payment, I incur additional daily interest on the amount due. Evaluate the right-hand side. Then, using either a spreadsheet or the power key $\boxed{y^x}$ on your calculator, raise the value of $(1 + i)$ to higher and higher powers n until you find the smallest value for n that makes the left-hand side larger than the right-hand side.)

12. (Spreadsheet helpful) Regarding the credit-card bill in Exercise 11: By paying \$10 each month, approximately how much interest would I have paid by the time I pay off the original \$500? (*Hint:* You won't be off by much if you estimate the last payment to be \$10.)

13. (Spreadsheet helpful) According to the regulations for the credit card discussed in Exercises 11 and 12, the minimum payment is supposed to be the greater of \$10 or 2% of the balance (rounded to the nearest dollar amount). Suppose that you have such a card and the balance is \$1500. Neglect the rounding down and assume that you pay exactly 2% of the current balance each month until the balance reaches \$500. Notice that if you make a payment of 2% and the bank charges daily interest of 0.04932% in each 30-day month, you in effect reduce the balance to $(1 - 0.02)(1.0004932)^{30} =$ 99.46043% of what it was the previous month. How many 30-day months will it take to reduce the balance of \$1500 to \$500? (Again, assume that you make payments right away.)

14. Exercises 9–13 explore how long it might take to pay off a credit-card balance. However, for a high enough interest rate, paying 2% of the balance due will not cover the interest, so the balance actually would increase (this is called *negative amortization*). How high would the APR have to be to make this happen? (Careful: It's not just 12 × 2%.)

15. (Spreadsheet helpful.) Because credit-card interest rates are above the rate of Exercise 14, the U.S. Treasury Dept. urged credit-card companies to raise minimum payments for all customers. Many banks now require payment of the "finance charge" (the amount of interest billed) plus 1% of the New Balance (which includes the interest billed). (Although this new policy sounds reasonable, the effect on some consumers was to double their monthly payments.) For a credit card of mine, the minimum payment is the New Balance if less than \$10; otherwise, it is the largest of the three amounts, rounded to the nearest dollar: \$10, 2% of the New Balance, or interest billed plus 1% of New Balance (plus late and over limit fees, which we disregard here). Under this policy, how long would it take to pay off a \$1500 balance with a daily interest rate of 0.04932%, as in Exercises 11 and 13? Include the rounding in your calculations.

16. Assume the same situation as in Exercise 15, but for an APR of 22.74%.

17. In January 2008 a dealership in Northern Illinois was offering a new 2008 Corolla CE car for \$14,462, about 10.5% lower than the list price. One option for financing was 2.9% over 36 months. What was the monthly payment?

(Ron Kimball/Kimball Stock.)

18. Assume the same situation as in Exercise 17, but for the second financing option of 3.9% over 48 months.

19. Assume the same situation as in Exercise 17, but for the third financing option of 4.9% over 60 months.

20. (Spreadsheet helpful.) Put off by even the lowest monthly payment in Exercises 17–19, you might have been attracted to the dealer's roster of "pre-owned" vehicle, including an older but fancier 2007 Toyota Corolla S ("loaded w/ extras, low miles, auto sunroof"). It listed for \$13,995 (after a down payment of \$1000) and a monthly payment of \$245 over 72 months. What was the APR? (Use a spreadsheet and try successive approximations in the amortization payment formula, or a calculator with a financial mode, or a spreadsheet's RATE function.)

21. A TV ad in January 2008 offered a 2006 Dodge Stratus car for \$170.75 per month for 84 months and cited a 7.3% APR.

(a) What was the price of the car?
(b) How much would you pay in interest over the course of the 84 months (7 years, an unusually long period for a car loan)?

22. Check the newspapers and pick a car that you would be interested in buying. Give the price, the interest rate, the term, and how much interest you would pay over the course of the loan.

23. Suppose that your Stafford loans plus accumulated interest total \$20,000 at the time that you start repayment, and that you elect the standard repayment plan of a fixed amount each month for 10 years, at 6.8% APR.

(a) What is your monthly payment?
(b) How much will you pay in interest?

24. Suppose that your Stafford loans plus accumulated interest total \$40,000 at the time that you start repayment. You could elect the standard repayment plan, as in Exercise 23, and your payments over the 10 years would be double the amount calculated there. Instead, since your accumulated outstanding federal loans total more than $30,000, you can instead elect to repay over 25 years. If you do that:

(a) What is your monthly payment?
(b) How much will you pay in interest?

Refer to the following for Exercises 25 and 26:

Your parents (if their credit rating qualifies) can take out a federal PLUS loan to pay for the total cost of your undergraduate education, less any other financial aid (such as a Stafford loan). The interest rate on a PLUS loan is fixed at 8.5% after July 1, 2006. (There are also fees, which we neglect here.) Unlike the Stafford loan, repayment starts right away (strictly speaking, within 60 days). The standard repayment plan is fixed monthly payments over 10 years.

25. Suppose that your parents take out a PLUS loan on your behalf on September 1 before your senior year for $10,000 and begin paying it back a month later. How much is their monthly payment?

26. If your parents instead take out a PLUS loan for $10,000 for each of your four years of college, how much is their monthly payment when you graduate?

Refer to the following for Exercises 27 and 28:

An alternative to saddling your parents with a PLUS loan is to augment your Stafford loan with a private federally-guaranteed direct student loan (to you, so you will be repaying). The terms and repayment options of such loans are: The interest is capitalized quarterly and when repayment begins, interest continues during the grace period. (The interest rate is not fixed but can change annually based on interest rate fluctuations.) A bank may charge an origination fee for such a loan, as a percentage of the loan amount; the fee depends on creditworthiness of the borrower and cosigner.

27. In December 2007 National City Bank advertised a loan for $10,000 with origination fee of $471.20 (so that the initial principal is $10,471.20), an interest rate of 8.21%, and repayment over 20 years beginning after 45 months in school and 6 months grace. Assuming that the interest rate remains 8.21% throughout the term of the loan:

(a) What is the principal of the loan when repayment begins?
(b) What is the monthly payment?
(c) How much interest would be paid over the course of the loan?

28. The bank of Exercise 27 offers a 0.25% reduction in the interest rate at the time repayment begins if the monthly payments are transferred electronically from a bank account, and an additional 0.25% reduction if the first 36 payments are made on time. If the borrower fulfills these conditions:

(a) What is the monthly payment for the first 36 months?
(b) What is the monthly payment after that?
(c) How much interest would be paid over the course of the loan?

29. Suppose that you have good credit and can get a 30-year mortgage for $100,000 at 6.5%. What is your monthly payment?

30. Assume the same situation as in Exercise 29, except that your credit is not as good and the rate that you are offered is 7.125%.

31. Assume the same situation as in Exercise 29, but you inquire about a 15-year loan instead. You are offered 6.125%. What is your monthly payment?

32. Assume the same situation as in Exercise 31, but your credit is not as good, and you are offered 6.75%. What is your monthly payment?

33. For the mortgage in Exercise 29, how much equity would you have after five years?

34. For the mortgage in Exercise 30, how much equity would you have after five years?

35. For the mortgage in Exercise 31, how much equity would you have after five years?

36. For the mortgage in Exercise 32, how much equity would you have after five years?

37. Despite a filter, lots of spam gets into my email. For a while I was getting mortgage offers, such as "$160,000 for less than $735 per month" (for a 30-year mortgage). What would be the corresponding interest rate?

38. Suppose that you and two friends decide to live off-campus in your senior year. One of them (who has wealthy parents) suggests that instead of renting an apartment, you could buy a house together, live in it for your senior year, then rent it out or else sell it. Assuming that (with the help of her parents and their good credit rating) you could get a mortgage for $180,000 to buy a house near the campus, what would be the monthly mortgage payment on a 30-year mortgage at 6.75%?

(amana images/Getty Images.)

For Exercises 39 and 40, refer to the following.

Payday lenders provide small loans until the borrower's next payday. The borrower receives the desired cash in exchange for a postdated check in the amount of the loan plus a fee, which is usually a percentage of the loan amount. In many states, there are now more payday loan offices than McDonald's fast food outlets. The average loan amount is $300.

39. For one payday lender, the fee for a $100 loan for up to two weeks is $15. What is the APR if the loan is for the full two weeks?

40. Another payday lender charges $18.62 for a $100 loan for 7 to 14 days. What is the APR if the loan is for 7 days?

For Exercises 41 and 42, refer to the following.

Many income tax preparation services, including the large national chains, offer refund anticipation loans (RALs), or "rapid refunds." These are similar to payday loans in providing an advance on anticipated income–in this case, a tax refund. The loan is repaid when the IRS pays the refund, usually about 7 to 17 days after the loan is made. The cost of the loan is deducted from the proceeds to the client, so this is a discounted loan. The RAL business takes in about $2 billion each year (including tax preparation and check-cashing fees) to arrange payment of the earned-income tax credit to working parents, about 7% of the total of this aid to poor families. A RAL for an anticipated refund usually is issued for a flat fee, often $88, and the average loan is $1500.

41. Suppose that the RAL speeds the refund by 7 days. What is the APR for the average RAL?

42. Suppose that the RAL speeds the refund by 17 days. What is the APR for the average RAL?

43. When interest rates drop, it may become attractive to refinance your home. Refinancing means that you acquire a new mortgage to borrow the current principal due on your home and use the proceeds to pay off your old mortgage. You then begin a new 15- or 30-year mortgage at the new, lower interest rate. A second factor that reduces your monthly payment is that the equity you accumulated under the old mortgage reduces the amount that you have to borrow under the new mortgage. Suppose that you have an existing 30-year $100,000 mortgage at 8.375%, on which you have been paying for five years, and you are considering refinancing at 7.0%.

(a) What is your payment under the old mortgage?
(b) How much equity do you have in the home?
(c) If you use all your equity to reduce the amount of the new mortgage, how much will your monthly payment be?
(d) How long is the payback period for the $2000 loan charge–that is, how many months will it take before you have saved $2000 in monthly payments?

44. One of the advantages of buying a home with a fixed-rate mortgage is that your payment stays the same but your earnings and the value of your home are likely to go up with inflation. You are paying back the loan with dollars of lesser value.

Consider the following scenario. Suppose that you buy a "starter" two-bedroom home for $105,000 under a special program for first-time home buyers that requires a down payment of only $5000. You have a 30-year fixed-rate mortgage for $100,000 at 7%, on which the monthly payment is $665.30. You also have a $2000 one-time expense in closing costs and annual costs of $200 for insurance and $2000 for property taxes.

You live in the home for five years and spend $10,000 on maintenance, upkeep, and improvements. You then sell the home for $125,000, pay a realtor $9000 to sell it, and pay closing costs of $500 (for title insurance and other costs). Finally, it costs $3000 to move.

(a) Make out a balance sheet of revenue and expenses. How did you make out on owning the home?
(b) Remember, you also got to live in the home without paying rent! Translate the cost of owning the home into an equivalent monthly rent.

22.4 Annuities

45. The largest amount won by an individual in a U.S. lottery was $314.9 million, by Jack Whittaker of West Virginia on Christmas Day 2002. (His subsequent life has been far from a fairy tale, as a Google search will reveal.) Instead of receiving $314.9 million in 30 equal annual payments, including one immediately, he chose a lump sum, which came to $170 million. What was the corresponding interest rate of the annuity?

46. Today, winners of the Powerball lottery can elect either an immediate lump sum (almost all do) or else an annuity. In the latter case, the advertised jackpot amount is paid in 30 annual payments, including one immediate payment. To keep up with inflation, each payment is 4% more than the previous year's; such an annuity is called a *graduated annuity*. On October 10, 2007, Eugene and Stanislawa Markiewicz took their prize as a $20 million annuity.

(a) What was the amount of their first payment, and how much will they receive in their last payment in October 2036?
(b) The winners could have chosen instead a lump sum of $9,402,914.90. What was the corresponding interest rate of the annuity?

47. Suppose that a man retires at age 65 and in addition to Social Security needs $2000 per month in income. Based on an expected lifetime of 16.6 more years (for men) or 19.6 more years (for women), how much would he have to invest in a life income annuity earning 4% to pay that much per year?

48. Assume the same situation as in Exercise 47, but for a woman at age 65, whose expected lifetime is 19.6 more years.

APPLET EXERCISES

To do these exercises, go to www.whfreeman.com/fapp8e.

There are two ways to buy a car: save up and pay cash or borrow the money. In the *Buying a Car: Cash vs. Loan* applet, you can explore just how much more expensive it is to borrow the money.

WRITING PROJECTS

1. In recent years, incentives from auto manufacturers to potential customers have taken the form of offering either a reduced interest rate on the loan for the car or else a rebate (reduction in price) on the cost itself. In fall 2004 you could buy a 2005 Toyota Camry for $13,570, with one of the following options for payment:

- $750 rebate
- 1.9% APR over 24 months
- 1.9% APR over 36 months
- 2.9% APR over 48 months
- 3.9% APR over 60 months

Suppose that you could afford a $2000 down payment and could get a loan from a credit union at 6.0% over 60 months if you opt for the $750 rebate.

(a) What was your monthly payment under each option?
(b) Suppose that the rate of inflation over the course of the loan was a steady 3% per year. How do the various options compare in terms of present value of the loan?
(c) Locate current advertised incentives for a car that you would like to buy and compare them in an essay of two to three pages.

2. A substantial proportion of new cars today are not sold but leased. Contact a local car dealer about a car that you are interested in and find out the details on leasing. Compare the cost of the lease and associated expenses with the cost of purchasing and owning the car. Include estimated maintenance, repair, and insurance costs for each option. Which seems like a better deal, and why? Write two to three pages describing and comparing the two options.

3. Banks often offer choices of mortgages with various combinations of interest rates and "points." A point is 1% of the mortgage amount. Points are paid to "buy down" the interest rate for the mortgage; they are paid upfront to the bank at the closing of the house sale. For example, you may have a choice between a mortgage at 6% with 2 points (2%) and a mortgage at 8% and no points. Which would you choose, and why? Does it make a difference how long you are planning to own the home? Or how expensive the home is? Write a page justifying your decision.

4. Explore actual costs of homes in your area, mortgages with local banks (including closing costs), and property taxes and insurance. Come up with data on a particular mortgage, and the costs and benefits of refinancing, and make out a corresponding balance sheet for five-year ownership.

SUGGESTED READINGS

KASTING, MARTHA. *Concepts of Math for Business: The Mathematics of Finance* (UMAP Modules in Undergraduate Mathematics and Its Applications: Module 370–372), COMAP, Inc., Arlington, Mass., 1980.

MILLER, CHARLES D., VERN E. HEEREN, and JOHN HORNSBY. Consumer mathematics. In *Mathematical Ideas,* 11th ed., Pearson Education/Addison Wesley, Reading, Mass. 2008.

YAREMA, CONNIE H., and JOHN H. SAMPSON. Just say "Charge it!" *Mathematics Teacher* 94 (7) (October 2001), 558–564. Shows how to apply the savings formula and the amortization formula and graph the results on the TI-83 calculator. Notes that the 78% of undergraduates in the United States who have credit cards carry an average debt of more than $2700, with 10% owing more than $7000.

SUGGESTED WEB SITES

www.lendingtree.com/stmrc/calculators1.asp
Java applet calculators (for any platform) to calculate payments and amortization schedules for conventional loans, adjustable-rate mortgages, auto loan vs. home-equity loan, and credit-card payoff. (*Note:* Lending Tree, Inc., is a loan broker; mention here of calculators at its Web site does not imply endorsement of its other services by this book's authors, editors, or publisher.)

www.leaseguide.com/ A guide to how car leasing works, including what "money factor" means, and how leasing cost is determined.

www.edmunds.com/apps/calc/Calculator Controller?pmtcalAction=apr_cash_calc
Commercial site offering a calculator to compare rebate vs. interest-rate offers for car purchase. (*Note:* Edmunds is a loan broker; mention here of calculators at its Web site does not imply endorsement of its other services by this book's authors, editors, or publisher.)

CHAPTER 23

The Economics of Resources

We use resources all the time: food, money, natural resources, labor, time. Some resources, such as annual flowers, are perishable, while others, such as money and standing timber, can be used now or saved for later use.

Our use of resources involves a complex intermixture of biological, ethical, practical, technical, and economic issues:

- How many people will there be in the world in another 20 or 40 years, and how will those numbers affect resources available to you and to them?
- What should we consume for our own use, give to others more needy, or leave for future generations, in terms of wealth, well-being, and wilderness?
- Will our standard of living keep getting better, or is it not maintainable, even at current levels, in the long run?
- How long will it be, at current patterns of use, until we exhaust a particular resource?
- Can we develop more efficient technology so that we can get more out of the resources that we use?
- How do we balance economics with other important considerations? How much would it cost–how much is it worth–to ensure that we do not let tigers, elephants, or rhinos go extinct?

On a personal level, we face similar but more immediate questions about the balance between consumption and conservation. How much must you save to be able to afford to retire? How do you take into account the fact that \$1000 after you retire won't be worth as much as \$1000 today? Once you retire, how much can you spend without exhausting your nest egg?

In Chapters 21 and 22, we explored mathematical models for saving, accumulating, and borrowing–the building up of resources. From Chapter 21, we use here only two formulas, specialized to an annual interest rate r and number of years n.

Compound interest formula: If a principal P is deposited into an account that pays interest at rate r per year, then after n years the account contains the amount

$$A = P(1 + r)^n$$

Savings formula: For a uniform deposit of d per year (deposited at the end of the year) and an interest rate r per year, the amount A accumulated after n years is

$$A = d\left[\frac{(1 + r)^n - 1}{r}\right]$$

A reader who can use these formulas can proceed in this chapter without first reading Chapters 21 and 22.

In this chapter, we model processes in the other direction–the use, decay, depletion, or spending down of resources, including some resources that replenish themselves regularly. Our models will lead us into the mathematics of dynamical systems and chaos. The models provide important insights into answers to the questions above.

23.1 Growth Models for Biological Populations

We encountered geometric growth models for savings accounts in Chapter 21. Growth is proportional to the amount present, and such growth is expressed in terms of compound interest and its formula. We now use a geometric growth model to make rough estimates about sizes of human populations. In addition to the **rate of natural increase**–the annual birth rate minus the annual death rate–we must take into account net migration. The sum of the two, in the terminology of financial models, is the effective rate.

Birth, death, and migration rates rarely remain constant for long, so projections must be made with care. In the short run, however, predictions based on the model may be useful. Let's apply this model to two questions about the population of the United States.

EXAMPLE 1 ■ Predicting the U.S. Population

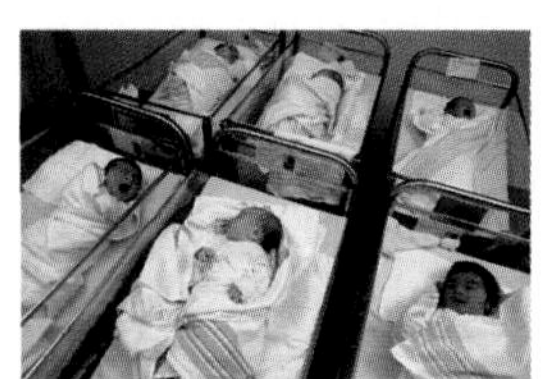

(*Blaine Harrington III/ Corbis.*)

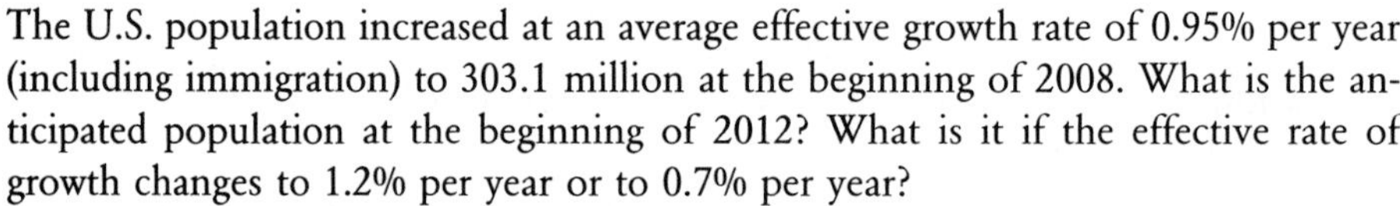

The U.S. population increased at an average effective growth rate of 0.95% per year (including immigration) to 303.1 million at the beginning of 2008. What is the anticipated population at the beginning of 2012? What is it if the effective rate of growth changes to 1.2% per year or to 0.7% per year?

SOLUTION We apply the compound interest formula with initial population size ("principal") 303.1 million. Using a year as the compounding period and the formula $A = P(1 + r)^n$, where $n = 4$, the projected population size in 2012 for a rate $r = 0.0095$ is

$$\begin{aligned} \text{population in 2012} &= (\text{population in 2008}) \times (1 + \text{growth rate})^4 \\ &= 303{,}100{,}000\ (1 + 0.0095)^4 \\ &= 303{,}100{,}000\ (1.0385) \\ &\cong 315{,}000{,}000 \end{aligned}$$

Because of the limited accuracy of the estimates of population and growth rate, we round off the final answer. The result of a calculation can't be more precise than the ingredients.

In the same way, with a growth rate of 1.2% per year, we predict a population of 318 million, while a growth rate of 0.7% per year yields 312 million.

So an uncertainty of one-fourth of one percentage point in the growth rate has major implications, even over fairly short time horizons. The presence or absence of 6 million people would have a significant impact on our social and economic systems! Indeed, much of the concern over long-range funding of Social Security programs results from uncertainties over birth and immigration rates. Figure 23.1a gives a graph of the U.S. population in 2007, structured by age and sex. Figure 23.1b shows possible futures for India.

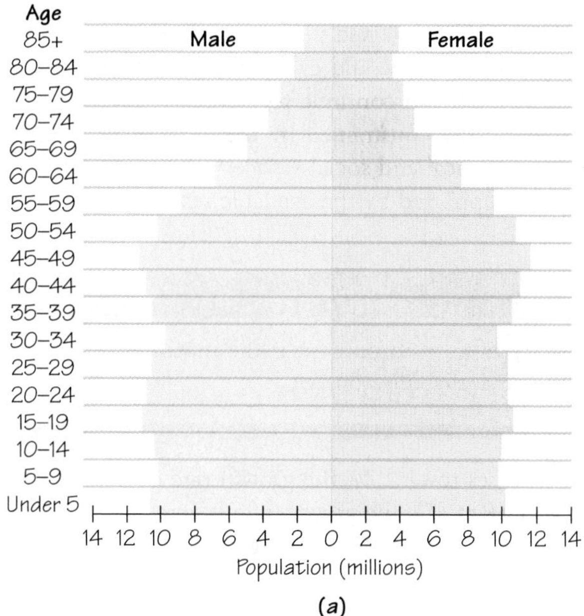

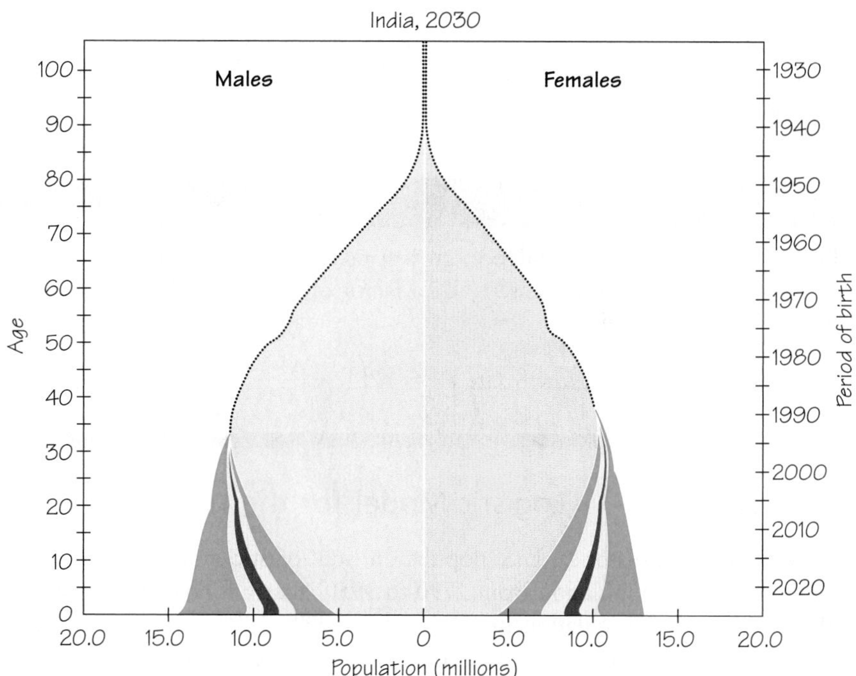

FIGURE 23.1 Graphs of the population of the United States in 2007 grouped by age and gender (a), and of scenarios for the population of India in 2030 (b).

Rates of increase in most developing nations are much higher than in industrialized nations. With a growth rate of 2.5%, Nigeria, Africa's most populous country, will grow from 144 million in mid-2007 to 254 million in mid-2030, an increase of three quarters. Such projections raise concern over providing sufficient food and resources for all people.

It is not just the number of people that is crucial, but also the **population structure**. In poorer countries, the proportion of the population over 60 years of age will be 20% by 2050, compared with 8% now; in Japan, it will be 40%.

Limitations on Growth

A population that keeps adding a fixed percentage each year, like a bank account accumulating compound interest, would eventually grow to astronomical numbers. But no biological population can continue to increase without limit (see Spotlight 23.1). Its growth is eventually constrained by the availability of resources such as food, shelter, and psychological and social "space." There may be a maximum population size that can be supported by the available resources, the **carrying capacity** of the environment.

Carrying Capacity — DEFINITION

The **carrying capacity** of an environment is the maximum population size that it can support with the available resources.

As the population increases toward M, the growth rate decreases. For a population at the carrying capacity ($P = M$), the growth rate is zero.

If the population ever exceeds the carrying capacity, the growth rate becomes negative (because $P > M$) and the population decreases. The carrying capacity is the long-range capacity to support the population, so the population could exceed it for short periods of time. This could happen either because the population grows very rapidly and surges above the carrying capacity or because of a sudden decrease in the food supply, thus temporarily lowering the carrying capacity, as happens to deer and other animals in winter. The **logistic model** is a simple model, but it provides excellent predictions for some populations.

Logistic Model — DEFINITION

The **logistic model** for population growth takes carrying capacity into account by reducing the natural increase rP by a factor of how close the population size P is to the carrying capacity M:

$$\text{growth rate } P' = rP\left(1 - \frac{P}{M}\right)$$

EXAMPLE 2 ■ Logistic Model for the U.S. Population

How well does the historical U.S. population fit a logistic model?

SOLUTION The U.S. population from 1790 to 1950 closely followed a logistic model with $r = 0.031$, $P =$ population in 1790 = 3,900,000, and $M = 201$ million. In the first decades after 1790, the population was a small fraction of this carrying capacity, and it grew at close to the rate r of 3.1% per year (a rate higher than in many

SPOTLIGHT 23.1 12 Billion by 2050–or Only 9 Billion?

How many people can the world hold? Are developing countries heading for a population disaster? Will falling fertility play havoc with Social Security in the United States? Will aging result in 50% of Japanese being over 60 in 2100?

Answers come from mathematical modeling of the future from predicted trends. The best analyses suggest a probability distribution over a range of estimates. They project separately by age, gender, education, and other characteristics. They try to factor in improvements in agriculture, spread of diseases (especially HIV), changes in urbanization, increases in economic aspirations, and the potential for climate change (for example, from global warming).

A basic concept is *total fertility rate* (TFR), the average number of births per woman. Absent catastrophes (such as war or disease), a rate of 2.1 continues a population at the same size. A model that assumes a value above 2.1 will predict an ever-increasing population; one that assumes a value below 2.1 will predict an ever-dwindling one. Most of the world's population growth will occur in the lesser-developed countries, whose TFR values are well above 2.1 (for example, it is 4.6 for Africa). Many countries in Europe have TFR values near 1; without immigration, they will lose population and struggle with fewer workers to provide social benefits to the elderly.

China's situation illustrates *demographic momentum*. Even though its fertility rate has been below replacement level for 20 years, the number of women in the childbearing years was (and still is) so large that China's population will continue to grow until 2040.

The most sophisticated models try to assess how the TFR will vary with changing social circumstances. The most important single factor impacting fertility is education of women. There is a strong negative association between level of education and TFR, and the effect can be very large: In India and China, women with some college education have on average only half as many children as women with no education. Hence, a country's policies about education, and their success, may directly impact future population levels.

(Will & Deni McIntyre/Corbis.)

As we have revised this book for successive editions, we have seen population projections change. The estimates have decreased, because fertilities have declined. The key questions are how to model such declines, whether they will continue, and how they will adapt to other world changes. In Spotlight 21.1 on p. 688, we saw Malthus's over-simplified prediction that mere arithmetic growth in food supplies would limit the geometric increase of human populations. Some demographers now think that population growth will remain a serious problem in some parts of the world but that global population may stabilize or even decline after 2050.

How many people the world will have depends on how well we as a world conserve the environment, distribute food, provide jobs, produce and consume energy, and make other critical decisions about our money and resources. The key concern is the quality of life of *all* people. Political and economic events in far corners of the world, and even natural disasters such as the tsunami at the end of 2004, impact us all. Neglecting problems faced by increasing numbers of poor people provides no security, peace, or moral refuge for anyone.

developing nations today). By 1920, the U.S. population had reached 106 million, and the growth rate had slowed by about one-half, to 1.5% per year (see Figure 23.2).

The 2008 U.S. population of 303.1 million far exceeds the hypothesized carrying capacity of 201 million. The structure of the U.S. population changed, from a large proportion of people making their livings on family farms to a highly urbanized society. The average number of children per family shrank. As the structure changed, the model based on the prior structure gradually became invalid.

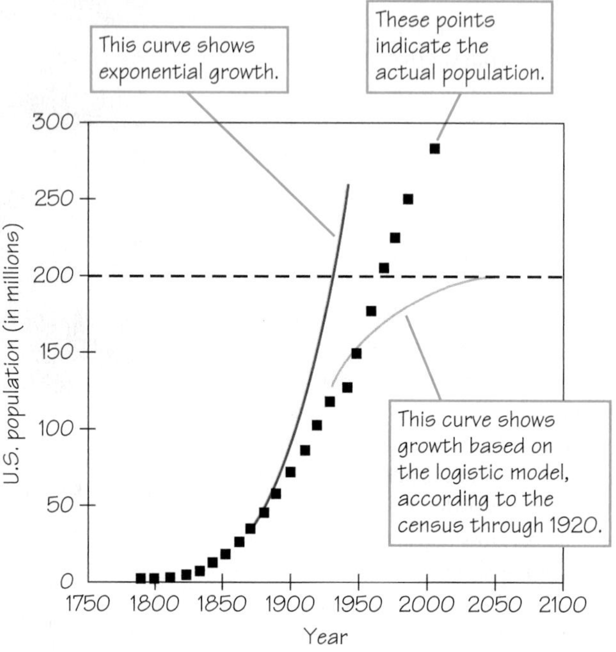

FIGURE 23.2 U.S. population by year, showing actual growth, exponential (geometric) growth, and logistic growth.

The logistic model applies not only to population growth limited by carrying capacity but also to modeling the spread of a technology or product, such as DVD players or flat-screen plasma TVs. Initially, sales are slow. Then they begin to climb rapidly. Finally, as the market gets saturated, new sales slow. We will see in the next section that the logistic model also can apply to exhaustion of nonrenewable resources.

23.2 How Long Can a Nonrenewable Resource Last?

People use resources, some of which are renewable, but others are not. In this section, we model depletion of **nonrenewable resources**. In the next, we treat renewable resources.

Nonrenewable Resource DEFINITION

A **nonrenewable resource** is one that does not tend to replenish itself.

Gasoline, coal, and natural gas are important examples, while lottery winnings and inheritances could be examples from personal affairs. There is no practical way to recover or reconstitute these resources after use. Some substances, such as aluminum

(*Joseph Baylor Roberts/ Getty Images.*)

or the sand used to make glass, are potentially recyclable, but to the extent that we do not recycle them, they, too, are nonrenewable.

For a nonrenewable resource, only a fixed supply S is available. Even without human population increases, we face dwindling nonrenewable resources. We are interested in the question: How long will the supply of a resource last?

As long as the rate of use of the resource remains constant, the answer is easy. If we are using U units per year and continue using U units per year, then the supply will last S/U years. This kind of calculation is the basis for statements such as those claiming that at the current rate of consumption, U.S. recoverable coal reserves will last 250 years or that the U.S. strategic reserve of gasoline (stored in underground salt domes in the South) will last 60 days.

However, the rate of use of resources tends to increase with population and with a higher standard of living. For example, projections for use of electric power often assume that use will increase by a fixed percentage each year. This is the simplest situation (apart from constant usage) and one that we can easily model.

Suppose that $U_1 = U$ is the rate of use of the resource in the first year (this year), and that usage increases $r = 0.05 = 5\%$ each year. Then the usage in the second year is

$$U_2 = U_1 + 0.05U_1 = 1.05U$$

and usage in the third year is

$$U_3 = U_2 + 0.05U_2 = 1.05U_2 = 1.05(1.05U) = (1.05)^2U$$

Generalizing, we see that usage in year i will be $(1.05)^{i-1}U$. Total usage over the next five years, for example, will be

$$U + (1.05)^1U + (1.05)^2U + (1.05)^3U + (1.05)^4U$$

This situation should remind you of the accumulation of regular deposits plus interest (see Chapter 21). Here the usage U corresponds to a deposit, and the increasing rate of use r corresponds to the annual interest rate. We may think of the situation as making regular withdrawals (with interest) from a fixed supply of the nonrenewable resource. The savings formula gives

$$A = d\left[\frac{(1 + r)^n - 1}{r}\right]$$

In the resource situation, A is the accumulated amount of the resource that has been used up at the end of n years, and U is the initial rate of use. We have

$$A = U\left[\frac{(1+r)^n - 1}{r}\right]$$

To find out how long the supply S will last, we set the supply S equal to the cumulative use A over n years and then determine what n has to be. We have

$$S = U\left[\frac{(1+r)^n - 1}{r}\right]$$

We perform some algebra to isolate the term involving n, to get

$$(1+r)^n = 1 + \frac{S}{U}r$$

At this point, to isolate n, we have to take the natural logarithm of both sides.

$$\ln[(1+r)^n] = n\ln(1+r) = \ln\left(1 + \frac{S}{U}r\right)$$

thus gives the final expression

$$n = \frac{\ln[1 + (S/U)r]}{\ln(1+r)}$$

which may look complicated but is quite easy to evaluate on a calculator. The expression S/U is called the **static reserve**, and n is called the **exponential reserve**.

Static Reserve and Exponential Reserve DEFINITION

The **static reserve** is how long the supply S will last at a particular constant annual rate of use U, namely, S/U years. The **exponential reserve** is how long the supply S will last at an initial rate of use U that is increasing by a proportion r each year, namely

$$\frac{\ln[1 + (S/U)r]}{\ln(1+r)} \text{ years.}$$

EXAMPLE 3 ■ U.S. Coal Reserves

Coal accounts for 30% of U.S. energy use, including 50% of U.S. electricity. Recoverable reserves of U.S. coal would last about 250 years at the current rate of use, so the static reserve is 250 years. How long would the supply last if the rate of use increases 1% per year, as it did from 2001 to 2007 (that is, at about the same rate that the U.S. population grew)?

SOLUTION The corresponding exponential reserve is

$$n = \frac{\ln[1 + (250)(0.01)]}{\ln 1.01} = \frac{\ln 3.5}{\ln 1.01} \approx 126 \text{ years}$$

That's quite a difference!

We must not take such projections as exact predictions. Estimates of supplies of a resource may underestimate how much is available, and previously unknown sources may be discovered or the technology improved to extract previously unavailable supplies. In addition, as supplies dwindle, the economic considerations of supply, demand, and price come into play. We will never completely run out of oil. It will always be available "at a price."

We must not take such projections lightly, either, because we are discussing resources that, once used, are gone forever. In any projection, it is very important to examine the assumptions, because small differences in the rate of increase of use can make big differences in the exponential reserve.

EXAMPLE 4 ■ Using Up Retirement Savings

Suppose that you begin retirement with \$1 million in savings, and you don't trust banks or the stock market, so you keep it all under your mattress. Suppose that it costs you \$50,000 per year to live at your accustomed standard of living and there is no inflation. How long will your retirement nest egg last? How long will it last if inflation is constant at 3%?

SOLUTION The static reserve is \$1,000,000/\$50,000 per year = 20 years. If, however, there is constant 3% per year inflation (as during 2001–2007), then it will cost you increasingly more per year to live, so you should realize that your savings will last only for the length of the exponential reserve, which is

$$n = \frac{\ln(1 + 20(0.03))}{\ln 1.03} = 15.9 \text{ years}$$

You have a fine strategy if you expect to live just 16 more years and want to die broke! ■

In our examples so far, we have assumed that the resource is just sitting there, waiting to be used up. For many natural resources, however, we have to find and develop new sources. As doing that becomes more difficult and more costly, at some point the exponentially increasing demand outstrips the ability to meet that demand.

Such a situation is modeled well by the logistic model famously applied to oil by M. King Hubbert, director of Shell Oil Company's research laboratory. Figure 23.3 shows data for cumulative U.S. oil production through 2001 compared with the logistic curve for the ultimate production of $M = 240$ gigabarrels (240 billion barrels). In 1956, Hubbert predicted that U.S. production would peak in the early 1970s (it did) and decline steadily thereafter (it did that, too, except for a blip from Alaska in the 1980s) (see Figure 23.4, whose curve is similar to but "heavier in the tails" than the normal distribution curve of Chapter 5).

If we rearrange the logistic equation on p. 734 by dividing both sides by P and doing a little algebra, we get

$$\frac{P'}{P} = r - \frac{r}{M}P$$

You can recognize this as the equation of a straight line $y = a - bx$, where P'/P takes the role of y and P takes the role of x. In other words, for a logistic model, if we graph P'/P against P, we get a straight line. Figure 23.5 shows that the data fit the Hubbert model well.

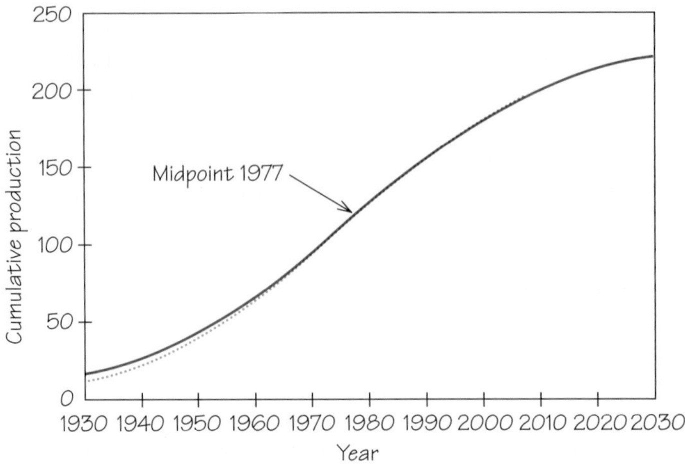

FIGURE 23.3 Logistic model (solid curve), assuming ultimate production of 240 billion barrels, and actual data (points) for cumulative U.S. crude oil production, in billions of barrels, through 2007.
Source: Adapted from Seppo A. Korpela, Oil depletion in the United States and the world, www.mecheng.osu.edu/files/u57/opmatalk.pdf, with revised and further data from the Energy Information Administration through 2007.

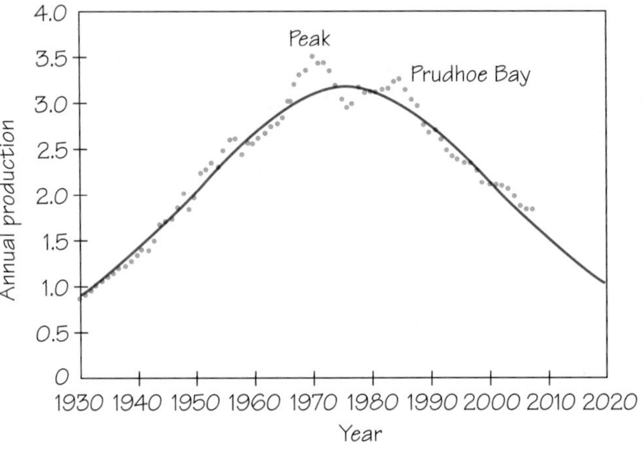

FIGURE 23.4 U.S. crude oil production, in billions of barrels versus year through 2007, and production as predicted by the logistic model, assuming ultimate production of 240 billion barrels.
Source: Adapted from Seppo A. Korpela, Oil depletion in the United States and the world, www.mecheng.osu.edu/files/u57/opmatalk.pdf, with revised and further data from the Energy Information Administration through 2007.

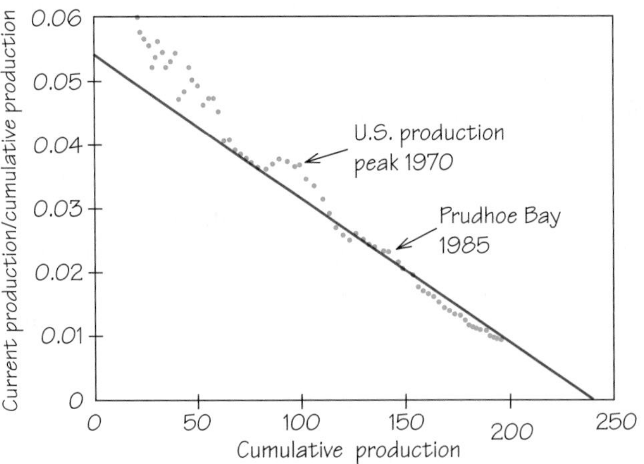

FIGURE 23.5 Display of fit of data to the logistic model (solid line) for U.S. oil production. The line follows the equation $y = 0.054 - 0.000225x$.
Source: Adapted from Seppo A. Korpela, Oil depletion in the United States and the world, www.mecheng.osu.edu/files/u57/opmatalk.pdf, with revised and further data from the Energy Information Administration through 2007.

The world's oil and gas are running out far faster than most people realize or than their governments are willing to acknowledge (see Exercises 11 and 12). The need for "affordable" fuels will likely soon dominate the political agendas of the entire world, particularly since experts have long predicted that world oil output would peak around 2006. In fact, production was level for 2005, 2006, and 2007.

23.3 Sustaining Renewable Resources

A **renewable natural resource** is a resource that tends to replenish itself, such as fish, wildlife, and forests. How much can we harvest and still allow for the resource to replenish itself?

Renewable Resource DEFINITION

A **renewable resource** is one that tends to replenish itself.

Other renewable resources are biological populations. We concentrate on the subpopulation with commercial value. For a forest, this might be trees of a commercially useful species and appropriate size. We measure the population size as its **biomass**, the physical mass of the population. For example, we measure a fish population in pounds rather than in number of fish, and a forest not by counting the trees but by estimating the number of board feet of usable timber.

Reproduction Curves

Our models for growth include many simplifications. Real populations may behave according to one of our models or another known model. Complicated factors that can affect populations, such as climatic or economic change, may mean that the only way to understand a population is to plot a graph of its size over time. Either from such a graph or from a model, we can construct a **reproduction curve**, which predicts next year's population size (biomass) based on this year's size. Although the precise shape of the curve varies from one population to another, the shape in Figure 23.6 is typical. For all possible sizes, it shows for the size this year (on the horizontal axis), the size next year (on the vertical axis), taking into account growth in size of continuing members and addition of new members, minus losses due to death and other factors.

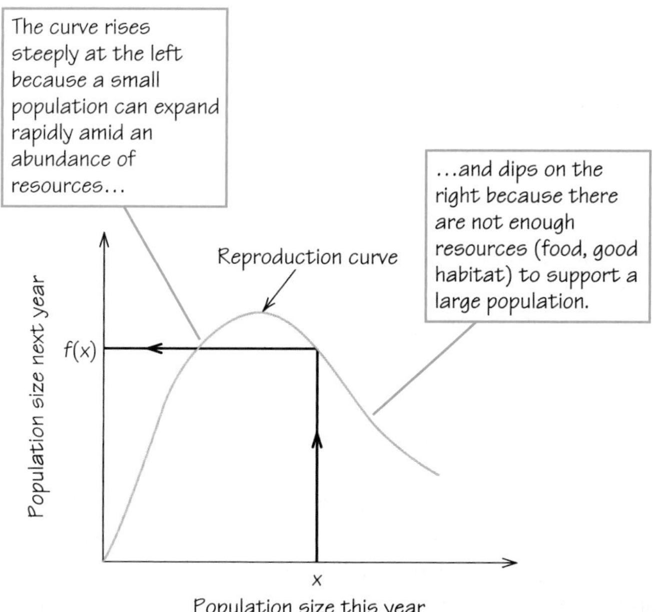
The curve rises steeply at the left because a small population can expand rapidly amid an abundance of resources...
...and dips on the right because there are not enough resources (food, good habitat) to support a large population.
Reproduction curve
Population size next year
f(x)
x
Population size this year

FIGURE 23.6 A typical reproduction curve.

Let x on the horizontal axis be a typical size of the population in the current year. The size *next* year is given by the height of the curve above the point marked x. This value is denoted by $f(x)$. (You can think of f as standing for "function of," or even as "forthcoming.")

Figure 23.7a shows the same reproduction curve, plus the broken line $y = x$ (which makes a 45° angle with the horizontal axis). You can trace what happens for various choices for x. For an x for which the curve is above the broken line, next year's size, $f(x)$, is larger than this year's, x. In Figure 23.7b, the **natural increase**, or gain in population size, is shown as the length of the green vertical line from the broken line to the curve, which in algebraic terms is $f(x) - x$. For an x for which the curve is below the broken line, next year's size is smaller than this year's and $f(x) - x$ is negative. For the size labeled x_e, for which the curve crosses the broken line, the size is the same next year as it was this year. This is the **equilibrium population size.**

Equilibrium Population Size

DEFINITION

An **equilibrium population size** does not change from year to year.

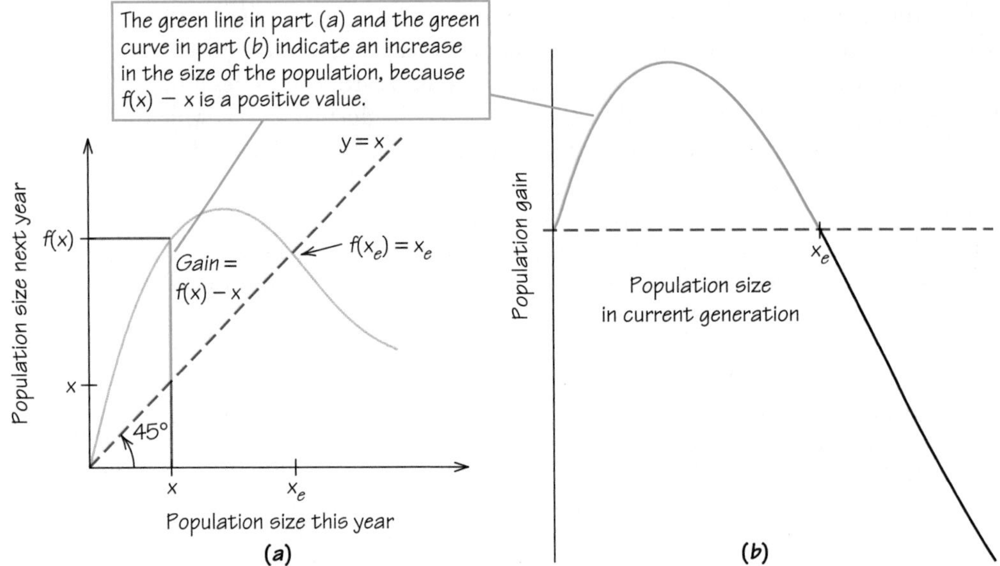

FIGURE 23.7 Depiction of the natural increase (gain) in population from one year to the next. The population size x_e is the equilibrium population size, for which the population one year later is the same, or $f(x_e) = x_e$.

Sustained-Yield Harvesting

In the case of fishing, for example, the harvest **yield** h depends on both the population x of fish and the amount of fishing effort (number of boats, hours of fishing). If the fishing fleet fishes the same banks with the same effort in a year when there are only half as many fish as usual, we can expect the fleet to catch only half as many fish. A simple model is that for any particular level of fishing effort, the harvest is proportional to the fish population, so that a graph of harvest versus population would be a straight line with a steeper slope corresponding to greater effort.

For each level of fishing effort, there is a level of **sustainable yield**, one that could be sustained year after year because the fish population would recover to the same level after each harvest. Figure 23.8 shows the gain curve and two possible harvest lines, with gain and harvest on the vertical axis and current population along the horizontal axis. We focus first on comparing the gain curve with the solid black harvest line. Where the curve lies above the harvest line, the gain exceeds the number harvested. Where the curve lies below the line, the harvest exceeds the annual growth. Where the curve and the line intersect, harvest equals growth; *next year the population will return to the same initial size and we can harvest the same amount.*

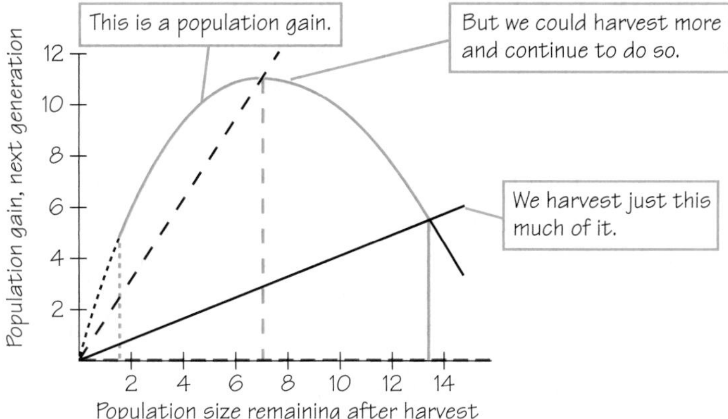

FIGURE 23.8 This population curve shows the effects of harvesting on a fish population. The scales for population and gain sizes are in millions of pounds.

The reason is that we harvest all of the gain and bring the population back to the same level as before the current season's growth, from which the population should generate the same gain next year. In the figure, the curve and the black harvest line intersect at 5 million pounds, where the harvest equals the population gain (indicated by height of the green line). The post-harvest population is 13.5 million pounds (on horizontal axis at foot of green line); the pre-harvest population was 13.5 + 5 = 18.5 million pounds. The 13.5 million pounds remaining after harvest will again grow by 5 million pounds (height of green line) to 18.5 million pounds by next fishing season, when we can again harvest the gain, 5 million pounds.

However, at a higher level of fishing effort, shown by the steeper dashed black harvest line, the fleet could harvest 11 million pounds, leaving a post-harvest population of 7 million pounds (at foot of dashed green line). The pre-harvest population was 11 + 7 = 18 million pounds, about the same as in the previous scenario. The blue population gain curve indicates that the 7 million pounds left after the harvest can be expected to grow by 11 million pounds (height of dashed green line) by the start of next fishing season–the same total pre-harvest level of 18 million pounds as the previous year. As before, the cycle can repeat, but this time with more than double the harvest each year.

Finally, we consider the dotted black harvest line at far left. It corresponds to harvesting 5 million pounds (height of the dotted green line) and leaving just 1.5 million pounds (at foot of dotted green line). The main difference between this and the first scenario is that here *if the fleet takes 6.5 million pounds instead of 5 million, it wipes out the fish altogether.* Even if it takes only 5.5 million pounds, the fish grow back the next year by only 3 million, to a total of 4 million pre-harvest; and trying to harvest 5 million then extinguishes the fish population.

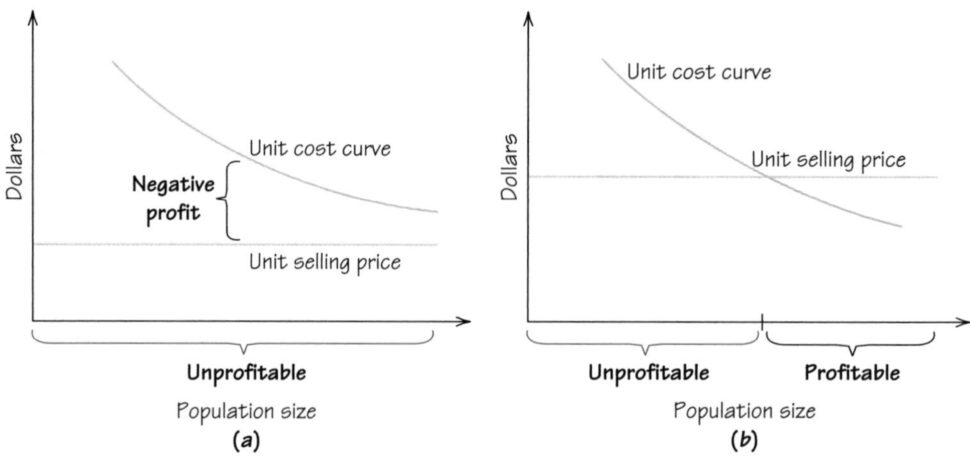

FIGURE 23.14 The unit cost, unit revenue, and unit profit of harvesting one unit, as a function of population size, for fishing or logging. (a) The market price is below the harvesting cost for all population sizes. (b) The operation is profitable for populations above a certain minimum size.

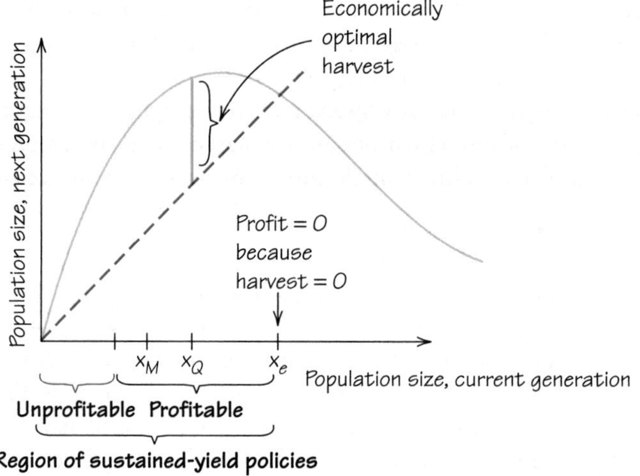

FIGURE 23.15 Reproduction curve showing regions of profitability for sustained-yield policy, with the economically optimal population size x_Q marked.

Our simple models fail to take into account a very critical feature of a modern economy: the *time value of money,* as measured by the interest that capital can earn. We now investigate why biological populations are susceptible to overexploitation and even extinction.

Why Eliminate a Renewable Resource?

Some species, such as the passenger pigeon, were harvested to extinction. In other cases, an entire ecosystem has been destroyed (see Spotlight 23.2). Why would anyone eliminate a renewable resource? Our approach helps explain why.

Sustained-yield policies involve revenues that will be received, year after year, in the future. The value of these revenues should be discounted to reflect the lost investment income that we could earn if instead we had the revenues today. For funds invested at a return of $100r\%$ per year, compounded annually, the present value P of an amount A to be received in n years in the future is related to A by the compound interest formula $A = P(1 + r)^n$.

SPOTLIGHT 23.2 The Tragedy of Easter Island

Easter Island
(Tom Till/Tony Stone Images.)

Easter Island is famous for its isolation—1400 miles to the nearest island—and for its hundreds of huge stone statues. For 30,000 years before the arrival of people in about 400, Easter Island maintained a lush forest, with several species of land birds. By the time of the first visit by Europeans in 1722, the island was barren, denuded of all trees and bushes over 10 feet high, and with no native animals larger than an insect. The 2000 or so islanders had only three or four leaky canoes made of small pieces of wood.

What happened? Careful analysis of pollen in soil samples tells the sad story. The settlers and their descendants cut wood to plant gardens, build canoes, make sledges and rollers to move the huge statues, and burn for cooking and warmth in the winter. In addition to crops they raised and chickens they had brought to the island and cultivated, they ate palm fruit, fish, shellfish, the meat and eggs of birds, and the meat of porpoises that they hunted from seagoing canoes. The population of the island grew to 7000 (or perhaps even 20,000).

By 1500, the forest was gone. Most tree species, all land birds, half of the seabirds, and all large and medium-sized shellfish had been extinguished. There was no firewood, no wood for sledges and rollers to transport hundreds of statues at various stages of completion, and no wood for seaworthy canoes. Without canoes, fishing declined and porpoises could not be taken. Stripping the trees exposed the soil, which eroded, so crop yields fell. The people continued raising chickens, but warfare and cannibalism ensued. By 1700, the population had crashed to 10–25% of its former size.

Why didn't the people notice earlier what was happening, imagine the consequences of keeping on as they had been, and act to avert catastrophe? After all, the trees did not disappear overnight.

From one year to the next, changes may not have been very noticeable. The forests may have been regarded as communal property, with no one charged with limiting exploitation or ensuring new growth. There was no quantitative assessment of the resources available and need for conservation versus the long-term needs of the "public works" program of erecting statues. Moreover, the religion of the people, the prestige of the chiefs, and the livelihood of hundreds depended on the statue industry. There was no perceived need to limit the population and no technology for birth control. Once the large trees were gone, there was no means for excess population to emigrate.

Adapted from Jared Diamond, "Easter's end," *Discover*, 16(8) (August 1995): 63–69.

The economic goal is to maximize the sum of the present values of all future receipts from harvesting. The optimal harvesting policy thus must depend on the expected rate of return r. We don't delve into the details of the calculations here but instead just give the results of the analysis.

Again, there are several cases to consider:

1. The unit cost of harvesting exceeds the unit price received, for all population sizes. Then it is impossible to make a profit.

2. For small r: for some population size x, the unit cost of harvesting equals the unit price received. Then there is a size between x and x_e (the equilibrium population size) for which the present value of the total return is maximized and the population and its yield are sustained.
3. For larger r, *the economically optimal policy may be to harvest the entire population, immediately extinguish the resource, and invest the proceeds.* The unit price exceeds the unit cost for *all* population sizes.

Let's put this in the simplest and starkest terms. Suppose that you own a resource, such as a forest, whose cost of harvesting is small relative to its value. If the rate at which the forest population grows is greater than what you can earn on other investments, it pays to let the forest keep on growing.

On the other hand, if the forest grows more slowly than the rate of return on other investments, the economically optimal harvesting policy is to cut down all the trees now and invest the money. You could then start raising cattle on the land–and right there you have the scenario that is resulting in deforestation all over the world.

The sobering fact is that *very few economically significant renewable resources can sustain annual growth rates over 10%.* Many, like whales and most forests, have growth rates in the 4% to 5% range. These values–even a growth rate of 10%–are far below the return that many investors expect on their investment. For example, over the long run, the U.S. stock market has yielded an average 11% return, but venture capital firms expect to exceed a 25% profit.

The concept of maximum sustainable yield is an attractive ideal if the expectations of investors are low enough. However, there are still difficult problems:

- One problem is "the tragedy of the commons," discussed by ecologist Garrett Hardin. Several hundred years ago, English shepherds would graze their flocks together on common land. The grass of the commons could support only a fixed number of sheep. Each shepherd could reasonably think that adding just one or two more sheep to his flock would not overtax the commons. Yet if each did so, there could be disaster, with all the sheep starving. Many natural-products industries, such as fisheries, are a form of commons. Small overexploitation by each harvester can produce disastrous results for all. Global warming may be a tragedy of a worldwide commons.
- How, in the presence of human needs or greed, can we anticipate and prevent overexploitation and possible extinction of a resource? By and large, it has been politically impossible to force a harvesting industry to reduce current harvests to ensure stability in the future.
- In some industries, such as a fishery, growth of the population may be abundant one year but meager another, so that a steady yield cannot be sustained without damaging the resource. A few good years in a row may provoke increased investment in fishing capacity. Then attempting to harvest at the same levels in succeeding normal or below-normal years results in overfishing. This exact scenario destroyed the California sardine fishery in the 1930s, the Peruvian anchovy fishery in 1972, and much of the North Atlantic fishery in the 1980s. The ocean off northwest Africa was nearly picked clean in the past decade.

Were the fishers and regulators mentioned above at fault for extinguishing these fisheries by overexploiting a dependable resource? Or were the extinctions due to chance variations of the fish stocks? In the next section, we examine a third possibility–that the fish stocks followed simple rules that nevertheless produced "chaotic" behavior of stocks, that is, wide variation from one year to the next. When we do not see the pattern, we interpret such behavior as randomness, much as the moves in a chess game may appear random and inexplicable to someone who does not know the rules of the game.

23.5 Dynamical Systems and Chaos

In this and the two previous chapters, we have considered systems that change over time: bank accounts, the amount due on a loan, and the size of a population. Other examples are a dripping faucet, a playground swing, a pinball play, the solar system, the business cycle, epidemics, the passage of a drug through the human body, and the weather. Some of these are very predictable (interest on a bank account), while others are notoriously unpredictable (the weather). Some involve no outside influences (the amount due on a loan, assuming that you don't get behind on payments!), while others are the result of many contributing factors (the business cycle).

In some systems (such as the population of a country), the state of the system may depend largely on its states at previous times (e.g., last year's population), while in other systems (such as an epidemic) chance may play a large role (e.g., in who and how many become infected).

We are interested in modeling systems, such as a fishery, as they operate without influence from outside or from chance. The applicable mathematical tool is a **dynamical system**.

Dynamical System DEFINITION

A **dynamical system** is a mathematical model for a system whose state evolves with time and whose future states depend deterministically on its present and past states.

To make this definition meaningful, we need to be explicit about what we mean by **deterministically**.

Deterministic DEFINITION

A system is **deterministic** if its changes through time depend only on natural and mathematical laws and are not substantially affected by what we consider to be chance or free will.

An example of a deterministic system is the path of a golf putt, which is governed by gravity, terrain, wind, and the force imparted by the golfer. A non-example is the outcome of a vigorous toss of a coin or a random number generator; although the result, like the golf putt, is determined by physical laws, we consider the result to be random. Another non-example is the outcome of an election, which involves choices by humans.

Mathematical Chaos

We think of **chaos** as referring to general confusion, unpredictability, and apparent randomness. Mathematicians and other scientists use the word to describe systems whose behavior over time is inherently unpredictable.

Chaos DEFINITION

A dynamical system exhibits **chaos** if it is:

1. *Near-periodic*–any state is near one that eventually will repeat;
2. *Transitive*–from any state you can eventually get close to any other; and
3. *Sensitive*–a small change in the initial state can produce widely diverging results later.

EXAMPLE 8 ■ Chaos in Manhattan

You may already know from experience that getting around Manhattan can be a chaotic experience, in the ordinary sense of the word. We show here that Manhattan's transit system is also chaotic in the mathematical sense.

Consider an urban area, such as Manhattan, where several modes of public transport are available. Subway trains and buses leave from set stops, and taxis wait at taxi stands; call these locations *transit points*. Trains and buses retrace their routes, hence are periodic; taxis are not. The system is closed: Vehicles do not leave the city nor do they travel through the fourth dimension.

This system fits the definition of chaos:

1. *Near-periodic*: Wherever you live, there is a transit point nearby.
2. *Transitive*: Wherever you live, and wherever you want to go, there is a transit point near you whose vehicle will take you near to where you want to go.
3. *Sensitive*: Vehicles at two different transit points near you can eventually take you to vastly different places.

If the system covers the city, then (1) is a consequence of (2). Also, anyone who has gotten on the wrong subway train or bus realizes that (3) is an inevitable consequence of (1) and (2). So in fact (1) and (3) both follow from (2)–a conclusion that is true not just of this Manhattan example but of a large class of dynamical systems. ■

The most noticeable property of a chaotic system is sensitivity–that a small change now can make a big difference later.

This feature is sometimes known as the **butterfly effect**, from the title of a 1979 talk by meterologist E. N. Lorenz: "Predictability: Does the Flap of a Butterfly's Wings in Brazil Set Off a Tornado in Texas?" (The phrase probably traces to a 1953 science fiction story by Ray Bradbury, "A Sound of Thunder," in which history is changed by a time-traveler who steps on and kills a prehistoric butterfly.)

We can get a feel for chaotic systems by playing with some **iterated function systems (IFS)**. The fancy name just means that we take an initial value, apply a function to it, then repeat over and over. This is exactly what we did earlier, geometrically, with reproduction curves for populations. (See section 19.5, pp. 631–635, for more about IFS and their connection to fractals.)

EXAMPLE 9 ■ Doubling on a "Stone Age" Calculator

Imagine that you have a calculator that keeps only the last two digits of a number. It has a special key marked DBL that doubles the number in the display and keeps *only* the last two digits. For example, DBL applied to 52 gives 04 (*not* 104).

Let's start with two numbers that are as close together as can be on this calculator, such as 37 and 38. As we push the DBL key over and over again, will the result stay close?

SOLUTION

37, 74, 48, 96, 92, 84, 68, 36, 72, 44, 88, 76, 52, 04, . . .
38, 76, 52, 04, 08, 16, 32, 64, 28, 56, 12, 24, 48, 96, . . .

Already, by the fourth iteration, the two sequences are far apart.

EXAMPLE 10 ▪ The Solar System

The American moon landings in 1969 and later, as well as all other space missions, were possible because of the predictability, or *determinism*, of the solar system. The moon and planets follow their orbits like clockwork. So how could the solar system be chaotic?

SOLUTION Over tens of millions of years, the orbit of each planet is chaotic, meaning that the slightest change in its position or velocity–due to, say, a comet passing nearby–could produce a huge difference later.

More down-to-earth examples of physical systems that can exhibit chaos include the fluttering of a falling autumn leaf, heart arrhythmias, and the Tilt-A-Whirl amusement park ride.

Chaos in Biological Populations

If we measure this year's population as a fraction x of the carrying capacity, and do the same for next year's population as a fraction $f(x)$, the logistic model takes the form

$$f(x) = \lambda x(1 - x)$$

where the Greek letter lambda $\lambda = 1 + r$ is the amount by which the population is multiplied each year. When expanded, the equation has the familiar form of a quadratic in x:

$$f(x) = -\lambda x^2 + \lambda x$$

EXAMPLE 11 ▪ The Logistic Population Model

What behaviors can occur in the logistic model?

SOLUTION For different values of the parameter λ and different starting values for the population fraction, each of the behaviors of Figure 23.11 on p. 746 can occur:

- $\lambda = 2.8$ and starting population fraction $x = 0.36$ produces Figure 23.11a.
- $\lambda = 3.1$ and starting population fraction $x = 0.235$ produces Figure 23.11b.
- $\lambda = 3.0$ and starting population fraction $x = 0.4$ produces Figure 23.11c.

In other words, for population growth rates (values of λ) that are fairly close together (2.8, 3.1, 3.0), the population evolves very differently. This is a surprising and nonintuitive conclusion.

But there is more. For $\lambda = 4$ and any starting population fraction, the population does not settle down into any of the patterns of Figure 23.11; year after year it wanders "unpredictably" all over the place (Figure 23.16). This is *chaotic behavior:* It is deterministic, complex, and–in the long run–unpredictable. In the short run, the behavior is completely predictable. For example, from this year's population fraction, the equation tells us exactly what next year's will be. Repeating the use of the equation, we can determine what it will be the following year. But as the years pass, any sense of pattern gets lost in the complexity.

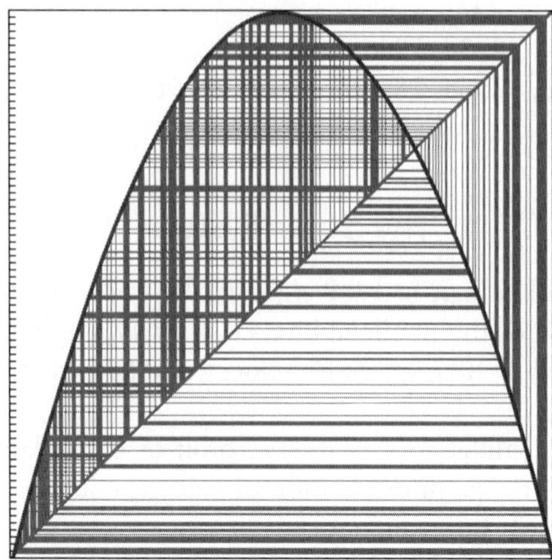

FIGURE 23.16 Chaotic behavior of a population.

This potentially chaotic behavior of a biological population is bad news for those who manage a ranch or any biological population in the wild or in captivity. In recent years, the lobster catch in Maine has been much higher than in previous years, reaching record levels, for no discernible reason. On the other hand, in the late 1950s the annual harvest of Dungeness crabs off the central California coast declined from 12 million pounds to less than 1 million pounds without any evidence of disease, heightened predation, or increased crabbing effort. Researchers who modeled that population in 1994 found that booms and busts are the rule. The population can remain nearly level for generations before suddenly exploding or crashing without warning.

Searching for an environmental cause for these fluctuations could be futile, because there may not be one. Moreover, observing the population over a few generations provides no help in predicting future behavior.

EXAMPLE 12 ■ Childhood Disease Epidemics

The incidence of childhood diseases such as chickenpox and measles varies greatly from year to year. Why?

SOLUTION There are three plausible explanations for the fluctuations:

- There is an underlying regular cycle that is perturbed and occasionally overwhelmed by random events ("noise").
- There is no pattern, because the fluctuations are due solely to chance.
- There is no discernible pattern, because such fluctuations are inherent in the epidemiology of the disease, a chaotic system.

The first explanation, a perturbed cycle, fits chickenpox, with a cycle of one year. For measles, either the second or the third explanation may be correct, depending on the size of the community. For small communities, chance is an adequate explanation. For large communities, historical data from before the era of mass immunization suggest that measles cases were chaotic. That doesn't mean that they occurred at random but rather that they were unpredictable. Research also shows that there is a critical community size above which a disease will not die out solely by chance. For measles, this size is about 250,000.

What you need to understand about chaos is that behavior that appears to be random can be produced even with very simple systems that are completely governed by deterministic rules. Just because the behavior appears chaotic does not mean a lack of underlying order and structure, though discovering that structure may be difficult.

Even if we discover the structure, prediction may elude us because of chaos. If we had an absolutely correct model of how weather behaves and measurements at every location on and above the Earth, we might still not be able to forecast the weather accurately a week ahead.

What about the fishery extinctions? Perhaps the fishers and the fish were victims not of greed or chance but of the chaotic nature of the reproduction curve for the fish.

REVIEW VOCABULARY

Biomass A measure of a population in common units of equal value. (p. 741)

Butterfly effect A small change in initial conditions of a system can make an enormous difference later on. (p. 752)

Carrying capacity The maximum population size that can be supported by the available resources. (p. 734)

Chaos Complex but deterministic behavior that is unpredictable in the long run. (p. 751)

Cobweb diagram A kind of graphical portrayal of the evolution of a dynamical system, such as a population. (p. 745)

Compound interest formula Formula for the amount in an account that pays compound interest periodically. For an initial principal P and effective rate r per year, the amount after n years is $A = P(1 + r)^n$. (p. 732)

Deterministic A system is deterministic if its future behavior is completely determined by its present state, past history, and known laws. (p. 751)

Dynamical system A system whose state depends only on its states at previous times. (p. 751)

Economy of scale Costs per unit decrease with increasing volume. (p. 747)

Equilibrium population size A population size that does not change from year to year. (p. 742)

Exponential reserve How long a fixed amount of a resource will last at a constantly increasing rate of use. A supply S, at an initial rate of use U that is increasing by a proportion r each year, will last

$$\frac{\ln\left(1 + \frac{S}{U} r\right)}{\ln(1 + r)} \text{ years. (p. 738)}$$

Iterated function system (IFS) A sequence of elements (numbers or geometric objects) in which the next element is produced from the previous one according to a function (rule). (p. 752)

Logistic model A particular population model that begins with near-geometric growth but then tapers off toward a limiting population (the carrying capacity). (p. 734)

Maximum sustainable yield The largest harvest that can be repeated indefinitely. (p. 744)

Natural increase The growth of a population that is not harvested. (p. 742)

Nonrenewable resource A resource that does not tend to replenish itself. (p. 736)

Population structure The division of a population into subgroups. (p. 734)

Rate of natural increase Birth rate minus death rate; the annual rate of population growth without taking into account net migration. (p. 732)

Renewable natural resource A resource that tends to replenish itself; examples are fish, forests, wildlife. (p. 741)

Reproduction curve A curve that shows population size in the next year plotted against population size in the current year. (p. 741)

Savings formula Formula for the amount in an account to which a regular deposit is made (equal for each period) and interest is credited, both at the end of each period. For a regular deposit of d and an effective interest rate r per year, the amount A accumulated after n years is

$$A = d\left[\frac{(1 + r)^n - 1}{r}\right]. \text{ (p. 732)}$$

Static reserve How long a fixed amount of a resource will last at a constant rate of use; a supply S used at an annual rate U will last S/U years. (p. 738)
Sustainable yield A harvest that can be continued at the same level indefinitely. (p. 743)
Sustained-yield harvesting policy A harvesting policy that can be continued indefinitely while maintaining the same yield. (p. 744)
Yield The amount harvested at each harvest. (p. 742)

✓ SKILLS CHECK

1. The carrying capacity of a population is
(a) the largest recorded population.
(b) the largest supportable population.
(c) the change in population.

2. The shape of the reproduction curve reflects that a small population has abundant resources and can grow quickly by ________________ steeply at the left.

3. The logistic curve is a model for a population that is growing
(a) linearly.
(b) exponentially.
(c) with a ceiling.

4. U.S. oil use has grown according to a(n) ________ model.

5. Management of a nonrenewable resource can be modeled by
(a) an annuity.
(b) a savings account.
(c) an add-on loan.

6. If we have enough reserves of a product to last 200 years at the current rate of use, but the rate of use increases by 10% per year, the supply will last ________ years.

7. The equilibrium population size is the same as
(a) the carrying capacity.
(b) the intersection point of the reproduction curve and the diagonal.
(c) the natural increase of the population.

8. If we have enough reserves of a product to last 1000 years at the current rate of use, but the rate of use increases by 1% per year, the supply will last ________ years.

9. If the starting population for a reproduction curve is changed, the subsequent population pattern
(a) will still drift to the same pattern.
(b) will change to a different pattern.
(c) will sometimes change to a different pattern.

10. For the logistic model $f(x) = 3x(1 - x)$, if the starting population fraction is 0.5, what is the next population fraction is __________ .

11. Economic considerations
(a) always work against the conservation of a resource.
(b) never interfere with the conservation of a resource.
(c) always affect in some way the conservation of a resource.

12. For the logistic model $f(x) = 3x(1 - x)$, if the starting population fraction is 0.4, the next population fraction is __________ .

13. The population size that leads to the maximum net profit under sustainable-yield harvesting is always
(a) the same as the maximum sustainable yield.
(b) smaller than the maximum sustainable yield.
(c) larger than the maximum sustainable yield.

14. For the logistic model $f(x) = 4x(1 - x)$, if the starting population fraction is 0.1, the next population fraction is __________ .

15. A system whose current state depends solely on its previous states is called
(a) a dynamical system.
(b) a stable system.
(c) an optimal system.

16. For the logistic model $f(x) = 4x(1 - x)$, a starting population fraction that will immediately lead to 0 population is __________ .

17. Chaotic behavior appears to be random
(a) but is actually not random.
(b) and is random.
(c) and is sometimes random.

18. The "butterfly effect" refers to a feature of chaos called ______________________ .

19. Pressing a digit and then repeatedly pressing the SIN key is a model of
(a) an iterated function system.
(b) chaos.
(c) randomness.

20. Elimination of a natural resource can be caused by ____ .

CHAPTER 23 EXERCISES

■ Challenge ◆ Discussion

23.1 Growth Models for Biological Populations

1. (Spreadsheet helpful) For many years, China has been the world's most populous country. However, India has been catching up, with 1132 million in mid-2007 and growing at 1.6% per year, versus China then with 1318 million and growing at only 0.6% per year. If these rates continue, when will India have more people than China?

2. The population of the less-developed countries (excluding China) in mid-2007 of 4.086 billion is expected to grow at 1.8% per year (this is an annual yield, so you may think of it as compounded annually). If this growth rate continues until mid-2025, what will be the size of the population then?

3. If the growth rate of the less-developed countries of Exercise 2 had changed suddenly to 1.7% in mid-2007, what would be the size of the population in mid-2025?

4. (Spreadsheet helpful) If the growth rate of the less-developed countries of Exercise 2 decreased by $\frac{1}{25}$ of a percentage point (0.04%) per year from 2007 through 2011, beginning in mid-2007, what would be the size of the population in mid-2012?

5. An advertisement for Paul Kennedy's book *Preparing for the Twenty-First Century* (Random House, 1993) asked: "By 2025, Africa's population will be: 50%, 150% or 300% greater than Europe's?" The population of Europe in mid-2007 of 729 million is expected to stay constant through 2025. The population of Africa in mid-2007 of 925 million is expected to increase at about 2.4% per year. What answer would you give to the question?

6. In its estimates for doubling times for populations in the world, the Population Reference Bureau uses a rule of 70, similar to (but slightly more accurate than) the rule of 72 used in banking and explained in Exercises 13 and 14 in Chapter 22. The rule of 70 says that if a country's population continues to grow at a constant rate of r% per year, then it will double in size every $70/r$ years. (As noted in Chapter 22, a rule of 69.3 would be even more accurate, but the difference between that and the rule of 70 is only 1%.) Apply the rule of 70 to estimate the doubling times for the following populations (figures are for mid-2007):

(a) Africa, 944 million, 2.5%
(b) United States, 302 million, 0.6% (not including immigration)

7. Do the calculations as in Exercise 6, but for:

(a) China, 1.318 billion, 0.5%
(b) The world as a whole, 6.625 billion, 1.2%

8. Wisconsin's electricity demand increased 3.03% per year for the 35 years from 1970 to 2005. If that trend continues, when will Wisconsin need to have twice as much generating capacity as it did in 2005?

(James Schnept/Getty Images.)

For Exercises 9 and 10, refer to the following.

Is Warren Sanderson right about world population growth slowing down (see Spotlight 23.1)? How much difference does it make in projections if we look at the world as a whole or break it down by countries or regions? In Exercises 9 and 10, we investigate this question, first projecting as a whole and then projecting by regions and adding.

9. The population of the world in mid-2007 of 6.625 billion is expected to increase 1.2% per year.

(a) Project the population to mid-2025 (by then you may have finished having children, if you have any) and to mid-2050 (by then you may be thinking about retiring).
(b) What are the assumptions involved in your projections?

10. Divide the countries of the world into three groups with differing rates of increase (see the table). (Why is this useful?)

Group	Population Mid-2007 (billions)	Rate of Growth (%)
More-developed countries	1.221	0.1
Less-developed countries (excluding China)	4.086	1.8
China	1.318	0.5

(a) Redo the projections in Exercise 9a for the years 2025 and 2050 by projecting each group separately and adding the totals. Is there a major difference from the results in Exercise 9a?
(b) Will the world be able to support the numbers of people that you project? What problems will these greater numbers of people cause? What could be done to avert those problems? Do you think that anything

will be done before there is some kind of worldwide crisis?

(In mid-1995, the world population was 5.7 billion and growing at 1.5% per year. Those figures led to projections as in Exercise 9 of 8.3 billion for 2020 and 11.1 billion for 2040. The corresponding growth rates for the groups of Exercise 10 were 0.2%, 2.2%, and 1.1%, which led to projections of 8.9 billion for 2020 and 12.7 billion for 2040.)

23.2 How Long Can a Nonrenewable Resource Last?

11. In 2005, world oil reserves totaled at most 2900 billion barrels, while daily consumption was 84.7 million barrels in 2006. The U.S. Geological Survey (USGS) projected then that world consumption would increase 1.9% per year through 2025.

(a) What was the static reserve for oil in 2005?
(b) What was the exponential reserve for oil in 2005?
(c) What considerations may affect the answers to parts (a) and (b) over time?

12. In 2008, world natural gas proven reserves totaled 6185 trillion cubic feet, while annual consumption was 105.5 trillion cubic feet in 2006. The USGS projected then that world consumption would increase 2.2% per year through 2025.

(a) What was the static reserve for natural gas in 2008?
(b) What was the exponential reserve for natural gas in 2008?
(c) What considerations may affect the answers to parts (a) and (b) over time?

13. Can our energy problems be solved by increasing the supply? [Thanks for the idea to Evar D. Nering of Arizona State University, in "The mirage of a growing fuel supply," *The New York Times* (June 4, 2001) Op-Ed page.]

(a) Suppose that we have a 100-year supply of a resource (such as oil, for which known world reserves will last less than 100 years at the current world rate of use). That is, the resource would last 100 years at the current rate of consumption. Suppose that the resource is consumed at a rate that increases 2.5% per year (this is the average increase in consumption for oil in the United States since 1973). How long will the resource last?
(b) Suppose that we underestimated the supply and actually have a 1000-year supply at the current rate of use. How long will that last if consumption increases 2.5% per year?
(c) Let's think big and suppose that there is 100 times as much of the resource as we thought–a 10,000-year supply. How long will that last if consumption increases 2.5% per year?

14. In this problem we explore the consequences of reducing the rate of growth of oil use. Suppose that we halve the growth rate from the 2.5% per year given in Exercise 13 to 1.25% per year. [Thanks for the idea to Evar D. Nering of Arizona State University, in "The mirage of a growing fuel supply," *The New York Times* (June 4, 2001) Op-Ed page.]

(a) How long will the 100-year supply last?
(b) How long will the 1000-year supply last?
(c) How long will the 10,000-year supply last?

15. We continue the ideas of Exercises 13 and 14, but with a more radical hypothesis.

(a) How long would the 100-year supply last if we reduced our consumption by just $\frac{1}{2}$% per year–that is, if we used $\frac{1}{2}$% less each year instead of 2% more?
(b) If we used 1% less each year?

16. By the time there is concern about using up a nonrenewable resource, it may be too late. Suppose that a resource has a static reserve of 10,000 years, but consumption is growing at 3.5% per year.

(a) How long will the resource last?
(b) How long before half the resource is gone?
(c) How much longer will the resource last if after half of it is gone, consumption is stabilized at the then-current level?
(d) What implications do you see to your answers?

17. Do a calculation to criticize the claim in the following quotation: "The United States holds 437 billion tons of known (coal) reserves, enough energy to keep 100 million large electric generating plants going for the next 800 years or so." [*Forbes* (December 15, 1975), p. 28; thanks to Albert A. Bartlett.]

For Exercises 18–20, refer to the following.

The formula for the average growth rate over a period of time is even simpler than the one for the exponential reserve of a resource. If usage at the beginning is N_0 and at the end of an interval of t years it is N, then the average annual rate of growth is

$$\frac{1}{t}\ln\left(\frac{N}{N_0}\right)$$

18. The U.S. population was 62.95 million in 1890 and 302 million in 2007 (at roughly the same times of the year). What was the annual rate of growth, to two decimal places?

19. The U.S. population was 3.93 million in 1790 and 62.95 million in 1890. What was the average annual rate of growth, to two decimal places?

20. (a) The average increase in oil consumption in the United States during 1993–2000 was nearly 2%. In fact, consumption in 1993 was 6.291 billion barrels and consumption in 2000 was 7.211 billion barrels. What is a more accurate (two-decimal-place) estimate of the average annual percentage increase in oil consumption? **(b)** Oil consumption in the United States was 19.701 million barrels per day in 2000 and 20.731 million barrels per day in 2004 (roughly one-fourth of world consumption). What was the average annual rate of growth? (From 2004 through 2007, growth was zero, no doubt due to much higher prices for oil.)

23.3 Sustaining Renewable Resources

23.4 The Economics of Harvesting Resources

For Exercises 21–25, refer to the following.

We suppose that a population has the reproduction curve shown in the following figure, with units of thousands of tons of biomass. The mathematical description is that the population in the following year, x_{n+1}, depends on the population x_n in the current year (after any harvest) according to

$$x_{n+1} = f(x_n) = \tfrac{1}{5}x_n(20 - x_n) = 4x_n(1 - 0.05x_n)$$

for x_n between 0 and 10, in units of millions of pounds. We start with a population this year of x_1 whose value we vary. (For these exercises, a spreadsheet or a programmable calculator is useful.)

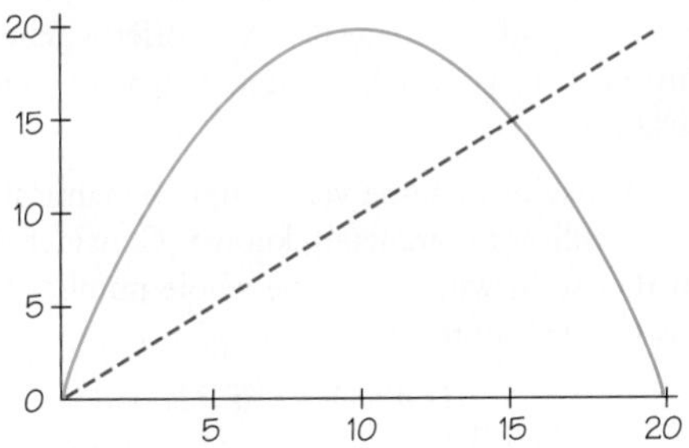

21. Start with $x_1 = 5$. Calculate numerically the population in the first few years, draw a cobweb diagram, and briefly describe the qualitative behavior of the population.

22. Repeat Exercise 21 with the starting value $x_1 = 10$.

23. Repeat Exercise 21 with the starting value $x_1 = 7$, going at least as far as x_{10}.

24. Try to find a starting value (besides 0, 5, 10, and 15) that leads to a stable population over time.

25. What is the equilibrium population size?

For Exercises 26–36, refer to the following.

We suppose that the population has the reproduction curve shown in the following figure, with units of thousands of tons of biomass, whose mathematical description is

$$x_{n+1} = f(x_n) = 3x_n(1 - 0.05x_n)$$

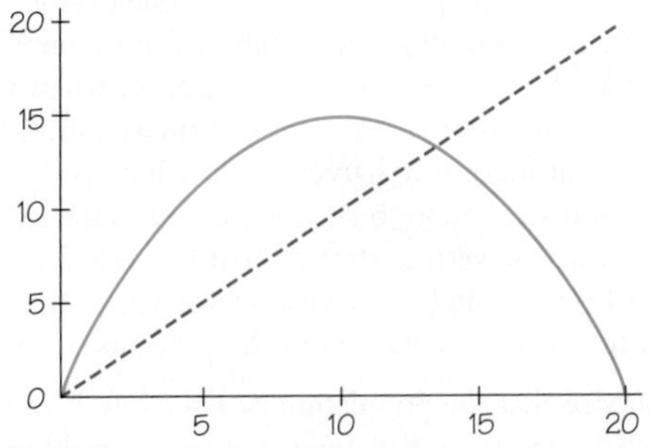

26. Start with $x_1 = 5$. Calculate numerically a population in the first 10 years, draw a cobweb diagram, and briefly describe the qualitative behavior of the population.

27. Repeat Exercise 26 with the starting value $x_1 = 10$.

28. Try to find a starting value (besides 0 and 20) that leads to extinction of the population.

29. What is the equilibrium population size?

◆ **30.** Which of the reproduction curves–the one for Exercises 21–25 or the one for Exercises 26–29–seems to you more realistic as a model of a biological population, and why?

31. What is the significance of the red dashed line in the preceding figures and its intersection with the blue curve?

32. The following figure shows the annual population gain in the absence of any harvesting. Determine the maximum sustainable yield to one decimal point.

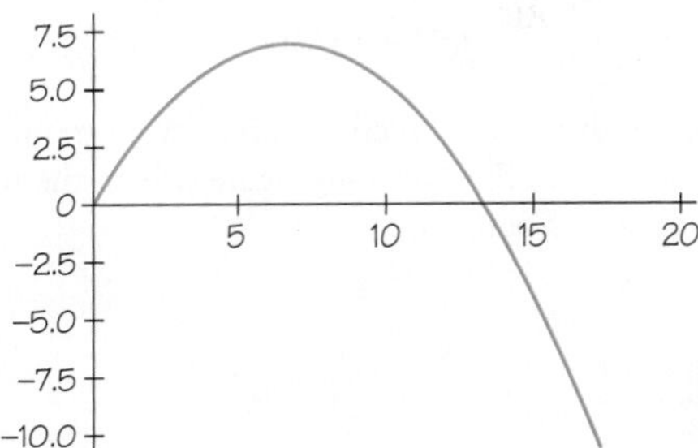

33. Suppose that the population of Exercises 26–32 starts with $x_1 = 11$ and each year we harvest half of the population. For example, in year 1, we harvest 5.5 million pounds, leaving the remaining 5.5 million pounds to reproduce (according to the reproduction curve) for the next year. Calculate numerically the population in the first 10 years, draw a cobweb diagram, and briefly describe the qualitative behavior of the population.

34. Repeat Exercise 33 but for a starting population $x_1 = 5$.

35. Harvesting a set proportion of a population is unrealistic for some situations, such as fishing, in which we can't know the size of the population or when we have harvested half of it. A more realistic situation for fishing is that increasing harvests attract increasing fishing effort (e.g., more boats). Repeat Exercise 33 with $x_1 = 11$ and a harvesting strategy that harvests 1 million tons the first year and every year harvests an extra 1 million tons (over the harvest of the previous year).

36. Suppose that the population of Exercises 26–35 has been overharvested to the point that only 1 million tons remain at the end of a particular year. If there is no harvesting at all until a year after a year with a population of 11 million tons, when can harvesting resume?

37. A reproduction curve for a population is shown in the following figure. Estimate the equilibrium population size and the maximum sustainable yield. (The units are in millions of pounds.)

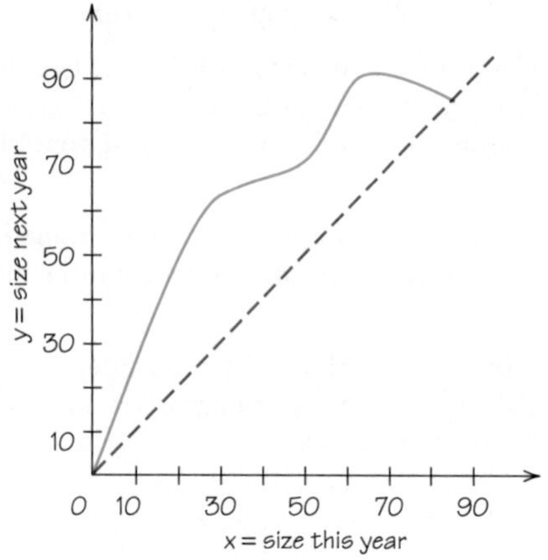

38. Suppose that a reproduction curve for a certain population is as in the following figure, where the units are in millions of pounds.

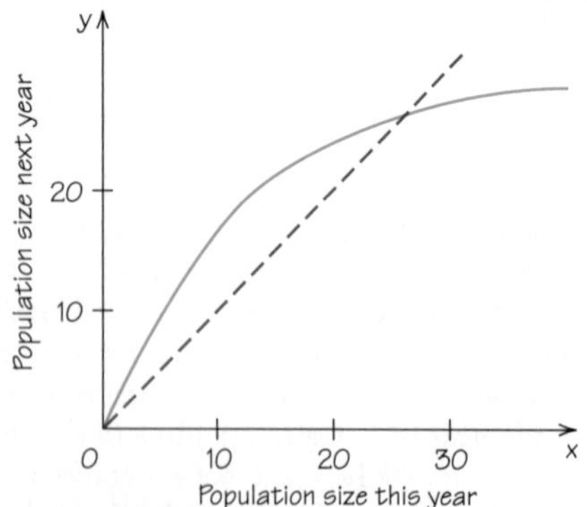

(a) Estimate the sustainable yields corresponding to a population of size 10 remaining after the harvest.
(b) Estimate the maximum sustainable yield.

23.5 Dynamical Systems and Chaos

39. In doubling on a "Stone Age" calculator (see Example 9):

(a) What do you notice about the two sequences that were produced?
(b) Suppose that we had started with a different seed, say 39. What would happen as we iterate the doubling?
(c) Explain what you observe in parts (a) and (b) and give a general argument about why it is true.

40. Explain why the logistic model on p. 734 is the same as the one on p. 753, where we have $f(x) = \lambda x(1 - x)$.

41. You saw in Figure 23.11 that a logistic model can result in a stable value, produce cycling between several values, or result in chaos. Other dynamical systems can exhibit similar behavior. Here we examine the system in which we start with a positive whole number n and iterate the following function:

$$f(n) = \text{sum of the squares of the digits of } n$$

For example, we have $f(133) = 1^2 + 3^2 + 3^2 = 19$.

(a) Calculate what happens as f is applied repeatedly, starting with 133. What do you observe?
(b) Pick a number different from 133 and different from 1, and iterate f repeatedly. What do you observe?
(c) Why did we exclude 1 in part (b)?
(d) Try some other values. Can you offer a general conclusion? Can you offer an argument why your conclusion is correct?

42. The behavior of some very simple dynamical systems is still not completely known. Consider the system that starts with a positive whole number n and gives as the next number

$$f(n) = \begin{cases} 3n + 1 & \text{if } n \text{ is odd} \\ n/2 & \text{if } n \text{ is even} \end{cases}$$

[This iterative function system was devised by Lothar O. Collatz, later of the University of Hamburg, during his student days before World War II. It is sometimes called the "$3n + 1$ problem" or the "Syracuse problem" (because it became popular at the Mathematics Department of Syracuse University), and the sequences generated are sometimes called "hailstone numbers."]

(a) Start with $n = 1$. What happens?
(b) Start with $n = 13$. What happens?
(c) Start with $n = 12$. What happens?

What you observe is known to happen for all $n < 10^{40}$, but after more than 60 years mathematicians have been unable to show that it happens for every n whatsoever.

43. (Requires programmable calculator, spreadsheet, or BASIC programming) A population model slightly different from the logistic model is given by the iterative function system

$$g(x) = x + rx(1 - x)$$

where x is a fraction of the limiting population and r is a growth rate.

(a) Set $r = 3$, start with $x = x_1 = 0.01$, and calculate the first 20 values $x_1, \ldots, x_{20}$.
(b) In part (a), you should have found $x_{10} = 0.722914$. Replace this value with the rounded-up value $x_{10} = 0.723$ and continue on to calculate x_{20}.
(c) Now replace x_{10} with the rounded-down value $x_{10} = 0.722$ and continue on to calculate x_{20}.

44. A dynamical system expressed as an iterated functional system $f(x)$ has an *equilibrium point* at a value x_0 if, once the system reaches x_0, it always stays at that value. In terms of an equation, an equilibrium point exists at x_0 if $f(x_0) = x_0$.

(a) For the dynamical system of Exercise 43, find all equilibrium points.
(b) For the logistic population model of Exercise 40, find all equilibrium points.

■ **45.** (Requires programmable calculator, spreadsheet, BASIC programming, or preferably use of software available under Suggested Web Sites) The behavior of the logistic population model $f(x) = \lambda x(1 - x)$ depends on the value of the positive parameter λ. As λ increases from 0 to 4, the system changes from one behavior to another, through the following possible states:

- The population simply dies out.
- The population tends toward a nonzero equilibrium point.
- The population oscillates between 2 points.
- The population oscillates between 4 points, then 8 points, then 16 points, and so on.
- The population oscillates between numbers of points that are not powers of 2, until at last . . . the population oscillates between 3 values.
- The population behaves chaotically.

Explore what happens for various values of λ between 0 and 4, trying to identify where the shifts in the system's behavior take place.

WRITING PROJECTS

1. Based on the calculations you did in Exercises 9 and 10, write a one- to two-page guest editorial for a newspaper. Describe your projections and how you arrived at them, how serious a problem you think population growth is, what problems it is likely to cause, what you think needs to be done, and what the implications are for your own life.

2. Identify a particular regional, national, or world nonrenewable primary resource (such as coal) or a secondary resource (one, such as electric power, that is produced from primary resources). Research how much of it is available now and what the current rate of consumption is. Determine the static reserve. Estimate the growth rate in consumption, taking into account human population increase, and determine the exponential reserve. What social and technological factors contribute to the increasing rate of consumption? Brainstorm how those factors could be changed. Write an essay of three to five pages.

3. Identify a particular regional, national, or world renewable resources (such as timber or clean drinking water). Research how much of it is produced now, how much is harvested now, and what the current rate of consumption is. Estimate the growth rate in consumption, taking into account human population increase. For how long can this resource continue to meet the demand? What social and technological factors contribute to the increasing rate of consumption? Brainstorm how those factors could be changed. Write an essay of three to five pages.

SUGGESTED READINGS

BARTLETT, ALBERT A. *The Essential Exponential! For the Future of Our Planet.* Center for Science, Mathematics, & Computer Education, University of Nebraska–Lincoln (126 Morrill Hall, Lincoln, NE 68588-0350), 2004.

CLOVER, CHARLES. *The End of the Line: How Overfishing Is Changing the World and What We Eat*, The New Press, New York, 2006.

COHEN, JOEL. *How Many People Can the Earth Support?* Norton, New York, 1995.

GLEICK, JAMES. *Chaos: Making a New Science*, Viking, New York, 1987.

PETERSON, IVARS. *Newton's Clock: Chaos in the Solar System*, W. H. Freeman, New York, 1993.

SCHWARTZ, RICHARD H. *Mathematics and Global Survival*, 4th ed., Ginn Press, Needham Heights, Mass., 1998.

 SUGGESTED WEB SITES

www.popin.org/ UN Population Division population statistics and estimates for all countries.

www.prb.org/ Population Reference Bureau population statistics and rates of growth by regions.

www.maths.anu.edu.au/~briand/chaos Downloadable Java applets to illustrate chaos.

www.ac.wwu.edu/~stephan/Animation/pyramid.html Animated display of population structures of individual countries from 1950 to 2050.

math.bu.edu/DYSYS/applets/ Downloadable Java applets designed to accompany Robert L. Devaney's *A Toolkit of Dynamics Activities,* but can be used independently.

staff.science.uva.nl/~alejan/dynamicstour.html Java applet for the logistic population model of Figures 23.11 and 23.16.

www.census.gov/ipc/www/idb International data base (IDB) of demographic data for 227 countries, with projections and capability to produce population pyramids for various years.

ANSWERS TO SKILLS CHECK EXERCISES

Chapter 1

1. a
2. 7; 8
3. a
4. 3
5. c
6. *B*; *E*
7. b
8. 6
9. c
10. 6
11. a
12. 4
13. b
14. 12
15. a
16. 3
17. b
18. digraph; graph; digraph
19. b
20. 8; 13

Chapter 2

1. c
2. 27
3. b
4. 26
5. b
6. 33
7. c
8. *V*
9. a
10. 54
11. b
12. 2600
13. c
14. 6; 8; 10
15. b
16. 9
17. a
18. 18
19. c
20. 16

Chapter 3

1. c
2. 14 min
3. b
4. 2 min
5. a
6. 4; 2; 3
7. c
8. 4; 2; 3
9. c
10. 10 min
11. b
12. 14
13. c
14. 3
15. b
16. 1
17. c
18. 2
19. a
20. 3

Chapter 4

1. a
2. 6; 2
3. a
4. 2; 5
5. b
6. 25; 6
7. c
8. $12
9. b
10. convex
11. c
12. 3; 2
13. c
14. 3
15. c
16. 3; 1
17. c
18. 57
19. c
20. −6

Chapter 5

1. a
2. 3
3. a
4. right
5. c
6. 14
7. c
8. 124.9
9. b
10. left
11. b
12. 50
13. c
14. 98, 120, 125.5, 132, 147
15. b
16. grams
17. a
18. standard deviation
19. c
20. 95

Chapter 6

1. a
2. positive
3. c
4. 70
5. c
6. −5
7. b
8. $500 + 100x$
9. b
10. 702.26
11. b
12. 0.86
13. c
14. $0.02x + 3$
15. b
16. 0.96
17. a
18. $-0.88 + 0.05x$
19. b
20. (8,3)

Chapter 7

1. b
2. 52
3. c
4. more
5. c
6. 25
7. a
8. 6694
9. b
10. 40
11. a
12. observational
13. b
14. placebo
15. c
16. 0.35
17. b
18. 0.065 (or 6.5%)
19. a
20. 0.06 (or 6%)

Chapter 8

1. a
2. 24
3. b
4. 0.85
5. a
6. $\frac{1}{18}$
7. b
8. $\frac{3}{10}$
9. b
10. 24
11. b
12. 0.3038
13. c
14. 0.4
15. b
16. less
17. a
18. 1511
19. c
20. 19.4

Chapter 9

1. b
2. if any two voters exchange ballots, the election outcome is unchanged
3. b
4. a switch in a ballot from being a vote for the loser to being a vote for the winner doesn't change the election outcome
5. c
6. defeats every other candidate in a one-on-one contest
7. d
8. sometimes produces no winner at all
9. c
10. receives the most first-place votes
11. a
12. has the highest Borda score
13. a
14. reverses the order in which this non winner and the winner were ranked
15. c
16. one-on-one contests take place according to an ordering of the candidates called an "agenda"
17. b
18. they are not monotone
19. c

20. satisfies the Condorcet winner criterion and independence of irrelevant alternatives, and always produces at least one winner in every election

Chapter 10

1. c
2. Borda
3. c
4. either an insincere ballot or a disingenuous ballot
5. a
6. monotonicity
7. b
8. treats both candidates equally and all voters equally
9. b
10. there are only three candidates
11. c
12. placing the additional j candidates at the bottom of each ballot (in any order whatsoever)
13. c
14. agenda manipulation
15. a
16. group manipulation
17. d
18. manipulable
19. a
20. the chair has the most power, but fares the worst

Chapter 11

1. c
2. 11
3. c
4. C
5. b
6. $\frac{1}{4}$
7. b
8. 720
9. c
10. the voters with weights 3 and 4
11. b
12. 256; The motion is defeated
13. a
14. 20
15. c
16. 8
17. a
18. 2^n
19. b
20. $\{A,B,C\}$, $\{A,B,D\}$, $\{A,C,D\}$

Chapter 12

1. b
2. B
3. c
4. just to the left or just to the right of M
5. c
6. C
7. a
8. only one candidate can win, and a median choice is not too far away from anybody, and two department stores at ends of a main street are closer to most consumers than one in the center
9. c
10. when he or she is one of the top two candidates identified by the poll
11. b
12. B
13. a
14. less power than voters in large toss-up states
15. b
16. about three times
17. c
18. it favors citizens who live in the largest states
19. b
20. the law mandates that the popular-vote winner wins if states with a majority of electoral votes pass it

Chapter 13

1. a
2. reflects the relative worth of each issue to that party
3. a
4. the transfer of items (or parts thereof) from one party to the other until points are equalized
5. d
6. the boat, car, and part of the land
7. b
8. cash only
9. b
10. never willingly choose his or her least-preferred item, and avoid wasting a choice on an item that he or she knows will remain available and can be chosen later
11. b
12. no other player received more than he or she did
13. a
14. each nondivider receives a portion that he or she has approved
15. b
16. leaves the game
17. a
18. one player separates the remainders into two portions and the other chooses
19. b
20. the first player

Chapter 14

1. a
2. 352; 44; 4.545; 2.273; 1.182
3. c
4. 14; 13; 17; 17; 19
5. b
6. 300
7. c
8. 0; 0; 100
9. b
10. 1; 1; 98
11. a
12. 0.5
13. a
14. 2; 3
15. a
16. Hill–Huntington
17. c
18. Jefferson
19. c
20. Webster

Chapter 15

1. c
2. third
3. b
4. third
5. a
6. Three
7. a
8. the value of the saddlepoint
9. c
10. prevent a player from being exploited by always choosing a pure strategy
11. a
12. each player's strategic choices are the same
13. c
14. to expect more fastballs than curves
15. b
16. more often kick side and break side
17. b
18. it yields greater payoffs
19. b
20. players must think ahead about what moves are optimal in the future in order to make optimal choices in the present

Chapter 16

1. a
2. 1
3. a
4. 9
5. b
6. 0
7. b
8. 10
9. b
10. 9
11. c
12. 9
13. b
14. 11
15. c
16. 100; 100
17. a
18. 20001-2800-7

19. c
20. 3765

Chapter 17

1. b
2. 1011
3. b
4. 3
5. a
6. one
7. a
8. 3
9. b
10. 0010100
11. a
12. 3
13. c
14. MATH
15. c
16. 29
17. c
18. either P or Q is true
19. a
20. 00111001

Chapter 18

1. c
2. 120
3. a
4. $6.40
5. b
6. $100^{1/3} \approx 4.6$
7. c
8. approximately 26.22 mi
9. c
10. 60
11. a
12. 19
13. c
14. 187
15. b
16. 4
17. c
18. 10
19. a (or b)
20. 1.86

Chapter 19

1. d
2. patterns
3. a
4. 12
5. c
6. 34
7. b
8. H; I; N; O; S; X; Z
9. b
10. translation; rotation
11. c
12. translation; reflection
13. b
14. half-turn rotation
15. c
16. translation
17. a
18. reflection
19. a
20. infinitely many

Chapter 20

1. b (or c)
2. 45°
3. c
4. 3; 4; 6
5. a
6. six
7. c
8. five
9. a
10. seven
11. b
12. Conway; translations; half-turns
13. b
14. translations and half-turns
15. b
16. quasi
17. b
18. the golden ratio
19. b
20. fivefold

Chapter 21

1. a (or c)
2. $15
3. c
4. $8396.19
5. c
6. $49.09
7. b
8. $1186.00
9. b
10. $182.08
11. c
12. 6.06%
13. a
14. 5.13%
15. b
16. $1491.82
17. c
18. a sinking fund
19. b
20. $9216

Chapter 22

1. c
2. $1219.89
3. c
4. $2543.20
5. a
6. annual percentage rate
7. c
8. 19.56%
9. b
10. 18%
11. c
12. $135.47
13. b
14. $456.33
15. c
16. $1\frac{1}{4}$ to $1\frac{1}{2}$
17. c
18. $500
19. b
20. $13.55 million if you get the first payment right away or $12.78 million if you have to wait a year for the first payment

Chapter 23

1. b
2. rising
3. c
4. exponential
5. a
6. 32
7. b
8. 241
9. c
10. 0.75
11. c
12. 0.72
13. b
14. 0.36
15. a
16. 0 or 1
17. a
18. sensitivity to initial conditions
19. a
20. greed, chance, or chaotic variation

ANSWERS TO ODD-NUMBERED EXERCISES

Chapter 1

1. (a) 8
(b) 12
(c) *A*: 3; *B*: 2; *C*: 3; *D*: 2; *E*: 4; *F*: 4; *G*: 3; *H*: 3
(d) *A*, *D*, and *F*
(e) *E*, *G*, and *H*

3. (a) No
(b) *EC*, *AD*, *BD*, and *AC*
(c) 5; 6

5. *E*: 0; *A*: 1; *H*, *D*, and *G*: 2; *B* and *F*: 3; *C*: 5

7. (a) *BCGDFB*
(b) **(i)** *BD*; *BFD*; **(ii)** *CBF*; *CGDF*; *CGDBF*
(c) *GDBCG*

9. (a) 4; 4
(b) 7; 6
(c) 10; 14

11. 2

13. Drawings can vary. Possible renderings for **(a)** and **(b)** include the following:
(a)

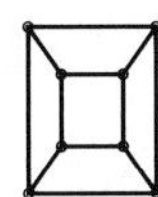

(b)

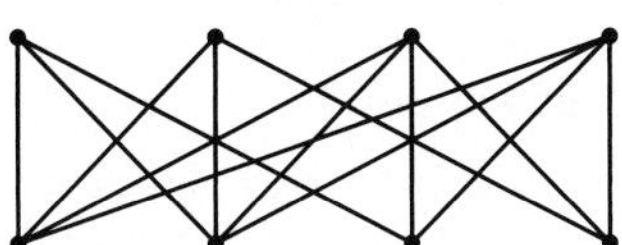

(c) Yes

15. Drawings can vary. Possible renderings include the following:
(a)

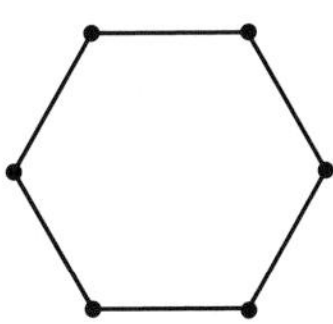

(b)

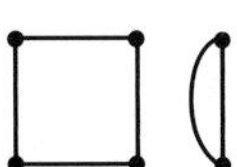

17. Drawings can vary. Possible renderings include the following:
(a)

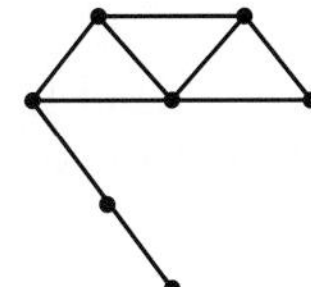

(b)

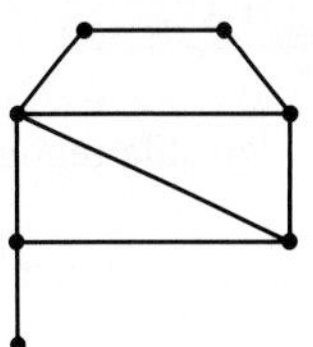

19. (a) Not all edges are traveled by worker; (b) end of route not the same as beginning of route; not realistic; no Euler circuit in graph

21. Since this graph is connected and even-valent, it has an Euler circuit.

23.

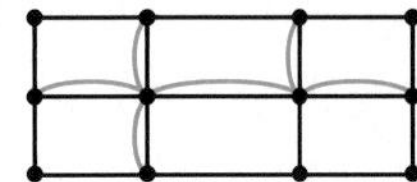

25. (a) 3; **(b)** and **(c)** Answers will vary.

27. Do not choose edge 2, but edges 1 or 10 could be chosen.

29. Answers will vary.

31. 2

33. Answers will vary; no

35. Answers will vary. Possible answers include *AECDABDCBEA*.

37. Drawings can vary. Possible renderings for **(a)**, **(b)**, and **(c)** include the following:
(a)

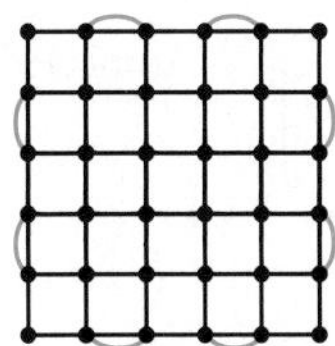

(b)

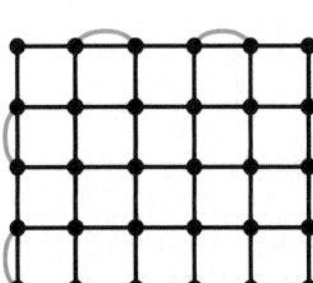

(c)

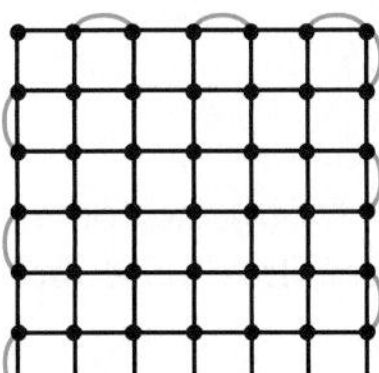

(d) Yes; no

39. (a) 2
(b) Yes

(c) 2
(d) No

41. Answers will vary.

43. Drawings can vary. Possible renderings include the following:

45. There are many circuits that achieve a minimum length of 44,000 feet.

47. (b) and (c); Additional answers will vary.

49. Answers will vary.

51. **(a)** Drawings can vary. Possible renderings include the following:

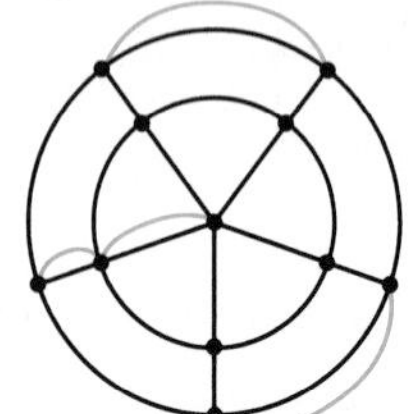

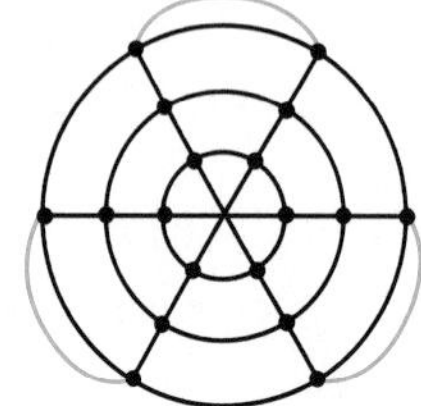

(b) The best eulerization for the four-circle, four-ray case adds two edges.
(c) Answers will vary.

53. Yes

55. Answers will vary.

57. 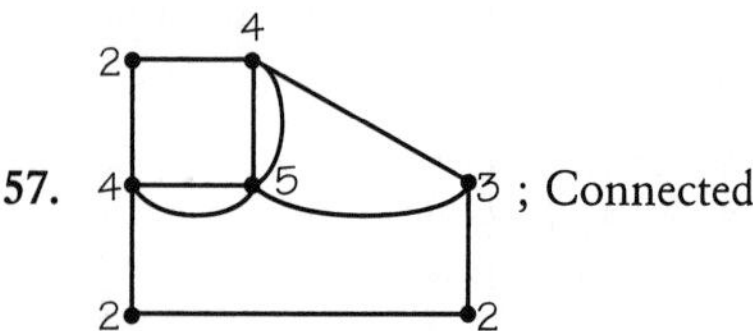

; Connected

59. Answers will vary.

61. Answers will vary. Possible answers include *ABDEFBEBFEDBACDCA*.

Chapter 2

1. **(a)** $X_5X_6X_1X_3X_4X_2X_5$
(b) $X_5X_4X_3X_2X_1X_6X_7X_8X_9X_{10}X_{11}X_{12}X_5$
(c) $X_5X_4X_3X_1X_2X_7X_6X_9X_8X_5$

3. **(a)** Yes
(b) Yes
(c) Yes

5. **(a)** No for (a); yes for (b); no for (c)
(b) No longer be possible to send messages between these two sites

7. Answers will vary.

9. **(a)** Yes for both.
(b) Answers will vary.
(c) Add edges X_2X_8, X_8X_6, X_6X_4, and X_4X_2.

11. Drawings can vary. Possible renderings include the following:

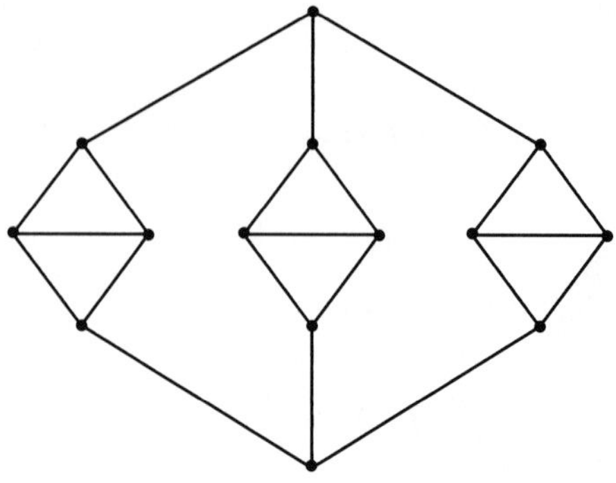

13. **(a)** No
(b) No

15. **(a)** Yes
(b) No
(c) No

17. **(a)** Yes
(b) No
(c) Answers will vary.

19. **(a)** Hamiltonian circuit: yes; Euler circuit: yes; Additional answers will vary.
(b) Hamiltonian circuit: yes; Euler circuit: yes
(c) Hamiltonian circuit: yes; Euler circuit: no
(d) Hamiltonian circuit: no; Euler circuit: yes; Additional answers will vary.

21. **(a)** Hamiltonian circuit: yes; Euler circuit: no
(b) Hamiltonian circuit: yes; Euler circuit: no
(c) Hamiltonian circuit: yes; Euler circuit: no
(d) Hamiltonian circuit: no; Euler circuit: no

23. **(a)** Drawings will vary.
(b) Drawings will vary.
(c) Answers will vary.

25. **(a)** 2520
(b) 16,807
(c) 16,800

27. **(a)** 15,120
(b) 17,576

29. Yes; 172

31. **(a)** 17,558,424
(b) Answers will vary.

33. 10,000,000; 900

35. Drawings will vary.; 6, 10, and 15 edges, respectively; $\frac{n(n-1)}{2}$ edges; 3, 12, and 60, respectively

37. **(a)** Possible drawings include the following:

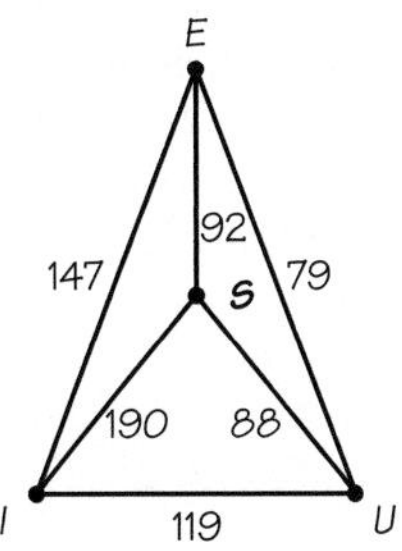

(b) Tour (1): *UISEU*: 480
Tour (2): *USIEU*: 504
Tour (3): *UIESU*: 446
(c) Tour (3)
(d) No
(e) Tour (1); yes; yes; no
(f) Tour (2); no

39. *FMCRF*

41. *MACBM*

43. A traveling salesman problem

45. Yes; Hamiltonian circuit; Chinese postman problem; Answers will vary and requires at least 9 reuses of edges.

47. Answers will vary.

49. The optimal tour is the same but its cost is now 4700.

51. Diagram (a): **(a)** There is a circuit and wiggled edges do not include all vertices. **(b)** The circuit does not include all the vertices of the graph.
Diagram (b): **(a)** The tree does not include all vertices of the graph. **(b)** Not a circuit
Diagram (c): **(a)** Not a tree **(b)** Not a circuit
Diagram (d): **(a)** Not a tree **(b)** Not a circuit

53. **(a)** 1, 2, 3, 4, 5, 8; Cost is 23.
(b) 1, 1, 1, 2, 2, 3, 3, 4, 5, 6, 6; Cost is 34.
(c) 1, 1, 1, 2, 2, 2, 2, 2, 3, 3, 3, 3, 4, 4, 4, 5, 5, 6, 7; Cost is 60.
(d) 1, 2, 2, 3, 3, 3, 4, 5, 5, 5, 6, 6; Cost is 45.

55. 27; 27; at least 26

57. Yes

59. Yes; Additional answers will vary.

61. Yes; yes

63. There are three different trees with the same cost.

65. **(a)** True
(b) False (unless all the edges of the graph have the same weight)
(c) True
(d) False
(e) False

67. **(a)** Answers will vary for each edge.
(b) 5; one less than the number of vertices in the graph
(c) No (*CD* must be included.)

69.

	A	B	C	D
A	0	16	13	5
B	16	0	19	11
C	13	19	0	8
D	5	11	8	0

71. **(a)** 22; $T_3T_2T_5$
(b) 30; $T_3T_5T_7$

73. T_1, T_5, and T_7; 28; T_1, T_4, and T_7

75. Answers will vary.

77. Drawings can vary. Possible renderings include the following:

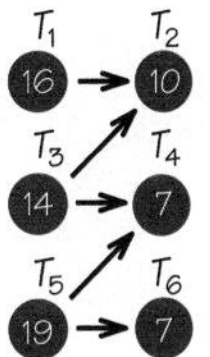

Chapter 3

1. Answers will vary.

3. Answers will vary.

5. **(a)** Processor 1: T_1, T_2, T_3, T_5, T_7; Processor 2: Idle 0 to 2, T_4, T_6, idle 4 to 5
(b) Processor 1: T_1, T_2, T_3, T_6, T_7; Processor 2: Idle 0 to 2, T_4, T_5, idle 4 to 5
(c) Yes
(d) No
(e) T_3 and T_5

7. **(a)** **(i)** Processor 1: T_1 from 0 to 13, T_3 from 13 to 25, T_6 from 25 to 45; Processor 2: T_2 from 0 to 18, T_4 from 18 to 27, T_5 from 27 to 35, idle from 35 to 45
(ii) Processor 1: T_1 from 0 to 13, T_3 from 13 to 25, T_4 from 25 to 34, T_5 from 34 to 42; Processor 2: T_2 from 0 to 18, T_6 from 18 to 38, idle from 38 to 42
(b) Yes
(c) T_2, T_6, and 38; Sum of the task times divided by 2 is 40.

9. **(a)** Yes
(b) No

11. **(a)** Processor 1: T_1, T_6, idle 15 to 21, T_7, idle 27 to 31; Processor 2: T_2, T_5, T_8; Processor 3: T_3, T_4, idle from 13 to 31
(b) Processor 1: T_1, T_6, idle 15 to 21, T_7, idle 27 to 31; Processor 2: T_3, T_4, idle from 13 to 21, T_8; Processor 3: T_2, T_5, idle from 21 to 31
(c) Processor 1: T_4, idle 10 to 11, T_6, idle 18 to 21, T_8; Processor 2: T_2, T_5, T_7, idle 27 to 31; Processor 3: T_1, T_3, idle 11 to 31

13. Answers will vary.

15. Yes

17. (a) No
(b) T_2 should have been scheduled at time 0.
(c) Use the digraph with no edges and the list: T_2, T_1, T_3, T_4, T_5.

19. (a) T_1, T_2, T_3, and T_6
(b) No tasks require that T_1 and T_6 be done before these other tasks can begin.
(c) T_6
(d) Processor 1: T_1, T_6; Processor 2: T_2, T_4, idle from 18 to 30; Processor 3: T_3, T_5, idle from 12 to 30
(e) No
(f) Processor 1: T_6, idle from 20 to 22; Processor 2: T_3, T_5, T_1; Processor 3: T_2, T_4, idle from 18 to 22
(g) Yes
(h) Yes

21. (a) 120
(b) No; T_1 must be assigned to the first machine at time 0.
(c) No; Since when 2 divides 31, there is a remainder of 1.
(d) No

23. Yes

25. Answers will vary.

27. No

29. (a) Task times: $T_1 = 3$, $T_2 = 3$, $T_3 = 2$, $T_4 = 3$, $T_5 = 3$, $T_6 = 4$, $T_7 = 5$, $T_8 = 3$, $T_9 = 2$, $T_{10} = 1$, $T_{11} = 1$, and $T_{12} = 3$. This schedule would be produced from the list: T_1, T_3, T_2, T_5, T_4, T_6, T_7, T_8, T_{11}, T_{12}, T_9, T_{10}.
(b) Task times: $T_1 = 3$, $T_2 = 3$, $T_3 = 3$, $T_4 = 2$, $T_5 = 2$, $T_6 = 4$, $T_7 = 3$, $T_8 = 5$, $T_9 = 8$, $T_{10} = 4$, $T_{11} = 7$, $T_{12} = 9$, and $T_{13} = 3$. This schedule would be produced from the list: T_1, T_5, T_7, T_4, T_3, T_6, T_{11}, T_8, T_{12}, T_9, T_2, T_{10}, T_{13}.

31. (a) (i) Processor 1: T_1, T_3, T_5, T_7, idle from 16 to 20; Processor 2: T_2, T_4, T_6, T_8
(ii) Processor 1: T_8, T_5, T_4, T_1; Processor 2: T_7, T_6, T_3, T_2
(b) Yes

33. Answers will vary.

35. In part (a), 33 is not exactly divisible by 4; In part (b) 56 is not exactly divisible by 5.

37. (a) (i) Machine 1: 12, 9, 15, idle from 36 to 50; Machine 2: 7, 10, 13, 20
(ii) Machine 1: 12, 13, 20; Machine 2: 7, 9, 15, 10, idle from 41 to 45
(iii) Machine 1: 20, 12, 9, idle from 41 to 45; Machine 2: 15, 13, 10, 7
(b) An optimal schedule is possible.
(c) The critical path list is T_6, T_5, T_4, T_1, T_7, T_2, T_3, using the first processor.

39. (a) Machine 1: 129; Machine 2: 129
(b) Machine 1: 123; Machine 2: 123
(c) Yes

41. (a) Processor 1: 12, 13, 45, 34, 63, 43, 16, idle 226 to 298; Processor 2: 23, 24, 23, 53, 25, 74, 76; Processor 3: 32, 23, 14, 21, 18, 47, 23, 43, 16, idle 237 to 298
(b) Processor 1: 12, 24, 14, 34, 25, 23, 16, 16, 76; Processor 2: 23, 23, 21, 63, 43, idle 173 to 240; Processor 3: 32, 23, 53, 74, idle 182 to 240; Processor 4: 13, 45, 18, 47, 43, idle 166 to 240
(c) Three machines: Processor 1: 76, 45, 43, 24, 23, 18, 16, 13; Processor 2: 74, 47, 34, 32, 23, 21, 14, 12, idle 257 to 258; Processor 3: 63, 53, 43, 25, 23, 23, 16, idle 246 to 248
Four machines: Processor 1: 76, 43, 24, 23, 16, idle 182 to 194; Processor 2: 74, 43, 25, 23, 16, 13; Processor 3: 63, 45, 32, 23, 18, 12, idle 193 to 194; Processor 4: 53, 47, 34, 23, 21, 14, idle 192 to 194
(d) Processor 1: 84, 45, 43, 25, 23, 23, 16, 12; Processor 2: 82, 55, 34, 32, 23, 18, 14, 13; Processor 3: 71, 61, 43, 24, 23, 21, 16, idle 259 to 271

43. Answers will vary.

45. Each task heads a path of length equal to the time to do that task.

47. 9; Number of bins would not change, but the placement of the items in the bins would differ.

49. (a) 17
(b) 16
(c) 16
(d) 13

51. Yes, both are acceptable.

53. (a) Answers will vary.
(b) It is possible.

55. No; yes

57. Answers will vary.

59. Answers will vary.

61. (a) 152 min; 124 min
(b) 155 min; 120 min
(c) Yes; five-processor decreasing-time schedule
(d) 11
(e) NFD: 13; WFD: 11
(f) An optimal packing with 10 bins exists.

63. (a) Answers will vary.
(b) Answers will vary.
(c) Packing rectangles of width 1 in an $m \times 1$ rectangle is a special case of the two-dimensional problem.
(d) Answers will vary.

65. Answers will vary.

67. (a) Graph (a) no; (b) yes; (c) no; (d) yes; (e) yes; (f) yes
(b) Graph (a) yes; (b) yes; (c) no; (d) yes; (e) yes; (f) yes
(c) Graph (a) 4; (b) 3; (c) 5; (d) 3; (e) 2; (f) 2

69. **(a)** Drawings can vary. Possible renderings include the following:

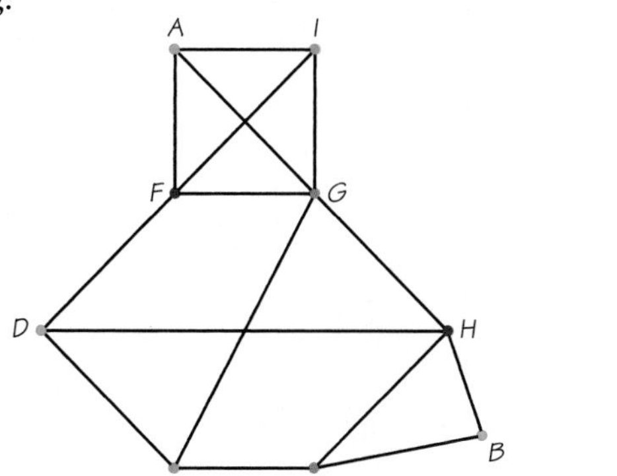

(b) 4
(c) The coloring in **(a)** indicates one possible arrangement.

71. **(a)** Drawings can vary. Possible renderings include the following:

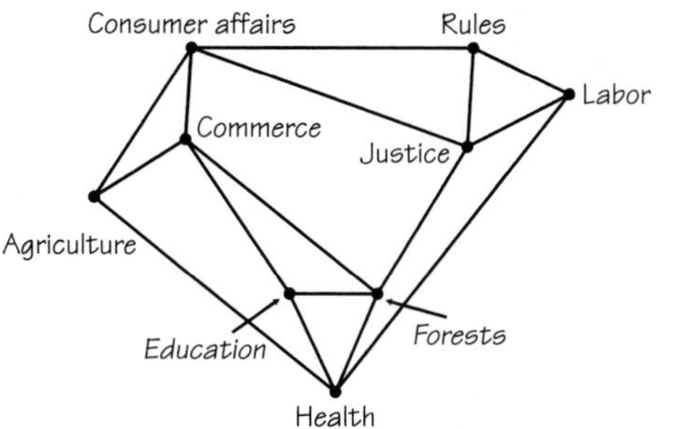

(b) 3
(c) 3; Additional answers will vary.

73. **(a)** Drawings can vary. Possible renderings include the following.

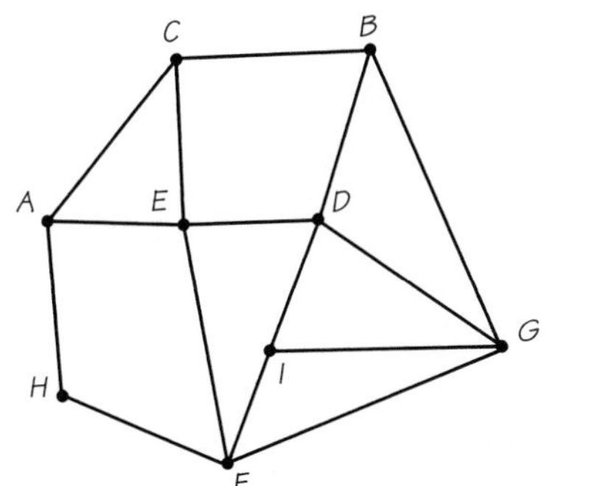

(b) 3
(c) 3

75. Answers will vary.

77. Graph (a) 6; (b) 8; (c) 6; (d) 3; (e) 3; (f) 4; Minimum is either the maximal valence of any vertex or one more than the maximal valence.

79. **(a)** Graph (a) 4; (b) 2; (c) 4; (d) 4; (e) 2; (f) 3
(b) Answers will vary.

81. 3

83. 3; only if each child formed his/her own play group

Chapter 4

1. **(a)**

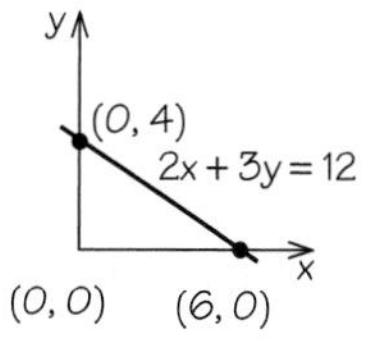

(b)

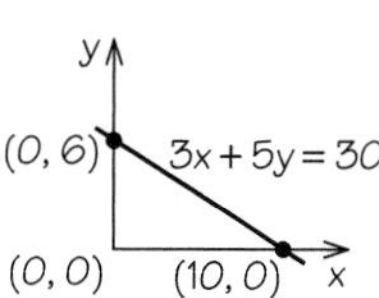

(c)

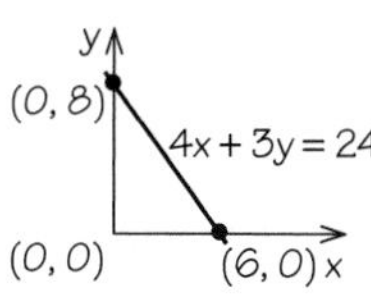

3. **(a)**

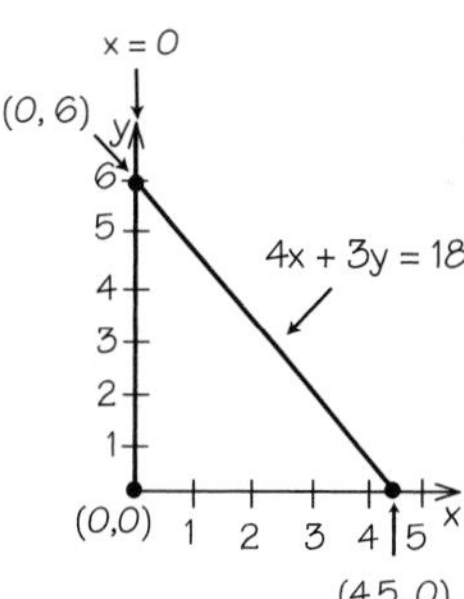

(b)

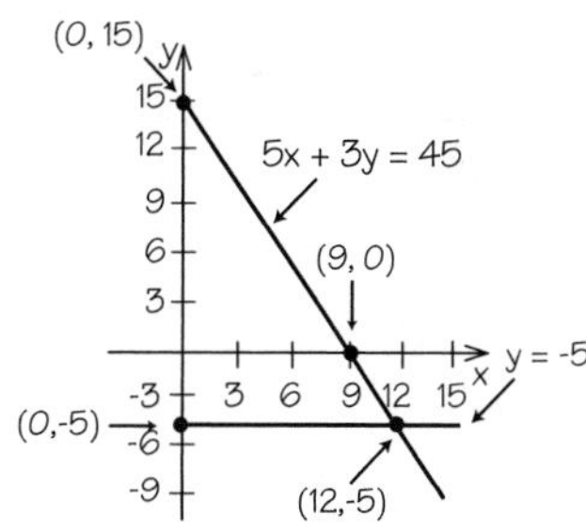

(c)

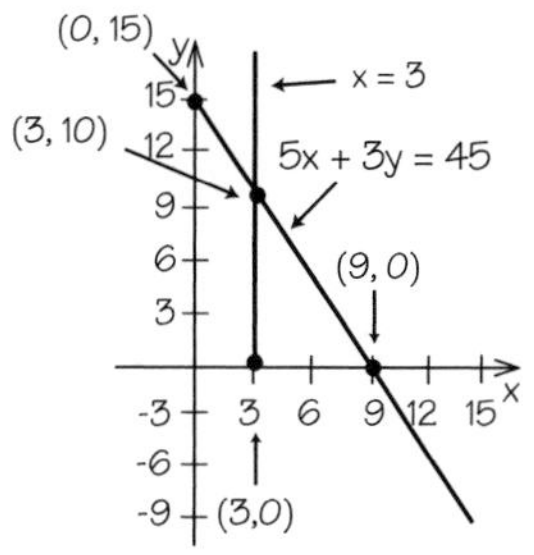

Note: For Exercises 5 and 7, first quadrant only is shown. Point of intersection is labeled.

5. (a)

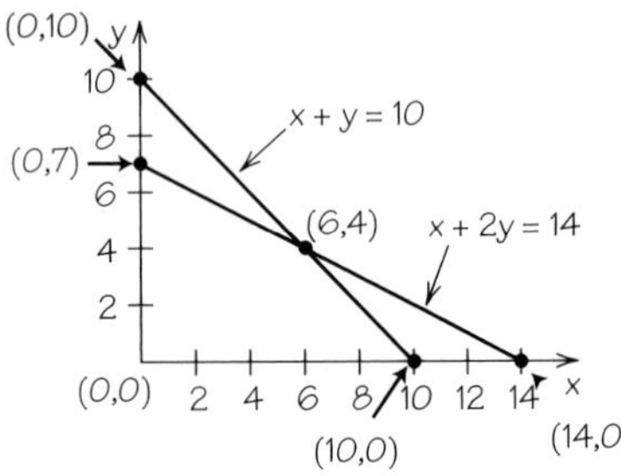

(b)

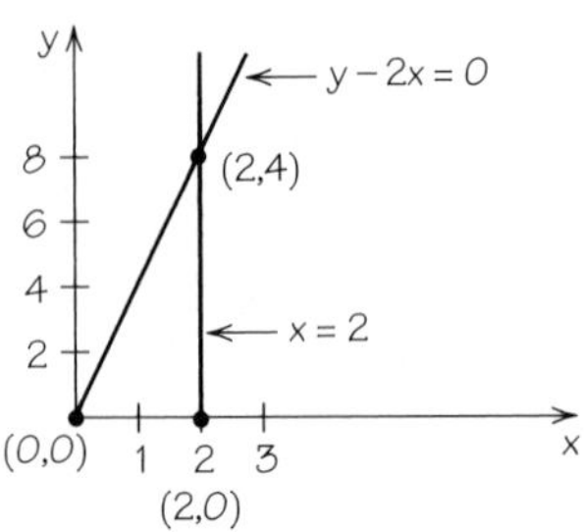

(c)

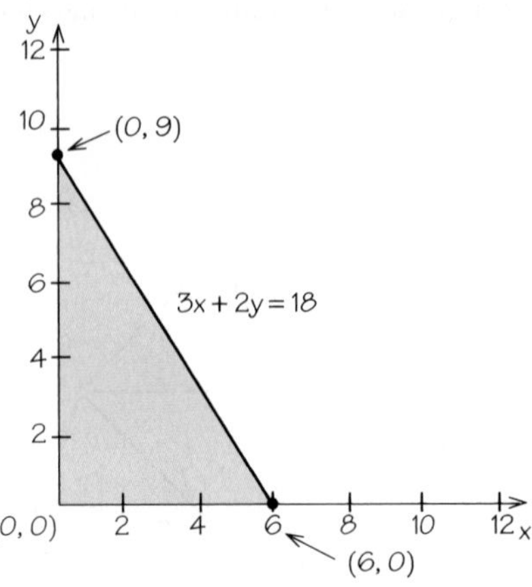

(d)

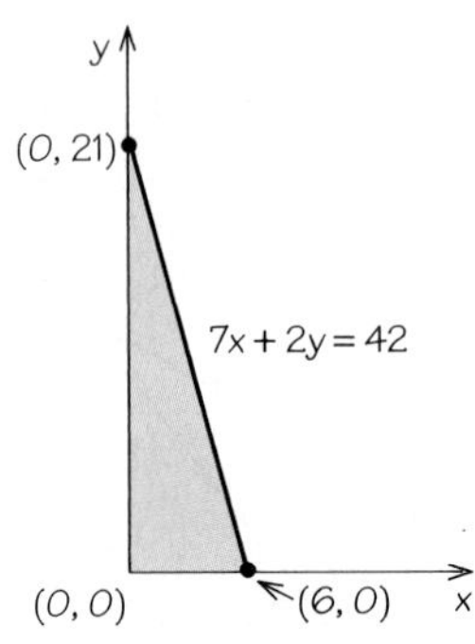

7. (a)

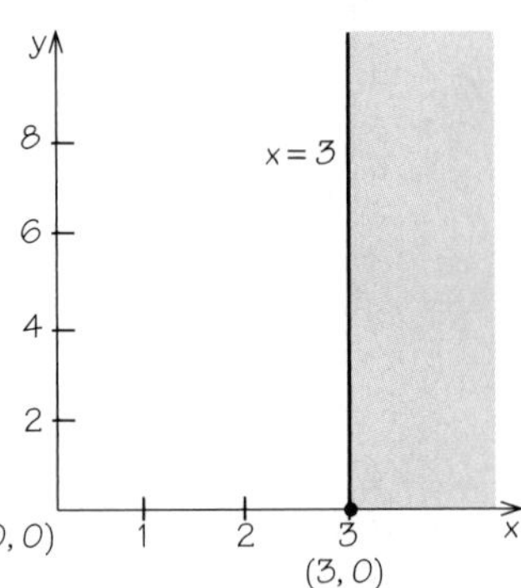

(b)

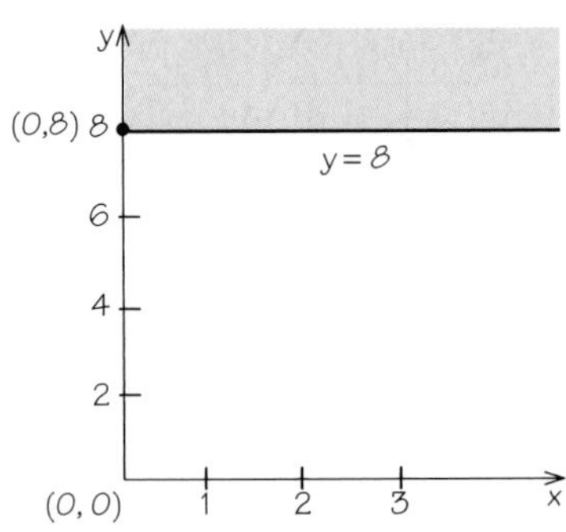

9. (a) $6x + 4y \leq 300$
(b) $30x + 72y \leq 420$

11.

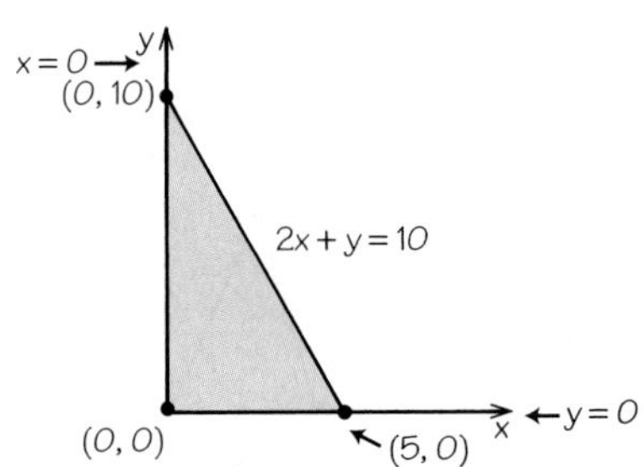

13.

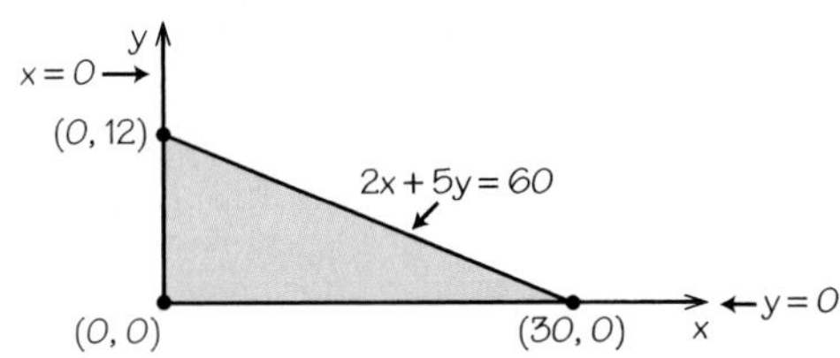

15.

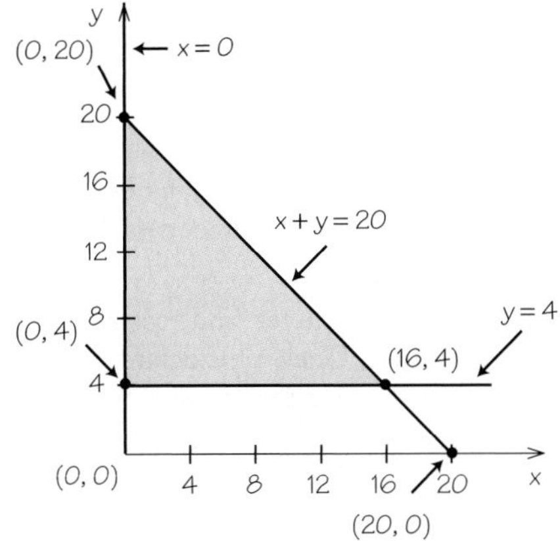

17. Exercise 11: (2, 4): yes; (10, 6): no; Exercise 13: (2, 4): yes; (10, 6): yes; Exercise 15: (2, 4): yes; (10, 6): yes

19. Make 0 skateboards and 30 dolls for a profit of \$111.

21. Note: These situations are shown only for the first quadrant.

(a)

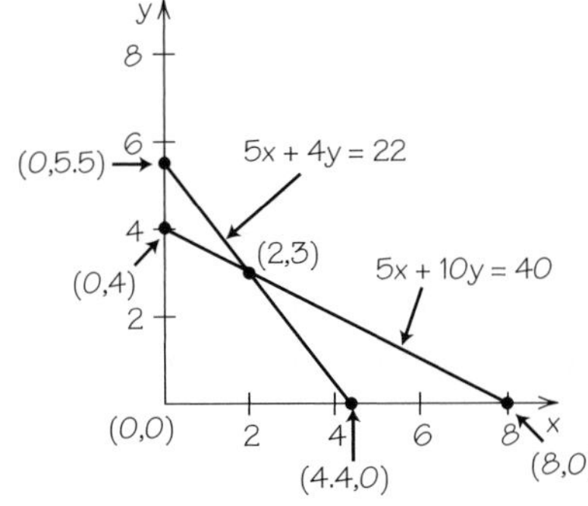

(b)

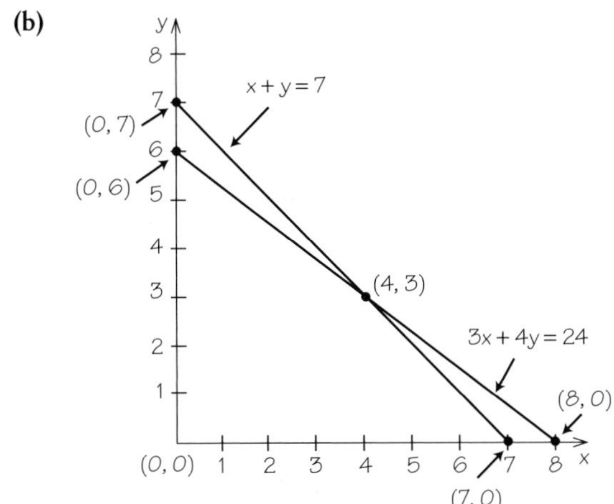

23.

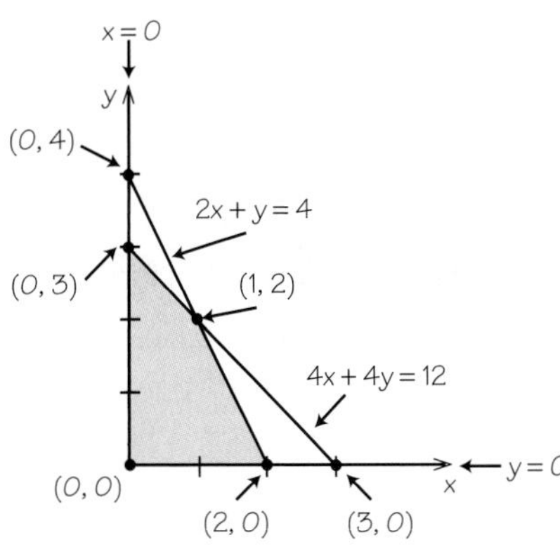

25.

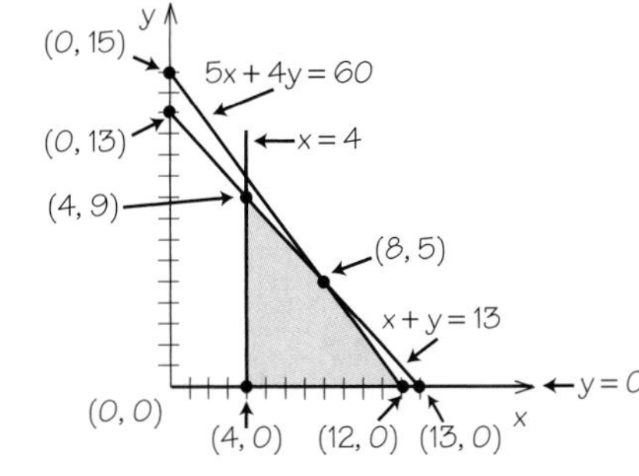

27. 28

29. 38

31. **(a)** (2, 0)
(b) It is not.
(c) Profit at (2, 0) greater than the profit at $\left(\frac{13}{7}, 0\right)$
(d) Yes; Profit at R is less than the profit at Q.
(e) Answers will vary.

33. Schedule 400 oil changes and no tune-ups.; Schedule 300 oil changes and 20 tune-ups.

35. Schedule 360 routine visits and no comprehensive visits.; Schedule 210 routine visits and 30 comprehensive visits.

37. Take four math courses and no other courses.; Take two math courses and 2 other courses.

39. Make 2 grade A and 5 grade B batches in both cases.

41. Make 3000 cartons of regular and 2000 cartons of diet in both cases.

43. Make no desk lamps and 1200 floor lamps.; Make 150 desk lamps and 1080 floor lamps.

45. Make 50 chairs, 10 tables, and no beds each month.

47. Make 470 pounds of Excellent, none of Southern, 2400 pounds of World, and 320 pounds of Special.

49. 43

51. Make 60 business and no charity calls.; Make 45 business and 10 charity calls.

53. Make 3 bikes and 2 wagons in both cases.

55. (a)

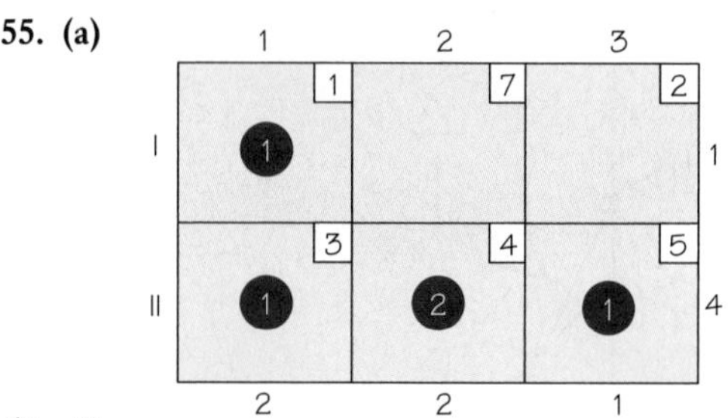

(b) 17
(c) (I, 2): 5; (I, 3): −1

57. (a) Connected and has no circuit
(b) Add edge joining Vertex I to Vertex 2.; Add edge from Vertex I to Vertex 3.
(c) Circuit 2, I, 1, II, 2 corresponds to the circuit of cells, (I, 2), (I, 1), (II, 1), (II, 2), (I, 2). Circuit 3, I, 1, II, 3 corresponds to the circuit of cells, (I, 3), (I, 1), (II, 1), (II, 3), (I, 3).

59. (a) (i)

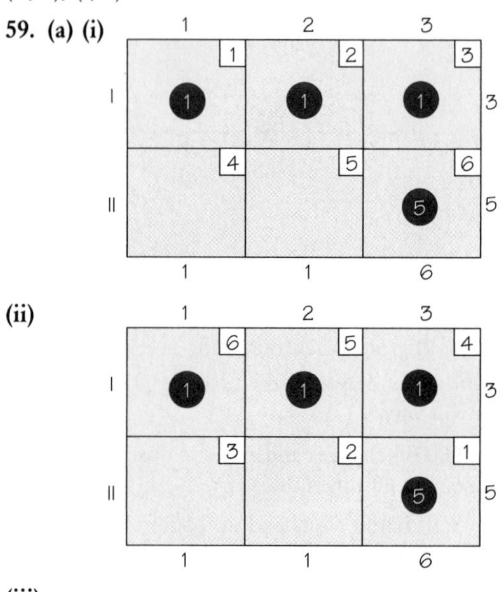

(ii)

(iii)

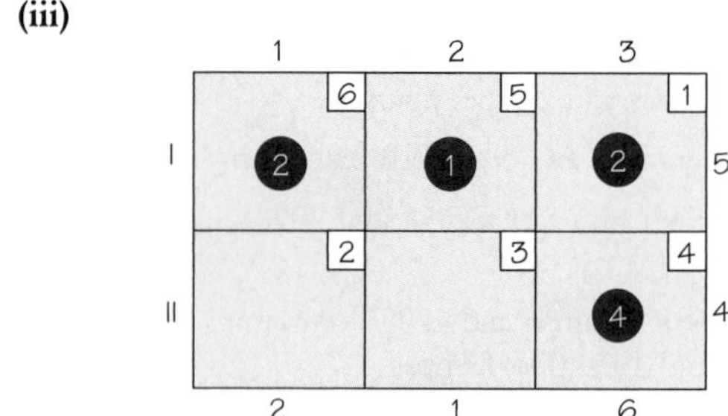

(b) For both **(i)** and **(ii)** the tableaux shown are optimal. However, there are also other optimal tableaux. For **(iii)** the tableau shown is not optimal. Using the stepping stone algorithm, the cost can be reduced to 16.

61. (i) (a) 33
(b) No
(c) Cost: 24
(ii) (a) 37
(b) No
(c) Cost: 28

Chapter 5

1. (a) Vehicle makes and models
(b) Vehicle type, transmission type, number of cylinders, city MPG, and highway MPG; cylinders (maybe), the two MPGs (certainly)

3. Given earlier years on the left and recent years on the right, skewed to the left; Draw a histogram skewed to the left because most coins were minted in recent years.

5. (a) Big countries would always top the list.
(b) Using class widths of 2 metric tons per person starting 0.0–1.9, the distribution is skewed to the right.; high outliers: Canada, Australia, and United States

7. (a) Alaska: 5.7%, Florida: 17.6%.
(b) Single-peaked and roughly symmetric, center near 12.7%, spread from 8.5% to 15.6%

9. There is one high outlier, 200. The center of the 17 observations other than the outlier is 137 (9th of 17). The spread is 101 to 178.

11. (a) Repeated stems break up the intervals further.
(b) Reasonably symmetric

13. (a) 141.1
(b) 137.6; High outlier pulls the mean up.

15. \$66,570; distribution strongly right-skewed

17. Examples will vary. One possible answer is 1, 2, 2, 2, 3, 3, 4, 17.

19. 5.7, 11.7, 12.75, 13.5, 17.6

21. 5799, 20000, 25942, 34986, 36700; two distinct clusters of values

23. 0.0, 0.75, 3.2, 7.8, 19.9; Q_3 and maximum are much farther from the median than Q_1 and minimum.

25. Median for bachelor's is greater than Q_3 for high school. Bachelor's distribution is very much more spread out, especially at the high-income end but also between the quartiles.

27. (a) Either histogram or stemplot will do.
(b) $\overline{x} = 48.25$, $M = 37.8$; Long right tail pulls the mean up.
(c) 2.0, 21.5, 37.8, 60.1, 204.9; Q_3 and maximum are much farther above the M that Q_1 and minimum are below it, showing that the right side of the distribution is much more spread out than the left side.

29. Arizona, California, Nevada, New Mexico, and Texas

31. (a) $\overline{x} = 5.448$, $s = 0.221$
(b) $M = 5.46$; yes

33. For both datasets, $\overline{x} = 7.50$ and $s = 2.03$. Data A have two low outliers and Data B have one high outlier.

35. (a) $s = 0$ is smallest possible. Examples will vary. One possible answer is 1, 1, 1, 1.

(b) 0, 0, 10, 10
(c) Yes
(d) No

37. About 27

39. Mean: *A*, median: *B*

41. **(a)** 327 to 345 days
(b) 16%

43. $Q_1 = 1381$, $Q_3 = 1641$

45. **(a)** −20.51% to 41.53%
(b) 20.51% or greater

47. **(a)** Median = 10% by symmetry
(b) 9.6% to 10.4%
(c) 9.866% to 10.134%

49. **(a)** 50%, 2.5%
(b) 0.37 to 0.43

51. Red: somewhat right-skewed (with no outliers), yellow: quite symmetrical (with no outliers)

53. Red: $\overline{x} = 39.71$, $s = 1.799$; yellow: $\overline{x} = 36.18$, $s = 0.975$; yellow distribution

55. Between 2.5% and 16%

57. Mode, mean, and median

Chapter 6

1. **(a)** Latter case; study time
(b) Relationship only
(c) Latter case; rainfall
(d) Relationship only

3. **(a)** Life expectancy increases with GDP in a curved pattern. The increase is very rapid at first, but levels off for GDP above roughly $5000 per person.
(b) Answers will vary.

5. Strong positive straight-line relationship

7. **(a)** Speed
(b) The regression line for Exercise 31 is included.

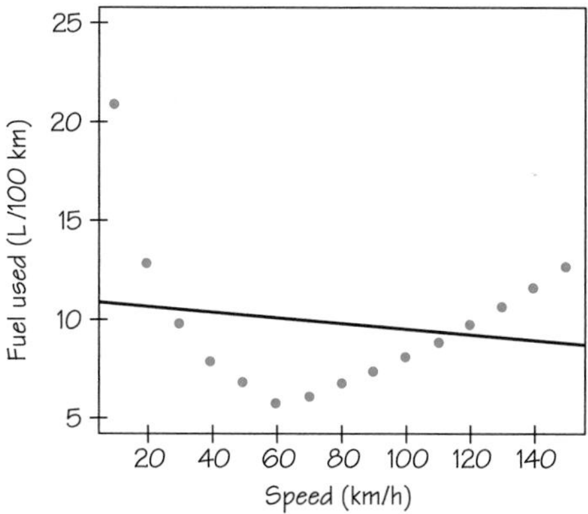

Relationship is curved; Additional answers will vary.
(c) None overall
(d) Quite strong; little scatter

9. 2.31

11. **(a)** *Hint*: Choose two values of weeks, preferably near 1 and 150.
(b) 5.42 and 4.64
(c) −0.0053; On average pH declined by 0.0053 per week during the study period.

13. 1

15. 0.9353; strong straight-line pattern

17. 0.9674; Point extends (strengthens) the straight-line pattern.

19. −0.1700; relationship strong but curved

21. 1

23. **(a)** Negative
(b) Negative
(c) Positive
(d) Small

25. **(a)** Dividend Growth: 0.98; Small Cap Stock: 0.81; Emerging Markets: 0.35
(b) No, just moved in the same direction

27. **(a)** Predicted highway mpg = 10.48 + 0.89 × (city mpg)
(b) 25.61 mpg
(c) Yes; Points follow a straight line.

29. Predicted highway mileage of a car that gets 18 mpg in the city is approximately 26 mpg or 27 mpg (26.5 mph).

31. See Exercise 7 answer for plot; Predicted values are approximately 10.91, 10.03, and 8.85, respectively.

33. Predicted height of husband = 33.67 + 0.54 × (height of woman); 69.85 in.

35. In $\hat{y} = mx + b$, substitute $b = \overline{y} - \left(r\frac{s_y}{s_x}\right)\overline{x}$, $m = r\frac{s_y}{s_x}$, and $x = \overline{x}$.

37. Answers will vary.

39. **(a)** All four have $r = 0.816$ and $y = 3.0 + 0.5x$.
(c) Data Set A; Additional answers will vary.

41. Answers will vary.

43. Answers will vary.

45. Lead level and reading score, respectively; negative; Answers will vary.; yes

47. **(a)** Positive; not
(b) $r = 0.3602$

49. **(a)** 0.42; For each additional inch of women's height, the height of the next person dated goes up by 0.42 inch on average.
(b) 69.22 in.

Chapter 7

1. **(a)** U.S. residents aged 18 and older
(b) 1027

3. Answers will vary.

5. **(a)** Answers will vary.
(b) Larger; bias due to voluntary response

7. Taylor, Brianna, Alexis

9. **(a)** 001 to 371
(b) 214, 235, 119

11. Repeated samples of the same size from the same population will always be the same.

13. **(a)** 35, 75, 115, 155, 195
(b) Answers will vary.

15. **(a)** All people aged 18 and over living in the United States
(b) 30%
(c) Answers will vary.

17. Answers will vary.

19. No imposed treatment; observational

21. Answers will vary.

23. Classes 1, 2, 7, 8, 9, 13, 14, 15, 17, 18, 19, 22, 25, 27, and 29 receive the treatment.

25. This is a randomized comparative experiment with four branches, similar to Figure 7.3 with two more branches. The "flow chart" outline must show random assignment of subjects to groups, the four treatments, and the response variable (healthcare spending). We can't show the group sizes because we don't know how many people or households are available to participate.

27. **(a)** There are six treatments:

	Discount Level		
	20%	**40%**	**60%**
50% on sale	1	2	3
100% on sale	4	5	6

(b) Label the subjects 01 to 60. The first group contains subjects labeled 7, 8, 10, 15, 25, 27, 54, 55, 58, 60.

29. **(a)** This is a randomized comparative experiment similar to Figure 7.3.
(b) Tea group contains rats 4, 6, 7, 8, 9, 11, and 12.

31. **(a)** This is a randomized comparative experiment with four branches. Best to use groups of size 216.
(b) 253, 296, 304, 470, 731
(c) Neither the subjects nor those working with the subjects know the contents of the pill each subject took daily.
(d) Differences could be due to the chance assignment of subjects to groups.
(e) Answers will vary.

33. Answers will vary.

35. Observational

37. Both are statistics.

39. **(a)** Approximately normal with mean 0.14 and standard deviation 0.0155.
(b) 0.109 to 0.171

41. **(a)** 60% of the digits 0–10 are 0–5.
(b) 29; 72.5%; 1; 6

43. 0.397 to 0.443

45. **(a)** 0.167 to 0.221
(b) It is likely that more than 171 ran a red light.

47. **(a)** 0.5
(b) $\frac{1}{\sqrt{n}}$

49. **(a)** 11.3%
(b) Answers will vary.

51. **(a)** No
(b) Yes

53. Smaller

55. 2% to 12.4%

57. About 0.16

Chapter 8

1. **(a)** Results will vary.
(b) Results will vary.

3. 0.105

5. **(a)** {0, 1, 2, 3, 4, 5, 6, 7, 8, 9, 10}
(b) {0, 10, 20, 30, 40, 50, 60, 70, 80, 90, 100}
(c) {Yes, No}

7. **(a)** {HHHH, HHHM, HHMH, HMHH, MHHH, HHMM, HMMH, MMHH, HMHM, MHHM, MHMH, HMMM, MMMH, MHMM, MMHM, MMMM}
(b) {0, 1, 2, 3, 4}

9. **(a)** 0.19
(b) 0.287

11. $\frac{1}{216}$

13. **(a)** Histograms should include grades 0–4, inclusive, on the horizontal axis and the associated probabilities on the vertical axis.
(b) 0.64

15. **(a)** No; Rule 2 violated
(b) Yes

17. The probability model for this pair of dice is the same as the one for regular dice.

19. $\frac{8}{15}$

21. **(a)** 1024
(b) $\frac{1}{512}$

23. **(a)** 0.919
(b) 0.377

25. **(a)** 6
(b) *asp, pas, sap, spa*
(c) 66.7%

27. **(a)** 4

(b) 2,598,960
(c) 0.00000154

29. (a) $\frac{1}{2} \times \text{base} \times \text{height} = \frac{1}{2}(2)(1) = 1$
(b) $\frac{1}{2}$
(c) 0.125

31. 0.25

33. 2.78; 0.8669

35. Owner-occupied units: 6.248; rented units: 4.321

37. Both models have mean 1, because both density curves are symmetric about 1.

39. $30

41. $0.45

43. (a) Loss of a $\frac{1}{4}$ point
(b) Yes

45. Between 0.11 and 0.19

47. (a) 5.77 mg
(b) 4; Answers will vary.

49. (a) Sketch not included in answers.
(b) 4580 to 4620
(c) 4588.46 to 4611.54

51. (a) About 0.16
(b) Mean: 20.8; standard deviation: 1.6
(c) About 0.0015

53. (a) 45,697,600
(b) 2600
(c) 0.0000569

55. (a) 0.18
(b) 0.39

57. (a) 1.156 days
(b) Between 0.54 day and 1.00 day

59. (a) 100%
(b) 80%
(c) 12%

Chapter 9

1. Answers will vary.

3. Answers will vary.

5. Each one-on-one score will have a winner because there cannot be a tie.

7.

Rank	Number of voters (4)			
	1	1	1	1
First	*A*	*B*	*C*	*D*
Second	*B*	*C*	*D*	*A*
Third	*C*	*D*	*A*	*B*
Fourth	*D*	*A*	*B*	*C*

9. (a) Yes
(b) Alfonse D'Amato (*D*)

11. (a) *C*
(b) *A*
(c) *D*
(d) *D*

13. (a) Five-way tie
(b) *C*
(c) Five-way tie
(d) *E*

15. (a) *C*
(b) *E*
(c) *E*
(d) *E*

17. (a) *E*
(b) Answers will vary. One possible answer is the following:

Rank	Number of Voters (2)	
	1	1
First	*A*	*C*
Second	*B*	*B*
Third	*C*	*A*

19. (a) If everyone prefers *B* to *D*, for example, then *D* has no-first place votes at all.
(b) Moving a winning candidate up one spot on some list neither decreases the number of first-place votes for the winning candidate nor increases the number of first-place votes for any other candidate.

21. (a) Condorcet winner always wins this kind of one-on-one contest.
(b) Moving a candidate up on some list only improves that candidate's chances in one-on-one contests.

23. In order to have one candidate's ranking be consistently higher than another candidate's would imply that only one candidate would be considered.

25. Answers will vary.

27. Answers will vary. One possible answer is the following:

Rank	Number of Voters (5)	
	3	2
First	*A*	*B*
Second	*B*	*C*
Third	*C*	*A*

29. (a) Since *D* has the least number of first-place votes, *D* is eliminated. Since *B* has the least number of first-place votes, *B* is eliminated. Thus, *A* is the unique winner.
(b) *B*

31. Answers will vary.

33. **(a)** A three-way tie with both methods
(b) That alternative is the sole winner with both methods.
(c) No; Either the situation in part (a) or the situation in part (b) must occur.

35. **(a)** *D*
(b) *A*, *B*, *D*, and *F*
(c) *B*, *D* and *F*
(d) *A*, *B*, *D*, and *F*

Chapter 10

1. Answers will vary. One example of two such elections is the following:

Rank	Election 1 Number of Voters (3)		
First	*A*	*A*	*B*
Second	*B*	*B*	*A*

Rank	Election 2 Number of Voters (3)		
First	*B*	*A*	*B*
Second	*A*	*B*	*A*

3. Answers will vary. One example of two such elections is the following:

Rank	Election 1 Number of Voters (3)		
First	*A*	*B*	*B*
Second	*B*	*A*	*A*

Rank	Election 2 Number of Voters (3)		
First	*B*	*B*	*B*
Second	*A*	*A*	*A*

5. **(a)** Doesn't treat all voters the same
(b) A dictatorship in which Voter #1 is the dictator
(c) A dictatorship in which Voter #2 is the dictator and a dictatorship in which Voter #3 is the dictator

7. Consider the leftmost voter changes his or her preference ballot to the following:

C
B
D
A

9. Consider the leftmost voter changes his or her preference ballot to the following:

A
C
D
B

11. Answers will vary.

13. Consider the leftmost voter changes his or her preference ballot to the following:

B
A
D
C

15. Consider the leftmost voter changes his or her preference ballot to the following:

A
C
B

17. Consider the leftmost voter changes his or her preference ballot to the following:

B
A
C

19. **(a)** Consider the agenda *D*, *A*, *C*, *B*.
(b) Consider the agenda *B*, *D*, *A*, *C*.
(c) Consider the agenda *B*, *A*, *C*, *D*.

21. Consider the voters in the 7% group to change their ballots to the following:

H
J
D

23. **(a)** To go from having a unique winner to a different unique winner occurs if the winning alternative in the first election has exactly two first-place votes, and one of these two voters changed his or her ballot by moving some other alternative into first place (yielding a worse outcome for this voter).
(b) Tie in second election
(c) Answers will vary.

25. **(a)** Answers will vary.
(b) Answers will vary.

27. 1 and 4

29. Answers will vary.

Chapter 11

1. **(a)** A winning or blocking coalition would be 50 senators plus the vice president, or more than 50 senators.
(b) At least 41 senators

3. Weight-5 and weight-4 voters; weight-3 voter

5. **(a)** 1958: *B*, *G*, *L*; 1964: *N*, *G*, *L*; later years: none
(b) 1958 and 1964: *N*, *B*, *G*, *L*; 1970 and 1976: none; 1982: *G*

7. The last juror in the permutation is the pivotal voter.

9. **(a)** $\left(\frac{5}{12}, \frac{1}{4}, \frac{1}{4}, \frac{1}{12}\right)$
(b) $\left(\frac{1}{3}, \frac{1}{3}, \frac{1}{6}, \frac{1}{6}\right)$
(c) $\left(\frac{1}{4}, \frac{1}{4}, \frac{1}{4}, \frac{1}{4}\right)$

11. Nevada

13. **(a)** NNNN, NNNY, NNYN, NNYY, NYNN, NYNY, NYYN, NYYY, YNNN, YNNY, YNYN, YNYY, YYNN, YYNY, YYYN, YYYY
(b) { }, {*D*}, {*C*}, {*C*, *D*}, {*B*}, {*B*, *D*}, {*B*, *C*}, {*B*, *C*, *D*}, {*A*}, {*A*, *D*}, {*A*, *C*}, {*A*, *C*, *D*}, {*A*, *B*}, {*A*, *B*, *D*}, {*A*, *B*, *C*}, {*A*, *B*, *C*, *D*}
(c) Each subset corresponds to a set of "yes" voters.
(d) (i) 1; (ii) 4; (iii) 6

15. *A* has a critical vote in 6 coalitions; *B*, *C*, and *D* each have critical votes in 2. The Banzhaf power index is (12, 4, 4, 4).

17. **(a)** 35
(b) 0
(c) 105
(d) 105

19.

Year	Banzhaf Index
1958	(32, 0, 0, 0, 0)
1964	(32, 0, 0, 0, 0)
1970	(32, 2, 2, 2, 2)
1976	(32, 2, 2, 2, 2)
1982	(28, 4, 4, 0, 4)

21. $\frac{5}{32}$

23. **(a)** {*A*, *C*, *D*} and {*A*, *B*}
(b) Each winning coalition includes *A*, so *A* has veto power. That makes {*A*} a minimal blocking coalition. There are two other minimal blocking coalitions: {*B*, *C*} and {*B*, *D*}.
(c) (10, 6, 2, 2)
(d) Answers will vary. Possible answers include [5 : 3, 2, 1, 1]
(e) $\left(\frac{7}{12}, \frac{1}{4}, \frac{1}{12}, \frac{1}{12}\right)$

25. Answers will vary.

27. **(a)** [4 : 2, 1, 1, 1]
(b) [6 : 2, 2, 1, 1, 1]

29. (24, 24, 24, 24, 20, 20, 20); faculty

31. If there were only two minimal winning coalitions, those two coalitions must overlap; there is a voter we'll call *V* who is in both. This *V* is therefore in every winning coalition, and hence has veto power.

33. **(a)** We will omit *E* from the list: One could include him in any winning or losing coalition without altering the vote total.

Winning Coalition	Extra Votes	Losing Coalition	Votes Needed
{*A*, *B*, *C*, *D*}	49	{*A*}	3
{*A*, *B*, *C*}	42	{*B*, *C*}	6
{*A*, *B*, *D*}	27	{*B*, *D*}	21
{*A*, *C*, *D*}	26	{*C*, *D*}	22
{*A*, *B*}	20	{*B*}	28
{*A*, *C*}	19	{*C*}	29
{*A*, *D*}	4	{*D*}	44
{*B*, *C*, *D*}	1	{ }	51

(b) 4
(c) 19
(d) 4

35. **(a)** [8 : 6, 1, 1, 1, 1, 1, 1, 1, 1]
(b) (492, 16, 16, 16, 16, 16, 16, 16, 16)
(c) $\left(\frac{2}{3}, \frac{1}{24}, \frac{1}{24}, \frac{1}{24}, \frac{1}{24}, \frac{1}{24}, \frac{1}{24}, \frac{1}{24}, \frac{1}{24}\right)$
(d) No

37. [9 : 4, 4, 4, 1, 1, 1, 1, 1]

39. The three weight-3 voters or two weight-3 voters and one weight-1 voter form minimal winning coalitions. The Banzhaf power index is (30, 30, 30, 6, 6, 6).

41. Case 1 (three weight-1 voters): $\left(\frac{17}{60}, \frac{17}{60}, \frac{17}{60}, \frac{1}{20}, \frac{1}{20}, \frac{1}{20}\right)$ and Case 2 (four weight-1 voters): $\left(\frac{9}{35}, \frac{9}{35}, \frac{9}{35}, \frac{2}{35}, \frac{2}{35}, \frac{2}{35}, \frac{2}{35}\right)$; In Case 1, the weight-3 voter is $5\frac{2}{3}$ times as powerful as the weight-1 voter and in Case 2 the weight-3 voter is $4\frac{1}{2}$ times as powerful as the weight-1 voter.

Chapter 12

1. Assume a distribution is skewed to the left. The heavier concentration of voters on the right means that fewer voters are farther from the median. Because there are fewer voters "pulling" the mean rightward, it will be to the left of the median. Likewise, a distribution skewed to the right will have a mean to the right of the median.

3. While there is no median position such that half the voters lie to the left and half to the right, there is still a position where the middle voter (if the number of voters is odd) or the two middle voters (if the number of voters is even) are located, starting either from the left or right. In the absence of a median, less than half the voters lie to the left and less than half to the right of this middle voter's (voters') position (positions).

Hence, any departure by a candidate from a position of a middle voter to the position of a non-middle voter on the left or right will result in that candidate's getting less than half the votes–and the opponent's getting more than half. Thus, the middle position (positions) is (are) in equilibrium, making it (them) the extended median.